THIN-LAYER CHROMATOGRAPHY

A LABORATORY HANDBOOK

BY

H. R. BOLLIGER · M. BRENNER · H. GÄNSHIRT
H. K. MANGOLD · H. SEILER · EGON STAHL · D. WALDI

EDITED BY

EGON STAHL

WITH 197 FIGURES AND 2 PLATES IN COLOR

1965
SPRINGER-VERLAG
BERLIN · HEIDELBERG · NEW YORK
ACADEMIC PRESS INC., PUBLISHERS · NEW YORK · LONDON

SPRINGER-VERLAG · BERLIN · HEIDELBERG · NEW YORK
Berlin 31 (Wilmersdorf), Heidelberger Platz 3

ACADEMIC PRESS INC., PUBLISHERS
111 Fifth Avenue, New York, New York 10003

Library of Congress Catalog Card Number 63—19323

Dünnschicht-Chromatographie.
Springer-Verlag, Berlin · Göttingen · Heidelberg 1962
English translation by
Cambridge Consultants Limited. Principal Editors —
A. N. Howard and L. J. Morris
In Cooperation with H. K. Mangold, University of Minnesota,
The Hormel Institute, Austin, Minnesota, U. S. A.

Title No. 1051

Printed in Germany

Preface

Thin-layer chromatography has become so widely known in the space of a few years that it has proved necessary to gather into book form and thus make generally accessible the experimental material previously only available in isolated publications. As thin-layer chromatography can be used both for organic and inorganic matter as well as on quantities ranging from the *nanogram* to the microgram, it is impossible for any one individual to possess sufficient laboratory experience or overall knowledge to produce a practical handbook that will be of real assistance to beginner and specialist alike. For this reason, an international group was formed, who made it their task to produce the best possible treatise. In view of the present stage of development reached by thin-layer chromatography, it seems specially apt that the authors should include yet unpublished work of their own.

As thin-layer chromatography is used in many different fields in natural science and medicine, the kind of brief description of materials intelligible only to the expert has been avoided. The short guides to the chemical properties of the groups to be separated, their names, and relevant bibliographic details should facilitate introductory studies and make possible a close acquaintance with the material in hand. It also seemed advisable to give brief details of the analytical classification of material, which is so often necessary. Although the classification used may appear unusual, it is in fact pre-eminently suitable to thin-layer chromatography. Thus the special section begins with the lipids and ends with the hydrophilic compounds. Each chapter brings the subject dealt with completely up to date.

The general section, and in particular the part dealing with instruments, their use, and working procedure, does not aim at a complete picture, since this would only be confusing. The instructions thus include techniques of proven value only, so that even the beginner should experience no difficulty in acquainting himself with them. A whole chapter on its own is devoted to the theoretical principles underlying chromatographic procedure, ranging from an empirical, intuitive approach to description and evaluation in mathematical terms.

In the hope that this handbook will prove as great a help in the laboratory as thin-layer chromatography itself, I would like to extend my special thanks to all those who have helped to make this book possible.

Saarbrücken, Summer 1962. EGON STAHL

Contributors

H. R. Bolliger, Messrs. F. Hoffmann-La Roche & Co., Basel, Switzerland.

M. Brenner, Institute for Organic Chemistry, University of Basel, 19 St. Johanns-Ring, Basel, Switzerland.

H. Gänshirt, Messrs. Karl Engelhard, Manufacturers of Pharmaceutical Preparations, 94 Sandweg, Frankfurt am Main, Germany.

Helmut K. Mangold, University of Minnesota, The Hormel Institute, Austin, Minn., U.S.A.

H. Seiler, Institute for Inorganic Chemistry, University of Basel, 51 Spitalstraße, Basel, Switzerland.

Egon Stahl, Institute for Pharmacognosy, University of the Saar, Stadtwald, Saarbrücken 15, Germany.

D. Waldi, Messrs. E. Merck AG., Darmstadt, Germany.

Table of Contents

General Section

Special Section

General Section

A. History of the Development of Thin-Layer Chromatography

By

Egon Stahl

It is remarkable that although the principle of thin-layer chromatography was described as early as 25 years ago, this method has become accepted as indispensable only during the last 3 years. A short summary of the development of thin-layer chromatography is, therefore, not only interesting in itself, but also will help towards an understanding of the technique. The first publication in the field dates back to the period when Tswett's column chromatography was enjoying a great success. It was at this time that constant efforts were being made to achieve "micro-chromatography" by reducing the diameter of the column to 1 mm. The inherent difficulties were summed up by Zechmeister [*74*] in 1938 as follows: "The main problem in 'micro-chromatography' is not the development of the right apparatus (i.e. for column chromatography), but the accurate identification of adsorbed substances." This problem was solved by the change-over from the "closed" to the "open" column, to the thin separating layer, in fact. This method, published some years ago, demonstrated the simplicity of the process and its wide applicability [*56*].

In 1938 Izmailov and Shraiber [*20*] described the basic principle underlying the process. They used the method for separating plant extracts reported in the Russian Pharmacopeia VII and described the method in a paper entitled "The application of analysis by drop-chromatography to pharmacy."

They slurried the adsorption medium (usually alumina) with water and spread this in a layer 2 mm thick, on to glass plates. Cracking was observed with thicker layers, while thinner layers led to difficulty in evaluating results. One drop of the alcoholic plant extract to be examined was applied to the center of the dried layer. In many cases satisfactory fractionations were achieved in concentric rings, and, where this did not occur, more alcohol was added until several zones appeared. They then compared their ring chromatogramms with the corresponding column chromatograms (Fig. 1, comp. Fig. 20) and pointed out the advantages of the new method.

Izmailov and Shraiber's paper, written in Russian, ends with the following summary:

"A method for chromatographic adsorption analysis is elaborated, based on the observation of the division of substances into zones on a thin layer of adsorbent, using one drop of the substance.

The results obtained by the method proposed are qualitatively the same as those obtained by the usual chromatographic adsorption method of analysis. The method enables one to obtain satisfactory results using one drop of the substance under test, very small quantities of the adsorbent and minimal time.

The method may be used for the evaluation of galenical preparations and their identification as well as for preliminary test of the adsorbent and the kind of the developer. Sixteen galenical preparations are studied using the method proposed."

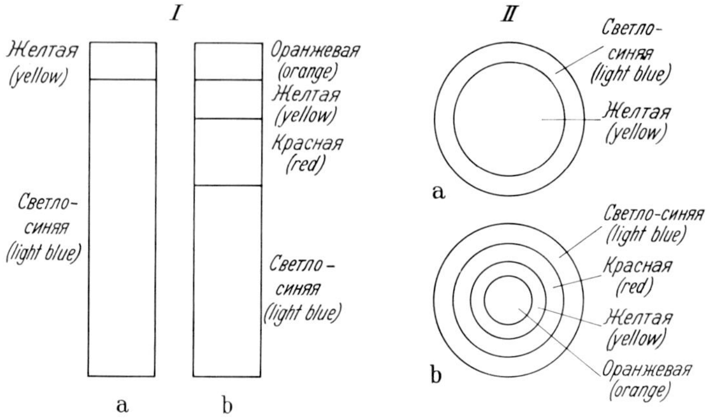

Fig. 1. Comparison of fluorescent dyes from a *Belladonna* tincture using an alumina column chromatogram (*I*) with a "drop chromatogram" (*II*) before development with alcohol (a) and after development (b). This diagram is taken from the work of IZMAILOV and SHRAIBER [*20*] with the translation of the fluorescent colors added

CROWE, in a review of this work, published in 1941 [*9*], reported that he and his colleagues had, for some time, been using a thin layer of adsorbent in a Petri dish and achieving similar results on such loose layers. These were, in any case, only preliminary tests prior to column chromatography.

We can, therefore, distinguish two processes at this stage, the one using adhesive layers and the other, using loosely applied layers. An improved version of the latter method was described by WILLIAMS in 1947 [*70*]. He protected the loose layer by covering it with a glass plate, which was perforated in the center to allow the mixture to be separated and the eluent to be added drop-wise.

Several years earlier MARTIN and SYNGE [*32*] developed a method of partition chromatography for separating amino acids and their derivatives. They fixed one phase (the stationary phase) on a support, e.g. powdered silica gel, and placed this in a column. As eluent, known in this case as "mobile phase", chloroform or a similar solvent was used.

In order to carry out partition chromatography on a microscale, CONSDEN, GORDON and MARTIN [*8*] (1944) started using filter paper, and thus, an "open" column. Their work in the field of amino acid analysis met with considerable response and their method was generally adopted; by 1956 over 10,000 papers had been published dealing

[1] In the last decade this method has been developed mainly by MOTTIER [*40, 41*] and has been in recent use in the countries of Eastern Europe [*10, 11, 39, 46*].

with the applications of this "Universal Method" [*15*]. It was natural that researchers, impressed by the results obtained, repeatedly attempted to eliminate the difficulties that still arose by altering the degree of impregnation, employing new solvent combinations, reversing phases, and by changing the chemical structure of the cellulose fiber. Glass fiber paper was also expected to be of great help in eliminating adsorption effects.

For separating lipophilic mixtures, attempts were made using adsorption chromatography on impregnated filter paper, and later glass fiber paper coated with silicic acid or alumina. KIRCHNER in 1950, was one of the first to do this [*25*]. It is not clear, however, why in collaboration with MILLER he later continued work done by MEINHARD and HALL [*34*] in 1948, who had called the technique developed by IZMAILOV and SHRAIBER, "surface chromatography", occurs. Perhaps the reason is that results using silica gel impregnated paper were already proving insufficient. KIRCHNER and MILLER proceeded to investigate methods of separating terpene derivatives on thin adsorbent plates; and published the results in several papers [*26, 36, 37, 38*]. They introduced their method under the name "Chromatostrip Technique" for use in separating terpene derivatives, and quite a large number of publications appeared subsequently. One of the authors (REITSEMA 1954 [*50, 51*]) later used plates (12.5 × 17.8 cm.) instead of narrow glass strips, and was thus able, as in paper chromatography, to separate several mixtures at a time. It is astonishing, that almost all authors abandoned the method, although this may be due to the fact that results were least satisfactory when separating essential oils, and considerable variation of *Rf* values occured because factors influencing them where yet unknown. It may also be that gas chromatography was considered to be the most suitable method at this period. Whatever the reason, however, the universal applicability and numerous additional advantages of surface chromatography remained unexplored. The clearest indication of this is apparent from an analysis of the investigations published annually on the subject.

Tswett's column chromatography has had a very similar history (Fig. 2). More than 20 years elapsed before RICHARD KUHN's group recognized the significance of this technique. Paper chromatography underwent an even longer "incubation period"; although a book by GOPPELSROEDER [*14*] appeared as early as 1906 with the title: "Suggestions for the study of capillary analysis based on capillarity and adsorption phenomena", it was only half a century later that a practicable technique was developed by CONSDEN, GORDON, MARTIN and SYNGE.

Ten years ago, I was occupied by the problem of separating the contents of plant cells, and was attempting to find a suitable method. As the capacity of the plant glandular fibres under investigation was less than 0.1 μg, I was soon forced to realize that the results I desired could not be achieved by any of the then current chromatographic techniques. Microscopic examination showed, in fact, that the glandular fibres were smaller than the individual particles of the common adsorbents and the cellulose fibers of the paper. I then changed to "open" columns of fine-grained material and used the well-known magnesia rods and grooves. Results were very good, but adsorption activity was too limited. GÄNSHIRT, who was in my group at that time, started working on the "chromatostrip" method, using it for separating essential oils. This technique was of no immediate help with my problems, and further tests with pressed plates of silica gel, roughened and sintered glass, clay plates and anodally oxidized aluminum foil were only partially successful. Most suitable were thin layers of very fine-grained silica gel. I thus gradually got to know the influencing factors and applicability of the process, and in 1956 I reported my findings in a fairly short publication [*55*]. This work too, remained ignored as that of my predecessors. The position at the time can best be summed up by the opinion that the method was "one of the usual chromatographic gimmicks".

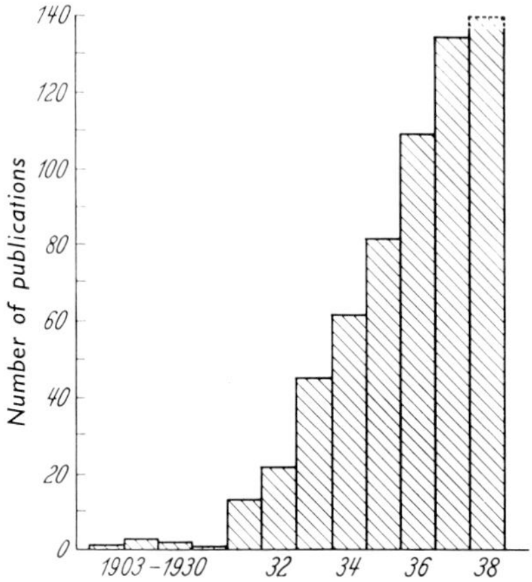

Fig. 2. Number of publications appearing annually between 1903 and 1938 in which column chromatography is discussed

As I was convinced this was not so, I tried to find reasons which might hinder its general application. The following seemed of importance:

1. Standardization of the method (thickness of layer, dimensions, method of production, separation chamber).

2. Design of suitable auxiliary equipment to form a basic kit.

3. The development of a multi-purpose adsorbent of constant composition which should be commercially available.

4. Determination of the applicability of the method.

The basic equipment for use in thin-layer chromatography, which was evolved from these inquiries, was presented to ACHEMA[1] in 1958 [*56*]. A great amount of interest was shown by industrial laboratories, particularly in the short separation time and the excellent separation effect of thin silica gel/gypsum plates. Thin-layer chromatography became generally known and was widely employed. In 1961/62 the first general surveys appeared, reviewing results obtained up to that time [*2, 12, 21, 42, 45, 54, 59, 61, 67, 69, 73*].

[1] ACHEMA: European Convention of Chemical Engineering, helt at Frankfurt-M., Germany

B. Instruments used in Thin-Layer Chromatography and their Operation

By

EGON STAHL

I. The application of thin-layers to carrier plates

There are various ways of applying thin-layers of powdered solids or their suspensions to carrier plates. Similar techniques have been known for a long time in the film and varnish industries, for example, and the different methods can be classed according to the method of application.

A. *Spreading*. 1. Using a mobile applicator and a stationary plate; 2. Using a stationary applicator and pushing or pulling the plate through accordingly.

B. *Pouring* of suspension or shaking of solids on to the plate, and then smoothing down.

C. *Dipping* of plates in a suspension.

D. *Spraying* with a thin suspension (spray technique).

At first method B was used [*20*], and it is still occasionally employed today as it requires no instruments. It is not possible, however, to obtain thin and uniform layers by this method. This is also true of methods C and D. In 1954 MILLER and KIRCHNER [*38*] described a fixed, funnel-shaped applicator (method A. 2.) through which narrow glass strips were pushed. The mixture flows through an adjustable outlet slit on to the strip being passed through. Their report included directions for construction, sketches, and a diagram. A modified design of this apparatus was used in 1961 for producing "chromatostrips" [*1*].

In 1961 WOLLISH and co-workers [*73*] described a device based on the same principle (A. 2.). The thickness of the layer can be adjusted by two "touching-screws". If several plates of varying thickness, but of the same width, have to be spread in a series, the plates should be placed in descending order of thickness. This apparatus requires that two plates of equal thickness be spread side by side; otherwise layers of uneven thickness will be produced. It is possible to obtain uniform layers by spreading glass strips of unequal thickness with an automatic thin-layer spreader described in 1958 [*56*] (Fig. 3). The plates are pressed against the applicator and transported by a small conveyer belt. The spreader is driven by an electric motor and is capable of applying a uniform layer to 1,000 and more plates per hour. Experience has shown, however, that an output of this order is unnecessary; it is normally sufficient to set up a row of plates of equal thickness and to coat them with the thin-layer spreader in one run. The apparatus, first described in 1956, and using method A. 1., is, in any case, only suitable for plates up to 5 cm. wide. Further development, however, produced a spreader for plates 20 cm. wide[1]. In the early stages the outlet slit was not adjustable, as every worker was then using plates of differing thickness; and it was necessary, when introducing the method, to begin with layers of a standard thickness.

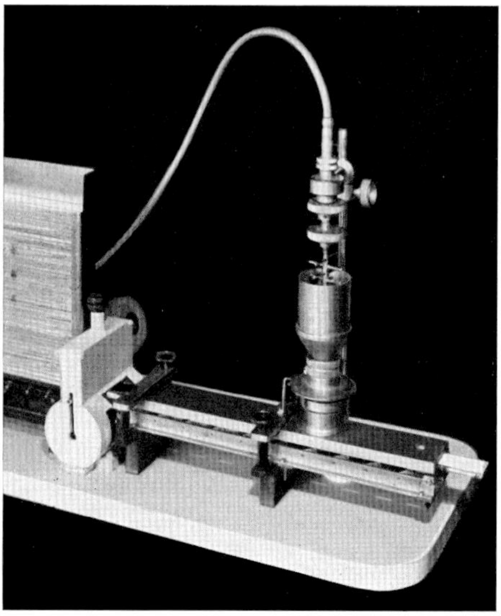

Fig. 3. Automatic coating device. On the left the magazine with glass plates, and under it the conveyer belt and drive. To the right the template with apparatus for accurate regulation of layer thickness. The coating mixture is fed through the funnel, and sedimentation is prevented by agitation [56]

After countless experiments, a layer thickness roughly corresponding to normal chromatographic paper, i.e. 150—250 μ, was accepted as most suitable for analytic purposes, and this thickness is now generally preferred. On occasion, as for the micro-preparation and isolation of single components, thicker layers are required. For this purpose a DESAGA thin-layer spreader was produced in 1962 with an outlet slit of max. 2 mm.; the layer thickness can easily be adjusted on this apparatus (Fig. 5).

With normal spreading mixtures, layers should not exceed 1 mm in thickness. In the author's laboratory layer thicknesses of 500 and 750 μ are employed for micropreparative work, with plates 40 cm wide by 20 cm. The mixtures (up to 500 mg) are applied as bands.

Thin-layer spreader

The apparatus developed by Stahl for producing thin adsorbent layers has been generally successful, and is, therefore, described in further detail. The unit consists of two parts: the aligning tray, on which the plates are set out in a line; and the *spreader*. This takes up the spreading mixture and applies it uniformly in a thin layer. The method of operation is shown in Fig. 4.

The spreader, which weighs nearly 2 kg., consists of a chromium-plated brass block into which is set a tube slotted on one side. This tube holds

[1] Unit forming part of the basic equipment for use in thin-layer chromatography. DESAGA, Hauptstrasse 60, Heidelberg, Germany. U. S. representative: C. A. Brinkmann, Cantiague Rd., Westbury, Long Island, N. Y.

the slurry, and can be revolved 180 degrees by means of a lever. This mechanism enables the adsorbent and water to be mixed directly and it prevents premature escape of the slurry. The unit has a capacity of approximately 150 ml and can, therefore, coat at least 10 plates (20 × 20 cm., thickness 250 μ).

Method of operation

a) Aligning tray. The aligning tray, 112 cm long, 22 cm wide, made of colored plastic, is set on a firm table with the long retaining edge nearest the operator. The short retaining edge is thus on the right-hand side. Foam-rubber strips on the bottom of the tray prevent its slipping. The rectangular plates, all of equal thickness, are then placed on the tray: either five 20 × 20 cm glass plates, ten 10 × 20 cm, or two large plates, 40 × 20 cm. The plates, which must be clean and free of grease, should lie flush to each other and to the long retaining edge. One 5 × 20 cm plate should then be placed at the left and right-hand end of the line of plates (starting and finishing plates).

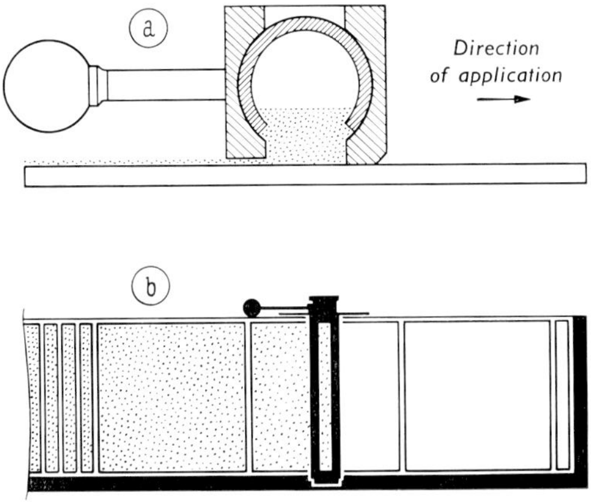

Fig. 4. a) Operation of thin-layer spreader. b) aligning tray with glass slide partially coated. (dots = adsorbent)

b) Positioning of the spreader. The spreader is then placed empty on the first plate. The red engraved arrow is now pointing to the right, in the direction

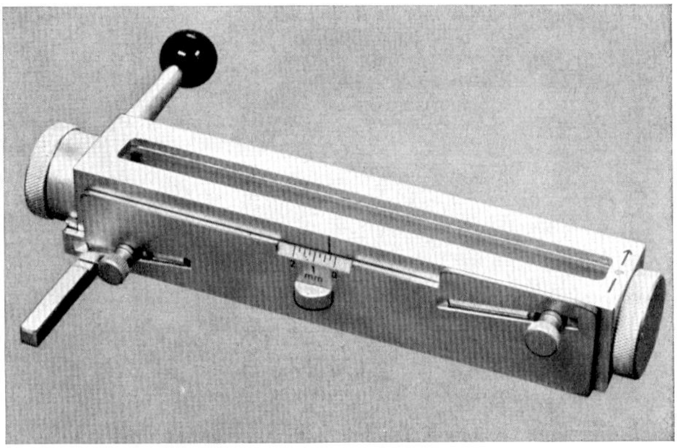

Fig. 5. TLC spreader with layer thickness adjustable from 0—2 mm. (photograph by DESAGA, Heidelberg)

of spreading. The guide bar of the spreader is flush to the aligning tray and can be moved effortlessly over the row of plates. If the operator is a novice, he should practice the movement of the spreader several times; this should be even, and neither too fast nor too slow. It is best to stand at the middle of the aligning tray, to grip the spreader with both hands, and to move it from the starting to the finishing plate without stopping and without applying pressure. The procedure for a line of plates 1 m long should take about 5—6 sec. This kind of preliminary test will also tell whether the plates in the row are all of the same thickness; there should, of course, be no catching as the spreader slides over them.

Fig. 6. Simple and clean coating of 5 plates in the laboratory. To the left, separation tank and chromatogram and behind, drying rack of light alloy (photograph by Merck AG, Darmstadt)

c) Adjustment of layer thickness. With the new DESAGA thin-layer spreader the problem of adjusting layer thickness is simple and readily solved. The two regulating screws on the face of the spreader are loosened, and the front plate can be raised or lowered by moving to right or left. The fixed mark indicates the layer thickness on the moving scale. Maximum depth of the outlet slit is 2 mm, which thus makes it possible to spread coarse-grained adsorbents of the type used in column chromatography in loose, but even, layers.

d) Filling of spreader and application. For filling, the lever arm must point in the direction of the red arrow engraved on the top of the spreader.

Water and adsorbent are then poured in, the jacket is closed by rotating the lever 90 degrees, and the contents mixed by shaking the whole unit. The mixture can also be made up free from grains and without air bubbles in a porcelain mortar, and then poured into the spreader. The spreader is then set on the 5 cm wide starting plate on the tray, and the lever rotated 180 degrees using the right hand. The left hand is used to press the spreader to the plate. The operator then waits for the mixture to come out through the outlet slit; while waiting, he may moisten the plate next to the spreader on the right with his wetted finger. This ensures smoother movement of the spreader, which is then drawn over the line of plates until the finishing plate is reached. The lever arm is now tipped back. If 10 plates (20 × 20 cm) are to be coated in one operation, a second aligning tray is necessary. In this case, a double amount of mixture is poured into the spreader, and when the finishing plate on the first row has been reached, it is lifted off together with the spreader itself and positioned as starting plate on a second aligning tray. The process then continues. The whole operation, from mixing adsorbent, containing gypsum, with water, to completion of spreading, should be finished inside 5 min.

e) Cleaning and storage of spreader. As soon as spreading is completed, the remainder of the mixture should be flushed out of the spreader by powerful water jet. The screw cover on the end of the tube is detached and the tube itself removed. A bottle brush of the right size will then clean the unit thoroughly. The rim of the outlet slit should be handled with particular care; it should be dead-smooth and without marks or scratches. The operator himself should, therefore, be responsible for cleaning and storing, preferably in a wooden box.

The same spreaders have been in use in my laboratory for several years and about 15,000 TLC-plates have been prepared without the spreaders showing any signs of wear.

It should finally be mentioned, that TLC-spreaders based on the same principle have been described in several recent reports [*3, 10, 11, 29, 46, 53*]. Their purpose has been to simplify the design, but they produce less uniform layers than the original spreader. Preliminary tests were carried out some years ago using similar apparatus, and these have led to the TLC-spreader as it is today.

II. Drying, storage, and handling of TLC-plates

The adsorbent is normally spread on the plates in the form of an aqueous suspension, and subsequently the water (6—8 ml per 20 × 20 cm. plate) must be removed. Water evaporation goes through the following stages:

1. First, the surface of the thin layer shines watery.
2. After several minutes this shine disappears and the layer becomes opaque.
3. When about 50% of the water has evaporated, the transparent effect fades and the layer becomes white.

Fig. 7 shows the evaporation of water during the drying of a Silica Gel G layer.

It is obvious that it saves time first to remove the main water deposit by an air jet (pre-drying) and then to complete the process by drying the plates at a higher temperature. For pre-drying, the plates, when coated, are left on the tray. The "Astron 2000" hot air drier is a multipurpose laboratory instrument and is suitable for producing a multiple-stage adjustable cold and hot air stream (Fig. 8)[1].

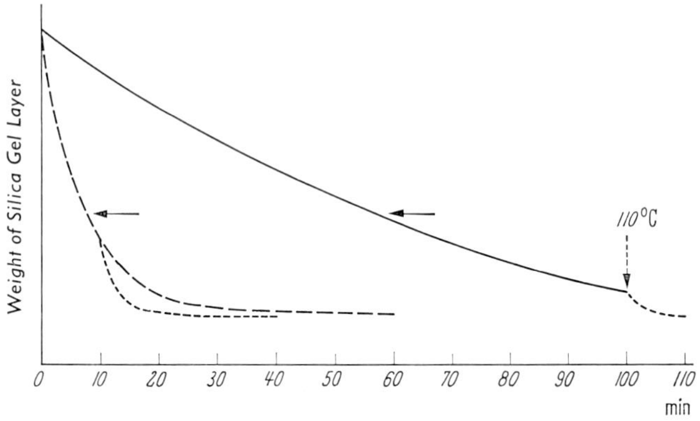

Fig. 7. Time taken by water evaporation from newly deposited Silica Gel G layers under differing conditions. —— = without drier; 22° C; ----- = with drier, 22° C; ····· = drying cabinet, 110° C. The two arrows indicate the point at which the plate is no longer transparent but white [63].

The drying is completed at a higher temperature. It is usually sufficient to heat the TLC-plates in a *vertical* position for 30 min. at 110° C. The most suitable units are rectangular circulating-air drying cabinets, at least 42 cm. wide, with an air-blower.

Fig. 8. "Astron 2000" hot air blower for quick predrying of layers (rear view). (Photograph by Sprenger KG).

To provide very active layers, silica gel and alumina plates should be heated to 150° C. for 3—4 h.

High activation is only worthwhile if application of samples is carried out in a correspondingly dry atmosphere, and if completely dry solvents are used. In thin-layer chromatography these conditions are seldom necessary, and only when dealing with hydrocarbon mixtures. With highly active adsorbents the danger of decomposition of substances is great.

Opinions as to suitable drying period and temperature differ somewhat. In TLC of polar compounds, e.g. amino acids, the plates are left to dry overnight at room temperature. Layers of this type have satisfactory adhesive strength and give better reproducible results than activated plates (see p. 394).

[1] A. Sprenger KG, St. Andreasberg, Harz, Germany.

As it is usual to coat 5 or 10 plates (20 × 20 cm.) in one operation, a compact and convenient drying rack of light alloy was evolved (Fig. 30). This is designed so that the plates can be inserted horizontally but left in a vertical position when the rack is tilted. This is of importance in assisting the escape of moisture.

Method of operation

After the layer has been applied, the plates remain on the tray; a suitable air blower (Fig. 8) is used for passing a current of hot or cold air over them. After about 5 min the plates, taken in turn, are inserted layer uppermost into the guide bars on the light-alloy drying rack. In this position the blower can be used for a few more minutes. The rack is then tilted so that the plates are vertical. The operator grips the handle with both hands and inserts the rack with plates into the drying cabinet, which has already been heated to 110° C. The door of the cabinet should be opened several times during the first ten minutes of drying to let the steam escape. It is even better to use a drying cabinet with air circulation and fresh air inlet. The hot rack is next removed with a cloth, cooled briefly (2—3 min), and tilted to the original position, when it is placed in a "Novus" vacuum desiccator (30 cm diameter) filled with indicating silica gel or phosphorus pentoxide. The cock of the desiccator should be left open for pressure equalization, and it is advisable to provide it with a drying tube.

The plates should always be protected against fumes in the laboratory. The desiccator will store up to 10 TLC-plates, while larger numbers can be kept available in air-tight portable cases (capacity 22 plates, 20 × 20 cm.), or in suitable cabinets.

III. Preparation of TLC-plates for use in chromatography

a) Checking dried thin-layers. Seen in transmitted light the layers should appear regular, without any streaks or rough patches. Under vertical light the layer surface should appear smooth and not show any coarse grains. Adhesive strength can be tested by lightly running the index finger over the layer, which should not become detached from the plate. A TLC-plate which has defects on the contact surface should not be chosen; however, such plates can be used for preliminary tests.

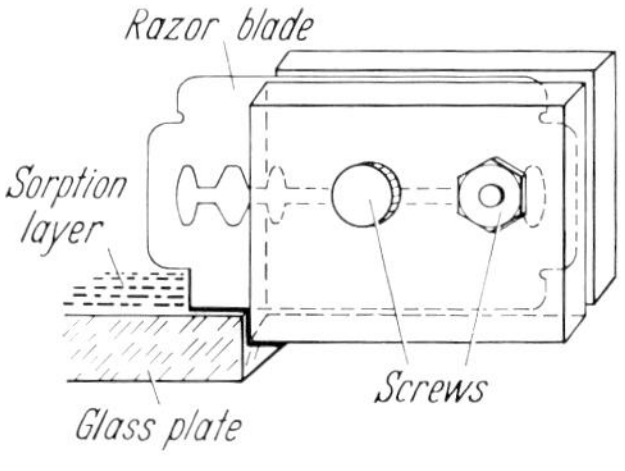

Fig. 9. Simple home-made edge stripper of flexible material, e.g., razor blade (stripping range 7 mm).

b) Stripping of thin-layer edge. The edge of the thin layer laid down by the contact surface of the spreader is normally quite thin, and it is best to remove it. It can be stripped off using the thumb of the right hand and the nail on the index finger as guide. Alternatively, the easily made edge stripper shown in Fig. 9 may be used.

c) Marking the layer. A sharp hard pencil or a dissection pin (from a biological dissecting set) can be used for this purpose. A flat bridge of transparent plastic, known as the template, for marking and sample application, is used to prevent damage to other parts of the layer. A scale is engraved on the top edge of the template to ensure proper marking of starting points at regular intervals. 15 mm from the edge there is engraved, a cross-line A and 100 mm beyond this, another line B.

Operation and use of the template for marking and sample application

The plastic template is laid like a bridge across the TLC-plate. The template is positioned so that cross-line A on it coincides with the bottom edge of the plate. Using the scale, the subsequent points of application are marked by small dots. The distance from the edge of the plate should be 20 mm and from point to point, 10—20 mm. The template is then moved to the second engraved cross-line. It is not advisable to draw a line across as "front line", but it is better to mark it with dots at regular intervals. At the same time, the substance to be separated or any particular remarks may be noted. It is also advisable to list solvents and spray reagents on each plate; the top third of each layer is normally available for this purpose.

d) Instruments for application of spots. Even when work is not quantitative, the use of graduated micro-pipettes is recommended; special pipettes of 10 mm³ capacity have proved particularly suitable for spot application. They are divided into ten sections with 1 mm³ between each graduation. Accuracy of calibration differs considerably from one manufacturing firm to the other, and for quantitative work checking of calibration is advisable. In general, 5 mm³ is as much as one applies at one time. Larger amounts should be applied in small portions. In the case of aqueous solutions intermittent drying should take place using a current of hot or cold air.

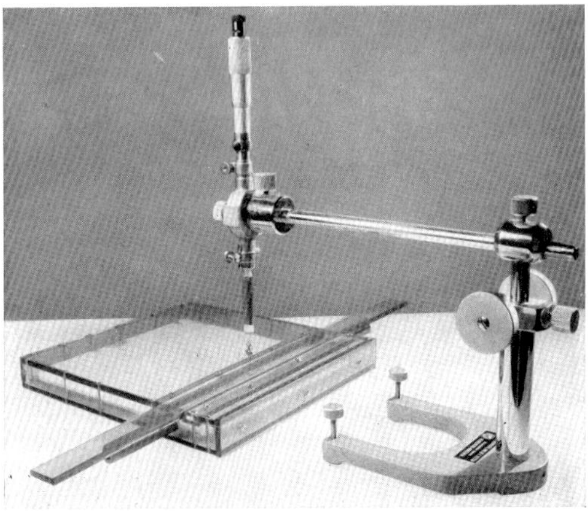

Fig. 10. Quantitative application of substances to a TLC-plate under inert gas and with UV-protection. To the left, DESAGA application tank[1] devised by HALPAAP [*16*]; right, Agla micro-syringe and support frame[2] (photograph by HALPAAP).

The easiest way of cleaning the pipette is as follows: connect pipette to the hose of the water-jet pump and flush through briefly with a suitable solvent, e. g. water → methanol → acetone, followed by air. In case of blockage, suck the impurity out through the head of the pipette.

---

[1] DESAGA, Hauptstrasse 60, Heidelberg, Germany. U. S. representative: C. A. Brinkmann, Cantiague Rd., Westbury, Long Island, N. Y.

[2] Burroughs Wellcome & Co., London, Great Britain.

Very small quantities of fluid can be applied using micrometer syringes, the best-known model being the English Agla micro-syringe[1]. In U.S.A., Hamilton micro-syringes[2] are widely used. The "micrometer dosage instrument" No. 550[1] operates on the same principle. An additional support frame is advisable (Fig. 10).

Note. Only transparent solutions and not "crystal suspensions" should be applied, as crystals dissolve very slowly in solvent, thus sometimes causing too marked a tail formation. The same danger arises from crystallization of compounds at the starting point during application.

e) Streak application of larger quantities of mixture for micro-preparative thin-layer chromatography. Up to the present "streak chromatograms" have been obtained by applying the dissolved mixture on the

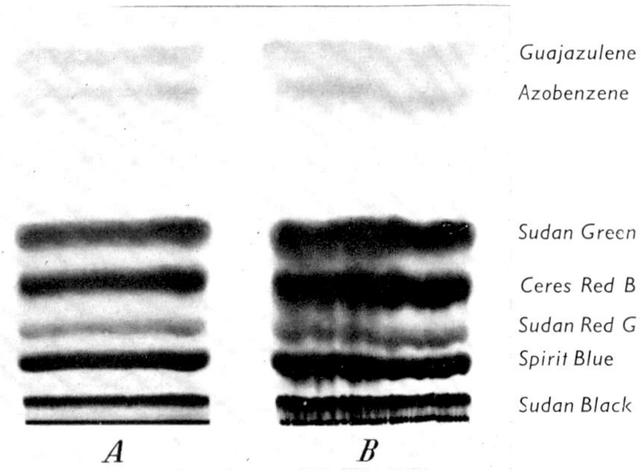

Fig. 11. Comparison of separation achieved after using different methods of application for "band chromatograms": *A* is sprayed and *B* spotted.

starting line in very close-set spots. This method is time-consuming and demands a certain skill; the layer is often damaged and, in any case, starting bands of this kind are too wide. After chromatography, the zones are correspondingly wide and often unequal in shape. This makes accurate scraping of the required substances difficult and sometimes impossible.

These difficulties can easily be solved by spraying the solution on in streak formation using a specially designed instrument [52]. In my laboratory a "micro spray gun"[3] using jets of 0.1—0.25 mm diameter has been used for this purpose. It is advisable to use an inert gas, such as carbon dioxide or nitrogen, for the spray-gun, and not compressed air. The required pressure must be accurately set beforehand and should be sufficient to vaporize the fluid; if too high, it will blow off the layer of adsorbent.

[1] See footnote 2, p. 12.
[2] Hamilton Company, Inc., P. O. Box 307, Whittier, California, U.S.A.
[3] See footnote 1, p. 12.

Method of operation

The TLC-plate is covered with two glass plates of requisite size leaving a starting band about 2 mm wide. The solution is then fed in through the aperture in the spray-gun (capacity 1 ml) and sprayed from a range of about 1 cm on to the uncovered starting band. The covering plates are then removed and chromatography follows [*63*].

With this method it is also possible to deactivate the starting band by first spraying it with water; this prevents decomposition on active adsorbents.

IV. Separation chambers and conditions of saturation

The replacement of column chromatography by "open" separation plates led to the development of several types of chambers. The success of separation, and consequently, the *Rf*-values are greatly influenced by the type and size of chamber used. This factor is often neglected and deserves more detailed comment.

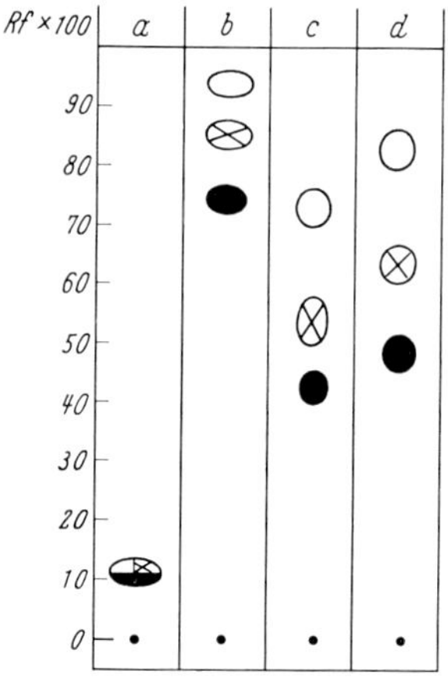

Fig. 12. The influence of different tanks on *Rf*-values of the three-color test mixture. *a* without tank, *b* trough tank with insufficient saturation, *c* the same with saturation, and *d* S-chamber. ○ = Butter Yellow, ⊕ = Sudan Red G, ● = Indophenol

Systems in general use were first similar to tanks used for paper chromatography and differed only in shape. They can be classed along the lines of separation methods then in use into:

1. Tanks for ascending development,
2. Tanks for descending development,
3. Tanks for horizontal development,
4. Tanks for thin-layer electrophoresis (thin-layer ionophoresis).

Glass, or occasionally, stainless steel, is most suitable for methods 1—3. Plastic tanks can only be used in special cases [*4*], as they are not resistant to all organic solutions. In tanks of the kind considered so far the *saturation condition of the tank atmosphere* is of particular importance [*60*]. The following four tests indicate the details in question.

The three-color test mixture (see p. 28) is applied at a distance of 1 cm from the bottom to four standard thin layer plates (10 × 20 cm), coated with Silica Gel G.

a) The first plate is placed in a flat basin filled to 0.5 cm with methylene chloride.

b) The test is repeated in the DESAGA trough tank (21 × 21 × 9 cm) using the next plate; once without special saturation of the tank but with the cover closed, and the second time,

c) with lining of filter paper and preliminary shaking.

d) The last plate is used in the S-chamber (Fig. 16).

Fig. 12 summarizes the results of these tests.

It is clear that, in the tests described, the extent of migration of substances depends on the rate of flow of the solvent. In test a) there is so much evaporation that the methylene chloride front rises only about 1 cm. As it runs steadily each dye soon migrates into the area of the solvent front. In test *b*) (Fig. 12) this evaporation effect is less marked but it takes the solvent 60 mins. to migrate across the 10 cm path.

In the S-chamber (*d*) (Fig. 12) where the volume is smaller, the evaporation effect is even more strongly suppressed; the time of run is only 21 mins, and as the rate of flow is less, the dyes have migrated correspondingly less far. There is, therefore, a characteristic value in the case of a usual tank for the relationship between the evaporation surface (cm^2) and tank volume (cm^3).

Type of chamber	Evaporation surface after 10 cm run	Volume of chamber for plates 20/20 cm
No tank (test a)	1	∞
Normal tank (test b)	1	~ 20
S-chamber (test d)	1	~ 0.5
BN-chamber	1	~ 0.1
Closed column	1	0

The method of chamber saturation (c) is not taken into consideration in this table.

1. Saturation and edge effects

When we turn from thin layers 1—2 cm wide to layers of standard size (10 × 20 and 20 × 20 cm), we occasionally find that the same substances migrate higher at the edge of TLC-plates than in the center (Fig. 13). This undesirable edge effect occurs in adsorption-TLC in standard tanks, particularly in the case of solvents composed of fairly strongly polar constituents. As Stahl has shown, this effect is caused by insufficient tank saturation and by the fact that the glass plate forms a partition in the tank. This, in turn, causes the solvent to evaporate more quickly at the edge, and the rear part of the tank to become saturated. It is also probable that in the case of mixtures such as hexane-ethyl acetate or chloroform-methanol, an increase in concentration to the more polar solvent occurs in the edge zone as a result of this evaporation effect. In the former mixture the weakly polar hexane is less strongly adsorbed and, therefore, evaporates more readily than the ethyl acetate which is more polar. In addition, the different vapor pressures of the solvents must be taken into account.

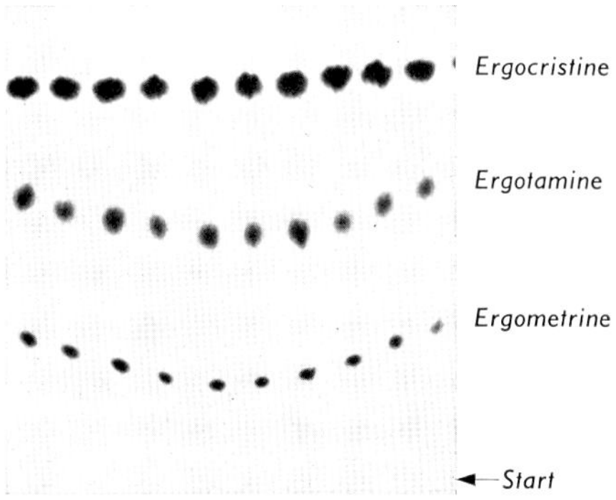

Fig. 13. "Edge effect" in TLC using a trough tank with insufficiently saturated atmosphere. Solvent: Chloroform + 5% Methanol [*60*]

It should also be pointed out that different *Rf* values for the same TLC-plate can be caused by fairly large variations in thickness in the thin layer itself. The effect of layer thickness on *Rf* values was first established in 1956, and is of particular importance in the case of thicknesses of less than 150 μ. More recently, BRENNER and co-workers have made a thorough study of the factors affecting *Rf* values [7].

This edge effect can be eliminated by uniform saturation of the tank with the solvent vapor before chromatography takes place. In the case of normal tanks (characteristic 1 : 10—1 : 50) uniform saturation may be achieved as follows:

Method of operation

A smooth (resin and grease-free) filter paper approx. 15 × 40 cm is placed in a U-shape into the glass trough and soaked in the solvent with developing solvent. The moistened paper is then pressed to the chamber wall, to which it usually adheres. It can, however, be fixed more securely by means of a flexible ring of thin steel wire which can be introduced into the chamber and pressed on to the paper. Before the plate is set in position the tank is tilted and the filter paper soaked in solvent again.

To differentiate this technique from that of normal saturation, it is known as chamber saturation (earlier called "chamber over-saturation"). Chamber saturation does not, however correspond to the equilibration which is often specified in paper partition chromatography.

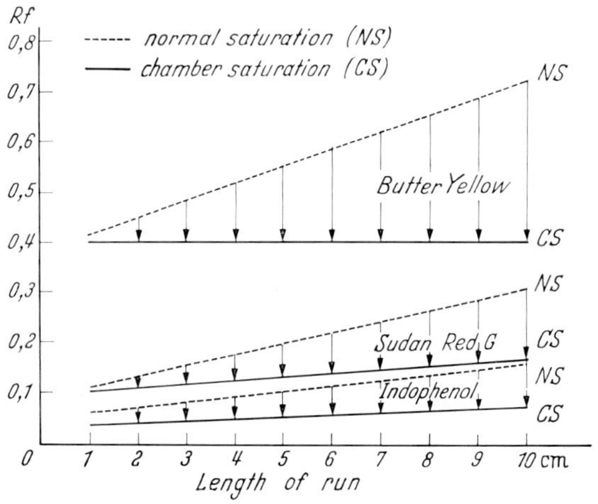

Fig. 14. Comparison of *Rf* values of test mixture using chamber with (CS) and without (NS) chamber saturation, in terms of the elevation of the hight front (Silica Gel G layer; Solvent: benzene)

It is only in partition-TLC and reversed-phase partition-TLC that chamber saturation corresponds to any great extent to "equilibration" as it is known in paper chromatography. Here, mutual saturation of the phases is called for.

Tests b and c, described above, indicate that, when working with tanks which are saturated, the time of migration is considerably shortened.

Study of the relation between *Rf* values and the length of run shows that there is a close connection in conditions of normal saturation between *Rf* values and the length of run (Fig. 14). It is only valid to refer to *Rf* values when working with chamber saturation or when using S- and BN Chambers as specially designed for thin-layer chromatography. The same length of run and variations from the 10 cm separation run (p. 41) should, however, always be specified. These observations show that the saturation of the chamber is of vital importance if satisfactory separation and comparable results are to be achieved.

2. Assembly of chambers (Temperature, light, protection against oxidation)

Work is usually carried out at room temperature (18—23° C). In adsorption chromatography, as distinct from partition chromatography, temperature has little effect on separation and its results. It is important, however, when setting up separation chambers to ensure against heating on one side only, as even small temperature variations inside the tank, due to light, sunlight, or a draught, can give rise to an undesired "oblique formation" of the solvent front.

In most cases work is done in diffuse daylight; sunlight should always be avoided. It is advisable to cover the inside of the window with anti-glare foil[1] which is impermeable to ultraviolet rays. A protective cover of black cardboard should be set in an inverted position over the chamber in work with substances sensitive to light. A black self-sealing plastic sheet (e.g. Decofix) may be stuck to the chamber. The solvent front can be observed by making a small inspection window in the plastic sheet about 12 cm above the bottom. Compounds that are extremely sensitive to light should be applied in red or green light.

Oxidizable compounds may decompose during application if they are left on the active layer without solvent. This can be avoided as follows: the TLC-plate is placed in a "developer bowl" which, in turn, is covered with a glass sheet. This sheet of glass is perforated and through the hole a tube is passed and carbon dioxide or nitrogen fed through. In order to apply the substance, the cover sheet is pushed back just enough to enable the starting point to be reached by the pipette. Some degree of protection against autoxidation can be obtained by previously moistening or spraying the starting zone with water (p. 14).

Halpaap [*16*] has designed a container of plastic which affords protection against ultra-violet rays and allows for the entry of a gas (Fig. 10) during application of the sample to the plate. Protection against oxidation during chromatography is best achieved by filling the tank with carbon dioxide. If a basic environment exists, however, (alumina, basic solvent), the chamber is first flushed with nitrogen and its cover only opened briefly in order to insert the TLC-plate.

[1] Fabriek van chemische Producten Vondelingenplaat N. V., Holland.

3. Chambers for ascending development

a) Rectangular trough chamber

In rectangular glass troughs in normal commercial use the edges are generally unevenly ground, which makes a tight seal impossible. These and other disadvantages are, however, eliminated in the specially designed DESAGA chamber ($21 \times 21 \times 9$ cm) which has a wide protruding edge and plain ground all-glass cover. The volume of this tank is approx. 4 liters.

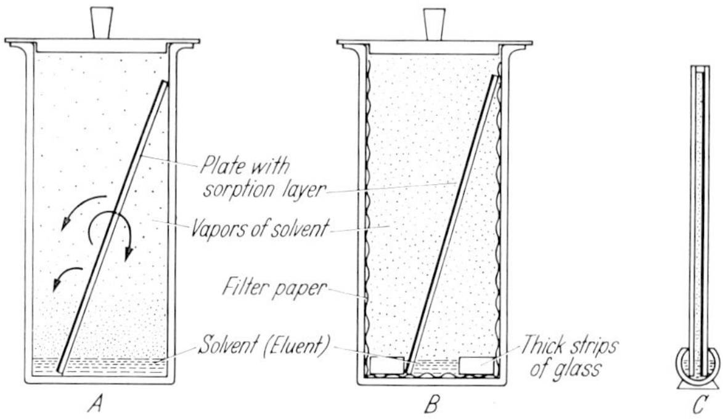

Fig. 15. Separation and saturation. *A* trough with normal saturation (= NS). The arrows pointing away from the layer of adsorbent represent the evaporation of the solvent and the dots the vapor density. *B* trough saturated with solvent by means of filter paper lining. *C* The chamber volume reduced using the S-chamber system

Method of operation

The chamber is first thoroughly cleaned and dried and then 10—100 ml of the solvent to be used are added. The chamber is saturated using a lining of filter paper, as described on page 16. The amount of solvent used can be reduced by placing two glass strips (0.5—0.8 cm thick, 20 cm long, and 2.5 cm wide) at the bottom of the chamber (Fig. 15B). Care should be taken to ensure that the TLC-plate is dipped evenly into the solvent when it is placed in the tank. Two plates (20×20 cm) can be inserted in a V, while, by means of a holding frame of stainless steel, up to five plates can be accommodated in a single chamber. The distance between plates should, however, be at least 0.5 cm. With this layout, it is important to fix a piece of filter paper to the uncoated back side of the plate so that it contacts the developing solvent. Brenner and Niederwieser [*6*] suggest a simple clamp-type holder for keeping five plates together; this can easily be made on the spot by bending a thin glass rod into wave shape. Round upright cylinders with ground glass stopper or preserving tubes of the right size may be used for plates 5 and 10 cm wide.

b) S-Chamber system [63]

In serial tests, with particular emphasis on quantitative comparison, as well as for micro-preparative work and evaluation of the *Rf* values of 20–35 substances, plates 40 cm wide and 20 cm high have proved suitable. Chambers adapted to this dimension of plate are not, however, available through commercial channels.

The volume of the chamber can most easily be reduced, by using the coated plate as the rear wall to the tank. A cover plate (frame plate)

of the same size, with glass strips 0.5 cm wide and 0.3 cm thick sintered along three edges, acts as the front wall (Fig. 15C and 16). Both plates are held together by two clamps. This S-chamber is placed in a trough filled with solvent.

The trough for the S-chamber consists of a double jacket of stainless steel. The outer jacket contains slits at varying distances which enables plates of differing width to be inserted and ensures a tight seal to the chamber. The "20 cm S-Chamber Trough" is designed for plates 10 and 20 cm wide, and the "40 cm S-Chamber Trough" for plates 20 and 40 cm wide. One cover plate (frame plate) is required for each plate dimension.

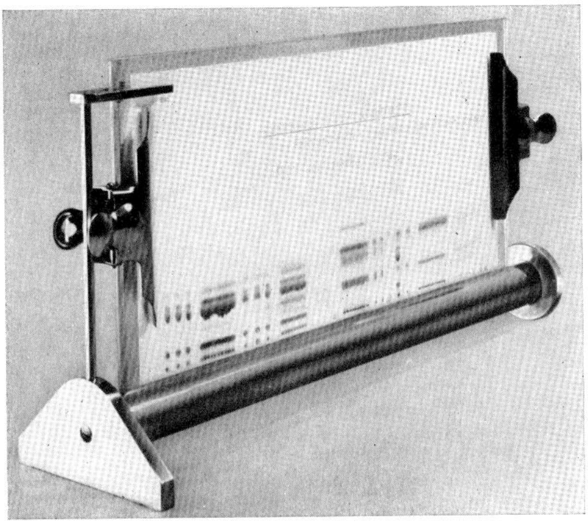

Fig. 16. S-chamber with a 40 cm. wide thin-layer chromatogram [*63*]

With the S-chamber system and the BN-chamber, saturation is easily achieved. The saving in solvent is also considerable; 15 ml are needed for the 20 cm S-chamber, and 30 ml for the 40 cm S-chamber. It should also be mentioned that with these chambers the separation process can be filmed without any distortion.

Method of operation

Before application, a 7—10 mm strip of adsorbent should be removed from each side of the plate. This can be done using a hand stripper (Fig. 9). This device can easily be made from thin spring steel (thick razor blade) or from some flexible plastic material. The edge must be free of graining or other irregularities as a close fit is essential. When the substances have been applied, the frame plate is placed on top, and fastened by two wide clamps at left and right. The chamber thus created is then placed in the trough (S-chamber trough) filled with the solvent. The chamber is kept upright by a fork-type holding arm inserted on the trough. The trough slit is narrowed by gentle rotation and withdrawal of the outer jacket, thus sealing the glass tank in the trough satisfactorily.

For cleaning the S-chamber the outer jacket is removed, whereupon the solvent can be easily tipped out.

4. Equipment for descending development

This technique has no advantages over the ascending method, described above, as far as separation and migration speed in TLC are concerned. There are, in fact, certain difficulties in providing suitable equipment.

Stanley and Vannier [*66*] have described a device for narrow TLC-strips ("Chromatostrips"), details of which are shown in Fig. 17. An attachable "rucksack tank" with a tongue of filter paper offers a possibility of using the wider TLC-plates. The solvent in the trough migrates through the tongue of filter paper onto the plate. The rate of flow can be regulated by varying the strength and type of filter paper. In order to avoid evaporation of the solvent, the whole system must be enclosed in a tight-fitting glass trough. During printing another system for use in descending development has been described [*4*]. "Continuous chromatograms" are also possible using the specially-designed BN-chamber described on pages 22—24.

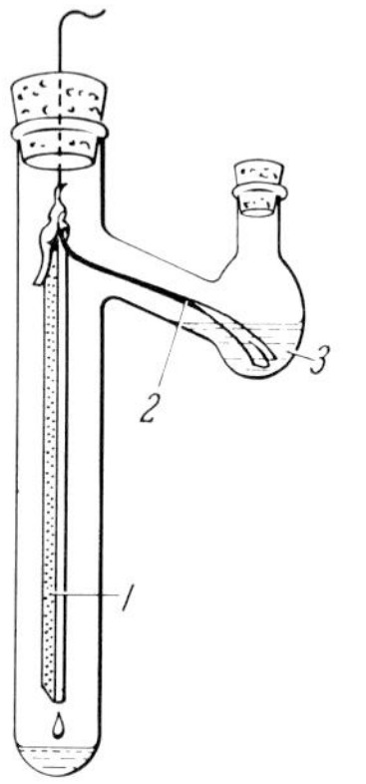

Fig. 17. Large test tube with side arm for descending development of narrow TLC-strips (STANLEY and VANNIER)

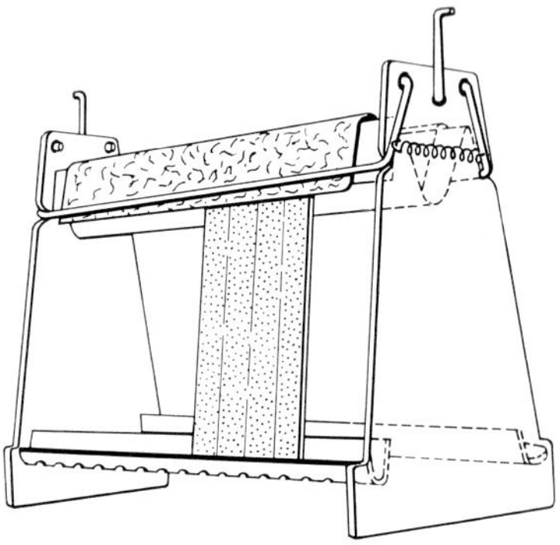
Fig. 18. Frame for preliminary washing of TLC-strips and plates. The eluent is in the trough (top) and is fed to the adsorbent layer via a filter paper. The frame presses the paper on to the plate. (STANLEY and co-workers)

This process is of value for the eluting of zones in the "continuous process". The device described by STANLEY, VANNIER and GENTILI [*65*] for preliminary washing of adsorbent plates is used for this purpose (Fig. 18). The objective is achieved more quickly, however, by scraping off the zones in question and then eluting them.

5. Equipment for horizontal development

While all the techniques described hitherto can be carried out with a layer of adsorbent that adheres to the plate, only development in a more or less horizontal position is suitable for thin layers spread loose on the plates. After the eluent has been fed in and the application technique established the following processes may be employed:

	Description	Method of applying solvent	Zones after separation
a	Circular technique (= ring chromatography — "round filter technique")	in spots (central)	ring-shaped
b	Horizontal technique (closed or "open" tank)	in streaks	spots or bands, depending on method of application
c	Flow technique	in streaks	spots or bands, depending on method of application

a) Circular technique

In its simplest form ring chromatography was already being used by IZMAILOV and SHRAIBER in 1938. They placed a drop of the solution they were investigating on the thin-layer and gradually added the solvent in drop form in the center. In this way the first ring chromatograms were produced (Fig. 1). MEINHARD and HALL[1] used the same technique, also without a tank. Their self-standing pipette for the solvent appears very suitable for this purpose (Fig. 19). The pipette is placed in the center of the spot of the

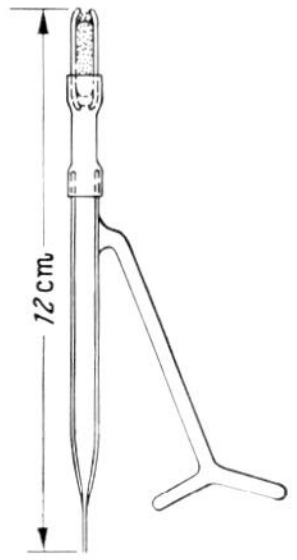

Fig. 19. Self-supporting pipette used for developing ring chromatograms (MEINHARD and HALL [34])

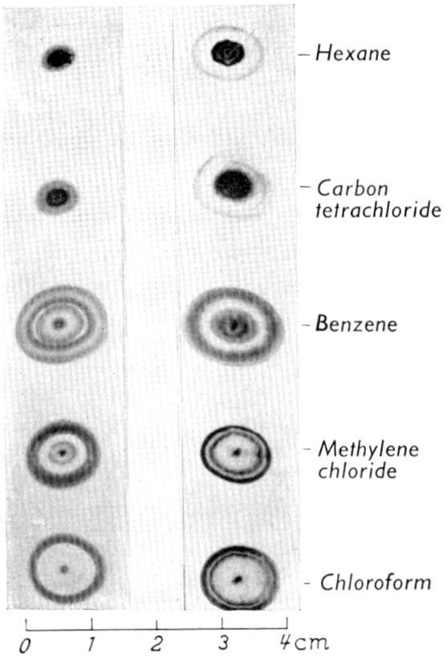

Fig. 20. Chromatograms obtained by the micro-circular method of two different dyestuff mixtures. Test to determine the most suitable solvent [56]

substance which has previously been applied in drop form. The micro-circular technique is eminently suited, as STAHL demonstrated in 1958, for rapid determination of a suitable solvent (Fig. 20).

Because of rapid evaporation, the ring chromatograms obtained are only 1—2 cm in diameter. With the circular technique, closed tanks, saturated with the solvent, should be used. Two easily assembled systems have proved their suitability for adhesive adsorbent plates (Fig. 21); both units (*A* and *B*) ensure satisfactory chamber saturation and are thus typical forerunners of the S- and BN-chambers.

Process A is characterized by the fact that the glass plate (1), with the layer side down (2), rests flush on a flat dish of suitable diameter filled with solvent (7). A

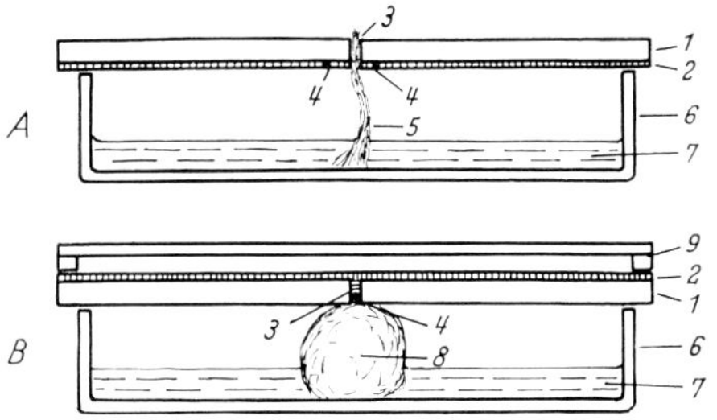

Fig. 21. Plan showing two layouts of equipment used for the circular method with adsorbent plates. *A* for separating up to 8 different mixtures. *B* with small pre-saturation column for mixture. *1* slide 20 × 20 cm; *2* layer of adsorbent; *3* hole 2 mm diameter; *4* point of application; *5* cotton wick; *6* cover to Petri dish; *7* solvent; *8* cotton wadding; *9* glass strips [57]

[1] LAGONI and WORTMANN [*27*, *28*] used a circular technique on loose layers.

piece of cotton (5) twisted into a wick acts as the lead from the solvent to the thin layer. The mixtures to be separated can be spotted around this hole at a distance of about 1 cm (4). It is thus possible to compare a series of substances on a single plate.

Process B, with its "pre-separation column," is more suited to the separation of larger amounts. The thin layer lies uppermost, and the solvent (7) rises past the wadding (8) through the pre-separation column (3) to the thin layer (2). The substance is previously applied to the uncoated side of the pre-separation column (4). It is advisable to cover the thin layer in order to prevent too strong an evaporation of the solvent as it spreads out in ring formation. A second 20 × 20 cm plate, without perforation, can be used for this purpose, with glass strips 5 mm wide and 3 mm thick stuck to its edge to form a kind of picture frame (9).

b) Horizontal method in closed tank

The process embodied in the circular method described above is only occasionally used in thin-layer chromatography. Although it is easy to execute and does provide distinctive separation, visual evaluation, i. e. comparison of zone size, is easier with "spot or streak chromatograms".

Fig. 22 offers a simple horizontal apparatus with linear migration of the solvent. The solvent is in a shallow layer in a glass "instrument dish". A thick strip or glass rod (1 cm diameter) tightly wrapped

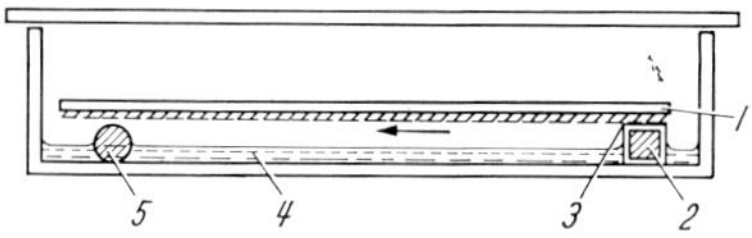

Fig. 22. Simple layout for horizontal method in a glass basin with cover.
1 TLC-plate, face down; *2* glass rod wrapped with filter paper; *3* point of application of the sample; *4* solvent; *5* support

several times in filter paper, acts as transfer for the solvent on to the thin layer (= bridge). A glass rod or tube serves as the second support. These and similar devices ensure satisfactory saturation. A layout similar in principle for use with loose layers has recently been described by MISTRYUKOV [*39*].

c) Continuous flow technique (BN-Chamber)

Substances which are not not resolved in a normal 10–18 cm run, may be separated by the following technique:

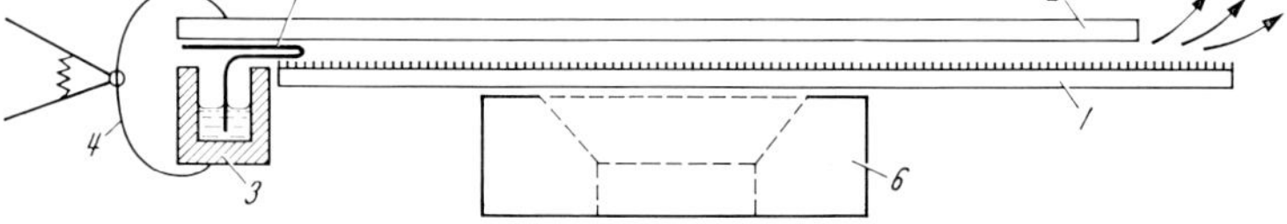

Fig. 23. Layout of BN-chamber, *1* plate with thin layer, *2* cover plate, *3* trough with solvent, *4* clamp (schematic representation) (Figs. 23 and 24 from BRENNER and NIEDERWEISER), *5* filter paper tongue and *6* cork ring as support for tank

The rate of flow can be increased quite simply by changing over after a certain length of run from a "closed tank" (characteristic 1 : 0.1)

to an "open tank" (1 : ∞). The solvent evaporates from the uncovered area and thus flows continuously (Fig. 23). The continuous flow method described by BRENNER and NIEDERWIESER [6] offers a simple solution.

In its layout the BN-chamber resembles the S-chamber used for the ascending method. A cover plate with two glass strips sintered on at the sides is required, and the edges of the carrier plate must be completely free of adsorbent up to a width of 6 mm. The plates are held together by strong clamps. The small trough, which is of stainless steel, has a plane-ground top and is used to hold the solvent. This solvent reaches the thin layer via a tongue of filter paper, migrates through the thin layer to the end of the bridge-shaped cover plate, and then evaporates so that a continuous chromatogram is obtained. Condensation of the solvent on the cover slide can occasionally be observed where there is a significant variation in room temperature. In this case the slide is warmed briefly by placing the hand on it.

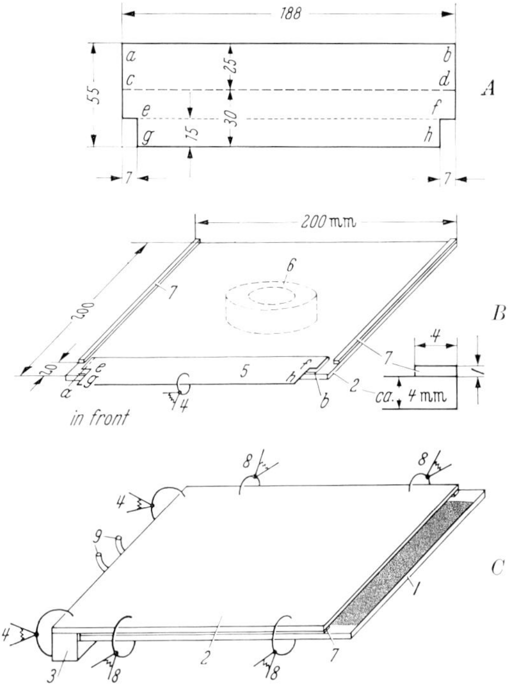

Fig. 24. BN-chamber. *A* dimensions of tongue of filter paper and fold lines. *B* tongue of filter paper set on cover slide. *C* Final layout. *7* sidefitted glass ledges, *8* side clamps (schematic representation), *9* tubes for pressure balance in trough. See also Fig. 23

With continuous chromatography it is advisable to apply an additional colored or fluorescent test mixture of similar properties for the purpose of checking progress.

Method of operation

The TLC-plate (20 × 20 cm) is prepared in the same way as for the S-chamber (see above). The resin-free filter paper (e.g. Whatman No. 1) is cut to the measurements given in Fig. 24*A*, folded once along line *cd* and placed with piece *abcd* on the front side of the cover plate (Fig. 24*B*). The paper tongue must not touch the ledges of the cover plate, and the distance between rim *a b* and the front edge of the slide should be only 1—3 mm (starting point 15 mm from edge). This position is normally fixed by means of a large paper clip (GK). The TLC-plate, with the substances already applied, is then placed on the ledges. The chromatoplate and the cover plate are held together by strong clamps at the sides. The paper clip used to fix the paper tongue is then removed and the paper strip folded (Fig. 24*B*, line *e f*) so that it can enter the trough at a later stage (Fig. 23). The trough unit is then tightly attached to the trough (Fig. 24*C*) and held together by means of two more clamps. The solvent is fed into the trough through a polyethylene tube (15 ± 5 ml). The tongue of paper can be moistened quickly and uniformly by tilting it gently. The BN-chamber is then placed on a cork ring (Fig. 24 *B*).

6. Apparatus for electrophoresis and ionophoresis in thin-layers of adsorbent

As early as 1946 CONSDEN, GORDON and MARTIN were using silica gel as stationary phase for separation of amino acids and peptides by electrophoresis. When the use of thin layers of fine grained adsorbents gave no further progress with specific problems, HONEGGER's method of thin-layer

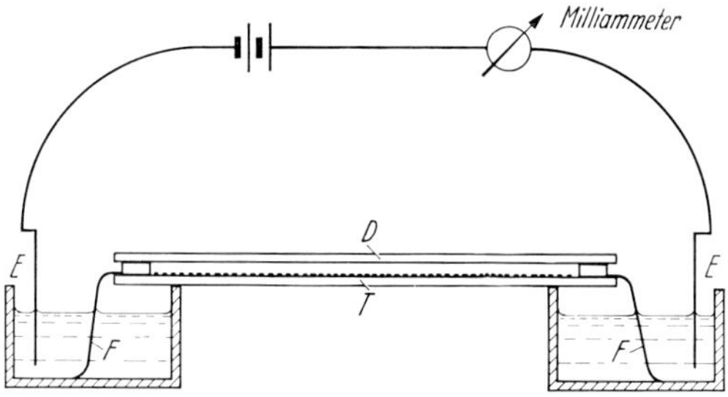

Fig. 25. Schematic representation of temporary equipment for use in thin-layer electrophoresis. *T* TLC-plate, *D* cover plate, *F* filter paper strips, and *E* electrodes. (PASTUSKA and TRINKS)

ionophoresis [*19*] and PASTUSKA and TRINK's technique of thin-layer electrophoresis [*43*, *44*] were adopted. It became evident that thin, inorganic fine-grained layers were superior to the cellulose paper which had previously been in general use (pp. 433—435). Fig. 25 shows a layout for thin-layer electrophoresis which, though not wholly satisfactory, is simple to construct (field intensity 20 V/cm). The layout described by HONEGGER would seem very suitable for higher loads. Thin layers (250 μ) are produced from one of the normal adsorbents. A buffer solution can be used instead of water when preparing the coating solution. The thin layers must be wetted evenly before use. Buffer solution can be sprayed on to the layer to form a transparent moist film. The starting points are marked on the dampened thin layers 5—10 cm from the edge nearest the anode, and the mixtures to be separated are applied (13—17). Fig. 26 shows a schematic representation of the layout described by HONEGGER.

Base M is a polished metal block 40 × 40 cm, cooled by an internal spiral system with water. It is insulated by means of a polyethylene foil wrapping about 0.2 mm thick. The wetted TLC-plate (T) is placed on the center of the block; to left and right glass plates of equal thickness (5 × 20 cm) are placed (G), and filter paper strips (= connecting strips U) 8 × 19 cm are saturated with a buffer solution and set on each of these. The paper strips must be of identical size, and should be pressed firmly on to strips G using a hand-roller. They should overlap the TLC-plate on both sides (anode and cathode) by 0.5—1 cm. Their other ends rest on the polyethylene foil. Flow between the electrode containers (E) and the connecting strip (U) is via two cellophane tubes containing gauze impregnated with the buffer (C). A plastic frame (which may have a cover pasted on) is attached to the cross strips (Q_1 10 mm, Q_2 7 mm wide) above the plates (UTU). These cross strips press on the "boundary points" and thus ensure satisfactory contact. The layout is covered with a glass plate measuring 35×25 cm.

HONEGGER separated amines and amino acids at 440 and 460 volts (19.6 and 12.6 mA) over periods of one hour. After the thin layers had been dried, further separation by either partition chromatography or adsorption chromatography could be carried out (see pp. 433—435), and the substances could be rendered visible in the usual way.

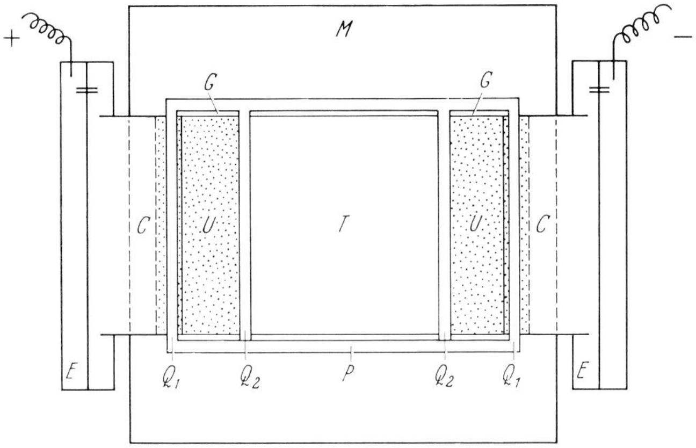

Fig. 26. Plan of HONEGGER'S ionophoresis equipment. Explanation in text

V. Spraying equipment and fume hoods

A particular advantage of pure inorganic layers is the opportunity they offer of using even the most corrosive spray reagents to visualize

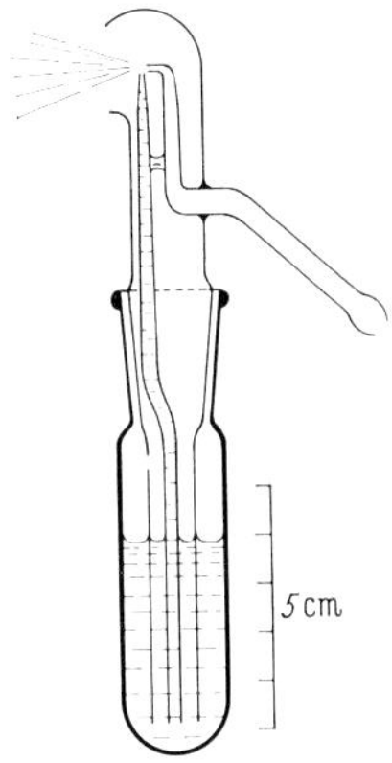

Fig. 27. Two-part DESAGA sprayer for atomizing of reagents by compressed air

Fig. 28. Diaphragm pump for producing oil-free compressed air, mounted on Multifix multi-purpose motor (Photograph by Schwinherr)

substances. With small spots, the reagent has to form a fine mist on the layer, but with the "throat sprayers" often used in paper chromatography, this is usually not possible. It is, therefore, advisable to employ compressed-air sprayers with fine nozzles. Of the many sprayers available a two-part sprayer[1] with container holding 10 ml has proved most satisfactory (Fig. 27). This permits one to keep a check on the amount of reagent sprayed.

If there is no compressed air connection available, an air cylinder with fine-adjustment valve can be used. Oil-free compressed air is best supplied by means of an efficient diaphragm pump[2] (Fig. 28). Other methods are described in a special chapter, pp. 483—502.

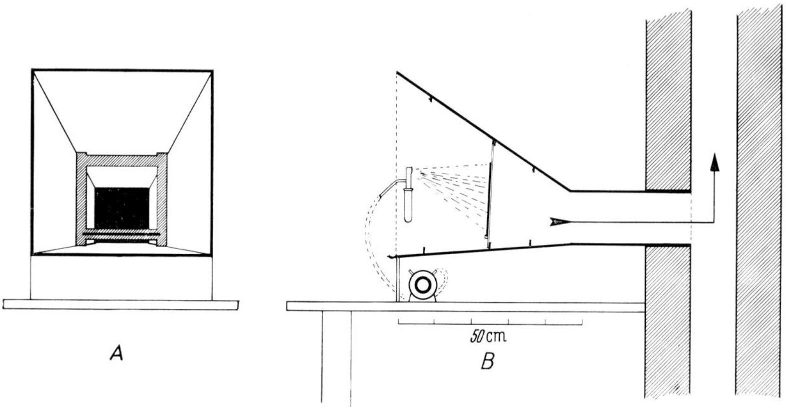

Fig. 29. Plastic spraying cabin connected to the ventilation flue. *A* front view (shaded section shows the frame for insertion of chromatograms). *B* Side view (Section); the sprayer for compressed air is also shown

Cleaning of sprayer. If reagents containing antimony chloride are used, the sprayer should be flushed out with chloroform or carbon tetrachloride and left in *aqua regia* (widemouthed bottle with ground-stopper) for several hours. Deposits can be removed by a 20% sodium hydroxide solution.

The spraying of corrosive reagents makes good ventilation imperative. Difficulties that may arise can be dealt with as follows:

a. If the available fume hood is large or draws badly, or if there are several openings leading into the flue, all except one opening is sealed off and a small welded spraying cabinet (Fig. 29) of acid-resistant plastic is connected to it. The cabinet contains a removeable frame of the same material on which the plate is set.

b. If there is no outlet available, a plastic tube (diameter approx. 25–30 cm) containing an acid-resistant and explosion-proof extractor should be provided and vented outside. The spraying cabinet depicted above should then be connected up.

[1] Manufacturer: DESAGA, Hauptstrasse 60, Heidelberg, Germany. U. S. representative: C. A. Brinkmann, Cantiague Rd., Westbury, Long Island, N. Y.

[2] Addition to the Multifix multi-purpose motor by A. Schwinherr, Schwäbisch Gmünd, Germany; max. pressure — 2.5 atmospheres, delivery — 1000 liters per hour.

VI. Standard conditions in thin-layer chromatography

Standardization has been of the greatest significance in the develop ment of thin-layer chromatography, establishing a basic procedure which should only be varied when absolutely necessary. The approved standards are as follows:

Dimensions of carrier plates:
200 × 200 mm; 100 × 200 mm (for preliminary tests)[1]; 400 × 200 mm (serial tests, micro-preparations).

Thickness of adsorbent layer (when spread, before drying)[2]: 250 μ for analytic separation; 500 μ and 750 μ for special cases and for micro-preparation.

Drying of layer:
Preliminary drying in air at room temperature, 10 min drying by hot air in a vertical position, 30 min at 110° C.

Storage:
Over blue silica gel and protected from laboratory fumes.

Starting points:
15 mm from the lower edge (distance between points 10—15 mm).

Length of run:
100 mm (start to front).

Separation Chambers:
a) Trough tank with plane-ground glass cover and tank saturation; b) S-chamber system; c) BN-chamber.

Depth of insertion of thin layer in solvent: approx. 5 mm.

Adsorbent:
Standardized adsorbents in accordance with directions (see pp. 29—34).

Test substances:
2—3 pure commercially available compounds are used; these should for preference be colored or fluorescent, and should, when separated satisfactorily lie in the *Rf* range 0.2—0.8, e. g. three-color test mixture (Butter Yellow, Sudan Red G, Indophenol).

The methods of operation given both above and below are shown in small print.

VII. Basic equipment for use in thin-layer chromatography

Taking into account the requirements of the modern laboratory, a list of basic equipment for use in thin-layer chromatography has been drawn

[1] The more recently recommended smaller dimensions, using specimen plates etc., are only suitable for certain separations.

[2] The thin layer contracts when it dries. The extent of shrinking depends upon grain size and the solvent. A Silica Gel G thin layer, 275 μ thick on application, contracts to approx. 150 μ after drying.

up comprising instruments that are needed but which are not normally at hand.[1]

The aim is to enable the worker to proceed immediately with a particular method without the necessity of preparation of single items, which can be difficult and expensive to make. It is also valuable because it serves to standardize techniques, which is of special importance in analytical processes.

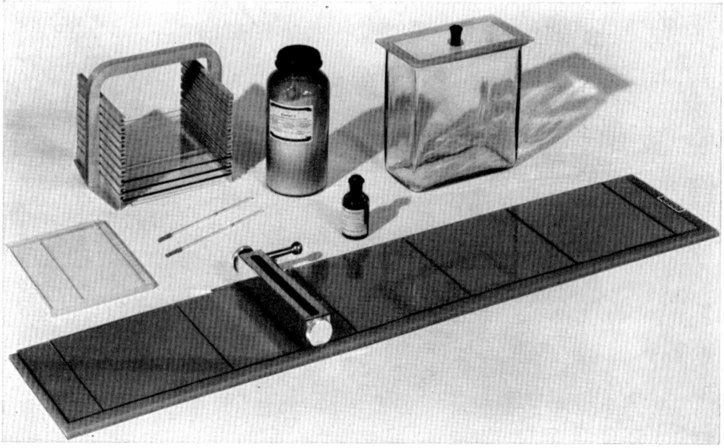

Fig. 30. Basic equipment for use in thin-layer chromatography[1]. The figures refer to the items detailed in the text (photograph DESAGA)

The DESAGA basic equipment kit No. 600 (Fig. 30) consists of the following items:

1. Thin-layer spreader DS 200/0–2, with regulation of thin layer thickness;

2. Plastic aligning tray for a row of plates is 1.1 m long (i. e. 5 plates 200 × 200 mm);

3. 10 glass carrier plates (200 × 200 mm) of identical thickness, plus one starting and one finishing plate (50 × 200 mm);

4. Rectangular chambers with glass edging and ground all-glass cover;

5. Plastic marking template with engraved scale, starting and front line;

6. Light-alloy drying rack for 10 TLC-plates, 200 × 200 mm, or 20 TLC-plates, 200 × 100 mm;

7. 2 Graduated special pipettes of 10 mm^3 capacity each;

8. Bottle of test mixture solution (Butter Yellow, Sudan Red G, and Indophenol) for determining the activity of Silica Gel G thin layers and Alumina G layers, as well as for the first trial of the method;

9. 500 g Silica Gel G (according to STAHL) for thin-layer chromatography, "Merck".

[1] Manufacturer: DESAGA, Hauptstrasse 60, Heidelberg, Germany, U. S. representive: C. A. Brinkmann, Cantiague Rd., Westbury, Long Island, N. Y.

Supplement to basic equipment

Special sprayer consisting of two parts, with ground joint, capacity 10 ml (Fig. 27), or with 50 ml Erlenmeyer flask,

S-Chamber according to Stahl for 10 and 20 × 20 cm plates, also for plates 20 and 40 cm wide, consisting of S-chamber and rod, cover plate with sintered glass strip and two clamps.

BN-Chamber according to Brenner and Niederwieser for 20×20 cm plates, consisting of stainless steel trough, cover plate and clamps,

Vacuum desiccator, model "Novus", with cock, 30 cm diameter, for storage of light-alloy drying rack.

DESAGA has designed a "small laboratory kit" for laboratories which use thin-layer chromatography only occasionally.

C. Coating Materials for Thin-Layer Chromatography

By

D. Waldi

The term coating materials, as used here, implies all solid materials which can also be employed in column chromatography. In addition to adsorbents, partition materials and ion exchangers are included.

It is often difficult to recognize whether a separation of compounds is due to ion exchange, adsorption or partition; in many cases these effects overlap. One should also never consider the adsorbent per se without taking into consideration the physico-chemical characteristics of the compounds to be separated, as well as that of the solvent [*68*].

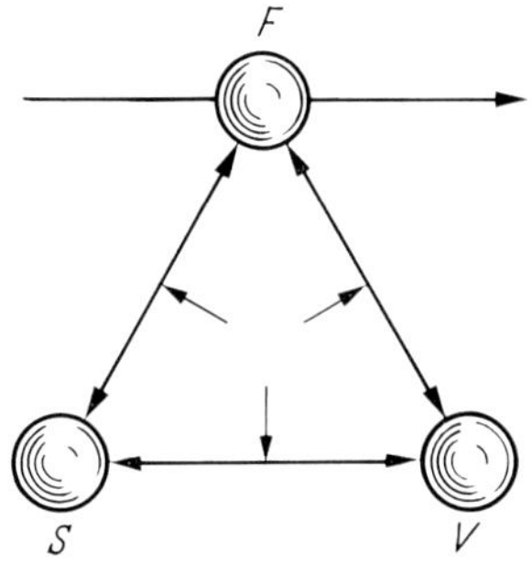

Fig. 31. Reciprocal influence of "van der Waals' forces" in a chromatogram. S = adsorbent, F = solvent, V = sample to be fractionated. cf. the diagram to Fig. 69

Fig. 31 illustrates how the adsorbent, the compounds to be separated and the solvent are reciprocally related and how the action of these three factors is interrelated, as shown by the arrows. The impelling forces (e.g. capillarity or gravity) which move the solvent, and the interacting forces of the three factors result in a characteristic rate of migration for each specific compound.

When choosing adsorbents, one will be guided by the characteristics of the compounds to be separated. To begin with, some thought must be given to the solubility of the substance to be chromatographed, i. e. whether it is hydrophilic (water-soluble) or lipophilic (fat-soluble), these being properties which, essentially, result from the number and nature of the functional groups (C/O ratio).

Once the solubility of a compound is known it must be ascertained whether it has basic, acidic or neutral properties, or whether it is amphoteric. Finally, it must be determined whether the compound is liable to react chemically with the adsorbent (or solvent), or whether chemical changes may occur due to the action of adsorbents or binders.

Bearing the above considerations in mind, three adsorbents were developed initially:

Silica Gel G according to STAHL for thin-layer chromatography[1],
Alumina G according to STAHL for TLC[1],
Kieselguhr G according to STAHL for TLC[1].

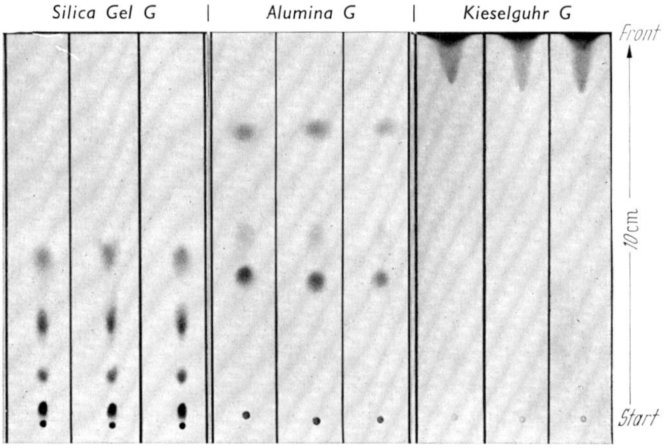

Fig. 32. Comparison of the activity of three of the most widely used inorganic adsorbents. A mixture of four fatty substances was developed with benzene. Silica gel shows greatest activity and the best separation

Silica Gel G, according to the directions for its preparation on pp. 7—9, produces a retentive, highly active acidic plate[2]: This layer is mainly designed to separate acidic and neutral substances which are not too hydrophilic. By choosing appropriate solvents (see p. 135) the chromatography of basic compounds (e. g. alkaloids), and also of hydrophilic compounds, can be carried out.

In addition to choosing an appropriate solvent (with acidic and basic additives), it is possible to influence the characteristics of the adsorbent by the manner in which it is dried (i. e. drying by air or otherwise, time exposed to the air, active nature of plates) and by the use of additives (see pp. 36—37).

Alumina G produces a retentive, active, weakly basic layer which is mainly designed for the fractionation of neutral and basic compounds

---

[1] Manufactured by E. Merck, Darmstadt, Germany. For further details see this firm's pamphlet.

[2] The designation "acidic" for silica gel or "basic" for alumina does not refer to their measured pH-values but to the property of the adsorbent of retaining a strong acid or base at the starting point in a neutral solvent.

(see p. 284). These adsorbent's characteristics can also be varied by appropriate choice of solvent, drying process and additives, and can then be suitably used for other substances.

Recently the firms Fluka and Woelm [*72*] have manufactured alumina and other adsorbents for TLC which have a considerably smaller grain size, with correspondingly longer migration times.

Kieselguhr G (Diatomaceous Earth) produces a practically inactive retentive thin layer which is mainly designed for the separation of strongly hydrophilic compounds and possibly also of amphoteric ions. KAUFMANN et al. [*22, 23*] have recently advocated the use of Kieselguhr G-layers impregnated with high-boiling petroleum fractions. MEYER [*35*] uses a mixture of Silica Gel G and Kieselguhr G (2 + 1 or 25 + 5) for the separation of antioxidants. We have been able to separate a number of compounds, e. g. barbituric acids, sugars and dyes, on a mixture of Silica Gel G and Alumina G[1] (1 + 1).

Further properties of adsorbents

The activity of Silica Gel G and Kieselguhr G depends on the active surface of the particles, i. e. the consistency of the granulation, and on the "channels" which are formed when silicic acid is precipitated. The smaller the particles, the larger the active surface. It is known from column chromatography that the rate of migration is a function of particle size. Small particles give long and large particles give short migration periods, but the latter yield inferior separation. The standardization of adsorbents is a rather difficult problem. One must, therefore, allow for certain variations in activity, so that all *Rf*-values can be considered only as a guide (for use of a test mixture see pp. 14. 16).

When using these adsorbents it must be remembered that binding material is also present and can interfere in the separation process. This binding material is calcium sulphate which will, for instance, retain free phosphoric acid at the starting point. In such cases, gypsum-free layers, or layers with organic binding materials are used[2].

Preparation of slurries of Silica Gel G, Alumina G or Kieselguhr G "Merck" for five 20 × 20 cm plates and a layer thickness of approx. 250 μ.

25 g adsorbent are put into a porcelain mortar of about 10 cm diameter, and 35 ml distilled water added, stirring slowly until a homogenous mass is obtained which should be devoid of iumps or air bubbles. Then an additional 15 ml of distilled water is added with stirring. The entire stirring period should not exceed 100 sec. The thin liquid suspension is immediately placed into the spreader and spread out on the plates.

The adsorbent can also be put into the spreader with double its quantity of water, rotating the tubular through 90° degrees and mixing the whole by cautious shaking.

---

[1] Designated as "Alusil."

[2] According to E. Merck AG., Darmstadt, Germany, the following adsorbents are now commercially available: Silica Gel H (which is adhesive without gypsum additive) and two silica gels with fluorescing indicators (Silica Gel GF_{254} and Silica Gel HF_{254}).

Additional adsorbents and adsorbent combinations

A number of substances, such as those described in the chapter on vitamins, undergo chemical changes when in contact with highly activated adsorbents. It is, therefore, advisable to use for vitamin A paraffin-impregnated Silica Gel G [*71*] or a (1:1) mixture of Silica Gel G with calcium phosphate. The only suitable agents for carotenoids are calcium hydroxide-Silica Gel G (6 : 1), *sec.* magnesium phosphate and calcium hydroxide [*71*].

For inorganic TLC it is necessary to use iron-free silica gel (see p. 475). Ferric oxide-hydrates as well as charcoal were used as adsorbents by Hesse et al. [*18*].

The following organic cellulose powders manufactured by Macherey, Nagel & Co., Düren, Germany are available:

a) Varieties without gypsum additive:		Capacity m val/g
MN 300	cellulose	—
MN 300 Ac	acetylated cellulose (about 10%)	—
MN 300 CM	carboxymethyl-cellulose	approx. 0.7
MN 300 P	phosphorylated cellulose	approx. 0.7
MN 300 DEAE	diethylaminoethyl-cellulose	approx. 0.7
MN 300 ECTEOLA	anion exchanger	approx. 0.35
b) Varieties with gypsum additive:		
MN 300 G	cellulose	—
MN 300 G/Ac	acetylated cellulose (about 10%)	—
MN 300 G/CM	carboxymethyl-cellulose	approx. 0.7
MN 300 G/P	phosphorylated cellulose	approx. 0.7
MN 300 G/DEAE	diethylaminoethyl-cellulose	approx. 0.7
MN 300 G/ECTEOLA	anion exchanger	approx. 0.35

Method of operation

The spreader described on p. 11 is used to produce chromatoplates. It is important to clean the glass plates thoroughly before applying the thin layer. It has been found advantageous to wash the plates in a concentrated soda solution followed by rinsing with cold distilled water. In order to obtain a completely smooth layer, it is advisable to stir the mixed powder thoroughly for about one min in an electric mixer.

Below are given instructions for preparing various layers. The amounts specified are sufficient to coat five 20×20 cm glass plates. Drying temperatures and drying periods are also given:

Cellulose

MN 300 or MN 300 G (cellulose, grade 300 G with gypsum additive)
15 g powder + 90 ml dist. water. Dry coated plates at 105° C, 10 min.

MN 300 Ac or MN G/Ac (acetylated cellulose, grade 300 G/Ac with gypsum additive)
10 g powder + mixture of 50 ml methanol and 5 ml dist. water. Dry at 60° C, 5—10 min.

Cellulose Ion Exchangers

MN 300 G/CM (Carboxymethyl-cellulose, with gypsum additive)
15 g powder + 90 ml dist. water. Dry at 50° C, 40 min.

MN 300 G/P (phosphorylated cellulose, with gypsum additive)
15 g powder + 70 ml dist. water. Dry at 50° C, 40 min. To obtain a completely smooth layer with this powder, stir the slurry thoroughly for 5 min. in an electric mixer.

MN 300 G/DEAE (diethylaminoethyl-cellulose, with gypsum additive)
10 g powder + 90 ml dist. water. Dry at 50° C, 40 min.

MN 300 G/ECTEOLA (anion exchanger, with gypsum additive)
10 g powder + 95 ml dist. water. Dry at 50° C, 40 min.

Cellulose ion-exchangers without gypsum do not yield an ideal thin layer. It is advisable to add a small quantity of cellulose powder MN 300 to the exchanger powders.

MN 300 CM (carboxymethyl-cellulose)
8 g powder MN 300 CM + 3 g powder MN 300 + 70 ml dist. water. Dry at 50° C, 40 min. To obtain a completely smooth layer with this powder, stir the slurry thoroughly for about 5 min in an electric mixer.

MN 300 DEAE (diethylaminoethyl-cellulose)
8 g powder MN 300 DEAE + 2 g cellulose powder MN 300 + 90 ml dist. water. Dry at 50° C, 40 min.

MN 300 ECTEOLA (anion exchanger)
8 g powder MN 300 ECTEOLA + 2 g cellulose powder MN 300 + 95 ml dist. water. Dry at 50° C, 40 min.

Table 1. *Coating Materials Used in TLC*

Adsorbent	Properties			Classes separated
	Acidity	Activity	Separating effect	
Silica gel	acidic	active	adsorption, partition (ion exch.)	Practically all
Alumina	basic	active	adsorption, partition (ion exch.)	Mainly bases and steroids
Kieselguhr	neutral	inactive	partition	Sugars, pharmaceuticals
Magnesium hydrogen phosphate		weakly active	adsorption	Carotenoids, Tocopherols
Magnesium trisilicate		weakly active	adsorption	Carotenoids, Tocopherols
Calcium hydroxide	basic	weakly active	adsorption	Carotenoids, Tocopherols
Calcium phosphate	weakly basic	weakly active	adsorption	Carotenoids, Tocopherols
Calcium sulphate [24]		weakly active	adsorption	Fatty acids and glycerides
Silica gel + alumina (1 + 1)	acidic + basic	active	partition (ion exch.)	Dyes, barbituatres
Ferric oxide hydrate		active	adsorption	Polar substances
Charcoal		active	adsorption	Non-polar substances
Cellulose powder	neutral	none	partition (ion exch.)	Dyes, amino acids
Polyamides	neutral	inactive	(ion exch. ?)	Flavones
Polyacrylonitrile	neutral	inactive	(ion exch., dispersion ?)	Authocyanins and similar substances
Ion Exchangers	acidic + basic		ion exch. (partition)	Nucleic acids and their derivatives

Excorna[1] produces a cellulose powder with gypsum additives for "analytical chromatography" which can also be used for TLC.

[1] Excorna, Mainz, Germany.

Polyamide[1] and polyacrylonitrile layers[2] are used for the separation of flavone glycosides and antioxidants [*4*, *10*, *11*, *13*]. Randerath [*47*, *48*] recommends the use of ion exchangers for the chromatography of nucleic acids.

Coating materials for TCL are listed in Table 1. Most separations can be achieved with Silica Gel G or Alumina G.

Storage and treatment of adsorbents

Coarse-grained silica gel is a well known drying agent, i. e., it combines with atmospheric moisture, and for this reason it is generally stored in closed vessels. Silica Gel G, Alumina G and Kieselguhr G contain gypsum, which is also hygroscopic. The vessels containing these substances must therefore always be kept closed and dry. Adsorbents whose binders are altered by atmospheric humidy produce layers of limited retentivity.

It is advisable to pre-dry the layers first and to activate them later (see p. 10). If they are heated immediately at above 100° C, without predrying, fissures may appear. If heating occurs for long periods at over 125° C, there is a danger of the gypsum becoming too strongly dehydrated; its binding power may thus be diminished. In all cases, the active adsorbent layers should be protected from atmospheric moisture and laboratory fumes since they gradually lose their activity. It is therefore necessary to store adsorbents over an efficient drying agent.

D. Special Techniques

By

Egon Stahl

1. Continuous flow, and multiple development techniques

If the substances to be separated have an *Rf*-value of less than 0.5, they may be further resolved by continuous flow chromatography in the BN-Chamber p. 22). Alternatively, the chromatogram may be developed again with the same solvent. This process may be repeated several times as the migration period is generally short. The multiple development technique is used when one solvent is incapable of separating some of the substances so that they are crowded together either in the lower (0—0.2) or upper (0.8—1) ranges. It is often possible to separate first with one and then with another solvent, each with different lengths of run. Two examples will make this clear: a) A mixture of lignanes and their glycosides (podophyllum-containing substances) was to be separated. In the first run (6 cm) the glycosides were separated from each other whereas the aglycones were crowded together in the solvent front. These substances were separated with a solvent of lower polarity (12 cm lenght of run) [*64*]. — b) A carotene mixture (β-carotene, lycopene,

[1] Polycaprolactam powder (polyamide or Perlon), Farbwerke Hoechst, Werk Bobingen, Germany, cf. also Woelm [*72*].

[2] Polyacrylonitrile powder, Farbenfabriken Bayer, Werk Domagen, Germany.

canthaxantin and bixin) were separated by chloroform after only a short length of run. However, β-carotene and lycopene migrated to the front. After 5 cm, the process was interrupted and chromatography continued with hexane. After this second run (10 cm) the two carotene hydrocarbones were separated from each other [*60*].

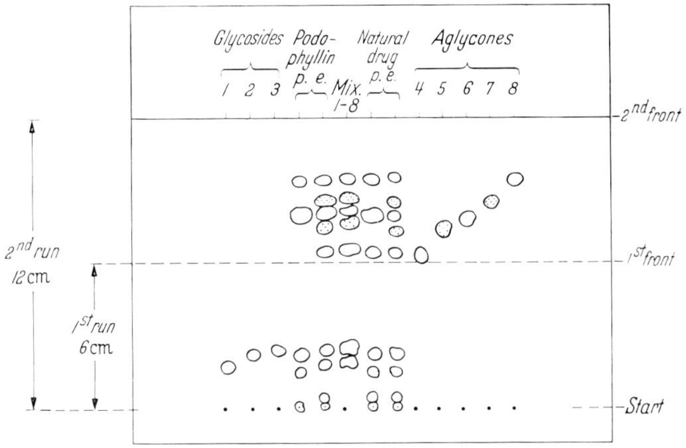

Fig. 33. Multiple development of podophyllum-containing substances on a Silica Gel G layer. Solvents, 1st step: chloroform + 10% methanol, 2nd step: chloroform + 35% acetone. *1* α-peltatin-β-glucoside, *2* podophyllotoxin-glucoside, *3* β-peltatin-β-D-glucoside, *4* 4′-dimethylpodophyllotoxin, *5* α-peltatin, *6* podophyllotoxin, *7* β-peltatin, *8* 1-deshydroxy-podophyllotoxin. 1.0 μg of the pure substances and 0.5 mm^3 of a 10% alcoholic extract of the podophyllines and the drug were applied. Visualization with sulphuric acid-acetic anhydride (1 : 3), 15 mins, 100° C. The stippled zones react also with diazonium salts. (*e.* = *P. emodi*, *p.* = *P. peltatum*) [*64*]

2. Wedged-tip technique

The "wedged-tip-technique" of paper chromatography, as described by Reindell and Hoppe [*49*], and by Matthias [*33*], led to improved separation. Owing to the wedged-shaped cut of the paper strip the substances are forced to assume an almost band-like path. It is very easy to apply this technique to thin adsorbent

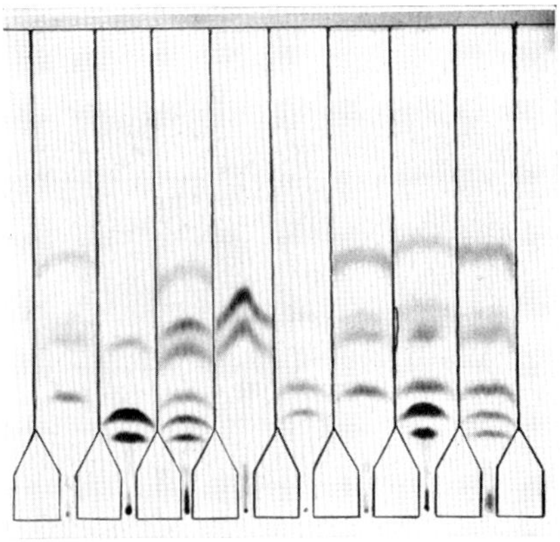

Fig. 34. TLC with wedged-tip divisions. The separation process assumes a band-like course. The $\wedge$-shape (middle) only occurs with large quantities

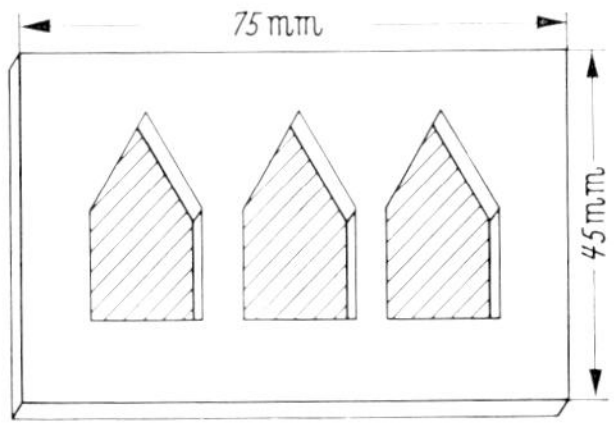

Fig. 35. Plexiglass stencil (2—3 mm thick) to draw dividing line. The shaded pentagons in the picture are cut out

layers. Dividing lines, 0.5—1 mm broad, are drawn in the layer with a narrow metal spatula.

The possible shapes of the wedges and their distances from each other were investigated. The shapes illustrated in Figs. 34 and 185 appear to give the best results. The scratching-in of the pentagons is facilitated by using a stencil made of transparent plastic material (Fig. 35). One simply places the lower edge of the stencil on the TLC plate and cuts around the five edges, after which the longitudinal lines are drawn. The substances are applied to the narrow portion of the wedge. The length of the run of the chromatograms is increased by the narrow band of solvent.

3. Two-dimensional separation, SRS-technique

The two-dimensional technique is a valuable tool for complex mixtures. Separation is first made in one direction, followed by rotation of the plate through 90° after which separation is carried out in the second dimension. This technique is successful only if the pattern of separation in the second dimension differs from that in the first dimension. Figs. 131, 149 and 168 illustrate the principle of this procedure adequately. The combination of adsorption-TLC (1st dimension) with partition-TLC (2nd dimension) as carried out by Kaufmann et al. [*22*] and Honegger's combination of adsorption and partition-TLC represent notable advances in two-dimensional-TLC (pp. 434—435).

Separation may be carried out in both directions under exactly identical conditions. The compounds, unless they decompose during development, must subsequently lie on a diagonal line bisecting the plate. This affords a simple way of determining whether a certain solvent affects one or more substances of the mixture. If a reaction is introduced after completion of chromatography in the first dimension, which leads to a structural change in one or more compounds which have already been separated, this will be recognized during subsequent chromatography with the solvent in the second dimension. This simple technique makes possible the rapid recognition of changes in individual components of a mixture following the action of radiation (γ-rays, X-rays, UV-light), gases, heat, etc. These techniques can be used for highly topical problems such as radiation protection, photochemistry and stability control. The SRS-technique, already known for a long time in paper chromatography, has been successfully used in the study of the inactivation of pyrethrines (Fig. 149) [*60*].

4. Variation in the separation characteristics of a layer

(see also pp. 29—34)

The experience gained in adsorption chromatography with "basic", "neutral" and "acidic" aluminas, as well as those gained with PC (buffering or impregnation of papers), may be applied to TLC. The Silica Gel G-layers used in TLC are weakly acidic. The Alumina G-layers have a slightly basic reaction. By mixing both adsorbents (1 : 1 by wt.) one can, according to WALDI, produce neutral plates, which, on occasion, are most adventageous. It is equally simple to produce acidic or basic layers: 0.1–0.5 N-oxalic acid or 0.1–0.5 N-KOH are added to the adsorbent in lieu of water. Similarly, buffered plates are made by adding suitable buffer solutions.

"Acidic" Silica Gel G-layers have been successfully used to separate acidic compounds such as phenols, acids. Similarly, basic plates have been used for alkaloids and amines. By consecutive use of such plates, it is possible to determine how an unknown compound reacts. If the substances form salts with the "impregnating agent" they do not migrate in weakly polar solvents, or at least slower than the free bases or phenols [*58*].

It may be advantageous to impregnate with bisulphite (for aldehydes), mercuric salts, silver salts (addition compounds) boric acid (formation of complexes) or with urea (inclusion compounds).

Rendering the adsorbent hydrophobic: In the TLC of lipids the technique of "reversed-phase partition chromatography", which is also used in PC, has proved useful. The adsorbent layer is rendered hydrophobic with a non-polar, lipophilic "oil", the mobile phase being a polar solvent which is immiscible, or only slightly miscible, with it. On such layers, the *Rf*-values of compounds from homologous series (e. g. C_5—C_{50}) decrease with rising C-numbers. Depending on the degree of "volatility" the impregnation is described as either temporary or permanent. MANGOLD [*30*, *31*], WINTERSTEIN [*71*] and, recently, also KAUFMANN [*23*] favor making the system permanently hydrophobic. The well-dried TLC-plate is dipped into a 5—10% solution of the impregnating agent in a volatile solvent. During impregnation it is important to see that both the plate and impregnating mixture are kept at the same temperature. Also, the plate must be dipped in slowly and the immersion time, which is usually given, must be adhered to. Plates rendered hydrophobic with silicone oil or paraffin are well suited for storing. At the present time, the following hydrophobic agents are in use:

Silicone oil 10 c St (Dow Corning, Midland, Mich., USA) 5% in ether.

Silicone oil 50 c St and 1.5 c St (Wacker-Chemie, München, Germany), 7% in petroleum hydrocarbon.

Liquid paraffin for IR-spectroscopy (E. Merck, Darmstadt, Germany) 5–10% in petroleum hydrocarbon.

Undecane (for making temporarily hydrophobic layers), 15% in petroleum hydrocarbon.

Tetradecane = petroleum fraction, B.P. 240–250° (Haltermann, Hamburg, Germany), 5% in petroleum hydrocarbon.

KAUFMANN [*23*] has described siliconizing and fixation of the layer. The TLC-plate is placed in a vacuum desiccator containing 3—5 ml dichlorodimethylsilane in a glass dish. Then, a vacuum of about 300 mm is applied for 15—30 min. The plate is left in air for 30 min. The siliconized layer thus obtained may be treated like a paper chromatogram, any reagent may be washed off with water.

Bibliography to Chapters A – D: General Section

[1] Applewhite, T. H., M. J. Diamond and L. A. Goldblatt: J. Am. Oil Chem. Soc. **38**, 609 (1961).
[2] Baehler, B.: Schweiz. Apotheker-Ztg. **99**, 543 (1961).
[3] Barbier, M., H. Jäger, H. Tobias u. E. Wyss: Helv. Chim. Acta **42**, 2440 (1959).
[4] Birkofer, L., Ch. Kaiser, H.-A. Meyer-Stoll u. F. Suppan: Z. Naturforsch. **17 b**, 352 (1962).
[5] Bolliger, A., and J. E. Bolliger: The Royal Australian Chemical Institute Proc. Feb. 1962, S. 78.
[6] Brenner, M., u. A. Niederwieser: Experientia (Basel) **17**, 237 (1961).
[7] Brenner, M., A. Niederwieser, G. Pataki u. A. R. Fahmy: Experientia (Basel) **18**, 101 (1962).
[8] Consden, R., A. H. Gordon and A. J. P. Martin: Biochem. J. **38**, 224 (1944).
[9] Crowe, M. O'L.: Anal. Chem. **13**, 845 (1941).
[10] Davídek, J., u. E. Davídková: Pharmazie **16**, 352 (1961).
[11] Davídek, J., a Ž. Procházka: Collection Czechoslov. Chem. Commun. **26**, 2947 (1961).
[12] Demole, E.: J. Chromatogr. **6**, 2 (1961).
[13] Egger, K.: Z. analyt. Chem. **182**, 161 (1961).
[14] Goppelsroeder, F.: Anregungen zum Studium der auf Capillaritäts- und Adsorptionserscheinungen beruhenden Capillaranalyse. Basel: Verlag Helbing und Lichtenhahn, 1906.
[15] Hais, I. M., u. K. Macek: Handbuch der Papierchromatographie. Jena: VEB Fischer Verlag 1958/1960. Engl. Ed. New York: Academic Press, Inc. 1964.
[16] Halpaap, H.: Private communication.
[17] Häusser, H.: Mitt.-Bl. Ges. Dtsch. Chem. Lebensmittelchem. gerichtl. Chem. **13**, 194 (1959).
[18] Hesse, G. u. M. Alexander: Abstract, «Journée internationale d'Étude des Méthodes de séparation immédiate et de Chromatographie», 13., 14., 15. June 1961. Paris.
[19] Honegger, C. G.: Helv. Chim. Acta **44**, 173 (1961).
[20] Izmailov, N. A., u. M. S. Shraiber: Farmazia **3**, 1 (1938).
[21] Jensen, A.: Tidsskr. Kjemi, Bergvesen Met. **1**, 14 (1961).
[22] Kaufmann, H. P., u. Z. Makus: Fette, Seifen, Anstrichmittel **62**, 1014 (1960).
[23] Kaufmann, H. P., Z. Makus u. T. H. Khoe: Fette, Seifen, Anstrichmittel **63**, 689, 807, 828 (1961).
[24] Kaufmann, H. P., u. T. H. Khoe: Fette, Seifen, Anstrichmittel **64**, 81 (1962).
[25] Kirchner, J. G., and G. I. Keller: J. Am. Chem. Soc. **72**, 1867 (1950).
[26] Kirchner, J. G., J. M. Miller and G. I. Keller: Anal. Chem. **23**, 420 (1951).
[27] Lagoni, H., u. A. Wortmann: Milchwissenschaft **10**, 360 (1955).
[28] Lagoni, H., u. A. Wortmann: Milchwissenschaft **11**, 206 (1956).
[29] Machata, G.: Mikrochim. Acta, **1960**, 79.
[30] Mangold, H. K.: Fette, Seifen, Anstrichmittel **61**, 877 (1959).
[31] Mangold, H. K.: J. Am. Oil Chemists Soc. **38**, 708 (1961).
[32] Martin, A. J. P., and R. L. M. Synge: Biochem. J. **35**, 91, 1358 (1941).
[33] Matthias, W.: Naturwissenschaften **41**, 18 (1954).
[34] Meinhard, J. E., and N. F. Hall: Anal. Chem. **21**, 185 (1949).
[35] Meyer, H.: Deut. Lebensm.-Rundschau **7**, 170 (1961).
[36] Miller, J. M., and J. G. Kirchner:Anal. Chem. **24**, 1480 (1952).
[37] Miller, J. M., and J. G. Kirchner: Anal. Chem. **25**, 1107 (1953).
[38] Miller, J. M., and J. G. Kirchner: Anal. Chem. **26**, 2002 (1954).
[39] Mistryukov, E. A.: Collection Czechoslov. Chem. Commun. **26**, 2071 (1961).
[40] Mottier, M.: Mitt. Gebiete Lebensm. u. Hyg. **47**, 372 (1956).
[41] Mottier, M.: Mitt. Gebiete Lebensm. u. Hyg. **49**, 453 (1958).
[42] Neubern de Toledo, T. A., e R. Wasicky: Correio do Mundo Farmaceutico (Brasilien) v. 15. Juli 1961.
[43] Pastuska, G., u. H. Trinks: Chemiker-Ztg. **85**, 535 (1961).
[44] Pastuska, G., u. H. Trinks: Chemiker-Ztg. **86**, 135 (1962).

[45] PEEREBOOM, J. W. C.: Chem. Weekblad **49**, 625 (1961).
[46] PROCHÁZKA, Ž.: Chem. Listy **55**, 974 (1961).
[47] RANDERATH, K.: Angew. Chem. **73**, 436 (1961).
[48] RANDERATH, K.: J. Chromatogr. **6**, 365 (1961).
[49] REINDEL, F., u. W. HOPPE: Naturwissenschaften **40**, 245 (1953).
[50] REITSEMA, R. H.: J. Am. Pharm. Assoc. **43**, 414 (1954).
[51] REITSEMA, R. H.: Anal. Chem. **26**, 960 (1954).
[52] RITTER, F. J., and G. M. MEYER: Nature (Lond.) **193**, 941 (1962).
[53] *RSCO Review* **3**, 18 (1961).
[54] SCHORN, P. J.: Glas-Instr.-Tech. **5**, 43 (1961).
[55] STAHL, E.: Pharmazie **11**, 633 (1956).
[56] STAHL, E.: Chemiker.-Ztg. **82**, 323 (1958).
[57] STAHL, E.: Parfümerie u. Kosmetik **39**, 564 (1958).
[58] STAHL, E.: Arch. Pharmaz. **292**, 411 (1959).
[59] STAHL, E.: Pharmaz. Rundsch. **1**, Nr. 2, 1 (1959).
[60] STAHL, E.: Arch. Pharmaz. **293**, 531 (1960).
[61] STAHL, E.: Z. analyt. Chem. **181**, 303 (1961).
[62] STAHL, E.: Angew. Chem. **73**, 646 (1961).
[63] STAHL, E.: Unpublished.
[64] STAHL, E., u. U. KALTENBACH: J. Chromatogr. **5**, 458 (1961).
[65] STANLEY, W. L., S. H. VANNIER and B. GENTILI: J. Assoc. Offic. Agr. Chemists **40**, 282 (1957).
[66] STANLEY, W. L., and S. H. VANNIER: J. Assoc. Offic. Agr. Chemists **40**, 582 (1957).
[67] TEIJGELER, C. A.: Pharmac. Weekblad **97**, 43 (1962).
[68] UMLAND, F., u. K. KIRCHNER: Z. anorg. Chem. **280**, 211 (1955).
[69] WAGNER, H.: Mitt. Gebiete Lebensm. u. Hyg. **51**, 416 (1960).
[70] WILLIAMS, T. I.: Introduction to Chromatography. Glasgow: Blackie & Son 1947.
[71] WINTERSTEIN, A., A. STUDER u. R. RÜEGG: Chem. Ber. **93**, 2951 (1960).
[72] WOELM, M.: Mitteilungen AL 10 (1961).
[73] WOLLISH, E. G., M. SCHMALL and M. HAWRYLYSHYN: Anal. Chem. **33**, 1138 (1961).
[74] ZECHMEISTER, L., u. L. VON CHOLNOKY: Die chromatographische Adsorptionsanalyse. 2. Aufl. Wien: Springer Verlag 1938.

Some recent noteworthy articles

BEKERSKY, I.: Anal. Chem. **35**, 261 (1963): Spray technique for the preparation of TLC-plates.
BRODASKY, T. F.: Anal. Chem. **35**, 343 (1963): Carbon layers, microbiological detection of antibiotics.
CERRI, O., and G. MAFFI: Boll. Chim. Parmac. **100**, 940 (1961): Review article.
DUNCAN, G. R.: J. Chromatog. **8**, 37 (1962): TLC-spreader.
FELTKAMP, H.: Dtsch. Apoth. Ztg. **102**, 1269 (1962): TLC on a glass rod.
GAMP, A., P. STUDER, H. LINDE and K. MEYER: Experientia **18**, 292 (1962): TLC on grooved plates.
HALPAAP, H.: Chem. Ing. Techn. **35**, 488 (1963): Preparative TLC.
HÖRHAMMER, L., H. WAGNER and G. BITTNER: Dtsch. Apotheker **14/4**, 1 (1962): Manually prepared TLC-plates.
HONEGGER, C. G.: Helv. Chim. Acta **46**, 1772 (1963): Preparative TLC.
HÜTTENRAUCH, R., L. KLOTZ and W. MÜLLER: Z. Chem. **3**, 193 (1963): Layers of ion-exchange resins.
KELEMEN, J., and G. PATAKI: Z. analyt. Chem. **195**, 81 (1963): Differences between layers prepared manually and by the standard method.
KOKOTI-KOTAKIS, E.: Chim. Chronika (Athens) **27** A, 59 (1962): Review article.
LIE, K. B., and J. F. Nyc: J. Chromatog. **8**, 75 (1962): TLC in test tubes.
LÜDY-TENGER, F.: Pharmac. Acta Helv. **37**, 770 (1962): Review article; spreader.
MALINS, D. C., and J. C. WEKELL: J. Chem. Educ. **40**, 531 (1963): Review article.
MORGAN, M. E.: J. Chromatog. **9**, 379 (1962): Apparatus for simultaneous application of several spots.

MORITA, K., and F. HARUTA: J. Chromatog. **12**, 412 (1963): Spraying method for the preparation of TLC-plates.
MUNTER, F.: Chemiker Ztg. **87**, 657 (1963): Choice of eluents.
PATAKI, G., and J. KELEMEN: J. Chromatog. **11**, 50 (1963): Differences between layers prepared manually and by the standard method.
PEIFER, J. J.: Mikrochim. Acta **1962**, 529: Microchromatoplates.
RUSIECKI, W., and M. HENNEBERG: Farmacja Polska **18**, 203 (1962): Review article.
RUSSEL, J. H.: Rev. Pure Appl. Chem. **13**, 15 (1962): Review article.
SEHER, A.: J. Soc. Cosm. Chem. **13**, 385 (1962): Review article.
SHELLARD, E. J.: Research and Development No. 21, 30 (1963): Review article.
SREPEL, B.: Farmac. Glasnik (Zagreb) **18**, 64 (1962): Review article.
STAHL, E.: Development and application of TLC, 11th commun., review article, Sci. Reports Istituto Superiore di Sanita, Roma, Elsevier Publ. Comp. Amsterdam, in press.
WALDI, D.: Arch. Pharmaz. **296**, Mitteilungen 1 (1963): Review article.
WALDI, D.: Glas Instrumenten Technik **7**, 477 (1963): Review article: adsorbents.

Books:

BOBBITT, J. M.: Thin-Layer Chromatography, Reinhold Publishing Corp., New York/London 1963.
RANDERATH, K.: Thin-Layer Chromatography (Engl. Edit.) Verlag Chemie, Weinheim 1963; (German Edit.: Dünnschicht-Chromatographie 1962).
STAHL, E.: Dünnschicht-Chromatographie, Ein Laboratoriumshandbuch, Springer, Berlin-Göttingen-Heidelberg 1962.
TRUTER, E. V.: Thin Film Chromatography, Cleaver-Hume Press, London 1963.

Films:

AHRENS, E. H.: TLC, ca. 20 min, colored with sound, 16 mm, The Rockefeller Institute, New York, U.S.A. 1963.
PRIVETT, O. S.: TLC, ca. 45 min, colored with sound, 16 mm, The Hormel Institute, Austin, Minn., U.S.A. 1963.
SEILER, H.: Short motion picture, ca. 10 min, Institut für Anorganische Chemie der Universität Basel, Switzerland.

E. Documentation of Thin-Layer Chromatograms

By

H. GÄNSHIRT

In contrast to paper chromatograms, thin-layer chromatograms are sensistive to mechanical damage and the stained spots, in general, fade more rapidly than in PC. For these reasons, and since the glass plates may be used again, it is not recommended to keep a developed thin-layer chromatogram unless it is in a specially prepared state, as to be described at the end of this section. It is not sufficient to record the location of the spots in a chromatogram merely by giving the *Rf*-values, as can be done with PC[1]. In TLC the *Rf*-values are regarded rather as guides which give the relative position of the spots within a chromatogram. It is, therefore, necessary to make very detailed notes of all experimental conditions when carrying out TLC and is convenient to use a printed form, as illustrated in Fig. 36.

[1] $Rf = \frac{\text{distance of center of spot from starting point}}{\text{distance of solvent front from starting point}}$.

THIN-LAYER CHROMATOGRAM NO.

Substance: ....

Literature reference: ....

Layer: .... Size: ....

Thickness of layer: ....

Production and drying: ....

....

Chamber, Type: .... Technique of separation: ....

State of saturation of chamber: ....

Length of run: .... Time of run: ....

Solvent mixture: ....

....

....

Solvents used for spotting and their concentration: ....

....

Quantities applied: ....

Special remarks: ....

....

Date: ....

Substance	Rf × 100	Rf × 100	Rf × 100	Sensitivity.	Substance detected.
1.					
2.					
3.					
4.					
5.					
6.					
7.					
8.					
9.					
10.					

Fig. 36. Proposed scheme for a printed form giving all necessary data for thin-layer chromatograms

All Rf-values in this book have been multiplied by a factor of 100. Thus, with a solvent front migration of 100 mm, the hRf-value is obtained directly as the distance from starting point to mid-point of the spot (zone) in millimeters. Although, in some cases, the utilization of the entire plate length makes possible the separation of complicated mixtures of substances, it is advisable, in general, to use a distance of separation of 100 mm. Experience shows that with greater distances spots with higher Rf-values are compressed towards the solvent front. When Rf-values are given it cannot be ascertained whether the substances are separated. When it becomes necessary for the writer to record sharply outlined spots this is, if possible, shown by a transverse stroke. If sharp separation is not possible we record this by placing the Rf-values in a bracket.

If a series of chromatograms is investigated, the use of reference substances is recommended for correlating spots in TLC. This method is already known from the PC of sugars. The R_{St}-value will thus be given by:

$$R_{St} = \frac{\text{Distance of sample spot from starting point}}{\text{Distance of reference material spot from starting point}} .$$

If one component from the mixture to be analyzed is used as reference material, it is preferable to express the R_{St} values thus obtained as Rf-values.

In addition to the plotting of results, thin-layer chromatograms are most suitable, by reason of their format, for the production of graphic or photographic copies. This is usually not convenient with paper chromatograms.

The simplest and most generally applied method consists of tracing the chromatograms on transparent paper (Fig. 37). Spots which appear only under UV-light are located by pricking with a needle and then transferred to the tracing paper. The spots may also be measured planimetrically by drawing them as rectangles of appropriate sizes, but having the same length of base, i.e. representing them semi-quantitatively (Fig. 37). When two or more spots lie close to one another, the scale must be enlarged in order to keep to the same base-edge length.

If a thin-layer chromatogram shows highly constrasted spots it is possible to use a photostat technique, e. g., the Agfa-Copyrapid procedure [*31*], in order to obtain a true black and white reproduction of the chromatogram (Fig. 37). With this procedure the negative is developed within a few seconds. Simultaneously, an automatically produced positive picture is obtained by means of a so-called "transfer sheet" [*14*]. With this method, the choice of spraying agent is of importance, corrosive reagents should be avoided. It may be possible to avoid damage to the printing paper by interposing a transparent plastic sheet. Several authors have already reported on this possibility of documentation [*10*, *13*].

The usual techniques are applicable for the photographic reproduction of chromatograms sprayed with corrosive reagents and for the documentation of chromatograms whose spots may be recognized only under UV-light by their fluorescence or by quenching on a fluorescent layer.

In order to do away with the need for subsequent magnification, a reflex camera with close-up lens attachment is most suitable. Black and white photographs of poorly-contrasted chromatograms are best produced with transmitted light. In this way, for instance, photographs were taken of lipid spots stained with antimony chlorides [*15*]. Faults in the layer become apparent with this procedure. Plates showing such faults can be photographed with a combination of incident and transmitted light. For highly contrasted chromatograms only incident light is used. For this purpose, the light sources must be placed on both sides and the incident angle must be below 45°. For black and white photographs of spots showing fluorescence an incident UV-light source with a wavelength of 254 mμ, is used.

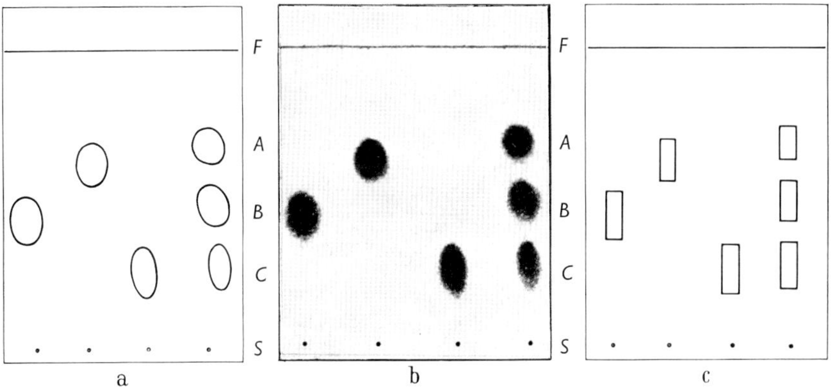

Fig. 37. Possible ways of documenting thin-layer chromatograms. a) Traced on transparent paper (transfered back to normal paper). b) Documentation by photostating. c) Documentation in the form of suitably-sized rectangles after planimetry. Example: Separation of bromine hypnotics with the solvent cyclohexane-chloroform-pyridine (20 + 60 + 5) on Silica Gel G layers. *S* start, *F* solvent front, *A* Adalin, *B* Abasin, *C* Bromural

Agfa-AGP or Agfa-JFF is a suitable film. In order to exclude diffracted and reflected primary light, a UV-filter and a medium-strength yellow filter are placed in front of the lens. To photograph the dark absorption spots on fluorescent-layers (p. 54) the same technique is used [*8*].

For color photographs of thin-layer chromatograms, the Agfa reversal film CT-18 can be used [*6*]. From the reversal diapositives any desired number of copies can be made with color reversal paper (Agfa-CT-copies). For color photographs electronic flashes may also be used as light sources. To photograph fluorescent spots the light source used is a low-pressure mercury lamp (Heraeus, Frankfurt, Germany) with UV-filter UG5 and a wave length of 254 mμ. In front of the camera lens a combination of filters, WG 2 (2 mm), GG, (13 mm) and GG 3 (4 mm) (Schott & Gen., Mainz, Germany) is attached with a socket in order to filter out diffracted and reflected UV-light. Precise indications concerning camera set-up and light can be found in the original paper [*6*].

As already stated at the beginning of this section, it is also possible to preserve the chromatograms by special treatment. Amino acid chromatograms were preserved by Barrolier [*2*] by imbedding the thin layer in a 4% collodion solution containing 7.5% glycerol. The adsorbent layer could then be stripped off the plate.

50 ml of the above-mentioned solution is carefully poured onto a 20 × 20 cm TLC-plate after staining. To prevent the solution running off, the chromatogram

is framed with a 1 cm-broad elastoplast strip. After drying, the collodion film is cut with a razor blade parallel to the elastoplast strip. The chromatogram then lifts off easily from the plate as a flexible film. For final documentation, the film is mounted on paper with transparent plastic adhesive tape.

Collodion films tend to curl and stick slightly. Preservation is improved by spraying a synthetic aerosol medium (polyacrylic acid ester, polyvinyl chloride or polyvinyl proprionate) on the developed chromatogram [*20*]. Such a dispersion is obtainable commercially under the name of Neatan (E. Merck, Darmstadt, Germany). To facilitate spraying, dilution of this aerosol medium with methanol is recommended.

Onto a dry 20×20 cm thin-layer chromatogram, about 10 ml Neatan-methanol (5+5) are sprayed until the chromatogram is well wetted. The plates are air-dried and the layer is stripped by following one of the following methods:

a) Press a thin, transparent adhesive tape on the entire chromatogram, dip the plate in water, and carefully strip off the layer.

b) The margins of the chromatogram are covered with thin, adhesive tape, the plate is dipped briefly into water, and the layer is carefully stripped off.

c) The chromatogram is briefly dipped into water, and the layer is very carefully removed with a knife or spatula.

After complete air-drying the chromatograms which are stuck on to cardboard may be easily preserved.

F. Quantitative Evaluation of Thin-Layer Chromatograms

By

H. Gänshirt

Thin-layer chromatograms must be standardized in order to permit their quantitative evaluation. This is achieved by a defined range of particle size, chemical purity of the adsorbent and reproducible thickness of the layer. The remaining prerequisites are already known from experience with paper chromatography. Suitable equipment must be available for spotting. This consists of calibrated capillary pipettes and glass-plunger syringes with screw-thread micrometer-advance (see p. 12).

Experiments were made to investigate the errors arising from the application of spots with a glass-plunger syringe. Solutions of benzoic acid esters were used as a test substance [*7*]. With volumes of 0.05 ml containing about 60 μg substances, the standard deviation was ±1.2%. This includes errors arising from spectrophotometric evaluation.

Whether application takes the form of spots or bands depends essentially on the quantity to be applied. The application of samples as streaks is relatively easy, without touching the silica gel surface, if a stencil is used (see p. 13).

For the quantitative estimation of substances separated by TLC the following methods have been described and are summarized in Tables 2 and 3:

Table 2. *Methods of Quantitative Evaluation of Thin-Layer Chromatograms without Extraction of Separated Substances*

Method I	Substances investigated	Quantities[2] used in μg	Error %	Literature
1. Visual comparison of stained spots with those of test substances[1]	D-Vitamins	0.1—1	semi-quantitative	[*16*]
	Ergot alkaloids of the clavin series	0.1 —30	semi-quantitative	[*18*]
	Retinene	0.05—0.1	approx. ± 30[3]	[*40*]
	Weakly polar steroids	0.1—150	semi-quantitative	[*5*]
2a. Production of photocopies and evaluation of spot area	Antioxidants (3-butyl-4-hydroxy-anisole)	20—150	approx. ± 5	[*32*]
	Tocopherols	10—60	approx. ± 5	(*33*]
	Synthetic aliphatic nitrogenous, phospho- and sulpholipids	10—130 acc. to substance	approx. ± 5	[*22*]
2b. Rendering chromatograms transparent. Production of photo-tracing and its spectrophotometric evaluation	Menthofurane content of peppermint oils	5—15	approx. ± 5	[*12*]
3. Photodensitometric evaluation	Mono-, di- and triglycerides	2—10	approx. ± 5	[*27*]
	Glycerides and their aldehydic degradation products	2—30	approx. ± 5	[*27*]
	Test mixtures of synthetic aliphatic nitrogenous lipids	5—35	approx. ± 10	[*22*]
	Cholesteryl esters	5—10	(± 1.5) —(± 5)	[*41*]
4. Autoradiographic evaluation of radioactively-labelled substances	Lipids	traces	—	[*21*]

[1] Only a few references are given here, as examples.
[2] In mixtures of substances, the values refer to individual substances, unless stated otherwise.
[3] Approximate values are given when no estimations of error were presented in the original publication.

Table 3. *Methods of Quantitative Evaluation of Thin-Layer Chromatograms after Extracion of Separated Substances*

Method II	Substances investigated	Quant. used[1]	Error %	Literature
1a) Weighing after location by fluorescence in UV-light	Neutral lipids from faeces and faecaliths	20.000 to 50,000 per plate	semi-quant.	[*39*]
1b) UV-spectrophotometry after location by fluorescence in UV-light	Rauwolfia alkaloids	50—70	± (3)—(5)	[*30*]
1c) Colorimetric estimation after location by fluorescence in UV-light	Ergot alkaloids	100—2000	± (0,5)—(4)	[*19*]
2a) Weighing after location in UV-light on fluorescent layers	α- and β-5-Cholestanol	10,000 per plate	approx ± 10[2]	[*4*]
2b) UV-spectrophotometry after location on fluorescent layers	p-Hydroxybenzoic acid ester Diphenyl (in citrus fruits and preparations)	20—70 2—120	± 3—5 —	Comp. Table 5 [*7*] [*17a*]
2c) Colorimetric estimation after location on fluorescent layers	Hydrocortisone and hydrocortisone acetate	10—100	± (2)—(16)	[*8*]
3a) Staining of radioactive substances and evaluation with Geiger-counter	Labelled derivatives of lipids	0,1—100	approx ± 5	[*21*, *24*]
3b) Location by iodine vapors Colorimetric determination of hydroxamic acid-iron complexes	Esters of fatty acids and hydroxy-fatty acids, also mono-, di- and triglycerides	100—1000	± 5	[*37*]

Continuation of Table 3

Method II	Substances Investigated	Quant. used[1]	Error %	Literature
3c) Staining with bromothymol blue and colorimetry after extraction with anthrone	Glycolipid components of cerebral lipids	10—40	± 8	[17]
3d) Staining with bromothymol blue and estimation of P-content after mineralization	Sphingomyelin from cerebral lipid	0.2—25 P	±0.01 P	[11]
	Plasma phospholipids	10—50	—	[17]
4a) UV spectrophotometry after location by means of reference chromatograms	Follicular hormones	approx. 5	± 10	[36]
4b) Location by reference chromatogram and colorimetric evaluation	Sugars	50	approx ± 5	[25]
4c) Location by spraying with water. Colorimetry after heating with sulphuric acid	Bile acids	10—60	±(1.5)—(3.0)	[9]

[1] In mixtures of substances the values refer to individual substances unless stated otherwise.

[2] Approximate values are given, when no estimates of error were presented in the original publication.

I. Determination without extraction of separated substances from the chromatogram (Method 1)

1. Method of visual comparison

With this method equal quantities of the solution to be analyzed and increasing quantities of a standard solution of precisely known content of the pure component to be estimated are alternately applied to the starting points and chromatographed. Then the spot sizes of the colored chromatogram are compared (with fluorescent spots this is done under a UV-lamp). The spot of the standard most nearly agreeing in intensity with the sample to be analyzed is chosen: this gives the result. The majority of analyses obtained by this method are only semi-quantitative (see Table 2).

This method, which, when applied in PC, often gives quite accurate results [*38*] frequently gives less accurate values when used with TLC. Fig. 38 shows the development of spot size in TLC as compared to PC.

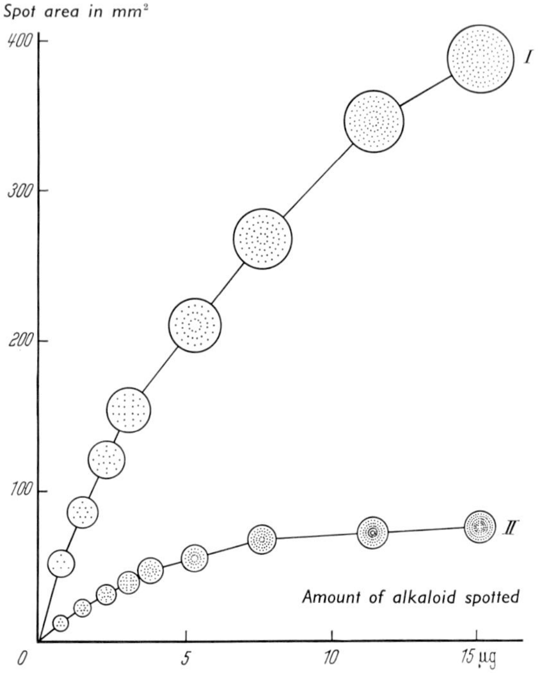

Fig. 38. Comparison between "spot areas" of increasing quantities of alkaloids with equal length of run. Curve I on formamide-impregnated paper. Curve II on a Silica Gel G layer. The density of stipling indicates the number of molecules [*35b*]

This method may be applied if the only staining methods known are those which produce rapid color-changes, or when the quantities of substances present are so small that accurate results could not be obtained with any other method.

2. Evaluation by photographic methods

Thin-layer chromatograms which after having been stained by a spraying reagent show highly contrasted spots, whose color remains fast for at least the duration of the analysis, may be evaluated objectively by means of the *Agfa "Copyrapid"* method. The dark spots obtained on a photostat may be measured by planimetry and can be evaluated by means of a calibration curve obtained by using known quantitites of the substance to be analyzed. A. Seher [*32*] used this method to determine the proportion of active agent in commercially available 3-butyl-4-hydroxy-anisole as well as for quantitative estimation of tocopherols [*33*] (cf. Tabel 2). From these investigations, it appears that it is necessary to prepare a calibration curve for each thin-layer chromatogram. This is to some extent due to the fact that the thickness of thin-layer plates varies slightly (cf. Fig. 39). Agreement between values within ±5% was obtained if, after several trials, the quantity of the component to be analyzed was chosen such that the results of the measurement fell within the range of the calibration curve.

An example of such a procedure is given: Determination of the degree of purity of 3-butyl-4-hydroxyanisole: 12 spots containing increasing quantities of the pure compound and in alternate sequence, spots containing a constant quantity of the sample are placed on the starting line of a thin-layer plate. The plate is developed with solvent and stained with phosphomolybdic acid solution. The areas of dark spots of a photostat are measured by planimetry. The mean of 6 planimetrically measured areas of the sample is then obtained. The areas, which were determined using the pure substance, provide the calibration curve from which the absolute quantity of the sample to be analyzed may be obtained.

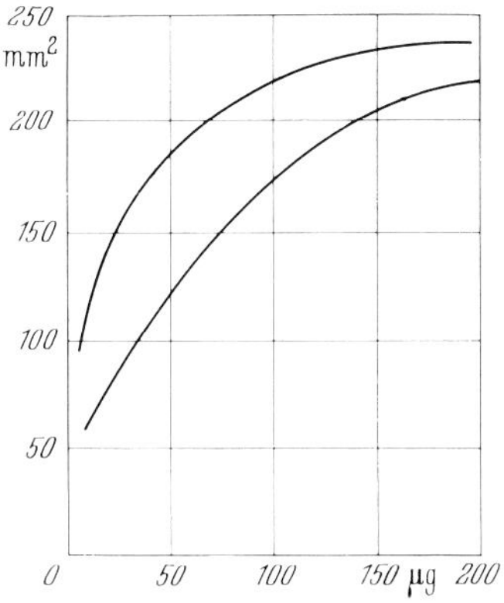

Fig. 39. Relationship of spot area to the quantity of substance applied (3-butyl-4-hydroxyanisole) on two different silica gel plates

PURDY and TRUTER [*28*], using the above data [*3, 33, 35b*], investigated the correlation between the planimetrically determined spot area and the quantity of substance applied. They described a graphical and arithmetical procedure for quantitative evaluation, and report that there exists a direct proportionality between the logarithm of the amount of the substance in grams and the square root of the spot surface area in square millimeters.

The "photoprint method" has been used also for the evaluation of thin-layer chromatograms. The spots are stained and the adsorbent layer is made transparent by means of an appropriate solvent. A phototracing is then made and, after dividing it into strips, it is evaluated in a spectrophotometer. This method, however, requires highly-contrasted spots which last for at least the period necessary for analysis, and also, that the adsorbent layer is made transparent without injuring the spots. This method of quantitative evaluation has, therefore, only a limited application. It was used by HEFENDEHL [*12*] to determine the menthofurane content of peppermint oils. When using this method, it is also recommended to make a calibration curve for each individual chromatogram.

3. Photo-densitometric determination after staining

PRIVETT and BLANK [*27*] converted mixtures of saturated and unsaturated glycerides by ozonisation and cleavage reactions into substances of varying polarity, separated them chromatographically on Silica Gel G-plates, 5 ×20 cm, and, after carbonization with aqueous sulphuric acid, evaluated them in a densitometer (cf. Fig. 40 and Tabel 2). In a further paper by these authors and LUNDBERG [*27a*], mono-, di- and triglycerides were separated and evaluated by densitometry (cf. Table 2). The mono-, di- and triglycerides were chromatographed individually with a solvent consisting of ether and petroleum hydrocarbon in varying proportions.

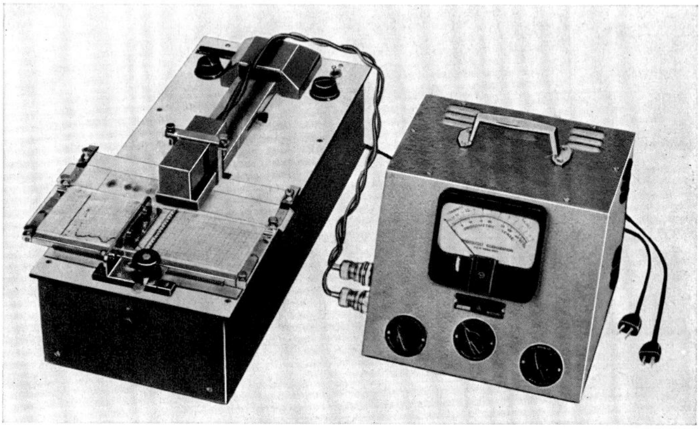

Fig. 40. Densitometer (Photovolt Corporation)

Method. After development, the plates were dried at room temperature, sprayed with 50% aqueous sulphuric acid and heated. The spots were evaluated in a densitometer [Photovolt Corporation, 52—C and 521 A (cf. Fig. 40)] which had accessories for semi-automatic measurement of the curves. The slit-width of the densitometer is 1 × 5 or 1 × 7.5 mm. Readings were made over the entire length of the chromatogram at intervals of 1 mm. No filters were used. Deviations due to the background were small. A linear relationship was found, over a certain range, between quantity of substance and spot density. However, the spot densities were not of equal size in all cases even where equal quantities of glycerides of identical or nearly identical carbon number, but of different structure, had been applied. Therefore, it was necessary to make calibration curves for each glyceride (cf. Fig. 41).

The densitometric method will lead to rapid results, if the "background" does not interfere very much and if there exists, over a certain range, a reproducible relationship between spot area and quantity of substance. Thus, thin-layer chromatograms of synthetic nitrogenous, as well as phosphorus and sulfur-containing aliphatic lipids, were evaluated densitometrically on plates, 20×20 cm after charring with chromic-sulphuric acid [*22*]. Test mixtures of saturated amines, amides and nitriles in quantities ranging from 5—35 μg, could be analyzed with an error of ±10% (cf. Table 2). It is possible to evaluate photocopies of such chromatograms in the manner described on p. 49, by a comparison of the spot surface areas. The error amounts to ±5% if 30—130 μg of substance are used [*22*].

ZÖLLNER et al. [*41*] separated mixtures of cholesteryl esters by TLC on silica gel layers into 5 main fractions which were stained with antimony trichloride and evaluated photometrically using various apparatus (an Electrophoresis scanner manufactured by Bender and Hobein, a chromatogram scanner for the Eppendorf-photometer and accessory for the Beckman DU G 4700). According to STAHL et al. [*35a*], the automatic recording and integrating double-ray reflection densitometer "CHROMOSCAN"[1] is best suited for rapid quantitative evaluation oft hin-layer chromatograms on plates 10×20 cm.

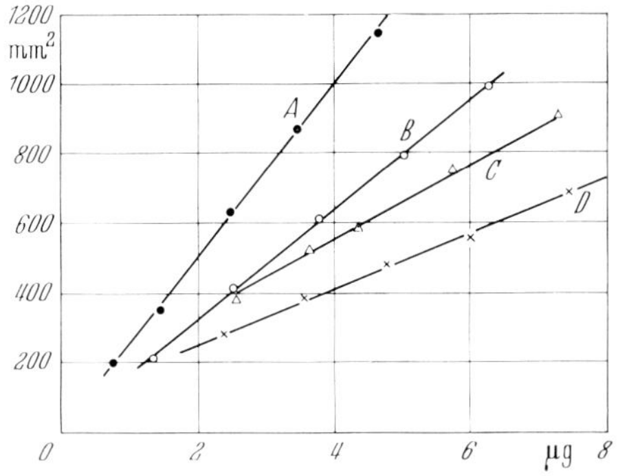

Fig. 41. Relationship of the spot area of various glycerides to the quantities of substance applied [*27a*]. *A* triolein, *B* tripalmitin, *C* dipalmitin, *D* monopalmitin

4. Autoradiographic evaluation

Radioactively-labelled lipids (see pp. 68—70) have been separated according to class, and the proportions of the various classes determined by autoradiometry [*21*] (see p. 60). A suitable film material was placed on the chromatograms in the darkroom, the films developed after an appropriate time and fixed. They were then evaluated densitometrically.

II. Determination of separated substances after extraction (Method II)

The evaluation of thin-layer chromatograms by spraying with a reagent followed by extraction and photometric evaluation is often fraught with difficulties. The ratio of concentration of reagent to the substance to be analyzed differs in the center of the chromatogram from that at the margin. The extent of reaction may thus be different. It also often happens that the stain is more difficult to extract than the separated substances themselves. Moreover, the background is frequently slightly stained by the reagent, and this leads to high blank values. On the other hand, substances separated by TLC may be further resolved and quantitatively determined by other methods, such as, for example, gas liquid chromatography (see p. 56). The spraying reagent should not interfere with GLC.

If, however, it proves possible to locate the separated substances, for instance by their fluorescence, or by their UV-absorption on layers impregnated with fluorescent material, without the risk of chemical changes taking place, and if the substances are present in sufficient

[1] Joyce, Loebl and Co., Gateshead-on-Tyne, Great Britain.

quantities to enable a physico-chemical estimation to be made after extraction, then this method should be used.

In many cases it is possible to carry out a photometric evaluation in UV-light or in the visible light after color development, as shown in Fig. 42, for the case of bile acids. It is convenient first to determine the blank value by elution of the adsorbent with an appropriate solvent; such methods are, for instance, described for Silica Gel G as adsorbent and methanol as extracting agent for evaluation in UV-light

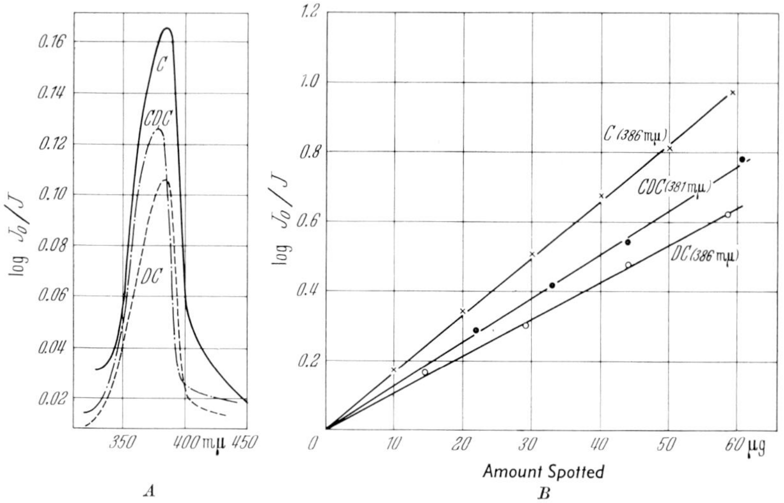

Fig. 42. Spectrophotometric determination of bile acids after extraction with diluted sulphuric acid which also served as reagent. *A* spectra of cholic acid (C), desoxycholic acid (DC) and chenodesoxycholic acid (CDC) after separation and staining. *B* Relationship between absorption and quantity applied to the plate

[7]. Hydrophilic solvents dissolve silica gel to some extent, but in some cases it can be removed by filtration of the extract through a short kieselguhr column. It must be borne in mind that the calcium sulphate binder is also soluble in water.

After these preliminary tests, it must be checked if the substance to be analyzed can be completely extracted with the chosen extracting solvent. Samples of the substance to be extracted are put on plates coated with the appropriate adsorbent. After using the same drying procedures to be used later in chromatography, the solvent is used to determine what proportion of the applied sample is recovered, and what errors occur on repeating this procedure (cf. Table 4).

In general, over 95% of the amounts applied can be re-extracted, with a suitable solvent. Once the spots have been applied to the starting point, the plates should not be exposed for any length of time before placing in the solvent, since, owing to the extraordinarily large contact area of sample and sorbent, atmospheric action may cause changes in the sample.

If after separation and extraction unexpectedly high and scattered blank values occur, it must not be forgotten that such values may be due to the incomplete removal of UV-absorbing impurities in the solvent.

Table 4.
Ethanol as Eluant for Hydrocortisone Alcohol and Acetate from Silica Gel G-Layers

	Quantity applied [μg]	Mean value of quantity recovered [μg]	Standard deviation, σ, or relative %
Hydrocortisone alcohol	100	99.1	1.4
Hydrocortisone acetate	100	101.2	1.4

for colorimetric estimation after extraction see p. 54.

$$\sigma\,\text{rel.} = \frac{\sigma \times 100}{MW} = \frac{\sqrt{\frac{\Sigma(f^2)}{N-1}}}{MW}$$

σ = Standard deviation
f = Deviation of individual measurements from mean
N = Number of measurements (in this case 10)
MW = Mean

1. Location by color or fluorescence

If the separated substances are colored or if they fluoresce when irradiated by UV-light, they may be easily recovered after chromatographic separation.

As far as possible, the wave length of the exciting radiation could always be stated when describing work with fluorescent substances.

Test mixtures of reserpine and rescinnamine were analyzed by this method [3] (cf. Table 3). After location by UV-illumination (254 or 366 mμ), the alkaloid spots were scraped from the plate, the non-fluorescent zones of corresponding size were treated similarly as blanks and their contents were extracted and submitted to UV-spectrophotometry. The same method was used for ergot alkaloid extracts of *Secale cornutum* [*19*]. In this particular case, the estimation of alkaloids was carried out by adding a suitable reagent, after extraction, followed by photometric evaluation in visible light. Similar methods were used to determine, quantitatively, the lipid content of faeces and faecoliths with a semi-micro procedure (20–50 mg) [*39*]. The lipids were recognized after separation by their fluorescence under UV-illumination, and then extracted. The residues were weighed after evaporation of the solvent (cf. Table 3).

2. Use of fluorescent layers

In the majority of cases, the substances are not immediately visible on the plate after separation. Organic compounds which absorb light in the UV range can be visualized by spraying the chromatograms with solutions of organic fluorescent dyes, UV-illumination, or by adding inorganic fluorescent material to the adsorbent. In either case, UV-absorbing organic substances can be recognized as dark spots on a fluorescent background when viewed in UV-light. Whereas spraying must be carried out most evenly to ensure a high sensitivity of detection, the addition of an inorganic fluorescent dye to the adsorbent is extremely simple and does not interfere with the extraction of the separated organic substances [*17a*]. The addition of 2–5% of ZS-Super "Riedel de Haen"[1] to the adsorbent, before the mixing of the material for spreading, has proved most satisfactory [*7*]. After location, the UV-absorbing substances may be extracted and estimated, for instance, by spectrophotometry. A detailed investigation into the possibility of location, followed by spectrophotometric estimation of the extracted components in the UV range, was undertaken with a test mixture of the methyl- and propyl-esters of p-hydroxybenzoic acid [*7*]. The extinction coefficient was given by the specific extinction at the absorption maximum of the substance to be estimated. It could be shown, experimentally, that the error is small if a test quantity of the substance to be estimated is placed on each plate, and the amount of unknown sample is then estimated from the extinctions observed (Table 5).

Table 5. *Quantitative Evaluation of p-Hydroxybenzoic Acid Methyl and Propyl Ester-Mixtures after TLC-Separation*
(Both substances applied in quantities of approx. 50 μg)

	Evaluation with constant extinction coefficient, recovered mean value %	Std. dev. σ rel (%).	Evaluation with test quantities of individual subst. which had also run on the plate. Recovered mean value %	Std. dev. σ rel. (%)
Methyl ester	96.0	±6.7	100.4	±3.2
Propyl ester	93.6	±5.7	98.0	±4.8

$$\sigma \text{ rel.} = \frac{\sigma \times 100}{MW} \qquad \sigma = \sqrt{\frac{\Sigma(f^2)}{N-1}}$$

σ = Standard deviation
f = Deviation of individual measurements from mean
N = Number of measurements (in this case 10)
MW = Mean

Application of method. 22 g of Silica Gel G, Merck, are mixed with 0.5 g fluorescent mineral ZS-Super (Riedel de Haen), the plates being prepared in the usual manner (see pp. 7—9). The mixtures of esters and test samples of the individual components are applied, as methanol solutions, in quantities of about 50 μg. After separation on the adsorbent, spots visible under short-wave UV-illumination (254 mμ) are marked with a needle. These spots, as well as adjacent silica gel serving as blanks are scraped off with a razor blade, placed into a small pointed tube (G. 5), according to GORBACH and extracted several times with about 2 ml of methyl alcohol using a magnetic stirrer. The extracts are filtered into a 10 ml glass flask through a G 4 glass filter rod (acc. to GORBACH), using a slight vacuum. After filling the flask with solvent, readings are taken at the absorption maximum using a spectrophotometer.

Essentially the same procedure may, of course, be applied after separation with a suitable solvent, location with UV-illumination and extraction, followed by photometric estimation in visible light. This method was used in the analysis of hydrocortisone alcohol and -acetate.

[1] Silica gels with fluorescence indicators, see footnote p. 31.

Staining was carried out after extraction with 3,3′-dianisyl-bis-4,4′-(3,5-diphenyl)-tetrazolium chloride [*8*] in basic solutions. Measurements were made at 520 mμ (cf. Table 3).

In a detailed investigation, quantitative estimations were also carried out of steroids on loose adsorbent layers of alumina or silica gel with added fluorescent substances, but without binding material [*4*]. The substances were separated by horizontal TLC, extracted after location in UV-light and weighed (cf. Table 2 and p. 257). Since loose adsorbent layers are very sensitive to shaking, and since they may only be sprayed with reagents when wet, their use is not recommended.

3. Staining of separated substances before extraction

As previously mentioned, it is not advisable to stain the separated substances on the chromatogram and to extract these colored derivatives and measure them photometrically. This method was used, however, with apparent success, by Barrolier [*2*], for the quantitative evaluation of amino acids which had been separated according to Brenner and Niederwieser [*3*]. No details were given regarding the reproducibility of this analytical method.

In certain cases, it is also possible to use spraying reagents to locate separated substances which, after the spots have been extracted, permit the use of a second staining reaction for colorimetric estimation, providing that chemical reaction between the two spraying reagents does not take place. With this procedure, a second reagent, which reacts with a different functional group in the substance being investigated, is used. Thus, fatty acid esters could be located with iodine vapors and estimated quantitatively after elution and conversion of the esters into colored iron complexes of hydroxamic acids [*37*] (see p. 157). The reproducibility of this method was checked with a mixture of methyl esters of palmitic acid, 6-hydroxystearic acid and 9,10-dihydroxystearic acid. When applying 100–1000 μg of the individual components in bands to the starting point, the standard deviation for 10 estimations was 4–7%. The same method was used for the analysis of mono-, di-, and triglycerides [*37*] (cf. Table 3). Glycolipid fractions of cerebral lipids were stained with ammoniacal bromothymol blue solution and were evaluated colorimetrically with anthrone sulphuric acid [*17*] [cf. Table 3). In order to be able to work reproducibly with minute quantities of solvents, calibrated micro-instruments of the "Ultramicro Analytical System", manufactured by Beckman Instruments, were used [*29*]. Lipid fractions of human plasma [11] and sphingomyelin fractions of cerebral lipids [*17*] were stained on thin-layer chromatograms with ammoniacal bromothymol blue, mineralized, and the phosphate content of the ash was estimated colorimetrically (cf. Table 3). When analysing plasma lipid fractions, Habermann et al. recovered 88–102% of the total phosphorous present. The limits of detectability were 0.2–25 μg P per spot with a standard deviation of 0.1 μg P. The silica gel blank value was 0.05–0.1 μg P/cm^2.

Evaluation of stained chromatograms by gas chromatography has also been described. By combining TLC and GLC, it proved possible to separate saturated from unsaturated fatty acids, and to evaluate them quantitatively [*23*] (see pp. 174—178 and Table 3).

When evaluating thin-layer chromatograms of radioactively-labelled substances, it is also possible to stain them first, and after extraction to measure their radiation with a Geiger-Müller counter. Lipid mixtures were analyzed with this method [*21, 24*]; 0.1–100 μg of the various lipids could be estimated with an error of $\pm$ 5% (cf. Table 3). If even smaller quantities have to be determined, direct autoradiographic evaluation of the chromatogram is preferable (see pp. 60—63).

4. Other methods of locating separated substances

The analysis of bile acid mixtures can be easily carried out (cf. Table 3). Following separation, water is sprayed on the chromatogram to make the bile acids visible. For quantitative estimation, the acids are extracted with 65% sulphuric acid which simultaneously serves as a reagent. After heating evaluation is made spectrophotometrically [*9*] (see p. 52).

As in PC, the separated substances may be located by comparison with a reference chromatogram. In addition to the sample, one puts, at a second starting point, a mixture of the substances to be estimated. After separation, the path of the sample chromatogram is covered and the reference chromatogram is stained. At the level where the test substances stain, the corresponding areas of the sample chromatogram are marked and evaluated quantitatively after extraction. This method can only be used if the qualitative composition of sample solution and test mixture is similar, since impurities in the sample may considerably alter the *Rf*-values of the various substances.

Completely smooth-layered plates must be used and all preliminary conditions must strictly be adhered to in order to ensure the most uniformly possible path of the solvent front (see p. 27). This method was used to separate and estimate, quantitatively, estrone, 17 β-estradiol and 16 α-, 17 β-estratriol [*36*] (cf. Table 3 and p. 258).

In order to prevent entry of the staining reagent (12% solution of antimony trichloride) into the sample, the chromatogram was not sprayed directly; instead, the reagent was sprayed on a blotter covered with filter paper which was then rolled on to the reference chromatogram. In the sample chromatogram, spots were removed at corresponding positions and then extracted and evaluated by UV-spectrophotometry.

Reference chromatograms were also used for the quantitative analysis of sugars [25] [cf. Table 3 and p. 467). Since the plates were produced by hand, and thus, the above preliminary conditions are not fulfilled, the method is reproducible only with difficulty. Silica gel layers, produced manually with starch as binder, were used to separate ipomeamarones from a naturally occurring mixture. The substance was located by means of a reference chromatogram, extracted and estimated colorimetrically after staining with 2, 4-dinitrophenylhydrazine [*1*].

With thin-layer chromatograms, substances may also be detected by means of biological methods:

a) Bacteriostatic and bactericidal compounds (see pp. 311—318).
b) Insecticides (see pp. 359—365).
c) Haemolysing compounds (see p. 388).
d) Bitter substances (see p. 387).

Mention must also be made of BAEHLER's [*1a*] method for location by direct sublimation from the layer onto a second glass plate.

Finally, mention must be made of HEYNS' and GRÜTZMACHER's [*12a*] combination of TLC and mass spectrometry. With this method the silica gel samples containing aromatic and heterocyclic compounds, steroids, derivatives of sugars, and amino acids were introduced directly into the mass spectrometer in a small glass capillary tube. At the operating temperature of 150–200° C, these compounds sublime directly into the electron beam.

Bibliography to Chapters E and F: General Section
(Documentation and quantitative evaluation)

[*1*] AKAZAWA, T., and K. WADA: Agr. Biol. Chem. (Japan) **25**, 30 (1961).
[*1a*] BAEHLER, BR.: Helv. Chim. Acta **45**, 309 (1962).
[*2*] BARROLLIER, J.: Naturwissenschaften **48**, 404 (1961).
[*3*] BRENNER, M., u. A. NIEDERWIESER: Experientia (Basel) **16**, 378 (1960).
[*4*] ČERNY, V., J. JOSKA a L. LÁBLER: Collection Czechoslov. Chem. Commun. **26**, 1658 (1961).
[*5*] DAM, M. J. D. VAN, G. J. DE KLEUVER and J. G. DE HEUS: J. Chromatogr. **4**, 26 (1960).
[*6*] EGGERS, J.: Phot. u. Wiss. **10**, 40 (1961).
[*7*] GÄNSHIRT, H., u. K. MORIANZ: Arch. Pharm. **293/65**, 1065 (1960).
[*8*] GÄNSHIRT, H., u. K. MORIANZ: unpublished.
[*9*] GÄNSHIRT, H., F. W. KOSS u. K. MORIANZ: Arzneimittel-Forsch. **10**, 943 (1960).
[*10*] GETZ, H. R., and D. D. LAWSON: J. Chromatogr. **7**, 266 (1962).
[*11*] HABERMANN, E., G. BANDTLOW u. B. KRUSCHE: Klin. Wschr. **39**, 816 (1961).
[*12*] HEFENDEHL, F. W.: Planta med. **8**, 65 (1960).
[*12a*] HEYNS, K., u. H. F. GRÜTZMACHER: Angew. Chem. **74**, 387 (1962).
[*13*] HILTON, J., and W. B. HALL: J. Chromatogr. **7**, 266 (1962).
[*14*] HORNUNG, W.: Handbuch der Agfa-Photopapiere. Düsseldorf: Karl Knapp Verlag 1955.
[*15*] HUHNSTOCK, K., u. H. WEICKER: Klin. Wschr. **38**, 1249 (1960).
[*16*] JANECKE, H., u. I. MAAS-GOEBELS: Z. analyt. Chem. **178**, 161 (1960).
[*17*] JATZKEWITZ, H.: Hoppe Seyler's Z. physiol. Chem. **326**, 61 (1961).
[*17a*] KIRCHNER, J. G., J. M. MILLER and R. G. RICE: J. Agr. Food Chem. **2**, 1031 (1954).
[*18*] KLAVEHN, M., u. H. ROCHELMEYER: Deut. Apotheker-Ztg. **101**, 477 (1961).
[*19*] KLAVEHN, M., H. ROCHELMEYER u. J. SEYFRIED: Deut. Apotheker-Ztg. **101**, 75 (1961).
[*20*] LICHTENBERGER, W.: Z. analyt. Chem. **185**, 111 (1962).
[*21*] MANGOLD, H. K.: Fette, Seifen, Anstrichmittel **61**, 877 (1959).
[*22*] MANGOLD, H. K., and R. KAMMERECK: J. Am. Oil Chemists' Soc. **39**, 201 (1962).
[*23*] MANGOLD, H. K., and R. KAMMERECK: Chem. & Ind. **1961**, 1032.
[*24*] MANGOLD, H. K., R. KAMMERECK and D. C. MALINS: Microchem. J., Symposium Vol. II. 697 (1962).
[*25*] PASTUSKA, G.: Z. analyt. Chem. **179**, 427 (1961).
[*26*] PETROWITZ, H. J.: Materialprüfung **2**, 309 (1960).
[*27*] PRIVETT, O. S., and M. L. BLANK: J. Lipid Research **2**, 37 (1961).

[27a] PRIVETT, O. S., M. L. BLANK and W. O. LUNDBERG: J. Am. Oil Chemists' Soc. **38**, 312 (1961).
[28] PURDY, S. J., and E. V. TRUTER: Chem. & Ind. **1962**, 506.
[29] SANZ, M. C.: Beckman Report **1**, 4 (1960).
[30] SCHLEMMER, F., u. E. LINK: Pharmaz. Ztg. **104**, 1349 (1959).
[31] SEHER, A.: Fette, Seifen, Anstrichmittel **61**, 345 (1959).
[32] SEHER, A.: Nahrung **4**, 466 (1960).
[33] SEHER, A: Mikrochim. Acta (Wien) **1961**, 308 [**2**].
[34] STAHL, E.: Chemiker-Ztg. **82**, 323 (1958).
[35] STAHL, E.: Naturwissenschaften **47**, 114 (1960).
[35a] STAHL, E.: Private communication.
[35b] STAHL, E.: Z. analyt. Chem. **181**, 303 (1961).
[36] STRUCK, H.: Mikrochim. Acta (Wien) **1961**, 634 [4].
[37] VIOQUE, E., and R. T. HOLMAN: J. Am. Oil Chemists' Soc. **39**, 63 (1962).
[38] WALDI, D.: Chromatographie, S. 67, E. Merck, Darmstadt 1959.
[39] WILLIAMS, J. A., A. SHARMA, L. J. MORRIS, and R. T. HOLMAN: Proc. Soc. Exptl. Biol. Med. **105**, 192 (1960).
[40] WINTERSTEIN, A., u. B. HEGEDÜS: Hoppe Seyler's Z. physiol. Chem. **321**, 97 (1960).
[41] ZÖLLNER, N., G. WOLFRAM u. G. AMIN: Klin. Wschr. **40**, 273 (1962).

Some recent noteworthy articles

AURENGE, J. et al.: Bull. Soc. Chim. France 8—**9**, 1732 (1963): Relation between weight of material and size of the spot.
BHANDARI, P. R. et al.: Pharmaz. Ztg. **107**, 1618 (1962): Preservation of chromatograms. Compare also "Neatan neu", prospectus E. Merck A.G., Darmstadt, Germany, and preservation lacquer "Fluka" for TLC, prospectus FLUKA A.G., Buchs, S. G., Switzerland.
BIRD, H. L. et al.: Analyt. Chem. **35**, 346 (1963): Spectrophotometry after extraction.
BRODASKY, T. F.: Analyt. Chem. **35**, 343 (1963): Determination of neomycine spots with a agar diffusion technique.
HANSBURY, E. et al.: J. Chromatog. **9**, 393 (1962): Polaroid-photographs.
MATTHEWS, J. S. et al.: J. Chromatog. **9**, 331 (1962): Localisation with iodine vapor. Spectrophotometry after extraction.
PURDY, J., and E. V. Truter: Analyst **87**, 802 (1962): Relation between weight of material and size of spot.
SEILER, N. et al.: Naturwiss. **50**, 643 (1963): Scanning of fluorescent spots (365 mμ).
SQUIBB, R. L.: Nature **198**, 317 (1963): Scanning of spots on plastic plates.
WALDI, D.: Arch. Pharm. **295**, 125 (1962): Visual comparison of spot areas.
WALZ, D. et al.: Experientia **19**, 213 (1963): Photographs, UV-photographs (365 mμ) UV-photostats.

G. Isotope Techniques

By

Helmut K. MANGOLD

The specific advantage of applying radioisotopes in chemical research is well known: it is the high sensitivity in detecting radioactive elements which is not affected by their chemical binding.

Chromatographic procedures have been used lately to separate and isolate chemically similar radioactive elements [*17*]. The same methods were also applied to fractionate radioactively-labelled organic substances. In a recent review article, ROCHE, LISSITZKY and MICHEL [*44*] have shown how significantly the different chromatographic separation methods have increased the application of isotopes, especially in biochemical research. Several authors have described special biochemical applications of radiochromatographic methods [*2*, *14*]. Of particular interest are CALVIN's studies [*13*] on the assimilation of radioactive carbon dioxide and the paper chromatographic analysis of the labelled primary products of photosynthesis in algae as well as in other green plants. Since FINK, DENT and FINK [*16*] described the photographic localization of radioactive substances on paper chromatograms, autoradiography has become an indispensable tool in investigations of the mechanism of photosynthesis [*5*, *6*, *13*] as in other biochemical fields.

Can thin-layer chromatography be equally adopted to fractionations of radioactively labelled compounds?

The author ventures to predict that in the near future, TLC will replace most other radioachromatographic techniques to a great extent. The following advantages of TLC should justify this presumption:

The technique of thin-layer chromatography is characterized as a micro-preparative method by the high capacity of the thin layer. On a single plate, 20 $\times$ 20 cm one can easily separate several mg of a mixture, and one may isolate radioactive compounds in quantities large enough for most subsequent chemical and biological studies.

Making use of TLC as an analytical method offers further advantages:

In contradistinction to all column chromatographic techniques, TLC, like paper chromatography, discloses the entire path of separation and one cannot fail to notice any nonvolatile substance.

Fractionations by TLC always appear much sharper than those obtained by column and paper chromatography. The sensitivity of detection is even greater than on paper chromatograms, because the separated substances are concentrated in much smaller areas. Therefore, substances with low radioactivity, as well as isotopes of low radiation energy, can be detected on thin-layer chromatograms. Possible quenching by the adsorbent layer seems to be of little significance.

The technique is simple to operate and most of the separations take less than one hour. These advantages prove especially valuable when working with very "hot" radioisotopes of a short half life. Long-lived radioisotopes, in most cases weak β emitters, are almost exclusively used for chemical and biological studies. Particularly important are radiocarbon, C^{14}, (half life 5,568 years, max. energy 0.155 MeV) and radioactive hydrogen, H^3, (Tritium), (half life 12.2 years, max. energy 0.018 MeV). Radioactive phosphorus, P^{32}, (14.3 days, 1.701 MeV), radioactive sulphur, S^{35}, (87.1 days, 0.167 MeV) and radioactive iodine, I^{131}, [80.4 days, 0.608 MeV (β)] are also often used.

I. Layers, solvents, and chemical methods of detection

The separation conditions specified in the various chapters of this book also apply when the different classes of substances are present in the form of radioactively labelled compounds. The fractionations may be effected by adsorption or partition on adsorbents, by partition in reversed phase on hydrophobic layers, by ion exchange, or by thin-layer ionophoresis and by thin-layer ionophoresis-TLC. The solvents are also the same.

Chemical reagents can be applied to thin-layer chromatograms bearing radioactive substances. In addition, various radiometric procedures may be used for the detection and for the quantitative determination of labelled compounds. Both chemical and radiometric detection methods should always be applied on the same thin-layer chromatogram in order to obtain information regarding both chemical and radio-chemical purity of a sample.

II. Methods of detecting radiation

Autoradiography is, above all, suitable for the detection of radioactive materials on plates. Detection by means of a Geiger-Müller tube is possible, but considerably more troublesome and generally less exact; therefore, it has scarcely been used so far. The same is true of proportional counting devices.

1. Autoradiography

Photographic emulsions are darkened by α-, β-, and γ-rays. For the photographic detection of radioactive substances on a plate, the dry chromatogram is pressed carefully against an x-ray film, and this "sandwich" is wrapped in black cloth and kept in the dark. It should be kept in mind that small amounts of chemical contaminants in the solvents may stick to the thin layer and cause formation of artifacts on the film. Hence, solvents should always be completely removed before the chromatogram is brought in contact with the photographic emulsion.

When working with several thin-layer chromatograms, it is advisable to mark the different plates with luminescent dye.

The film most commonly employed is "No-Screen Medical X-Ray Safety Film" from the Eastman Kodak Corporation[1]. The emulsion of this material is rather sensitive and the resolution [*49*] is adequate for the requirements of TLC. Similar films are also manufactured by several other firms[2,3,4,5]. "Supermix developer" from the General Electric Corporation[6] or any other x-ray developer[1,2,3,4,5] is suitable for processing the films. The developing is carried out in the dark or in red safety light. The developing time depends upon the temperature and on the age of the developer, ranging between 3 and 6 min at 20° C. X-Ray fixing salt may be obtained from the firms mentioned[1,2,3,4,5,6]. The films are fixed for from 10 to 30 min and, subsequently, rinsed for one-half to one hour with running tap water.

The length of exposure is governed by the activity of the separated substances, by the kind and energy of the radiation of the radioisotope used for labelling, and by the desired effect. Blackening of the film is obtained, when, depending on the type of isotope, from 1 to 10 million β particles appear per cm^2 during the time of exposure [*23*, *71*]. When working with radioactive carbon, one can assume that amounts showing an activity of at least twice the background in a Geiger-Müller-counter, will produce distinct blackening of the x-ray film after an exposure time of 2 days or more. The detection of compounds of radioactive carbon by autoradiography is from 10 to 100 times less sensitive on paper chromatograms.

Identical chromatograms of C^{14} labelled compounds are usually left in contact with x-ray film for 1, 2, 4 and 8 days. Some of the autoradiographs thus obtained will reproduce the components of the separated mixture in proportion to the quantities present. Other autoradiographs will show photographically over-saturated spots caused by the major substances, but, at the same time, minor contaminants will appear. In this way, one will get a complete picture of the qualitative composition of the separated mixture.

The chromatogram and the film must be kept in a refrigerator if an exposure of several weeks is necessary.

Thin-layer chromatograms of tritiated substances have to be exposed for more than a week, whereas compounds containing P^{32} or I^{131} can often produce good autoradiograms within 30 min exposure, and in most cases, within less than 6 hr.

It is sometimes necessary to detect compounds containing different radioisotopes and double labelled substances on the same chromatogram [*5*, *27*]. This is easily possible through autoradiography if the two isotopes differ widely in their half life and/or in the energy of the β-radiation. Chromatograms containing both C^{14} and S^{35} labelled components are exposed to x-ray film at once and after several months. The first autoradiograph shows spots caused by both isotopes, whereas the latter only detects the isotope with the longer half life; in this case C^{14}. As a rule, the second autoradiograph should not be taken before 5 to 10 times the half life of the short-lived isotope has elapsed.

A method of distinction, based on different radiation energies of the isotope, is illustrated by the classical example presented by Benson and Calvin [*5*]. Two x-ray films are placed on top of each other to cover a paper chromatogram carrying C^{14} and P^{32} labelled substances. After

[1] Eastman Kodak Corp., Rochester 4, N. Y., U.S.A.

[2] Agfa-Photofabrik, Leverkusen-Bayerwerk, Germany.

[3] Ansco, Vestal Pkwy East, Binghamton, N. Y., U.S.A.

[4] E. I. du Pont de Nemours & Co., Inc., Photo Products Division, Wilmington 98, Delaware, U.S.A.

[5] Dr. C. Schleussner Fotowerke G.m.b.H., Frankfurt a. M., Germany.

[6] General Electric X-Ray Department, Milwaukee 1, Wisconsin, U.S.A.

developing, the film closer to the paper chromatogram shows black areas caused by both isotopes. The second film, not having been in immediate contact with the paper chromatogram, shows only black spots caused by P^{32}. The radiation energy of C^{14} is so weak that it cannot pass through the first film to have any effect upon the second. P^{32}, however, with its 10 times higher energy, penetrates through the first film and causes spots on both the first and the second.

Similarly, I^{131} can be detected in the presence of S^{35} or C^{14} [*27, 28, 63, 65*]. An aluminum foil of < 0.1 mm thickness is used to screen off radiation of the radioactive sulphur or radio-carbon. The harder radiation of radioactive iodine passes through the foil and reaches the film.

Autoradiographs of both paper and thin-layer chromatograms can be quantitatively evaluated by photodensitometry or by measuring the spot size. The possibilities of these analytical procedures will be discussed later.

Inactive compounds can be detected by autoradiography after being activated by neutrons or in the form of labelled derivatives. This technique is elaborated upon in the section, "Analysis by means of radioisotopes".

2. Counting tubes and scintillation counters

The quantitative evaluation of paper chromatograms of radioactive substances by a Geiger-Müller-counting tube, by means of a proportional counter, or by scintillation counting, has often been described. Devices for the simultaneous detection and recording of material on an automatically transported paperstrip or on a two-dimensional chromatogram are commercially available. Pocchiari and Rossi [*42*] have described the best known apparatus in a recent review; less detailed descriptions can be found in various textbooks [*19, 26, 49*].

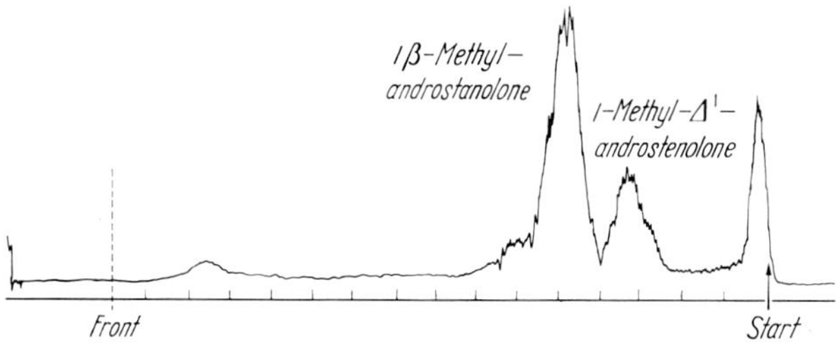

Fig. 43. Isolation of two tritiated steroids by thin-layer chromatography [*51*]. Adsorbent: Silica Gel G (0.9 mm). Solvent: cyclohexane-ethyl acetate, 60 : 40. Time: 40 minutes. Amount: 8 mg of the reaction mixture. The distribution of activity on the plate was measured with a special counting tube. Zones containing radioactive material were scraped off and eluted

Similar instruments for use with plates are not yet known, but several teams are occupied with the direct quantitative evaluation of radio thin-layer chromatograms.

Schulze and Wenzel [*51*] have constructed a counting tube for measuring tritiated compounds on plates. Fig. 43 shows a curve recorded by this instrument.

The quantitative evaluation of such curves is still rather inaccurate. So far, the preferred method has been to elute the radioactive substance from the adsorbent and to measure it in a conventional counter in order to eliminate self-absorption. The use of a liquid scintillation counter is essential if the different fractions contain different amounts of "cold" carrier substance. This will always be the case in biological investigations.

Several solvents used in scintillation counters are also suitable as eluents according to SNYDER and STEPHENS [*53*]. This facilitates the quantitative analysis of radio thin-layer chromatograms. Two frequently used "cocktails" have the following composition:

a) 5 g of 2,5-diphenyloxazol ("PPO") as primary scintillator, and 0.3 g of dimethyl-1,4-bis [2(5-phenyloxazol)-benzene] ("Dimethyl-POPOP") as secondary scintillator are dissolved in 1 liter of toluene.

b) 104 g naphthalin, 6.5 g PPO and 0.130 g POPOP in 5 ml of toluene, 500 ml dioxane and 300 ml methanol.

In order to avoid the formation of precipitates in the scintillation solvent, the vials should be counted at + 10° C. Moreover, according to SNYDER and STEPHENS [*53*], the activities of labelled components can be determined on Silica Gel G without prior elution. The adsorbed radioactive substance is scraped off the plate together with the Silica Gel G. The adsorbent is held in suspension by a gel of the counting solution (a) (Table 1) described above, i.e., a mixture of PPO, Dimethyl-POPOP in toluene and 4% of "Cab-O-Sil"[1] [*7*]. The cocktail is homogenized by vigorous shaking (c). A very high counting efficiency can be obtained. No quenching is observed with radio-carbon and up to 100 mg Silica Gel G, which equals app. 10 cm^2 of a standard thin layer, per counting vial containing 15 ml of cocktail (c).

c) 5 g PPO, 0.3 g Dimethyl-POPOP and 40 g Cab-O-Sil in 1 liter toluene.

The method of SNYDER and STEPHENS is, in the author's opinion, superior to all other radiometric methods used in TLC.

The direct measurement of the radioactivity on adsorbent layers is not very sensitive because of the "softness" of the glass background. The self absorption of the glass plate is too high for this method to be reliable or quantitatively reproducible.

The quantitative evaluation of autoradiographs by means of a photodensitometer is rather time-consuming. Substances occurring in small amounts are very hard to determine on autoradiographs in the presence of the main components of a complex mixture. This is due to the fact that the latter will have over-saturated the emulsion long before the effect of the radiation of the by-products will become visible.

The photodensitometrical or planimetrical evaluation of autoradiographs is completely impossible if the various radioactive fractions are diluted with different amounts of "cold" material.

III. Preparation of radioactively labelled substances

A molecule can be radioactively labelled in one or more positions by means of chemical synthesis. Procedures for the synthesis of tagged organic substances are described in several publications [*2, 14, 66*]. The monograph by MURRAY and WILLIAMS is the most complete [*41*].

[1] Packard Instrument Company, Inc., Box 428, La Grange, Ill., U.S.A.

Non-specifically or randomly labelled compounds can often be obtained through biosynthesis [*2, 14, 74*]. For example, the leaves of the plant *Canna indica* produce C^{14} labelled carbohydrates by assimilation of "hot" CO_2 for 10 to 20 hrs [*2, 14*]. In a similar manner, radioactive starch can be isolated from tobacco leaves [*14, 34*]. Radioactive peptides and amino acids are obtained from the yeast *Torula utilis* [*25*], more commonly, however, from algae such as *Chlorella vulgaris* [*33*] or *Chlorella pyrenoidosa* [*46*], which are grown in an atmosphere of radioactive CO_2 for a few days or weeks. If the latter alga is grown in a nitrogen deficient medium, it produces mostly lipids. This organism is, therefore, most useful for synthesizing randomly labelled fats and fatty acids [*36*]. The fungus *Phycomyces blakesleanus* has also been used for the biosynthesis of C^{14} labelled fatty acids [*8*]. "Hot" nucleic acids, nucleotides and nucleosides can be isolated from *Torula utilis* [*25*]. Soybeans are also suitable for the production of radioactive natural products [*15*].

A host of specifically labelled compounds is commercially available. Radioactive yeast and algae as well as protein hydrolyzates, saccharides and lipid extracts of these organisms are also on the market. A brochure published by the International Atomic Energy Commission lists the commercially available products and their specific activities as well as the addresses of the manufacturers [*25*].

The most commonly available products are frequently impure and can only be utilized in chemical and biochemical investigations after careful purification.

The classical methods for purification of the crude products of chemical syntheses are often not satisfactory when applied to a small amount of a radioactive preparation. In this connection reference is made to the criterion given by RAPPORT and LERNER [*43*] to prove the success of fractional crystallization. Recrystallization of an impure component is usually connected with great loss because the procedure has to be continued until some of the physical constants of the main fraction no longer alter. The above mentioned authors recommend performing the fractional crystallization in such a way that only 1 to 5% of the material remains in the mother liquor. By comparison of the composition of the *residue* with that of the main fraction, a far better picture of the purification success is gained than through the criteria used until now.

TLC is especially suitable for purifying radioactive compounds. Such preparative uses are described in a special section.

WILZBACH [*69*] found an elegant method of labelling organic compounds with tritium: The substance is exposed to tritium gas in a sealed glass container for several days or weeks. During this time, part of the H-atoms of the exposed molecules are replaced by H^3-atoms, but, at the same time, chemical side-reactions may take place. For instance, unsaturated compounds are often partly or completely hydrogenated. The Wilzbach method has just been described in a detailed monograph [*50*]. Several firms are doing this kind of labelling on requisition [*25*].

The difficulties of the Wilzbach method lie in the working up of the tritiated substances as well as in the purification of the compounds which are decomposed through radiolysis during prolonged storage. SCHULZE and WENZEL [*51*] have pointed out that TLC can be applied for this purpose with good success. Examples of the purification of tritium labelled compounds are mentioned in the following section.

IV. Isolation of radioactive compounds by thin-layer chromatography

Usually only a few mg. of a labelled compound are necessary for a radiochemical investigation. Such small amounts can be isolated by TLC without much effort. MANGOLD, KAMMERECK, and MALINS [*39*] applied TLC to the analysis and purification of C^{14}-, H^{3}-, and I^{131}-labelled lipids. Fig. 44 shows the autoradiograph of a thin-layer chromatogram of various commercially available fatty acids, labelled in the carboxyl group with C^{14}.

Every one of these preparations contains several contaminants. Conspicuously, most of these substances – intermediary products of the synthesis – are very polar, and this facilitates the isolation of the pure acids by TLC.

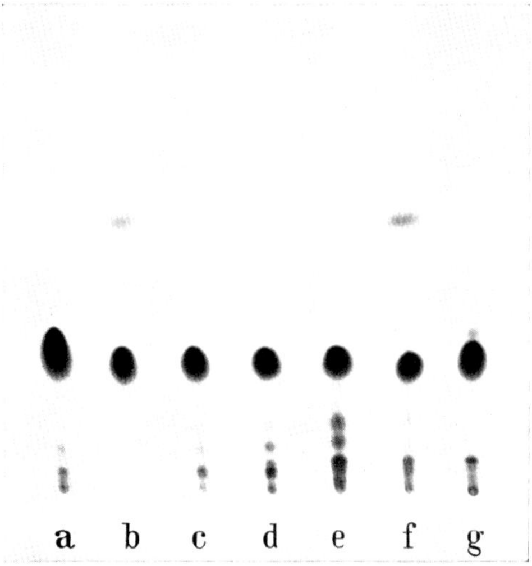

Fig. 44. Autoradiograph of a thin-layer chromatogram of labelled fatty acids. Adsorbent: Silica Gel G. Solvent: Petroleum hydrocarbon (B.P. 60—70° C) — diethyl ether — acetic acid (90 + 10 + 1). Film: Eastman Kodak "No-Screen Medical X-ray Safety Film". *a*) Lauric acid-1-C^{14}, *b*) Myristic acid-1-C^{14}, *c*) Palmitic acid-1-C^{14}, *d*) Stearic acid-1-C^{14}, *e*) Oleic acid-1-C^{14}, *f*) Linoleic acid-1-C^{14}, *g*) Linolenic acid-1-C^{14}

Appropriate solvents for elution of chromatographically separated radioactive compounds can be found by applying several polar solvents for chromatographing these substances. Solvents which carry a compound with the solvent front, are suitable for eluting this compound. The adsorbent is scraped from the plate and extracted with at least three portions of solvent. The clear solutions are then decanted and filtered over a small sintered glass funnel. The use of filter paper usually causes great losses of material.

As an example, the purification of tritiated stearic acid is described here. This commercial preparation contained at least seven impurities [*39*].

4.4 mg Stearic acid-9,10-H^{3} was applied to a plate, 20 × 20 cm, coated with Silica Gel G, in a line of spots and developed with a solvent consisting of petroleum hydrocarbon (B.P. 60—70° C), -diethyl ether-acetic acid, (80 + 20 + 1), for 1 hr. The separated substances were made visible by iodine vapors. The radioactive stearic acid was scraped off the plate, together with the adsorbent, and eluted with 3 portions, of 10 ml each, of diethyl ether-methanol-acetic acid) 80 + 20 + 1). The eluted fractions were combined, filtered, washed, and the solvent evaporated. Yield: 2 mg (45%).

If a compound is contaminated with substances of similar polarities, it may help to chromatograph it as a derivative. Acids, for instance, can be purified as methyl esters; alcohols and amines as acetyl compounds.

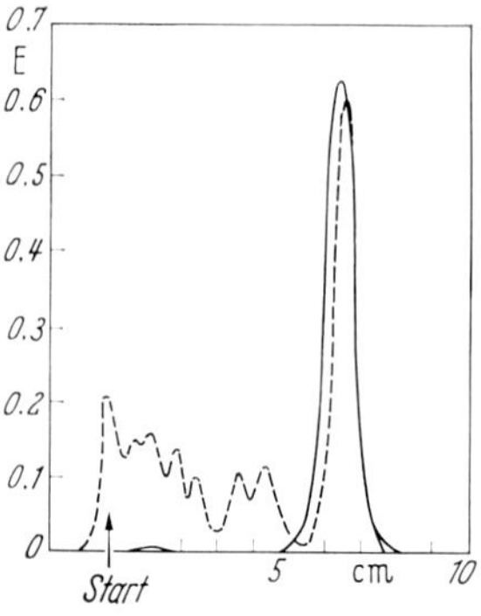

Fig. 45. Purification of radioactively labelled tripalmitin by thin-layer chromatography [*39*]. Densitometer curves were taken of the autoradiograph of a thin-layer chromatogram of commercial (-----) and purified tripalmitin (——) (for details see text page 66)

Derivatives like these are also much easier to elute than the polar starting material.

Figure 45 shows the photodensitometer curve of the autoradiograph of a thin-layer chromatogram of C^{14}-labelled tripalmitin before and after purification [*39*].

It was found advisable to use deactivated plates for the chromatography of substances containing ester groups. Deactivating the adsorbent layer can be accomplished by exposing the plates to air for at least one day. Fresh and highly activated silica gel hydrolyzes esters during chromatography.

Within an hour, 7.8 mg tripalmitin-1-C^{14} was separated on a Silica Gel G plate with the solvent petroleum hydrocarbon (B. P. 60—70° C)-diethyl ether-acetic acid (90 + 10 + 1). The triglyceride was located by means of autoradiography and was then eluted with 3 portions of 10 ml diethyl ether-methanol, (90 + 10) each. The eluted fractions were filtered through a small sintered glass funnel. The filtrate was evaporated and twice rechromatographed. Yield: 3,8 mg (48%).

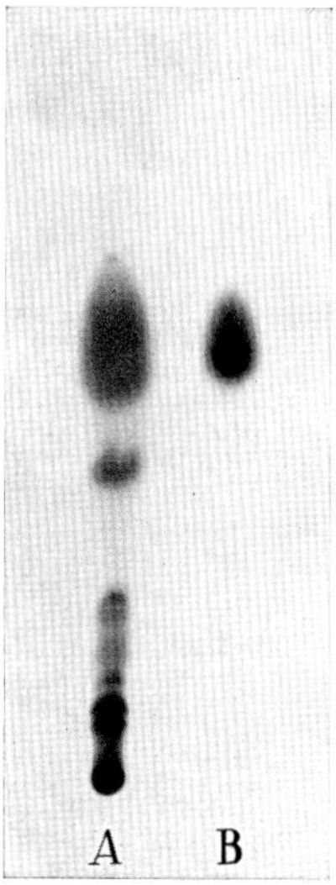

Fig. 46. Autoradiograph of a thin-layer chromatogram of commercial (A) and purified (B) "Radioiodinated-I^{131}-triolein" [*61*]

Amounts of several grams are best purified by column chromatography. Fig. 46 shows the autoradiograph of a plate of "radioiodinated triolein" (glyceryltri-diiodo-9,10-I^{131}-stearate) before (*A*) and after (*B*) having been purified by column chromatography.

The commercially available product is contaminated with iodinated mono- and diolein, as well as the corresponding acetyl compounds; also iodinated oleic acid, and a little methyl oleate [*61*]. Such material, as well as vegetable oils labelled with radioactive iodine, have been routinely used clinically for studying "fat" absorption!

Figure 43 demonstrates the separation of tritiated steroids on Silica Gel G by SCHULZE and WENZEL [*51*]. The authors determined the distribution of activity on the plate with a special counting tube. The areas containing activity were scraped off and the individual substances were eluted.

It is absolutely essential to verify the homogeneity of a substance purified by TLC by means of other methods. Compounds having been isolated by adsorption-TLC should be analyzed by partition TLC, paper chromatography and/or gas liquid chromatography.

V. Analysis by means of radioisotopes

The following radiochemical analysis procedures can be distinguished [*2, 10, 11, 26, 49*]:

Indicator analysis,
isotope-dilution analysis,
activation analysis,
and the isotopic derivative method.

These techniques are often used in combination with paper chromatography. To date there are only a few instances known of the application of these techniques in combination with TLC.

1. Indicator analysis

Indicator analysis proved valuable in testing the sharpness of chromatographic separations. This method has often been used in paper chromatography for this purpose [see *63, 65*].

TUNA, KAMMERECK, and MANGOLD [*62*] have proved, by indicator analysis, that fractions of naturally occurring lipid compounds separated by TLC are not grossly contaminated by one another. A small amount of "hot" tripalmitin was mixed with shark liver oil and separated by adsorption TLC. An autoradiograph of the chromatogram showed the total radioactivity to be in the triglyceride fraction. The glyceryl ether diesters were not contaminated with the triglycerides, although they are of a very similar structure (see Fig. 72, p. 152).

2. Isotope dilution analysis

The isotope dilution method is often used in biochemistry for the quantitative analysis of small amounts of compounds, which otherwise can only be measured very inaccurately. The substance to be determined must be available in radioactive form and its radioactivity must be known. A known quantity of the radioactive compound is mixed with the material to be analyzed and then isolated in pure form. The amount (M_1) in the original mixture can be calculated from the quantity (M_2) and the activity (A_2) of the added substance, and from the specific activity (A_3) of the isolated compound, by the following formula:

$$M_1 = M_2 \left(\frac{A_2}{A_3} - 1\right)$$

The quantity of radioactive material added is usually so little that it can be neglected in relation to the amount of "cold" substance. This means that the proportion of the activity of the added and of the isolated material indicates the yield of the separation and thus makes the determination of the unknown quite easy.

The isotope dilution method should be suitable for quantitative determination of substances giving a rather small yield if isolated by TLC.

In inverse isotope dilution, the radioactive sample is mixed with a known amount of a "cold" substance to be determined.

3. Activation analysis

Some elements can be converted into radioactive isotopes by neutron bombardment in a reactor, in a cyclotron or in a van de Graaf generator. The radioactive isotopes can be detected by radiometric methods. In this manner, phosphorus [*59*], sulphur [*59*] and chlorine [*71*] compounds have been detected on paper chromatograms. For the quantitative determination of traces of a compound, a known amount of this substance is irradiated on the same chromatogram.

Applications of activation analysis in connection with TLC are not yet known. Because calcium and sulphur are activated by neutrons, calcium sulphate is excluded as a binding agent in the adsorbent. For the same reason, one would have to use plastic plates instead of glass plates.

4. Isotopic derivative method

The isotopic derivative technique will often be helpful if classical analytical methods fail, either because too small amounts of the sample are available or because the mixture contains too small a concentration of the substance which has to be determined. The isotopic derivative method has proved useful even in cases where the technique of isotope dilution can not be applied because the substance to be determined is not available in radioactive form.

The principle of the method lies in the circumstance that an inactive element, radical or molecule reacts with a radioactive reagent. In this way it becomes detectable as a "hot" derivative with radiometric methods and it can be quantitatively determined.

The analysis can be performed according to one of the following principles:

a) Fractionation before radioactive labelling. The inactive sample is separated and the resolved fractions are made accessible to radiometric measurement by reaction with a radioactive reagent. For instance, amino acids have been located as radioactive derivatives on paper chromatograms by reacting them with methyl iodide, $CH_3\ I^{131}$, [*71*] or copper acetate, $Cu^{64}\ (AcO_2)$ [*68*]. Methyl bromide, CH_3Br^{82}, has been applied also [*70*]. Many unsaturated compounds can be detected by addition of radioactive iodine [*12*]. This technique has not yet been applied in connection with TLC.

b) Separation of radioactive derivatives. One mg or less of a mixture of compounds is first reacted with a radioactive reagent and the "hot" derivatives are then fractionated. The following agents were recommended as radioactive reagents for labelling inactive hydroxy and amino compounds: 3-chloroanisoyl chloride-3-Cl^{36} [*54*], 4-iodobenzyl chloride-I^{131} [*57*] and especially acetic anhydride-1-C^{14} or -2-H^3 [*3, 7, 24, 29, 38, 39, 67*]. Procedures for the acetylation of sterols and aliphatic lipids are given on page 71. Acids can be reacted with 4-chloroaniline-Cl^{36} [*55*] or with C^{14} labelled diazomethane ($C^{14}H_2N_2$) [*4, 37, 47, 48, 52, 58*]. Tritiated diazomethane cannot be used [*4, 32*]. A procedure for the esterification of long chain fatty acids is given on page 71.

The author has previously described a scheme for the analysis of radioactive lipid derivatives by means of adsorption-TLC or column chromatography and reversed-phase paper chromatography. Mixtures of fatty alcohols, mono- and diglycerides, and other lipids which can be acetylated, are reacted with tagged acetic anhydride. The acetyl derivatives are separated according to classes of compounds by adsorption-TLC or by column chromatography on silica gel. The relative amounts of the acetylated lipid classes are determined by radiometry of the eluate. Each of these groups of compounds is then further separated by reversed-phase partition on siliconized paper. Paper chromatograms of radioactive lipid derivatives can be quantitatively evaluated by means of a proportional counter provided with an attachment for automatic transport of the paper strips and for recording of the results.

MANGOLD, KAMMERECK, and MALINS [*39*] have analyzed the fatty acids of castor oil in a similar way. The acids were esterified with radioactive diazomethane. Adsorption-TLC was first used to separate the methyl esters of the common fatty acids from those of the monohydroxy and dihydroxy acids (Fig. 47). The three types of methyl esters were eluted and their ratios determined by measuring the respective radioactivities. The mixture of esters of non-oxygenated acids was further fractionated by reversed-phase paper chromatography and its composition was quantitatively determined by measuring the distribution of radioactivity on the paper strip with a proportional counter. It might have been better to determine the proportion of the monohydroxy and the dihydroxy fatty acids as "hot" acetylated acids or esters. The small amount of dihydroxy fatty acids would, in this case, have been labelled with two radioactive acetyl groups per molecule. This would have meant a doubling of the specific activity and the results would, thereby, have become even more exact.

Fig. 47. Autoradiograph of a thin-layer chromatogram of methyl-C^{14} esters from castor oil [*39*]. Adsorbent: Silica Gel G. Solvent: Petroleum hydrocarbon (B.P. 60 to 70° C). - diethyl ether - acetic acid (70+30+2) Time: 40 minutes. Film: Eastman Kodak "No-Screen Medical X-Ray Safety Film". *a*) esters of common fatty acids, *b*) esters of monohydroxy fatty acids, *c*) esters of dihydroxy fatty acids

Analyses of this kind can, in principle, be carried out with all classes of substances containing reactive groups [compare, e. g., *3, 29, 48, 67*]. The methyl esters of acids and the acetyl derivatives of alcohols and amines are less polar than the substances to be analyzed. It is, therefore, easy to elute them quantitatively from the adsorbent.

c) Fractionation after adding a radioactive derivative to the mixture of non-labelled derivatives. The sample to be analyzed is reacted with a "cold" reagent and a small amount of the radioactively labelled derivative of the compound to be determined is added to the mixture of derivatives. The amount of this substance in the original material can be calculated after isolating and determining the mixture of its cold and hot derivative by using the formula from the isotope dilution method (p. 67) [*48*, *54*, *55*]. This procedure has not yet been used in conjunction with TLC.

d) Separation after adding an inactive derivative to the mixture of radioactively labelled derivatives of the compound to be determined. This procedure represents an inversion of the isotope dilution method. The derivative of the substance to be analyzed is isolated in radiochemical purity and from the activity of this product, as well as from the activity of the added derivative, the concentration of the unknown in the original material can be calculated. This method was developed in 1946 [*27*] and has proved to be of special value in the analysis of protein hydrolyzates [*60*]. Uses of this procedure have not yet been described in conjunction with TLC.

e) Application of two different radioactive isotopes. The materials to be analyzed are reacted with a radioactive reagent. The substance to be determined is added to the reaction products in the form of a compound, chemically identical with the derivative present in the mixture, but labelled with a different isotope. After fractionation of the mixture, the effectiveness of the separation and the original amount of substance can be calculated from the results of separate measurements of the activities of both radioactive isotopes in one fraction and measurements of the total activity of the same fraction. This procedure was developed for the analysis of protein hydrolysates. I^{131}-labelled p-iodophenylsulphonylchloride and S^{35}-labelled p-iodophenylsulphonylchloride (I^{131}- and S^{35}-"Pipsylchloride") [*27*, *28*, *63*, *64*, *65*], as well as H^3 and C^{14} labelled acetic anhydride [*3*, *67*] are frequently used as radioactive reagents.

One can assume that this principle of analysis also can be combined favorably with the thin-layer chromatographic technique.

VI. Procedures for radioactive labelling

The procedures given below are useful for labelling small amounts of inactive acids and alcohols by "hot" esterification. Acids are reacted with radioactive diazomethane; alcohols with radioactive acetic anhydride.

1. Esterification of acids with diazomethane, $C^{14}H_2N_2$

Radioactive diazomethane can be obtained from nitrosomethyl-C^{14} urea [*52*][1]. The more stable p-tolylsulphonyl-methyl-C^{14}-nitrosamide [*4*, *37*, *58*] (radioactive "Diazald") is often given preference. A detailed procedure for the synthesis of this compound from methyl-C^{14}-amine has been described by STOLL and coll. [*58*, see also *9*]. Labelled Diazald is

[1] International Chem. and Nuclear Corp., 13332 E. Torch Street, City of Industry, Calif., U.S.A.

now also commercially available[1]. A test tube with a sidearm attached [*47*] or a micro gas-generator [*45*] are used for the preparation and distillation of radioactive diazomethane.

Procedure

A solution of 10 mg (0.05 mMole) p-tolysulphonylmethyl-C^{14}-nitrosamide, specific activity about 0.6 m Curie/mMole, in 1 ml diethyl ether, is mixed with 2 ml of a cold solution of 0.1 g sodium hydroxide in ethanol-water, 10 + 1, and heated in a stream of nitrogen on a water bath at 60—70° C. The diazomethane distills and is collected in two connected test tubes each containing 1—2 ml ice-cold diethyl ether. The two diazomethane solutions are combined and *at once* used to esterify the fatty acids in diethyl ether-methanol, 9 + 1 [*39, 47, 58*].

Diazomethane easily forms polymethylene and substances of as yet unknown structures [*40*]. Pure methyl ester can be separated quickly from these by-products by means of adsorption-TLC [*62*].

It is often favorable to reduce the mixture of acids or esters to be analyzed with lithium aluminum hydride quantitatively to alcohols and then to label them, radioactively, with acetic anhydride.

2. Acetylation of alcohols with acetic anhydride $(C^{14}H_3CO)_2O$ or $(CH_3^3CO)_2O$

The acetylation reaction is especially easy to perform. Acetic anhydride labelled with either C^{14} or H^3 can be obtained from several firms.

Procedure

10—20 mg of a lipid mixture containing compounds with free amino or alcohol groups are sealed in a glass ampoule, 6/150 mm, with a 20% excess of a solution of acetic anhydride-1-C^{14}, specific activity of 0.6 mCurie/mMole, and pyridine 1 + 10. The tube is heated in a water bath for 30—60 min at 100° C. After cooling, the tube is opened and the mixture is diluted with 10 ml N-sulphuric acid. The acetylated lipids are extracted with diethyl ether, washed neutral with several portions of water and dried over anhydrous sodium sulphate [*7, 24, 38*].

Compounds containing ester groups, such as mono- and diglycerides, are subject to acetolysis to a small extent. Hence, the acetylation of such substances should be carried out with utmost care. Reaction with radioactive ketene [*1*] might yield purer acetylation products.

A device for acetylation of amino acids has been described by WHITEHEAD [*67*].

VII. Application of thin-layer chromatography in chemical and biochemical investigations with radioisotopes

MAHADEVAN and LUNDBERG [*35*] used autoradiography of thin-layer chromatograms to prove the purity of synthetically produced cholesteryl oleate-C^{14}. Further applications of TLC in synthetic work have already been mentioned [*39, 51*].

[1] New England Nuclear Corp., 575 Albany Street, Boston 18, Mass., U.S.A.

Several laboratories have used TLC for biogenetic studies with radiocarbon. GRIESEBACH and coll. [*20, 21, 22*] traced the formation of isoflavones in different plants by means of radioactively labelled precursors. These authors used paper chromatography and two-dimensional TLC to separate the isoflavones, biochanine A and formononetine, as well as other metabolic products. KRATZL and PUSCHMANN [*30*] injected coniferine-3-C^{14} in the twigs of a birch tree and isolated two radioactive products by TLC which had been formed in the plant by methoxylation. KRATZL [*31*] has described the application of autoradiography of thin-layer chromatograms in a comprehensive review on the biogenesis of lignins.

FULCO and MEAD [*18*] made use of TLC in investigations concerned with the biosynthesis of hydroxy acids in the rat brain. STEIN and STEIN [*56*] traced the incorporation of radioactive fatty acids in the epididymal fat of the rat by the same technique. Reference is made to two recent publications by DHOPESHWARKAR and MEAD, which are discussed in the chapter "Aliphatic Lipids".

WINTERSTEIN, STUDER and RÜEGG [*72*] have isolated a series of carotenoids from natural products. In several cases, the preparation of uniform carotenoid fractions on a micropreparative scale was only possible by TLC. WINTERSTEIN [*73*] stressed, in a summarising report, that they succeeded only due to substantially improved methodology, especially due to the application of TLC acc. to STAHL and through the use of radioactive derivatives.

Bibliography to Chapter G: Isotope Techniques

[*1*] ALDERHOUT, J. J. H., G. K. KOCH and A. H. W. ATEN jr.: Rec. trav. chim. **76**, 7/2 (1957).

[*2*] ARANOFF, S.: Techniques of Radiobiochemistry. Ames: The Iowa State University Press 1956.

[*3*] AVIVI, P., S. A. SIMPSON, J. F. TAIT and J. K. WHITEHEAD: 2nd Radioisotope Conference, Vol. I, p. 313. London: Butterworth's Scientific Publications 1954.

[*4*] BAUMGARTNER, W. E., L. S. LAZER, A. M. DALZIEL, E. V. CARDINAL and E. L. VARNER: J. Agr. Food Chem. **7**, 422 (1959).

[*5*] BENSON, A. A., and M. CALVIN: Cold Spring Harbor Symposia Quant. Biol. **13**, 6 (1948).

[*6*] — J. A. BASSHAM, M. CALVIN, T. C. GOODALE, V. A. HAAS and W. STEPKA: J. Am. Chem. Soc. **72**, 1710 (1950).

[*7*] BERLINER, D. L.: Proc. Soc. Exp. Biol. Med. **94**, 126 (1957).

[*8*] BERNHARD, K., M. ROTHLIN, J. P. VUILLEUMIER u. R. WYSS: Helv. Chim. Acta **41**, 1017 (1958).

[*9*] DE BOER, TH. J., and H. J. BACKER: In Organic Syntheses **34**, 96 (1954), W. S. JOHNSON, Editor. John Wiley & Sons, Inc., New York, New York and Chapman & Hall, Ltd., London.

[*10*] BRODA, E., u. T. SCHÖNFELD: In Handbuch der mikrochemischen Methoden Band II, p. 130, edited by F. HECHT and M. K. ZACHERL. Wien: Springer-Verlag 1955.

[*11*] — — Die technischen Anwendungen der Radioaktivität. Berlin: VEB-Verlag Technik; München: Porta-Verlag 1957.

[*12*] BUDZYNSKI, A. Z., Z. J. ZUBRZYCKI and I. G. CAMPBELL: Nature **182**, 178 (1958).

[13] CALVIN, M.: Nobel-Lecture, 1961, see Angew. Chem. **74**, 165 (1962).
[14] — CH. HEIDELBERGER, J. C. REID, B. M. TOLBERT and P. E. YANKWICH: Isotopic Carbon. New York: John Wiley & Sons, Inc.; London: Chapman & Hall, Ltd. 1949.
[15] CHORNEY, W., N. J. SCULLY, L. H. MASON and H. J. DUTTON: In preparation.
[16] FINK, R. M., C. E. DENT and K. FINK: Nature (Lond.) **160**, 801 (1947).
[17] FINSTON, H. L., and J. MISKEL: Ann. Rev. Nuclear Sci. **5**, 269 (1955).
[18] FULCO, A. J., and J. F. MEAD: J. Biol. Chem. **236**, 2416 (1961).
[19] GLASSTONE, S.: Sourcebook on Atomic Energy, 2nd ed. Princeton, Toronto, London, New York: D. van Nostrand Company, Inc. 1958.
[20] GRISEBACH, H., u. N. DOERR: Z. Naturforsch. **15**b, 284 (1960).
[21] — L. PATSCHKE: Chem. Ber. **93**, 2326 (1960).
[22] — G. BRANDNER: Z. Naturforsch. **16**b, 2 (1961).
[23] HERZ, R. H.: Nucleonics **9**, 24 (1951).
[24] HOLLANDER, V. P., and J. VINECOUR: Anal. Chem. **30**, 1429 (1958).
[25] *International Directory of Radioisotopes*, Vol. I and II. The International Atomic Energy Agency, Wien 1959.
[26] KAMEN, M. D.: Isotopic Tracers in Biology, 3rd edition. New York: Academic Press Inc., Publishers 1957.
[27] KESTON, A. S., S. UDENFRIEND and R. K. CANNAN: J. Am. Chem. Soc. **68**, 1390 (1946).
[28] — — and M. LEVY: J. Am. Chem. Soc. **72**, 748 (1950).
[29] KLIMAN, B., and R. E. PETERSON: J. Biol. Chem. **235**, 1639 (1960).
[30] KRATZL, K., u. G. PUSCHMANN: Holzforschung **14**, 1 (1960).
[31] — Holz, Roh- u. Werkstoff **19**, 219 (1961).
[32] LEITCH, L., P. E. GAGNON and A. CAMBRON: Can. J. Research **28**B, 256 (1950).
[33] LIEBSTER, J., M. DOBIÁŠOVÁ, J. KOPOLDOVÁ i J. EKL: Collection Czechoslov. Chem. Commun. **26**, 1700 (1961).
[34] LIVINGSTONE, L. G., and G. MEDES: J. Gen. Physiol. **31**, 75 (1947).
[35] MAHADEVAN, V., and W. O. LUNDBERG: J. Lipid Res. **3**, 106 (1962).
[36] MANGOLD, H. K., and H. SCHLENK: J. Biol. Chem. **229**, 731 (1957).
[37] — J. L. GELLERMAN and H. SCHLENK: Federation Proc. **17**, 268 (1958).
[38] — Fette, Seifen, Anstrichmittel **61**, 877 (1959).
[39] — R. KAMMERECK and D. C. MALINS: Microchem. J., Symposium Vol. II 697 (1962).
[40] MORRISON, W. R., T. D. V. LAWRIE and J. BLADES: Chem. & Ind. (London) **1961**, 1534.
[41] MURRAY, A. III and D. L. WILLIAMS: Organic Syntheses with Isotopes, Vols. I and II. New York and London: Interscience Publishers, Inc. 1958.
[42] POCCHIARI, F., and C. ROSSI: J. Chromatog. **5**, 377 (1961).
[43] RAPPORT, M. M., and B. LERNER: J. Biol. Chem. **232**, 63 (1958).
[44] ROCHE, J., S. LISSITZKY et R. MICHEL: In Chromatographie en chimie organique et biologique, Vol. I, p. 321, sous la direction de E. LEDERER. Paris: Masson et Cie. Editeurs 1959.
[45] ROPER, R., and T. S. MA: Microchem. J. **1**, 245 (1957).
[46] SCHIELER, L., L. E. MCCLURE and M. S. DUNN: J. Biol. Chem. **203**, 1039 (1953).
[47] SCHLENK, H., u. J. L. GELLERMAN: Anal. Chem. **32**, 1412 (1960).
[48] — H. K. MANGOLD, J. L. GELLERMAN, W. E. LINK, R. A. MORRISSETTE, R. T. HOLMAN and H. HAYES: J. Am. Oil Chemists' Soc. **37**, 547 (1960).
[49] SCHMEISER, K.: Radioaktive Isotope, ihre Herstellung und Anwendung. Berlin-Göttingen-Heidelberg: Springer-Verlag 1957.
[50] SCHULZE, E., u. M. WENZEL: Tritium-Markierung. Herstellung, Messung und Anwendung nach WILZBACH ^{3}H-markierter Verbindungen. Berlin: Walter de Gruyter Verlag 1962.
[51] — — Angew. Chem. **74**, 777 (1962).
[52] SMIRNOV, B. P., R. A. POPOVA u. R. A. NISKANEN: Biokimiya **25**, 368 (1960).
[53] SNYDER, F., and N. STEPHENS: Oak Ridge Institute of Nuclear Studies Report No. 41 (1962). Anal. Biochem. **4**, 128 (1962).
[54] SORENSEN, P.: Anal. Chem. **27**, 388 (1955).

[55] — Anal. Chem. **28**, 1318 (1956).
[56] STEIN, Y., and O. STEIN: Federation Proc. **20**, 271 (1961).
[57] STOKES, W. M., F. C. HICKEY and W. A. FISH: J. Am. Chem. Soc. **76**, 5174 (1954).
[58] STOLL, A., J. RUTSCHMANN, A. v. WARTBURG u. J. RENZ: Helv. Chim. Acta **39**, 993 (1956).
[59] STRICKLAND, E. H., and A. A. BENSON: Arch. Biochem. Biophys. **88**, 344 (1960).
[60] THOMPSON, E. O. P., u. A. R. THOMPSON: In Fortschritte der Chemie organischer Naturstoffe, Band XII, p. 297, edited by L. ZECHMEISTER. Wien: Springer-Verlag 1955.
[61] TUNA, N., H. K. MANGOLD and D. G. MOSSER: J. Lab. Clin. Med. **61**, 620 (1963).
[62] — R. KAMMERECK and H. K. MANGOLD: unpublished.
[63] UDENFRIEND, S.: J. Biol. Chem. **187**, 65 (1950).
[64] — and S. F. VELICK: J. Biol. Chem. **190**, 733 (1951).
[65] — T. C. CLARK u. E. TITUS: Experientia **8**, 379 (1952).
[66] WEYGAND, F., u. H. SIMON: In Methoden der organischen Chemie, Band IV/2, p. 529, edited by E. MÜLLER, 4th. ed. Stuttgart: Georg Thieme Verlag 1955.
[67] WHITEHEAD, J. K.: Biochem. J. **68**, 662 (1958).
[68] WIELAND, T., K. SCHMEISER, E. FISCHER u. H. MAIER-LEIBNITZ: Naturwiss. **36**, 280 (1949).
[69] WILZBACH, K. E.: J. Am. Chem. Soc. **79**, 1013 (1957).
[70] WINTERINGHAM, F. P. W.: J. Chem. Soc., Suppl. **1949**, 416.
[71] — A. HARRISON and R. G. BRIDGES: Analyst **77**, 19 (1952).
[72] WINTERSTEIN, A., A. STUDER u. R. RÜEGG: Chem. Ber. **93**, 2951 (1960).
[73] — Angew. Chem. **72**, 902 (1960).
[74] WOODRUFF, N. H., and E. E. FOWLER: Nucleonics **7**, 26 (1950).

Some recent noteworthy articles

BROWN, J. L., and J. M. JOHNSTON: J. Lipid Res. **3**, 480 (1962): Radioassay of lipids on thin-layer chromatograms.
CSALLANI, A. S., and H. H. DRAPER: Anal. Biochem. **4**, 418 (1963): Scanning of layers that have been peeled off a plate.
DAVIDOFF, F., and E. D. KORN: J. Biol. Chem. **238**, 3199 (1963): Application in a study of the biosynthesis of fatty acids.
ERWIN, J., and K. BLOCH: J. Biol. Chem. **238**, 1618 (1963): Application in a study of the biosynthesis of fatty acids.
FRAY, G., and J. FRAY: Bull. Soc. Chim. Biol. **45**, 1201 (1963): Autoradiography of lipids.
GOLDRICK, B., and J. HIRSCH: J. Lipid Res. **4**, 482 (1963): Device for recovering labeled fractions from a chromatoplate.
HARLAN, W. R., and S. J. WAKIL: J. Biol. Chem. **238**, 3216 (1963): Application in a study of the biosynthesis of fatty acids.
KARLSON, P., R. MAURER and M. WENZEL: Z. Naturforsch. **18**b, 219 (1963): Purification of a steroid labeled by the Wilzbach technique.
MEYER, F., and K. BLOCH: J. Biol. Chem. **238**, 2654 (1963): Application in a study of the biosynthesis of fatty acids.
RICHARDSON, G. S., I. WELIKY, W. BATCHELDER, M. GRIFFITH and L. L. ENGEL: J. Chromatog. **12**, 115 (1963): Autoradiography of lipids.
RIVLIN, R. S., and H. WILSON: Anal. Biochem. **5**, 267 (1963): Technique for recovering labeled steroids from the liquid scintillation phosphor.
ROSENBERG, J., and M. BOLGAR: Anal. Chem. **35**, 1559 (1963): Scanning of chromatoplates.
SCHILDKNECHT, H., u. O. VOLKERT: Naturwissenschaften **50**, 442 (1963): Scanning of chromatoplates.
SHEPPARD, H., and W. H. TSIEN: Anal. Chem. **35**, 1992 (1963): Autoradiography of tritiated compounds.
SNYDER, F.: Anal. Biochem., in press. Instrument for removing small sections of thin layers for radiometry.

VAHOUNTY, G. V., C. R. BORJA and S. WEERSING: Anal. Biochem. 6, 555 (1963): Determination of free and esterified cholesterol.

Several firms offer instruments for scanning thin-layer chromatograms bearing labeled substances (see pp. 506—508, Nos. 4, 41, 50, 51).

H. Theoretical Aspects of Thin-Layer Chromatography

By

M. BRENNER, A. NIEDERWIESER, G. PATAKI, and R. WEBER

General Remarks

In principle, all chromatographic separation methods are alike: a mobile phase passes over a stationary phase and thereby transports different substances with different speeds in the direction of flow. On the basis of experimental technique, it is common practice to distinguish between elution chromatography, displacement chromatography, and frontal analysis. Thin-layer chromatography, as usually performed, belongs to the first type, i.e., elution chromatography. Therefore, displacement chromatography and frontal analysis will not be discussed in this chapter.

Under ideal conditions of elution chromatography, each migrating substance travels independently of other substances present. During migration, the individual substance particles constantly move back and forth between the two phases constituting the chromatographic system. After each entry into and before subsequent withdrawal from the mobile phase, any one of them resides there for a certain time interval. Adding up all such intervals, beginning at the onset of chromatographic migration, we obtain a sum of time intervals which may be defined as the particles "residing time." It corresponds, of course, to a fraction of the total time elapsed during the experiment. For identical particles, this "residing time" fluctuates around a mean which depends on their physical properties and the experimental conditions.

All particles are convectively transported only while residing in the mobile phase. They propagate, therefore, stepwise and in each step at an average rate which is identical for all of them and which corresponds to the flow rate of the mobile phase.

Actual differences in the propagation of various substances result only from differences in the "residing times" of the corresponding particles. The distance of migration of any one substance is obtained by multiplying the mean convective velocity with the average "residing time." Here again, we are evidently dealing with means, the deviations resulting from both the deviation from the average longitudinal velocity and the average "residing time."

It is understood, thus, that not all the particles of a substance S will migrate on a chromatogram the same distance in the direction of flow. Instead, they form a zone or band on the chromatogram with maximum concentration at the center of the band, or in case of perturbation[1], a band with asymmetric substance distribution. If there is a possibility for transverse movement of the particles, e.g., in paper and thin-layer chromatography due to diffusion perpendicular to the direction of flow, the bands acquire the shape of "spots." The effects caused by transverse diffusion and by the fluctuations of the parameters which determine the movement in longitudinal direction are more pronounced the longer the duration of the experiment. The bands of spots broaden, becoming diffuse with increasing time. The dispersion of the bands or spots is even observed when – relative to the stationary phase – small amounts of substance are used. In fact, it is qualitatively independent of the solvent or adsorbent capacity of the stationary phase[2].

In the past, a number of authors have attempted the mathematical description of the processes occurring during chromatography. By introducing simplifying assumptions, various theories of varying validity were obtained. They all have – though expressed in different terms depending on the underlying theoretical model – common features, and all of them are, in principle, applicable to all chromatographic separation methods, including thin-layer chromatography.

In the first section of this chapter, a brief account of a quantitative treatment is given which is based on a perceptual experimental model, not impaired by a priori assumptions (e.g., the theoretical plate concept). As long as partition or adsorption phenomena may be considered as being independent of concentration, this treatment is general and mathematically exact. We shall see how the distribution of a substance in a moving band is arrived at, how the distortion of a band can be explained, and how the rate of transport of a band – the *Rf*-value – is related to the adsorption or partition coefficient.

In a second section, the *empirically* observed relationship between chromatographic behavior (*Rf*-value) and chemical structure is discussed. It is the purpose of that discussion to rationalize these correlations which, in turn – by comparison of *Rf*-values – can, to some extent, be evaluated to aid in structure determination. A *theoretical* understanding, however, of the interdependence between partition coefficient and chemical structure still requires intensive further investigation and is not considered in this section.

In a third section, we shall deal, in detail, with the characteristics of thin-layer chromatography.

---

[1] As caused, for example, by concentration dependent on partition or adsorption coefficients, overloading of the chromatogram, irregularities in the packing or in the chromatography column.

[2] Ideally, the zone diameter increases with the square root of the time. Cf. Einstein-Smoluchowsky formula on particle displacement by Brownian movement and [2].

I. General theory of chromatography

Chromatography represents a continuous process. When theoretically treated, the continuous overall process is generally divided into a discontinuous series of successive elementary processes.

At first glance at least, the procedure adapted by MARTIN and SYNGE deserves the credit of being particularly conceivable. MARTIN and SYNGE arbitrarily divide the chromatographic column into segments, so-called theoretical plates, and then represent the development of the chromatogram as a consequence of binomial distribution of the chromatographed substance [3]. The analogy to the behavior of the solute in the Craig counter current distribution[1] seems to be perfect. It indeed applies until, in the course of further treatment, the binomial passes into a Poisson distribution. A reader not familiar with statistics is hardly prepared to appreciate the meaning of that step and to arrive at a full understanding of the final result. He may find it bewildering that substance distribution differs from that observed in a countercurrent distribution machine (cf. Figures 48 and 49). In other papers [5 to 17], the development of the chromatogram is derived from pure mathematical arguments.

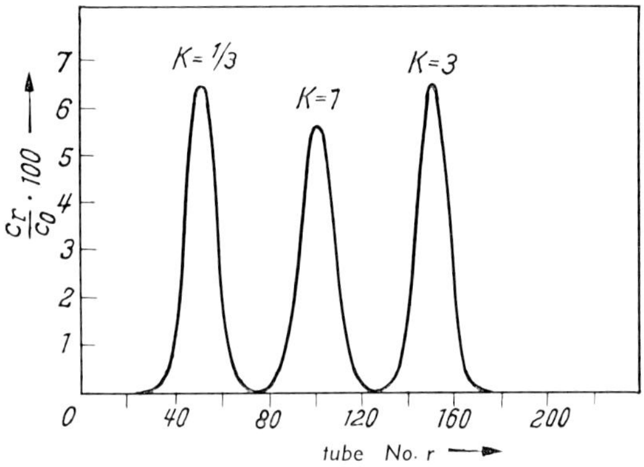

Fig. 48. Solute distribution caused by intermittent solvent flow (Craig), assuming 201 tubes, 200 transfers, equal volumes of mobile and stationary phases, and partition coefficients $K = \frac{1}{3}$, 1 and 3. c_0 = initial concentration in first tube (No. 1); c_r = final concentration in tube No. r (binominal expansion of $\left(\frac{1}{1+K} + \frac{K}{1+K}\right)^{200}$. (From LINSKENS, H. F., Papier-Chromatographie in der Botanik, p. 4, Springer-Verlag, Berlin 1959)

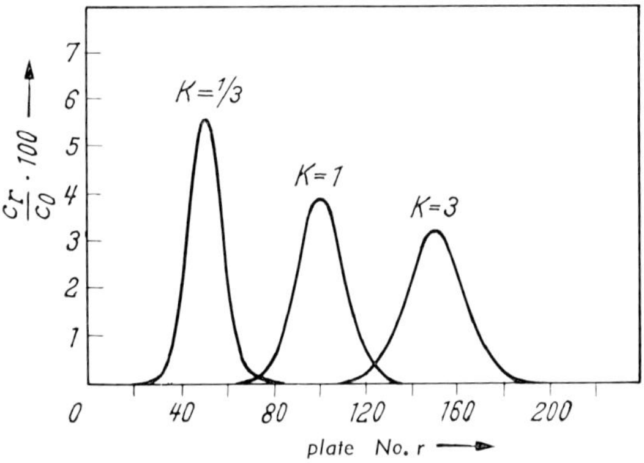

Fig. 49. Solute distribution caused by continuous solvent flow (MARTIN-SYNGE), assuming 201 "theoretical plates", equal volumes of mobile and stationary phases, and partition coefficients $K = \frac{1}{3}$, 1 and 3. c_0 = initial concentration in first plate (No. 1); c_r = final concentration in plate No. r after arrival of the solvent front in plate No. 201. Poisson distribution, graph of equation (9), with $p = K/(1+K)$ and $n = 201$. (From LINSKENS, H. F., Papier-Chromatographie in der Botanik, p. 4, Springer-Verlag, Berlin 1959)

To a laboratory worker, the principles of chromatography are preferably explained using a real model. In order to be adequate, that model should involve a continuous process. With this reasoning in mind, the authors of this chapter became aware of some

[1] L. C. CRAIG and D. CRAIG in [4].

work of R. SIGNER, of the Institute of Organic Chemistry at the University of Berne, Switzerland. SIGNER had been interested in the construction of a continuous countercurrent distribution machine, and in that context he had performed a number of elucidating model experiments. Enlarging on SIGNER's ideas and results, one may, in a very direct way, demonstrate the phenomena of chromatography and offer their quantitative description in a conceivable manner.

1. An introductory experiment

Forgetting for a while all about chromatography, consider a tank (volume v_m) filled with a liquid, e.g., water, and fitted with an inlet for filling, on outlet for discharging, and a very efficient stirrer ensuring thorough mixing within the liquid. If we wish to replace the water in the tank by some colored liquid, e.g., ink, we can do so by first pouring out the water and then pouring in the ink. The volume of ink required will obviously be equal to the volume of the tank, v_m. In an equally feasible procedure, ink can be introduced slowly at one end of the tank while water will simultaneously leave at the other end (Fig. 50a). However, due to the action of the stirrer, the outflow will contain some ink. Therefore, in this second procedure, a volume of ink greater than v_m will be required for complete water replacement. Now, what will happen if we divide the tank by one, two or more dividing walls into compartments, as indicated in Figs. 50b, c? From a series of experiments, we learn that the volume of ink required for complete water replacement decreases as the number of compartments increases. When this number becomes very large, the volume approaches the value v_m. This is again evident: when the compartment volume becomes comparable to the volume of the inflow, transmission and outflow openings, mixing in the direction of flow is no longer possible, except by longitudinal diffusion.

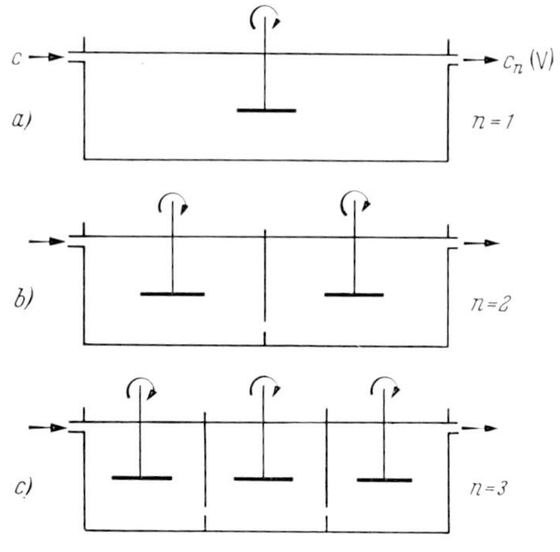

Fig. 50a—c. Tanks of volume v_m divided by separating walls into compartments of volume v_m/n. Each compartment is equipped with an efficient stirring device, and a small opening in each separating wall allows for liquid passage in the direction of flow

We conclude that the water is more efficiently replaced by ink if we more effectively prevent mixing of the two liquids in the direction of flow. In a successful mathematical treatment, the subject was indeed considered to represent a *mixing problem*, and the following equation was derived[1,2]:

$$c_n(V) = c - c \cdot e^{-\frac{V}{\frac{v_m}{n}}} \left[1 + \frac{1}{1!} \frac{V}{\frac{v_m}{n}} + \frac{1}{2!} \left(\frac{V}{\frac{v_m}{n}} \right)^2 + \cdots + \frac{1}{(n-1)!} \left(\frac{V}{\frac{v_m}{n}} \right)^{n-1} \right] \quad (1)$$

$c_n(V)$: substance concentration in outflow from a tank divided in n compartments, as a function of V
c : substance concentration in inflow
n : number of compartments of the tank
V : total volume of inflow = total volume of outflow, at the time of observation of $c_n(V)$
v_m : total volume of liquid in the tank.

When $n > 25$, equation (1) may be replaced by the more convenient approximate equation (1a):

$$c_n(V) = \frac{c}{2} \left\{ 1 + 2\Phi \left[\left(\frac{V}{v_m} - 1 \right) \sqrt{n} \right] \right\} \quad (1a)$$

wherein the term

$$\Phi \left[\left(\frac{V}{v_m} - 1 \right) \sqrt{n} \right] = \Phi(u), \text{ with } \Phi(-u) = -\Phi(u),$$

represents the familiar error integral[3]

$$\Phi(u) = \frac{1}{\sqrt{2\pi}} \int_0^u e^{-\frac{\tau^2}{2}} d\tau$$

Fig. 51 represents a plot of concentration $c_n(V)$ versus volume ratio V/v_m for different values of n, calculated according to (1). It is seen that with $n = 5$, a volume V of ink equal to about $3\ v_m$ is required to displace the water from the tank, whereas with $n = 25$ or 125, V decreases to about $1.6 \cdot v_m$ or $1.3 \cdot v_m$, respectively.

In the derivation of (1), mixing of the contents of a given compartment with the solution entering that compartment was assumed to be immediate and complete, thus effecting, at any time in any compartment, a homogeneous concentration, identical with the concentration of the solution leaving that compartment. Are such ideal conditions practically

[1] H. Hadwiger and P. Glur, Experientia **19**, 270 (1963).

[2] The present authors are aware that many equations in this chapter could be considerably simplified by introducing new auxiliary variables in place of some of the more complex terms. It was felt, however, to be preferable not to disguise the physical meaning of the parameters involved. Thus, for instance, $\frac{v_m}{n}$ is always seen to represent the volume of a single tank compartment, depending on v_m and n. The term $\frac{v_m}{n \cdot p}$ (p. 85) recalls to the reader the increase in compartment capacity due to a second phase present, and the dependence of this increase on the partition coefficient of the substance involved.

[3] Areas of the normal curve of error, c.f. Handbook of Chemistry and Physics ([*19*] pp. 210—213).

attainable? What is the result if they are not met in practice? By experiment, using a tank with 29 compartments, SIGNER found values of $c_n(V)$ shown in Fig. 51 by circles and crosses, respectively. Points indicated by circles refer to optimal conditions with regard to flow-rate and

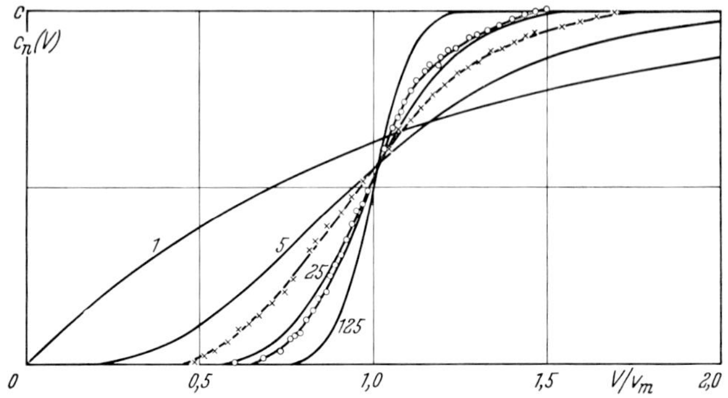

Fig. 51. Substance concentration in outflow from tanks with n = 1, 5, 25 or 125 compartments with respect to substance concentration, c, in inflow, as a function of V [18]. Full-line curves: calculated, using equations (1) and (1a). Curves with crosses or circles: experimental data, cf. text

speed of the stirrers. Observations indicated by crosses were made after a 4-fold increase of stirrer speed. The circles coincide with a theoretical curve illustrating the outflow from a tank with n = 30 ± 2 compartments. Evidently, experimental conditions in this case were such as to very nearly correspond to the ideal assumptions underlying equation (1). On the other hand, the curve drawn through the crosses suggests a tank with n = 13 ± 2 compartments. This last result may be expressed in two ways:

a) the tank has an efficiency of 45 ± 7%,

b) the tank operates with n′ = 13 ± 2 effective compartments.

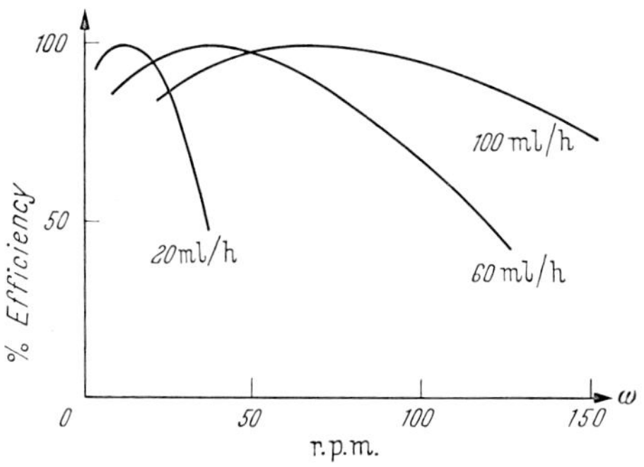

Fig. 52. Efficiency of a tank with 29 compartments as a function of flow rates, u, and of stirring speed, ω; v_m = 550 ml; apertures in dividing walls 2 mm; liquid contained in the tank: 7.5% ammonium benzoate solution; inflow: water. Concentration in outflowing solution determined by refractive index measurements [18]. Abscissa: revolutions per minute; Ordinate: % efficiency

In Fig. 52, the efficiency dependence upon experimental conditions is represented somewhat more in detail. On the basis of Figs. 51 and 52, the following conclusions may be drawn:

1. Ideal mixing within compartments in conformity with the assumptions made to develop a theory is, in practice, attainable (efficiency 100%) if flow-rate u and stirring speed ω are correctly balanced.

2. Imperfect mixing within compartments, which is due to ω being too small or to u being too large or to turbulent a flow (both ω and u large), reduces the performance of a tank in such a way that it operates like a tank with n′ effective compartments, n′ being smaller than n, the number of constructed compartments.

3. Partial mixing of the contents of neighboring compartments, which is due to u being too small (longitudinal diffusion), or to u as well as ω being too large (turbulent flow), also reduces tank performance as described sub 2.

4. The size of the transmission-openings (apertures) in the dividing walls between compartments is a further important parameter.

To summarize, the substance concentration, c(V), in a solution coming out of a SIGNER *tank depends, under otherwise identical conditions, on the degree of mixing between the ingoing solution and the liquid held by the tank. Lowering the degree of mixing in the direction of flow causes an increase of the slope of the c(V) versus* V/v_m *curve. Thus, the slope increases when the number of lateral dividing walls becomes larger. On the other hand, given a certain attainable degree of non-mixing by a certain number of tank compartments, the slope may decrease as a result of unappropriate operation favoring, e.g., longitudinal diffusion or turbulent flow*[1], *and thus stochastically*[2] *impairing the mixing process. The number of effective compartments, n′, has no physical meaning. It is, however, a convenient parameter characterizing tank performance, and its magnitude is easily determined from the relative position of the curve obtained on plotting observed values of c(V) against* V/v_m *as shown in Fig. 51. A tank composed of n compartments operated under non ideal conditions will behave like a tank composed of n′ compartments operated under ideal conditions, and n′ is always smaller than n.*

Now, imagine a vertically held glass tube filled with glass beads or some other *inert* material. Suppose you displace the remaining air by water, connect the top of the tube to a reservoir containing ink, and let ink slowly penetrate into the tube. The slope of the curve obtained on plotting ink-concentration c(V) versus V would increase when, under otherwise identical conditions (e.g., same volume v_m of liquid in the column, same rate of flow, same temperature), bead size were made smaller. It would also respond to the homogeneity of bead packing: irregularities should cause a flattening of the curve. Provided v_m were known, one could calculate or graphically estimate the number of the tube's effective compartments n′, which, in such a case, usually is called the number of theoretical plates. Actually, the envisaged arrangement

[1] Turbulent flow will cause certain elements of the liquid to go astray whilst others will travel through one or several compartments by the shortest possible route. Think of logs floating down a turbulent river!

[2] Systematic deviations from the assumptions made, e.g., layer formation in the compartments as a result of large density differences between solute and solvent, produce c(V) curves cutting across the family of curves shown in Fig. 51. Such cases can no longer be treated as pure mixing problems and our definition of the number of effective compartments becomes invalid.

corresponds to the arrangement used in "Frontal Analysis" [1], except for the fact that glass beads do not appreciably adsorb ink. However, this difference has no qualitative effect on the nature of the elution curve[1]. The quantitative effect of adsorption will be discussed later on.

2. Another experiment

The apparatus illustrated by Fig. 50 can serve us to perform a further experiment. Starting at left, give a number, 1, 2 . . . r . . . n, to each compartment. Put ink into the first compartment, water into all the others and connect to a water reservoir delivering water at a constant flow rate to compartment No. 1. Pass the outflow from compartment No. n through a colorimeter, recording ink concentration c (V) as a function of V/v_m.

For a while, we expect to observe an increase in c (V) similar to that noticed in the introductory experiment. Later, however, as ink in No. 1 is gradually replaced by water, c (V) must pass through a maximum, and then it will decrease until the outflow consists of practically pure water. Thus, the elution curve might reasonably be expected to exhibit some degree of symmetry. Indeed, it becomes symmetrical, at least for all practical purposes, when n is large.

A mathematical expression for $c_n(V)$ can be derived[2] from a general mathematical theory of continuous countercurrent extraction [21]; it is of course again assumed that mixing in each compartment is immediate and complete. For a tank with n compartments (numbering from 1 to n), HADWIGER's formula (40) reads in our notation:

$$c_n(V) = c_0 \frac{\left(\frac{V}{\frac{v_m}{n}}\right)^{n-1} \cdot e^{-\left(\frac{V}{\frac{v_m}{n}}\right)}}{(n-1)\,!} \qquad (2)^3$$

$c_0 = \frac{M_0}{\frac{v_m}{n}}$: initial concentration in compartment No. 1

V : total volume of outflow at the time of obversation of $c_n(V)$,

v_m : total volume of liquid in the tank.

[1] "Break through curve" would be a more appropriate term, putting emphasis on the fact that the column functions as on obstacle.

[2] In this context, the reader may also consult the paper of HADWIGER and GLUR referred to in footnote 1 on p. 79.

[3] Equation (2) may also be derived from equation (1) by differentiation with respect to $\frac{V}{\frac{v_m}{n}}$.

In this case, substance distribution in the outflow is of the Poisson type. Graphs of the function with n = 1, 5, 10, 20 and 29, $v_m = 1000$ ml. and $c_0 = 1$ g. per $\frac{1000}{n}$ ml. are shown in Fig. 53. With increasing n, the substance distribution takes the form of a symmetrical band with the concentration maximum in the center.

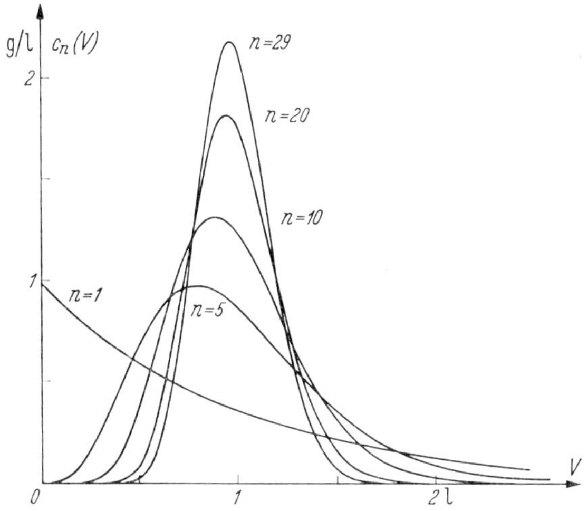

Fig. 53. *Graphs of equation (2), taken from* [22]
$v_m = 1$ liter, $M_0 = 1$ gram, $c_0 = n$ grams per liter. Areas under curves represent weights, i.e., $\int_0^\infty c_n(V)\,dV = 1$ gram. $c_n(V)$ is at a maximum when $V = V_{max} = \frac{n-1}{n} v_m$

Such c(V)-curves can, in fact, be obtained experimentally. In line with our experience gathered from the introductory experiment, we expect the effective number of compartments, n′, to depend on experimental conditions. The same conclusion is borne out from the mathematical considerations of HADWIGER and GLUR, cf. footnote 1 on p. 79.

Now, let us revert to the idea of the vertically held glass tube filled with glass beads. Suppose that we displace most of the air by water, displace the air remaining in the uppermost part by a small volume of ink, and then connect the top of the tube to a reservoir delivering water at a constant flow rate. Plotting ink concentration in the outflow versus time would yield an elution curve of the type observed in actual chromatography. This type of curve would be obtained despite the fact that adsorption or partition phenomena are not involved in the envisaged process.

Band formation is, in such a case, due exclusively to mixing phenomena within the liquid passing through the bed of glass beads. The number of theoretical plates n′, under a given set of operating conditions may be estimated by comparison of the observed elution curve [plot of c(V) versus $\frac{V}{v_m}$] with calculated curves [graphs of equation (2)]. The parameter n′ is a measure for column performance, expressed in terms of the column's capacity to reduce longitudinal mixing within the moving liquid. Otherwise, it has no physical meaning. – The setup just described corresponds to the setup commonly used in elution chromatography, except for the fact that glass beads do not appreciably adsorb ink. The influence of a non inert stationary phase on the elution curve will be discussed in the next chapter.

3. The model

The Signer tank, as used in the proceeding experiment, did not contain a *stationary* phase capable of extracting or adsorbing a solute from the moving liquid. Therefore, it was not yet a true model of a chromatographic column. We are now going to eliminate that imperfection. We enlarge the tank, raising its walls, and take advantage of the volume increase, v_s, by introducing a corresponding amount of a suitable liquid or solid forming a heavy phase henceforth called the stationary phase. The originally present liquid of volume v_m forms a layer above the stationary phase and is henceforth called the mobile phase.

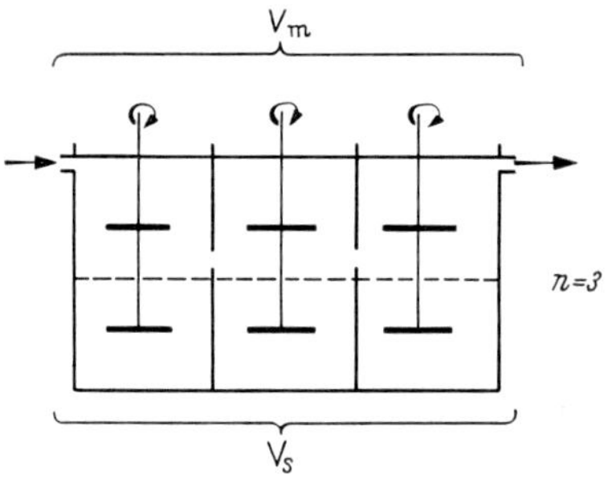

Fig. 54. Tank as in Fig. 50c modified to contain an additional stationary phase of volume v_S. The dotted line represents the interface between the mobile (upper) and the stationary (lower) phase. Equilibration within each compartment is, as nearly as possible, effected by stirring in both phases. With a solid phase in v_S, the tank will not actually work. For the sake of discussion, however, its functioning is supposed to be retained

Dissolve M_0 grams of a solute in compartment No. 1, connect that compartment to a reservoir containing mobile phase, and start passing mobile phase through the tank. What equation will describe the solute concentration in the outflow, $c_n(V)$?

In equation (2), for a given value of n, $c_n(V)$ was a function of $\frac{M_0}{v_m}$ and of $\frac{V}{v_m}$. V and v_m were quantities of the same phase. At every instant, their ratio determined the position (with respect to the tank) of greatest solute concentration. This position was found within the tank, when $V \leqq \frac{n-1}{n} \cdot v_m$, and outside, when $V > \frac{n-1}{n} \cdot v_m$.

In the modified tank, v_m is the volume of a phase in contact with another phase, of volume v_s. A solute present in either phase will partly pass to the other phase, until equilibrium is attained. Therefore, the volume v_m now no longer represents the total solute holding capacity of the tank. If the general form of equation (2) is to be retained, the term v_m of that equation must obviously be replaced by a corrected term, $v_m + \Delta v_m$ accounting for the total solute holding capacity, which is represented by the sum $v_m + v_s$, and the partition coefficient, K. It follows from well known relationships[1], that $\Delta v_m = \frac{1}{K} \cdot v_s$ and $v_m + \Delta v_m = v_m \cdot \frac{1}{p}$.

[1] cf. Footnote 1 on p. 85.

Substituting v_m in equation (2) by $v_m \cdot \frac{1}{p}$ yields

$$c_n(V) = c_o\, p \frac{\left(\frac{V}{\frac{v_m}{n \cdot p}}\right)^{n-1} \cdot e^{-\left(\frac{V}{\frac{v_m}{n \cdot p}}\right)}}{(n-1)!} \tag{3}$$

$$\text{wherein } p = \frac{K \cdot \varrho}{1 + K \cdot \varrho} \text{ and } c_o = \frac{M_o}{\frac{v_m}{n}}$$

We call the term $\frac{v_m}{p}$ the effective tank volume, and the term $\frac{v_m}{n \cdot p}$ the effective compartment volume.

Application of equation (3) requires a full understanding of its meaning. Such understanding will also help in deriving another equation (5) which is particularly useful for approximate calculations of $c_n(V)$. In fact, with that latter expression at hand, numerical application of the theory becomes an easy matter.

To begin with, let us look at Fig. 55. There are plots of $c_n(V)$, calculated according to equation (3), against outflow volume, V, representing elution curves from modified Signer tanks. It may be important to point out that $c_n(V)$, while denoting solute concentration in the outflow from compartment No. n, mathematically represents the solute concentration in the mobile phase within compartment No. n. In practice, when continually sampling and analyzing mobile phase from compartment No. n, we in fact analyze for solute contained within that compartment. The argument is based on the assumption that withdrawal of a volume dV from a vessel does neither appreciably affect the volume of the solution nor the amount of material remaining in the vessel.

Now, suppose, for a moment, that all mixing phenomena in the tank were missing. Under such conditions, the total solution originally present in compartment No. 1 were displaced to compartment No. n by a volume V equal to $(n-1)$ times the effective compartment volume, $v_m/n \cdot p$. This particular situation is, by the way, most easily visualized if $p = 1$. Throughout the displacing process, the solution would evidently retain its original concentration, c_0, and the outflow from compartment No. n

---

[1] $p + q = 1$; $\frac{p/v_m}{q/v_s} = \frac{p}{q} \cdot \frac{1}{\varrho} = K$ $\qquad \frac{1/\Delta\, v_m}{1/v_s} = K$

K	:	partition or adsorption coefficient, supposed to be independent of concentration. For the latter, this is only true at very high dilution.
v_s	:	volume (or weight) of liquid (or solid) stationary phase
p	:	fraction of solute in mobile phase, at equilibrium
q	:	fraction of solute in stationary phase, at equilibrium
ϱ	:	phase ratio (v_m/v_s)

would, of course, reflect this situation: all solutes were eluted in a volume fraction appearing between $V = (n-1)\frac{v_m}{n \cdot p}$ and $V = \frac{v_m}{p}$.

In practice, mixing phenomena do occur, and it is precisely their effect which is expressed by equation (3) and shown in Fig. 55 [cf. also equation (2). The volume V_{max} required to displace the concentration maximum from compartment No. 1 to No. n still equals $(n-1)\frac{v_m}{n \cdot p}$.

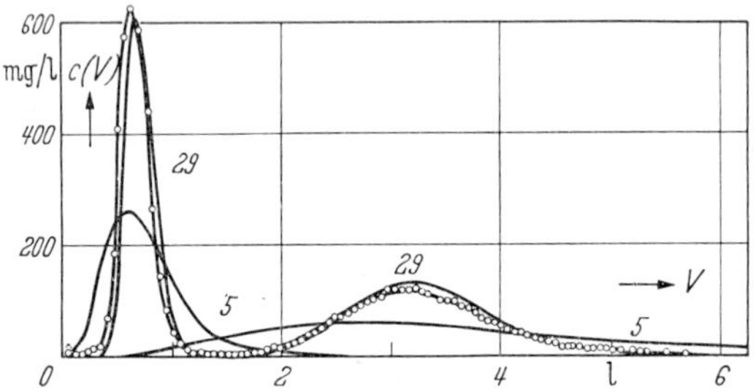

Fig. 55. Separation of a mixture of 200 mg each of benzamide and nicotinamide by a tank with 29 compartments. Mobile phase ethyl acetate, v_m = 580 ml, flow rate 60 ml per hour, diameter of apertures in separating walls 2 mm; stationary phase water, v_s = 580 ml. Stirring speed 25 r.p.m.; $K_{benzamide}$ = 3.59; $K_{nicotinamide}$ = 0.216. The dotted curve is experimental [*18*], and the full line curves are plots of data calculated from equation (3), with n = 5 and 29, respectively

This may be seen from the graph and is easily verified by differentiation of $c_n(V)$ with respect to V, setting the derivative equal to zero, and solving for $V\Big/\frac{v_m}{n \cdot p}$:

$$\frac{V_{max}}{\frac{v_m}{n \cdot p}} = n - 1 \qquad V_{max} = \frac{n-1}{n} \cdot \frac{v_m}{p}.$$

In spite of mixing, then, V_{max} retains its value. However, $c_n(V_{max})$ now equals $c_0 \cdot p/\sqrt{2\pi(n-1)}$ instead of $c_0 \cdot p$[1]. This strong degree of dilution is due to spreading of the solute within the tank. Evidently, when the volume ratio, $V\Big/\frac{v_m}{n \cdot p}$, has become equal to $V_{max}\Big/\frac{v_m}{n \cdot p} = n-1$, then only a fraction of all solute particles originally contained in compartment No. 1 has moved to compartment No. n. The displacement of this fraction is the result of (n−1) transfers undergone by each individual particle present in the fraction, one transfer being defined as the transition from compartment No. r to No. r+1. At volume ratios larger or smaller than (n−1), the particles assembled in compartment No. n, after (n−1) transfers of each, constitute still smaller fractions of the total solute. It follows that the value (n−1) of the volume ratio, while not occurring

[1] Solve equation (3) for $\frac{V}{\frac{v_m}{n \cdot p}} = (n-1)$, and $(n-1)! \approx \sqrt{2\pi(n-1)}\left(\frac{n-1}{e}\right)^{n-1}$ (stirling's formula).

exclusively, at least occurs more frequently than any other value securing (n−1) transfers of individual solute particles.

The number of particles, dN, eluted while the volume ratio changes from $x = V \Big/ \frac{v_m}{n \cdot p}$ to (x + dx) must be proportional to the number, N, of all particles initially present, to the magnitude, g(x), of the fraction present in compartment No. n at the time of elution, and to the width of the interval dx. It reads:

$$dN = N \cdot g(x) \cdot dx, \text{ with } g(x) = \frac{x^{(n-1)} \cdot e^{-x}}{(n-1)!} \,.$$

Consequently

$$\frac{dN}{dV} = N \cdot g(x) \cdot \frac{dx}{dV} = c_0 \cdot p \cdot g(x) = c_n(V), \text{ and } \Sigma\, dN = N \cdot \int_0^\infty g(x) \cdot dx = N.$$

The Poisson function in equation (3), equal to g(x), obviously represents a frequency distribution within the population of volume ratios, x, potentially effecting (n—1) transfers of an individual particle. There exists a mean, $\bar{x}$, of this population. It is defined[1] by

$$\bar{x} = \int_0^\infty g(x) \cdot x \cdot dx\,,$$

and it is found that $\bar{x} = n$.[2]

The variance of the distribution, σ^2, is defined by

$$\sigma^2 = \int_0^\infty g(x) \cdot (x-n)^2 \cdot dx$$

and it is also found that $\sigma^2 = n$.

We see that the mean of the distribution does not coincide with (n—1), i.e., the volume ratio occurring at greatest frequency. This finding is in line with the asymmetry of the distribution (cf. Fig. 55). However, with increasing n, the distribution tends to become symmetrical, and incidentally, the value of (n—1) approaches the value of $\bar{x} = n$. At very large values of n, the distribution becomes bellshaped, the location of the distribution maximum now practically coincides with the mean, $\bar{x} = n$, and $(n-1) \approx n$ now indicates the volume ratio observed at 50% solute elution[3].

If n is large, the frequency distribution of the volume ratios securing (n—1) transfers assumes the general form

$$\varphi(x) = \frac{1}{\sigma\sqrt{2\pi}} \cdot e^{-\frac{(x-\bar{x})^2}{2\sigma^2}}$$

Substituting $\frac{1}{c_0 \cdot p} \cdot c_n(V)$ for $\varphi(x)$, $V \Big/ \frac{v_m}{n \cdot p}$ for x, (n−1) for $\bar{x}$, and $\sqrt{n-1}$ for σ, gives

$$c_n(V) \approx \frac{c_0 \cdot p}{\sqrt{2\pi(n-1)}} \cdot e^{-\frac{\left[\left(V \Big/ \frac{v_m}{n \cdot p}\right) - (n-1)\right]^2}{2(n-1)}} \qquad (4)^4$$

[1] Page 12—13 in O. L. DAVIES, Statistical Methods in Research and Production, published by Oliver and Boyd, London 1958.

[2] Note: the mean outflow volume, after complete solute elution, is $V = \frac{U_m}{p}$.

[3] Note: the outflow volume at 50% solute elution now is $V_{50} = \bar{V} = \frac{v_m}{p} = V_{max}$.

[4] A mathematical derivation of (4) from (3) is given in [23], p. 116—117.

Expanding the exponent of e by $\left(\frac{v_m}{n \cdot p}\right)^2$, substituting V_{max} for $\frac{n-1}{n} \cdot \frac{v_m}{p}$, $\frac{M_0}{\frac{v_m}{p}}$ for c_0, and τ for $\frac{(V - V_{max})\sqrt{n-1}}{V_{max}}$, gives

$$c_n(V) \approx \frac{M_0 \cdot \sqrt{n-1}}{V_{max}} \left[\frac{1}{\sqrt{2\pi}} \cdot e^{-\frac{\tau^2}{2}}\right], \qquad (5)^1$$

wherein the expression in brackets represents ordinates of the Normal curve of error[2].

Note: If n gets very large, equations (4) and (5) become rather exactly valid. Otherwise, they represent useful approximations. Their use should be avoided if $n < 25$. A comparison between plots of equations (3) and (5) for $n = 29$ is offered in Fig. 56 in the Appendix; another one, for $n = 100$, is given in [*3*].

Equation (5) represents a Normal or Gaussian distribution with the mean $\mu_{(3)} = \frac{n-1}{n} \cdot \frac{v_m}{p}$, the standard deviation $\sigma_{(3)} = \frac{\sqrt{n-1}}{n} \cdot \frac{v_m}{p}$, and the ordinate $c_n(V_{max})_{(3)} = 0.3989 \frac{M_0}{\sigma_{(3)}}$, the subscript (3) referring to the derivation from equation (3). By the same sort of reasoning, equation (2) can be approximated by a Normal distribution with

$$\mu_{(2)} = \frac{n-1}{n} \cdot v_m; \; \sigma_{(2)} = \frac{\sqrt{n-1}}{n} \cdot v_m; \; c_n(V_{max})_{(2)} = 0.3989 \frac{M_0}{\sigma_{(2)}}.$$

While the terms μ and $c_n(V_{max})$ require no further explanations, a short comment on the meaning of the parameter σ appears to be justified. From a table[3] on data of the Normal curve of error, we learn that the areas under the curve between $\mu-\sigma$ and $\mu+\sigma$, between $\mu-2\sigma$ and $\mu + 2\sigma$, and between $\mu-3\sigma$ and $\mu+3\sigma$ comprise 68.26%, 95.46% and 99.74%, respectively, of the total area. Therefore, the width of the curve at its base line extends essentially from $\mu - 3\sigma$ to $\mu+3\sigma$. For most practical purposes, a width of 4 σ-units is considered to represent an acceptable approximation of peak extension along the base line. For an example, cf. Fig. 57 in the Appendix.

Examining the expressions for $c_n(V_{max})_{(2)}$, $c_n(V_{max})_{(3)}$, $\sigma_{(2)}$, $\sigma_{(3)}$, $\mu_{(2)}$ and $\mu_{(3)}$, we immediately see the effect of a stationary phase in a Signer tank on both the shape and V_{max} of the elution curve: peak height is diminished by a factor p, peak width is enlarged by a factor $\frac{1}{p}$, and V_{max} is also increased by a factor $\frac{1}{p}$. For a given phase ratio, ϱ, the value of p is uniquely determined by the value of the partition coefficient, K.

[1] An expression of this form may also be obtained from equation (1a) by differentiating with respect to $V \Big/ \frac{V_m}{n}$.

[2] Tabulated, e.g., in Handbook of Chemistry and Physics [*19*], p. 209—213.

[3] cf. Footnote 3 on p. 79.

Consequently, if K-values of two different solutes introduced simultaneously to compartment No. 1 of a modified Signer tank are different, then the individual values of V_{max} will also be different, and if the difference is sufficiently large, then one solute will still be inside the tank while the other has already completely emerged.

This prediction has been verified using a modified tank with 29 compartments (cf. Fig. 55) [*22*]. Again, it was possible to empirically find operating conditions affording a tank efficiency of nearly 100%. Under less favorable conditions, departure from ideal behavior may now be due to incomplete attainment of partition equilibrium or to non linearity of the partition isotherm, in addition to causes previously discussed. There is practically no change in V_{max} if $n' < n$, but the peaks broaden, and separation may become incomplete. This is readily seen from a comparison of theoretical curves, assuming a given tank is divided in $n = 5$ or 29 compartments, respectively (cf. Fig. 55).

Note: σ gets larger, V_{max} practically remains the same, and $c_n(V_{max})$ becomes smaller if operating conditions on a given tank reduce n to n'. On the other hand, σ gets smaller, V_{max} gets smaller, and $c_n(V_{max})$ becomes larger, if n is reduced because the length of the tank is reduced $\left(\frac{v_m}{n} \text{ and } M_0 \text{ remaining constant}\right)$.

So far, we considered elution from the modified Signer tank. In chromatographic practice, one is not necessarily concerned with elution. Especially in TLC and in PC, the chromatographic process is stopped when the solvent front arrives at the end of the layer or paper. Notwithstanding certain important differences to be discussed in chapter III, the experimental arrangement in TLC and PC is, in many ways, analogous to a modified Signer tank operated in the usual manner until $V = v_m - \frac{v_m}{n}$. By this time, solvent inflow is stopped. The mobile phase, i.e., the solvent originally present in the first compartment, now has as a "front" arrived in the nth compartment; in fact, its movement corresponds to the movement of ink as illustrated above. Where is the solute now?

Reverting to equation (3), we substitute the variable V by the constant term $\left(v_m - \frac{v_m}{n}\right)$, and the constant term $(n-1)$ by the variable $(r-1)$, r representing any number 1, 2, 3 ... n, and now get an expression for $c_V(r)$, the solute concentration in the upper phase of compartment No. r, when $V = v_m - \frac{v_m}{n}$.

It reads:

$$c_V(r) = c_0 \cdot p \frac{[(n-1) \cdot p]^{r-1} \cdot e^{-(n-1) \cdot p}}{(r-1)!} \tag{6}$$

and is valid as long as $p + 3\sqrt{\frac{p}{n-1}} < 1$[1].

[1] This restriction is borne out from a consideration on peak position and peak width outlined below. A peak moving closer to the end of the tank causes solute to be dammed up at that end, and the solute distribution is no longer in line with equation (6).

If L is the length of the tank, then the total amount of solute per unit length, at a distance $d_r = r \cdot \frac{L}{n}$ from the origin, is expressed by

$$M(d_r) = M_0 \cdot \frac{n}{L} \cdot \frac{c_V(r)}{c_0 \cdot p} \tag{7}$$

What is the meaning of equation (6)? Multiplying both sides of (6) with $\frac{v_m}{n \cdot p}$, substituting N_r for $c_V(r) \cdot \frac{v_m}{n \cdot p}$, N for $c_0 \cdot p \cdot \frac{v_m}{n \cdot p}$ and $f(r-1)$ for $\frac{[(n-1)p]^{(r-1)} \cdot e^{-(n-1)p}}{(r-1)!}$, gives

$$N_r = N \cdot f(r-1).$$

This expression tells us that the number of particles, N_r, starting from compartment No. 1 and arriving after (r−1) transfers in compartment No. r, is proportional to the number, N, of particles initially present in compartment No. 1, and to the relative frequency, f(r−1), of occurrence of (r−1) transfers. The location of the maximum of f(r−1) cannot be found by differentiation because the variable (r−1), unlike the variable V in equation (3), changes stepwise. However, we can guess by analogy that f(r−1) will become greatest when (r−1), i.e., the number of transfers, becomes equal to (n−1)p. Indeed, if $(n-1) \cdot p$ is an integer number, it can easily be shown that

$$f([n-1] \cdot p - 2) < f([n-1]\, p-1) = f([n-1]\, p) > f([n-1] \cdot p + 1).$$

By convention, we set $(r_{max}-1) = (n-1) \cdot p$, and we may, therefore, state that N_r is greatest in compartment No. $r = r_{max} = (n-1) \cdot p + 1$.

The numbers $(r_{max}-1)$ and $(n-1)$ correspond to distances traveled by the solute and the "front" of the mobile phase. Their ratio is usually called Rf, and we see that $Rf = p$.

The mean, $\overline{(r-1)}$, and the variance, σ^2, of f(r−1) are found to be

$$\overline{(r-1)} = \sum_{r=1}^{r=\infty} (r-1) \cdot f(r-1) = (n-1) \cdot p$$

$$\sigma^2 = \sum_{r=1}^{r=\infty} [(r-1)-(n-1)\, p]^2 \cdot f(r-1) = (n-1)\, p,$$

and, in this case, unlike the mean and the variance of g(x) on p. 87, are seen to coincide with $(r_{max}-1) = (n-1) \cdot p$. When $(n-1) \geqq 30$, a plot of f(r−1) against (r−1) approaches a smooth and almost symmetrical curve amenable to approximation by a Gaussian curve. Therefore,

$$f(r-1) \approx \frac{1}{\sqrt{2\pi(n-1) \cdot p}} \cdot e^{-\frac{[(r-1)-(n-1)p]^2}{2(n-1) \cdot p}}, \text{ if } n \geqq 30.$$

Expanding the exponent by $\left(\frac{L}{n}\right)^2$, substituting the resulting expression for $\frac{c_V(r)}{c_0 \cdot p}$ in equation (7) and rearranging, we obtain, for large values of n

$$M(d_r) \approx \frac{M_0}{d_\sigma \sqrt{2\pi}} \cdot e^{-\frac{(d_r - d_{r,\,max})^2}{2\,d_\sigma^2}} \qquad (8)$$

which is valid as long as

$$d_{r,\,max} + 3\,d_\sigma < L, \text{ and wherein}$$

$$d_r = \frac{L}{n} \cdot r, \quad d_{r,\,max} = \left(\frac{n-1}{n} \cdot p + \frac{1}{n}\right) L \approx L \cdot p,$$

$$d_\sigma = \frac{\sqrt{(n-1) \cdot p}}{n} \cdot L \approx L \cdot \sqrt{\frac{p}{n}}\,.$$

From the parameters of equation (8), we can directly see the influence on solute distribution caused by changing either the length of the tank, or the number of compartments, or both. *Rf* is independent of such changes. This statement still holds, of course, if operating conditions of the tank are not ideal and n', therefore, becomes smaller than n. Peak height, however, which is equal to $M_0/0.3989 \cdot d_\sigma$, will decrease under such conditions, and peak width will correspondingly increase. Any change in the phase ratio, $\frac{v_m}{v_s}$, will affect $p = Rf$ as well as the shape of the peak. The parameters also help us to estimate the degree of separation of two solutes with partition coefficients, K_1 and K_2, corresponding to values p_1 and p_2, respectively, of p. The solutes in the tank no longer overlap if the tail of the faster moving peak, located at $d_{r2} = d_{r2,\,max} - 2 \cdot d_{\sigma 2}$, has moved a longer distance than the front of the slower peak, situated at $d_{r1} = d_{r1,\,max} + 2 \cdot d_{\sigma 1}$, that is, when $d_{r2} - d_{r1} > 0$.

We conclude our considerations on the model represented by the modified Signer tank with the statement that a stationary phase added to the mobile phase simulates a solute specific variation of the volume, v_m, of the mobile phase, and thus, generates a possibility for separation of solutes differing from each other in the solute-solvent interactions expressed by the partition coefficient.

4. The chromatographic column

a) Comparison with the model

A chromatographic column is a tube filled with an adsorbent or with a carrier supporting a stationary liquid phase, or it is a strip of filter paper or a thin layer of some material capable of reversibly extracting a solute from a solvent. Under certain restricting conditions explicitly stated below, a solute transported by a moving solvent through a properly packed chromatographic column behaves very much like a solute transported through a Signer tank containing a stationary phase. Indeed, solute distribution during development of the chromatogram and upon elution looks like solute distribution effected by the tank. We should, as a matter of fact, expect such behavior. It is true that a column is not divided into compartments and is devoid of any device for stirring. However, it is just as obvious that the column filling presents an obstacle to spreading of the solute in the direction of flow, and that equilibration of concentrations in the direction perpendicular to the column axis is, owing to shortness of distances involved, efficiently achieved by diffusion. There is a difference to the Signer tank only in so far as the efficiency of the column cannot be predicted from a known number of compartments, and that the other parameters involved in determining solute distribution, p and v_m, are, in general, also unknown. However, in the case of solute distribution by the tank, n, p and v_m may be estimated from the outcome of experiments, i.e., from an analysis of elution and distribution curves. The same procedure can be adapted to chromatographic results. These can be interpreted in terms of:

a number, n', of "effective compartments",

a volume, v_F, of mobile phase within the column, which generally corresponds to the difference between the volume of the column and the volume occupied by the stationary column filling, and is equivalent to v_m of the tank, and

a factor p' simulating a solute specific variation of v_F, thus effecting a solute specific solute retention by the column.

The number n' is called the number of theoretical plates. It is a measure of column performance in the same way as n' was a measure of tank performance. The factors affecting the magnitude of n' are listed in Table 6, both for the tank and the column, in order to demonstrate the analogy of the two arrangements. In practice, however, there is one important difference between the tank and the column. All our mathematical considerations involving n and n' were based on the assumption that the solute was initially contained in compartment No. 1 of the tank. In the case of the column, we apply the solute to the top of the column, avoiding, of course, any unnecessary spreading. Nevertheless, we do not know the number of theoretical plates thus filled with solution prior to onset of the chromatographic distribution. If these plates get too numerous, there will be an increase of spread in the final solute distribution. This is equivalent to an impairment of column performance, and the situation can be expressed in terms of a decrease of the number n'.

The factor p' is an equivalent to the factor p accounting for the effect of a stationary phase in the tank. Like p, it varies between the limits 0 and 1, depending on the solute, the solvent, the stationary phase and the phase ratio. Unlike p, on the other hand, p' usually is rather ill defined with regard to the underlying partition mechanism. In a column this mechanism is often unknown. That is the reason why we distinguish between the two quantities. In fact, p' represents a generalized p. It expresses solute retention in a general sense, and, in principle, there is no difference whether such retention is due to reversible adsorption on a solid surface, enrichment at an interface between liquid phases, partition between liquid phases, solubility of a vapor in a liquid phase, or to a combination of some of these or to eventually still other mechanisms. Even such phenomena as solute exclusion from a stationary phase, observed in the case of gel filtration on Sephadex or of ion exclusion by an ion exchange resin, may be expressed in terms of a factor p'.

Table 6. *Comparison of Mechanical Factors Responsible for Solute Distribution Observed in a Signer Tank and in a Chromatographic Column*

Tank	Chromatographic Column	General Effect	Effect on n′ Increase or no decrease + Decrease —
compartment walls	column packing	obstacle to longitudinal mixing	+
compartments	void spaces	no obstacle to lateral mixing	+
stirrers	convection and diffusion within and between void spaces[1]	lateral mixing	+
apertures in walls compartment	void spaces	longitudinal transport	±
mobile phase	solvent	longitudinal transport, lateral equilibration	
a) optimal flow rate		deviations from ideal behavior at minimum	±
b) flow rate too fast		lateral mixing and equilibration incomplete, turbulent flow, irregular longitudinal transport	—
c) flow rate too slow		noticeable longitudinal diffusion	—
Stirring speed optimal	diffusion optimal due to sufficiently fine and regular packing	uniform lateral mixing	±
stirring speed too fast (turbulence)	convection due to coarse or irregular packing	non uniform lateral distribution, irregular longitudinal transport	—
stirring speed too slow	column too wide[1]	lateral mixing and equilibration incomplete	—

[1] Note the significance of a small column cross section!

We arrive at the following conclusion, supported by experience:

The mathematical expressions derived for solute distribution effected by our model tank can be applied to solute distribution effected by a chromatographic column, provided:

a) p' is independent of solute concentration, and

b) the flow rate of the solvent is the same in all parts of the column and constant throughout the experiment.

Consequences of concentration dependence of p' and of variation of the flow rate will be discussed in paragraph e.

b) Developing a chromatogram

Solute distribution within a chromatographic column is the process that "develops" a chromatogram. For instance, an initial band of mixed solutes, while moving along the column, is gradually split up into individual bands, representing single components of the original mixture. The process can, for each component, be described by an expression of the type of equation (8), L representing, in each case, the position of the solvent front. The value of p′ equals $Rf = \frac{d_{r,\,max}}{L}$, and the value of n′ may be estimated by a method outlined in paragraph d of this chapter.

Note: Equation (8) refers to the tank containing both stationary and mobile phases. Consequently, equation (8) is applicable only to a "wet" column, i.e., a column of stationary phase embedded in mobile phase prior to application of the solute. Common technique of PC and TLC is different, the column being dry at the onset of chromatography. As a consequence, flow rate decreases with time, and the phase ratio becomes a function of time and of distance from the origin[1]. Neither $n'/_L$ nor p′ are constant[2]. Therefore, equation (8) is valid only in a qualitative sense when applied to TLC.

c) Elution

The elution curve from a tank with a large number of compartments is represented by equation (5). This equation also applies to elution from a column, provided the conditions mentioned in paragraph a) are met. $V_{max} = \frac{n-1}{n} \cdot \frac{v_m}{p}$ now becomes $V_R = \frac{v_F}{p'}$. The quantity V_R is called "retention volume" and is, for a given set of conditions, specific for each solute. Nearly symmetrical elution curves, as postulated by theory, are observed, e.g., in gas chromatography [*23*], ion exchange chromatography of amino acids [*25*] and in gel filtration [*26*, *27*].

The value of the retention volume is directly obtained from the elution curve. Separate estimation of v_F, and consequently of p′ of any solute, would require chromatography of a standard solute with p′ = 1. Estimation of n′ is outlined below.

d) Estimation of the number of theoretical plates and of the height equivalent to a theoretical plate (HETP)

The theoretical plate number n′ and the length L of a chromatographic column determine the quantity H = L/n′. H denotes the vertical section of a column which has the same effect as one tank compartment. H is called the "height equivalent to a theoretical plate" (HETP). If H or n′ is known, the separation efficiency of a chromatographic column can to some extent be predicted. An exact prediction is not possible as n′ and H do not depend on column operation alone, but also on the rate of solute equilibration between solvent and stationary phase. This rate is not the same with all solutes. Thus, n′ and H, in principle, have individual values

[1] cf. Chapter III, p. 106.

[2] cf. Summary and Concluding remarks, p. 96.

for each solute. However, for solutes with similar values of p′, the rate differences are probably small and calculations of separation effects may be based on a value of n′ common to chromatography of both solutes.

There are several possibilities for the estimation of n′. Visual comparison of the experimentally obtained elution curve with a family of calculated elution curves (cf. Fig. 53) is not a generally practical method. A calculation of n′ on the basis of a comparison of peak width and $d_{r,\max}$ or V_R, respectively, is easier[1]. Apply equation (9) to a developed chromatogram, and equation (10) to an elution curve.

$$H \approx \frac{(4\,d\sigma)^2}{16 \cdot d_{r,\max}} = \frac{b^2}{16 \cdot d_{r,\max}}, \tag{9}$$

wherein, b is the visible peak width and $d_{r,\max}$ is the distance from the top of the column to the peak maximum.

$$n' \approx \left(\frac{4 \cdot V_R}{4\sqrt{\frac{1}{n}}\,V_R}\right)^2 = \left(4\,\frac{V_R}{V_B}\right)^2, \tag{10}$$

wherein, V_R is the retention volume, and V_B is the peak width of the elution curve, measured at the base line in units of volume. V_B, as defined by equation (10), can be determined quite accurately by drawing the inflection tangents to the peak. V_B is then equal to the distance between the intersection points of the tangents with the base line.

The derivation of equations (9) and (10) is easily understood on the basis of comments to equations (5) and (8).

e) Gross deviations from ideal behavior

Deviations occur whenever one or more of the parameters in equation (5) or (8) exhibit a systematical drift during the chromatographic process. There are numerous reasons such as overloading the column with solute, uneven column packing, a change in the phase ratio, a change in temperature, demixing of a multicomponent phase, a change in flow rate, uneven solute distribution perpendicular to the column axis and concentration dependence of adsorption or partition coefficients, K′[2].

Appropriate equipment and careful column operation allow the elimination of all but one of them. There is no simple way to correct for the concentration dependence of the partition parameter, K′, except dilution. Especially in case of adsorption, K′ may not become constant before solute concentration is extremely low. In other words, an adsorbent exposed to increasingly concentrated solutions very soon, and long before saturation is reached, adsorbs less than proportional quantities of the solute. An adsorption isotherm is usually curved towards the

[1] There are limitations to this method. They are discussed in [*23*], p. 61—65, 113—114, 183.

[2] K′ denotes a generalized partition coefficient K in the same way as p′ denoted a generalized retention factor p (cf. p. 94). Consequently, K′ refers to liquid-liquid partition as well as to adsorption phenomena. K′ will henceforth be called the partition parameter.

concentration coordinate. In the case of liquid-liquid partition, the isotherm may be curved to either coordinate, usually due to solute association being more favored in one of the phases. Solute concentration in a chromatographic column assumes all values between a maximum and ~ zero. Therefore, a curved isotherm has an inavoidable effect on solute distribution in the chromatographic process.

For example, if $^1/_{K'}$ decreases as the solute concentration is raised, there will be an acceleration of the migration rate of the peak maximum as compared to the migration rates of the front and the tail of the peak. The front will gradually be overrun by the center, and the tail will be left further and further behind, the net result being a very asymmetric solute distribution no longer amenable to interpretation on the basis of the equations discussed earlier in this chapter. Fortunately, tailing can often be checked by discontinuously or continuously changing the chromatographic system. An appropriate concentration gradient in the mobile phase, for instance, can increase the eluting effect in a way which becomes equivalent to a reduction of isotherm curvature. This special technique is called gradient elution [*32*].

5. Summary and concluding remarks

Our model, the Signer tank, allows a straightforward interpretation of the overall effect of a chromatographic column.

The migration rate of a solute, equal to

$$\frac{L}{V_R/u} = \frac{d_{r,\max}}{v_F/u} = L \cdot p \cdot \frac{u}{v_F},$$

proves to be an intelligible function of the adsorption capacity of the stationary phase, the phase ratio and the ratio of the flow rate to the solvent volume embedding the stationary phase.

The formation of bands with bellshaped solute distribution is obviously due to the column filling acting as an obstacle to mixing processes in the direction of solvent flow. Concentration equilibration perpendicular to the column axis is effected by diffusion. Stochastic aberrations from such behavior merely cause a flattening of peaks. Systematic deviations, however, give rise to an eventual solute distribution which is asymmetric instead of bellshaped.

On the basis of the model, we may, therefore, consider chromatography as a combination of processes of mixing and transportation.

Under the ideal conditions prevailing in a properly operated tank we can predict the actual overall behavior of a solute. In case of certain experimental deviations from expected behavior, we can empirically adjust the theory, viz., by introducing the concept of the theoretical plate number, n'. With a tank, n' still has some bearing to the idea of compartments. With a column, n' is a quantity exclusively defined by experiment. In both cases, a change of n' due to a change in operating conditions, while qualitatively understood, is on the basis of our theory not amenable to quantitative prediction. If the solute retention capacity of a tank or of a column depends on solute concentration, the theory reduces again to a rule providing but qualitative information.

The theory, inter alia, is based on the assumption that the flow rate of the mobile phase is constant throughout the experiment. The outflow volume, V, thus constitutes a measure of time. Explicitly, however, time does not show up as a parameter in any of the expressions derived.

Let us now divide both sides of equation (9) by the time, t, elapsed during chromatographic migration of a solute. Substituting Δ for b/2 and rearranging, we obtain

$$\frac{2 \cdot d_{r,\max} \cdot H}{t} = \frac{\Delta^2}{2\,t} \tag{11}$$

The left hand side of equation (11) now has the dimension of a diffusion coefficient and the equation, as a whole, recalls the Einstein diffusion equation. Thus, Δ can be considered as a solute displacement in either direction from the center of a band, caused by diffusion. Band broadening in chromatography can, in fact, be treated a priori as a diffusion problem [*2*, *33*]. Such an approach fits physical reality better than the theoretical plate concept deduced from our model. It has been possible, e.g., in gas chromatography, to correlate individual parameters to stochastic irregularities in column packing, to longitudinal diffusion, and to delay in equilibration, and to evaluate the contribution of each of these factors to the eventual solute distribution, thus affording an interpretation of the height equivalent to a theoretical plate, H, on the basis of molecular kinetics [*11*, *16*, *23*, *34*–*36*]. A discussion of chromatography from the point of view of such advanced concepts is beyond the descriptive scope of the present article.

We may state, in conclusion, that the time factor must not be overlooked in the evaluation of chromatograms. Apart from being involved in migration rates of solutes and solvents, time is implicitly connected with the phenomenon of band broadening. A simple mathematical relation between migration rates and peak shape is only to be expected when the rate of solvent flow is constant and identical at all distances from the origin of the chromatographic column.

Appendix

A comparison of Gaussian with Poisson curves and a method for approximate calculation of band purity in separation experiments.

Fig. 56 gives

a) an example, based on calculation using equation (3), of incomplete separation of equal amounts of two closely related solutes emerging from a tank with 29 compartments,

b) an idea of the fit of Gaussian approximations of equation (3), when n = 29, and

c) the results of cutting the outflow at certain selected values of V into an A and a B fraction.

When studying the figure, remember that the approximation improves if the compartment number, n, is increased. The same is true for conclusions drawn from the approximation with regard to optimal fraction cutting and achieved band purity. As a matter of fact, our example is close to the limit where the approximation becomes questionable.

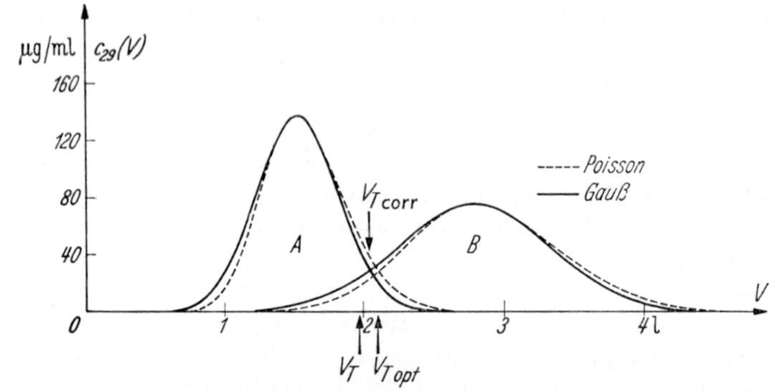

Fig. 56. *True and approximate concentrations of two closely related solutes, Leucine (A) and Valine (B), in the outflow from a tank with 29 compartments, as calculated from equations (3) and (5)*

$M_{0,Leu} = M_{0,Val} = 100$ mg. Mobile phase n-butanol, $v_m = 750$ ml; stationary phase water, $v_s = 150$ ml. $\varrho = 5$; $n = 29$; $K_{Leu} = 0.18$; $K_{Val} = 0.07$. Cuts at $V_T = 1976$, $V_{T,corr} = 2022$, $V_{T,opt} = 2083$ ml, respectively, yield fractions as listed below, the figures indicating true weights estimated by planimetry of Poisson areas. Figures in parentheses refer to calculated weights (see text):

	Leucine fraction				Valine fraction			
	V_T	$V_{T,corr}$	$V_{T,opt}$		V_T	$V_{T,corr}$	$V_{T,opt}$	
Leu . . .	92.0	92.6	94.7	(93.9)	8.0	7.4	5.3	(6.1)
Val. . . .	3.1	3.9	5.3	(6.1)	96.9	96.1	94.7	(93.9)
Total. . .	95.1	96.5	100.0	(100.0)	104.9	103.5	100.0	(100.0)

If two overlapping bands contain equal amounts of material, a cut is considered to be optimal when it yields fractions containing equal weights. The position of the optimal cut can of course be derived from a planimetric evaluation of areas under the elution curves. For instance, in the case of the example of Fig. 56, the position found by planimetry is located at $V_{T,opt}$. This procedure is, however, tedious. An approximate determination of $V_{T,opt}$ based on areas under the Gaussian curves is much easier. It gives V_t instead of $V_{t,opt}$. The change in fraction composition effected by cutting at V_T is small and becomes still smaller (cf. caption to Fig. 56) when the cut is placed at $V_{T,corr} = V_T + \Delta V_T$, with

$$\Delta V_T = \frac{V_{max,A} - V_{max,B}}{2} \cdot \frac{Z^2}{4(n-1)} \quad [1]$$

The meaning of Z will presently be explained.

How is the approximation of $V_{T,opt}$ arrived at? Fig. 57 represents once more the Gaussian curves of Fig. 56. The shaded areas of both

[1] This somewhat empirical expression is based on a correction formula mentioned in [*3*]. $V_{T,corr}$ is supposed to approach $V_{T,opt}$ if $n > 100$.

peaks are identical, because the cut is placed at V_T which is on the respective σ-scales equidistant from $V_{max,A}$ and $V_{max,B}$. The condition for equality of the shaded areas, therefore, is

$$V_{max,A} + Z \cdot \sigma_A = V_{max,B} - Z \cdot \sigma_B ,$$

wherein Z denotes the distance between $V_{max,A}$ and $V_{max,B}$ measured in units of the sum $(\sigma_A + \sigma_B)$. Rearranging and substituting give

$$Z = \sqrt{n-1}\ \frac{p_A - p_B}{p_A + p_B} = \sqrt{n-1}\ \frac{K_A - K_B}{K_A + K_B + 2K_AK_B\,\varrho} \qquad (12)$$

with $K_A > K_B$. It follows further that

$$V_T = V_{max,A}\left(1 + \frac{Z}{\sqrt{n-1}}\right) = V_{max,B}\left(1 - \frac{Z}{\sqrt{n-1}}\right) \qquad (13)$$

Now, apart from affording the desired approximation of $V_{T,opt}$, the Gauss function provides a simple method for an approximate calculation of fraction composition. The shaded area of each peak in Fig. 57 represents a certain percentage of the respective total peak area. This percentage is a function of Z and reads

$$f(Z) = 100\,[0.5 - \Phi(Z)] ,$$

wherein, $\Phi(Z)$ is the error integral (cf. Table 7). Inasmuch as Gauss areas are proportional to amounts of solute – which is only approximately true as seen in Fig. 56 – an A fraction produced by a cut at V_T will contain $[100 - f(Z)]$ % of A and $f(Z)$ % of B.

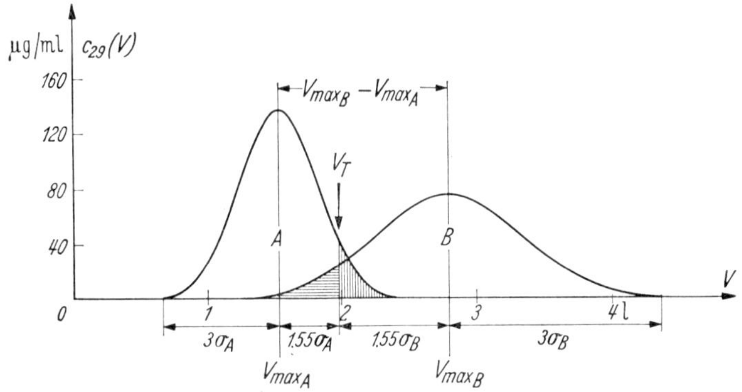

Fig. 57. *Redrawing of Gaussian curves of Fig.* 56

On the respective σ-scales, V_T is equidistant from $V_{max,A}$ and $V_{max,B}$, with Z = 1,55. A cut at V_T separates from each peak f(Z) = 6.1% of its total area. Band separation becomes satisfactory, if Z = 2, with f(Z) = 2.2%, and practically complete, if Z = 3, with f(Z) = 0.1%

With respect to Figs. 56 and 57, this corresponds to a Leucine fraction containing 9.39 mg of Leucine and 6.1 mg of Valine. On a weight basis the calculated band purity is 93.9%.

The general argument is not restricted to cases where the bands contain equal weights of material. f(Z) still represents percentages cut off from each peak, but on a weight basis, f(Z) is no longer identical with percentage contamination. V_T may, therefore, no longer denote the optimal position for a cut.

Giving names to Z and to f(Z), we might denote Z as "separation parameter" and f(Z) as "overlap." It is seen from equation (12) that the band separation is improved if the phase ratio ϱ is reduced, and is, for a given difference $(K_A - K_B)$, better with small rather than with large values of K_B. Table 7 illustrates the effect of a reduction of ϱ in a convincing manner.

Table 7. *Influence of the Phase Ratio ϱ on the Separation Parameter Z and the Overlap f(Z) of Leucine-Valine Separations in a Tank with 29 Compartments*

Fractions cut at V_T. Parameters are the same as in the example of Fig. 56, except v_m and v_s which now are defined by $v_m + v_s = v_m + \frac{v_m}{\varrho} = 900$ ml.

ϱ	$V_{max, Leu}$	V_T	Z	Φ(Z)*	f(Z)**
12	1175 ml	1405 ml	1.05	0.3539	14.6
5	1530 ml	1976 ml	1.55	0.4392	6.1
1	2848 ml	3985 ml	2.12	0.4828	1.7
0.2	4170 ml	5965 ml	2.28	0.4887	1.1

* Error-integral: $\Phi(Z) = \frac{1}{\sqrt{2\pi}} \cdot \int_0^Z e^{-1/2\,\tau^2} \cdot d\tau$.

Table in Handbook of Chemistry and Physics ([*19*] p. 210—213).

** $f(Z) = 100 \cdot [0.5 - \Phi(Z)]$ %.

Equation (12) may be rearranged to indicate the number of compartments, n, necessary to achieve any desired degree of separation of two solutes. We obtain

$$(n-1) = \left(Z \cdot \frac{\alpha + 1 + 2\,K_A \cdot \varrho}{\alpha - 1}\right)^2 \tag{14}$$

wherein, $\alpha = K_A/K_B$. If $K_A \cdot \varrho$ is small compared to $(\alpha + 1)$, and n is large, the expression reduces to

$$n \approx \left(Z \cdot \frac{\alpha + 1}{\alpha - 1}\right)^2.$$

In the literature, α is usually termed "separation factor."

Equations (12), (13) and (14) refer to elution. In case of chromatogram development, we find from equation (8)

$$Z = \frac{d_{r,max,A} - d_{r,max,B}}{d_{\sigma,A} + d_{\sigma,B}} = \sqrt{n-1}\left(\sqrt{Rf_A} - \sqrt{Rf_B}\right). \tag{15}$$

II. Chromatographic behavior and chemical structure

1. Relations of the partition coefficient and phase ratio to the Rf-value and retention volume

In a given chromatographic system[1], the chromatographic behavior of a compound is best described by the *Rf*-value (development of a chromatogram) or the retention volume V_R (elution). This has been shown in the previous chapter. Both quantities are functions of the partition parameter K′ in the respective system. For the case of the partition between two liquid phases, the following relations apply using the definitions on p. 84 and the notations on pp. 91, 92, 95:

$$p' = \frac{K' \cdot \varrho}{1 + K' \cdot \varrho} \qquad \frac{1}{K'} = \varrho\left(\frac{1}{p'} - 1\right).$$

Here it is of practical importance that K′ does not need to be the same in a chromatographic column as in separatory funnel. Because of the possible participation of the support of the stationary phase, the situation is much more complicated in a chromatographic column than in a beaker [*38*], (see pp. 10–11 of [*39*]), [*40*].

For adsorption, relations similar to the ones given above can be derived from Freundlich's adsorption isotherm and the definitions on p. 84, together with $p' + q' = 1$. There, the phase ratio ϱ is substituted by a function of the ratio between the amount of adsorbent and the volume of the mobile phase. K′ will depend strongly on the concentration. This, however, will not change the overall analogy of partition and adsorption.

Now $p' = Rf$ (pp. 91, 93) and likewise

$$p' = \frac{v_F}{V_R} \quad \text{(see p. 94).}$$

Therefore we have:

$$\frac{1}{K'} = \varrho\left(\frac{1}{Rf} - 1\right) \tag{16}$$

when developing a chromatogram and

$$\frac{1}{K'} = \varrho\left(\frac{V_R}{v_F} - 1\right) = \varrho\,\frac{V_R - v_F}{v_F} \tag{17}$$

in case of elution,
where $V_R - v_F$ is the "apparent retention volume" often referred to and easily determined, e.g., in a gas-chromatography[2].

2. Qualitative Rules

When comparing substances of known structure, it is often possible to make qualitative statements about the ratio of the respective K′ values by

[1] A "chromatographic system" is defined by the environmental conditions like chromatographic technique, stationary phase, solvent, temperature a.s.o. The term was coined by R. Pohloudek-Fabini and H. Wollmann [*37*].

[2] See p. 16 in [*23*]. Keulemans uses the notation V'_R.

using well-known solubility rules[1]. According to (16) or (17), respectively, a larger K′ corresponds to a larger *Rf*-value or a smaller retention volume. In this way, useful qualitative predictions can be made regarding the chromatographic separation.

The chromatographic adsorption also follows certain rules. Polar adsorbents prefer polar solutes, and vice versa. The solvent will be adsorbed, too, and, therefore, will compete with the solute for the available surface. Moreover, experience shows that the affinity towards the adsorbent increases with increasing number of "unsaturated" sites of the solute's molecules.

3. Quantitative relations: the Martin-relation

In many cases, quantitative relations between structure and K′ become apparent if the quantities $\log\left(\frac{1}{\mathrm{Rf}} - 1\right)$ or $\log(V_R - v_F)$ of members of a homologous series are plotted in a coordinate system versus the number of "homologous structural elements"[2,3] or its logarithm [*43*]. The points often lie on a straight line! The behavior of other compounds belonging to the same homologous series may then be predicted by intra- or extrapolation.

a) Definition of the Rm-value

The linear relationship to the number of homologous structural elements may be understood by following considerations first put forth by MARTIN [*39*]. The coefficient K′ (partition between two liquid phases or between a liquid and a gas) is an equilibrium constant. As such, according to the easily derived thermodynamic relation

$$A_{\mathrm{rev}} = \mathrm{RT}\left[\ln \frac{c'_m}{c'_s} - \ln \mathrm{K}'\right], \tag{18}$$

it is a measure of the reversible work which has to be performed to transfer 1 mole of substance from the stationary phase at concentration c'_s into the mobile phase at concentration c'_m while the volume and the absolute temperature T are kept constant. Confining ourselves to the partition between two liquid phases, and denoting the reversible work by ΔF for the case $c'_m = c'_s$, we have:

$$\Delta \mathrm{F} = -\mathrm{RT} \ln \mathrm{K}' \tag{19}$$

Now we can take $\frac{1}{\mathrm{K}'}$ from (16) or (17) and insert it in (19). Equ. (17) gives a result which will not be discussed here any further because it pertains

[1] See, for instance, "the solubility classes" in [*42*]. See also pp. 8—11 in [*39*], and [*41*], [*43*], [*44*].

[2] For instance, the number of CH_2-groups in a n-paraffine.

[3] $\log\left(\frac{1}{\mathrm{Rf}} - 1\right)$: [*45—49*]. $\log(V_R - v_F)$: [*50—52*].

to the case of elution[1]. Equ. (16) gives:

$$\Delta F = RT \ln \left[\varrho\left(\frac{1}{Rf} - 1\right)\right] \quad \text{or}$$

$$\Delta F = 2{,}3\, RT \log \left[\varrho\left(\frac{1}{Rf} - 1\right)\right] \tag{20}$$

Equ. (20) may now be applied on a substitutable substance S_0; we write therefore ΔF_0 instead of ΔF. By introducing a substituent X into S_0 the free transfer energy ΔF_0 will naturally be changed by an amount ΔF_X. Another substitute Y will cause a change ΔF_Y. We assume now that by simultaneously introducing X and Y the changes are additive, and that the additivity holds when introducing several X's (e.g. $m \cdot X$) or Y's (e.g. $n \cdot Y$). Then we have the following general expression for the transfer energy of a derivative of S_0:

$$2.3\, RT \log \left[\varrho\left(\frac{1}{Rf} - 1\right)\right] = \Delta F_0 + m\Delta F_X + n\Delta F_Y + \ldots \tag{21}$$

and from there by transforming

$$\log\left(\frac{1}{Rf} - 1\right) = \frac{\Delta F_0}{2.3\, RT} + \frac{m\Delta F_X}{2.3\, RT} + \frac{n\Delta F_Y}{2.3\, RT} + \ldots\ldots + \log\frac{1}{\varrho} \tag{21a}$$

resp.

$$Rm = G_0 + mG_X + nG_Y + \cdots + G \tag{22}$$

where

$Rm = \log\left(\frac{1}{Rf} - 1\right)$ [45, 54, 55] (A table to convert Rf into Rm and vice versa may be found at the end of this book.)

$G_0 = \Delta F_0/2.3\, RT$: constant of the parent group S_0[2] [54, 55]

$G\mu = \Delta F\mu/2.3\, RT$: constant of group μ ($\mu = X, Y \ldots$)[2] [54, 55]

$G = \log 1/\varrho$: fundamental constant.[3]

b) Application of the Rm-value

The Rm-value is, according to (22), a linear function of the number of equal groups, while the number of the other groups remains constant. The contribution of a certain group to the Rm-value of a given substance would, therefore, prove independent from the rest of the molecule and from the position of the group in the molecule. This, of course, can only approximately be true. Whenever individual groups influence each other, deviations must occur. Good agreement may only be expected for homologous series, and even then, only if the spatial arrangement offers the

---

[1] In the form of $\Delta G = RT \cdot \ln K'$ (ΔG = Gibb's free energy), (19), in combination with (17), is of great importance in gas chromatography. See, for instance, [41, 53].

[2] If, for instance, S_0 = benzene and X = Cl (chlorobenzene) then the parent group S_0 is benzene, resp. phenyl. Benzene and phenyl are not distinguished from each other.

[3] With paper and thin-layer chromatograms, the phase ratio ϱ depends not only on the solvent but also on the paper or the layer. Hence, G is also called paper constant or layer constant [56].

same possibility for solvatation to each homologous structural element and if, at the same time, the rest of the molecule behaves identically in each member of the series. Thus, we are referring to homologous series in a restricted sense, e.g., to the series of straight chain carboxylic acids, alcohols, esters, amines, amino acids a.s.o.; also to benzene, toluene and ethyl benzene, but not to benzene, toluene and xylene [*56*]. Likewise, oxalic acid, malonic acid, succinic acid a.s.o. do not form a homologous series according to the above stated criterion: here, not only the chain length changes but also the mutual influence of the carboxyl groups. Experimental verification often proved the additivity implied in Equ. (22), as has been mentioned above[1]. Systematic evaluation of (22) yields empirical constants, the knowledge of which makes the computation of Rm or *Rf* values possible. REICHL [*55*], PARDEE [*46*], and MOORE and BAKER [*57*] found excellent agreement between measured and computed *Rf*-values. SCHAUER and BULIRSCH [*58–60*] finally determined from the inversion function of (22), with certain practical limitations, the nature and number of functional groups in unknown compounds with known *Rf*-values.

All constants in (22) depend on the used "chromatographic system"[2]. A change of the system, therefore, implies a change of Rm. This Δ Rm can be different for isomeric and epimeric compounds. MACEK [*61*] showed, for instance, with reference to 5 + 4 examples, that

$$\Delta\, \mathrm{Rm}_{\text{fomamide/formamide-ammonium formiate; paper}}$$

is 0.71–0.86 for reserpoids of the normal series, for reserpoids of the iso-series, however, 0.38–0.48, and that Δ Rm thus may be used to determine the configuration (epimery at carbon atom no. 3). The relative constancy of Rm within a steric series meets the scheme of Equ. (22) to a large extent. It ought to be the more pronounced, the more distinctly the epimery influences the shape of the molecule.

Isomers, i.e., compounds with identical structural elements, are often chromatographically separable[3]. This is no contradiction to (22), but an expression of more differentiated interactions with the environment, which were neglected in (22). Further theoretical investigations will not be possible until more knowledge will be obtained about these interactions and the mutual influences of equal but differently arranged structural elements. To empirical studies, however, the problem is already amenable today. One may add experimentally determined isomeric constants to Equ. (22) and thus bring a certain amount of order into the present data [*62*].

Finally, Equ. (22) offers the possibility to calibrate layer material or paper of different quality by means of layer or paper constants. Any differences arising may be computationally eliminated and standardized *Rf*-values may be obtained [*56, 63*].

---

[1] Footnote 3, p. 102.
[2] Footnote 1, p. 101.
[3] Examples in [*62*].

c) Exceptions

There are cases where Equ. (22) strangely does not apply. As mentioned, sometimes Rm changes within a homologous series proportionally to the logarithm of the number of homologous structural elements. There the influence per structural element decreases with increasing molecular weight. We would like to put up for discussion the opinion that, under such circumstances, adsorption chromatography prevails over partition chromatography and that, therefore, the validity of (22) was perhaps wrongly assumed. Among about 20 examples investigated in our laboratory, only one case was found where the log-log-relation was fulfilled. It is the series of DNP-amines (C_1-C_5) on silica gel in benzene as solvent[1]. On the other hand we could prove Equ. (22) to hold whenever chemical considerations were in favor of partition instead of adsorption chromatography. This could not be done, however, before we became aware that in thin-layer chromatograms multicomponent solvents penetrating a dry layer of silica, formed several solvent fronts because of unequal adsorption of the different components (frontal analysis). Under such circumstances, *Rf*-values may have to be determined with respect to another than the first solvent front. Taking into account this situation, we were able to reduce those examples also to the Rm-relation (22) which previously [*62*] were thought to obey an Ra-relation [*56*].

From what has been said above, it is clear that before planning to treat the *Rf*-values theoretically, one must be in a position to form an opinion about their reliability. Especially for data reported in the older literature, this might be almost impossible. It is, therefore, well advisable to be cautious when drawing conclusions based on the interpretation of data from experiments performed elsewhere[2]. In the next section, we will consider more thoroughly the problems of determining the *Rf*-value.

III. Particularities of thin-layer chromatography

A chromatographic process corresponds to the process of separation in a Signer tank only if the experimental arrangement is identical in principle. This condition is not fulfilled for thin-layer chromatography in its usual form. At the beginning of a run, the column does not yet contain the solvent. Moreover, there is lateral spreading of solute due to free diffusion perpendicular to the direction of solvent flow. This difference must not be overlooked in theoretical considerations. Of course, the same problem exists in paper chromatography. The following discussion is based on experiences from both fields of activity. Besides parallel features, there are also differences between thin-layer and paper chromatography. We shall briefly mention those, too, and finally we shall point out a relation of practical importance between thin-layer chromatography and a new kind of column chromatography.

---

[1] Horizontal technique, "covered plate", see [*64*].

[2] See, for instance, Franc and Jokl [*43*]. Despite the doubt indicated this paper deserves careful attention.

1. Features analogous to paper chromatography

a) Migration and distribution of the solvent, *Rf*- and *Rm*-values

α) Single component solvents

As has already been mentioned in the paragraph on the development of a chromatogram (section I, p. 94), the transport of solvent and solute along the column is different for thin-layer or paper chromatograms and Signer's tank. The reason is the "dry" column. Here, we find, for the motion of the solvent front, a particular dependence on time. A variable distribution of the solvent along the path of migration leads to a non-uniform relative speed of migration of the solute. This means that the *Rf*-value of a given solute changes during the chromatographic process.

Movement of the solvent front. The time dependence of the motion of the solvent front is determined by the chromatographic technique employed in a given chromatographic system (ascending, horizontal, and descending). For paper – in the case of horizontal and ascending chromatography – a relatively simple relation holds between the length of run of the front, z_f, and the elapsed time t, after a short starting period:

$$(z_f)^2 = a + b \cdot t, \text{ where a and b are constants } [14, 65-68]$$

We were able to confirm this relation on silica gel for the ascending and horizontal technique (covered plate), (see Fig. 58a and b).

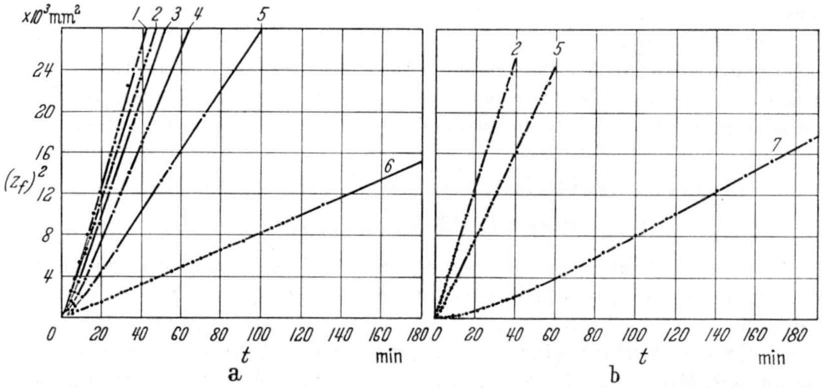

Fig. 58. Speeds of migration of solvent fronts at 20° C on air-dried layers of Silica Gel G. Fig. 58a: horizontal technique (covered plate [*64*]. Fig. 58b : ascending technique (Desaga separation chamber, gas phase saturated). The smaller flow rate observed with the horizontal technique is probably due to a decelerating effect of the filter paper connecting the solvent to the layer
1 Ether, *2* ether-glacial acetic acid (95+5), *3* ethyl acetate, *4* benzene, *5* carbon tetrachloride, *6* n-butanol, *7* phenol-water (75+25, w/w)

As a practical outcome, we have the fact that equations which assume a uniform motion of the front cannot be used to determine the number of theoretical plates in the presently used versions of thin-layer chromatography. The HETP is not constant, but it increases with increasing length of run. Accordingly, the power of separation is highest in the beginning. – In this connection, we would like to mention once more that during the run the spreading of the spot occurs proportionally to $\sqrt{t}$ (see section I, pp. 96, 97).

Profile of solvent concentration. Like the motion of the front, the distribution of the solvent along the length of run also depends on the chromatographic technique [*67*, *68*]. GIDDINGS and collaborators [*68*] coined the term "concentration profile" for this distribution. They determined a number of profiles on horizontal[1] paper chromatograms with radial and rectangular migration of the solvent. In the case under discussion, especially the rectangular migration is of interest where the solvent moves from a starting line perpendicular to the direction of flow. Fig. 59 shows this situation at various stages of chromatogram development. Of interest and significance is the fact that the profiles consecutively occurring in the course of a development may be reduced to a common basic profile which is characteristic for the given chromatographic system[2]. For this purpose, we substitute the distance z of Fig. 59 by the reduced distance z/z_f. Here z denotes distances from the starting line (dip level) to the points of observation and z_f denotes the distances between the starting line and the respective solvent fronts (Fig. 59).

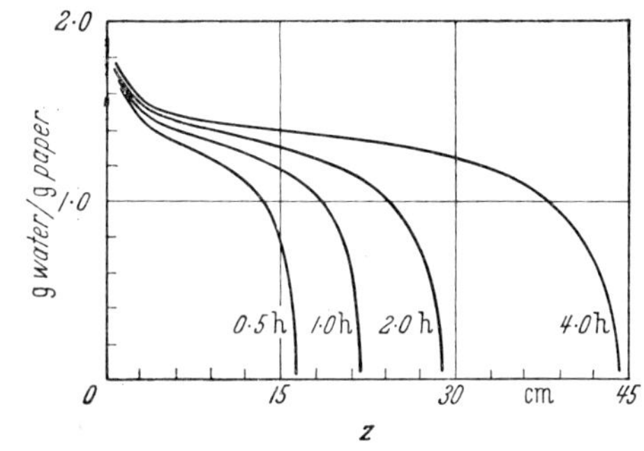

Fig. 59. Distribution of solvent (concentration profile) as a function of time and the distance z between the immersion line (dip level) and the point of observation. (Water, WHATMAN 3 MM paper, acc. to GIDDINGS et al. [*68*])

Fig. 60 shows that there is a zone observable throughout the development which, although migrating, keeps its relative position with respect to the front. A solute of a given arbitrary *Rf*-value starting simultaneously with the solvent from a *common* starting line will, therefore, encounter the same phase ratio ϱ throughout the whole chromatographic process. Solutes with different *Rf*-values encounter different phase ratios ϱ according to the shape of the basic profile. If the starting line of the solvent (dip level) does *not coincide* with the starting point of the solute

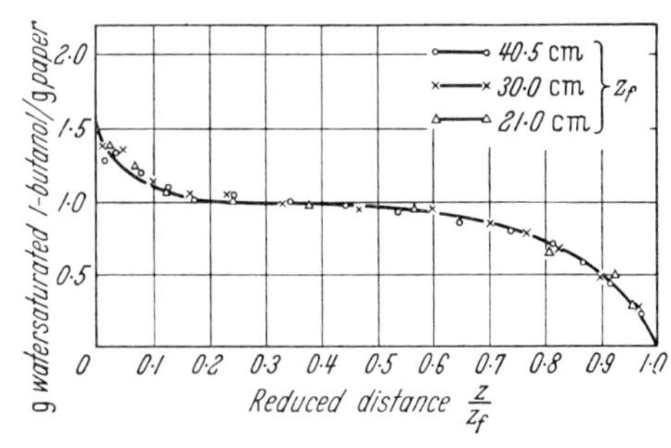

Fig. 60. Reduced solvent distribution (basic profile), acc. to GIDDINGS et al. [*68*]. (Water-saturated n-butanol, WHATMAN 3 MM paper)

[1] For ascending and descending chromatography, the situation is not clear at all (see [*68*]).

[2] See footnote 1, p. 101.

(conventional version of thin-layer and paper chromatography), the operative ϱ-value is not constant anymore during the development: depending on the distance *dip level-starting point*, the shape of the basic profile and the substance, ϱ increases more or less rapidly from 0 to a final constant.

Speed of migration of the solvent behind the front: true and observed Rf-value. Not only the phase ratio changes along the basic profile but also — in interaction with ϱ — the local speed of the migration of the solvent. The product of both, i.e., the amount of the solvent traversing a cross-section perpendicular to the direction of flow, decreases with increasing value of $z:z_f$ [*68*]. Therefore, the phase ratio ϱ and the speed of solvent migration, although following opposing trends both spatially and temporally, are not simply inversely proportional to each other. The convefitionally measured Rf-value[1] ($=\mathrm{Rf}_{\mathrm{observed}}$, abbreviated $\mathrm{Rf}_{\mathrm{obs.}}$) of a solute will be influenced by this fact in a very complicated way. $\mathrm{Rf}_{\mathrm{obs}}$ *is always smaller than the "true" Rf-value defined by means of our model (Signer's tank) and used in deriving equation (22) in section II.* The difference is about 10—20%.

To understand this, it is necessary to consider the sections ds traversed by the migrating spot. Their sum is the total distance s covered by the spot of the solute, and we have

$$\mathrm{Rf}_{\mathrm{obs}} = \frac{\int ds}{z_f - s_0} \qquad (23)$$

where s_0 is the distance between the *starting line of the solvent* and the *starting point of the spot*, and z_f is the total length of run of the solvent from the starting line to the front. — According to section I, p. 91, Rf is defined as being the ratio p of two distances. If we divide both distances by the time elapsed since the beginning of the experiment, we obtain *Rf* as being the ratio p of two velocities. Even if this ratio of velocities is not constant during the migration, the statement

$$\mathrm{Rf} = \frac{\text{momentary velocity of the substance}}{\text{velocity of solvent at the position of the spot}},$$

referring to a section ds, still agrees with the previous definition of Rf. — According to section II, p. 101, the relation $\mathrm{Rf} = \frac{K' \cdot \varrho}{1 + K' \cdot \varrho}$ holds [see (16)]. Because of the variability of ϱ, we write it in the form of $\mathrm{Rf}_{\mathrm{int}} = \frac{K' \cdot \varrho'}{1 + K' \cdot \varrho'}$ · $\mathrm{Rf}_{\mathrm{int}}$ denotes the "intrinsic" Rf-value which is valid at the moment of observation and which is defined by K′ and ϱ'. Evidently, $\mathrm{Rf}_{\mathrm{int}}$ and the above mentioned true Rf-value, defined in section I, are identical. Therefore, the following is also true:

$$\mathrm{Rf}_{\mathrm{int}} = \frac{\text{momentary velocity of the substance}}{\text{velocity of solvent at the position of the spot}}.$$

Let z be the position of a solute behind the front, u_z its velocity at z and v_z the velocity of the solvent at z. Then we may write:

$$u_z = v_z \cdot \mathrm{Rf}_{\mathrm{int},\,z}. \qquad (24)$$

v_z is different from the velocity v_f of the solvent front. Following GIDDINGS we have:

$$v_z = v_f \cdot \frac{A_z}{W_z}, \qquad (25)$$

[1] Ratio of the distances *start-center of spot* and *start-solvent front.*

where A/W denotes a spatially variable factor; it is always less than unity. W denotes the local solvent concentration (= mass of solvent per unit mass of paper), as can be seen in Fig. 59. A is a variable factor of proportionality and connects the flux of solvent at a given point with the frontal velocity v_f; the flux of solvent at position z is, therefore, $A_z \cdot v_f$. GIDDINGS has calculated A for the case of the reduced profile of Fig. 60 and tabulated the quantity A/W for z/z_f running from 0 to 1 (see Fig. 4 in GIDDINGS [*68*], uppermost curve with index c = 0). Inserting (25) into (24) we obtain

$$u_z = v_f \cdot \frac{A_z}{W_z} \cdot Rf_{int,z} \tag{26}$$

or by multiplying with dt

$$ds_z = \frac{A_z}{W_z} \cdot Rf_{int,z} \cdot dz_f \tag{27}$$

The total distance s of migration corresponds to the integral

$$s = \int_{\text{starting point of solute}}^{\text{solvent front}} \frac{A_z}{W_z} Rf_{int,z} \cdot dz_f \tag{28}$$

The total length of run of the solvent from the starting point of the substance to the solvent front is

$$\int_{\text{starting point of solute}}^{\text{solvent front}} dz_f = z_f - s_0$$

and (23) yields now

$$Rf_{obs} = \frac{\displaystyle\int_{\text{starting point of solute}}^{\text{solvent front}} \frac{A_z}{W_z} \cdot Rf_{int,z} \cdot dz_f}{z_f - s_0} \tag{29}$$ [1]

It is important to realize the significance of the integrand of (29). Together with the shifting dz_f of the front, a shifting of the solvent occurs at the position z by an amount of

$$\frac{A_z}{W_z} \cdot dz_f \quad [\text{see (25)}]$$

and, therefore, a shifting of the spot by

$$Rf_{int,z} \cdot \frac{A_z}{W_z} \cdot dz_f \quad [\text{see (27)}].$$

Because of (26) the integrand corresponds to the ratio

$$\frac{u_z}{v_f} = \frac{\text{momentary velocity of the substance}}{\text{front velocity}}.$$

[1] As can be shown by some algebraic manipulation, the integrand $(A_z/W_z \cdot Rf_{int,z}$ is identical to the quantity $A_z/(W_z + c)$ used by GIDDINGS. He computed it for the case of the reduced profile of Fig. 60 for all values z/z_f and several values of c (see graph of Fig. 4 in GIDDINGS [*68*]). The parameter c is the partition function used in this paper. It is equivalent to the ratio

$$\frac{\text{grams of stationary phase per gram supporting material}}{\text{partition coefficient } K'}.$$

Its magnitude is given by the shape of the profile at point z, disregarding constant quantities like K' and the characteristics of the chromatographic system[1].

At the moment the traveling solvent front hits the spot of the substance at the starting point the velocity of the substance u_z is 0 because of $\varrho'_z = 0$ and $Rf_{int,z} = 0$. The substance starts to move immediately ($\varrho'_z > 0$, $Rf_{int,z} > 0$). With $u_z > 0$ also $u_z/v_f > 0$ follows. With further increasing z_f, we find that the slope of u_z/v_f decreases until u_z/v_f reaches a limiting value which stays constant despite the continued spreading of the solvent. This may be explained in the following way:

The linear measure, with unit z_f, changes during the spreading of the solvent in the same way as a yardstick made of rubber of length z_f, when stretched with velocity of stretching v_f. A marker fixed at an arbitrary point thereby changes its absolute but not its relative position. If the latter is defined by a given ratio z/z_f, the velocity of displacement of the marked point is $(z/z_f) \cdot v_f$. A solute spot located at the same point moves according to (26) with the velocity

$$Rf_{int,z} \cdot \frac{A_z}{W_z} \cdot v_f .$$

If

$$\frac{A_z}{W_z} \cdot Rf_{int,z} < \frac{z}{z_f},$$

then the absolute movement of the spot is different from the absolute movement of the marked point (point z/z_f in the reduced profile). The spot travels slower and, therefore, shifts to a position of a smaller value of z/z_f. If this smaller figure is still different from the new value $(A_z/W_z) \cdot Rf_{int,z}$ now valid, a further shifting results. This will continue tending toward the state

$$\left(Rf_{int,z} \cdot \frac{A_z}{W_z}\right)_{final} = \left(\frac{z}{z_f}\right)_{final} .$$

With further migration of the solvent, a marker fixed at the point $(z/z_f)_{final}$ travels, from now on, with the same velocity as the solute spot. The latter does not change its position in the reduced profile any more. The shape of the profile in the neighborhood of the spot, characterized by $(z/z_f)_{final}$, remains constant, likewise the integrand of (29) and the velocity ratio u_z/v_f. Hence, a stationary state has been reached.

Let us look once more at Equ. (29). With the integrand increasing from zero to $\left(\frac{A_z}{W_z} \cdot Rf_{int,z}\right)_{final}$, it follows

$$\lim_{z_f \to \infty} Rf_{obs} = \left(\frac{A_z}{W_z} \cdot Rf_{int,z}\right)_{final} \tag{30}$$

As GIDDINGS has shown, this limiting value will be approached mostly under practical chromatographic conditions (distance between *dip level* and *starting point* $\ll$ distance between *starting point* and *solvent front*). If $(A_z/W_z)_{final}$ can be determined numerically, the stationary true Rf-value is

$$Rf_{true} \equiv Rf_{int,final} = \frac{\lim Rf_{obs}}{\left(\frac{A_z}{W_z}\right)_{final}}, \tag{31}$$

or simplified

$$Rf_{true} \approx \frac{Rf_{obs}}{\left(\frac{A}{W}\right)_{Rf_{obs}}}, \tag{32}$$

where

$$\left(\frac{A}{W}\right)_{Rf_{obs}} \approx \left(\frac{A_z}{W_z}\right)_{final} .$$

[1] See footnote 1, p. 101.

Of course, each value of $Rf_{int,z}$ is a "true" Rf-value. For the definition of Rf_{true} by means of Equ. (31), the quantity $Rf_{int,final}$ should be preferred, however, because it relates to a phase ratio ϱ which is at least approximately the same for all substances traveling stationarily in the central portion of the profile (Fig. 60). If $(A_z/W_z)_{final}$ can be determined in this region[1], Rf_{true} according to (31) forms a base from where the chromatographic behavior of very many solutes may be compared quantitatively with a good degree of accuracy. Below we will encounter an empirical relation similar to Equ. (31) resp. (32).

For direct chromatographic comparison of two solutes, the difference between Rf_{obs} and the true Rf-value is without significance. However, if we want to establish a relation between actual chromatographic behavior and chemical structure (see section II), the observed Rf-value should be inserted into Equ. (16) only with caution (see below), since it is no characteristic measure for the partition coefficient K′ and the phase ratio ϱ. In view of this shortcoming, it is surprising that Rf-values (Rf_{obs}) of members of homologous series' observed in paper chromatography satisfy[2], in many cases, Martin's Rm-relation. This is also true for Rf_{obs} of amino acids on thin-layer chromatograms, as we were able to show (see Fig. 61) [*56*].

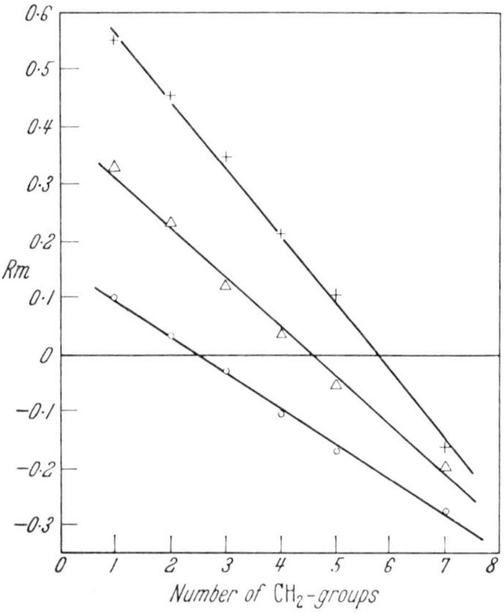

Fig. 61, Linear relation between Rm and the number of CH_2-groups in the series of the unbranched α-amino acids. The regression lines are computed [*56*]

○ Methanol/water, △ ethanol/water, + n-propanol/water: 70 + 30; Silica Gel G, ascending (cf. [*71*])

Each point represents the average of 18 observations. Each of the 6 amino acids was chromatographed twice on the same plate (arbitrary order), and the experiment was repeated nine times. The addition of 2,4,8 or 16% glacial acetic acid did not influence the linearity, but altered the slope and position of the lines

How can this discrepancy between theory and experiment be explained? A first indication may be found by comparing the Rf_{obs} which yield satisfactory Rm-values, with those which do not satisfy Equ. (22). As a rule, application of Equ. (22) demands $0.1 < Rf_{obs} < 0.7$ [*69*]. This is just a region where the concentration profile of the solvent may run more or less horizontally (see for instance Fig. 60). Preliminary studies of our own seem to prove that similar profiles form on layers of silica gel with both horizontal and vertical techniques [*56*, *70*]. Solutes traveling in the "horizontal" region of the profile encounter approximately the

---

[1] In the example (Fig. 60) treated numerically by GIDDINGS, the quantity $\left(\frac{A}{W}\right)_{Rf_{obs}}$ varies between 0.85 and 0.9 for Rf_{obs} lying between 0.2 and 0.7. Hence, it is practically constant.

[2] See footnote 3, p. 102.

same phase ratio ϱ. Thus, an essential condition is met for the applicability of (22). – A second indication comes from the following observation: If we prevent the further advancement of the solvent front on silica gel by drawing a separating line after a length of run of 10–15 cm, the flow of solvent does not stop instantly. The profile does not level out even after 24 hours (ascending technique), but it flattens considerably. At the same time all the Rf_{obs} increase by a factor of about 1.1. Thus they tend to a limiting value. The ideal limiting value would be reached after complete leveling of the profile. Obviously, it would correspond to the true Rf-value defined in (16) (ϱ everywhere the same).

Relations of Rm_{obs} to Rm′ and Rm; additivity of Rm_{obs}. Let us now assume that an observed Rf-value in the region 0.1 to 0.7, and especially an average value $\overline{Rf}_{obs}$ calculated from many individual observations, Rf_{obs} follows the relation

$$\overline{Rf}_{obs} \cdot \xi = Rf_{true} \qquad (33)^1$$

Taking for ξ a value between 1.1 and 1.2, it can be shown that

$$Rm_{obs} = \log\left(\frac{1}{\overline{Rf}_{obs}} - 1\right)$$

is often a fair estimate of

$$Rm' = \log \xi \left(\frac{1}{Rf_{true}} - 1\right)$$

and, therefore, may quite well reflect the additivity of Rm. For this reason, Rm_{obs} is in many cases of practical use [*56*].

Form (21a) together with (33) and adding log ξ on both sides, one obtains:

$$\log \xi \left(\frac{1}{\overline{Rf}_{obs} \cdot \xi} - 1\right) = \frac{\Delta F_0}{2.3\,RT} + \frac{m\Delta F_X}{2.3\,RT} + \cdots + \log \frac{\xi}{\varrho}, \qquad (21\,b)$$

where

$$\log \xi \left(\frac{1}{\overline{Rf}_{obs} \cdot \xi} - 1\right) = \log\left(\frac{1}{\overline{Rf}_{obs}} - \xi\right) \qquad (34)$$

and

$$\log\left(\frac{1}{\overline{Rf}_{obs}} - \xi\right) = \log\left(\frac{1}{\overline{Rf}_{obs}} - 1\right) + \delta \qquad (35)$$

or

$$Rm' = Rm_{obs} + \delta, \text{ with } Rm' = Rm + \log \xi \qquad \text{[see (22)]}$$

δ is the error we make if we replace the term on the left hand side of (21b) by $\log\left(\frac{1}{\overline{Rf}_{obs}} - 1\right)$. It is easy to compute. In the most unfavorable case of $\xi = 1.2$, the value of δ, a function of ξ and $\overline{Rf}_{obs}$, may be taken from the difference between Rm′ and Rm_{obs} from Fig. 62. Rm′ and Rm_{obs} are represented as bands to show the spread of the experimentally measured Rf-values. The standard deviation of $\overline{Rf}_{obs}$ in thin-layer and

[1] Note the formal agreement with relation (32) based on theory. The empirical ξ corresponds even quantitatively with the factor $1\Big/\left(\frac{A}{W}\right)_{Rf_{obs}}$ which is nearly constant in the central region of the profile.

paper chromatograms is on the average: ${}^{s}\overline{\mathrm{Rf}}_{\mathrm{obs}} \approx 0.0037$ (18 observed values). A good interval of confidence is, therefore, given by $\overline{\mathrm{Rf}}_{\mathrm{obs}} \pm 0.011$ [*71*].

As one may see, Rm′ and $\mathrm{Rm}_{\mathrm{obs}}$ for $0.1 < \overline{\mathrm{Rf}}_{\mathrm{obs}} < 0.3$ coincide within the error margins. δ increases steadily up to $\overline{\mathrm{Rf}}_{\mathrm{obs}} = 0.6$, but increases rapidly beyond $\overline{\mathrm{Rf}}_{\mathrm{obs}} = 0.7$. In the region, $0.1 < \overline{\mathrm{Rf}}_{\mathrm{obs}} < 0.6$, $\mathrm{Rm}_{\mathrm{obs}}$ is indeed an approximation for Rm′. Its usefulness is apparent in Fig. 63. There the values of the middle curve in Fig. 61 are compared with Rm′ ($\xi = 1.2$) in a slightly different scale. For all points, the dispersion interval,

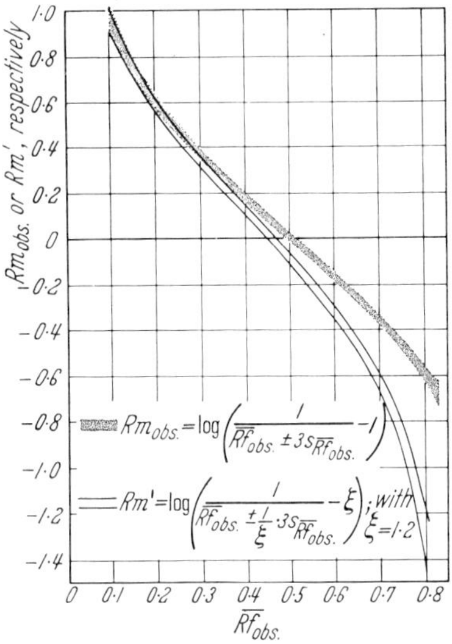

Fig. 62. Comparison of $\mathrm{Rm}_{\mathrm{obs}}$ with Rm′ for different values of $\mathrm{Rf}_{\mathrm{obs}}$, with experimental errors. See text and the Equ. (21b), (34), (35), [*56*]

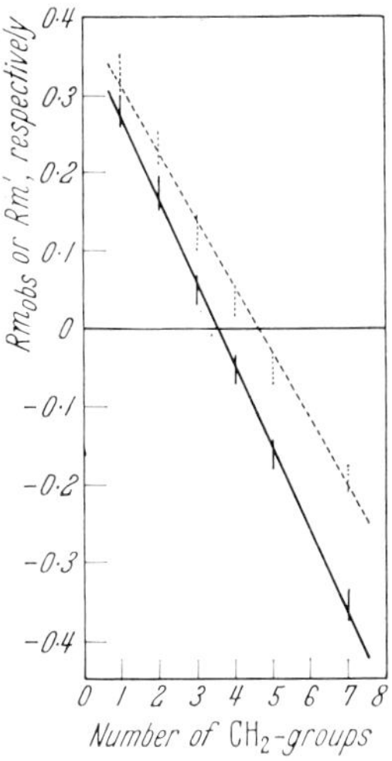

Fig. 63. Comparison of the computed regression line through $\mathrm{Rm}_{\mathrm{obs}}$ with the visually estimated regression line through Rm′. Homologous series of the unbranched α-amino acids. Experimental data from Fig. 61, solvent ethanol/water. $\mathrm{Rm}_{\mathrm{obs}}$ broken line, Rm′ solid line [*56*]

$\pm\, 3\, s_{\overline{\mathrm{Rm}}}$, is shown. The approximation of Rm′ by $\mathrm{Rm}_{\mathrm{obs}}$ is satisfactory concerning additivity and good, concerning the slope of the regression line. The vertical displacement δ is of no consequence. These considerations show that $\mathrm{Rm}_{\mathrm{obs}}$ is for intermediate Rf-values additive within the experimental error and that $\mathrm{Rm}_{\mathrm{obs}}$, therefore, may replace Rm′ or Rm, respectively. Thus they confirm the empirical evidence.

Comparing $\mathrm{Rm}_{\mathrm{obs}}$ with Rm′ in Fig. 63, one may be tempted always to substitute $\log\left(\frac{1}{\overline{\mathrm{Rf}}_{\mathrm{obs}}} - 1\right)$ by $\log\left(\frac{1}{\overline{\mathrm{Rf}}_{\mathrm{obs}}} - \xi\right)$, to choose an appropriate ξ (best linearity) and thus take optimal advantage of (22), resp. (21b). In view of our problematic assumptions (horizontal profile, $\mathrm{Rf}_{\mathrm{obs}} \cdot \xi = \mathrm{Rf}_{\mathrm{true}}$), we consider this as wrong. The purpose of the above discussion was to defend the empirical use of Rm-values against doubts based on the exact theory. Thin-layer chromatography will hardly ever exploit Martin's relation (22) as fully as gas chromatography does. It would be wrong, however, because of perfectionism, to give up all possibilities which may arise from the actual additivity of Rm_{obs}.

This additivity may be found sometimes even in such cases where it does not seem to exist. For instance, in the examples of Fig. 64, the values of Rm_{obs} of members of homologous series clearly do not lie on straight lines. The reason is not a failure of the theory, but the

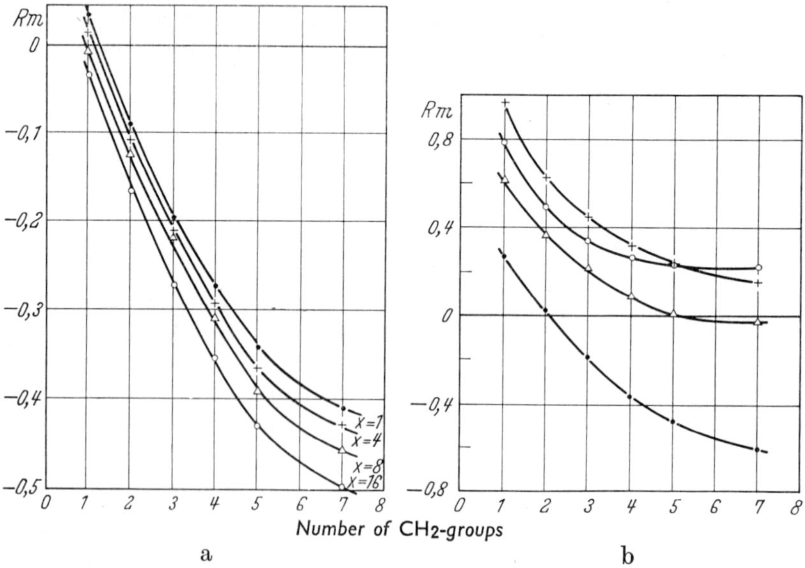

Fig. 64. Rm_{obs} is not additive, because Rf_{obs} has conventionally been related to the first solvent front, despite of demixing of the solvent [*56, 70*]

a) Homologous series of the unbranched α-amino acids (glycin to α-amino caprylic acid) in methanol/water/aqueous ammonia 34% (70 + 30 + x). Silica Gel G, ascending technique [*71*]

b) Homologous series of the unbranched N-2,4-dinitrophenyl amino acids (glycin to α-amino caprylic acid) in 4 different multicomponent solvents:
+ : chloroform/ethyl acetate/glacial acetic acid (80 + 20 + 1),
○ : chloroform/methanol/glacial acetic acid (95 + 5 + 1),
△ : toluene/glacial acetic acid (90 + 10),
● : diethyl ether/glacial acetic acid (95 + 5).
Silica Gel G, horizontal technique (covered plate) [*64*] see also [*71*]

neglect of an interference due to chromatographic demixing (frontal analysis) of the solvent used. The conventionally measured Rf-values (Rf_{obs}) in such a case, do not even approximately satisfy the assumption

$$Rf_{true} = 1.2 \cdot Rf_{obs}$$

[see (33)].

β) Multicomponent solvents
Chromatographic demixing of the solvent

Solvents consisting of several components undergo partial demixing while penetrating a dry adsorbent. This phenomenon is well known[1]. An equivalent process forms the basis of a special chromatographic

[1] See also [*73*], [*74*].

analyzing technique, originated by TISELIUS, and named "frontal analysis" [*75–78*].

Several solvent fronts. The demixing occurs because the advancing solvent deposits its components on the adsorbent according to their adsorption coefficients and concentrations. Hence, the front changes its composition. After some time – maybe very soon – a first component disappears completely (formation of the last front), later a second one (formation of the second to last front), a.s.o., till finally only one component remains which travels alone forming the foremost or first front. Fresh solvent is supplied continuously from the solvent reservoir. From the dip level until close to the last front, which of course keeps moving, the wetted adsorbent is in equilibrium with the solvent of the original composition. In the following zone up to the second to the last front, the adsorbent is in equilibrium with a solvent mixture lacking one component. Two components are lacking in the following zone, a.s.o. As a rule, a solvent consisting of n components forms n different zones separated by (n – 1) fronts. We denote these, starting with the first one, as α-, β-, γ-, ... ω-front. Let ${}^{\alpha}z_f$ (= z_f), ${}^{\beta}z_f$, ${}^{\gamma}z_f$ a.s.o. be the distances traveled by the respective fronts up to the moment of observation, and $d\,{}^{\alpha}z_f/dt$, $d\,{}^{\beta}z_f/dt$ a.s.o. be their velocities of migration. Then one may write

$$\frac{d\,{}^{\beta}z_f}{dt} = k_{\beta} \cdot \frac{d\,{}^{\alpha}z_f}{dt} \quad \text{resp.} \quad {}^{\beta}z_f = k_{\beta} \cdot {}^{\alpha}z_f$$

$$\frac{d\,{}^{\gamma}z_f}{dt} = k_{\gamma} \cdot \frac{d\,{}^{\alpha}z_f}{dt} \quad \text{resp.} \quad {}^{\gamma}z_f = k_{\gamma} \cdot {}^{\alpha}z_f \quad \text{a.s.o.}$$

where k_{β}, k_{γ}, ... k_{ω} are retardation factors which, in connection with the velocity of the α-front, determine the velocities of the other fronts.

A solvent consisting of two components with little interaction may be considered an ideal solution of one component in the other. The relation between the concentration c_M of the (more strongly) adsorbed component, its adsorption coefficient and the factor k_{β} may be derived from Freundlich's adsorption isotherm [*70*]. It is

$$\log\left(\frac{1}{k_{\beta}} - 1\right) = \varkappa - \zeta \cdot \log c_M \tag{36}$$

where $\varkappa$ and ζ are constants. $\varkappa$ contains Freundlich's adsorption coefficient and the phase ratio. We were able to confirm Equ. (36) experimentally in the systems benzene/pyridine (range of concentration 1—42 mole-percent pyridine) and toluene/glacial acetic acid (range of concentration 2—32 mole-percent glacial acetic acid) [*70*]. Using horizontal 0.3 mm thick air-dry (relative humidity 40%) layers of Silica Gel G, the following values for the constants at 20° C were obtained:

	$\varkappa$	ζ
Benzene/pyridine	0.65	0.84
Toluene/glacial acetic acid	0.91	0.82

For c_M = 5 mole-percent k_{β} was 0.32 in toluene/glacial acetic acid and 0.47 in the system benzene/pyridine! Thus, depending on polarities, a solvent component may contribute to a β-front with high k_{β}-value despite its small percentual concentration. If the difference in the polarities of the components is less pronounced than in the above mentioned examples — for instance, in a solvent consisting of alcohol and water — the k_{β}-values become greater, eventually approaching $k_{\beta} = 1$.

The fronts formed in the demixing process are directly visible only in a few cases (Fig. 65). They are easily recognized, however, by the chromatographic behavior of suitable solutes or mixtures of solutes which will now be discussed.

The solvent chloroform/methanol/glacial acetic acid (95 + 5 + 1)[1] forms, on silica gel an α-, a β- and a γ-front. Hence, chromatograms with this solvent show an α-zone ($CHCl_3$), a β-zone ($CHCl_3/CH_3OH$) and a γ-zone [$CHCl_3/CH_3OH/CH_3COOH$ (95 + 5 + 1)][1]. The distance traveled by a substance depends now on the zone with which it starts to move, how

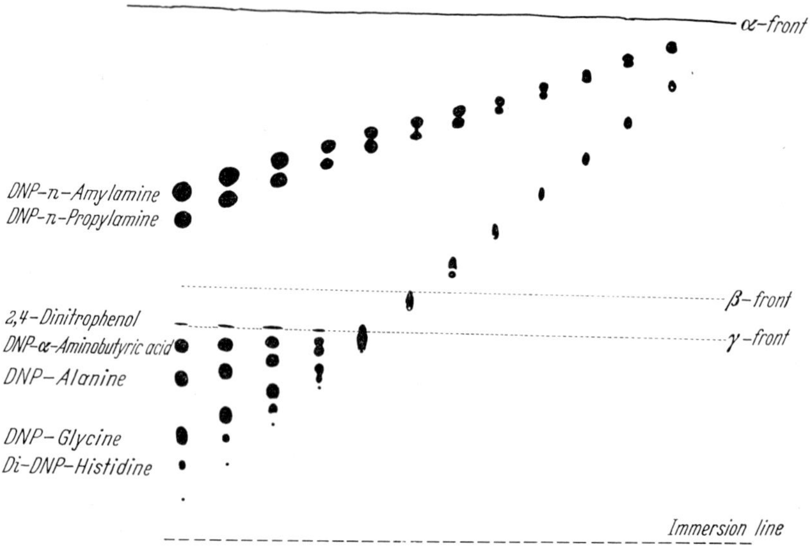

Fig. 65. Demonstration[1] of solvent demixing in horizontal thin-layer chromatography (covered plate [*64*]) of a mixture of 2,4-dinitrophenol and several 2,4-dinitrophenyl-(DNP-) compounds in chloroform/methanol/glacial acetic acid (95 + 5 + 1) [*70*]. The substance mixture was applied to several starting points located on a straight line running from the lower left to the upper right corner. The α-, β-, and γ-fronts are directly visible in UV-light. The zones divided by the fronts have different solvent compositions: pure chloroform between the α- and the β-front (α-zone), chloroform/methanol between the β-and the γ-front (β-zone), chloroform/methanol/glacial acetic acid between the γ-front and the dip level (γ-zone)

[1] See also [*73*], [*74*].

rapidly and how long it travels with the zone, and whether it will be overtaken by one or more zones during its migration. Fig. 65 gives some examples. The particular arrangement of the starting points[2] and an appropriate choice of solutes help to demonstrate the dramatic effect of solvent demixing.

Polar substances (dinitrophenol, DNP-amino acids) practically do not move until the γ-front has reached the respective starting points. Hence, they travel almost exclusively in the γ-zone. Their Rf_{obs}-values must, therefore, correctly be related to the distance *respective starting-point-γ-front* (reference distance). We will call such Rf-values $^{\gamma}Rf_{obs}$. If the starting points are located on an oblique straight line (oblique

[1] Unless indicated otherwise, volume parts are used (ml). Parts of weight are denoted by g (grams).

[2] See also [*73*], [*74*].

starting line), the end points of the traveling distance, determined for each substance by the product ${}^{\gamma}Rf_{obs}$ · *reference distance*, must also lie on an oblique straight line (connecting line)[1]. At the crossing point between starting line and connecting line, the traveling distance is zero. This point, therefore, must lie on the front of the γ-zone. It defines the location of the γ-front. Indeed, in Fig. 65, the γ-front lies on the approximately[2] common crossing point of six straight lines: oblique starting line, connecting line of the spots of DNP-histidine, connecting line of the spots of DNP glycine a.s.o.

The less polar substances (DNP-amines) already start to move when the α-front crosses the starting points. Because these spots move solely in the α-zone, their connecting lines converge on the crossing point of the oblique starting line and the α-front with good approximation (Fig. 65). Here it is permissible to use, as customary, the distance *starting-point-α-front* as reference distance when determining the Rf-values. We call such an Rf-value ${}^{\alpha}Rf_{obs}$ and retain this notation even if the distance to the α-front was wrongly chosen as reference distance. — The actual solvent in the α-zone of the chromatogram of Fig. 65 is pure chloroform. This can easily be confirmed by chromatographing DNP-amylamine and DNP-propylamine with pure chloroform as solvent: The values of Rf_{obs} are identical to ${}^{\alpha}Rf_{obs}$ obtained by the above procedure.

The β-front in Fig. 65 is of no importance because all compounds contained in the sample move either in front of or behind the region of the β-zone. A starting point in this region will not be reached by the γ-front; hence, the polar compounds remain stationary. The same is true for the polar compounds on starting points between the β- and α-front.

From all this, we can conclude that the conventional Rf-value measurement with the distance *starting point-α-front* as reference distance may lead to physically meaningless Rf-values. It is wrong to relate the migration of a substance to the α-front and to report ${}^{\alpha}Rf_{obs}$ if the substance moves with a zone other than the α-zone[3]. The correct reference distance for the Rf-value measurement is the distance *starting-point-β-front* or *starting-point-γ-front* if the substance moves in the β-zone or in the γ-zone. The Rf-values measured in this way — we call them ${}^{\beta}Rf_{obs}$ or ${}^{\gamma}Rf_{obs}$, as already stated — must, however, only be used if, during the total developing time, the solutes moved almost exclusively in the respective zone.

Distinct fronts[2] may be expected particularly if the leveling of concentration differences via the vapor phase[4] is prevented. In this connection, the covered plate offers an advantage. It has, therefore, been used in the test shown in Fig. 65. With the

---

[1] A condition is that ${}^{\gamma}Rf_{obs}$ does not noticeably depend on the distance *dip level-starting point*. It is very well fulfilled (see p. 120).

[2] The quality of the approximation depends, among other things, also on how sharply the front is defined, i.e., on the magnitude of the gradient of the concentration at the transition between two zones.

[3] We will see, however, that the error is sometimes too small to be of any practical consequence.

[4] See also [*74*].

common ascending technique, the fronts are the more blurred the more spacious the gas chamber is. The amount of solvent in the vapor phase, favoring equalization of solvent composition along the length of run, is proportional to the volume of the gas chamber. For large reserves of solvent stored in the vapor phase, the composition of the solvent is defined only behind the ω-front. Then only the values of ${}^{\omega}\mathrm{Rf}_{\mathrm{obs}}$ are useful for theoretical considerations.

Additivity of ${}^{\beta}$Rm and ${}^{\gamma}$Rm. Consider the data plotted in Figs. 66a and b from this new point of view. The points connected by curves should lie on straight lines. They fail to do so because the $\mathrm{Rf}_{\mathrm{obs}}$-values, on which

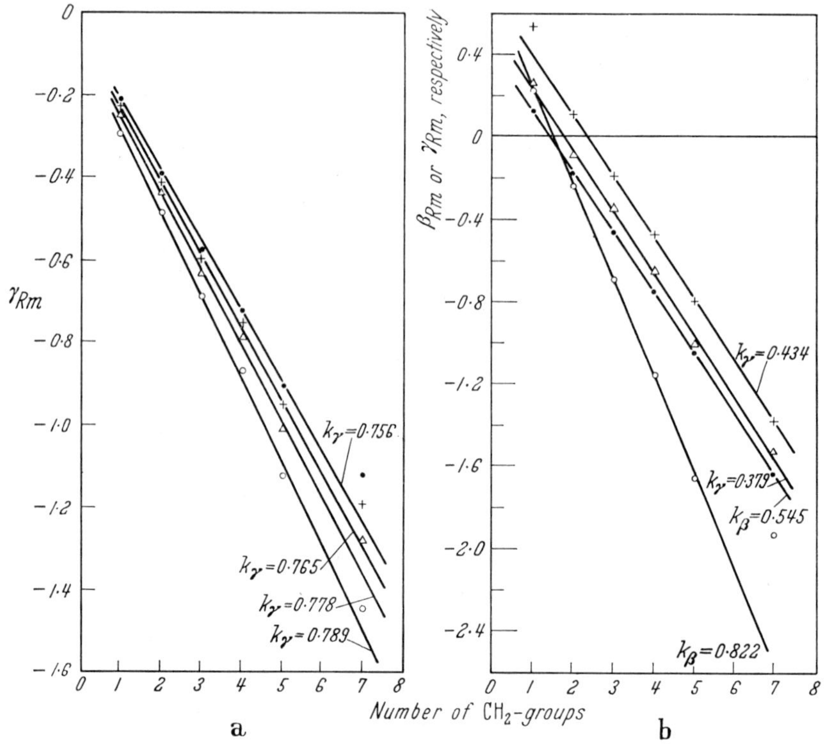

Fig. 66. Rm is additive, because solvent demixing, was taken into account when measuring $\mathrm{Rf}_{\mathrm{obs}}$. (See text) [56], [70]
a) Test as in Fig. 64a. Solvent consisting of 3 components: only the α- and the γ-front are visible. $\mathrm{Rf}_{\mathrm{obs}} = {}^{\gamma}\mathrm{Rf}_{\mathrm{obs}}$.
b) Test as in Fig. 64b. Solvent consisting of 2 components: both fronts are visible. $\mathrm{Rf}_{\mathrm{obs}} = {}^{\beta}\mathrm{Rf}_{\mathrm{obs}}$. Solvent consisting of 3 components: all fronts are visible. $\mathrm{Rf}_{\mathrm{obs}} = {}^{\gamma}\mathrm{Rf}_{\mathrm{obs}}$.

the Rm-values were based, were conventionally related to the α-front. If we substitute, however, ${}^{\alpha}\mathrm{Rf}_{\mathrm{obs}}$ by ${}^{\beta}\mathrm{Rf}_{\mathrm{obs}}$ and ${}^{\gamma}\mathrm{Rf}_{\mathrm{obs}}$ respectively, and if we calculate

$$ {}^{\beta}\mathrm{Rm} = \log\left(\frac{1}{{}^{\beta}\mathrm{Rf}_{\mathrm{obs}}} - 1\right) $$

resp.

$$ {}^{\gamma}\mathrm{Rm} = \log\left(\frac{1}{{}^{\gamma}\mathrm{Rf}_{\mathrm{obs}}} - 1\right) $$

and plot the resulting values versus the respective numbers of CH_2-groups, then straight lines are obtained in the range C_1–C_6 (Figs. 65a and b) [*56, 70*]. The deviation at 7 CH_2-groups may be attributed to the uncertainty of the Rf measurement in the region Rf $\approx$ 1.

Provided the values of Rf_{obs} are measured in a correct way, Martin's relation (22) is found to be useful also in certain cases where demixing of the solvent occurs.

Putting $^{\omega}\overline{\mathrm{Rf}}_{\mathrm{obs}} \cdot \xi = {}^{\omega}\mathrm{Rf}_{\mathrm{true}}$, according to (33), one often finds that ξ must have a magnitude close to 1. Otherwise $^{\omega}\overline{\mathrm{Rf}}_{\mathrm{obs}}$ values up to nearly 0.95 could not yield – as they do – additive $^{\omega}$Rm-values. Here, obviously, the decrease of the phase ratio[1] does not begin until close to the front, whereas it starts relatively early in the case of α-fronts.

Admissible and inadmissible use of conventionally measured Rf-values. We refer again to Fig. 65. Measuring the values $^{\alpha}\mathrm{Rf}_{\mathrm{obs}}$ of the spots lying in the γ-zone one notices that they change with the ratio *dip level – starting point*: *dip level – α-front*. It is instructive to consider more closely this dependence[2] and the relation between $^{\alpha}\mathrm{Rf}_{\mathrm{obs}}$ and the Rf related to

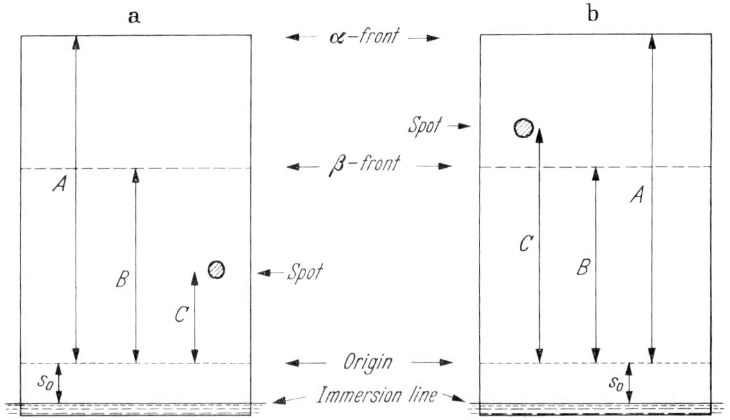

Fig. 67. Demixing of a solvent consisting of 2 components. The spot travels behind the β-front (67a) or in front of the β-front (67b), See discussion in the text [*70*]

the correct front. The situation is clearest in the case of demixing of a solvent consisting of only two components. Figs. 67a and b are schemes of corresponding chromatograms. In Fig. 67a the substance has traveled behind, in Fig. 67b, in front of the β-front, up to the moment of observation.

Let s_0 be the distance *dip level-start*, A the distance *start-α-front*, B the distance *start-β-front* and C the distance *start-center of the spot*. Then we have for Fig. 67a [*70*]:

[1] See Figs. 59 and 60, respectively.
[2] It has nothing to do with the variability of $\mathrm{Rf}_{\mathrm{obs}}$ according to Equ. (30) (see p. 110).

Total length of run of the solvent: ${}^{\alpha}z_f = z_f = A + s_0$, length of run of the β-front: ${}^{\beta}z_f = k_\beta \cdot z_f = B + s_0$, further

$$\frac{{}^{\alpha}Rf_{obs}}{{}^{\beta}Rf_{obs}} = \frac{\frac{C}{A}}{\frac{C}{B}} = \frac{B}{A} = \frac{k_\beta \cdot z_f - s_0}{z_f - s_0}, \text{ and}$$

$$^{\alpha}Rf_{obs} = \frac{k_\beta \cdot z_f - s_0}{z_f - s_0} \, {}^{\beta}Rf_{obs} \tag{37}$$

${}^{\alpha}Rf_{obs}$ assumes its maximum value at $s_0 = 0$; therefore, ${}^{\alpha}Rf_{obs}$ cannot become larger than

$$^{\alpha}Rf_{max} = k_\beta \cdot {}^{\beta}Rf_{obs} \tag{37a}$$

For a given ${}^{\beta}z_f$, ${}^{\beta}Rf_{obs}$ is practically independent of s_0. A spatial variation of the degree of wetting (see Fig. 59 and 60) is, of course, not excluded, but it seems to be likely that the phase ratio changes less near the demixing lines[1] than near the α-front.

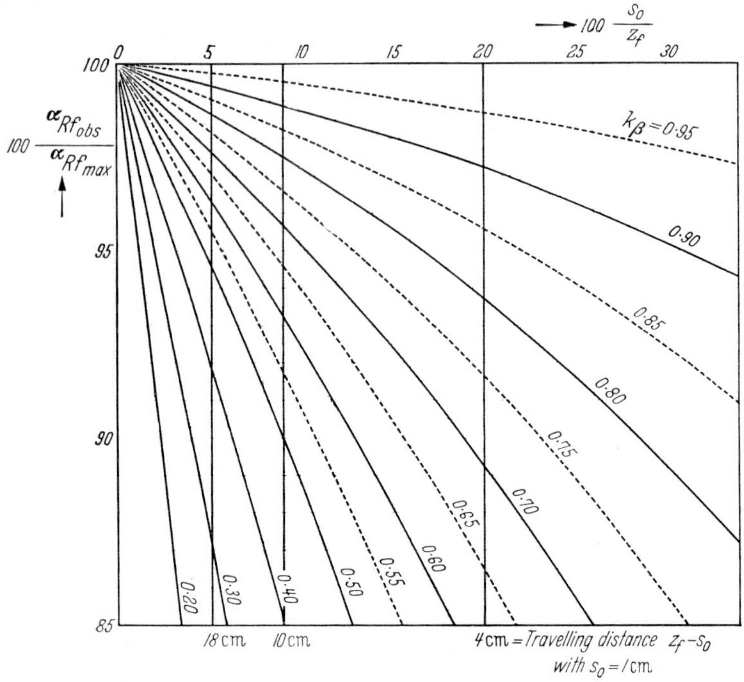

Fig. 68. Influence on ${}^{\alpha}Rf_{obs}$ of the ratio s_0/z_f of the distances dip level-starting point dip level-front, in case of demixing of a solvent consisting of 2 components and a solute with ${}^{\alpha}Rf_{obs} < k_\beta$. k_β is the ratio of the migration speeds of the α- and the β-front [70]

The difference between ${}^{\alpha}Rf_{obs}$ and ${}^{\alpha}Rf_{max}$ may be clearly expressed by the ratio of these two quantities. We have

$$\frac{{}^{\alpha}Rf_{obs}}{{}^{\alpha}Rf_{max}} = \frac{\frac{k_\beta \cdot z_f - s_0}{z_f - s_0} \cdot {}^{\beta}Rf_{obs}}{k_\beta \cdot {}^{\beta}Rf_{obs}}$$

[1] Each front except the α-front represents a demixing line.

From there, we obtain by transforming

$$\frac{1 - \dfrac{{}^{\alpha}\mathrm{Rf}_{\mathrm{obs}}}{{}^{\alpha}\mathrm{Rf}_{\mathrm{max}}}}{\dfrac{1}{k_\beta} - \dfrac{{}^{\alpha}\mathrm{Rf}_{\mathrm{obs}}}{{}^{\alpha}\mathrm{Rf}_{\mathrm{max}}}} = \frac{s_0}{z_f} \tag{38}$$

According to (38), the ratio ${}^{\alpha}\mathrm{Rf}_{\mathrm{obs}} - {}^{\alpha}\mathrm{Rf}_{\mathrm{max}}$ is a function of the parameter k_β and the variable s_0/z_f. This is shown graphically in Fig. 68. From there, ${}^{\alpha}\mathrm{Rf}_{\mathrm{obs}}$ may be read directly as percentage of ${}^{\alpha}\mathrm{Rf}_{\mathrm{max}}$. Only the range 5 to 20 on the abscissa is of practical importance (lengths of run, 18 cm to 4 cm at $s_0 = 1$ cm).

First inference from Fig. 68.

For a transition on the abscissa from 5.2 to 9.1 (Fig. 68) and for $k_\beta > 0.5$, the percentual change in ${}^{\alpha}\overline{\mathrm{Rf}}_{\mathrm{obs}}$ is of the order of the percentual error of measurement[1] (Table 8) and may, therefore, be neglected. If the region on the abscissa extends from 5.2 to 20, this is only true for $k_\beta > 0.9$! Beyond these limits ${}^{\alpha}\overline{\mathrm{Rf}}_{\mathrm{obs}}$ depends more or less heavily on the ratio of the distances *dip level-start* and *dip-level-α-front.*

Second inference from Fig. 68.

For the region on the abscissa between 5.2 and 9.1 and for $k_\beta \geqq 0.85$, the values of ${}^{\alpha}\overline{\mathrm{Rf}}_{\mathrm{obs}}$ and ${}^{\alpha}\overline{\mathrm{Rf}}_{\mathrm{max}}$ become identical within the error of measurement (percentual error $> 2\%$) and independent of s_0/z_f. Because of

$${}^{\alpha}\overline{\mathrm{Rf}}_{\mathrm{obs}} \approx {}^{\alpha}\overline{\mathrm{Rf}}_{\mathrm{max}} \text{ and } {}^{\alpha}\overline{\mathrm{Rf}}_{\mathrm{max}} = k_\beta \cdot {}^{\beta}\overline{\mathrm{Rf}}_{\mathrm{obs}} \tag{37a}$$

there holds the approximation

$${}^{\alpha}\overline{\mathrm{Rf}}_{\mathrm{obs}} \cdot \frac{1}{k_\beta} \approx {}^{\beta}\overline{\mathrm{Rf}}_{\mathrm{obs}}.$$

If

$${}^{\beta}\overline{\mathrm{Rf}}_{\mathrm{obs}} = \frac{1}{\xi} \cdot {}^{\beta}\mathrm{Rf}_{\mathrm{true}} \qquad \text{(p. 119)},$$

then follows

$${}^{\alpha}\overline{\mathrm{Rf}}_{\mathrm{obs}} \cdot \frac{1}{k_\beta} \approx \frac{1}{\xi} \cdot {}^{\beta}\mathrm{Rf}_{\mathrm{true}} \tag{39}$$

or

$${}^{\alpha}\overline{\mathrm{Rf}}_{\mathrm{obs}} \cdot \xi' \approx {}^{\beta}\mathrm{Rf}_{\mathrm{true}}. \tag{39a}$$

Table 8. *Comparison of Two Averages Rf-Values: Uncertainty of Averages Obtained from 8 Observed Values of Rf* (see text)

${}^{\alpha}\overline{\mathrm{Rf}}_{\mathrm{obs}}$ *	possible percentual error $\pm \dfrac{0.022 \cdot 100}{{}^{\alpha}\overline{\mathrm{Rf}}_{\mathrm{obs}}}\ \%$ *
0.9	2.5
0.8	2.8
0.7	3.2
0.6	3.8
0.5	4.5
0.4	5.6
0.3	7.5
0.2	11.3

* When comparing the values of this table with those of Fig. 68, note that ${}^{\alpha}\mathrm{Rf}_{\mathrm{obs}}$ is always smaller than k_β!

ξ', contrary to ξ, is related to a chromatographic system[2] with demixing of the solvent [*70*]. As long as ξ', as ξ, lies in the range between 1 and

[1] The standard deviation $s_{\overline{\mathrm{Rf}}}$ of the mean $\overline{\mathrm{Rf}}_{\mathrm{obs}}$ is roughly $0.015/\sqrt{n}$, where n is the number of observations of Rf [*71*]. According to the statistical rules for the comparison of two average values, the interval of dispersion of $\overline{\mathrm{Rf}}_{\mathrm{obs}}$, for $n = 8$, is $\pm\sqrt{2} \cdot 3 \cdot s_{\overline{\mathrm{Rf}}} = \pm\, 0.022$ units of Rf. This "experimental error" is fairly independent of the magnitude of $\overline{\mathrm{Rf}}_{\mathrm{obs}}$; if Fig. 68 is used, it must be converted into the "percentual error" based on the magnitude of $\overline{\mathrm{Rf}}_{\mathrm{obs}}$ (Table 8).

[2] See footnote 1, p. 101.

1.2, the same conclusions must be drawn from (39a) as from (33)! Indeed, it is very likely that the additivity of Rm in Fig. 61, resp. Fig. 63, is based on the validity of (39a).

There is no doubt that the solvents used in those experiments undergo demixing. The β-fronts are not visible, however, but may be detected by the method illustrated in Fig. 65. Because of the similarity between alcohols and water, they are located near the α-front. For instance, for methanol/water (70 + 30) k_β was determined to be 0.95 [*56*]. The addition of glacial acetic acid, mentioned in the caption to Fig. 61, causes, without doubt, the formation of an invisible γ-front, which also may have a very high k_γ value.

Finally, let us consider the situation shown in Fig. 67b. In the moment of observation we have ${}^{\alpha}Rf_{obs} = C/A$ – see p. 120 – and, therefore, also

$$ {}^{\alpha}\overline{Rf}_{obs} \cdot \xi = Rf_{true} \tag{33}$$

Continuing the development the ratio $s_0 : z_f$ decreases. Thus, three possibilities arise:

1. The β-front cannot catch up with the spot, because

$$u_{spot} > u_{\beta\text{-front}} \qquad (u = \text{velocity})$$

$${}^{\alpha}Rf_{obs} \cdot \frac{dA}{dt} > k_\beta \cdot \frac{dA}{dt},$$

$${}^{\alpha}Rf_{obs} > k_\beta .$$

The ${}^{\alpha}\overline{Rf}_{obs}$ observed in the composed solvent is, and remains till the end of the experiment, equal to $\overline{Rf}_{obs}$ found upon chromatography of the solute with only the faster moving component of the solvent.

2. The spot travels at a constant distance from the β-front, because:

$$u_{spot} = u_{\beta\text{-front}},$$

$${}^{\alpha}Rf_{obs} = k_\beta .$$

As above, there is no difference to chromatographing with the respective pure solvent component.

3. The distance between the spot and the β-front decreases, because

$$u_{spot} < u_{\beta\text{-front}},$$

$${}^{\alpha}Rf_{obs} < k_\beta ,$$

and the front catches up with the spot if

$$A = \frac{s_0 (1 - k_\beta)}{k_\beta - {}^{\alpha}Rf_{obs}} .$$

Now the relative velocity of migration $\frac{u_{spot}}{u_{\alpha\text{-front}}}$ of the spot changes. In an extreme case, it corresponds to the relative velocity of migration of the β-front (${}^{\beta}Rf = 1$). Usually it will be smaller (${}^{\beta}Rf < 1$). In any case, by the time the experiment is stopped, it will not be easy to define a sensible *Rf*-value of such a spot.

Practical rules. *In summary, we can make the following statements about measuring the Rf-value in the case of demixing solvents* [*70*]. Substances traveling ahead of the β-front are not influenced in their chromatographic behavior if the solvent demixes. The solvent, however, does not have the composition present in the solvent container. ${}^{\alpha}\overline{Rf}_{obs}$ is determined in the conventional way. ${}^{\alpha}Rm$ is additive.

If the substance travels behind the β-front, we can distinguish several cases.

1. $k_\beta > 0.85$. The β-front may be neglected and ${}^{\alpha}\overline{Rf}_{obs}$ is determined in the conventional way. ${}^{\alpha}Rm$ is additive.

2. It is a "lower limit" $< k_\beta < 0.85$, where the "lower limit" may be located between 0.2 and 0.4. If the substance does not start to travel until the β-front crosses the starting point, ${}^{\beta}\overline{Rf}_{obs}$ is to be determined. ${}^{\beta}Rm$ is additive. However, any reasonable Rf-measurement is impossible if there is already a noticeable migration in the α-zone.

3. It is $k_\beta <$ a "lower limit". Under such circumstances, one may determine values of ${}^{\beta}Rf_{obs}$, but the corresponding Rm-values are not additive.

4. If there is a γ-front besides the β-front the situation may become confusing. Tentatively, rules 1 to 3 may be applied.

5. If the spots travel behind the ω-front, the rules 1 to 3 with k_ω instead of k_β apply; first, k_ω is to be determined and eventually ${}^{\omega}\overline{Rf}_{obs}$. The situation is insofar clear as the composition of the solvent corresponds to that in the jar.

b) Quantitative evaluation of thin-layer chromatograms

The quantitative evaluation of column chromatograms offers no difficulties. If the chromatographed solutes are separated on the column and if we succeed in measuring their concentration in the eluate any time without errors, it is sufficient to plot the concentration versus the volume and to find the area under the resulting curves. With Gaussian curves, this can simply be done by measuring two distances. The area F is

$$F = w_{1/2} \cdot h \cdot {}^1/_2 \cdot \sqrt{\frac{\pi}{\ln 2}}$$
$$= w_{1/2} \cdot h \cdot 1{,}0645\ ,$$

where h is the height of the maximum of the concentration curve and $w_{1/2}$ is the width of the curve measured at half of the height of the maximum. The areas under the curves are proportional to the quantities of the solutes.

With thin-layer and paper chromatograms, the situation is more complicated. Of course, we can divide the chromatogram into segments, wash out each segment with a solvent, determine the concentration in each eluate and proceed in a similar way as above. If there are colored spots, the procedure may be simplified by cutting out the entire spots instead of segments. Colorless solutes must be made visible. If this revelation is performed by a chemical reaction, producing, for instance, a dye amenable to colorimetry, special precautions are necessary to ensure a stoichiometric or at least a proportional reaction[1]. All this is

[1] See for instance [79].

very troublesome and can hardly be compared with the elegance of the qualitative application of thin-layer and paper chromatography. Therefore, it is often preferred to evaluate the spots *in situ*, i.e., to try to correlate quantitatively their magnitude and intensity with the amounts of solute present.

We wish to consider briefly the principles of the methods employed. We assume proportionality between the amount of color and the amount of solute, which is self evident only for inherently colored substances. In the ideal case, the spots on thin-layer and paper chromatograms are elliptical [*80*]. The solute concentration is not only variable in the direction of flow as in column chromatograms, but also perpendicular to it. This is due to the possibility of cross diffusion which is lacking in "closed columns".

Starting from an idealized assumption about the distribution of the color density in the spot, we can discuss the kind of results to be expected if the concentration is to be evaluated either

from the color intensity, or

from the area of the spot.

Actual deviations from the idealized assumptions are sometimes easily recognized and can be taken into account when judging the quality of the method of evaluation.

In the case of a thin-layer or paper chromatogram with elliptically shaped spots (after chromatography) [*80*], we assume that:

1. the color extinction of the spot is proportional to the density of the substance,

2. the density Ψ of the substance along the axes x and y of the ellipse follows a Gaussian distribution of the form

$$\Psi = k \cdot e^{-\frac{x^2}{2\sigma^2_x}} \quad \text{and}$$

$$\Psi = k \cdot e^{-\frac{y^2}{2\sigma^2_y}}$$

3. and that the time t in the parameters

$$\sigma_x = \sqrt{2 D_x t} \;;\; D_x : \text{"diffusion coefficient along x"}$$

$$\sigma_y = \sqrt{2 D_y t} \;;\; D_y : \text{"diffusion coefficient along y"}$$

is identical for spots to be compared.

Evaluation by densitometry. Densitometers have a probing slot or point. It is important that the color density to be measured does not vary excessively over this aperture. For the densitometry of spots on chromatograms, especially, a point-like probe is suitable (for instance, a very small circle). Running this point along one axis of the ellipse of the spot, we should find that the area under the resultant extinction curve is proportional to the quantity of substance, if the conditions 1–3 are strictly fulfilled.

Comparing spots on chromatograms that have been developed for different lengths of times, condition 3 is not fulfilled anymore (see above). For elliptical spots, mathematical arguments now suggest proportionality between the substance quantities and the expression $\frac{F_x \cdot F_y}{E_{max}}$; F_x and F_y are the areas under the extinction curve obtained by probing the spots along both axes of the ellipse. E_{max} is the extinction in the common vertex of the curves (center of the ellipse).

In some cases, the spot is not elliptical. Nevertheless, the proportionality may still exist if the probing line is properly chosen. For the frequently occurring crescent shaped spots, one chooses the axis of symmetry. Completely irregular spots may be measured better with a probing slot whose length approximately corresponds to the spot diameter.

A careful calibration is essential. It should cover the whole range of concentration to be encountered.

Evaluation by measuring spot diameters (spot area). The edge of a spot, which can be observed on the plate or on a photograph, obviously corresponds to the transition between a not discernible and a just discernible everywhere equal color density. The area of the marginal ellipse thus defined is proportional to the logarithm of the quantity of substance present if the conditions 1–3 are fulfilled [*80*]. The factor of proportionality remains constant for a given substance in a given chromatographic system[1]. A measure for the area of the ellipse is the product of the two axes [*80*]. With irregular spots, such measurements do not lack a certain amount of arbitrariness. In such cases, it is especially important to fix the procedure by calibration and to account for the respective error of measurement.

It was found empirically that over a fairly large range of weights, the logarithm of the quantity present is proportional to the square root of the spot area[2]. In practice, the spot area thus increases more rapidly with increasing substance quantity than is predicted by the theory. This deviation can easily be explained. It is probably a consequence of a delay in equilibration of the solute between the mobile and stationary phase. The larger the quantity of the solute, the longer it takes until all of it starts to migrate.

Measuring the spots is advantageous over densitometry because the color yield needs to be reproducible only at the spot edges where the excess of staining reagent is largest.

It should be emphasized that the discussion in this section is confined to general directions, the validity of which depends upon the aforementioned conditions. A critical examination is, in any case, indispensable. A paper by GIDDINGS [*80*] contains a list of references about practical experiences in measuring spots on paper chromatograms. Results obtained in thin-layer chromatography are described in chapter F, p. 49 (see also[2]).

[1] See footnote 1, p. 101.
[2] S. J. PURDY and E. V. TRUTER, Chem. & Ind. **1962**, 506.

2. Differences between thin-layer chromatography and paper chromatography

The practical advantages of thin-layer chromatography on silica gel are well known. As far as the chromatographic process of separation is concerned, they are mainly based on lesser spreading of the spots (see Figs. 39, 159, 160, 178). Because of this:

a) the detection thresholds of the chromatographed substances are lowered,

b) the separation is possible over a shorter length of run, and

c) as a consequence of b), the time of development is substantially reduced.

One might be inclined to explain the reduced spreading tendency of the spots on the basis of differences between silica gel and cellulose. Processed cellulose powder[1], however, gives thin layers with properties more similar to the properties of silica gel than to those of filter paper. This suggests that the differences between silica and paper are more of a quantitative than of a basic nature. In our opinion, the characteristics of filter paper are to be ascribed primarily to its fibrous structure. It is well known that substance spreading at interfaces is almost instantaneous. The driving force is a change of the surface tension. In contrast to powder, fibers have long continuous boundaries. Along these, a solute might spread much faster than by diffusion alone. Only the space between the fibers must be bridged by diffusion. With powders, diffusion is the only possible spreading mechanism, except convection by the solvent. Upon this phenomenon, there may be superimposed an increased adsorptive capacity due to a larger surface of the pulverized material.

A difference, in the sense that thin-layer chromatography is based more on adsorption, paper chromatography more on partition between liquid phases, seems unlikely to us. In our own experience, the advantages of thin-layer chromatography are preserved if one changes over from solvents favoring adsorption to those that indicate a partition mechanism.

3. Relations to column chromatography

It is known that the behavior of a substance on paper chromatograms gives an approximate idea of its behavior on cellulose columns[2]. A first, easily fulfilled requirement is to employ comparable ratios of solute to adsorbent (paper, cellulose powder).

A second requirement more difficultly met would be an equal speed of migration and an equal distribution of the mobile phase along the

---

[1] The raw material is very pure cellulose with a high α-content. The powder, which is not acid treated, gives residual ashes of about 1000 ppm. 90% of it has a grain size of less than 20 μ. Such a high grade material is necessary for the layers to become smooth and perfect. A large specific surface, more than 15000 cm^2/g (acc. to Blaine), is of special importance with regard to the separating power. Ordinary cellulose powder has a substantially smaller specific surface and hence, yields layers of poorer separating power. (Private communication from Messr. Macherey, Nagel & Co., Düren, Germany).

[2] See for instance [*79*].

path of migration[1]. This condition is certainly not fulfilled if one uses a column which has been wetted in advance by the solvent, as is commonly done in column chromatography [*79*]. Basically, the same is true if experiences from thin layer chromatography are to be applied to silica columns. Recently, a new version of column chromatography has been developed [*81*]. It allows, to a certain extent, to meet both requirements mentioned above. This method is characterized by the fact that the solvent penetrates the silica gel exclusively because of capillarity, as on a horizontal thin layer [*64*]. After the solvent has wetted the column completely, it still keeps moving on because of evaporation at the end of the column in a fashion similarly to the continuous – flow – technique with a covered plate [*64*]. According to DAHN and FUCHS [*81*], experience shows that the Rf-values on the covered plate are, in many cases, comparable to those on the column. The reason for the similarity of the situations on the column and the covered plate might be seen in the fact that the interference from chamber saturation, which is unimportant on the covered plate, is completely absent on the column.

Addendum: Displacement and ion exchange on thin-layer chromatograms

Substances with similar Rf-values, whose retention is based on equal mechanisms, may influence each other's partition between the mobile and the stationary phase by the phenomenon of displacement. This is beautifully demonstrated by chromatographing a substance A in increasing quantities to which always the same amount of substance B is added ($Rf_A \lesssim Rf_B$). The A-spots get larger and push the B-spots ahead in such a way as to increase their measured Rf-values.

Silica gel has, to a limited extent, the character of an ion exchanger. This is discussed in greater detail in the chapter on inorganic thin-layer chromatography. When chromatographing salts of organic compounds, one may often find that they split into acids and bases which travel independently of each other. The dilution by the solvent, of course, favors this dissociation. Therefore, one can observe the same Rf-values for hydrochlorides as for free bases. The same is true for alkali salts of carboxylic acids. The Cs-salt of 2,4-dinitrophenylglycine, e.g., forms the same spot as free 2,4-dinitrophenylglycine and leaves a Cs spot at the starting point (revelation by autoradiography of ^{134}Cs) [*70*]. This holds for acidic, basic, and neutral solvents.

Salts of polyvalent acids or bases may form several spots because of the possibility of partial dissociation. This phenomenon (multiple spots) is, in principle, always to be expected if a substance exists in several mutually interconvertible forms (different states of ionization, formation of complexes, tautomery). The problem has been thoroughly discussed by JÄGER, RAMEL and SCHINDLER [*82*], as well as by KELLER and GIDDINGS [*13*].

---

[1] See the discussions of p. 106ff. on the spreading of the solvent and the shape of the solvent profile on thin-layer and paper chromatograms.

List of symbols

A	distance start-α-front	119, 122
A_{rev}	reversible work	102
A	see page	108
$\mathsf{A/W}$	factor of proportionality between the local velocity of the solvent and the velocity of the front	108, 109, 110, 111
a and b	coefficients	106
B	distance start-β-front	119
b	width of the band, in cm	95
C	distance start-center of spot	119
c	concentration	79
c	partition parameter used by GIDDINGS	109
c_M	concentration in mole percent	115
$c_V(r)$	concentration in compartment no. r after the flowing out of the constant volume $V = v_m - v_m/n$	89, 90
c_o	initial concentration in compartment (tube, plate, chamber) no. 1	77, 82
c_r	concentration in tube (plate, chamber) no. r	77
c (V)	concentration of the solution flowing out of Signer's tank after the flowing out of the volume V	86
c_n (V)	c(V) for n chambers	85—88
c'_m	arbitrary concentration in the mobile phase	102
c'_s	arbitrary concentration in the stationary phase	102
D_x	coefficient of diffusion in the x-direction	124
D_y	coefficient of diffusion in the y-direction	124
d	distance of migration of the maximum of the band, in cm	95
d_r	distance from the origin to compartment No. r of a Signer tank	90, 91
$d_{r,max}$	distance from the origin to compartment No. r_{max} of a Signer tank	91, 94—97
d_σ	standard deviation from $d_{r,max}$, in cm	91, 95, 100
E_{max}	extinction in the maximum of a spot	125
e	base of the natural logarithm	85, 87, 88, 89, 91, 100, 124
F	area under the c(V)-curve (GAUSS)	123
F_x, F_y	areas under the extinction curve	125
f(r—1)	Poisson distribution, discontinuous	90
f (Z)	degree of overlapping	99, 100
G, G_o, G_μ ...	constants in Martin's relation	103
g(x)	Poisson function (continuous)	87
H	HETP (height equivalent to a theoretical plate)	94, 106
h	height of the maximum of the c(V)-curve (GAUSS)	123

K	partition- resp. adsorption-coefficient, independent of the concentration (the latter can be regarded as independent of the concentration only at small concentrations!)	84, 85, 88, 99—101
K′	K for not completely definable mechanisms of partition (chromatographic column)	95, 101—103, 108, 109
$k_\beta, k_\gamma \ldots k_\omega$	delaying factors	115, 118—123
L	length of a Signer tank or of a chromatographic column	90, 91, 94
M_o	amount of substance initially present in compartment No. 1	85, 88—91
$M(d_r)$	amount of substance per unit length, at a distance d_r from the origin of a Signer tank, after the flowing out of the volume $V = V_m \frac{V_m}{n}$	90, 91
N	number of solute particles initially present in compartment No. 1 of a Signer tank	86, 87, 90
N_r	number of solute particles present in compartment No. r of a Signer tank	90
n	number of compartments	85—92, 98—100
n′	number of effective compartments (theoretical plates)	92, 94—96
p	fraction of the total amount of substance present in the mobile phase in a compartment (theoretical plate, segment of a column)	85—92, 96, 99
p′	p for not completely definable mechanisms of partition (chromatographic column)	92—95, 101
q	fraction of the total amount of substance present in the stationary phase in a compartment (theoretical plate, segment of a column)	85
R	gas constant	102
Ra	Rm-value for association	105
Rf	definition on page	91, 94, 102, 108, 109
Rf_{obs}	conventionally measured Rf-value $\left(\frac{\text{distance of migration of the substance}}{\text{distance of migration of the front}}\right)$	108—114
$\overline{Rf}_{obs}$	average value of Rf_{obs}-observations	112, 113
Rf_{int}	intrinsic Rf-value	108—110
$Rf_{int, final}$	final value of the variable Rf_{int}	110
Rf_{true}	$= Rf_{int, final}$	111—114, 122
${}^\alpha Rf_{obs}, {}^\beta Rf_{obs} \ldots {}^\omega Rf_{obs}$:	conventionally measured Rf-values related to the α, β, ω-front	116—123
${}^\alpha\overline{Rf}_{obs}, {}^\beta\overline{Rf}_{obs}, \ldots {}^\omega\overline{Rf}_{obs}$:	average values of ${}^\alpha Rf_{obs}$, ${}^\beta Rf_{obs}$, ${}^\omega Rf_{obs}$-observations	119, 121—123
${}^\alpha Rf_{max}$	upper limit of ${}^\alpha Rf_{obs}$ for vanishing s_0	120, 121
${}^\beta Rf_{true}, {}^\gamma Rf_{true}, \ldots {}^\omega Rf_{true}$:	true Rf-values related to the β, γ and ω-front	119, 121, 122

Rm	$= \log \left(\frac{1}{Rf} - 1\right)$	103, 112, 118
Rm_{obs}	Rm from Rf_{obs}	112—114
Rm′	$= Rm + \log \xi$	112—114
$^{\alpha}Rm, {}^{\beta}Rm, \ldots {}^{\omega}Rm$:	Rm-values computed from $^{\alpha}Rf, {}^{\beta}Rf \ldots {}^{\omega}Rf$.	118, 119
r	compartment (chamber, plate, segment) number . . .	89
r_{max}	r with highest concentration of substances	90
S_0	substitutable substance	103
s	distance covered by the solute	108, 109
s_0	distance immersion line (dip level)-start.	108, 109, 119—122
$s_{\overline{Rf}\ obs}$	standard deviation of $\overline{Rf}_{obs}$	113,121
$s_{\overline{Rm}}$	standard deviation of $\overline{Rm}$	113
T	absolute temperature in ° K	102, 103, 112
t	time	97, 106, 115, 122, 124
u	flow rate	96
U	$= \left(\frac{V}{v_m} - 1\right) \sqrt{n}$	79
u_z	velocity of a substance spot at position z	109, 110
V	volume flown out since the beginning of the experiment	84—89
V_B	width of band in units of volume.	95
V_R	retention volume (chromatographic column)	95, 96
V_T	volume flown out at the time of fraction cutting (description of c(V) after GAUSS)	98—100
$V_{T,\,corr}$	corrected V_T	98
$V_{T,\,opt.}$	volume flown optimal moment for fraction cutting . .	98
V_{max}	volume flown out (Signer's tank) when the solute concentration has arrived at its maximum, identical with retention volume.	86—88, 94
v_F	volume of the solvent within a chromatographic column	92, 94, 96
v_f	velocity of the solvent front	108—110
v_m	total volume of the mobile phase in Signer's tank . . .	82, 83
v_s	total volume of the stationary phase in Signer's tank. .	85
v_z	velocity of the solvent at position z	108
W	see pages	108, 109
$w_{1/2}$	width of a Gaussian curve at one half of the maximum height	123
x	volume ratio $\dfrac{V}{\frac{V_m}{np}}$	86
$\bar{x}$	average of all values of x	87
Z	separation parameter	98—100

z	distance between the starting line of the solvent (immersion line, dip level) and the point of observation along the length of run of the solvent	107—110
z_f	distance of migration of the solvent front	106—110, 120, 122
$^{\alpha}z_f$, $^{\beta}z_f$, ... $^{\omega}z_f$	length of run of the α-, β-, ... ω-front	115, 120
α-, β-, γ-, ... ω-front:	notation of fronts formed by demixing solvents	115—120, 122, 123
Δ	half-width of the "visible" band in Signer's tank	97
ΔF	reversible work (energy of transfer) under standard conditions	102, 103
ΔF_o	ΔF for a substitutable substance S_0	103
ΔF_X, ΔF_Y	change of the energy of transfer ΔF_0 due to a substituent (X, Y ...) in S_0	103, 112
$\Delta F\mu$	change of the energy of transfer ΔF_0 due to a substituent μ in the S_0	103
ΔG	Gibb's free energy for a substance changing from the liquid to the gaseous phase	103
ΔRm	change of Rm due to a change in the chromatographic system	104
Δv_m	correction to the volume on page	85
ΔV_x	correction to the volume on page	98
δ	error in Rm′, due to the use of Rf_{obs}	112
ζ	coefficient	115
$\varkappa$	coefficient	115
ξ	factor of correction for Rf_{obs}	112, 113, 119, 121, 122
ξ'	ξ for demixing of the solvent	121
π	= 3.14159...	87, 123
ϱ	phase ratio	85, 99—101, 103, 108, 111
ϱ'	variable phase ratio	108
σ	standard deviation of the normal distribution function	87, 88, 90, 99
τ	independent variable in the normal distribution function	88, 100
$\Phi(u)$	error integral	99, 100
$\varphi(x)$	Gauss function, Normal distribution	87
Ψ	density of the substance	124
ω	speed of stirring	80
μ	mean of the normal distribution function	88

Bibliography to Chapter H: Theoretical Aspects of Thin-Layer Chromatography

[1] MARTIN, A. J. P.: Endeavour **6**, 21 (1947).
[2] GIDDINGS, J. C.: J. Chromatog. **2**, 48 (1959).
[3] MARTIN, A. J. P., and R. L. M. SYNGE: Biochem. J. **35**, 1358 (1941).
[4] WEISSBERGER, A.: Technique of organic chemistry, Vol. 3, p. 529—311. Interscience Publishers Inc. 1950.

[5] WILSON, J. N.: J. Am. Chem. Soc. **62**, 1583 (1940).
[6] DE WAULT, D.: J. Am. Chem. Soc. **65**, 532 (1943).
[7] GLÜCKAUF, E.: Discussions Faraday Soc. **7**, 12 (1949).
[8] SMIT, W. M.: Discussions Faraday Soc. **7**, 38 (1949).
[9] GLÜCKAUF, E.: Trans. Farady Soc. **51**, 34 (1955).
[10] — Trans. Faraday Soc. **51**, 1540 (1955).
[11] VAN DEEMTER, J. J., F. J. ZUIDERWEG and A. KLINKENBERG: Chem. Eng. Sci. **5**, 271 (1956).
[12] GIDDINGS, J. C.: J. Chromatog. **2**, 44 (1959).
[13] KELLER, R. A., and J. C. GIDDINGS: J. Chromatog. **3**, 205 (1960).
[14] RUOFF, A. L., and J. C. GIDDINGS: J. Chromatog. **3**, 438 (1960).
[15] BOGUE, D. C.: Anal. Chem. **32**, 1777 (1960).
[16] HAMILTON, P. B., D. C. BOGUE and R. A. ANDERSON: Anal. Chem. **32**, 1782 (1960).
[17] GIDDINGS, J. C.: J. Chromatog. **5**, 46 (1961).
[18] SIGNER, R., K. ALLEMANN, E. KÖHLI, W. LEHMANN, H. MEYER u. W. RITSCHARD: Dechema-Monograph. **27**, 32—44 (1956).
[19] Handbook of Chemistry and Physics: 38. Aufl. Cleveland, Ohio: Chemical Rubber Publishing Co. 1956—1957.
[20] Biochemisches Taschenbuch. Berlin-Göttingen-Heidelberg: Springer 1956.
[21] HADWIGER, H.: Experientia **9**, 391 (1953).
[22] Diss. H. MEYER: Universität Bern 1956.
[23] KEULEMANS, A. I. M.: Gas Chromatography. New York: Reinhold Publishing Corp. 1957.
[24] HECKER, E.: Verteilungsverfahren im Laboratorium. S. 87. Weinheim: Verlag Chemie 1955.
[25] SPACKMAN, D. H., W. H. STEIN and S. MOORE: Anal. Chem. **30**, 1190 (1958).
[26] z. B. STEPANOV, V., D. HANDSCHUH u. F. A. ANDERER: Z. Naturforsch. **166**, 626 (1961).
[27] BENNICH, H.: Biochim. et Biophys. Acta **51**, 265 (1961).
[28] WIELAND, TH., u. H. MERZ: Chem. Ber. **85**, 731 (1952).
[29] CLAESSON, S.: Arkiv Kemi Mineral. Geol. A **24**, 16 (1947).
[30] HAGDAHL, L.: Acta Chem. Scand. **2**, 574 (1948).
[31] — Sci. Tools **1**, 21 (1954).
[32] ALM, R. S., R. J. P. WILLIAMS u. A. TISELIUS: Acta Chem. Scand. **6**, 826 (1952).
[33] KAUMAN, W. G., and A. BAK: Nature **182**, 743 (1958).
[34] KHAN, M. A.: Nature **186**, 800 (1960).
[35] KIESELBACH, R.: Anal. Chem. **32**, 880 (1960).
[36] GIDDINGS, J. C.: J. Chromatog. **5**, 61 (1961).
[37] POHLOUDEK-FABINI, R., u. WOLLMANN: Pharmazie **15**, 590 (1960).
[38] MOORE, S., and W. H. STEIN: Ann. N. Y. Acad. Sci. **49**, 265 (1948).
[39] MARTIN, A. J. P.: Biochem. Soc. Symposia (Cambridge, Engl.) **3**, 4 (1950).
[40] — Ann. Rev. Biochem. **19**, 518—520 (1950).
[41] KOVÁTS, E.: Helv. Chim. Acta **41**, 1915, 1928 (1958).
[42] SHRINER, R. L., R. C. FUSON and D. Y. CURTIN: The systematic identification of organic compounds, 4. Aufl. Kapitel 6. New York: John Wiley & Sons Inc. 1956.
[43] FRANC, J., and J. JOKL: J. Chromatog. **2**, 423 (1959) und frühere Arbeiten.
[44] KOVÁTS, E.: Helv. Chim. Acta **42**, 2709 (1959).
[45] BATE-SMITH, E. C., and R. G. WESTALL: Biochim. et Biophys. Acta **4**, 427 (1950).
[46] PARDEE, A. B.: J. Biol. Chem. **190**, 757 (1951).
[47] GASPARIC, J., u. M. VECERA: Collection Czechosl. Chem. Commun. **24**, 1822 (1959).
[48] ROUX, D. G., and S. R. EVELYN: J. Chromatog. **1**, 537 (1958).
[49] GREBENOVSKÝ, E.: Z. anal. Chem. **185**, 290 (1962).
[50] JAMES, A. T., and A. J. P. MARTIN: Biochem. J. **50**, 679 (1959).
[51] RAY, N. H.: J. Appl. Chem. (London) **4**, 21 (1954).
[52] JAMES, A. T.: Biochem J. (London) **52**, 242 (1952).
[53] KOVÁTS, E.: Z. anal. Chem. **181**, 351 (1961).

[54] FRENCH, D., and G. M. WILD: J. Am. Chem. Soc. **75**, 2612 (1953).
[55] REICHL, E. R.: Monatsh. Chem. **86**, 69 (1955).
[56] Diss. G. PATAKI, Universität Basel, 1962.
[57] MOORE, T. B., and C. G. BAKER: J. Chromatog. **1**, 512 (1958).
[58] SCHAUER, H. K., u. R. BULIRSCH: Z. Naturforsch. **10**b, 683 (1955).
[59] — — Naturwiss. **43**, 34 (1956).
[60] — — Z. Naturforsch. **13**b, 327 (1958).
[61] MACEK, K.: Experientia **17**, 162 (1961).
[62] BRENNER, M., u. G. PATAKI: Helv. Chim. Acta **44**, 1420 (1961).
[63] — — Helv. Chim. Acta (im Druck).
[64] — u. A. NIEDERWIESER: Experientia **17**, 237 (1961).
[65] MÜLLER, R. H., and D. L. CLEGG: Anal. Chem. **23**, 296 (1951).
[66] KOWKABALY, G. M., and H. G. CASSIDY: Anal. Chem. **22**, 817 (1950):
[67] WOOD, S. E., and H. H. STRAIN: Anal. Chem. **26**, 260 (1954).
[68] GIDDINGS, J. C., G. H. STEWART and A. L. RUOFF: J. Chromatog. **3**, 239 (1960).
[69] FRANC, J., u. J. JOKL: Collection Czechoslov. Chem. Commun. **21**, 1161 (1956).
[70] Diss. A. NIEDERWIESER, Universität Basel, 1962.
[71] BRENNER, M., A. NIEDERWIESER, G. PATAKI u. A. R. FAHMY: Experientia **18**, 101 (1962).
[72] BRENNER, M., u. A. NIEDERWIESER: Experientia **16**, 378 (1960).
[73] BOMAN, H. G.: Nature **170**, 703 (1952).
[74] BUNGENBERG DE JONG, H. G., u. J. TH. HOOGEVEEN: Proc. Acad. Sci. (Amsterdam, Series B **64**, 1, 18, 167, 183 (1961); **63**, 228, 243, 383 (1960).
[75] TISELIUS, A.: Arkiv Kemi, Mineral. Geol. **14** B, Nr. 22 (1940).
[76] CLAESSON, S.: Arkiv Kemi, Mineral. Geol. **23**A. Nr. 1 (1946).
[77] — Arkiv Kemi, Mineral. Geol. **24** A, Nr. 7 (1947).
[78] GRIFFITHS, J., D. JAMES and C. PHILLIPS: Analyst **77**, 897 (1952).
[79] BRENNER, M., u. A. VETTERLI: Helv. Chim. Acta **40**, 943—949 (1957).
[80] CALVIN, J., J. C. GIDDINGS and ROY A. KELLER: J. Chromatog. **2**, 626 (1959).
[81] DAHN, H., u. H. FUCHS: Helv. Chim. Acta **45**, 261 (1962).
[82] JÄGER, H., A. RAMEL u. O. SCHINDLER: Helv. Chim. Acta **40**, 1310 (1957).

Some recent noteworthy articles

HALMEKOSKI, J., and H. HANNIKAINEN. Suomen Kemist **36**B, 24 (1963): Martin relation.
THOMA, JOHN A., and D. FRENCH: Analyt. Chem. **29**, 1645 (1957): Selection of Solvent Components and Solvent Proportions.
— Talanta **8**, 829 (1961): Parameters involved in paper chromatography.
— Analyt. Chem. **35**, 214 (1963): Application and Theory of Unidimensional Multiple Chromatography.
— A Quantitative Approach to Unidimensional Multiple Chromatography (in press) Partition Functions in Carbohydrate Chromatography. Their Measurement and Significance.

Special Section

Introduction

By

EGON STAHL

The most common and easiest separation process is based on the fact that substances dissolve in solvents to varying degrees. A complex mixture may be readily resolved into a lipophilic and a hydrophilic fraction. With the aid of further isolation procedures, it is possible to obtain subfractions, i.e. mixtures of similar compounds. Chromatography is often used for efficient separation in place of more cumbersome and time-consuming processes.

Thin-layer chromatography (TLC) is being increasingly employed as a rapid method, first for exploratory work and subsequently, for more exact analysis. The possibility of varying both the thin layer and the solvent, means that many difficulties arising in separation can be solved with amazing rapidity. It should, however, be made clear that even TLC is not a universal panacea. In the case of complex mixtures, twenty fractions is about the greatest number that can be resolved on a single chromatogram, for reasons of space. Prefractionation of complex mixtures by other techniques may be necessary.

It seemed most suitable, to arrange the groups according to their lipophilic or hydrophilic character in the special section that follows, i.e. to begin with lipids and to end with hydrophilic compounds[1].

Within the groups themselves, the less polar compounds are usually listed first and the more polar compounds later. This classification follows naturally from the method itself, and in following it the reader will be constantly reminded of the close relationship between the three main elements in chromatography as shown in Fig. 69. This figure is, in fact, based on data gathered from adsorption chromatography, which may be summarized as follows:

Mixture to be separated

a) Saturated hydrocarbons are adsorbed only slightly, if at all, and thus, migrate fastest. Unsaturated hydrocarbons are more strongly adsorbed the

[1] In certain chapters, e.g. on vitamins, this principle has been broken in order to give precedence to other factors.

more double bonds they contain. Hydrocarbons containing conjugated double bonds are more strongly adsorbed than those having a system of isolated double bonds.

To separate hydrocarbons, an active adsorbent and an only slightly polar solvent should be used. Alternatively, "phase reversal" may be employed, whereby the stationary phase is impregnated with a lipophilic substance and a hydrophilic solvent is used as the mobile phase.

b) If functional groups are introduced into a hydrocarbon, the adsorbtion affinity is increased in the following sequence: $-CH_3 \rightarrow$ O-Alkyl $\rightarrow$ $\rightarrow > C = O \rightarrow -NH_2 \rightarrow -OH \rightarrow -COOH$.

When benzene, for instance, is used as solvent on silica gel or alumina layers, the ethers and esters are found in the top part of the chromatogram, and ketones and aldehydes approximately in the center, the alcohols below them, while the acids remain at the starting point. The separation sequence thus follows the polarity of the compounds. Organic acids and strong bases can be chromatographed either in the form of their less polar derivatives, or by means of acidic or basic solvents.

Solvent

Solvents can be arranged according to their "eluting" effect in an "eluotropic series" [*128*]. Here too, a close relationship between polarity and elution effect is evident. The dielectric constant may be taken as an indication of polarity, and dielectric constants are, for this reason, included in Table 9.

Table 9. *Eluotropic Series of Solvents*

Solvent[1]	Boiling point (B. P. 760 mm in ° C)	Dielectric constant (20°)
n-Hexane	68.7	1.890
Heptane	98.4	1,924
Cyclohexane	81.4	2.023
Carbon tetrachloride	76.8	2.238
Benzene	80.1	2.284
Chloroform	61.3	4.806
Diethylether	34.6	4.34
Ethyl acetate	77.1	6.02[2]
Pyridine	115.3	12.3[2]
Acetone	56.5	20.7[2]
Ethanol	78.5	24.30[2]
Methanol	64.6	33.62
Water	100	80.37

[1] Additional solvents and their properties can be found in Biochemical Handbook, Springer-Verlag, Berlin-Göttingen-Heidelberg, 1956; Handbook of Chemistry and Physics, Chemical Rubber Publishing Co., Cleveland, Ohio, 1959.
[2] At 25°.

Adsorbent

Adsorbents can also be classified, the difference here being between those with a marked degree of adsorption activity and those which show little adsorbtivity. Details are given in the section on adsorbents on pp. 29—34.

The beginner can obtain a clear idea of the various relationships from a study of the diagram, which is of assistance in all types of chromatography[1]. The shaded triangle can be rotated. If, for example, a lipid mixture is involved, one of the apices of the triangle is pointed to "lipophilic" under the heading "mixture to be separated". The other two points of the triangle then show that a non-polar solvent is required in conjunction with an active adsorbent. If, however, the triangle is set for a polar mixture, the two other apices show that a hydrophilic solvent and a non-active adsorption layer are called for.

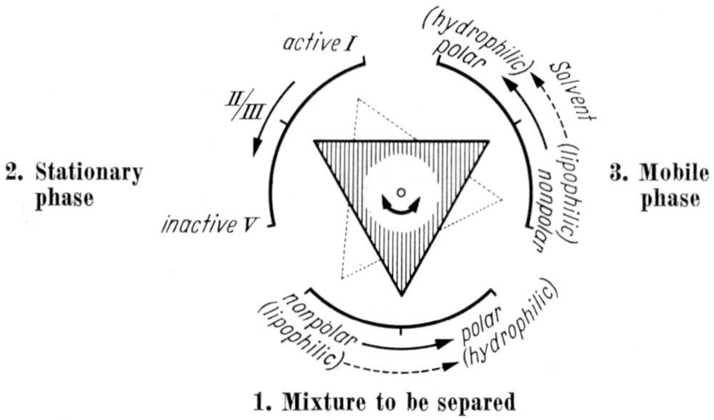

Fig. 69. Diagram showing the close relationship between the three main variable elements in chromatography as illustrated from adsorption chromatography [*124*]. For explanation see text

It is not possible, in the light of present knowledge, to lay down optimum separation conditions for the majority of mixtures, as the method itself is still in the stage of development, and new data are constantly being obtained. The experimental material contained in the chapters that follow will, however, serve to indicate various general principles, both for normal and for special separation problems.

As thin-layer chromatography, like paper chromatography, is also used for identification of single compounds, *Rf*-values are given, and these multiplied by 100 to give h*Rf*. It must, however, be emphasized that these should be taken as a guide only. For the sake of simplicity, even-number volume parts are given for solvent mixtures, if not specially indicated to the contrary, and these add up to 100 ml (amount used in the standard separation chamber).

---

[1] The diagram can readily be used for other chromatographic methods. If "active" at I is replaced by "lipophilic" or "non-polar" and "hydrophilic" or "polar" substituted at V, and under the heading "mobile phase", "non-polar" and "lipophilic" substituted for "polar" and "hydrophilic", etc., the diagram is ready for use with phase reversal.

A. Aliphatic Lipids

By

HELMUT K. MANGOLD

I. Introduction

1. Neutral lipids and their hydrolysis products

Long-chain hydrocarbons, alcohols, aldehydes, and acids occur in great abundance in the vegetable and animal kingdoms:

$H_3C—(CH_2)_X—CH_3$

Hydrocarbon

$H_3C—(CH_2)_{X'}—CH_2OH$

Alcohol

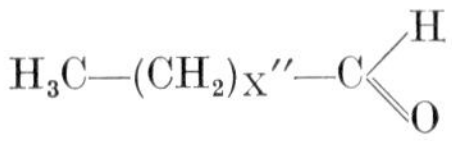

Aldehyde

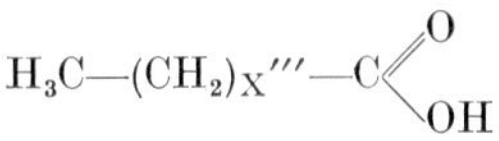

Acid

x, x′ . . . > 8

In each of these four lipid classes, saturated, mono- and polyunsaturated, straight-chain, and branched-chain components occur. Bifunctional compounds, such as epoxy and hydroxy acids, also are known.

Alcohols, aldehydes, and acids occur mostly in bound form in nature. Alcohols and acids form (ester-) waxes:

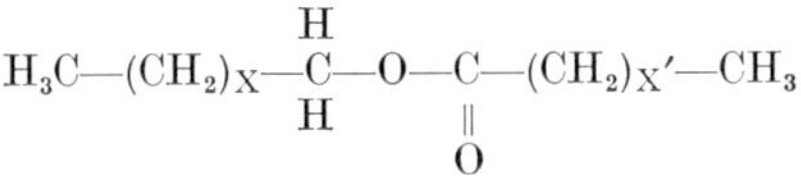

Wax Ester

Compounds containing aliphatic alcohols, aldehydes, and acids bound to glycerol are very common. Long-chain alcohols and glycerol form glyceryl ethers; aldehydes are bound to glycerol as semi-acetals or vinyl ethers; and acids are esterified with glycerol to form mono-, di-, and triglycerides:

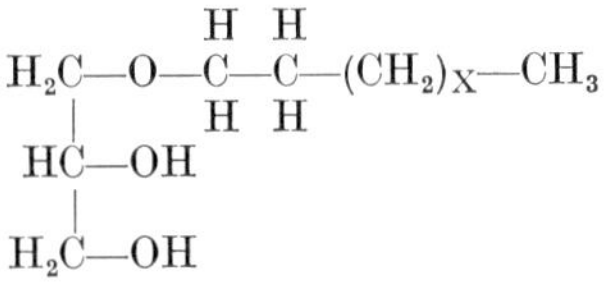

Glyceryl Ether

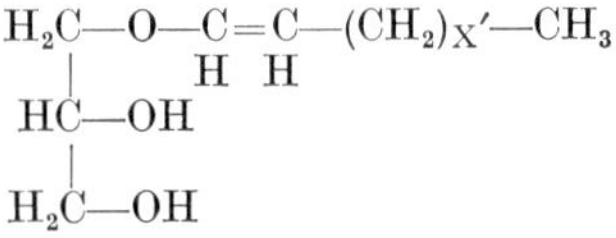

Vinyl Ether

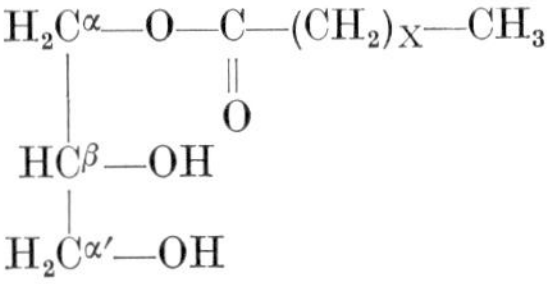

α-Monoglyceride

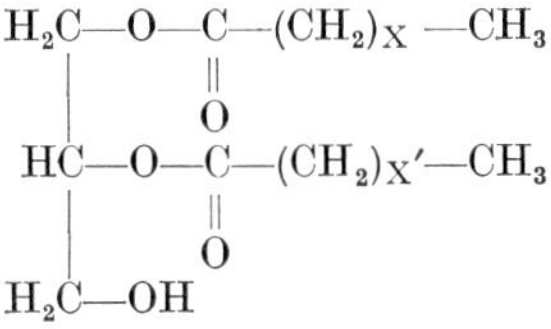

α, β-Diglyceride

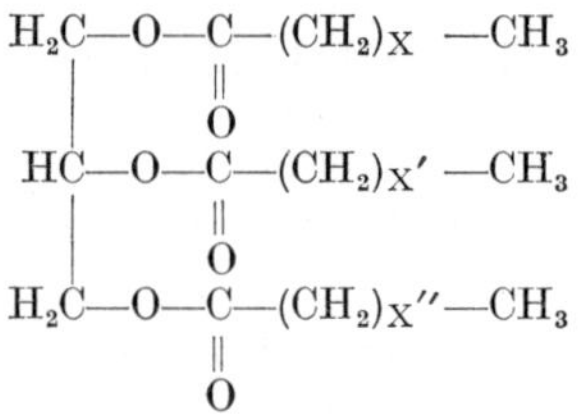

Triglyceride

Glyceryl ethers, vinyl ethers, and monoglycerides may occur in α- or β-forms. Diglycerides are built either symmetrically (α, α') or asymmetrically (α, β).

Glyceryl ethers and vinyl ethers occurring in nature usually are esterified with acids in the form of glyceryl ether diesters (alkoxydiglycerides) and "aldehydogenic triglycerides" ("neutral plasmalogens"). The triglycerides of long-chain (fatty) acids represent the major part of fats and oils; they are the "fats", in the common sense of the word.

2. Phospholipids, sulpholipids, and glycolipids

Phospholipids are lipid classes in which one of the hydroxy groups of the glycerol molecule is esterified with phosphoryl choline, phosphoryl ethanolamine, or phosphoryl serine. Ethanolamine phospholipids exist in the following three forms:

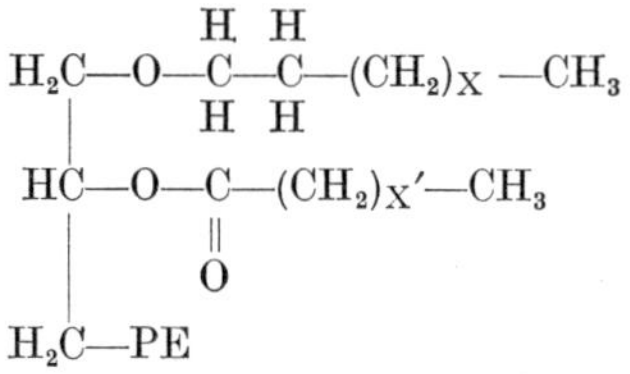

Ether-Ester Phosphatide

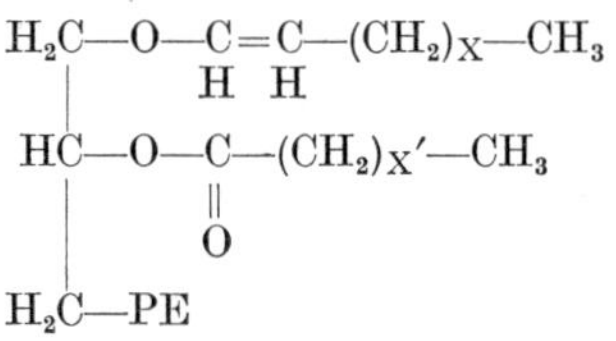

Vinyl Ether-Ester Phosphatide (Plasmalogen)

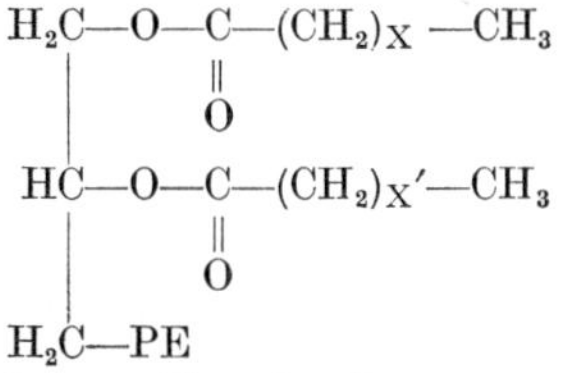

Diester-Phosphatide (Cephalin)

PE = Phosphoryl Ethanolamine:

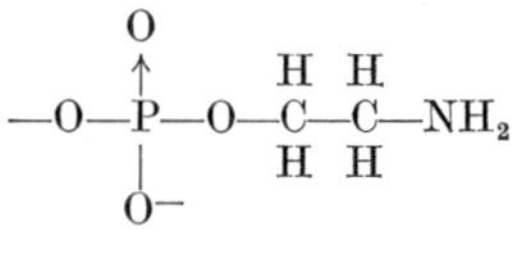

One can assume that choline phospholipids and serine phospholipids occur also in the ether-ester, vinyl ether-ester, and diester forms.

Glyceryl phosphoryl choline, glyceryl phosphoryl ethanolamine, and glyceryl phosphoryl serine, esterified with only one fatty acid moiety, and thus containing one free hydroxy group, are called lyso compounds. Lysolecithin is shown here as an example:

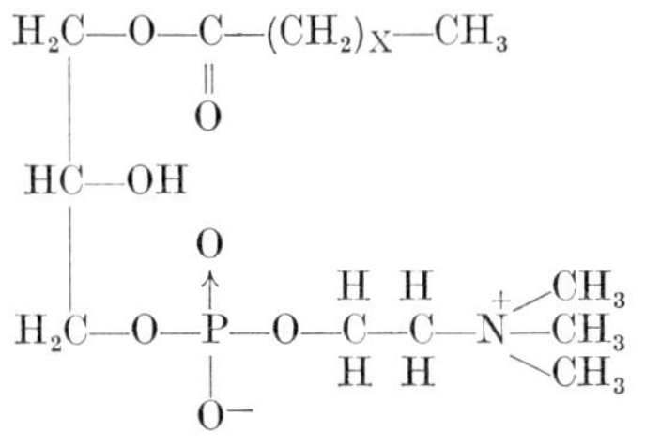

Lysolecithin

The phosphatidic acids are a group of phospholipids containing no bases. They occur in nature, possibly as hydrolysis products of choline, ethanolamine, and serine-phosphatides:

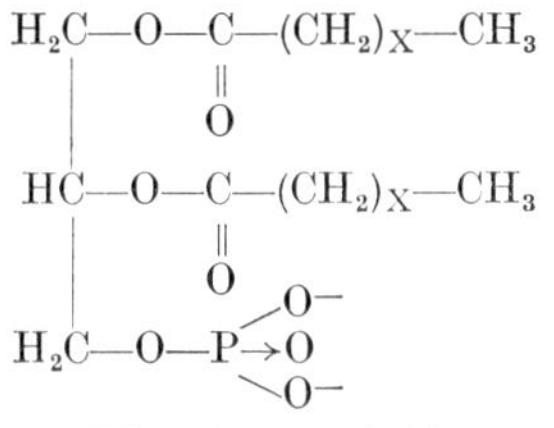

α-Phosphaticic Acid

Phosphatidyl glycerol, cardiolipin, and phosphoinositides also do not contain bases:

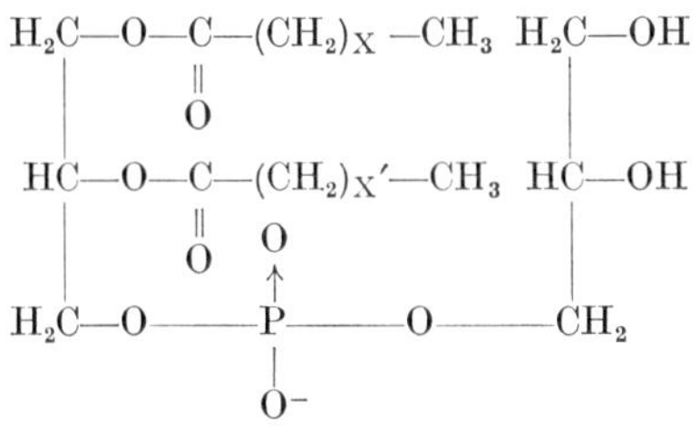

Phosphatidyl Glycerol

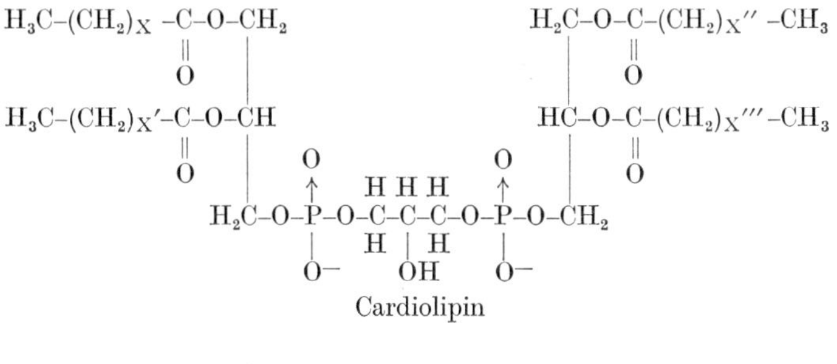

Cardiolipin

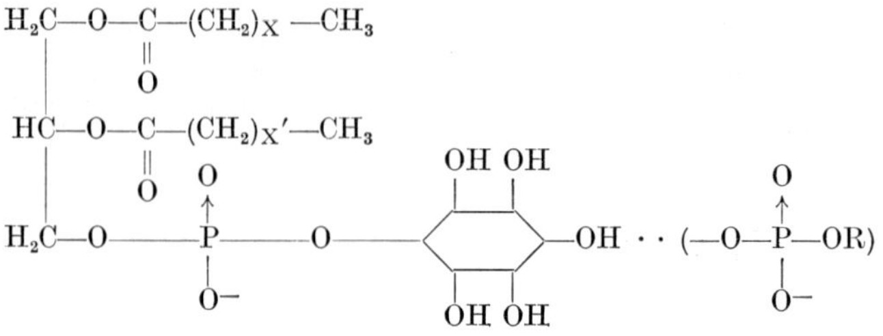

Phosphatidyl inositol and Diphosphatidyl inositol
(monophosphoinositide and diphosphoinositide)

A further group of phospholipids, the sphingomyelins, are composed of a polyvalent amino alcohol (i. e. sphingosine), phosphoryl choline, and a fatty acid.

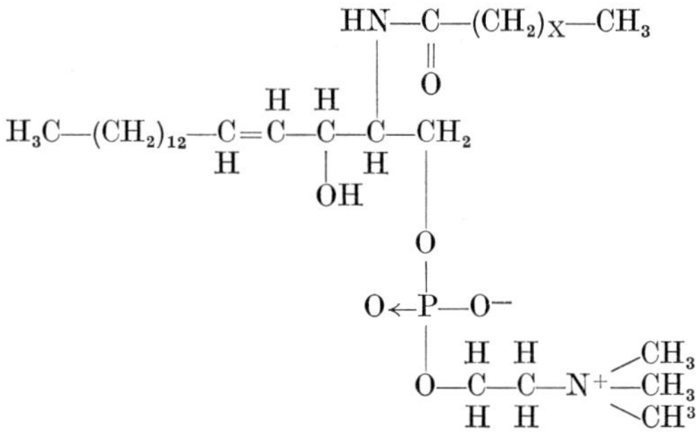

Sphingomyelin

Glycolipids, such as cerebrosides and gangliosides, contain sphingosine and a sugar:

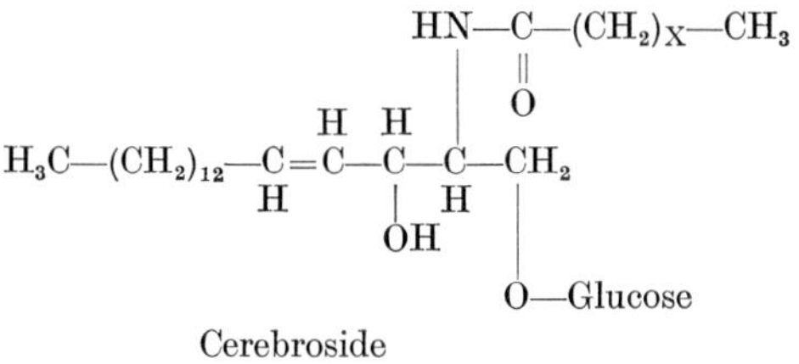

Cerebroside

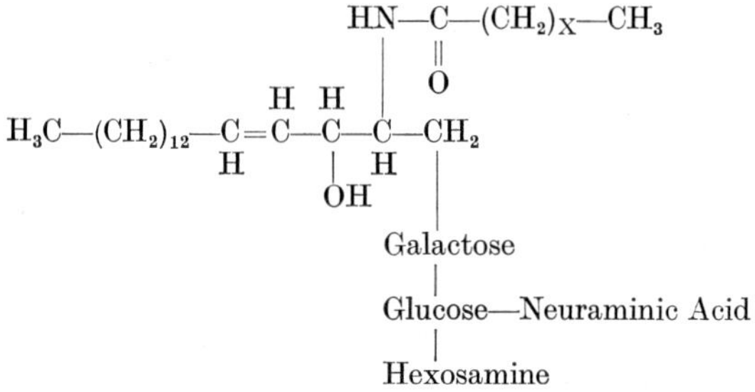

Ganglioside

The structures of the different sulpholipids are not yet elucidated. The following formulae have been proposed:

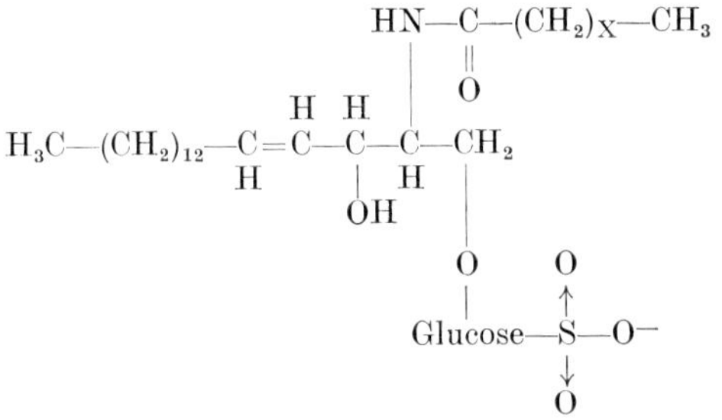

Cerebroside Sulphate

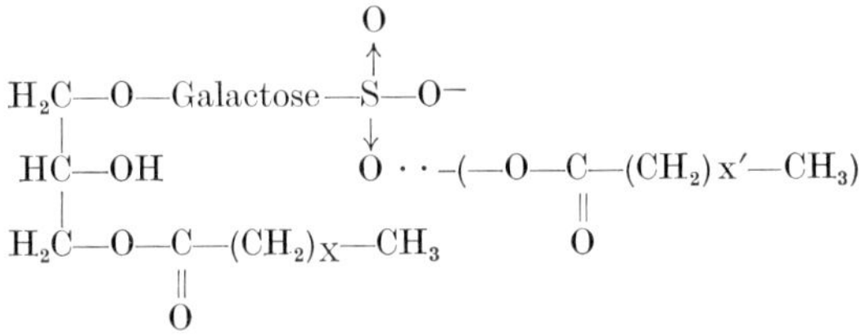

Plant sulpholipid

3. Older methods of lipid analysis

The extremely complex composition of naturally-occurring lipid mixtures makes it an impossible task to analyze, quantitatively, a group of these substances, or even an individual compound in such mixtures, by chemical methods only. Until about 10 years ago, constants like the acid, saponification, iodine, and diene "numbers" were given to characterize a lipid mixture. The determination of the amount of "unsaponifiable" material by alkaline hydrolysis was a standard method of fat analysis. Phospholipids and sulpholipids were analyzed quantitatively as inorganic phosphate and sulphate after mineralization. These methods were complemented by color reactions for the detection of minor fat constituents, such as lipochromes, sterols, and resin acids.

Physical methods for the characterization of lipids covered determinations of melting and freezing points, density, hardness, viscosity, surface and interfacial tension, and others. The classical methods of fat analysis are comprehensively described in the books of T. P. Hilditch [*33*] and H. P. Kaufmann [*56*].

During recent years, a number of procedures were developed for the fractionation of lipid mixtures. Low-temperature crystallization and urea adduct formation proved to be the most effective for the separation of saturated from unsaturated fatty acids and their methyl esters. Vacuum distillation was used for the isolation of mixtures of methyl esters of fatty acids of uniform chain length. Molecular distillation was applied for the separation of mono-, di-, and triglycerides. Countercurrent distribution between two immiscible solvents was used for fractionating fatty acids according to chain length or according to their degree of unsaturation, and for separating mono-, di-, and triglycerides, as well as mixtures of phospholipids. Segregations of neutral and acidic lipids were accomplished by dialysis through rubber membranes.

The application of these separation methods usually required from 1 g to 100 g of lipid. In most cases, these procedures made possible the enrichment of a component or a semi-quantitative fractionation. Ultraviolet and infrared spectroscopy were used to monitor the efficiency of such fractionations.

4. New procedures for the fractionation of lipids

Chromatographic techniques have replaced all other methods for fractionating lipids on an analytical or micro-preparative scale. Adsorption and partition chromatography on silica gel columns, on silicic acid-impregnated filter (cellulose) paper and on glass fiber paper are used to separate complex mixtures according to classes of compounds. Reversed-phase partition chromatography is used to separate the members of a homologous series on hydrophobic column packings or on hydrophobic paper. Gas-liquid chromatography especially, is utilized as a partition procedure for separating the methyl esters of fatty acids. The fractionation of lipid mixtures, according to unsaturation, is possible by chromatographing the mercuric acetate addition compounds of unsaturated lipids on silicic acid. Chromatography on ion-exchange columns and on ion-exchange paper is used for isolating acids and for fractionating strongly polar lipids. Straight-chain fatty acids can be resolved from branched-chain fatty acids on urea columns or on paper impregnated with urea. A number of comprehensive reviews describe the various methods for the chromatography of lipids [*23, 86, 96, 100*].

Each of these principles of chromatographic separation can be applied in thin-layer chromatography as well. In most cases, TLC results in better resolutions in a shorter period of time, and, therefore, this technique has become indispensable in laboratories where lipid work is carried on. However, the great advantages of TLC are fully utilized only if this tool is combined with other chromatographic techniques. It is advisable to apply absorption-TLC and reversed-phase partition-TLC, as well as paper and gas-liquid chromatography, consecutively, to the same sample. Groups of compounds which cannot be separated by adsorption chromatography may well be resolved by reversed-phase partition chromatography, or vice versa.

The following types of fractionation of lipids were accomplished by TLC: separations according to chemical classes, fractionations of homologous series, separations according to degree of unsaturation, and resolutions of stereo and positional isomers.

5. Preparation of the sample for analysis

The best separation method is worthless if the mixture to be analyzed is spoiled by incorrect treatment. Hence, a few methods for the isolation of plant and animal lipids are described here, and some rules for the handling of lipid extracts are given.

a) Homogenization and extraction

The problem of extraction of lipids is currently the subject of intensive investigations. The standard procedures described here have been in use for many years. Through the introduction of TLC by STAHL, and the application of this method to lipids, it has been shown that the classical extraction procedures lead to the formation of artifacts and show other deficiencies [*103*, *129*]. Unfortunately, improved procedures have not been published so far.

Preparation of vegetable lipids. Vegetable lipids are extracted from seeds and fruits by grinding and pressing and by extracting with solvents.

When working with plant material having a high oil content, it is advisable first to squeeze out the major part of the oil and to grind the residue further to avoid formation of a pulp mush, and then to press it. Appropriate presses and mills are described in handbooks of chemical technology, along with the addresses of their manufacturers [*56*, *115*]. In the laboratory, a simple coffee grinder, or, for soft plant material, a kitchen mixer, will serve one's purpose just as well as an expensive, special apparatus. Processing the material with abrasive sea sand in a mortar is also a very efficient method.

Treating the residue with solvent and extracting it exhaustively in a Soxhlet apparatus, or a similar device, will procure the remainder of the lipids. Extractors and their use are described adequately in several handbooks; commercial suppliers are listed therein also [*14*, *49*, *56*]. Pulverized plant particles usually are extracted on a water bath with anhydrous solvents, such as hexane, petroleum hydrocarbon, B.P. 60 to 70° C, or benzene, for 4 to 6 hrs. Equally suitable are chloroform, carbon tetrachloride, trichloroethylene, and diethyl ether. Ethanol [*121*], n-propanol, and isopropanol [*51*, *144*], as well as water-saturated n-butanol [*90*], sometimes are useful for the isolation of plant lipids. It is advisable to extract, successively, with two or three solvents of different polarities, e.g., diethyl ether, benzene, and petroleum hydrocarbon. It is essential to extract always in an atmosphere of nitrogen, in order to protect unsaturated lipids against autoxidation.

KATES [*51*] has shown that aliphatic ethers, ketones, and esters activate the enzyme phosphatidase C. This enzyme hydrolyzes phospholipids rapidly. For this reason, KATES recommends the use of n-propanol or isopropanol for the extraction of plant lipids, instead of diethyl ether, acetone, or ethyl acetate.

"Total lipid extracts" usually contain, also, free amino acids and peptides, sugars, and other hydrophilic natural products which are carried into the extract through the action of lecithin and other solubilizers. It is very difficult to remove these contaminants completely. Most suitable for separating lipids from non-lipids are chromatography on cellulose columns according to the method of LEA and RHODES [*70*], preparative paper chromatography with the solvents described by WESTLEY, WREN, and MITCHELL [*139*], but, above all, countercurrent distribution [*9*, *53*, *105*].

Procedures for the extraction of lipids, and directions for purifying lipid extracts, have been very conscientiously worked out for human and animal tissues. The methods described here, for working up animal tissues, apply essentially to plant material as well.

Extraction of animal lipids. Very thorough descriptions and comparisons of the methods used for extracting human and animal tissue lipids have been presented by SPERRY [*121*] and by ENTENMAN [*20*]. The study of ENTENMAN's critical discussion is especially recommended.

The following eight principles, as given by ENTENMAN, should be considered:

1. Carry out all procedures under an atmosphere of nitrogen.

2. Use purified solvents. Methanol and ethanol should be distilled over potassium hydroxide to remove aldehydes. Chloroform should always be freshly distilled. Diethyl ether should be distilled over hydroxylamine or ferrous sulphate and stored over iron or sodium wire. It is advisable to keep both chloroform and diethyl ether in an explosion-proof refrigerator. Petroleum hydrocarbon should be distilled over conc. sulphuric acid.

3. Remove the tissues rapidly after sacrificing the animal.

4. Finely subdivide the tissue immediately.

5. Use the proper solvent-to-tissue ratio.

6. Use heat only when absolutely necessary.

7. Remove non-lipid impurities without loss of lipids.

8. Store lipids under conditions that minimize their alteration.

SPERRY [*121*], as well as ENTENMAN [*20*], described apparatus for homogenizing tissues, and gave the addresses of manufacturers. Best known is the homogenizer designed by POTTER and ELVEHJEM [*106*], which allows complete disintegration of the cells. This apparatus consists of a pestle, which fits in a ground thick-walled test tube so as to leave a gap of less than 1 mm. The tissue is put into the glass tube in suspension, then the pestle is introduced and, by means of an electric motor, rotated at high speed. The tissue is crushed and completely ground between the pestle and the glass tube.

Extraction of lipids according to BLOOR *[4]*

One part of tissue homogenate is added to 20 (to 30) parts of a mixture of ethanol-diethyl ether (3 + 1) and left overnight under an atmosphere of nitrogen, at room temperature. The lipid extract is filtered from the residual protein and concentrated at less than 50° C in vacuo under nitrogen, but never evaporated to complete dryness. The lipids are re-extracted with three portions of petroleum hydrocarbon. Material not soluble in petroleum hydrocarbon is discarded. The petroleum hydrocarbon solution of lipids is dried over anhydrous sodium sulphate.

Complete extraction of lipids from very fatty tissues requires slight heating of the homogenate with ethanol-diethyl ether. Brain lipids are only partially extracted with the Bloor mixture.

Extraction of lipids according to FOLCH, ASCOLI, LEES, MEATH *and* LEBARON *[21]*

This method gives very good yields of complex lipids, such as proteolipids and gangliosides.

The ground tissue is extracted with 20 parts of a mixture of chloroform-methanol (2 + 1) at room temperature. Non-lipids are removed by the diffusion method developed by FOLCH and co-workers [*22*]:

A small glass beaker is placed at the bottom of a beaker of ten-fold capacity, which is filled with distilled water. The small, submerged beaker is filled two-thirds full with the chloroform-methanol lipid extract by means of a pipette. While standing over night, the methanol and all water-soluble contaminants of the lipid extract diffuse into the surrounding water. The lipids thus purified remain in the chloroform layer and also in a flocky layer at the interface of the two liquids. The aqueous solution is siphoned off as completely as possible. The small beaker is filled with the same amount of methanol that was present originally. The flakes dissolve again on mixing the methanol with the chloroform layer. Modifications of this method have been worked out for the isolation of proteolipids, phosphatidolipopeptides, gangliosides, sulphatides, and triphosphoinositides [*72*, *73*].

Extraction of whole animals by boiling with hydrochloric acid in ethanol is not satisfactory and, therefore, is rarely used. This method has the advantage that the tissue is finely distributed, and that the lipids can be extracted easily thereafter, but it will cause drastic alterations of many lipids.

b) Saponification and esterification

Very fatty tissues from plant or animal sources can be decomposed and, simultaneously, the lipids hydrolyzed by alkali. But usually the total lipids are first extracted by one of the above-described procedures and then saponified.

Classes of compounds are isolated by chromatographic methods, e. g., by TLC; their fatty acid composition and content of neutral substances are determined after alkaline hydrolysis. Most lipids can be eluted and recovered from Silica Gel G with diethyl ether or with diethyl ether-methanol (9 + 1). Mixtures of chloroform and methanol are used to isolate phospholipids.

The alkaline hydrolysis ("saponification") of fats produces water-soluble alkali salts of the fatty acids bound in these lipids and non-saponifiable neutral compounds, e.g., hydrocarbons, alcohols, and aldehydes. The "nonsaponifiables" are removed from the alkaline aqueous-alcoholic phase with non-polar solvents. The salts of the fatty acids then are liberated by acidification of the aqueous-alcoholic solution, and extracted with organic solvents.

Alkaline hydrolysis of lipids

One part, by weight, of potassium hydroxide, is dissolved in one part by weight of distilled water, and, after cooling, diluted with 3 to 4 parts of methanol. The lipid material to be saponified is treated with ten times the amount of this potassium hydroxide solution, and left overnight at room temperature. This "cold saponification" is especially recommended for hydrolyzing lipids containing vitamins or conjugated-unsaturated fatty acids. Heat usually is applied for saponifying ester-lipids. The sample is refluxed under nitrogen for 1—12 hrs. The length of boiling depends upon the nature of the lipids to be saponified. The temperature of the saponification mixture may be raised by adding 10—20% of toluene or xylene, if lipids are encountered that are hard to hydrolyze.

Adsorption-TLC is better suited for checking the course of hydrolysis reactions than any other method. Small samples of the reaction mixture are applied, at half-hour intervals, to narrow Silica Gel G plates, and developed with petroleum hydrocarbon (B.P. 60—70° C) — diethyl ether — glacial acetic acid (70 + 30 + 2). At the beginning of the saponification, mainly spots of rather non-polar substances can be seen above the fatty acid spot on the chromatogram. After complete saponification, only the spot of acids appears. Non-polar non-saponifiable constituents (hydrocarbons) show up above the fatty acid spot, whereas, highly polar compounds (alcohols and diols) remain below.

After the saponification is completed, the major part of the methanol is evaporated at 50° C *in vacuo* and the aqueous residue diluted with twice its volume of water. The solution becomes turbid if large amounts of non-saponifiable consituents are present. The non-saponifiable fraction is extracted with several portions of diethyl ether, each being one-half the volume of the aqueous solution. Emulsions can be broken by adding alcohol or by repeated cooling to approximately 5° C. Completeness of extraction should be verified by adsorption-TLC. The combined ether extracts are washed neutral with distilled water, which has been boiled and then cooled by bubbling a stream of nitrogen through it. After drying the solution and evaporating the ether, the non-saponifiable lipid fraction is sealed in an ampoule under nitrogen or dissolved in a suitable solvent and stored in a refrigerator. The aqueous solution of the potassium salts of fatty acids is acidified with 5 N-sulphuric acid under a layer of diethyl ether. The combined ether extracts are washed with distilled and oxygen-free water and then dried over anhydrous sodium sulphate. The ether is evaporated and the fatty acids sealed under nitrogen or stored, dissolved in petroleum hydrocarbon, in a refrigerator.

Until recently, it was customary to saponify fats with aqueous-ethanolic potassium hydroxide or with potassium ethylate in ethanol solution. Lately, methanol is given preference over ethanol. The reason is that fatty acids usually are analyzed in the form of their methyl esters by means of gas-liquid chromatography and small amounts of ethyl esters, which might occur if ethanol were used, will simulate methyl esters of odd-numbered and branched-chain fatty acids.

Esterification of fatty acids

Three methods have been proven to be especially valuable:

1. Reaction of fatty acids with diazomethane, which is generated from N-nitrosomethyl urea [*114*], or from p-tolylsulphonylmethylnitrosamide [*5*].

2. Reaction of acids with 2,2-dimethoxypropane in methanol and aqueous hydrochloric acid [*113*].

3. Esterification of acids with methanol in the presence of concentrated sulphuric acid or boron trifluoride [*91*].

Fatty acids esterified in glycerides or other lipids can be converted into methyl esters without prior saponification:

1. Trans-esterification ot the lipids with methanol in diethyl ether, in the presence of potassium hydroxide [*68*].

2. Alcoholysis of the lipids with methanol containing hydrochloric acid [*126*].

These procedures are described in detail in easily accessible original publications and in several handbooks. A procedure for the esterification of small amounts of fatty acids with diazomethane is given on page 71. Directions for acetylating lipids containing hydroxy and/or amino groups can be found in the same chapter (see p. 71).

It is of great importance that esters of fatty acids are purified by TLC prior to gas chromatographic analysis. A solvent mixture of petroleum hydrocarbon (B. P. 60–70°C) -diethyl ether (95 + 5) on Silica Gel G layers is recommended. The esters are eluted with petroleum hydrocarbon-diethyl ether (50 + 50) immediately following chromatography [*129*].

Acetyl derivatives of lipids containing hydroxy and amino groups also can be rapidly purified by TLC [*80*].

II. Thin-layer chromatography of lipids

1. Separation of lipids according to classes of compounds

More than twenty years ago, TRAPPE [*128*, and subsequent publications] showed that lipids can be eluted as classes of compounds in fractions from an adsorption column by applying consecutively various solvents of different polarities. He arranged different solvents according to their eluting effect and named this list "the eluotropic series of solvents". TRAPPE's eluotropic series, and similar series of solvents, are given on p. 135.

SCHROEDER [*120*] found that, instead of a series of individual solvents, mixtures of petroleum hydrocarbon and diethyl ether in various ratios could be used. MEAD and FILLERUP [*20*, *88*, *89*] as well as HIRSCH and AHRENS, JR. [*34*], carefully described methods for using these two solvents in the adsorption chromatography of complex lipid mixtures on silica gel columns. Both procedures were tested later in other laboratories, and were routinely applied as standard methods for the analysis of human and animal lipid extracts [*86*, *96*].

After the introduction of TLC by STAHL [*122*–*124*], several authors applied this new method to lipids [*46*, *80*, *137*] and found it exquisitely suitable for the separation of naturally-occurring lipids. VIOQUE [*130*] and MORRIS [*96*], in comprehensive reviews, dealt with the history of TLC in the lipid field.

Within a few years, TLC and gas-liquid chromatography have almost replaced all other chromatographic methods for the separation of lipids on an analytical scale.

Fats, oils, waxes, and other neutral lipids usually are separated by adsorption-TLC on Silica Gel G according to classes of compounds. Combinations of petroleum hydrocarbon and diethyl ether are used most often as solvents. Alcoholic and aqueous solvents must be chosen for separating very polar lipids on Silica Gel G layers. Thus, the fractionation of highly polar lipids, according to classes of compounds, is more often based on partition than on adsorption. Reversed-phase partition chromatography, on hydrophobic layers, also has been used in some cases for separating lipids according to classes of compounds (see Fig. 88, p. 166).

a) Neutral lipids and their hydrolysis products

Long-chain hydrocarbons, alcohols, aldehydes, acids, monoglycerides, diglycerides, triglycerides, and similar lipids can be separated by adsorption-TLC, according to the kind and number of functional groups, into classes of compounds of different polarities. Great differences in chain-length and degree of unsaturation of components of one class can lead, in rare cases, to subfractionations within classes. Such subfractionations are never so pronounced that they could interfere with the separation according to classes.

Fig. 70 depicts schematically the separation of some typical lipids according to classes of compounds. As was to be expected, the separation of various types of compounds on plates follows the rules given by Brockmann and Volpers [*6*]: hydrocarbons are not adsorbed, aldehydes

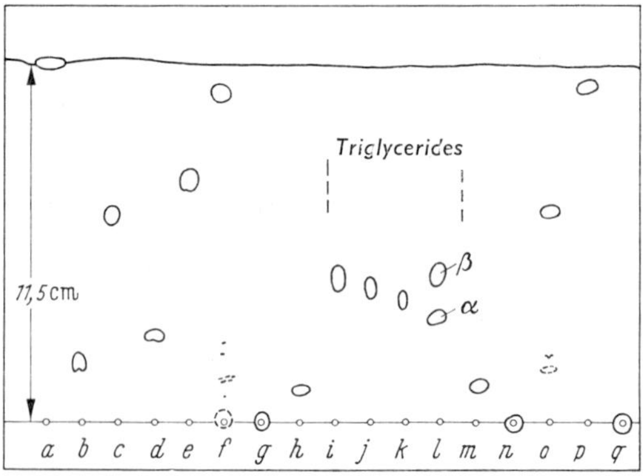

Fig. 70. Separation of neutral lipids and their hydrolysis products by adsorption-TCL [*76*]. *a* Octadecene-9, *b* oleyl alcohol, *c* oleyl aldehyde, *d* oleic acid, *e* methyl oleate, *f* cholesteryl oleate, *g* monoolein, *h* diolein, *i* triolein, *j* trilinolein, *k* trilinolenin, *l* tricaproin (α) and tristearin (β), *m* cholesterol, *n* selachyl alcohol, *o* selachyl diolein, *p* oleyl oleate, *q* dioleoyl lecithin. Adsorbent: Silica Gel G. Solvent: Petroleum hydrocarbon, B.P. 60—70° C, —diethyl ether— glacial acetic acid (90 + 10 + 1) Time: 40 minutes. Indicator: 2′,7′-Dichlorofluorescein in ethanol. Amounts: 20 μg

precede alcohols and acids, short-chain compounds and unsaturated compounds are more strongly adsorbed than longer chain and saturated compounds, respectively (see p. 135). As an example of the extent of subfractionation within the class of triglycerides, see Fig. 70.

Experimental conditions

Adsorbents. Of all adsorbents, silicic acid is most widely used for fractionating lipids. Wren [*145*] has described the characteristic features and advantages of this adsorbent. On a plate, 20 × 20 cm., coated with Silica Gel G, one can separate 10–20 mg of a complicated lipid mixture. If only a few substances of very different polarities are to be fractionated, up to 50 mg can be applied to a standard size plate. Alumina is rarely used, since it will hydrolyze [*128*] and isomerize [*119*] lipids. Florisil®, a synthetic magnesium silicate which is often used in column chromatography of lipids [*8*], can likewise be applied in TLC. *Sec.*-Magnesium phosphate also is a very suitable adsorbent for lipids (see p. 230). Sugar was used by Kathen [*52*] for separating lipids according to classes of compounds. This adsorbent has very good qualities but very low capacity.

Being water-soluble, it makes the recovery of adsorbed lipophilic substances very easy. For this reason, the applicability of sugar as an adsorbent for TLC of lipids should be investigated.

Solvents. The choice of solvents depends upon the polarities of the components of the lipid mixture and on the desired separation. A list of solvents for the separation of neutral lipids by adsorption-TLC is given in Table 10. Mixtures of petroleum hydrocarbon with 1–5% benzene or diethyl ether are especially suited for separating hydrocarbons, alkyl esters, sterol and polyenol esters according to classes of compounds [*26*, *75*, *80*, *81*]. Components containing one or more free functional groups remain at the origin when solvents like these are used.

Table 10. *Solvents for Separating Neutral Lipids and Their Hydrolysis Products by Adsorption-TLC on Silica Gel G*

Solvents	Ratio v/v	Literature
Petroleum hydrocarbon[1]-benzene	95+5	[*85*]
Carbon tetrachloride	—	[*47*, *58*]
Hexane-tetralin[2].	75+25	[*59*]
	50+50	[*59*]
Petroleum hydrocarbon	95+5	[*26*, *81*, *84*]
	90+10	[*2*, *76*, *95*]
	80+20	[*26*, *83*, *93*]
	70+30	[*2*, *47*, *109*]
	50+50	[*2*, *108*]
Petroleum hydrocarbon-diethyl-ether-acetic-acid	90+10+1	[*77*, *81*, *97*]
	80+20+1	[*84*, *140*]
	70+30+2	[*81*, *84*, *103*]
Benzene-diethyl ether	50+50	[*2*]
Dichloroethane	—	[*47*, *58*]
Chloroform	—	[*47*, *103*]
Diethyl ether	—	[*47*, *58*]
Diisopropyl ether	—	[*58*, *59*]
Chloroform-acetic acid	96+4	[*47*]
Diisopropyl ether-acetic acid	100+1,5	[*58*, *59*]
Benzene-methanol	85+15	[*2*]
n-Propanol-conc. ammonium hydroxide . . .	66+33	
followed by chloroform-benzene	60+40	[*37*, *137*, *147*]
followed by carbon tetrachloride[3]	—	

[1] Usually petroleum hydrocarbon, B.P. 60—70° C, i.e. mainly hexane.
[2] For the subfractionation of cholesteryl esters
[3] Stepwise elution.

For fractionating alcohols and aldehydes, mono-, di-, and triglycerides mixtures of petroleum hydrocarbon with 10–50% diethyl ether, are used (e.g. [*81*, *95*, *108*, *109*]). Non-polar lipids will migrate near the front with these solvents and will hardly be separated from one another. For TLC of lipid mixtures containing free fatty acids, the above-mentioned solvents should be used with 1–2% glacial acetic acid, to avoid formation of streaks and tailing of the fatty acids [*76*, *81*, *95*, *140*].

It is often impossible to obtain, with a single solvent, complete separations according to classes of compounds because most of the naturally-occurring lipid mixtures contain substances of widely different polarities. A better picture of the composition of a complicated lipid mixture can be obtained when it is chromatographed on separate plates with two or three solvents of different polarities. These solvents can also be used in the same direction on the very same plate. Two-dimensional TLC will likewise yield better separations [*58*]. TLC with a solvent gradient can be done for more complete separations of lipid classes of greatly different polarities[1].

Plant and animal waxes are usually very well fractionated on Silica Gel G with the solvent system petroleum hydrocarbon-diethyl ether (95 + 5). The system of petroleum hydrocarbon-diethyl ether-acetic acid (90 + 10 + 1) is the combination of solvents most often used for the TLC of lipids. It is very well suited for fractionating animal fats (Fig. 71 and 73). Vegetable oils, which often contain epoxy and/or hydroxy constituents and, therefore, are more polar than most of the animal fats, usually are fractionated with mixtures of petroleum hydrocarbon-diethyl ether-acetic acid (80 + 10 +1) or (70 + 30 + 2) (Fig. 74). The latter solvent system is very suitable also for TLC of alcohols and diols in the nonsaponifiable matter of vegetable and animal fats, and for TLC of unsubstituted fatty acids, epoxy, hydroxy and dihydroxy fatty acids and their methyl esters.

Detection methods. Almost all indicators used in the paper chromatography of lipids [*54*, *78*] can be applied to visualize lipids on plates. Furthermore, corrosive spray reagents can be used for charring all organic substances.

Neutral lipids are usually visualized on the plate by iodine vapors [*80*, *81*], by 2′, 7′-dichlorofluorescein [*76*] or Rhodamine B [*58*, *135*], and by chromic-sulphuric acid solution [*85*, *129*].

Iodine will stain all unsaturated lipids, and some nitrogenous saturated lipids, brown on a light yellow or white background. It is possible to detect less than 1 μg of a monounsaturated compound with iodine vapors; most saturated lipids will be only faintly colored. The brown spots disappear in a few minutes, but they can always be reproduced by re-exposure to iodine vapors.

Almost every lipid can be recognized, after spraying the chromatogram with an ethanolic 0.2% solution of 2′,7′-dichlorofluorescein, in U.V.light (270 mμ) as light green fluorescent spots on a dark violet background. With this reagent, 1–5 μg of a compound can be detected. Spraying with a 0.05% solution of Rhodamine B in 96% ethanol also can be used for observing lipids in U.V.light. Generally, as little as 1 μg of a lipid can be detected as a dark violet spot on a pink background. Chromic sulphuric acid solution is very satisfactory for charring all nonvolatile organic compounds on a plate by heating on an electric hot plate (Fig. 71). Less than 1 μg of a lipid can be recognized as a grey or black spot on a white background. During heating, several color changes can be observed; e.g., the spots of cholesterol and cholesteryl esters are

[1] See e.g. S. M. RYBICKA [*162*].

first red, then brown, and finally black; vitamin A and its esters are first blue, and at higher temperatures, they turn grey and black.

Iodine, 2′, 7′-dichlorofluorescein, and chromic sulphuric acid solution can be applied successively to the same plate. The successive use of three indicators increases the probability that all substances will be detected.

Less common spray reagents (pp. 483—502) are: solutions of bromothymol blue, phosphomolybdic acid, antimony trichloride, antimony pentachloride, α-cyclodextrine-iodine vapors, fluorescein followed by bromine vapors, and hydroxylamin-ferric chloride. Lipids which absorb U.V. light due to a system of conjugated double bonds are best developed on plates coated with Silica Gel G containing a fluorescent mineral. The spots of lipids are detected in U.V. light. Lipids containing hydroxy or amino groups can be reacted with radioactively-labelled acetic anhydride and separated as acetyl derivatives. Free fatty acids are converted into their labelled methyl esters. The radioactive lipid derivatives are localized by autoradiography (see p. 60). Most of the indicator reactions for lipids are 10—100 times more sensitive on thin-layer chromatograms than on paper chromatograms. Some of the above-mentioned reagents, such as 2′,7′-dichlorofluorescein and Rhodamine B, do not alter the lipids, which, therefore, can be isolated in the original state by means of TLC.

Applications and results

α) Fats, oils, and waxes. Adsorption-TLC was applied by Mangold and Malins [*81*] for fractionating vegetable fats and waxes, fish oils and their hydrolysates, i.e., fatty acids and nonsaponifiable matter, according to classes. Figs. 71 and 74 show two typical chromatograms.

Particularly noteworthy is the fact that, by TLC, classes of lipids can be separated which are closely related, e.g., alkyl esters, steryl esters, and polyenol esters. Even glycerylether diesters can be well resolved from triglycerides. Separations of comparable efficiency cannot be achieved by column or paper chromatography, nor by any other technique. For micro-biological and entomological studies, TLC is far superior to column chromatography, because much smaller amounts of material are required. Patel, Haydak and Lovell [*102*] report the analysis of "royal jelly" of the honey bee by TLC. Thin-layer chromatography should become the method of choice for separating lipophilic constituents of bacteria and viruses.

Tuna, Kammereck and Mangold [*129*] applied radioactively-labelled compounds to check whether the resolutions achieved by TLC are as good as they appear to be, and to determine whether some classes are contaminated by each other. Labelled tripalmitin-1-C^{14} was mixed with the total lipids from human depot fat and with dogfish liver oil. In the same manner, cholesteryl palmitate-1-C^{14} was added to the total lipid extract from atherosclerotic plaques of a human aorta. These lipid mixtures then were chromatographed on the same plate. The spots were first stained with iodine vapors and, subsequently, an autoradiograph was taken to determine the distribution of radioactivity. Both pictures are shown in Fig. 72.

It is easy to recognize that in both the chromatogram of dogfish liver oil and human depot fat the total activity is located in the triglyceride

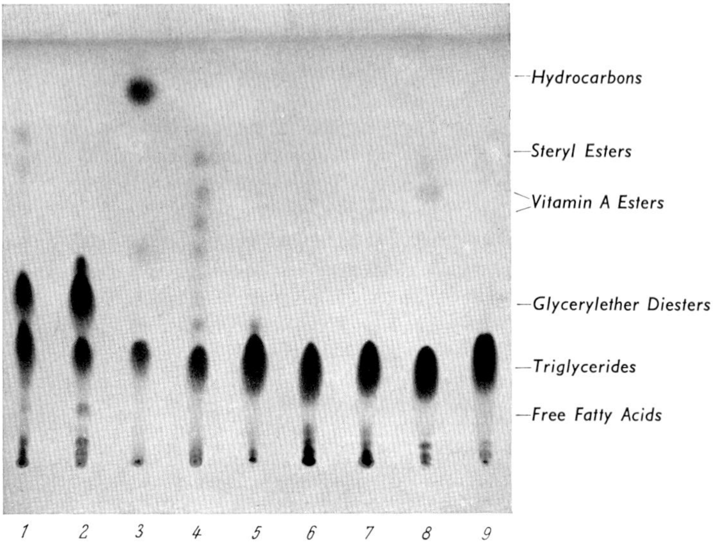

Fig. 71. Thin-layer chromatogram of marine oils [see *81*]. *1 Squalus acanthias* liver oil, *2 Hydrolagus colliei* liver oil, *3 Cetorhinus maximus* liver oil, *4 Galeorhinus galeus* liver oil, *5 Physter macrocephalus*, *6 Engraulis mordax*, *7 Thunnus thynnus*, *8 Onorhynchus gorbuscha* eggs, *9 Gadus morrhua* liver oil. Adsorbent: Silica Gel G. Solvent: Petroleum hydrocarbon, B.P. 60—70° C. — diethyl ether — glacial acetic acid (90 + 10 + 1). Time: 1 hour. Indicator: Charring with chromic sulphuric acid solution. Amounts: 200—300 μg

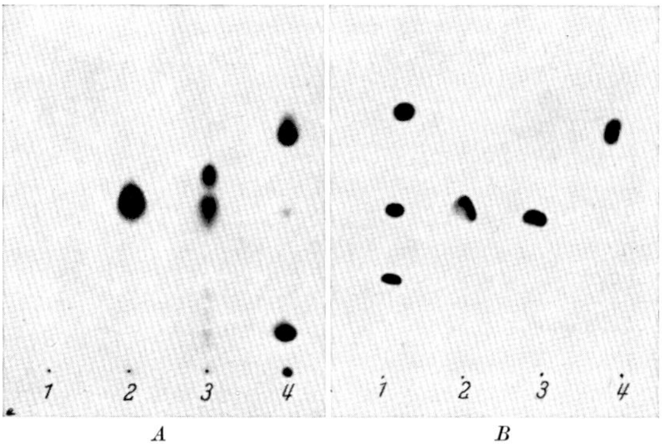

Fig. 72. Test for the sharpness of separations [*129*]. *1* Cholesteryl palmitate-1-C^{14}, tripalmitin-1-C^{14}, and palmitic acid-1-C^{14}. *2* Human depot fat and tripalmitin-1-C^{14}. *3* Dogfish liver oil and tripalmitin-1-C^{14}. *4* Total lipids of atherosclerotic plaques of a human aorta and cholesteryl palmitate-1-C^{14}. Adsorbent: Silica Gel G. Solvent: Petroleum hydrocarbon, B.P. 60—70° C, — diethyl ether — glacial acetic acid (80 + 20 + 1). Indicators: *A* Iodine vapors, *B* Autoradiograph. Amounts: About 200 μg of natural lipid mixtures

fraction. The spots above triglycerides are not a result of subfractionation of the triglycerides. The lipid extract from human aortic plaques shows the total activity in the spot representing cholesteryl esters.

These results prove that mutual contamination of the various lipid fractions occurs only to a minimal extent, if at all. It is especially astounding that the glycerylether diesters, having a chemical structure and physical properties similar to those of the triglycerides, are not contaminated by triglycerides. Diacetyl glycerylethers and diacetyl monoglycerides also can be separated by TLC [*80*]. The much more polar unesterified glycerylethers and the equivalent monoglycerides, however, cannot be separated on Silica Gel G. The extremely sharp separations obtained by TLC make it possible to isolate various classes of lipids of an extract and to determine their fatty acid composition. MALINS [*77*] fractionated the liver oil of dogfish by TLC and isolated the triglycerides and glycerylether diesters separately. He saponified both classes of lipids, isolated the mixtures of fatty acids of each of the two classes, and analyzed them by gas-liquid chromatography after esterification. MANGOLD and TUNA [*82*], as well as TUNA, KAMMERECK and MANGOLD [*129*], determined the fatty acid composition of cholesteryl esters, triglycerides, and phospholipids from the lipid extracts of human tissues.

TLC has proved to be of special value for making routine analysis of lipids from human and animal sera and organs. WEICKER was the first to analyze human serum lipids by TLC [*137, 138*]. Applications of WEICKER's method by ZÖLLNER and co-workers [*146–148*] are described on p. 257. WILLIAMS, SHARMA, MORRIS and HOLMAN [*140*] investigated the lipids of faecaloid deposits in the human caecum and compared their composition with that of faeces. Several investigators found that lipid extracts of human and animal tissues contain artifacts, e.g., fatty acid methyl esters [*12, 103, 104, 129*] and, in several cases, trimeric aldehydes (1,3,5-trioxanes) [*129*]. Several classes of compounds in human tissues, which could not be found by other methods, were detected by TLC. Among these are glycerylether diesters and very small amounts of "aldehydogenic triglycerides" (which apparently are present in almost any human organ), as well as sterols, less polar than cholesterol, which are found in the human aorta [*129*]. A chromatogram of human tissue lipid extracts is reproduced in Fig. 73.

The method of HIRSCH and AHRENS, Jr. [*34*] for the chromatographic fractionation of complex lipid mixtures on silicic acid columns, according to these authors, is not suited for separating glycerylether diesters from triglycerides. However, TLC of fractions of the eluate of a HIRSCH and AHRENS column demonstrates partial separation of these two lipid classes. Fig. 75 shows a chromatogram of fractions of the "triglyceride" eluate from human depot fat. The presence of two less polar classes of lipids in the first fractions of the triglyceride eluate can be recognized on the plate, but not by the corresponding weight curve.

Clinical applications of TLC are described in chapter 7. Reference is made here to an interesting publication which appeared when this monograph was already in press. HONEGGER [*154*] studied the lipids from the brains of normal healthy persons and compared their composition with the brain lipids of patients suffering from multiple sclerosis.

Dhopeshwarkar and Mead [*18a*] described experiments giving strong evidence that methyl esters of higher fatty acids occur as natural constituents in animal tissues and are not necessarily only artifacts

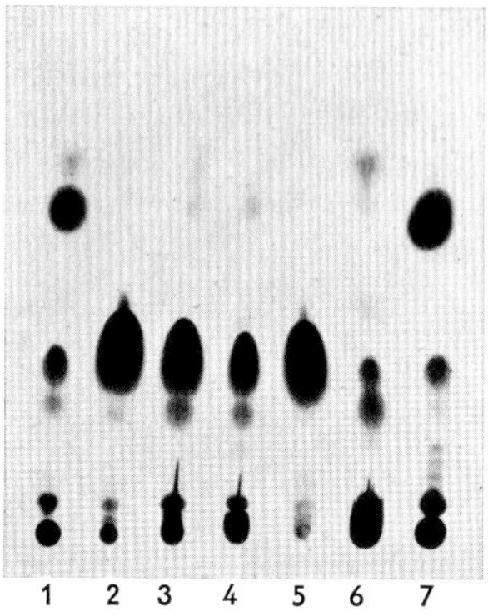

Fig. 73

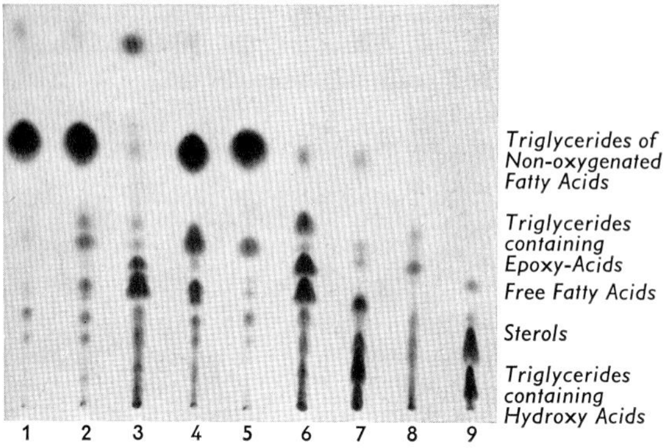

Fig. 74

Fig. 73. Thin-layer chromatogram of human tissue lipids [*129*]. *1* serum, *2* bone marrow, *3* liver, *4* kidney, *5* perinephric fat, *6* spleen, *7* aortic plaque. Adsorbent: Silica Gel G. Solvent: Petroleum hydrocarbon, B.P. 60—70° C, — diethyl ether - glacial acetic acid (90+10+1). Time: 1 hour. Indicator: Charring with chromic sulphuric acid solution

Fig. 74. Thin-layer chromatogram of seed oils [*97*]. *1 Olea europaea*, *2 Malope trifida*, *3 Vernonia anthelmintica*, *4 Artemisia absinthium*, *5 Ceiba pentandra*, *6 Cephalocroton cordofanus*, *7 Dimorphotheca auranthiaca*, *8 Onguekoa Gore*, *9 Ricinus communis*. Adsorbent: Silica Gel G. Solvent: Petroleum hydrocarbon, B.P. 60—70° C, — diethyl ether — glacial acetic acid (70+30+2) Time: 1 hour. Indicator: Charring with chromic sulphuric acid solution. Amounts about 200 μg

produced by extracting lipids with solvent mixtures containing methanol. The same authors were able to demonstrate that methyl esters can be absorbed by the animal body without having been hydrolyzed in the

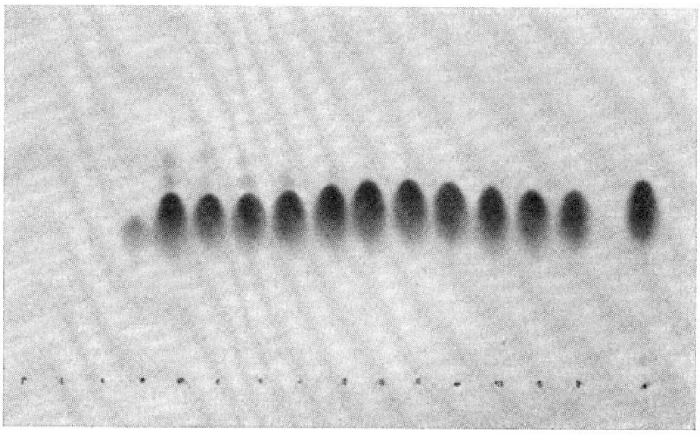

Fig. 75. Thin-layer chromatogram of the forerun of the triglyceride fraction from a column chromatographic separation of human depot fat [*129*]. Note the appearance of two classes of lipids less polar than the triglycerides

intestine [*18a, 18b*]. The following experiment is of interest: Dhopeshwarker and Mead fed radioactively-labelled methyl elaidate-1-C^{14} to guinea pigs. The animals were sacrificed after four hours and the lipids from blood, liver, kidney, lung, heart, spleen, and depot fat were extracted according to the method of FOLCH, LEES and SLOAN-STANLEY [*22*]. The cholesteryl esters, triglycerides, sterols, and phospholipids of the various tissues were isolated by silicic acid column chromatography, and the radioactivities of each of these classes were determined. Nearly all of the total activity was found to be in the cholesteryl ester fraction. The authors could prove, however, by TLC on Silica Gel G, that the "cholesteryl ester fraction" was contaminated with methyl elaidate and that only the latter was radioactive. These results are summarized in the following table, which was taken from one of the authors' publications:

Table 11. *Incorporation of Elaidic Acid Into Lipid Components of Pooled Tissues as shown by Silicic Acid Chromatography*

Eluent	Component Identified by Thin-Layer Chromatography	Specific Activity (dps/mg)
2% Diethyl ether in pentane	Cholesteryl esters	0.5
	Methyl esters[1]	708
5% Diethyl ether in pentane	Triglycerides	4.1
20% Diethyl ether in pentane	Sterols	0.4
Methanol	Phospholipids	3

[1] Cholesteryl esters and methyl esters were separated by TLC on Silica Gel G with the solvent hexane-diethyl ether, 98.5+1.5.

Fig. 76 shows the weight curve of the eluate of a column chromatogram of ratfish liver oil. The purity of the various fractions of the eluate

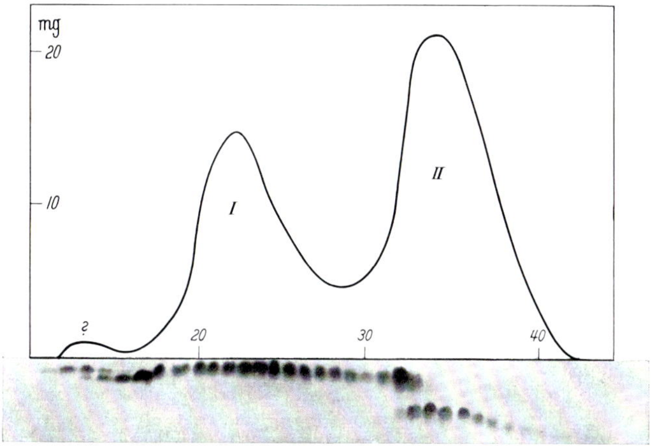

Fig. 76. Analysis of the eluate of a column chromatographic fractionation of *Hydrolagus colliei* liver oil. I Glycerylether diesters, II Triglycerides (compare with Fig. 71)

was determined by TLC. The application of TLC for monitoring column chromatographic separations has been mentioned by several authors [*2, 35, 92, 129*].

A thin-layer chromatogram of unusual vegetable oils is depicted in Fig. 74. The separation of triglycerides of the common fatty acids from those containing epoxy and/or hydroxy acids is remarkable. MORRIS and

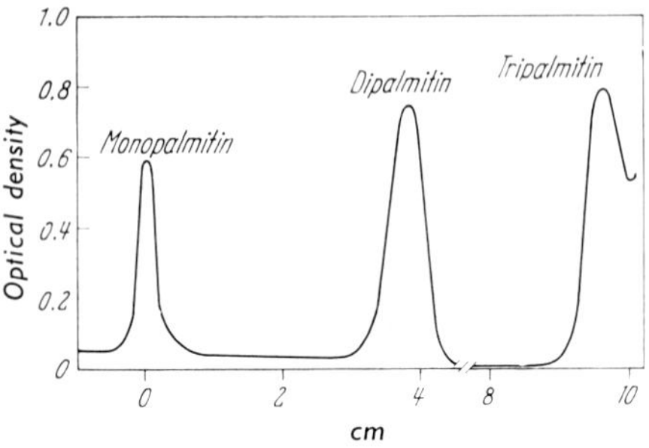

Fig. 77

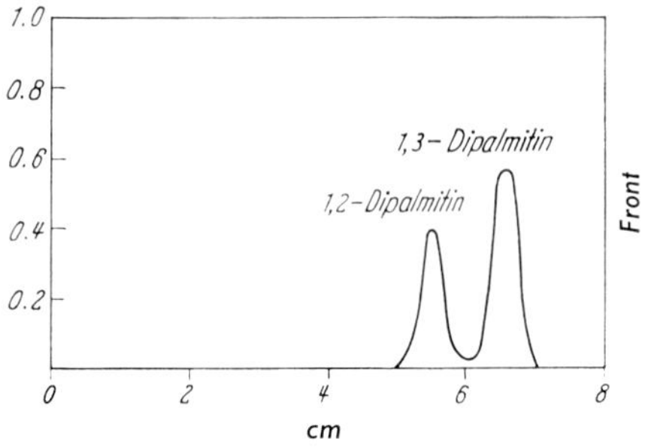

Fig. 78

Fig. 77. Densitometer curve of a thin-layer chromatogram of mono-, di-, and tripalmitin [*109*]. Adsorbent: Silica Gel G. Solvent: Petroleum hydrocarbon, B.P. 40—60° C, — diethyl ether, 70+30. Time: 40 minutes. Indicator: Charring with 50% aqueous sulphuric acid

Fig. 78. Densitometer curve of a thin-layer chromatogram of 1,2-dipalmitin and 1,3-dipalmitin [*109*]. Adsorbent: Silica Gel G. Solvent: Petroleum hydrocarbon, B.P. 40—60° C. — diethyl ether, 70+30. Time: 40 minutes. Indicator: Charring with 50% aqueous sulphuric acid

MANGOLD [*97*] applied TLC to determine the nature of the constituents causing color reactions, such as in the Halphen test and other tests frequently used in the analysis of fats.

NELSON [*99*] used TLC to analyze wheat lipids. The nonsaponifiable matter of methanolic extracts from wheat germs was investigated by JANISTYN [*41*].

PRIVETT, BLANK and LUNDBERG [*109*] described a photodensitometric method for the quantitative analysis of mixtures containing mono-, di-, and triglycerides. A characteristic curve is presented in Fig. 77.

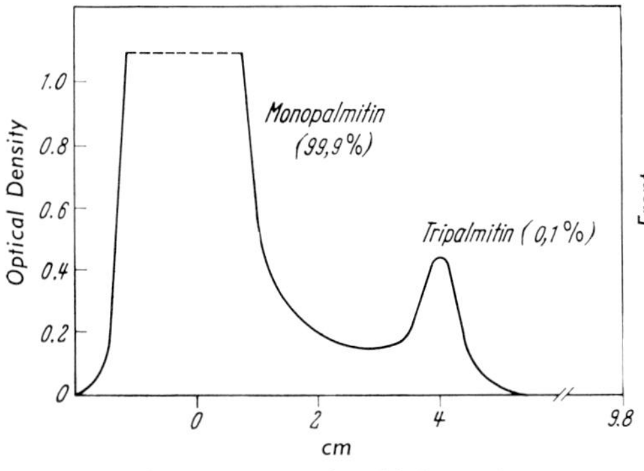

Fig. 79. Densitometer curve of a thin-layer chromatogram of monopalmitin containing 0.1% of tripalmitin [*109*]. For experimental conditions, see Fig. 78

Symmetrical (β) and asymmetrical (α) monoglycerides can be separated, and quantitatively determined, after cleaving the latter with either periodic acid or lead tetraacetate [*108*]. Symmetrical diglycerides can be directly resolved from asymmetrical diglycerides on Silica Gel G (Fig. 78), but not α- and β-monoglycerides. HOFMANN [*36*] succeeded in separating mixtures of α- and β-monoglycerides on thin layers of hydroxyl apatite.

As little as 0.1% of triglyceride can be detected in a monoglyceride by means of photodensitometry (Fig. 79).

PRIVETT, BLANK and LUNDBERG charred the lipids on the plate after spraying with 50% aqueous sulphuric acid. The plates were evaluated with the photodensitometer shown in Fig. 40. Lipids of one and the same class, having different degrees of unsaturation, yield different standard curves. This means that natural mixtures can be quantitatively evaluated only after having been hydrogenated. The method for the determination of ester lipids, described by VIOQUE and HOLMAN [*134*], is independent of the degree of unsaturation. This method is based upon the reaction of hydroxylamine with esters. The red hydroxamic acid-ferric ion complexes are quantitatively determined by colorimetry.

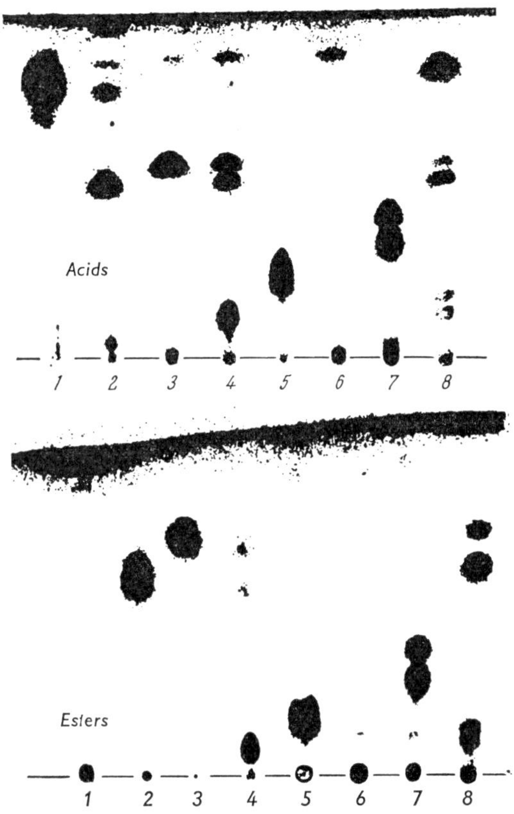

Fig. 80. Thin-layer chromatogram of naturally occurring epoxy and hydroxy fatty acids and their methyl esters [*95*]. *1* palmitoleic plus oleic acids and their methyl esters. *2* *cis*-9,10-epoxystearic acid and its ester. 3 *cis*-12,13-epoxyoleic acid and its ester. *4* *cis*, *cis*-9,10; 12,13-diepoxystearic acid and its ester, also two contaminants. *5* 12-hydroxyoleic acid and its ester. *6* *threo*-12,13-dihydroxyoleic acid and its ester. *7* *threo*-12,13-chlorohydroxyoleic acid plus *threo*-13,12-chlorohydroxyoleic acid and the corresponding esters. *8* acids and esters from *Artemisia absynthium* oil. Adsorbent: Silica Gel G. Solvents: Petroleum hydrocarbon, B.P. 60—70° C. — dietyl ether (90+10) (for esters). Petroleum hydrocarbon, B.P. 60—70° C. — diethyl ether-glacial acetic acid (90+10+1) (for fatty acids). Time: 40 minutes. Indicator: Charring with 50% aqueous sulphuric acid. Reproduced by photocopying. Amounts: About 50 μg

β) Free fatty acids and simple fatty acid derivatives. Applications of TLC for separating common fatty acids and their methyl esters have been described by several authors. MANGOLD and MALINS [*81*] fractionated the non-oxygenated acids from the hydroxy and dihydroxy acids of castor oil on Silica Gel G layers; and the same authors used alumina to separate the non-oxygenated acids from the keto acids of oiticica oil. The three types of acids in castor oil were quantitatively analyzed as radioactively-labelled methyl esters [*84*]. The esters of non-oxygenated acids were eluted and further fractionated by gas-liquid chromatography [*81*] and paper chromatography [*84*]. Linolenic acid was thus detected in castor oil for the first time.

VIOQUE and HOLMAN [*134*] determined the quantitative ratios of the methyl esters of the non-oxygenated acids to the mono- and the dihydroxy acids of castor oil by the hydroxamic acid method, after TLC (see p. 46).

MORRIS, HOLMAN and FONTELL described in several papers [*93, 94, 95*] the application of TLC for the analysis of epoxy and hydroxy acids in seed oils. These investigations led to the discovery of two isomeric hydroxy acids, the 9-hydroxy-*trans*-10-*cis*-12- and the 13-hydroxy-*cis*-9-*trans*-11- octadecadienoic acids.

By the use of TLC, MORRIS, HOLMAN and FONTELL [*93*] found three different epoxy acids in each of six seed oils. Fig. 80 depicts two chromatograms of acids and methyl esters of such oils.

VIOQUE, MORRIS and HOLMAN [*132*] found *trans*-9,10-epoxy oleic acid in orujo oil, a by-product of the olive oil industry. CHALVARDJIAN, MORRIS and HOLMAN [*11*], DHOPESHWARKAR and MEAD [*18*], as well as FULCO and MEAD [*25*], applied TLC to trace substituted fatty acids in nutritional and metabolic investigations.

Some of the separations reported by HOLMAN and collaborators were later repeated and confirmed with model mixtures by KAUFMANN and MAKUS [*58*]. APPLEWHITE, DIAMOND and GOLDBLATT [*2*] also described the use of TLC for the analysis of epoxy and hydroxy fatty acids.

MORRIS, HOLMAN and FONTELL [*94*] could prove by TLC that gas-liquid chromatography isomerizes the methyl esters of triple conjugated-unsaturated fatty acids and dehydrates the esters of vicinally unsaturated hydroxy acids. The primary products of autoxidation of unsaturated methyl esters, the hydroperoxides, are dehydrated during GLC and thus simulate esters of unsaturated fatty acids.

MORRIS, HAYES and HOLMAN [*92*] used TLC to follow the success of preparative methods for the isolation of epoxy acids.

Adsorption-TLC is very well suited for controlling synthetic procedures. Since all kinds of adsorption chromatography separate lipids chiefly according to classes of compounds, it is not necessary to develop a synthesis with a pure acid, e.g., oleic acid; a mixture of higher fatty acids can be used instead. GRUGER, Jr., MALINS and GAUGLITZ, Jr. [*29*] applied TLC to prove the purity of acetylated mono- and diglycerides, which they had prepared from herring oil, menhaden oil, and the oil of salmon eggs. GAUGLITZ, Jr. and MALINS used TLC for analyzing intermediate products in the synthesis of aldehydes from fish oils [*26*]. The authors condensed the methyl esters obtained from the acids of menhaden oil. They reduced the acyloins to diols and cleaved these to aldehydes. GAUGLITZ, Jr. and MALINS observed that the mixtures of methyl esters, and also of acyloins, and of aldehydes, show very definite subfractionations in adsorption-TLC, due to the big differences in chain length and in degree of unsaturation (Fig. 81). The strongly polar diols do not exhibit these subfractionations.

The application of TLC for the analysis of starting material, intermediates, and end products of chemical syntheses of lipids is also mentioned by other authors [*69, 74, 133*]. In many cases, analysis by thin-layer chromatography will help the investigator achieve a better evaluation of the mechanisms of preparative methods and a more realistic account of the purity of the end products of organic syntheses. APPLEWHITE, DIAMOND and GOLDBLATT [*2*] found that the catalytic isomerization of methyl 12-hydroxyoleate (methyl ricinoleate) to methyl 12-ketostearate, with Raney nickel, leads to the formation of methyl 12-hydroxystearate and at least one, not further characterized, by-product. The reduction of methyl 9-hydroxy-10,12-octadecadienoate (methyl dimorphecolate) with hydrogen and platinum dioxide results, according to the same authors, in the formation of 9-hydroxystearate, stearate, and ketostearate by hydrogenolysis, isomerization, and reduction reactions. APPLEWHITE, DIAMOND and GOLDBLATT used three indicators: a fluorescent mineral in the adsorbent layer for recognizing conjugated-unsaturated compounds in UV. light; the fluoresceine-bromine test for staining substances with isolated double bonds; and 2′,7′-dichlorofluorescein, which was used to detect all lipids, saturated and unsaturated, in U.V. light. The application of various indicators for detecting saturated and unsaturated lipids aids in studying reactions involving hydrogenations, because adsorption-TLC does not separate saturated and unsaturated compounds of a certain class from one another. VAN DEENEN and coworkers [*150*, *153*] utilized TLC as an analytical tool for checking syntheses of lecithins. DE HAAS and VAN DEENEN [*152*] studied the enzymatic hydrolysis of lecithins by TLC.

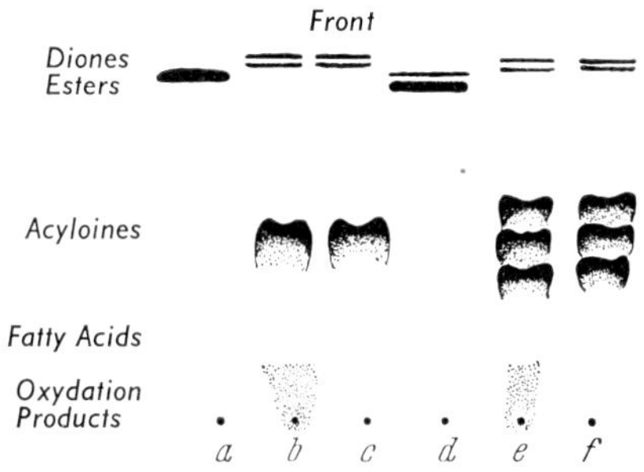

Fig. 81. Thin-layer chromatogram of intermediates of an aldehyde synthesis [*26*]. *a* methyl linolenate; *b* linolenoin, crude; *c* linolenoin, purified; *d* methyl esters from menhaden oil; *e* acyloins from Menhaden oil, crude; *f* acyloins from Menhaden oil, purified. Note the subfractionation of methyl esters and acyloins from Menhaden oil. Adsorbent: Silica Gel G. Solvent: Petroleum hydrocarbon, B. P. 40 — 60° C. — diethyl ether, 90+10. Time: 30 minutes. Indicator: Iodine vapors. Amounts: About 100 μg

It is becoming apparent that the results of classical methods of organic analysis have to be cross-checked by newly developed procedures. MORRIS [*98*] analyzed, by TLC, a preparation of monoacetyl ricinoleyl alcohol, a recently discovered insect sex attractant. He found that the compound, which gave good elemental analyses, and a sharp melting point, consisted of about equimolecular amounts of ricinoleyl alcohol, its monoacetyl and diacetyl derivatives. This example demonstrates that even multi-component mixtures can yield elemental analyses which fit perfectly the sum formula of one of their constituent compounds.

MANGOLD and KAMMERECK [*85*] developed procedures for separating, by TLC, mixtures of technically important lipids; for example, primary, secondary and tertiary amines, quaternary ammonium bases, amides and nitriles (Fig. 82). Extremely polar lipids, such as alkyl sulphates and sulphonates, alkyl phosphates and phosphonates, as well as other detergents, can be separated by TLC on plates coated with a mixture of

9 parts of Silica Gel G and 1 part of ammonium sulphate. Solvent mixtures consisting of chloroform and aqueous sulphuric acid-methanol are used for developing the plates. TLC will facilitate the analysis of detergents, which, so far, has been a difficult and circumstantial task.

Studies on the mechanism of autoxidation of unsaturated lipids led PRIVETT [*107*] to use TLC for elucidating the reaction mechanisms of autoxidation. It is generally assumed that hydroperoxides trigger the

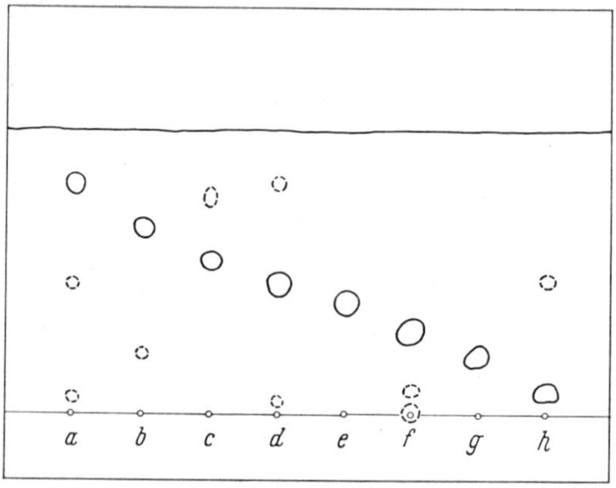

Fig. 82. Thin-layer chromatogram of technically important fatty acid derivatives [*85*]. *a* trialkyl amine, *b* acetylated N-2-hydroxyethylamide, *c* acetylated primary amine, *d* dialkylamine, *e* amide, *f* alkyldimethylamine, *g* hydroxyethylamide, *h* alkylamine. Adsorbent: Silica Gel G. Solvent: Chloroform-N-aqueous ammonia (10 + 1) with methanol (07 + 3). Time: 40 minutes. Indicator: Charring with chromic sulphuric acid solution. Ammounts: About 40 μg

autoxidation of unsaturated lipids. PRIVETT and BLANK [*110*] were able to demonstrate that, during autoxidation of methyl linoleate, autoxidation products of unknown structures can be found *prior* to the formation of hydroperoxides. They concluded that existing theories of autoxidation will have to be revised. The same authors have developed new ideas about the role of metals in the autoxidation of lipids, and they also point at a possible mode of action of antioxidants and synergists. Attention is drawn to two other papers related to the oxidation of lipids. Both RIECHE and co-workers [*161*] and VIOQUE and HOLMAN [*164*] utilized TLC in studying the oxidation of unsaturated lipids.

Lacquers and varnishes from drying oils contain oxygen in the form of various functional groups. DE FREITAS [*24*] reduced monomolecular films of drying oils with lithium aluminium hydride and analyzed by TLC the alcoholic compounds thus formed. Applications of TLC in this field of fat research promise significant new results with relatively little investment of effort and time.

b) Phospholipids, sulpholipids, and glycolipids

Various methods for the chromatographic fractionation of these compounds were described by HANAHAN [*32*]. Very polar natural lipids, as well as synthetic derivatives of similar character, can be separated according to classes of compounds by TLC. The fractionation of such compounds is accomplished by adsorption and by partition. Pronounced adsorption will lead to tailing. Good separations are obtained with layers which have been deactivated to suppress adsorption effects.

Plasmalogens cannot be completely separated from the corresponding diester compounds. Aldehydogenic phospholipids like these are only a little less polar than the diester phospholipids and, therefore, always are concentrated at the upper edge of the corresponding spot on thin-layer chromatograms. Little is known about the chromatographic behavior of polar lipids of very complicated structure. It may be expected that the intensive application of TLC will contribute greatly to the elucidation of the structures of these compounds.

Phospholipids are usually applied to the plates as 0.1 or 1% solutions in chloroform or chloroform-methanol (1:1).

Experimental conditions

Adsorbents. The various classes of phospholipids, sulpholipids and glycolipids are chromatographed on silica gel containing calcium sulphate (Silica Gel G) or on plain silica gel (Silica Gel H). Partition-TLC on "wet" Silica Gel G will produce much better separations than chromatography on dried and activated plates of this adsorbent, according to CHALVARDJIAN [*11*]. The addition of ammonium sulphate to the adsorbent will also give deactivated plates. A maximum load of 10 mg of lipid can be fractionated on one plate by partition-TLC, i.e. substantially less than by adsorption chromatography.

Silica Gel G plates containing ammonium sulphate are prepared from 25 g of Silica Gel G and a solution of 2.5 g of ammonium sulphate in 60 ml of water. It is necessary to coat the plates very quickly, as this mixture dries even faster than the usual pure aqueous slurry. The plates are first air dried, and then are dried further in an oven at 120° C. SKIPSKY, PETERSON and BARCLAY [*163*] recommend adding sodium acetate or sodium carbonate to Silica Gel G.

Plates coated with cellulose (see p. 32) should be suitable for partition-chromatographic separations of phosphatides and other polar lipids. It should be possible also, to apply thin layers of ion exchangers for fractionating such compounds (see pp. 451—456).

Solvents. Mixtures of chloroform and methanol, with a little water, are suitable solvent systems for the TLC of phospholipids. These solvents have been used for years to separate phospholipids on cellulose paper treated with silicic acid [*87*], on glass paper [*31*], and on silicic acid columns [*71*].

WAGNER [*135*], as well as WAGNER, HÖRHAMMER and WOLFF [*136*], recommend the solvent chloroform-methanol-water (65 + 25 + 4) for separating ester phosphatides according to classes of compounds. Fig. 83 shows a typical chromatogram published by these authors (see also Table 13).

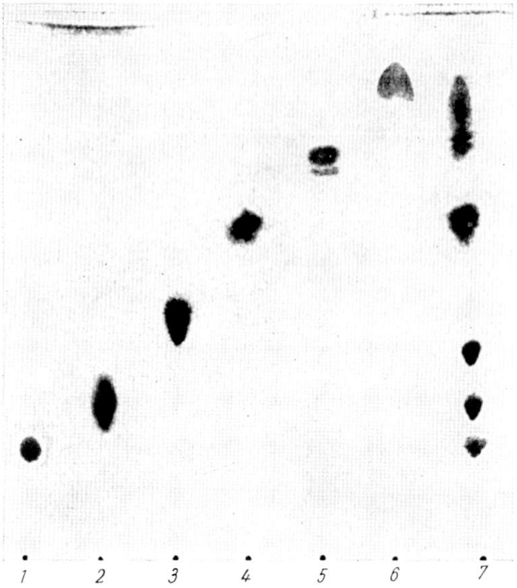

Fig. 83. Thin-layer chromatogram of polar lipids [*135, 136*]. *1* lycolecithin, *2* sphingomyelin, *3* lecithin, *4* ethanolamine phosphatide, *5* cerebroside, *6* cardiolipin, *7* mixture of *1*—*6*. Adsorbent: Silica Gel G. Solvent: Chloroform-methanol-water (65+25+4). Time: 2 hours. Indicator: Rhodamine B in ethanol and Dragendorff reagent. Amounts: 50—100 μg

SCHLEMMER [*116*] also used chloroform-methanol-water for separating phospholipids. He recommends ratios of 75+22+3 and 65 + 30 + 5. DAIN, WEICKER, SCHMIDT and THANNHAUSER [*15, 138*] prefer aqueous-ammoniacal mixtures of chloroform-methanol, such as chloroform-methanol, 25 + 75 or 75 + 25, each containing 40 ml of conc. ammonia per liter.

JATZKEWITZ was the first to apply TLC for separating phospholipids, sulpholipids, and glycolipids [*46, 47*]. He recommended a mixture of ammonia and n-propanol as a solvent system. JATZKEWITZ and MEHL [*47*]

Table 12. *Solvents for Separating Polar Lipids on Silica Gel G*

Solvent	Ratio v/v	Literature
Chloroform-methanol-water	75+22+3	[*116*]
	65+30+5	[*116*]
	65+25+4	[*30, 135, 136*]
	60+35+8	[*135, 136*]
Benzene-N-aqu. ammonia¹	10+1	[*85*]
Alkaline chloroform-methanol²	75+25	[*15, 36*]
(40 ml conc. (14 N) aqu. ammonia/l) . . .	25+75	[*15, 36*]
Acidic chloroform-methanol³	97+3	[*85*]
(Methanol containing 5% of 0.1 N-sulphuric acid)	80+20	[*85*]
Acetone-conc. (14 N) aqu. ammonia	10+1	[*85*]
n-Propanol-water	70+30	[*67*]
n-Propanol-12 N aqu. ammonia	80+20	[*46, 47*]
n-Propanol-conc. (14 N) aqu. ammonia . . .	70+30	[*47*]
n-Propanol-12 N aqu. ammonia	80+20	
followed by ethylene chloride-methanol .	98+2	[*47, 48*]
followed by chloroform-acetic acid (*96*%) .	95+5	
n-Butanol-pyridine-water	60+40+20	[*63*]

¹ The aqueous layer is discarded.

² A similar solvent system can be prepared as follows: Chloroform is equilibrated with N-aqu. ammonia, 10 + 1. A mixture of 97 vols. of this solution and 3 vols. of methanol serves as the developing solvent [*85*].

³ On Silica Gel G containing 5% of ammonium sulphate (see page 161).

used three solvent systems, successively, in the stepwise procedure. Table 12 summarizes these solvent systems, as well as others that have been described for the separation of polar and ionic lipids.

Detection methods. The spray reagents applied in paper chromatography of phospholipids can be used in TLC also: ninhydrin solution

Table 13. [*136*]

Substance	R_f-Value	Ninhydrin[1] Nr. 108	Dragendorff Reagent Nr. 60	Ammonium-molybdate-Perchloric Acid
Amino Acids	0 — 0.10	+	—	—
Lysolecithin	0.21 ± 0.037	—	+	+
Lecithin	0.39 ± 0.055	—	+	+
Sphingomyelin	0.29 ± 0.055	—	+	+
Colamine-Cephalin. . . .	0.57 ± 0.075	+	—	+
Cerebroside.	0.78 ± 0.075	—	—	+
Cardiolipin	0.92 ± 0.015	—	—	+

[1] + positive reaction; — no reaction.

serves for the detection of amine phosphatides [*136*], the Dragendorff reagent is used for choline phosphatides [*136*], and the ammonium molybdate-perchloric acid reagent can be applied for all phospholipids [*136*]. "Universal" indicators for lipids are alcoholic Rhodamine B solution [*136*], slightly alkaline bromothymol blue solution [*47*], and iodine vapors. The charring technique, i.e. spraying with 50% aqueous sulphuric acid or chromic sulphuric acid solution, followed by heating of the plate, applies to phospholipids, as well as to most other organic compounds. The phospholipids show up as black spots. The composition and uses of these and other spray reagents are discussed comprehensively on pp. 483—502 of this volume.

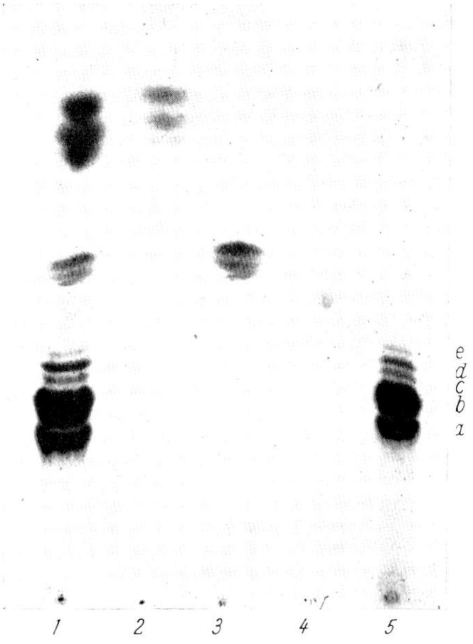

Fig. 84. Thin-layer chromatogram of Sphingolipids [*136*]. *1* mixture of sphingolipids from beef brain. *2* cerebrosides, *3* cerebroside sulphuric acid esters, *4* sphingomyelins, *5* gangliosides (a, b, c, d, e), Adsorbent: Silica Gel G. Solvent: Chloroform-methanol-water (60 : 35 : 8). Time: 2 hours. Amounts: 50—100 μg

It is always essential that several reagents are applied to one chromatogram, so no substance remains undetected. This is demonstrated in Table 13, which is taken from a publication by Wagner, Hörhammer and Wolff [*136*].

The most useful and reliable technique for showing organic material is charring (see Fig. 71, 73, 74).

Applications and results. JATZKEWITZ [*46*] accomplished the separation of cerebrosides and their sulphuric acid esters of the phrenosine type from those of the kerasine type, by TLC on Silica Gel G. He analyzed the cerebroside sulphuric acid esters of normal and pathological brains and assumed that they were identical. Shortly thereafter, WAGNER, HÖRHAMMER and WOLFF [*136*] confirmed, by TLC, that all cerebrosides, including nervone, occur in normal brains in the form of their sulphuric acid esters. Fig. 84 shows the separation of cerebrosides and their esters in three zones on a Silica Gel G plate, using chloroform-methanol-water 60 + 35 + 8) as the developing solvent.

KOCHETKOV, ZHUKOVA and GLUKHODED [*65*] also reported the application of TLC to the analysis of cerebrosides.

HABERMANN, BANDTLOW and KRUSCHE [*30*] made use of the solvent recommended by WAGNER, HÖRHAMMER and WOLFF [*136*] for separating phospholipids of human serum by TLC. The individual fractions were determined, colorimetrically, as phosphate after charring. JATZKEWITZ [*48*] worked out a method for the quantitative determination of various sphingolipids. WAGNER [*166*] reported methods for the quantitative analysis of lecithins and cephalins by TLC.

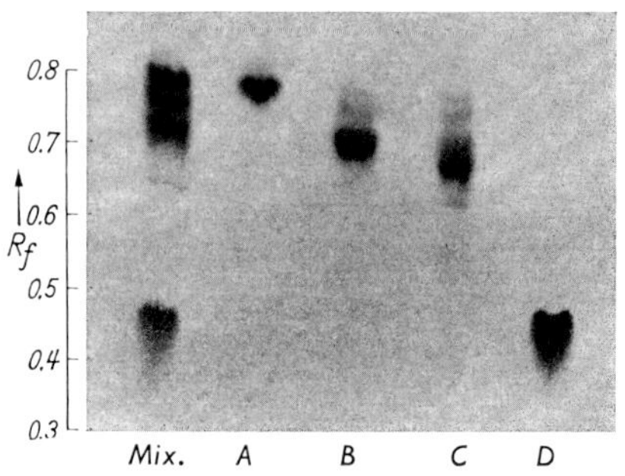

Fig. 85. Thin-layer chromatogram of a mixture of gangliosides and four individual gangliosides (*A*, *B*, *C*, *D*) from human brain [*64*]. Adsorbent: Silica Gel G. Solvent: n-Butanol-pyridine-water (3 + 2 + 1). Time: $2^1/_2$—3 hours. Indicator: Bial's reagent

TLC was applied, with great success, in the isolation and determination of the structures of gangliosides. WAGNER, HÖRHAMMER and WOLFF [*136*] separated, on Silica Gel G, a ganglioside fraction from beef brain into two main components and three zones of less intensity. The authors assumed that the two main components (Fig. 84a and b) were identical with two gangliosides, previously isolated by KUHN [*66*] as well as by KLENK and GIELEN [*63*]. KUHN, WIEGANDT and EGGE [*67*] and KLENK and GIELEN [*64*] were able to detect, isolate, and determine the structures of four gangliosides by TLC. The solvents applied by KUHN and coworkers and KLENK and coworkers are listed in Table 12. Other solvents for separating simple sphingolipids and complex glycolipids can be found there also.

c) Applications of thin-layer chromatography for structural analysis of lipids

PRIVETT and BLANK [*108*] described a method for the analysis of saturated and unsaturated α- and β-monoglycerides. This "Ozonization-

Reduction-TLC-Method" allows one to determine the following four types of monoglycerides (I–IV), quantitatively, in the presence of each other:

I II III IV

S = Saturated fatty acid
U = Unsaturated fatty acid

Mixtures of monoglycerides cannot be completely resolved by TLC or by any other method. Privett and Blank fractionated α- and β-monoglycerides, after glycol splitting, as glycol aldehyde esters and unchanged β-monoglycerides. The unsaturated glycol aldehyde esters and monoglycerides then were converted to aldehydic fragments by reductive ozonolysis:

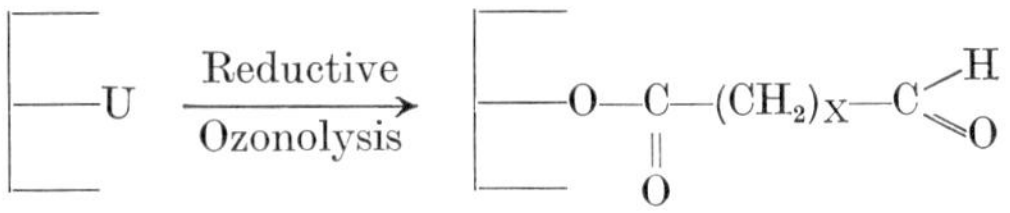

+ Short-chain aldehyde

These "aldehydic residues" are more polar than the glycol aldehyde esters and monoglycerides of saturated fatty acids, which are not attacked by ozone. It is easy to separate these groups of substances by adsorption-TLC (Fig. 88).

Privett and Blank [*108*] were able also to determine, quantitatively, by means of their ozonization-reduction-TLC method, six of the seven possible diglyceride types, and four of the six possible types of triglycerides. Fig. 87 illustrates the analysis of a mixture of 1,3-diolein, 1,2-distearin, and 1,3-distearin.

The aldehydic residues from mono-, di- and triglycerides give the same standard curve in densitometry as saturated mono-, di-, and triglycerides, if correction is made for the carbon lost by ozonolysis. This facilitates the quantitative analysis of complicated mixtures containing glycerides of saturated acids and glycerides of unsaturated acids of widely different degrees of unsaturation (see pp. 45, 51).

Privett and Blank [*111*] also applied their method to the quantitative determination of the four types of lecithins:

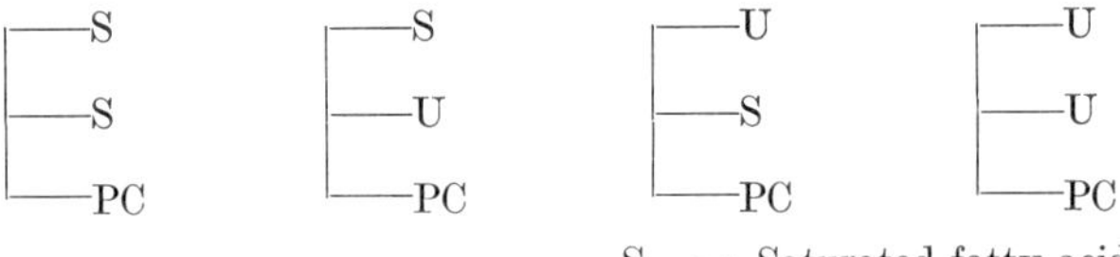

S = Saturated fatty acid
U = Unsaturated fatty acid
PC = Phosphatidyl choline

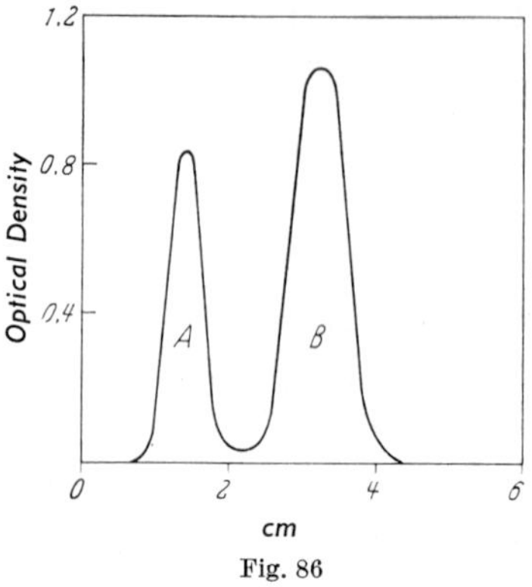

Fig. 86

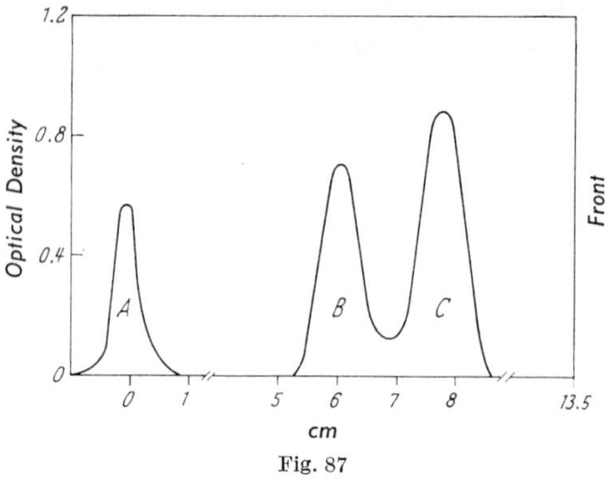

Fig. 87

Fig. 86. Densitometer curve of a thin-layer chromatogram of a mixture of monopalmitin and monoolein after reductive ozonolysis [*108*]. *A* "aldehydic core" from monoolein. *B* monopalmitin. Adsorbent: Silica Gel G. Solvent: Petroleum hydrocarbon, B.P. 40—60° C. — dietyhl ether, 10 + 90. Time: 40 minutes. Indicator: Charring with 50% aqueous sulphuric acid

Fig. 87. Densitometer curve of a thin-layer chromatogram of a mixture of 1,3-diolein, 1,2-distearin, and 1,3-distearin after reductive ozonolysis [*108*]. *A* "aldehydic core" from 1,3-diolein. *B* 1,2-distearin. *C* 1,3-distearin. Adsorbent: Silica Gel G. Solvent: Petroleum hydrocarbon, B.P. 40—60°. — diethyl ether (60+40). Time: 40 minutes. Indicator: Charring with 50% aqueous sulphuric acid

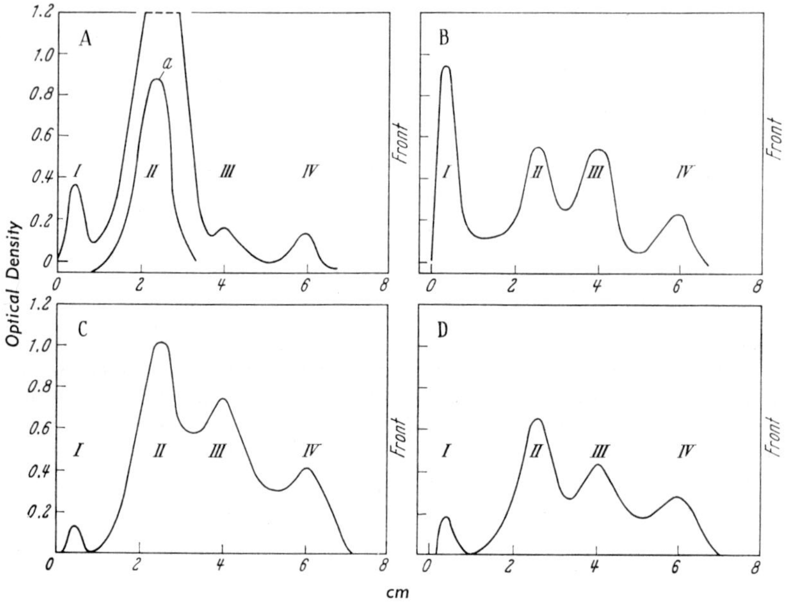

Fig. 88. Densitometer curves of the "aldehydic cores" obtained from natural mixtures of lecithins [*111*]. *A* egg lecithin, *B* lecithins of beef spinal cord, *C* soybean lecithins, *D* wheat lecithins; *I* saturated lecithins, *II* aldehydic residues from saturated (α) — unsaturated (β) lecithins, *III* aldehydic residues from saturated (β) — unsaturated (α) lecithins, *IV* aldehydic residues from lecithins containing two unsaturated acid moieties. Stationary phase: Silicone on Silica Gel G. Solvent: Glacial acetic acid-water (80+20). Time: $1^1/_2$ hours. Indicator: Charring with 50% aqueous sulphuric acid

The "aldehydic cores" obtained by reductive ozonolysis of lecithins were resolved on siliconized Silica Gel G [*76*]. Fig. 88 demonstrates the curves obtained by densitometry of thin-layer chromatograms of the aldehydic fragments of egg lecithins, the lecithins from beef spinal cord, soybean lecithins, and wheat lecithins.

Privett and Nickell [*112*] proved, by TLC, the formation of two isomeric ozonides of methyl oleate, methyl elaidate, and other *cis*- or *trans*-unsaturated methyl esters of fatty acids. At The Hormel Institute, ozonolysis of unsaturated lipids is always monitored and controlled by TLC. Over-ozonization can be recognized easily by this technique.

2. Fractionation of homologous series

Volatile compounds of a homologous series can be fractionated by adsorption-TLC in the form of their derivatives. Compounds of longer chain-lengths are resolved, in most cases, by reversed-phase partition-TLC, according to chain length and degree of unsaturation.

a) Procedures for fractionating short-chain compounds

Volatile aldehydes and methyl ketones occur in nature as a result of enzymatic action and autoxidative decomposition of lipids. These volatile materials are partially responsible for the flavors and odors of fats and oils. Traces of alcohols and acids, even amines and mercaptans, are found,

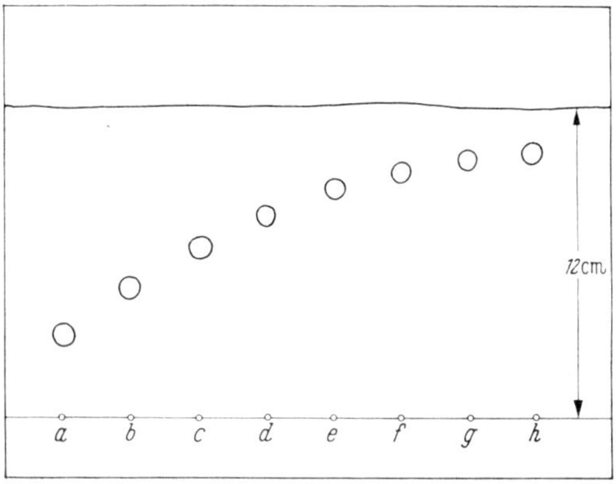

Fig. 89. Separation of derivatives of short-chain members of a homologous series by adsorption-TLC on Silica Gel G. Dinitrophenyl derivatives of *a* methylamine, *b* ethylamine, *c* propylamine, *d* butylamine, *e* amylamine, *f* hexylamine, *g* heptylamine, *h* octylamine. Adsorbent: Silica Gel G. Solvent: Petroleum hydrocarbon, B.P. 60—70° C. — diethyl ether, 70+30. Time: 40 minutes. Indicator: 1% of a fluorescent mineral in the layer. Amounts: 5 μg

especially in fish oils. It is a difficult task to detect and identify by chemical means, small amounts of these odoriferous compounds. TLC, especially in combination with GLC, seems to be an ideal tool for analyzing these unpleasant and unwanted minor constituents of fats and oils.

TLC is well suited for analyzing fragments formed while determining the chemical structures of unsaturated lipids. Reductive ozonolysis of unsaturated fatty acids results in the formation of short-chain aldehydes and aldehyde-acids. Both classes of compounds can be separated in the form of derivatives by TLC, according to chain-length. The acidic fragments from oxidative cleavage of unsaturated lipids also can be resolved easily by TLC.

The short-chain members of a homologous series are first converted to non-volatile derivatives and then separated on a plate. Fig. 89 shows, as an example, the resolution on Silica Gel G of the dinitrophenyl derivatives of primary amines of carbon numbers 1 to 8.

Alcohols are best fractionated as 3,5-dinitrobenzoates, and highly volatile aldehydes and ketones, as 2,4-dinitrophenylhydrazones. Acids are separated as anilides or N-methyl anilides. Mercaptans of chain-lengths 1 through 6 are resolved as 3,5-dinitrobenzoates.

Experimental conditions

Adsorbents. Thin layers of Silica Gel G, or of Alumina G, are useful for separating the derivatives of short-chain compounds.

Solvents. Mixtures of hexane or petroleum hydrocarbon, B.P. 60 to 70° C, and diethyl ether or ethyl acetate, are suitable solvent systems.

Detection methods. Aromatic compounds are either yellow and/or absorb U.V. light. Hence, no particular spray reagent is required for staining. They are most advantageously chromatographed on Silica Gel G containing 0.5–2% of a fluorescent mineral. After chromatography, the plates are viewed under a U.V. lamp. Non-volatile contaminants can be visualized as grey or black spots by the charring technique. The derivatives themselves, are usually quite resistant to charring; e.g. the yellow color of 2,4-dinitrophenyl hydrazones of aldehydes and ketones fades when heated after spraying with 50% aqueous sulphuric acid. On cooling, the spots appear as before.

In most cases, TLC of derivatives of the short-chain members of a homologous series is satisfactory only up to a chain length of 6 or 7 carbon atoms. Stepwise development, or two-dimensional TLC, yields very little improvement. Derivatives of compounds of 10 or more carbon atoms cannot be well resolved by adsorption chromatography; they migrate as classes.

Detailed descriptions of TLC of various short-chain aliphatic compounds and their derivatives are to be found in several chapters of this volume.

b) Procedures for separating long-chain compounds

Mixtures of long-chain members of a homologous series can be fractionated by partition-TLC in reversed phase.

The application of the reversed-phase technique requires that the homologous substances to be analyzed be completely free of other classes of compounds. Only rarely can complicated mixtures of compounds, which differ in the nature and number of their functional groups and in chain length and degree of unsaturation, be completely resolved by reversed-phase partition-TLC only. The consecutive application of adsorption- and reversed-phase partition-TLC is, therefore, an ideal combination for separating complex lipid mixtures. Groups of compounds which are not resolved by adsorption, are fractionated further by reversed-phase partition. GLC is used for determining the fatty acid composition of lipids isolated by adsorption- and reversed-phase partition-TLC.

The application of these three methods in combination allows "three-dimensional lipid analysis" [*129*]. The principle of this kind of analysis is summarized as follows:

A complicated mixture of lipids is first resolved into classes of compounds, e.g. cholesteryl esters, triglycerides, and acids, by means of adsorption-TLC on Silica Gel G.

Each class of compounds is fractionated further into simple mixtures, e.g. tripalmitin, palmitodiolein, and triolein, by reversed-phase partition-TLC on hydrophobic layers.

These groups of compounds, or individual compounds, are hydrolyzed, and their constituent fatty acids are identified and quantitatively determined as methyl esters, by GLC, after methylation.

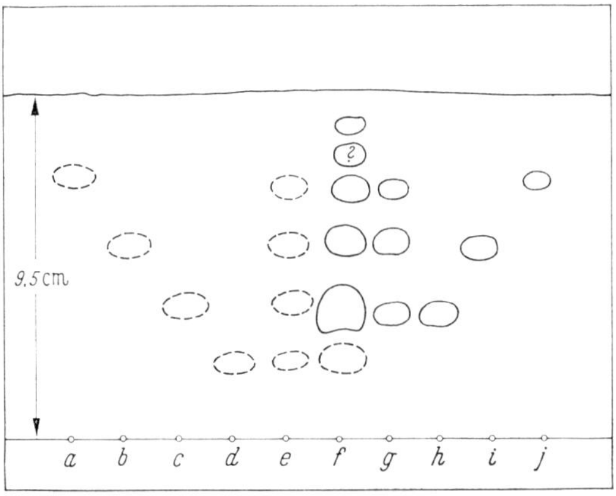

Fig. 90. Fractionation of a lipid class by reversed-phase partition-TLC [*76*]. *a* methyl laurate, *b* methyl myristate, *c* methyl palmitate, *d* methyl stearate, *e* mixture of saturated methyl esters, *f* C_{18}-fraction of menhaden oil methyl esters, g) mixture of unsaturated methyl esters, *h* methyl oleate, *i* methyl linoleate *j* methyl linolenate. Stationary phase: Silicone on Silica Gel G. Solvent: Acetonitrile-glacial acetic acid-water 70+10+25. Time: 40 minutes. Indicators: Iodine vapors (——) followed by α-cyclodextrin-iodine (-----). Amounts: 20 μg of individual esters

Fig. 90 shows the separation of the methyl esters of higher fatty acids by reversed-phase partition-TLC. Silicone was used for impregnating a Silica Gel G plate. It is evident that, by reversed-phase partition-TLC, it is possible to separate homologous compounds of the same class of compounds and vinylogues of the same chain-length: the methyl esters of unsaturated C_{18} acids are resolved according to their degree of unsaturation. Polyunsaturated lipids of a class of compounds migrate further than mono-unsaturated or saturated compounds of the same chain-length. Also, the different members of a homologous series can be fractionated by this method: saturated methyl esters are separated according to chain-lengths. The *Rf* values decrease with increasing chain-lengths. It is remarkable that the *Rf* values are reproduced better in reversed-phase partition-TLC than in adsorption-TLC.

Certain saturated and unsaturated compounds migrate together in TLC on hydrophobic plates. The same phenomenon is encountered in other procedures, such as countercurrent distribution [*1*], reversed-phase partition chromatography on columns [*118*] or on impregnated paper [*117*]. For example, methyl oleate (C_{18}, monounsaturated) and methyl palmitate (C_{16}, saturated) migrate together. Methyl linoleate (C_{18}, double-unsaturated) and methyl myristate (C_{14}, saturated) are also inseparable. The pairs of the corresponding free fatty acids, or of the corresponding aldehydes, also form "critical pairs". Triolein overlaps with tripalmitin; trilinolein cannot be separated from trimyristin. As a rule, the lipids of a certain class of compounds, which differ by 2 methylene groups and 1 double bond, will migrate together. It is possible to separate critical pairs by low-temperature chromatography [*57, 117*], by TLC on layers containing silver nitrate, or by chromatography with peroxidic solvents [*79*]. These special methods are further described below.

Experimental conditions

The stationary phase. The author prefers silicone for impregnating Silica Gel G plates. "Dow Corning 200 Fluid" (see p. 37) has been useful in paper chromatography of lipids. Hydrocarbons, e.g. undecane [*58*], tetradecane [*62*], and squalane [*10*], or mixtures of high molecular weight paraffins [*141*] can be used instead of silicone as hydrophobic agents.

High-molecular weight compounds are, in the author's experience, superior to the more volatile hydrocarbons. Plates impregnated with silicone, paraffin, or squalane do not change their qualities even after weeks of storage. In contrast, the plates will constantly lose "stationary phase" if impregnated with undecane, and, consequently, the results of separations will remain uncertain, unless the plates are kept in an atmosphere saturated with undecane.

Directions for impregnating Silica Gel G or Kieselguhr G layers are given on p. 37. The addresses of manufacturers of suitable silicones and paraffins are also given.

Recently, polyethylene, acetylated cellulose, and polyamide powders for TLC have been made commercially available (see p. 33). These synthetic products have proved very valuable for fractionating fatty acids [*28*] and fat-soluble vitamins [*143*] on columns. Other synthetic products, such as Teflon [*3*] might be suitable for separation of lipids by partition-TLC in reversed phase. Several papers on the application of polyamide powders for TLC have been published [*16, 17*].

Solvents. The behavior of a certain class of lipids on adsorption plates gives information about its polarity, and thus, facilitates the choice of suitable solvents for the further fractionation of these compounds by reversed-phase partition-TLC.

Most of the solvent mixtures used for paper chromatography of lipids are also suitable for TLC on hydrophobic layers [*58*, *76*]. Triglycerides and less polar lipids (Fig. 70) usually can be separated according to chain length and degree of unsaturation with glacial acetic acid or acetonitrile, each containing up to 10% water. Also, chloroform-methanol with 5% water and mixtures of acetone with acetonitrile, or methylethylketone with acetonitrile, or methylethylketone with acetonitrile are good solvents for fractionating classes of less polar lipids. Mixtures of glacial acetic acid-acetonitrile and mixtures of glacial acetic acid or acetonitrile or tetrahydrofuran with 10 to 50% water, are suitable for separating polar lipids according to chain length and degree of unsaturation. Very polar lipids, for example, alkylsulphates or quaternary ammonium bases, can be fractionated on cellulose with aqueous ethanol.

MALINS and MANGOLD [*76*] used glacial acetic acid-water (85 + 15) in order to separate higher fatty acids on siliconized Silica Gel G. The same authors also chromatographed methyl esters of higher fatty acids on silicone as stationary phase, with the solvent acetonitrile-glacial acetic acid-water (70 + 10 + 25). CHAKRABARTY [*10*] found that the same solvents could be applied when the Silica Gel G was impregnated with squalane instead of silicone.

KAUFMANN and MAKUS [*58*] worked with Silica Gel G impregnated with undecane. They made use of mixtures of glacial acetic acid-water (96+4) and glacial acetic acid-acetonitrile (50+50) as solvents for the separation of higher fatty acids. The same authors have separated fatty alcohols on Silica Gel G impregnated with undecane using glacial acetic acid-acetonitrile (25+75). They fractionated epoxy and episulphido fatty acids with glacial acetic acid-water (80+20).

KAUFMANN and MAKUS [*58*] described the separation of di- and triglycerides by partition-TLC in reversed phase. Chloroform-methanol-water (25+75+5) served for fractionating the diglycerides. Acetone-acetonitrile (70+30) was used to separate synthetic triglycerides on Silica Gel G impregnated with undecane.

KAUFMANN and his collaborators later switched to the use of hydrophobic agents of higher molecular weights. Tetradecane was used instead of undecane [*62*]; then they chose a fraction of petroleum hydrocarbon (b. p. 240–250° C), and finally silicone, for impregnation [*61*]. KAUFMANN, MAKUS and DEICKE [*59*] were able to separate 15 cholesteryl esters by two-dimensional TLC (see p. 257). KAUFMANN, MAKUS and DAS [*61*] separated mixtures of synthetic triglycerides, and even triglyceride fractions obtained from naturally-occurring fats and oils. The solvents applied were acetone-acetonitrile, (80+20), or methanol-acetonitrile-propionitrile (50+40+15). Recently, KAUFMANN and DAS [*156*] reported the separation of triglycerides by TLC on impregnated Kieselguhr G. Reference is made to two publications on the hydrogenation and bromination of unsaturated lipids on plates [*157*], and the separation of "critical" mixtures of fatty acids and of triglycerides [*158*]. KAUFMANN and KHOE [*158*] mention the use of gypsum as an adsorbent for coating plates.

Critical pairs can be fractionated by low-temperature chromatography or by chromatography with peroxidic solvents. These methods are further described in the following sections. Other Methods are based on the formation of mercuric acetate addition compounds of unsaturated lipids and on the formation of π-complexes of unsaturated substances (see pp. 176—179).

SCHLENK and collaborators [*117*] and, shortly afterward, KAUFMANN and MOHR [*57*], showed that critical pairs of saturated and unsaturated lipids can be separated by low-temperature chromatography on siliconized paper or on paper impregnated with undecane. MALINS and MANGOLD [*76*] utilized this method to fractionate a mixture of palmitic and oleic acids on siliconized Silica Gel G. The solvent, glacial acetic acid-formic acid-water (40+40+20) resolved oleic acid (*Rf* 0.10) from palmitic acid, which remained at the origin after 8 hrs at 4–6° C.

For the separation of critical pairs of saturated and unsaturated lipids on siliconized paper, peroxidic solvents can be utilized, according to MANGOLD, GELLERMAN and SCHLENK [*79*]. This method has subsequently been applied in other laboratories also [*7*, *80*, *85*]. MALINS and MANGOLD [*76*] described TLC with oxidizing solvents. In aqueous solvents, part of the water is replaced by perhydrol, and in solvents containing glacial acetic acid, part of the acid is replaced by peracetic acid. These solvents will oxidize all the unsaturated lipids during chromatography. The very polar reaction products migrate together with the solvent front, whereas, the saturated lipids migrate unchanged and are well separated from each other in the useful region of the chromatogram. Typical peroxidic solvents are glacial acetic acid-peracetic acid-water (75+10+15) and glacial acetic acid-water-perhydrol (85+5+10).

Three to 4 hrs are required for partition-TLC in reversed phase. Only solvents containing acetonitrile migrate faster.

Table 14 gives a summary of solvents applied for TLC in reversed phase.

Detection methods. Unsaturated lipids can be made visible on hydrophobic plates by iodine vapors [*80*]. Most organic substances can also be detected on siliconized Silica Gel G layers by charring with chromic sulphuric acid solution [*111*]. KAUFMANN and collaborators [*58*, *59*] prefer a solution of Rhodamine B and alcoholic phosphomolybdic acid as spraying reagents.

2′,7′-Dichlorofluorescein is not suitable for staining lipids on hydrophobic plates [*76*].

Applications and results

Fig. 90(*f*) shows, as an example of the application of hydrophobic layers, the separation of fatty acids from menhaden oil in the form of

Solvent[1]	Ratio v/v	Stationary Phase	For fractionation of (Lipid Class)	Literature
Methanol	—	Paraffin[2]	Carotenals	[*142*] (see pp. 214—219)
Acetic acid-water . .	96+4	Undecane[2]	Fatty acids	[*58*]
	75+25	Silicone[2]	Fatty acids and their methyl esters	[*76*]
	85+15	Silicone[2]	Fatty acids and their methyl esters	[*76*]
			„Aldehydic cores" from lecithines[5]	[*111*]
	85+15	Squalane[2]	Fatty acids and their methyl esters	[*10*]
Acetic acid-acetonitrile	50+50	Undecane[2,3]	Fatty acids	[*58, 60*]
Acetic acid-acetonitrile-water	10+70+25	Silicone[2]	Methyl esters of higher fatty acids	[*76*]
	10+70+25	Squalane[2]	Methyl esters of higher fatty acids	[*10*]
Acetone-ethanol-water	60+10+30	Polyethylene[4]	Methyl esters of higher fatty acids	[*129*]
Acetic acid-formic acid-water[6]	40+40+20	Silicone[2]	Fatty acids saturated/unsaturated	[*76*]
Acetic acid-peracetic acid-water	75+10+15	Silicone[2]	Fatty acids saturated/unsaturated	[*76*]
Methylethyl ketone-acetonitrile	70+30	Paraffin[2]	Colesteryl esters of higher fatty acids	[*59*]
Acetic acid-water . .	80+20	Undecane[3]	Keto fatty acids, lactones	[*62*]
	80+20	Paraffin[3]	Keto fatty acids, lactones	[*62*]
Acetic acid-water . .	90+10	Tetradecane[3]	Hydroxy fatty acids	[62]
	75+25	Paraffin[3]	Hydroxy fatty acids	[*62*]
Chloroform-methanol-water	15+75+5	Undecane[2]	Diglycerides	[*58*]
Methanol-acetonitrile	50+40	Silicone[3]	Triglycerides	[*61*]
Methanol-acetonitrile-propionitrile . . .	50+40+15	Silicone[3]	Triglycerides	[*61*]
Acetone-acetonitrile	80+20	Paraffin[3]	Triglycerides	[*60, 61*]
	70+30	Undecane[2]	Triglycerides	[*58*]

[1] All solvents must be saturated with stationary phase.
[2] On Silica Gel G.
[3] On Kieselguhr G.
[4] Directly on glass plate.
[5] Separation acc. to classes.
[6] Chromatographed at 4—6° C.

their methyl esters. The identity of the various esters was proved by chromatographing known standard substances, and by the U.V. spectra of the unsaturated esters, after alkali isomerization. An unknown substance ("?") was isolated but could not be identified by alkali isomerization nor by gas chromatography [*76*].

KAUFMANN, MAKUS and KHOE [*60*] fractionated the triglycerides from linseed oil, soybean oil, and cocoa butter substitutes by partition-TLC in reversed phase. KAUFMANN, MAKUS and DAS [*61*] described the separation of triglycerides from corn oil. The same authors also fractionated the triglycerides from sunflower oil, sesame seed oil, olive oil, lard, and beef tallow by the reversed-phase technique. They obtained several sharply defined spots of simple mixtures of triglycerides. Corn oil, for instance, could be separated into 8 triglyceride fractions. Linseed oil gave as many as 10 spots.

In this connection, reference is made to the work of WINTERSTEIN and collaborators [*141, 142*] on the separation of carotenoids (see p. 217).

3. Separation of lipids according to their degree of unsaturation

Adsorption chromatography separates lipids mainly into classes of compounds. Partition chromatography in reversed phase subfractionates the components of one class according to chain length and degree of unsaturation. The application of partition chromatography is limited by the fact that one double bond equals two methylene groups, e.g., methyl oleate and methyl palmitate migrate together. By means of low-temperature chromatography, sometimes it is possible to resolve such "critical pairs". Mixtures of saturated and unsaturated lipids of one class can also be separated after conversion of the unsaturated components into a derivative. The most simple procedure is to oxidize the unsaturated lipids. The oxidation of unsaturated lipids may be accomplished by chromatographing with peroxidic solvents (see p. 172). This method is quantitative, but it has the disadvantage that the unsaturated components cannot be regenerated from their oxidation products.

No effort has been spared to find derivatives of unsaturated lipids from which the original compound can be recovered after separation. Easily accessible are the acetoxymercuri-methoxy derivatives [*13*]. These addition compounds are formed by the reaction of mercuric acetate with unsaturated lipids in methanol solution.

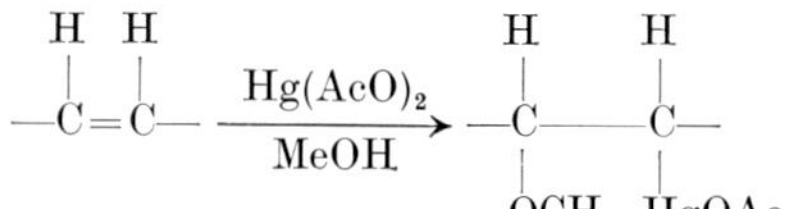

The addition reaction is quantitative if carried out at room temperature [*43*]. *Cis* compounds react 10 to 20 times faster than the corresponding *trans* isomers [*43*]. The unsaturated compounds can be regenerated by exposing their derivatives to hydrochloric acid. Shifting of double bonds or *cis-trans* isomerization does not occur [*42*, *45*].

INOUYE and collaborators separated, by reversed-phase paper chromatography, the mercuric acetate derivatives of methyl esters of long-chain fatty acids [*38*], as well as the derivatives from lecithins [*39*] and triglycerides [*40*, *101*]. KAUFMANN and POLLERBERG [*55*] paper-chromatographed the derivatives of allyl esters of long-chain fatty acids.

JANTZEN and ANDREAS [*42*, *44*] and JANTZEN, ANDREAS, MORGENSTERN and ROTH [*45*] separated the methyl esters of saturated fatty acids from the mercuric acetate derivatives of methyl esters of unsaturated fatty acids by adsorption chromatography on silicic acid columns. The same investigators were able to show that the derivatives of unsaturated fatty esters separate strictly according to degree of unsaturation, i. e., only by the number of acetoxymercuri-methoxy groups per molecule.

It was logical to apply this kind of fractionation to thin-layer plates. Fig. 91 shows, schematically, the possibilities TLC offers for the chromatography of mercuric acetate derivatives.

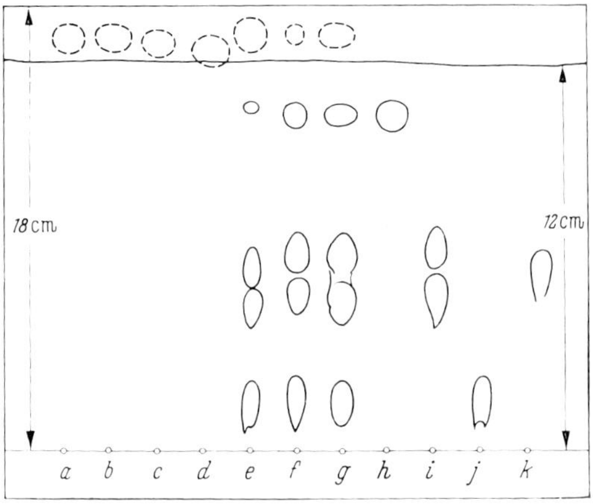

Fig. 91. Thin-layer chromatography of mercuric acetate derivatives [*83*]. *a* methyl stearate, *b* methyl oleate, *c* methyl linoleate, *d* methyl linolenate, (*e*—*j*) mercuric acetate derivatives of: *e* methyl esters of C_{16} acids from *Chlorella pyrenoidosa*, *f* methyl esters of C_{18} acids from *Chlorella*, *g* total methyl esters from *Chlorella*, *h* methyl oleate, *i* methyl linoleate, *j* methyl linolenate, *k* mercuric acetate. Adsorbent: Silica Gel G. Solvents: First, petroleum hydrocarbon, B.P. 60—70° C. — diethyl ether (80 : 20); Second, n-propanol-glacial acetic acid (100 : 1). Times: First solvent: $1^1/_2$—2 hours; second solvent: 3—4 hours. Indicators: s-Diphenylcarbazone in ethanol, followed by iodine vapors. Amounts: 20 μg of methyl esters, 50—100 μg of derivatives

Even mixtures of different classes of lipids can be separated by TLC in the form of their mercuric acetate derivatives, according to the degree of unsaturation, provided the types of compounds do not differ too much in polarity. Hydrocarbons, wax esters, and alcohol acetates have so little polarity that their chromatographic behavior is completely governed by the number of acetoxymercuri-methoxy groups per molecule. Hence, the derivatives of 9-octadecene, cetyl oleate and oleyl acetate migrate together, as do the derivatives of 9,12-octadecadiene cetyl linoleate and linoleyl acetate.

a) Preparation of mercuric acetate derivatives

A procedure described by JANTZEN and ANDREAS [*43, 44*] proved to be satisfactory.

The reagent is a solution of 14 g of mercuric acetate, dissolved in 250 ml of methanol, containing 2.5 ml of water and 1 ml of glacial acetic acid. For lipid mixtures of small iodine numbers, 40 ml of the reagent is sufficient per 1 g of sample. For lipids with iodine numbers exceeding 100, 0.4 times the iodine number in ml of reagent solution per 1 g of sample must be taken. The reaction mixture is kept under nitrogen in the dark.

The reaction is completed after 12 hrs at room temperature. If *trans* isomers are to be converted, it takes at least 2 days.

After the reaction is concluded, most of the methanol is evaporated *in vacuo* at 30° C. The residue is dissolved in 10 to 20 ml of chloroform and washed several times with water to remove excess mercuric acetate. The chloroform extract is dried over sodium sulphate.

A modification of this procedure was described by GOLDFINE and BLOCH [*27*].

The mercuric acetate derivatives of unsaturated lipids are clear, colorless oils. For separations by TLC, the derivatives are applied as chloroform solutions.

b) Experimental conditions

Adsorbents. Silica Gel G is suitable for the fractionation of mercuric acetate derivatives of lipids containing up to 3 double bonds [*83*]. Other types of adsorbents have not been investigated yet. It is desirable to find plates allowing the separation of derivatives of lipids with more than 3 double bonds.

The derivatives are usually chromatographed on a preparative scale. It is possible to resolve up to 10 mg of derivatives on a standard plate of 20 × 20 cm (see Fig. 92).

Solvents. A mixture of petroleum hydrocarbon (B. P. 60–70° C-diethyl ether (80+20) fractionates the methyl esters of saturated fatty acids from the derivatives of unsaturated ones within 1.5–2 hrs. During this time, the solvent front migrates 15–18 cm high. A second solvent system, n-propanol-glacial acetic acid (100 + 1) serves for the resolution of the derivatives of unsaturated fatty esters. Three to 4 hrs are required to move the solvent front to a height of 12–14 cm.

Saturated and unsaturated hydrocarbons, wax esters, alcohol acetates, nitriles, and other lipids of low polarity are separable under the same conditions. Surprising is the fact that even rather polar compounds, such as alcohols, monoglycerides, and fatty acids, can be fractionated in a similar manner.

Detection methods. The mercuric acetate derivatives turn purple when sprayed with a solution of 0.1% s-diphenylcarbazone in 96% ethanol. The derivatives can be recognized as violet-purple spots on a pink background. The saturated esters are visualized with 2′,7′-dichlorofluorescein in U.V. light or with iodine vapors (Reagents N°. 42 and 72).

The derivatives of monoenes (*Rf* 0.85), dienes (two spots, *Rf* 0.55 and *Rf* 0.45) and trienes (*Rf* 0.15) are separated very distinctly. The chain lengths of the various esters have little or no influence on the chromatographic behavior of the derivatives. The abovementioned *Rf* values are easy to reproduce. Obviously, the partition effect outweighs the adsorption effect in chromatography with the system n-propanol-glacial acetic acid.

The saturated esters of different chain lengths are located between the two solvent fronts.

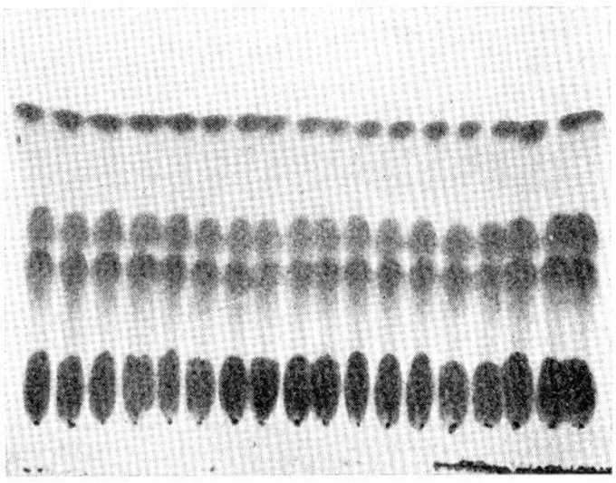

Fig. 92. Thin-layer chromatogram of the mercuric acetate derivatives of the total methyl esters derived from the alga *Chlorella pyrenoidosa* (compare with Figs. 91 and 93)

c) Recovery of derivatives

The 3 groups of derivatives (monoenes, dienes, and trienes) are scraped off the plate together with the silica gel in bands 2—3 cm wide. Each of the 3 fractions is put into a test tube containing 10 ml methanol

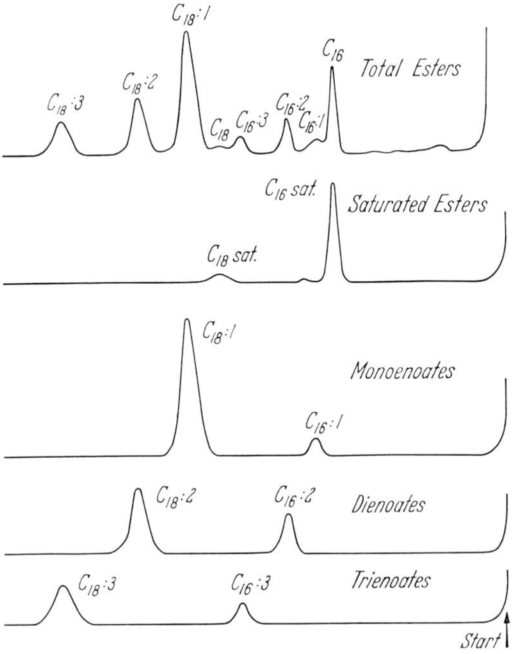

Fig. 93. Gas-chromatographic separation of methyl esters of uniform degree of unsaturation (compare with Fig. 92) [*83*]

and 0.5 ml conc. hydrochloric acid. After shaking, the adsorbent is allowed to settle, and the clear supernatant solution is decanted and filtered. The residue is treated a second time with methanol-hydrochloric acid. The combined filtrates are diluted with 25 ml of water, and extracted once with 25 ml of diethyl ether and 4 times with 10 ml of diethyl ether. The ether extracts are dried over sodium sulphate. The solvent is evaporated and the groups of lipids containing 1, 2, or 3 and more double bonds are recovered.

Saturated lipids are eluted with a mixture of petroleum hydrocarbon-diethyl ether (50+50).

d) Application and results

The separation of mercuric acetate derivatives of lipids by TLC was most often applied to mixtures of methyl esters [*83*]. This method complements gas-liquid chromatography in an ideal manner: complicated mixtures of esters of higher fatty acids, which cannot be resolved by GLC, are first separated by TLC in the form of their mercuric acetate derivatives. The groups of saturated methyl esters, and those containing 1, 2 or 3 and more double bonds, are obtained in more than 98% purity by this method. Each group is gas-chromatographed separately according to chain length. GLC of such groups of esters facilitates the identification of the components, and allows one to trace minute amounts of esters which could hardly be found if the total mixture were analyzed. Fig. 93 illustrates, schematically, the principle of the above-described method.

The combined application of TLC and GLC leads to more reliable quantitative analyses than could ever be achieved by GLC alone.

4. Separation of *cis-trans* isomers

MORRIS, HOLMAN and FONTELL [*93*] have shown that, in a few cases, it is possible to fractionate the esters of *cis* and *trans* isomeric fatty acids by TLC on Silica Gel G. A general method for the separation of steroisomers was developed very recently. Unsaturated components undergo complexing with silver ions. These complexes show differences in solubility.

In 1952, NICHOLS [*160*] pointed out that it should be possible to fractionate methyl oleate (*cis*-octadecenoate) and methyl elaidate (*trans*-octadecenoate) by partitioning between two immiscible solvents containing silver nitrate. DUTTON, SCHOLFIELD and JONES [*151*] performed this separation a few years later by counter-current distribution.

DE VRIES [*165*] was the first to show that *cis-trans* isomers can be resolved on silicic acid impregnated with silver nitrate. He fractionated mixtures of methyl stearate, methyl elaidate, and methyl oleate, as well as mixtures of elaidodipalmitin and oleodipalmitin, on silicic acid columns impregnated with silver nitrate. Furthermore, he accomplished the fractionation of methyl esters of acids of different degrees of unsaturation, e.g., methyl oleate, methyl linoleate, and methyl linolenate, as silver ion complexes. He achieved a sharp resolution of tristearin, oleodipalmitin, stearodiolein, and triolein.

At the same time, MORRIS [*159*] applied this principle to the thin-layer fractionation of the methyl esters of mono- and polyunsaturated acids. He also separated olefinic unsaturated esters from those containing a triple bond. The *cis-trans* isomers, methyl oleate and methyl elaidate, and the corresponding epoxy and hydroxy esters, are well separated on Silica Gel G layers impregnated with silver nitrate. The solvent mixture was petroleum hydrocarbon, B.P. 60–70° C, -diethyl-ether (60+40) and 2′,7′-dichlorofluorescein was used as the visualizing agent.

MORRIS [*159*] demonstrated the separation of the *threo-* and *erythro* forms of saturated dihydroxy esters, in the form of their borate complexes, on Silica Gel G impregnated with aqueous boric acid solution. The same investigator applied silver nitrate and boric acid on the same plate coated with Silica Gel G for fractionating saturated and unsaturated *threo-* and *erythro*-dihydroxy esters. A mixture of petroleum hydrocarbon, (B.P. 60–70° C)-diethyl ether (60+40) served as solvent, 2′,7′-dichlorofluorescein (Reagent No. 42) was used as indicator.

BARRETT, DALLAS and PADLEY [*149*] reported the fractionation of synthetic mixtures of triglycerides, and of lard, cocoa butter, cottonseed oil, and other fats and oils by TLC on Silica Gel G impregnated with silver nitrate.

5. Discussion

In the field of lipid analysis, TLC is mostly used as an adsorption method for the separation of complicated mixtures into classes of compounds. By means of this elegant and efficient method, one person can easily fractionate ten times more samples in one day, than could possibly be fractionated by column chromatography in an entire week. The resolutions achieved by TLC are much sharper than those obtained on columns. For the quantitative evaluation of the fractionated lipids, a number of procedures are known [*84, 108, 109, 139, 140*, see also pp. 45—47]. The individual classes of lipids can be eluted and isolated. It is not dificult to hydrolyze these lipids to obtain the constituent fatty acids, which can be converted to methyl esters and quantitatively analyzed by GLC [*76, 77, 82, 129*].

Adsorption-TLC is complemented by partition-TLC on water-containing deactivated layers, and by partition-TLC in reversed phase on hydrophobic layers. Partition-TLC, on deactivated plates, allows the separation of strongly polar lipids (see p. 160). By reversed-phase partition-TLC, lipids of a certain class can be fractionated (see p. 169).

The procedure of fractionating lipids according to degree of unsaturation by TLC of their mercuric acetate adducts complements the various methods. By means of this last-mentioned technique and on layers containing silver nitrate, it is possible to resolve mixtures of *cis* and *trans* isomers (see pp. 176, 178).

The great advantages of chromatography on thin layers, i.e. simplicity, speed, excellent resolution, sensitivity of detection of the separated substances, and high capacity of the plates, are fully utilized only if the various kinds of TLC are applied in combination with each other. In particular, adsorption-TLC and reversed phase partition-TLC should be applied successively; mixtures of lipids which cannot be separated by adsorption usually can be resolved by partition in reversed phase, and *vice versa*.

TLC of lipids on ion exchangers has not been described yet. Ion exchange-TLC is probably well suited for the fractionation of strongly polar lipids. Thin-layer ionophoresis and thin-layer ionophoresis-TLC might also be suitable for separations of similar types of compounds (see p. 433). Chromatography on thin layers of urea and other substances forming inclusion complexes is promising. It should be possible, by TLC on layers of urea, to separate straight-chain lipids from multiple branched-chain lipids, and, possibly, saturated lipids from unsaturated ones.

Only a substance, which is still uniform after the application of *various* principles of chromatography, can be considered to be "chromatographically pure". Page 65 shows the autoradiograph of a thin-layer chromatogram of radioactively-labelled fatty acids. The purity of each of these fatty acids was "proven" by GLC, by the manufacturers. Fig. 44 shows that each of the commercial products is a mixture. Almost all contaminants, intermediates in the syntheses of these acids, migrate less than the fatty acids. These contaminants will stay on the column in GLC and, therefore, are not registered by the detector.

The fact that in TLC the separation path lies open constitutes one of the main advantages of this method. In contrast to chromatography on columns, no substance can escape observation unless the material is volatile. Mixtures of volatile compounds, even gases, can be fractionated by TLC in the form of derivatives (see p. 189).

MAHADEVAN [*75*] investigated whether unsaturated lipids undergo oxidation during TLC, but even the most sensitive chemical detection methods could not show evidence of autoxidation of highly unsaturated fatty acids and their cholesteryl esters. MAHADEVAN found TLC to be an excellent technique to purify autoxidized lipids.

Two-dimensional TLC is another proof that no autoxidation occurs under normal conditions during the time of separation. If highly unsaturated lipids are chromatographed in two dimensions in the same solvent system, the resolved substances line up diagonally [*84*]. If the plate, after chromatography in one direction, is not developed in the second direction within 10 min, polar oxidation products, which migrate only a little, may be observed [*129*].

If TLC is used as a preparative tool, it is important that the spots are scraped off the plate while the layer is still wet. The scrapings should be transferred directly into the eluting solvent. Autoxidation occurs rapidly if the chromatogram is not protected by a solvent film. For this reason, unsaturated lipids are much more stable on hydrophobic layers than on dry adsorbent layers.

Appropriate solvents for elution of lipids are found by chromatographing the substances in various polar solvent systems. Solvents carrying a compound to the solvent front are suitable for elution (see p. 52). Diethyl ether and mixtures of diethyl ether and 10 to 20% methanol are usually satisfactory to elute neutral lipids. Very polar lipids tend to form complexes with the calcium ions of the gypsum used as a binder. Nevertheless, it is possible to elute them quantitatively [*111*]. It is often advisable to chromatograph strongly polar lipids in the form of less polar derivatives, which are eluted without difficulty [*80, 84*].

In GLC of methyl esters of higher fatty acids, substances are often recorded which are interpreted, based on retention time exclusively, as esters of branched-chain fatty acids. Actually, these substances are either aldehydes occurring in the animal body as such or as aldehydogenic lipids which are associated with ester lipids, or are dimethyl acetals which were formed from aldehydes during the esterification of the fatty acids with hydrochloric acid-methanol. The presence of such contaminants in methyl esters is easily proved by adsorption-TLC. The pure methyl esters can be isolated by the same method prior to GLC analysis [*129*]. In the opinion of the author, pre-purification by TLC of the material to be analyzed by GLC is an absolute necessity. Only adsorption chromatography guarantees that only *one* class of compound is subjected to GLC.

TLC is also an outstanding tool for the analysis and purification of radioactively-labelled lipids. Some typical applications are discussed on pp. 65—66.

Separations by TLC of alicyclic compounds occurring as minor constituents in fats are reported in several sections of this book. Attention is drawn to the contributions on steroids (p. 249), carotenoids (p. 214) fatsoluble vitamins (p. 186), and terpenes and resins (p. 210).

Bibliography to Chapter A: Aliphatic Lipids

[*1*] Ahrens, E., H., jr., and L. C. Craig: J. Biol. Chem. **195**, 299 (1952).
[*2*] Applewhite, T. H., M. J. Diamond and L. A. Goldblatt: J. Am. Oil Chemists' Soc. **38**, 609 (1961).
[*3*] Arcus, A. C., and G. G. Dunckley: J. Chromatog. **5**, 272 (1961).
[*4*] Bloor, W. R.: J. Biol. Chem. **77**, 53 (1928).
[*5*] de Boer, Th, J., and H. J. Backer: In Organic Syntheses, Vol. 34, p. 96, W. S. Johnson, Editor. New York: John Wiley & Sons, Inc.; London: Chapman & Hall, Ltd., 1954.
[*6*] Brockmann, H., u. F. Volpers: Chem. Ber. **82**, 95 (1949).
[*7*] Buchanan, M. A.: Anal. Chem. **31**, 1616 (1959).
[*8*] Carroll, K. K.: J. Lipid Res. **2**, 135 (1961).
[*9*] Carter, H. E., R. H. McCluer and E. D. Slifer: J. Am. Chem. Soc. **78**, 3735 (1956).
[*10*] Chakrabarty, M.: Private communication, 1961.
[*11*] Chalvardjian, A., L. J. Morris and R. T. Holman: J. Nutrition **76**, 52 (1962).
[*12*] — Private communication, 1961.
[*13*] Chatt, J.: Chem. Revs. **48**, 7 (1951).
[*14*] Craig, L. C., and D. Craig: In Technique of Organic Chemistry, Vol. III, p. 171, A. Weissberger, Editor. New York: Interscience Publishers Inc. 1950.
[*15*] Dain, J. A., H. Weicker, G. Schmidt and S. J. Thannhauser: In Symposium on Sphingolipidoses and Allied Diseases p. 289, Academic Press, New York 1961.
[*16*] Davidek, J., and E. Davidekova: Pharmazie **16**, 352 (1951).
[*17*] — u. J. Pokorny: Z. Lebensm.-Untersuch. u. -Forsch. **115**, 113 (1961).
[*18*] Dhopeshwarkar, G. A., and J. F. Mead: J. Am. Oil Chemists' Soc. **38**, 297 (1961).
[*18a*] — — J. Lipid Res. **3**, 238 (1962).
[*18b*] — — Proc. Soc. Exptl. Biol. Med. **109**, 425 (1962).
[*19*] Entenman, C.: J. Am. Oil Chemists' Soc. **38**. 534 (1961).
[*20*] Fillerup, D. L., and J. F. Mead: Proc. Soc. Exptl. Biol. Med. **38**, 574 (1953).
[*21*] Folch, J., I. Ascoli, M. Lees, J. A. Meath and F. N. le Baron: J. Biol. Chem. **191**, 833 (1951).

[22] FOLCH, J., M. LEES and G. H. SLOAN-STANLEY: J. Biol. Chem. **226**, 497 (1957).
[23] FONTELL, K., R. T. HOLMAN and G. LAMBERTSEN: J. Lipid Res. **1**, 391 (1960).
[24] DE FREITAS, A.: Thesis, Univ. of Minnesota, 1962.
[25] FULCO, A. J., and J. F. MEAD: J. Biol. Chem. **236**, 2416 (1961).
[26] GAUGLITZ, E. J., jr., and D. C. MALINS: J. Am. Oil Chemists' Soc. **37**, 425 (1960).
[27] GOLDFINE, H., and K. BLOCH: J. Biol. Chem. **236**, 2596 (1961).
[28] GREEN, T., F. O. HOWITT and R. PRESTON: Chem. & Ind. (London) **1955**, 591.
[29] GRUGER, E. H., jr., D. C. MALINS and E. J. GAUGLITZ, jr.: J. Am. Oil Chemists' Soc. **37**, 214 (1960).
[30] HABERMANN, E., G. BANDTLOW u. B. KRUSCHE: Klin. Wschr. **39**, 816 (1961).
[31] HAMILTON, J. G., and J. E. MULDREY: J. Am. Oil Chemists' Soc. **38**, 582 (1961).
[32] HANAHAN, D. J.: Lipide Chemistry. New York-London: John Wiley & Sons, Inc., 1960.
[33] HILDITCH, T. P.: The Chemical Constitution of Natural Fats, 3rd Edition. London: Chapman & Hall, Ltd., 1956.
[34] HIRSCH, J., and E. H. AHRENS, jr.: J. Biol. Chem. **233**, 311 (1958).
[35] — Federation Proc. **20**, 269 (1961).
[36] HOFMANN, A. F.: J. Lipid Res. **3**, 391 (1962).
[37] HUHNSTOCK, K., u. H. WEICKER: Klin. Wschr. **38**, 1249 (1960).
[38] INOUYE, Y., M. NODA and O. HIRAYAMA: J. Am. Oil Chemists' Soc. **32**, 132 (1955).
[39] — — Arch. Biochem. Biophys. **76**, 271 (1958).
[40] — — J. Agr. Chem. Soc. Japan **33**, 452 (1959).
[41] JANISTYN, H.: Parfümerie u. Kosmetik **41**, 371 (1960).
[42] JANTZEN, E., u. H. ANDREAS: Angew. Chem. **70**, 656 (1958).
[43] — — Chem. Ber. **92**, 1427 (1959).
[44] — — Chem. Ber. **94**, 628 (1961).
[45] — — K. MORGENSTERN u. W. ROTH: Fette, Seifen, Anstrichmittel **63**, 685 (1961).
[46] JATZKEWITZ, H.: Hoppe-Seyler's Z. physiol. Chem. **320**, 134 (1960).
[47] — u. E. MEHL: Hoppe-Seyler's Z. physiol. Chem. **320**, 251 (1960).
[48] — Hoppe-Seyler's Z. physiol. Chem. **326**, 61 (1961).
[49] JÜBERMANN, O.: In Methoden der organischen Chemie, Bd. I/1, S. 223. Herausgegeben von E. MÜLLER, 4. Aufl. Stuttgart: Georg Thieme Verlag 1958.
[50] KAMMERECK, R.: Unpublished.
[51] KATES, M.: Can. J. Biochem. and Physiol. **35**, 127 (1957).
[52] KATHEN, H.: Arch. Mikrobiol. **14**, 602 (1949).
[53] KAUFFMANN, T., u. R. KÖDDING: Z. Naturforsch. **12**, 735 (1957).
[54] KAUFMANN, H. P., u. W. H. NITSCH: Fette, Seifen, Anstrichmittel **56**, 154 (1954).
[55] — u. J. POLLERBERG: Fette, Seifen, Anstrichmittel **59**, 815 (1957).
[56] — (Herausgeber). Analyse der Fette und Fettprodukte. Berlin-Göttingen-Heidelberg: Springer-Verlag 1958.
[57] — u. E. MOHR: Fette, Seifen, Anstrichmittel **60**, 165 (1958).
[58] — u. Z. MAKUS: Fette, Seifen, Anstrichmittel **62**, 1014 (1960).
[59] — — u. F. DEICKE: Fette, Seifen, Anstrichmittel **63**, 235 (1961).
[60] — — u. T. H. KHOE: Fette, Seifen, Anstrichmittel **63**, 689 (1961).
[61] KAUFMANN, H. P., Z. MAKUS u. B. DAS: Fette, Seifen, Anstrichmittel **63**, 807 (1961).
[62] — u. Y. SU KO: Fette, Seifen, Anstrichmittel **63**, 828 (1961).
[63] KLENK, E., u. W. GIELEN: Hoppe-Seyler's Z. physiol. Chem. **323**, 126 (1961).
[64] — — Hoppe-Seyler's Z. physiol. Chem. **326**, 144 (1961).
[65] KOCHETKOV, N. K., I. G. ZHUKOVA i I. S. GLUKHODED: Akad. Doklady Nauk. Sci. S. S. S. R. **139**, 605 (1961).
[66] KUHN, R.: Naturwiss. **46**, 43 (1959).
[67] — H. WIEGANDT u. H. EGGE: Angew. Chem. **73**, 580 (1961).
[68] KURZ, H.: Fette und Seifen **44**, 144 (1937).

[*69*] LAWSON, D. D., and H. R. GETZ: Chem. & Ind. (London) **1961**, 1404.
[*70*] LEA, C. H., and D. N. RHODES: Biochem. J. **54**, 467 (1953).
[*71*] — — and R. D. STOLL: Biochem. J. **60**, 353 (1955).
[*72*] LE BARON, F. N., and E. E. ROTHLEDER: In Biochemistry of Lipids, p. 1, G. POPJÁK, Editor. New York-Oxford-London-Paris: Pergamon Press 1960.
[*73*] LEES, M., J. FOLCH, G. H. SLOAN-STANLEY and S. CARR: J. Neurochem. **4**, 9 (1959).
[*74*] MAHADEVAN, V., and W. O. LUNDBERG: J. Lipid Res. **3**, 106 (1962).
[*75*] — Private communication, 1961.
[*76*] MALINS, D. C., and H. K. MANGOLD: J. Am. Oil Chemists' Soc. **37**, 576 (1960).
[*77*] — Chem. & Ind. (London) **1960**, 1359.
[*78*] MANGOLD, H. K., B. G. LAMP and H. SCHLENK: J. Am. Chem. Soc. **77**, 6070 (1955).
[*79*] — J. L. GELLERMAN and H. SCHLENK: Federation Proc. **17**, 268 (1958).
[*80*] — Fette, Seifen, Anstrichmittel **61**, 877 (1959).
[*81*] — and D. C. MALINS: J. Am. Oil Chemists' Soc. **37**, 383 (1960).
[*82*] — and N. TUNA: Federation Proc. **20**, 268 (1961).
[*83*] — and R. KAMMERECK: Chem. & Ind. (London) **1961**, 1032.
[*84*] — R. KAMMERECK and D. C. MALINS: *Proceedings — 1961 International Symposium on Microchemical Techniques.* Microchem. J., Symposium Vol. II, (697) 1962.
[*85*] — and KAMMERECK R.: J. Am. Oil Chemists' Soc. **39**, 201 (1962).
[*86*] — J. Am. Oil Chemists' Soc. **38**, 708 (1961).
[*87*] MARINETTI, G. V., J. ERBLAND and J. KOCHEN: Federation Proc. **16**, 837 (1957).
[*88*] MEAD, J. F., and D. L. FILLERUP: Proc. Soc. Exptl. Biol. Med. **86**, 449 (1954).
[*89*] — — J. Biol. Chem. **227**, 1009 (1957).
[*90*] MECHAM, D. K., and A. MOHAMMAD: Cereal Chem. **32**, 405 (1955).
[*91*] METCALFE, L. D., and A. A. SCHMITZ: Anal. Chem. **33**, 363 (1961).
[*92*] MORRIS, L. J., H. HAYES and R. T. HOLMAN: J. Am. Oil Chemists' Soc. **38**, 316 (1961).
[*93*] — R. T. HOLMAN and K. FONTELL: J. Am. Oil Chemists' Soc. **37**, 323 (1960).
[*94*] — — — J. Lipid Res. **1**, 412 (1960).
[*95*] — — — J. Lipid Res. **2**, 68 (1961).
[*96*] — In Chromatography p. 428. E. HEFTMANN, Editor. New York: Reinhold Publishers 1961.
[*97*] — and H. K. MANGOLD: Unpublished.
[*98*] — Private communication, 1961.
[*99*] NELSON, J.: Thesis, Univ. of Minnesota, 1961.
[*100*] NODA, M.: J. Agr. Chem. Soc. Japan **33**, 417 (1959).
[*101*] — and O. HIRAYAMA: Yukagaku **10**, 24 (1961).
[*102*] PATEL, N. G., M. H. HAYDAK and R. LOVELL: Nature London **191**, 362 (1961).
[*103*] PEIFER, J. J.: Mikrochim. Acta **1962**, 529.
[*104*] — R. A. MUESING jr., and F. W. JANSSEN: J. Am. Oil Chemists' Soc. **39**, 292 (1963).
[*105*] POLONOVSKI, J. and L. DOUSTE-BLAZY: In Proceedings of the 2nd International Conference on Biochemical Problems of Lipids, p. 64. G. POPJÁK und E. LE BRETON, Editors. New York, London: Interscience Publishers Inc., and Butterworth Scientific Publications 1956.
[*106*] POTTER, V. R., and C. A. ELVEHJEM: J. Biol. Chem. **114**, 495 (1936).
[*107*] PRIVETT, O. S.: In Proceedings of the Flavor Chemistry Symposium, p. 147, Camden, New Jersey: Campbell Soup Company 1961.
[*108*] — and M. L. BLANK: J. Lipid Res. **2**, 37 (1961).
[*109*] — M. L. BLANK and W. O. LUNDBERG: J. Am. Oil Chemists' Soc. **38**, 312 (1961).
[*110*] PRIVETT, O. S., and L. M. BLANK: J. Am. Oil. Chemists' Soc. **39**, 465 (1962).
[*111*] — — Private communication, 1961).
[*112*] — and CH. NICKELL: J. Am. Oil Chemists' Soc. **39**, 414 (1962).
[*113*] RADIN, N. S., A. K. HAJRA and Y. AKAHORI: J. Lipid Res. **1**, 250 (1960).

[*114*] ROPER, R., and T. S. MA: Microchem. J. **1**, 245 (1957).
[*115*] RUMPF, H., TH. LANGE u. W. STROH: In Methoden der organischen Chemie, Bd. I/2, S. 3. Herausgegeben von E. MÜLLER, 4. Aufl. Stuttgart: Georg Thieme 1959.
[*116*] SCHLEMMER, W.: Boll. Soc. ital. Biol. Sperm. **37**, 134 (1961).
[*117*] SCHLENK, H., J. L. GELLERMAN, J. A. TILLOTSON and H. K. MANGOLD: J. Am. Oil Chemists' Soc. **34**, 377 (1957).
[*118*] — — J. Am. Oil. Chemists' Soc. **38**, 555 (1961).
[*119*] SCHMID, M. E., u. H. R. BOLLIGER: Helv. Chim. Acta **37**, 884 (1954).
[*120*] SCHROEDER, W. A.: Ann. N. Y. Acad. Sci. **49**, 204 (1948).
[*121*] SPERRY, W. M.: In Methods of Biochemical Analysis, Vol. II, p. 83. D. GLICK, Editor. New York: Interscience Publishers Inc. 1955.
[*122*] STAHL, E.: Chemiker Ztg. **82**, 323 (1958).
[*123*] — Parfümerie u. Kosmetik **39**, 564 (1958).
[*124*] — Pharm. Rdsch. **1**, Nr. 2 (1959).
[*125*] STEIN, Y., and O. STEIN: Biochim. Biophys. Acta **54**, 555 (1962).
[*126*] STOFFEL, W., F. CHU u. E. H. AHRENS jr.: Anal. Chem. **31**, 307 (1959).
[*127*] STRICKLAND, E. H., and A. A. BENSON: Arch. Biochem. Biophys. 88, 344 (1960).
[*128*] TRAPPE, W.: Biochem. Z. **305**, 150 (1940).
[*129*] TUNA, N., R. KAMMERECK and H. K. MANGOLD: Unpublished.
[*130*] VIOQUE, E.: Grasas y Aceites (Seville, Spain) **11**, 223 (1960).
[*131*] — Private communication, 1961.
[*132*] — L. J. MORRIS and R. T. HOLMAN: J. Am. Oil Chemists' Soc. **38**, 489 (1961).
[*133*] — and R. T. HOLMAN: Anal. Chem. **33**, 1444 (1961).
[*134*] — — J. Am. Oil Chemists' Soc. **39**, 63 (1962).
[*135*] WAGNER, H.: Fette, Seifen, Anstrichmittel **62**, 1115 (1960).
[*136*] — L. HÖRHAMMER u. P. WOLFF: Biochem. Z. **334**, 175 (1961).
[*137*] WEICKER, H.: Klin. Wschr. **37**, 763 (1959).
[*138*] — J. A. DAIN, G. SCHMIDT and S. J. THANNHAUSER: Federation Proc. **19**, 219 (1960).
[*139*] WESTLEY, J., J. J. WREN and H. K. MITCHELL: J. Biol. Chem. **229**, 131 (1957).
[*140*] WILLIAMS, J. A., A. SHARMA, L. J. MORRIS and R. T. HOLMAN: Proc. Soc. Exptl. Biol. Med. **105**, 192 (1960).
[*141*] WINTERSTEIN, A., u. B. HEGEDÜS: Hoppe-Seyler's Z. physiol. Chem. **321**, 97 (1960).
[*142*] — A. STUDER u. R. RÜEGG: Chem. Ber. **93**, 2951 (1960).
[*143*] WISS, O., u. U. GLOOR: Hoppe-Seyler's Z. physiol. Chem. **310**, 260 (1958).
[*144*] WREN, J. J., and H. K. MITCHELL: J. Biol. Chem. **234**, 2823 (1959).
[*145*] — J. Chromatog. **4**, 173 (1960).
[*146*] ZÖLLNER, N., K. KIRSCH u. G. AMIN: Verh. dtsch. Ges. inn. Med. **66**, 677 (1960).
[*147*] — u. G. WOLFRAM: Klin. Wschr. **39**, 817 (1961).
[*148*] — — u. G. AMIN: Klin. Wschr. **40**, 273 (1962).
[*149*] BARRETT, C. B., M. S. J. DALLAS and F. B. PADLEY: Chem. & Ind. (London) **1962**, 1050.
[*150*] DEENEN, L. L. M. VAN, J. DE GIER u. G. H. DE HAAS: Koninkl. Ned. Akad. Wetenschap. Proc. Ser. B **64**, 528 (1961).
[*151*] DUTTON, H. J., C. R. SCHOLFIELD and E. P. JONES: Chem. & Ind. (London) **1961**, 1874.
[*152*] HAAS, G. H. DE, et L. L. M. VAN DEENEN: Rec. trav. chim. **80**, 951 (1961).
[*153*] — — Koninkl. Ned. Akad. Wetenschap., Proc. Ser. C **64**, 592 (1961).
[*154*] HONEGGER, C. G.: Helv. Chim. Acta **45**, 281 (1962).
[*155*] JENSEN, R. G., and J. SAMPUGNA: J. Dairy Sci. **45**, 435 (1962).
[*156*] KAUFMANN, H. P., u. B. DAS: Fette, Seifen, Anstrichmittel **64**, 214 (1962).
[*157*] —, Z. MAKUS u. T. H. KHOE: Fette, Seifen, Anstrichmittel **64**, 1 (1962).
[*158*] KAUFMANN, H. P. u. T. H. KHOE: Fette, Seifen, Anstrichmittel **64**, 18 (1962).
[*159*] MORRIS, L. J.: Chem. & Ind. (London) **1962**, 1238.
[*160*] NICHOLS, P. L. jr.: J. Am. Chem. Soc. **74**, 1091 (1952).
[*161*] RIECHE, A., M. SCHULZ, H. E. SEYFARTH u. G. GOTTSCHALK: Fette, Seifen, Anstrichmittel **64**, 198 (1962).

[162] RYBICKA, S. M.: Chem. & Ind. (London) **1962**, 308.
[163] SKIPSKI, V. P., R. P. PETERSON and M. BARCLAY: J. Lipid Res. **3**, 467 (1962)
[164] VIOQUE, E., and R. T. HOLMAN: Arch. Biochem. Biophys. **99**, 522 (1962).
[165] VRIES, B. DE: Chem. & Ind. (London) **1962**, 1049.
[166] WAGNER, H.: Fette, Seifen, Anstrichmittel **63**, 1119 (1961).

Some recent noteworthy articles

Books and Reviews

BOEKENOOGEN, H.: Editor, Analysis and Characterization of Oils, Fats and Fat Products, Vol. 1, John Wiley & Sons, London, 1964.

ENTENMAN, C.: Methods of Analysis of Lipids and Lipid Constituents, Academic Press, Inc., New York and London, 1964.

HOFMANN, A. F.: In Biochemical Problems of Lipids, A. C. FRAZER, Editor, Elsevier Publishing Company, Amsterdam-London-New York, 1964. This article also contains much unpublished work.

JAMES, A. T.: Analyst 88, 572 (1963): Methods for analyzing fatty acids.

LINES, J. G.: In Biochemical Problems of Lipids, A. C. FRAZER, Editor, Elsevier Publishing Company, Amsterdam-London-New York, 1964. Quantitative TLC

MANGOLD, H. K.: J. Am. Oil Chemists' Soc. in press). A compilation of recent literature.

MORRIS, L. J., and A. T. JAMES: New Biochemical Separations, D. van Nostrand Co., Ltd., London, 1964. with contributions on TLC of lipids by HOFMANN, A. F.; TLC on hydroxyl apatite. MORRIS, L. J.; TLC on layers containing silver nitrate. NICHOLS, B. W.; TLC of plant phospholipids and glycolipids.

PADLEY, F. B.: In Thin-Layer Chromatography, G. B. MARINI-BETTOLO, Editor, Elsevier Publishing Company, Amsterdam-London-New York, 1964.

PASCAUD, M.: In Advances in Lipid Research, Vol. 1, R. PAOLETTI and D. KRITCHEVSKY, Editors, Academic Press, New York and London, 1963.

SCHLENK, H.: In Fatty Acids, Their Chemistry, Properties, Production and Uses, Second Edition, K. S. MARKLEY, Editor, Part 3, p. 2125, J. Wiley & Sons, Inc., New York, 1964.

Original Articles

ANKER, L., and D. SONANINI: Pharm. Acta Helv. **37**, 360 (1962): Reversed-phase partition TLC of fats and oils.

BARRETT, C. B., M. S. J. DALLAS and F. B. PADLEY: J. Am. Oil Chemists' Soc. **40**, 580 (1963): Quantitative TLC by densitometry.

BERGELSON, L. D., E. DYATLOVSKAYA and V. V. VORONKOVA: Izvest. Akad. Nauk. SSSR, Otdel. Khim. Nauk., **1963**, 954: TLC and GLC of isomeric monunsaturated fatty acids.

DOBIÁŠOVÁ, M., P. HAHN and O. KOLDOVSKÝ: Biochim. et Biophys. Acta **70**, 713 (1963): Changes of lipid patterns during postnatal development of the rat.

ENG, L. F., Y. L. LEE, R. B. HAYMAN and B. GERSTL: J. Lipid Res. **5**, 128 (1964): Separation of methyl esters and dimethylacetals prior to GLC.

GRÉEN, K., and B. SAMUELSSON: J. Lipid Res. **5**, 117 (1964): TLC of prostaglandins.

HORROCKS, L. A.: J. Am. Chemists' Soc. **40**, 235 (1963): TLC of phospholipids.

JATZKEWITZ, H., and K. SANDHOFF: Biochim. et Biophys. Acta **70**, 354 (1963): TLC of the lipids from normal and diseased human brains.

KAUFMANN, H. P., and B. DAS: Fette, Seifen, Anstrichmittel **64**, 214 (1962): TLC of triglycerides.

— — Fette, Seifen, Anstrichmittel **65**, 398 (1963): TLC of wax esters.

KOCHETKOV, N. K., I. G. ZHUKOVA and I. S. GLUKHODED: Proc. Acad. Sci. SSSR., Chem. Sect. **139**, 608 (1962): TLC of cerebrosides.

— — — Proc. Acad. Sci. SSSR., Chem. Sect. **147**, 376 (1962): TLC of sphingosine derivatives.

KOMAREK, R. J., R. G. JENSEN and B. W. PICKETT: J. Lipid Res., **5**, 268 (1964): Quantitative TLC.

MORRIS: L. J.: J. Lipid Res. **4**, 357 (1963): Fractionation of cholesteryl esters on layers containing silver nitrate.

— J. Chromatog. **12**, 321 (1963): TLC of isomeric polyhydroxy acids.

MÜLDNER, H. G., J. R. WHERRETT and J. N. CUMINGS: J. Neurochem. **9**, 607 (1962): TLC of brain lipids.

NICHOLS, B. W.: Biochim. et Biophys. Acta **70**, 417 (1963): TLC of photosynthetic tissues.
PURDY, S. J., and E. V. TRUTER: Analyst **87**, 802 (1962): Quantitative TLC.
RIEZEBOS, G., J. C. GRIMMELIKHUYSEN and D. A. VAN DORP: Rec. trav. chim. **82**, 1234 (1963): Isomeric ozonides.
SKIDMORE, W. D., and C. ENTENMAN: J. Lipid Res. **3**, 471 (1962): Two-dimensional TLC of phospholipids.
SKIPSKY, V. P., R. F. PETERSON, J. SANDERS and M. BARCLAY: J. Lipid Res. **4**, 227 (1963): TLC of phospholipids.
SVENNERHOLM, E., and L. SVENNERHOLM: Biochim. et Biophys. Acta **70**, 432 (1963): TLC of glycolipids from serum.
SVENNERHOLM, L.: J. Neurochem. **10**, 613 (1963): TLC of brain gangliosides.
TUNA, N., and MANGOLD, H. K.: In Evolution of the Atherosclerotic Plaque, R. J. JONES, Editor, The University of Chicago Press, Chicago and London, 1963, pp. 85—108: Human tissue lipids.
DE VRIES, B., and G. JURRIENS: Fette, Seifen, Anstrichmittel **65**, 725 (1963): TLC on layers containing silver nitrate.

B. Terpene Derivatives, Essential Oils, Balsams, and Resins

By

EGON STAHL and H. JORK

This chapter discusses possible ways of separating lipophilic mixtures of natural products, constituent parts of which can be theoretically derived from isoprene and thus correspond to Ruzicka's isoprene rule [*58*]. WALLACH [*81*] has divided the hydrocarbons and those of their derivatives considered here into the following groups:

Hemiterpenes	C_5H_8	and derivatives	steam-volatile: chief constituents of the essential oils including the phenyl propane group
Monoterpenes b.p. 140—180° C	$C_{10}H_{16}$	and derivatives	
Sesquiterpenes b.p. over 200° C	$C_{15}H_{24}$	and derivatives	
Diterpenes b.p. around 300° C	$C_{20}H_{32}$	and derivatives	little or not at all steam-volatile: constituents of balsams, resins, waxes, caoutchouc
Triterpenes	$C_{30}H_{48}$	and derivatives	
Polyterpenes[1]	$(C_{10}H_{18})_x$	and derivatives	

The important group of naturally-occurring C_9 compounds, which can formally be considered as phenyl propane derivatives, should be included under the hemiterpenes.

The numerous compounds of each group are distinguished by the type of cyclization (open-chain, mono-, bicyclic, etc.), the number and position of the double bonds, the asymmetric centers, and the type and number of functional groups. Thus, it is convenient to classify the terpenes according to functional groups, in the order of increasing polarity: hydrocarbons → esters → aldehydes and ketones → alcohols → acids.

[1] A further sub-division into tetra-, penta- and higher polyterpenes is omitted here.

The structural differences and the physical and chemical properties of the terpenes and phenylpropane derivatives occurring in essential oils are described in the standard works of GILDEMEISTER-HOFFMANN [*14*], GUENTHER [*17*], SIMONSEN [*60*], W. KARRER [*31*] and MORITZ [*42*]. HAAGEN-SMIT [*20*] deals with sesquiterpenes and diterpenes. STEINER and HOLTZEM [*72*] and also JONES and HALSALL [*29*] have published comprehensive treatises on triterpenes.

I. Separation of lipophilic, steam-volatile mixtures

In many cases it is necessary to concentrate the terpenes and derivatives from a reaction mixture, or from some plant or animal source before they are further fractionated. As already mentioned, the C_5 to C_{15} compounds considered here are steam-volatile, and may be isolated using appropriate apparatus [*18, 42a*]. For quantitative isolation of small amounts of such mixtures (1 to 1000 mg) the "Karlsruhe" apparatus (Fig. 94) has proved most valuable. It enables the lipophilic part of the steam-distillate to be obtained free from water in a microflask, for subsequent weighing and further use. In addition, it is possible – although of course less accurate – to measure it volumetrically. The original paper should be consulted for details [*69*].

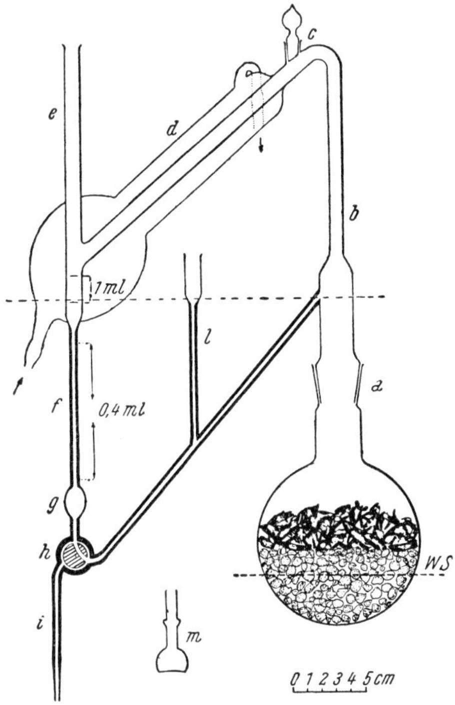

Fig. 94. "Karlsruhe" apparatus for isolation of small amounts of steam-volatile lipids. *a* Standard joint B 29, *b* ascending tube, *c* inlet for washing, *d* condenser, *e* pressure compensator with 1 ml. graduation, *f* graduated capillary with divisions of $^1/_{200}$, *g* bulb, *h* three-way stopcock, *i* outlet, *l* filling-tube, *m* microflask for gravimetric estimation of collected lipids, *WS* water level in flask [*69*]

II. Chromatographic separation of lipophilic, steam-volatile mixtures

RIGANESIS [*55*] has compiled an extensive summary of work on the column chromatography of terpene mixtures. Alumina, and occasionally silicic acid, have been the most widely used adsorbents. A as rule, separation of such mixtures by adsorption chromatography is likely to be successful only if the individual components havewidely differing polarities. An approximate criterion is the dielectric constant; this is of limited use however, since dielectric constants have been determined for only a few compounds of this class, and much of the earlier data should be considered with caution. The dielectric constants of seven monoterpene hydrocarbons investigated lie between 2.24 and 2.76, and it is not unexpected that this group can be separated from the other more polar groups without difficulty. Consequently, the dielectric constants are given in the special section, as far as they are known. The results of a large number of experiments (TLC on Silica Gel G) show that many compounds within a "polarity-group"[1] cannot be separated by adsorption chromatography. Thus, nearly all the terpene and sesquiterpene alcohols have the same *Rf* value when benzene or chloroform is used as the mobile phase. By using mixtures of solvents it is possible, in some cases, to separate various members of one group. These difficulties do not apply to the analysis of simple mixtures, containing only one member of the above-mentioned groups, but do arise with many natural mixtures. However, it is possible to overcome these difficulties by using different techniques of chromatographic separation. Relevant experiments have shown, that under certain conditions, "polarity-groups" which cannot be separated by adsorption chromatography can be resolved by gas-chromatography [*68*]. If amounts of about 1 g of mixture are available, they are separated into fractions ("polarity groups"[1]) on columns of silica gel or alumina, whichever is more suitable, and these fractions are then examined by gas-chromatography. With both methods, the fractions are controlled by thin-layer chromatography. To assign compounds to their groups, not only *Rf* values but also color-reactions are useful. MILLER and KIRCHNER have shown that reactions for identifying functional groups, such as dehydration, oxidation, reduction and saponification, can be carried out directly on the chromatographic plates.

One should however, always attempt to identify the compounds by established micro-methods such as derivative formation, boiling-point, optical rotation, and UV and IR spectra.

1. Mono- and sesquiterpene hydrocarbons

As already mentioned, dielectric constants give a certain guide to the adsorption affinity of the compounds to be separated; for the monoterpenes theylie between 2.24 and 2.76. Table 15 (taken from MILLER and KIRCHNER [*39*]) shows that separation can be expected only with solvents that have even lower dielectric constants ($\varepsilon = 1.844-2.023$). Moreover, highly active layers of silica gel are necessary. After the usual drying they must be stored in a vacuum dessicator at 3 mm Hg over phosphorus pentoxide. Since MILLER and KIRCHNER worked with narrow strips and thick layers, the danger of inactivation by atmospheric moisture during spotting was not as great as with the large plates with thinner layers currently in favor. Attention should be drawn, in parti-

[1] Compounds with the same kind and number of functional groups are considered here together under the expression "Polarity groups".

cular to the possibility of decomposition or rearrangement of these substances on the highly activated layers.

Table 15. *hRf-Values (= Rf × 100) of some Mono- and Sesquiterpenes, according to* MILLER *and* KIRCHNER [*39*]

Monoterpenes	*Rf* values × 100			
	I[1]	II[1]	III[1]	IV[1]
p-Cymene (ε = 2.24)	38	41	62	69
Limonene (ε = 2.3)	41	55	54	59
Terpinolene	64	67	60	65
β-Pinene	80	75	80	85
α-Pinene (ε = 2.7)	83	85	84	90
Camphene (ε = 2.33)	84	76	79	80
Sesquiterpenes				
α-Caryophyllene	50	35	47	33
β-Caryophyllene	62	60	52	65
γ-Caryophyllene	80	82	87	90
Cedrene	82	80	83	85

[1] Solvent I: Hexane; II: 2,2-Dimethyl butane; III: Cyclohexane; IV: Methylcyclohexane.

These *hRf*-values should be treated with caution, since at the time they were determined little was known about the factors influencing them. A new determination is required. Commercial azulene (Guaiazulene, Dragoco, Holzminden/Weser, Germany) is suitable as color standard.

If more strongly polar solvents are used, such as benzene (ε = 2.284) or chloroform (ε = 4.806) the mono- and sesquiterpenes appear together in a higher region of the chromatogram (*hRf* 80–100). As previous work has shown, it is preferable to analyze mixtures of this group by gas chromatography [*68*].

Visual **identification** is possible using conc. H_2SO_4, if necessary with the addition of aldehyde; or alternatively, the fluorescein-bromine test (Reagent No. 101 A) or antimony pentachloride (Reagent No. 13).

Low molecular-weight, volatile olefins. A study by PREY, BERGER and BERBALK [*49*] on the separation of unsaturated low molecular-weight olefins is relevant to the terpene hydrocarbons. Thin-layer chromatography of mercuric acetate addition compounds (also referred to in the chapter on lipids, p. 174) can possibly be applied successfully to terpene derivatives.

Since the first members of the homologous series of alkenes, up to butylene, are gases, and higher homologs are readily volatile, they are chromatographed as their involatile mercury addition compounds. Most stable are the mercuric acetate derivatives.

Preparation. Gaseous olefins are passed into a 1% methanolic solution of mercuric acetate, while liquids are added directly. Reaction is quantitative and almost instantaneous. This methanolic solution can be used directly for chromatography. 5—10 μg of addition compound should be applied.

Separation. The olefin derivatives are separated on a layer of Silica Gel G prepared in the standard way (p. 7), using the solvent system n-propanol-triethylamine-water (50 + 25 + 25). The following *hRf*-values are given: Ethylene 7; propylene 13; butylene-(1) 17; amylene-(2) 22; amylene-(1) 29; hexene-(1) 31 (90 min run).

Detection. The chromatogram is sprayed with a 2% alcoholic diphenyl carbazone solution, and then it is heated to 80° C for a short time in a drying-oven. The olefin derivatives are blue-violet in color.

2. Oxides and peroxides

Oxides and peroxides can occur in many essential oils. The two best-known compounds are 1,8-cineol (eucalyptol) and the peroxide ascaridole, which is readily accessible by a photochemical reaction from α-terpinene. Ascaridole is the chief constituent of poisonous wormseed oil. These substances cannot be separated on a layer of Silica Gel G using benzene (*hRf* 13 and 15), but they are easily resolved with chloroform; the *hRf* values are: ascaridole 63; 1,8-cineol 54. The antimony chloride reagent gives a grey color. For the detection of peroxides on the plate, the potassium iodide-acetic acid-starch test (Reagent No. 85) has usually proved better than ferrous thiocyanate [*65*].

Menthofuran is separated and detected in essential oils by chromatography on Silica Gel G layers using petroleum ether, or better, hexane, and with guaiazulene (see p. 189) run alongside as a standard [*64*]. Both substances appear in the same region. For visual identification the EP-reagent (No. 47) can be used, as for the azulene.

3. Esters of terpene alcohols

Essential oils often contain esters of terpene alcohols; most common are the acetates[1], but formates[1], propionates[1], butyrates[1], valerates[1], and higher acid esters occur occasionally.

The results of experiments with more than 20 different esters show that, using benzene, chloroform and similar solvents on Silica Gel G layers, the *Rf*-values for esters are significantly higher than those of the corresponding alcohols. The greatest differences in *Rf*-values are observed using the chromatography tank without special saturation (NS). With saturation (CS), the *Rf*-values of esters in chloroform are some 2.5 times higher than those of the alcohols. Interestingly, the acetates[1] migrate slower than the corresponding formates, propionates and butyrates, and thus, can be readily separated from them.

Table 16. *Separation of Terpene Alcohols on Silica Gel G Layers*

Esters of terpene alcohols	$Rf \times 100$	
	Benzene (CS)	Chloroform (CS)
Acetates[1] (9 different)	25—33	63—65
Formate[1], *propionate*[1], *butyrate*[1] (11 different)	42—47 (linalyl formate 37)	71—74
Alcohols (see Table 20)	5—8	26—33 (linalool 38)
Butter Yellow, Sudan Red G, Indophenol (= test substances)	40; 15; 5	73; 63; 58

[1] This nomenclature is common, but incorrect. These compounds should, to be precise, be called, for example, the acetic acid ester of geraniol, instead of geranyl acetate.

It remains to be seen whether the corresponding hydroxamic acids can be separated by thin-layer chromatography.

Detection

a) Reaction. For determining whether an ester is involved, the following possibilities supplement the methods described above.

1. Saponification with alcoholic potassium hydroxide, giving the free terpene alcohol and the salt of the acid.

2. Reduction with lithium aluminum hydride. This yields the terpene alcohol and the alcohol corresponding to the acid. The reaction can be carried out directly on the thin layer [*39*]. The ester is applied at the starting point and 1 drop of a 10% ethereal solution of $LiAlH_4$ is added. On subsequent chromatography, the ester spot should no longer appear, instead, a spot is seen in the alcohol region.

3. Demole [*7*] describes the following method, which gives good, reproducible results, after some practice.

After development, the dried thin layer is sprayed with water until transparent and a filter paper freshly impregnated with hydroxylamine laid on it while still wet. It is pressed down with a glass plate and the whole heated to 35—40° C on a hot-plate. The esters evaporate and react on the paper to give potassium hydroxamate. After 15—30 min the hydroxamate is detected visually by spraying the paper with 5% ferric chloride in 0.5 N-hydrochloric acid solution. For impregnating the filter paper, a mixture of 3.5 g of hydroxylamine hydrochloride in 50 ml of water and 16 g of potassium hydroxide in 50 ml of methanol is used.

b) Color reactions: All the color reactions described for terpene alcohols (pp. 195, 196) as well as the phosphomolybdic acid reagent, can serve for the visual detection of their esters. The sensitivity of detection, and the colors are much the same. No differences have been observed in the colors with esters of formic, acetic, propionic or butyric acids.

4. Aldehydes and ketones

Besides terpene alcohols, numerous aldehydes and ketones occur in the essential oils, and are distinguished by characteristic odors. Some of them are easily accessible by synthesis and are frequently used in the perfume industry. Cinnamaldehyde, smelling of cinnamon, vanillin, smelling of vanilla, and carvone, responsible for the odor of caraway, are known well enough to deserve mention.

The studies cited in Table 24 make it apparent that aldehydes and ketones are often separated on thin layers of adsorbent as their 2,4-dinitrophenylhydrazones (2,4-DNP).

As early as 1952, Onoe [*45*] was using 60 μ-thin layers of silica gel dried at 60° C. 2.5% Aqueous polyvinyl alcohol was used as binder. Acrolein, crotonaldehyde, citral, and a series of 14 aliphatic aldehydes from C_2—C_{10} were developed using water-saturated benzene.

Dhont and de Rooy [*9*] chromatographed a series of 2,4-dinitrophenylhydrazones in the standard way on Silica Gel G layers and referred the values to butter yellow run as standard (R_B-values, see p. 196). Their solvent systems were A: benzene-petroleum ether b.p. 60–80° C (75 + 25), and the more polar mixture B: benzene-ethyl acetate (95 + 5).

Mixtures of vanillin, veratraldehyde and ethyl vanillin could be separated by the two-dimensional technique, running with solvent A in one direction and solvent B in the other (Table 17).

Table 17. *R_B-Values of 2,4-DNP's of Aromatic Aldehydes* [9]

2,4-DNPH of	R_B values[1] A	R_B values[1] B
Vanillin	0,06	0,17
Veratraldehyde	0,0	0,45
Ethyl vanillin	0,0	streaks
Salicylaldehyde	0,50	0,85
Cinnamaldehyde . . .	0,83	1,04
Benzaldehyde	1,06	1,03
α- and β-Ionones . . .	1,40	1,14
Anisaldehyde	—	0,88

[1] Silica Gel G layer. A = Benzene-petroleum hydrocarbon (75 + 25), B = Benzene-Ethyl acetate (95 + 5); saturated chamber (CS).

Fig. 95 (by the same authors) shows that the 2,4-DNPH's of aliphatic aldehydes are best separated as a function of their chain length by solvent system A.

PETROWITZ [*48*] used thicker layers of Silica Gel G, which he prepared by hand, for the separation of some hydroxyaldehydes; but he gives no details of the conditions of chromatography. The separation distance was 12 cm and the solvent used was benzene containing 5% of methanol. The following *Rf*-values were recorded: protocatechu aldehyde 0.06; syringaldehyde 0.14; p-hydroxy benzaldehyde 0.17; vanillin 0.27; veratraldehyde 0.45; o-vanillin 0.53; anisaldehyde 0.65; salicylaldehyde 0.91.

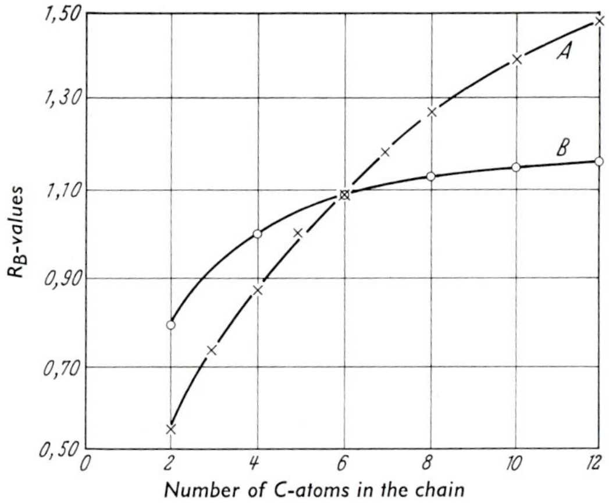

Fig. 95. Dependance of R_B-values of 2,4-dinitrophenylhydrazones of aliphatic aldehydes on chain length. Solvents A: Benzene-petroleum ether (75+25), *B* Benzene-ethyl acetate (95+5) [*9*]

The values given in Table 18 allow aldehydes and ketones to be compared under standard conditions with other "polarity groups".

The *hRf*-values of commonly occurring aldehydes and ketones run in chloroform on Silica Gel G lie between 55 and 65. The acetates of terpene alcohols also fall in this region (63—65) (Table 16). Aldehydes and ketones in essential oils can be further characterized by the SRS-technique (p. 36). A mixture is first chromatographed with chloroform in direction 1. Then the remainder of the chromatogram is covered while a 0.5% solution of 2,4-dinitrophenylhydrazine in methanol containing 1% of 25% hydrochloric acid, is applied in the shadowed region (see Fig. 149). Aldehydes and ketones react on the layer to give a yellow-orange color. After 10–15 min the chromatogram is developed with benzene in direction 2. In this way some, but not all, 2,4-dinitrophenylhydrazones can be separated. For solving particular problems the use of "reactive" layers should be considered, for example, layers containing sodium hydrogen sulphite.

Table 18. h*Rf-Values and Detection of Aldehydes and Ketones*

Substances	*Rf*-values × 100		Color reactions (5 μg) with	
	Benzene[1]	Chloroform[1]	2,4-DNPH	o-Dianisidine
Vanillin	5	29	orange	yellow-orange
Asarylaldehyde	6	39	orange brown	orange
Furfural	10	50	orange	violet
Citral	10	56	yellow	yellow-brown
Anisaldehyde	13	55	orange	yellow
Piperonal	14	55	orange	—
Cinnamaldehyde	15	60	orange	orange-brown
Benzaldehyde	25	59	yellow	—
Cuminic aldehyde	25	65	orange	yellow
Piperitone	8	43	—	from 20 to 50 μg near grey to brown
Fenchone	9	42	—	
Carvone	10	56	orange-yellow	
Methyl heptenone	13	56	yellow	
Methyl hexyl ketone	14	57	yellow	
Methyl nonyl ketone	16	57	yellow	
Menthone	20	60	—	
Butter Yellow	38	75		
Sudan Red G	13	65		
Indophenol	5	58		

[1] Silica Gel G layer 250 μ; trough tank with saturation.

Detection: An acid solution of 2,4-dinitrophenylhydrazine (Reagent No. 53) is most commonly used for the visual detection of aldehydes and ketones. The corresponding hydrazones appear as colored spots after a short time at room temperature. Acyclic aldehydes and ketones usually give yellow spots, while those from cyclic compounds are orange to orange-brown. The sensitivity varies but usually amounts of 1–5 μg give well-defined spots. Menthone, fenchone and piperitone require larger amounts. However, compounds other than aldehydes or ketones may give a yellow or orange color (examples are asarone, apiol, myristicin etc.) and this must be borne in mind when using the reagent.

The phosphomolybdic acid reagent (No. 120) does not give good results with 5 μg quantities of aldehydes and ketones. However, after employing the 2,4-DNP reagent, one can spray with the antimony tri- and penta-chloride mixture without disturbing the layer.

WASICKY and FREHDEN [82] have described the use of a saturated solution of o-dianisidine in acetic acid for the visual detection of aldehydes.

Already in the cold, aldehydes give the colors listed in Table 18; furfural reacts particularly readily. On heating, the colors are intensified and the layer darkens. Certain ketones, indicated in Table 18, react only when present in larger amounts. The reagent is non-specific; colors are obtained also, for example, with ascaridole, 1,8-cineol-hydroxyphenyl-propane derivatives and several terpenes.

5. Mono- and sesquiterpene alcohols

When a large number of alcohols of these groups are chromatographed with chloroform on Silica Gel G layers, it becomes apparent that they show very similar *Rf*-values. In a mixture, the C_{10} alcohols are for the most part separated from the C_{15} alcohols. For this reason, *Rf*-variations within the group are omitted.

Table 19. *Separation of Terpene Alcohols by Adsorption Chromatography on a Silica Gel G Layer using Chloroform as Solvent*

Number of alcohols investigated	*Rf*-values × 100		
	NS[1]	CS[1]	S-Chamber
11 Terpene alcohols	30—45	25—35	20—25
5 Sesquiterpene alcohols	45—55	35—40	25—30

[1] chromatography chamber; NS = without, CS = with saturated atmosphere.

The separation of the stereoisomeric menthols [16, 24, 47] show, that some problems can be resolved in this way, in spite of the limitations mentioned.

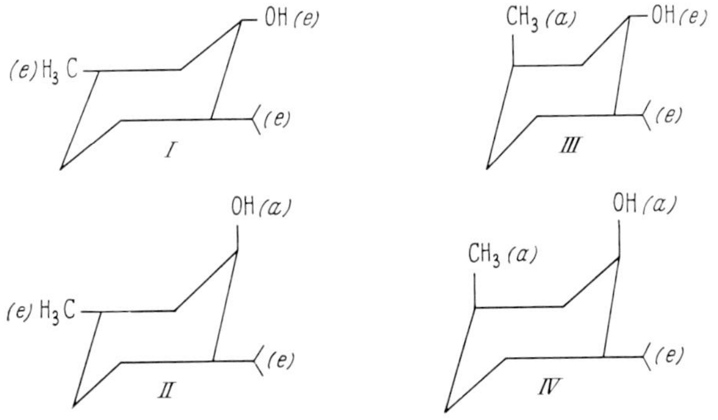

Fig. 96. Conformational formulae of the stereoisomeric menthols. *I* Menthol, *II* Neomenthol, *III* Isomenthol, *IV* Neoisomenthol (see text)

As early as 1957, ITO [25] found that the epimeric menthols with axial (*a*) hydroxyl group could be separated clearly from those with the hydroxyl group equatorial (*e*) on silica gel layers. ITO used hexane-ethyl acetate (85 + 15) as the solvent. PETROWITZ [*47*] preferred benzene-methanol (95 + 5). The following values were obtained by us. Under standard conditions on Silica Gel G layers.

Rf-Value × 100:	Chloroform[1]	Hexane-ethyl acetate (90 + 10)[2]
Menthol (I).	32	47
Isomenthol (III) . . .	29	41
Neomenthol (II) . . .	40	64
Neoisomenthol (IV) . .	36	68

[1] Chromatography chamber saturated (CS).
[2] Chromatography chamber not saturated (NS).

Work by GRAF and HOPPE [*16*] has confirmed these results.

With anisaldehyde-sulphuric acid (Reagent No. 9b) a blue violet color is obtained; antimony tri-and pentachloride (Reagent No. 11 and 13) gives a grey-brown coloration with menthol and grey-green with neomenthol.

Thus, a series of solvents, nearly isoeluotropic with chloroform, with dielectric constants between 3 and 5, can be tried in special cases. The corresponding esters and hydrocarbons travel closely behind the solvent front. If a less polar solvent such as benzene ($\varepsilon = 2.24$) is used the alcohols migrate too slowly (S-chamber, *hRf* = 10–15; saturated chamber h*Rf* = 5–10).

Phase-reversal: It is possible to achieve a good separation of alcohols, belonging to groups with the same number of carbon atoms (Tab. 20), by impregnating the Silica Gel G layer with paraffin or silicone oil and using a hydrophilic solvent ("reversed-phase technique"). Compared with the

Table 20. *Separation of Alcohols on a Paraffin-Impregnated Silica Gel G-Layer with 70% Methanol as a Solvent*

Number of carbon atoms	Alcohols	*Rf* × 100[1]
20	Phytol, Isophytol	2—5
15	Farnesol, Nerolidol, Cedrol, Guaiol, Elemol (*hRf* 25)	17—20
10	Geraniol, Nerol, Linalool, Menthol, Terpineol, Fenchol, Borneol, Isoborneol, Cuminic alcohol (*hRf* 50)	42—48
9	Cinnamyl alcohol	55 separated
7	Benzyl alcohol	59 from a
5	Furfuryl alcohol	63 mixture

[1] The *Rf*-values decrease with increasing numbers of carbon atoms; they should be regarded only as a guide.

untreated plate there is a good chance of making the distinction, so difficult with micro-samples, between mono-, sesqui- and diterpene alcohols. However, one must first make the assumption that the mixture under investigation consists of alcohols only. Terpene hydrocarbons give long streaks under these conditions. Phenol ethers show much the same "reversed" *Rf*-values as those obtained on untreated Silica Gel G layers; i.e. safrol travels slower than the C_{15}-alcohols, eugenol is found in the region of the $C_{10}-C_7$ alcohols; and those aldehydes investigated travel between the two.

In a separation on Silica Gel G impregnated with a lipophilic phase, it can be assumed that, apart from the partition effect, the adsorption on the silica gel still plays an important role. Thus, if one proceeds in the usual way, but using the inactive Kieselguhr G as the stationary phase, the alcohols run together to a higher region of the chromatogram (hRf 80).

Method: The Silica Gel G plate is dried and allowed to cool to room temperature, then the layer is immersed for 1 min in a 5% solution of paraffin or silicone oil DC 550 in petroleum ether (see p. 37); the light petrol (b.p. 40—60° C) is allowed to evaporate for 15 min in with the plate in the horizontal position, and the plate subsequently handled in the normal manner.

The preferred **solvent** is prepared from 70 ml of AR methanol mixed with 30 ml of distilled water. This mixture is saturated with liquid paraffin or silicone oil DC 550.

The *hRf*-values given in the above table were obtained in a 40 cm S-chamber. All the compounds (5 μg of each) were applied next to each other, and to mixtures of alcohols with different numbers of carbon atoms.

Alcohols and phenols can also be characterized by reaction with 3,5-dinitrobenzoyl chloride (see pp. 341, 342) and thin-layer chromatography of the resulting dinitrobenzoates (DNB). DHONT and DE ROOY [*10*] showed that the DNB-esters mentioned below can be separated on Silica Gel G layers using benzene-petroleum ether (50 + 50). Butter Yellow (p-dimethylamino azobenzene) was chosen as standard and R_B-values derived from the relation.

$$R_B = \frac{\text{Rate of migration of the substance}}{\text{Rate of migration of butter-yellow}}$$

Dinitrobenzoates of alcohols gave the following R_B-values: furfuryl alcohol-DNB 0.71, benzyl alcohol-DNB 0.76, maltol-DNB 0.92, geraniol-DNB 1.26, citronellol-DNB 1.69.

Menthol- and isomenthol-DNB's can be separated on Silica Gel G layers under standard conditions using petroleum (b.p. 105–120° C)-isopropyl ether (95 + 5) and a migration-distance of 15 cm [*16*].

For phenols: eugenol-DNB 0.55, isoeugenol-DNB 0.56, thymol-DNB 1.42. Detection is achieved by spraying with 1% ethanolic naphthylamine solution; yellow to orange-yellow spots appear.

Detection: Very sensitive but non-specific detection is possible using the phosphomolybdic acid reagent (No. 120). After spraying and heating, amounts as small as 0.05–1 μg can be detected as blue spots on the silica gel layers. The antimony tri- and pentachloride reagents (No. 11 and 13) are rather less sensitive. When they are used, the chromatogram should be examined under daylight and long-wave UV-light both before and after the heat treatment. Of the alcohols investigated (5 μg), the following gave grey to violet colors: geraniol, nerol, linalool, terpineol, nerolidol, farnesol, guaiol, and phytol. These mostly turned brown on heating; the other alcohols (Table 20) became brown only on heating. In contrast to the phenol ethers (q.v.) most alcohols and their esters show a brown-red fluorescence in long-wave UV-light.

Higher sensitivity and a certain color differentiation can be attained by using the anisaldehyde-sulphuric acid reagent (No. 9b). The following compounds give blue to violet colors on heating: menthols, guaiol, phytols; geraniol, nerol, nerolidol and farnesol go grey-violet; citronellol, terpineol and cedrol red-violet; cinnamyl alcohol and linalool grey; and cuminic alcohol turn reddish pink. Borneol, isoborneol and fenchol exhibit especially striking color changes; they turn, after some time, from brown to green.

6. Phenylpropane and phenol derivatives

In this group, the separation of the hydroxyphenylpropane derivatives, used in medicine and the perfume industry, is of particular interest. Many useful drugs can be obtained from plants; a few important ones for drugs are cloves, pimento, aniseed, fennel, parsley, dill, calamus and sassafras. A series of these compounds could be successfully separated by chromatography on 250μ thick Silica Gel G layers prepared by the standard method and using benzene as the solvent. Fig. 97 shows a thin-layer chromatogram of important phenol ethers from this group.

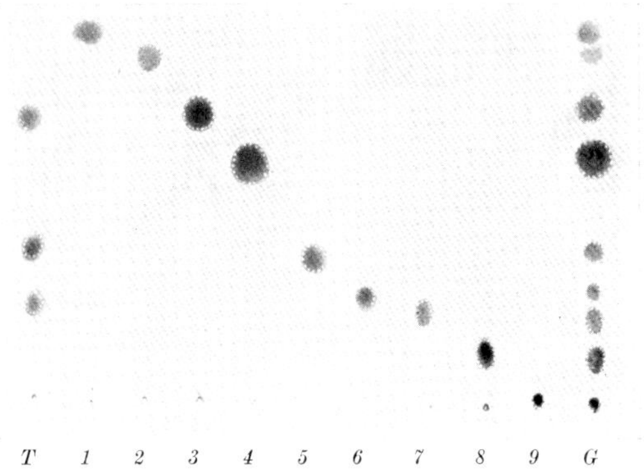

Fig. 97. Separation of hydroxyphenylpropane derivatives (4 μg each) on Silica Gel G layer with benzene, S-chamber, 10 cm run, 35 mins., reagent phosphomolybdic acid solution (Nr. 120 A). *T* DESAGA test mixture, *1* safrol, *2* methyl chavicol, *3* myristicin, *4* apiol, *5* engenol methyl ether, *6* asarone, *7* allyl tetramethoxy benzene, *8* elemicin, *9* pyrocatechin, *G* mixture (2 μg each)

Since the type of chromatography tank and its state of saturation have a considerable effect on the separation by adsorption chromatography, the data in Table 21 are given both for the S-chamber and for the tank (with and without saturation). Otherwise conditions are standard (pp. 7—12).

A variety of spray-reagents can be used for the **detection** of the substances in Table 21. All give a blue coloration after heating with the phosphomolybdic acid reagent (No. 120a) (Fig. 97). The clearest color-distinction is obtained by using the mixture of antimony tri- and pentachlorides 1 + 1 (Reagents No. 11 and 13,), or anisaldehyde-sulphuric acid (No. 9b). The addition of antimony pentachloride to antimony trichloride solution causes the colors to be intensified, but has the disadvantage that the spots do not show fluorescence in long-wave UV-light immediately; this develops after about one day. Some compounds can be clearly distin-

Table 21. *Phenol- and Hydroxyphenylpropane Derivatives, Rf-Values and Color Reactions* (Silica Gel G[1] layer; standard methods; solvent: benzene)

Substance	Rf-value × 100			Color reactions			
	chamber		S-chamber	Antimony (III)- + (IV)-chloride reagent 1+1 (No. 11; 13)			Anisaldehyde H_2SO_4; (No. 9b); after heating[3]
	CS	NS		room temp.	after heating[2]	after 24 hrs under UV (!)	
Safrole	57	90	89	—	grey-violet	red-brown, bright edge	blue-green-grey
Isosafrole	57	90	89	grey-violet	blue-violet	grey-violet, bright edge	grey-violet
Anethole	56	87	83	—	grey-violet	pink-violet	grey-red, blue edge
Methyl chavicol	55	85	80	—	olive-green-grey	pink	violet-grey (aft. 15′)
Myristicin	46	70	61	light-brown	grey-brown	dark, reddish edge	brown-green-grey
Iso-myristicin *(trans)*	46	70	61	—	brown-violet	brown-violet	grey-brown
Resorcinol dimethyl ether	40	72	56	grey-brown	brown-green	dark	red
Apiol	35	58	45	light-brown	olive-green	dark	violet-grey
Iso-Apiol *(trans)*	35	58	45	violet	brown-violet	dark	blue-violet
Hydroquinone dimethylether	31	54	45	—	yellow-green to brown	dark	weak violet-grey
Carvacrol	25	47	33	—	red to brown	red	light red-red-brown
Thymol	23	47	33	—	blue-red	red-violet	red
Pyrocatechol dimethyl ether	22	47	32	—	blue-green	dark	red
Guaiacol	18	40	27	grey	grey-black	dark	brown-grey
Eugenol	18	40	27	violet-brown	brown-grey	violet-grey	dirty green
Eugenol methyl ether	15	31	22	violet-green	brown-violet	beige-greenish edge	brown, blue edge
Iso-Eugenol methyl ether	15	31	22	—	violet	reddish-brown edge light	dark violet
Veratrol	13	29	17	—	blue-black	dark	weak red lilac
Allyl tetramethoxy benzene	11	20	12	—	olive-green	dark	beige-brown (aft. 15′)
Phenol	10	19	13	—	grey-brown	dark	red-orange
Asarone *(trans-Iso-)*	10	20	13	grey	grey-green	dark	red-lilac
Elemicin	8	11	7	grey-violet	brown-black	brown, light edge	dark violet
Catechol	0	0	0	—	light red-lilac	dark	red
Homocatechol	0	0	2	brown-grey	brown	dark	red
Resorcinol	0	0	0	—	brown	dark violet	red-orange
Hydroquinone	0	0	0	—	yellow-brown	dark	brown-grey
Butter Yellow	32	60	53				
Sudan Red G	10	24	17				
Indophenol	3	10	8				

CS = Chamber saturation, NS = Without chamber saturation (p. 18).
[1] Batch No. 62541. [2] 10 min at 100—105° C. [3] No color reaction before heating
Running times for 10 cm migration distance: NS = 35 min, CS = 50 min. S-chamber = 40 min.

guished by their different fluorescence colors. Examples are eugenol methyl ether, which gives a greenish-yellow fluorescence, and isoeugenol methyl ether (*cis-trans* mixture) which shows up reddish-brown. If antimony trichloride is used the fluorescence appears immediately on heating [*65*].

These color reactions depend generally on the amounts of substance and reagent used, as well as on the temperature and duration of the heating. A common fault is to use too little spray-reagent; 10–15 ml should be used for a plate of 20 cm square. An alkaline solution of a diazonium salt (Reagents No. 37; 61) can be sprayed on to determine whether a compound with a free phenol group is present. Phenols, as well as a number of other compounds, azulenes for example, couple and give colored products.

It is also possible to prepare 3,5-dinitrobenzoates and separate them by thin-layer chromatography (see p. 196). Chromatography on "*alkaline*" Silica Gel G layers is another way of identifying acid-reacting phenols. Phenols, presumably because they are in the phenolate form, show markedly lower *Rf*-values compared with the phenol ethers. On more acid layers on the other hand, *Rf*-values are higher [*66*].

The following *rules of thumb* are valuable for the separation and identification of phenylpropane and phenol derivatives under the conditions described above:

1. Free phenol groups affect the adsorption affinity strongly. While monophenols still migrate when benzene is used as solvent, diphenols remain at the origin.

2. Phenol ethers show distinctly lower adsorption affinity than the corresponding phenols. The *hRf*-value[1] decreases with increasing methoxy content. (Anethole, 1 OCH_3, $hRf = 83$; eugenolmethyl ether, 2 OCH_3, $hRf = 22$; asarone, 3 OCH_3, $hRf = 13$).

3. If two phenolic hydroxyl groups are connected by a methylene bridge, their influence on the *Rf*-value is almost annulled. (Catechol $hRf = 0$, safrol $hRf = 89$).

4. The introduction of an aliphatic side chain (methyl, ethyl, allyl or propenyl group) on the nucleus has a negligible effect on adsorption-affinity and therefore on the *Rf*-value.

5. When functional groups are situated closely together the adsorption affinity of each is decreased and thus the *Rf*-value of the compound increased.

Applications: Thin-layer chromatography was the basis for the rapid identification of the active substances in "chemical races" of medicinal plants containing hydroxyphenyl propane derivatives. For instance, the composition of the essential oils of calamus (*Acorus calamus L.*) varied sharply with genotype (Fig. 99) [*87*]. An investigation of the essential oils of parsley seeds from 17 different sources showed that they could be separated according to their chief constituents into three "chemical races":

1. Myristicin-race; 2. Apiol-race. 3. Tetramethoxy benzene-race.

Fig. 98 shows a 40 cm broad thin-layer chromatogram of the various parsley oils, developed in the S-chamber. Variations of this sort have

[1] $hRf = Rf\text{-value} \times 100$.

also been established by thin-layer chromatography in other plant families which contain phenols or phenylpropane derivatives as their most important active constituents. The remarkably significant quantitative

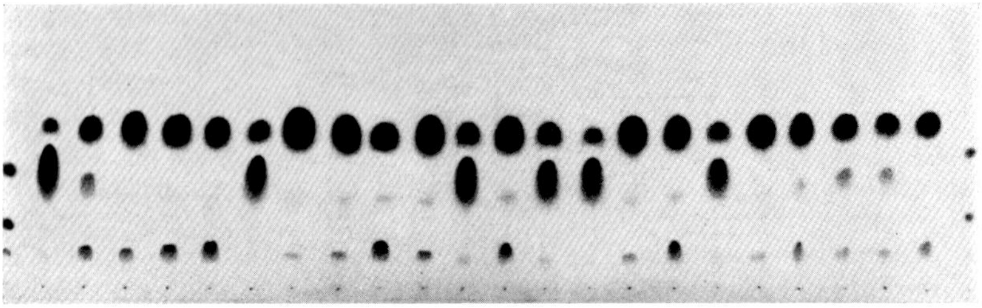

Fig. 98. Comparison of 22 different oils from parsley seeds by thin-layer chromatography on a 40 cm. TLC-plate. Spots in the upper region: Myristicin with Apiol below; lower region: Allyltetramethoxy benzene; on left and right, the DESAGA test mixture. Silica Gel G layer, S-chamber, solvent mixture trichloroethylene-chloroform (90+10). Distance of run 12 cm

and qualitative differences between oils from different parts of the same plant, for example, from roots, rhizomes and leaves, have frequently been observed (Fig. 99).

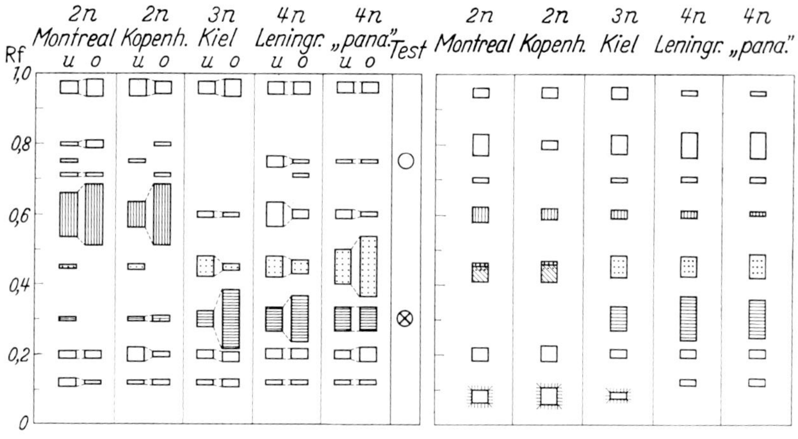

A. Leaf oils *B. Rhizome oils*

Fig. 99. Schematic representation of variation in essential oils (each 400 μg). *A* from the leaves and *B* from the rhizomes of different races of calamus (*2n* = diploid, *3n* = triploid, and *4n* = tetraploid plants from different sources. *u* = lower, *o* = upper half of leaf)
Silica Gel G-layer, solvent phase benzene; tank; normal saturation, visual detection by spray-reagent No. 11. [vertically hatched box] = Geranyl acetate, [dotted box] = Isoeugenol methyl ether, [horizontally hatched box] = Asarone, ○ = Butter Yellow, and ⊗ = Sudan Red G

7. Diterpene derivatives

DEMOLE and LEDERER [8] have reported the separation of four diterpene derivatives. These substances were chromatographed on silica gel layers, prepared essentially by the method of REITSEMA [52] using a mixture of n-hexane and ethyl acetate (85 + 15). The "*Rf*-values" × 100 given below can only be regarded as approximate; they have been taken from a schematic diagram: phytol 35, isophytol 50, geranyl-linalool 44, phytyl acetate 66. Visual detection was achieved by spraying with 0.25–0.5% aqueous potassium permanganate solution. After drying, brown spots developed on a white background. The compounds mentioned can be detected better using anisaldehyde-sulphuric acid reagent (No. 9b) or antimony trichloride (Reagent No. 11).

8. Triterpenes and derivatives

Relatively little work has been done on the paper-chromatographic separation of the triterpene mixtures commonly occuring in plants. [77] TSCHESCHE and coworkers [77] have dealt in detail with the thin-layer chromatography of the triterpene acids and a series of neutral triterpene derivatives. Even simple solvent mixtures like benzene and methylene chloride give very good separations of neutral triterpenoids on Silica Gel G layers. As expected, more strongly polar solvents, like di-isopropyl ether-acetone (75 + 30), must be used for the separation of the triterpene acids. With some acids, tailing develops (§), but this can be prevented by the addition of pyridine, or better, diethylamine.

It should be noted that the "*Rf*-values" quoted are obtained on layers prepared with a spatula, which were, therefore, uneven and thicker; also, no information is given regarding the type and state of saturation of the tank. It is advisable to run oleanolic acid as a standard. With neutral triterpenoids, the DESAGA test mixture of three dyes can also be used for this purpose.

Table 22. *Rf-Values of Triterpene Acids on Silica Gel G Layers* [77]

Triterpene acids	*Rf* × 100		Triterpene acids	*Rf* × 100	
	I[1]	II[2]		I[1]	II[2]
Acetylursolic acid	76		Chinovasic acid	55	
Ursolic acid	68	50	Masticadienonic acid	47	
Oleanolic acid	68	50	Isomasticadienonic acid	47	
Betulinic acid	68		Guaiiavolic acids	35	
Oleanonic acid	68		Bayogenin	31	
Siaresinolic acid	66		Acantholic acid	29	
Morolic acid	59		Medicagenic acid	29§*	35
Bredemolic acid	59§*		Machaerinic acid	23§*	25
Emmolic acid	59		Cochalic acid	18§*	45

* = Indicates tailing.
[1] Solvent mixture I: di-isopropyl ether-acetone (75 + 30).
[2] Solvent mixture II: ethyl acetate-methanol-diethyl amine (70 + 20 + 15).

Table 22 shows that the isomeric oleanolic, ursolic and betulinic acids are not separated under these conditions. Their separation can, however, be achieved by using anion exchange paper and various solvents, for instance, methyl cyclohexane-chloroform (80 + 20) saturated with 99% formic acid [77]. The spots are long-shaped. Conceivably better separations are obtained by using fine-grained ion exchange thin layers.

Table 23. *Rf-Values of Neutral Triterpenoids on Silica Gel G Layers* [77]

Neutral triterpenoids	*Rf*-values × 100		
	III[1]	IV[1]	V[1]
β-Amyrin		38	12
β-Amyrin acetate			45
α-Amyrenone			31
Lanosterol	75	40	14
Dihydrolanosterol acetate			43
Ursolic acid methyl ester acetate		77	26
Oleanolic acid methyl ester acetate		77	24
Ursolic acid methyl ester	85	51	
			Solvent I[1]
Crataegolic acid methyl ester monoacetate	40	13	80
Crataegolic acid methyl ester diacetate	72	37	92
Dehydrocrataegolic acid methyl ester diacetate	69	39	
11-keto crataegolic acid methyl ester monoacetate			77
Acantholic acid methyl ester monoacetate			73
Echinocystic acid methyl ester	59	15	
Emmolic acid dimethyl ester	73		

[1] Solvent I, see Table 22; III: Di-isopropyl ether, IV: Methylene chloride, V: Benzene.

Detection: Chlorosulphonic acid (Reagent No. 33) is mentioned as particularly suitable [77]. As little as 0.02 μg of oleanolic acid can be detected using it. Otherwise, the antimony tri- and pentachloride mixture (Reagents 11 and 13) or stannic tetrachloride (Reagent No. 156) can be used. The colored regions are generally red-violet to brownish.

Applications: Silica Gel G layers have been used for the separation of triterpene acids from *Bredemeyera floribunda* WILD [76] and of dihydroxy acids and their esters from *Crataegus oxyacantha L.* [76a]. THOMAS and MÜLLER [74] used them to follow the purification of methyl esters of triterpene acids from *Commiphora glandulosa* SCHINZ. They worked with Silica Gel G layers and with chloroform-ethyl acetate (80 + 20) as solvent. Conc. H_2SO_4 was used for visual detection. HUNECK [23] separated triterpenes from the bark of *Sorbus torminalis L.* on layers of fibrous alumina with ether-ethanol (98 + 2). Further applications are listed in Table 24.

9. Polyterpenes

The possibilities of thin-layer chromatographic separation of polyterpene mixtures, particularly carotenes and carotinoids, as well as ubiquinones, are described in the following chapter C (p. 210).

10. Essential oils (natural mixtures of terpene derivatives)

Essential oils are steam-volatile, lipophilic mixtures of natural products from plant sources that are distinguished by a special odor. As far as the physiology of the plant is concerned, they are excreta, which are often formed in particular cells, e.g. the glandular hairs of the *Labiatae*

Table 24. *Separations of Essential Oils and Terpene Derivatives on Thin Adsorption Layers, 1938—1962*

Year	Author	Ref.	Separation of:
1938	IZMAILOV and SHRAIBER	[*24*]	Tincture of peppermint and cinnamon
1951	KIRCHNER, MILLER and KELLER	[*32*]	*Rf*-values of 14 terpene derivatives; separation of simple mixtures
1952	MILLER and KIRCHNER	[*38*]	Investigation of fractions from columns
1952	ONOE	[*45*]	2,4-Dinitrophenylhydrazones
1953	FUKUSHI	[*12*]	Azulenes
1953	GÄNSHIRT	[*13*]	Essential oils from *Aristolochia clematitis L.*
1953	LABAT and MONTES	[*34*]	2,4-Dinitrophenylhydrazones and 3,5-Dinitrobenzoates
1953	MILLER and KIRCHNER	[*39*]	*Rf*-values of 38 terpene derivatives in various solvents
1954	GRÜNER and SPAICH	[*17*]	Essential oils from *Arnica montana* and triterpenes
1954	ITO, WAKAMATSU and KAWAHARA	[*26*]	Constituents of Japanese peppermint oil
1954	REITSEMA	[*52*]	*Rf*-Values of terpene derivatives Carvone-containing essential oils
1954	REITSEMA	[*53*]	Essential oils from *Mentha* species and strains
1955	BRYANT	[*4*]	Circular chromatography of *Eucalyptus* oils on magnesia layers
1955	COVENEY, MATTHEWS and PICKERING	[*5*]	Essential oils from *Strobilanthopsis linifolia*
1955	GOGRÖF	[*15*]	Sesquiterpenes from *Pogostemon Patschuli*
1955	KAISER	[*30*]	Essential oils and resins from *Grindelia* species
1955	DE NADAL	[*43*]	Terpenes, sesquiterpenes and terpenoid compounds
1955	RIGBY and BETHUNE	[*56*]	Fractions from hop-oils; colortable of 22 chromatograms
1955	WAGNER	[*80*]	Some volatile phenols
1955	WOTHERSPOON and BEDOUKIAN	[*85*]	Determination of oxidized compounds in essential oils
1956	DEMOLE	[*6*]	Jasmine oil; separation from isophytol
1956	LUYENDIJK	[*36*]	Essential oils of some *Umbelliferae*
1956	SPIEGELBERG	[*61*]	Essential oils of tansy and sage
1956	SPRECHER	[*62*]	Essential oil (methyl nonyl ketone) from *Ruta graveolens*
1956	STAHL	[*63*]	Essential oils, proazulenes and chamazulene derivatives
1957	FRYDMAN, MONTES and TROPAREVSKI, 3rd commun.	[*11*]	Numerous essential oils and pure substances
1957	ITO	[*25*]	Separation of menthol, neo-, iso-, and neoiso-menthols
1957	REITSEMA, CRAMER and FASS	[*54*]	Composition of the essential oils in the different organs or particular plants
1957	STANLEY and VANNIER	[*70*]	Citrus oils

Table 24. Continued

Year	Author	Ref.	Separation of:
1957	Stanley et al.	[71]	Coumarins from *citrus* oils; diphenyl.
1958	Demole and Lederer	[8]	Jasmine oil, diterpene derivatives
1958	Klohr-Meinhardt	[33]	Essential oils from *Petroselinum*- and *Levisticum*-grafts
1958	Naves	[44]	Essential oils
1958	Pryor and Bryant	[50]	*Eucalyptus* oils on magnesia layers
1958	Stahl	[65]	Calamus and valerian oils, peroxides etc.
1958	Stahl	[64]	Applications to the perfume industry; essential oils, resins, balsams
1959	Stahl	[66]	Phenols and phenol ethers on acidic and basic layers
1959	Winkler and Lunau	[84]	Essential oils from *Curcuma xanthorriza* and *C. longa*
1960	Brieskorn and Wenger	[3a]	Analysis of essential oils from sage
1960	Hefendehl	[21]	Peppermint oil; quantitative estimation of menthofuran
1960	Petrowitz	[48]	Isomeric menthols (see Ito, 1957)
1960	Stahl	[67]	Pyrethrine
1960	Stahl and Trennheuser	[68]	Hydroxyphenylpropane derivatives, and terpenes; combined with gas chromatography
1960	Tschesche and Sen Gupta	[76]	Triterpene acids
1960	Tschesche, Lampert and Snatzke	[77]	Triterpene acids and neutral triterpenoids, *Rf*-values
1960	Wenger	[83]	Dalmatian and Spanish sage oil; TLC and gas chromatography
1960	Wulff and Stahl	[86]	Essential oils of *Calamus* races
1961	Akazawa and Wada	[1]	Ipomeamarone from fungus-infected sweet potatoes
1961	Battaile et al.	[2]	Biosynthesis of Peppermint oil constituents
1961	Brieskorn, Klinger and Polonius	[3]	β-Sitosterol, ursolic acid
1961	Dhont and de Rooy	[9]	2,4-Dinitrophenylhydrazones
1961	Dhont and de Rooy	[10]	3,5-Dinitrobenzoates
1961	Hörhammer, Wagner and Lay	[22]	Oleanolic acids, β-sitosterol from Radix Panax Ginseng
1961	Jaspersen-Schib	[28]	Constituents of *Mentha* oils
1961	Neubern de Toledo and Wasicky	[44a]	Some essential oils
1961	Paris and Godon	[46]	Various essential oils, comparison with paper chromatography
1961	Zanini, Dal Pozzo and Dansi	[87]	Essential oils from *Pinus pumilio*
1962	Graf and Hoppe	[16]	Isomeric menthols
1962	Gabel, Müller and Schoknecht	[12a]	Determination of Thymol and Carvacrol
1962	v. Schantz	[59]	TLC of 21 essential oils

and *Compositae*, or the excretory canals of the *Umbelliferae*. The cultivation of these plants and the extraction of the essential oils is a major industry. The very expensive isolation procedures account for the relative high price, and explain why adulterants are so common. Simple and reliable methods are necessary for detecting these adulterants, apart from aroma analysis and the measurement of physical constants, and these are of interest for establishing the composition of essential oils[1].

The following points should be considered at the outset of any attempt to characterize essential oils by means of thin-layer and gas chromatography.

1. Commercial essential oils have almost invariably undergone rectification, to conform with demands of odor and/or taste. This frequently leads to removal of high and low boiling fractions. Essential oils carefully freed from terpenes and sesquiterpenes are particularly valued. Oils of differing quality may be mixed.
2. The composition of essential oils depends on a number of factors:
(a) on the botanical uniformity of the plant sources used ("Chemical races".).
(b) on the part of the plant used (e.g. stem, leaf, root, etc.), and on the place of origin, climate, harvesting and storage of the material.
(c) on the methods of extraction, and conditions used, as well as chemical modifications during distillation.
3. Only a combination of several analytical methods can yield a conclusive identification and quantitative estimation of any particular component.

It is outside the scope of this chapter to discuss in detail the work summarized in Table 24, nor would this be particularly useful, since most cases involve special problems. Of more recent work, JASPERSEN-SCHIB's [*28*] paper on the *mentha* oils deserves attention. It confirms the experience of others, that work on practically every essential oil needs thorough foundation before anything generally useful can be stated. Such extensive microanalytical methods of separation are not available for most oils. This is partly because of the difficulty of obtaining pure samples for purposes of comparison. Thus, one ought always to aim at establishing the chief constituents and characteristic minor components, as well as at a quantitative determination. The previous section should be of use towards this end.

11. Resins and balsams

Like the essential oils, resins and balsams are excreted from plants. They are usually formed in the cortical tissue as a result of some external stimulus, such as an injury. Unlike the essential oils, they are only slightly volatile or involatile in steam. The resins are yellow to brown solid, brittle amorphous masses which soften only at high temperatures and show no sharp melting points. They are insoluble in water and partially soluble in ethanol; the best solvents are chloroform or ethyl acetate. While only small amounts of essential oils can be present in resins, balsams contain much larger amounts, and as a result are mostly viscous liquids.

Resins and balsams consist of mixtures of terpene derivatives (in the widest sense) that are complex and very difficult to analyze. It is not

[1] Apart from one or more chief constituents characteristic of a particular oil, many other compounds can be present, of types dealt with in sections 1—6.

surprising that neither botanical nor chemical (Tschirch [*78*]) systems are satisfactory. The latter distinguished ester resins, resino-acid resins, and gum- or color-resins.

Attempts have been made to characterize resins and balsams by chromatography. Rothenheimer [*57*] and Stock [*73*] used capillary analysis, Valentin [*79*] used chromatography on alumina columns, and both Mills and Werner [*40, 41*] and also Rawlings and Werner [*51*] used the reversed-phase technique on paper. These authors had only partial success and none of the methods is very practicable. The first thin-layer chromatogram of resins and balsams, published some years ago, showed that this gave more chance of differentiation [*64, 66a*]. The subsequent extension of this investigation to a large number of commercially available resins and balsams has confirmed these findings.

Separation conditions: On chromatography under standard conditions using 250 μ thick Silica Gel G layers, distinctly better separations were obtained with the S-chamber (p. 18) than with the normal tank. Resins and balsams give narrow, elliptical, and partly-banded spots (Fig. 100).

The best **solvent** is benzene with 5% of methanol, for TLC in both the S-chamber and the trough-tank. For good distribution over the 10 cm chromatographic run, the chromatogram was always developed twice, i.e. the solvent phase passed twice (Fig. 100). No other solvent system tried gave such good separation, whether developed once or twice. This procedure is also better than experiments using reversed-phase chromatography on paraffin-impregnated Silica Gel G layers.

The best way of comparing resins and balsams is to apply 1 mm³ (30 µg) of a 3% ethyl acetate or chloroform solution. In each case, an authentic sample of the same concentration should be chromatographed, as well as the dye test-mixture.

Detection: The antimony trichloride reagent (No. 11) is very suitable for the visual detection of resins and balsams. The addition of antimony pentachloride to this spray reagent may cause the colors to be intensified to grey, but an undesirable weakening of fluorescence occurs.

The chromatogram is examined, before and after spraying, in daylight and long-wave UV light, and the zones are marked. Then it is heated to 110° C for about 10 min. Fig. 100 shows a thin-layer chromatogram of some resins and balsams taken in visible light, after heating. The zones are mostly brown, violet or grey. The yellow spots from synthetic Peru balsam (Perugen) are especially striking. Most resins and balsams can be characterized readily by the number, size and color of the zones. Examination of the chromatogram under long-wave UV-light allows further characterization. Many zones will fluoresce as shown in Fig. 101. Amongst others, a good differentiation of Sumatra and Siamese benzoin resins has been possible.

Another series of reagents are used for visual detection. However, they are not as sensitive as the method proposed. The Halphen-Hicks test, the Noller reaction, the "Natural-Product Reagent" (No. 55) and concentrated acids with the addition of aldehyde, have all been tried.

Note. The chromatogram of resins and balsams (Fig. 100) shows a selection of products popular in the German drug trade in the past five years. Usually 5—10 different specimens were examined. It is possible to state whether materials are really authentic only in cases where the path of the product can be traced all the way back to the plant source. When comparing them, one must consider that the collection procedure, the grading, the age of the sample, and many other factors, can influence the composition. Thin-layer chromatography now makes it possible to investigate these factors.

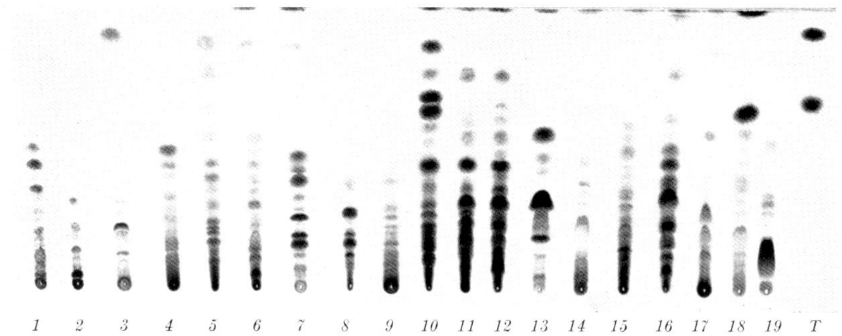

Fig. 100. Comparison of the most important resins and balsams by TLC on a 40 cm. plate (for details see text)

1 Siam-benzoin
2 Sumatra-benzoin
3 Styrax
4 Peru balsam
5 Perugen
6 Tolu balsam
7 Asa foetida
8 Galbanum
9 Myrrh
10 Olibanum
11 Mastic
12 Dammar
13 Terebinthina
14 Rosin
15 Canada balsam
16 Synthetic Canada balsam
17 Sandarac
18 Copaiba balsam
19 Gamboge

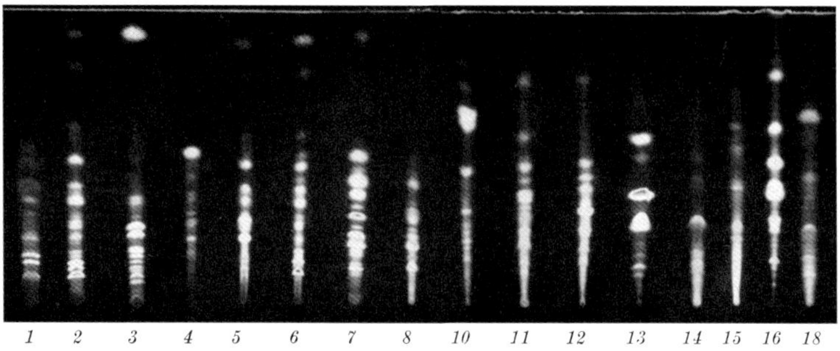

Fig. 101. Thin-layer chromatogram of resins and balsams taken in UV light. For separation conditions and notation see Fig. 100

It is further apparent from the chromatogram reproduced, that Tschirch's classification of the resins can no longer be regarded as satisfactory.

A recent report [*37a*] has dealt with the hydrolysis products of synthetic resins. The degradation products can be separated on Silica Gel G layers using methanol-glacial acetic acid (90 + 10).

Bibliography to Chapter B: Terpene Derivatives

[*1*] Akazawa, T., and K. Wada: Agr. Biol. Chem. Japan **25**, 30 (1961).
[*2*] Battaile, J., et al.: Biochem. Biophys. Acta **51**, 538 and 545 (1961).
[*3*] Brieskorn, Ch., H. Klinger u. W. Polonius: Arch. Pharm. **294**, 389 (1961).
[*3a*] — u. E. Wenger: Arch. Pharm. **293**, 21 (1960).
[*4*] Bryant, L. H.: Nature (Lond.) **175**, 556 (1955).

[5] Coveney, R. D., W. S. A. Matthews and G. B. Pickering: Colonial Plant and Animal Prods. (Gt. Brit.) **5**, 150 (1955).
[6] Demole, E.: Compt. rend. **243**, 1883 (1956).
[7] — Thèses de Doctorat, Paris, Ser. A. n° 844, N° d'Ordre 870 (1958).
[8] — et E. Lederer: Bull. soc. Chim. France **1958**, 1128.
[9] Dhont, J. H., and C. de Rooy: Analyst **86**, 74 (1961).
[10] — — Analyst **86**, 527 (1961).
[11] Frydman, B. J., A. L. Montes y A. Troparevski: Anales asoc. quim. (arg.) **45**, 248, 257, 261 (1957).
[12] Fukushi, S., and Y. Obata: J. Agr. Chem. Soc. Japan **27**, 353 (1953); ref.: Chem. Abstr. **50**, 15027 (1956).
[12a] Gabel, E., K. H. Müller u. I. Schoknecht: Dtsch. Apotheker-Ztg. **102**, 293 (1962).
[13] Gänshirt, H.: Pharm. Ind. **15**, 177 (1953).
[14] Gildemeister, E., u. Fr. Hoffmann: Die ätherischen Öle. Berlin: Akademie-Verlag 1956—1962.
[15] Gogröf, G.: Pharmazie **12**, 38 (1957).
[16] Graf, E., u. W. Hoppe: Dtsch. Apotheker Ztg. **102**, 393 (1962).
[17] Grüner, S., u. W. Spaich: Arch. Pharm. **287**, 243 (1954).
[18] Gstirner, F.: Prüfung und Verarbeitung von Arzneidrogen, Bd. I. Berlin-Göttingen-Heidelberg: Springer 1955.
[19] Guenther, E.: The Essential Oils. New York: D. van Nostrand Company 1952—55.
[20] Haagen-Smit, A. G.: In L. Zechmeister, Fortschritte der Chemie organischer Naturstoffe, Bd. XII. Wien: Springer 1955.
[21] Hefendehl, F. W.: Planta Med. **8**, 65 (1960).
[22] Hörhammer, L., H. Wagner u. B. Lay: Pharm. Ztg. **106**, 1308 (1961).
[23] Huneck, S.: J. Chromatog. **7**, 561 (1962).
[24] Izmailov, N. A., i M. S. Shraiber: Farmazia (Russian) No. 3, 1 (1938).
[25] Ito, M.: J. Chem. Soc. Japan, Pure Chem. Sect. **78**, 172 (1957); ref.: Chem. Zbl. **42**, 11595 (1957).
[26] — Sh. Wakamatsu and H. Kawahara: J. Chem. Soc. Japan, Pure Chem. Sect. **75**, 413 (1954); ref.: Chem. Abstr. **48**, 13172 (1954); Schimmel Ber. (1956) 91.
[27] Janistyn, H.: Parfümerie u. Kosmetik **41**, 371 (1960).
[28] Jaspersen-Schib, R.: Pharm. Acta Helv. **36**, 141 (1961).
[29] Jones, E. R. H., u. T. G. Halsall: In L. Zechmeister: Fortschritte der Chemie organischer Naturstoffe, Bd. XII. Wien: Springer 1955.
[30] Kaiser, H.-H.: Dissertation Karlsruhe, 1955.
[31] Karrer, W.: Konstitution und Vorkommen der organischen Pflanzenstoffe. Basel/Stuttgart: Birkhäuser 1958.
[32] Kirchner, J. G., J. M. Miller and G. I. Keller: Anal. Chem. **23**, 420 (1951).
[33] Klohr-Meinhardt, R.: Planta Med. **6**, 203, 208 (1958).
[34] Labat, J., y A. L. Montes: Anales asoc. quim. arg. **41**, 166 (1953).
[35] Luoms, W. D.: Biochem. Biophys. Acta **51**, 538 (1961).
[36] Luyendijk, E. N.: Dissertation Leiden 1956.,
[37] Marbet, R., and G. Saucy: Chimia **14**, 362 (1960).
[37a] Meckel, L., H. Milster u. U. Krause: Textil-Praxis **16**, 1032 (1961).
[38] Miller, J. M., u. J. G. Kirchner: Anal. Chem. **24**, 1480 (1952).
[39] — — Anal. Chem. **25**, 1107 (1953).
[40] Mills, J. S., and A. E. Werner: Nature (Lond.) **169**, 1064 (1952).
[41] — — J. Chem. Soc. **1955**, 3132.
[42] Moritz, O.: In W. Ruhland, Handbuch der Pflanzenphysiologie Bd. X. Berlin-Göttingen-Heidelberg: Springer 1958.
[42a] — in K. Paech u. M. V. Tracey: Moderne Methoden der Pflanzenanalyse, Bd. III. Berlin-Göttingen-Heidelberg: Springer 1955.
[43] de Nadal, G. N.: Am. Perf. Ess. Oil. Rev. **65**, No. 6, 17 (1955).
[44] Naves, Y.-R.: France et ses parfums **1**, 10 (1958); **2**, 18 (1959); according to: M. R. Paris and M. Godon.
[44a] Neubern de Toledo, T. A., u. R. Wasicky: Tribuna Farm. No. 7, 44 (1961).

[45] ONOE, K.: J. Chem. Soc. Japan, Pure Chem. Sect. **73**, 337 (1952).
[46] PARIS, M. R., et M. GODON: Ann. pharm. franç. **19**, 86 (1961).
[47] PETROWITZ, H.-J.: Angew. Chem. **72**, 921 (1960).
[48] — Z. analyt. Chem. **183**, 432 (1961).
[49] PREY, V., A. BERGER u. H. BERBALK: Z. analyt. Chem. **185**, 113 (1962).
[50] PRYOR, L. D., and L. H. BRYANT: Proc. Linnean Soc. N. S. Males **83**, 55 (1958).
[51] RAWLINGS, F. I. G., and A. E. WERNER: Endeavour **13**, 140 (1954).
[52] REITSEMA, R. H.: Anal. Chem. **26**, 960 (1954).
[53] — J. Am. Pharm. Assoc. Sci. Ed. **43**, 414 (1954).
[54] — F. J. CRAMER and W. E. FASS: Agr. and Food Chem. **5**, 779 (1957).
[55] RIGANESIS, M. D.: Essenze deriv. agrumari **26**, 29 (1956).
[56] RIGBY, F. L., and J. L. BETHUNE: Proceedings of Americ. Soc. of Brewing Chemists, p. 174 (1955).
[57] ROTHENHEIMER, C.: Pharm. Ztg. **74**, 712 (1929).
[58] RUZICKA, L.: Experientia **9**, 357 (1953).
[59] v. SCHANTZ, M.: Farmaseuttinen Aikakauslehti **2**, 52 (1962).
[60] SIMONSEN, J. L.: The Terpenes, Bd. I—V. Cambridge: University Press 1953—1957.
[61] SPIEGELBERG, CH.: Dissertation Berlin, 1956.
[62] SPRECHER, E.: Dissertation Karlsruhe, 1956.
[63] STAHL, E.: Pharmazie **11**, 633 (1956).
[64] — Parfümerie u. Kosmetik **39**, 564 (1958).
[65] — Chem. Ztg. **82**, 323 (1958).
[66] — Arch. Pharm. **292**, 411 (1959).
[66a] — Pharmaz. Rundschau **1**, Heft 2, 1 (1959).
[67] — Arch. Pharm. **293**, 531 (1960).
[68] — u. L. TRENNHEUSER: Arch. Pharm. **293**, 826 (1960).
[69] — Mikrochim. Acta **40**, 367 (1953).
[70] STANLEY, W. L., and S. H. VANNIER: J. Am. Chem. Soc. **79**, 3488 (1957).
[71] — — B. GENTILI: J. Ass. Offic. Agr. Chemists **40**, 282 (1957).
[72] STEINER, M., u. H. HOLTZEM: In K. PAECH u. M. V. TRACEY: Moderne Methoden der Pflanzenanalyse, Bd. III. Berlin-Göttingen-Heidelberg: Springer 1955.
[73] STOCK, E.: In K. DIETERICH u. E. STOCK: Analyse der Harze, Balsame und Gummiharze, 2. Aufl. Berlin: Springer 1930.
[74] THOMAS, A. F., u. J. M. MÜLLER: Experientia **16**, 62 (1960).
[75] TRENNHEUSER, L.: Dissertation Saarbrücken, 1962.
[76] TSCHESCHE, R., u. A. K. SEN GUPTA: Chem. Ber. **93**, 1903 (1960).
[76a] —, u. G. POPPEL: Chem. Ber. **92**, 320 (1959).
[77] — F. LAMPERT and G. SNATZKE: J. Chromatog. **5**, 217 (1961).
[78] TSCHIRCH, A.: Die Harze. Leipzig: Bornträger 1933/36.
[79] VALENTIN, H.: Chromatographische Adsorptionsanalyse in der Pharmazie I. Pharm. Ztg. **80**, 469 (1935).
[80] WAGNER, G.: Pharmazie **10**, 302 (1955).
[81] WALLACH, O.: Terpene und Campher, 2. Aufl. Leipzig 1914.
[82] WASICKY, R., u. O. FEHDEN: Mikrochim. Acta **1**, 55 (1937).
[83] WENGER, E.: Dissertation Tübingen ,1960.
[84] WINKLER, W., u. E. LUNAU: Pharm. Ztg. **104**, 1407 (1959).
[85] WOTHERSPOON, P. A., and P. Z. BEDOUKIAN: Am. Perfumer Essent. Oil Rev. **66**, No. 5, 17 (1955).
[86] WULFF, H. D., u. E. STAHL: Naturwiss. **47**, 114 (1960).
[87] ZANINI, C., A. DAL POZZO e A. DANSI: Boll. Chim. Farm. **100**, 83 (1961).

Some recent noteworthy articles

BADINGS, H. T., and J. G. WASSINK: Ned. Melk Zuiveltijdschr. **17**, 132 (1963): Dinitrophenylhydrazones of aliphatic aldehydes and ketones.
GMELIN, R.: Arzneimittel-Forsch. **13**, 771 (1963): Cucurbitacin glycosides.
JONG, K. DE, K. MOSTERT and D. SLOOT: Rec. trav. chim. **82**, 837 (1963): Unsaturated dinitrophenylhydrazones.

Klouwen, M. H., and R. ter Heide: Parfümerie Kosmetik **43**, 195 (1962). Phenols and phenol ethers.
Knappe, E., u. D. Peteri: Z. anal. Chem. **190**, 386 (1962). Organic peroxides.
Marcuse, R.: J. Chromatog. **7**, 407 (1962). Alkanals and alkanones.
Mehlitz, A., K. Gierschner and Th. Minas: Chem. Ztg. **87**, 573 (1963). Differentiation of dinitrophenylhydrazones.
Nigam, I. C., M. Sahasrabudhe and L. Levi: Can. J. Chem. **41**, 1535 (1963). Use of TLC in conjunction with GLC.
Schulte, K. E., F. Ahrens and E. Sprenger: Pharmaz. Ztg. **108**, 1165 (1963). TLC and PC of natural polyacetylenes.
Sundt, E., and A. Saccardi: Food Technol. **16**, 89 (1962). Aldehydes and coumarins.
Tyihák, E., D. Vágujfalvi and P. L. Hágony: J. Chromatog. **11**, 45 (1963). Stereoisomeric farnesols and derivatives.

C. Vitamins

By

H. R. Bolliger

I. Introduction

Although several vitamins can be estimated biologically, this method of assay is inferior in precision, time and cost. Physicochemical methods (such as colorimetry and spectrophotometry) are consequently being increasingly used. For these methods, however, the vitamins, after extraction from plant or animal tissues, foodstuffs and drugs, must be freed from substances which would interfere with the assay.

Methods of purification using column chromatography (adsorption, partition, ion-exchange), paper chromatography, (adsorption and partition), and electrophoresis, have proved to be satisfactory and are discussed in detail in the literature [*23, 34, 65*]. These methods are complemented by thin-layer chromatography, which has the advantages of being time-saving, and sometimes of increased sensitivity. The versatility of the method is such that it can also be used for the solution of other problems, such as stability tests of natural and synthetic vitamins, and the study of chemical reactions. New possibilities of vitamin estimation using thin-layer chromatography[1] are described below.

II. Method (General observations)

The method used to obtain the extract to be chromatographed depends on the properties of the original material, on the type of application, and on the quantity of the vitamins present. In natural products, the vitamins are not usually present in a free state, but are bound chemically. In man-made preparations the vitamins are frequently enclosed in gelatine to provide greater stability. The vitamins are extracted from the uniform sample (whether solution or powder) by the usual methods, either directly or after hydrolysis. The extracts obtained in this manner are then concentrated and subjected to further purification, for example,

[1] This includes our own unpublished studies.

by freezing-out or by column chromatography, and are subsequently applied to the coated plate and chromatographed in one or two dimensions with suitable solvents. One detects the vitamins on the plate by observing it in light of various wave lengths or by spraying with suitable reagents[1]. For a quantitative evaluation, one can compare with standards applied on the same plate, or, alternatively, an area carrying the spot can be scraped off, eluted and estimated by physicochemical methods. Bioautography, or the microbiological method for the vitamins, can also be used; a method for the paper chromatography of the B vitamins has been described by MARTEN [in *34*].

Sources of error during this process are the sensitivity of some of the vitamins to atmospheric oxidation (particularly vitamins A, D, E, K, and C), to light (vitamins A, D, E, K, B_1, B_2, B_6, and C), to active adsorbents (vitamins A, D and K), to heat and to solvents (vitamin D). The effects produced by these factors can be adequately and simply established by thin-layer chromatography and can be minimized or eliminated by suitable precautionary measures.

In our own experiments, it proved advisable to conduct vitamin analyses in rooms maintained at constant temperature (20° — 23° C), and in subdued light, covering the windows with transparent UVEX-foil[2] which absorbs short wavelength actinic light. Sunlight was excluded by venetian blinds. Only pure redistilled solvents were used, since the absolute ethanol commercially available, and also "pure" petroleum ether, contained traces of impurities which became concentrated during the processing of the extracts. These traces fluoresce on the adsorption layer, may take up some stain, and consequently cause interference.

Standard thin-layer chromatography (p. 7) was used. 2% of a fluorescent substance "ZS-Super"[3] added to the adsorbent caused the plates to fluoresce in ultraviolet light of 254 mμ. The coated plates must be of uniform activity and thickness. Unevenness in the thickness of the layer may lead to distortion, deformation and shifting of the spots. Similar effects can be produced by accompanying substances, or by too highly concentrated solutions causing overloading of the layers. To secure standard conditions, the separating tanks lined with filter paper were freshly filled with solvents for each new chromatogram. As the *Rf*-values (which should be regarded only as a guide) may be subject to considerable variation, the pure vitamin, alone and added to the extract, was run for the purpose of comparison. Before the chromatograms were sprayed they were always examined under short-wave ultraviolet light (254 mμ) for absorption or fluorescence, under long-wave ultraviolet light (365 mμ) for fluorescence, and in daylight for their own color.

If the extracts contain only small quantities of the vitamins in the presence of large quantities of fat and other substances, separation can be achieved by applying relatively large quantities of the extract (0.1—1.0 ml) in bands on thicker layers (0.4—2.0 mm). These layers can

[1] A list of reagents will be found at the end of the book (pp. 483—502).
[2] Special foil UVEX supplied by Cellpack A.G., Wohlen/AG, Switzerland.
[3] Supplied by Riedel-de Haën, Seelze/Hannover, Germany

be produced very evenly with the aid of an applicator which allows the layer thickness to be controlled[1] (see, for example, under vitamin D).

It has appeared to be advantageous to isolate the water-soluble and fat-soluble components of a vitamin-complex separately as this involves a simultaneous purification. In this way, overloading of the adsorbent layer during chromatography can be avoided and both groups of vitamins can be separated in less time and without losses on the plate.

III. Thin-layer chromatography of fat-soluble vitamins

Thin-layer chromatography has been found to be extremely useful (and probably superior to paper chromatography) in the separation of these highly lipophilic substances. Relatively large quantities of extract can be investigated chromatographically in a short time, and the active components can be visualized with corrosive and specific reagents. In the preparation of chromatograms it must be remembered that purified extracts frequently still contain fat and oil particles, which make the interpretation more difficult by fluorescing under ultraviolet light, and by staining readily. Moreover, carotenoids and the vitamins A, D and K are very unstable on the dry, active layers, and must therefore be treated with care during the application and subsequent scraping-off of the zone for quantitative analysis. In other words, the total time on the dry layer must be as short as possible. The decomposition of the vitamins on the plate can be detected by chromatography after applying a number of spots to the plate at different time intervals.

1. Mixed fat-soluble vitamins

Mixed fat-soluble vitamins can be separated simultaneously by various systems depending on the problem concerned[2]. Separation of various vitamins can thus be achieved in sixty minutes on Silica Gel G (Merck) with an 80 + 20 cyclohexane-ether mixture, and on alumina (Fluka or Merck) with benzene (see Table 26 and Fig. 102). In the solvent mixture cyclohexane-ether, different separation effects can be achieved by varying the proportions of the mixture. Spots of vitamin A alcohol and vitamin D which run almost together in the cyclohexane-ether mixture, may be made to migrate at different rates on Silica Gel G plates by the use of cyclohexane-ethyl acetate (80 + 20) or (70 + 30); at the same time, substances with higher *Rf*-values may merge. These separa-

[1] Supplied by C. Desaga GmbH., Heidelberg, Germany.

[2] Mixtures of β-carotene and vitamin A can be separated by the method of LAGONI et al. [*32, 33*]; α-tocopherol and vitamin A by the method of FONTELL et al. [*14*]; and mixtures of α-tocopherol, β-carotene, vitamin K and ubiquinone (50) can be separated by the method of WAGNER et al. [*64*]. Recently, DAVIDEK et al. [*7*] carried out chromatographic investigations on vitamin A (alcohol, acetate and palmitate), vitamin D_2, vitamin E (acetate), K_1, K_2, K_3 and the provitamins α- and β-carotene, on loose alumina layers without a binding agent. Model mixtures were separated with fourteen solvents and the *Rf*-values were reported.

tions also succeed on thicker layers (0.4–2.0 mm) and with larger quantities of vitamins and/or extracts (0.1–1.0 ml). Petroleum ether and chloroform are also suitable solvents. Carriers used in column chromatography, such as calcium phosphate also affect separation with certain solvents. Silica gel layers impregnated with liquid paraffin may be used in reversed phase chromatography.

Table 25. *Color Reactions of Fat-Soluble Vitamins with Perchloric Acid and Sulphuric Acid on Alumina Layers [3, 7]*

Vitamins	70% Perchloric acid	98% Sulphuric acid
A	violet	blue-violet
D_2	orange-brown	orange-red
E	brown	brown
K_1	yellow-brown	yellow-brown
K_2	yellow-brown	yellow-brown
K_3	yellow-brown	yellow-brown
Provitamins A .	blue	blue

Fat-soluble vitamins can be recognized in small quantities by mere inspection in light of various wave lengths, by their own color (the carotenoids), by absorption and/or fluorescence (see Table 26). These findings are valuable for subsequent elution and estimation. Exposure to ultraviolet light must, however, be brief to avoid photochemical decomposition.

Traces of all the vitamins can be rendered visible as grey-blue spots with phosphomolybdic acid (Reagent No. 120b) or as white spots on a grey-blue background with o-tolidine-potassium iodide reagent (No. 32) after treatment with chlorine. Excellent specific spray-reagents will be mentioned later. Various active substances can be stained individually

Table 26. *Approximate Values for Rf × 100, and the Direct Detection of Fat-Soluble Vitamins in Different Types of Light, on Layers to which Fluorescent Indicator has been added* (layer: Silica Gel G (Merck), Alumina (Fluka)

Vitamins	*Rf* × 100		Visibility in			Smallest detectable quantity (in μg)
	on silica gel with cyclohexane-ether (80 + 20)	on alumina with benzene	ultraviolet light (254 mμ)	ultraviolet light (365 mμ)	daylight	
β-Carotene	84	89	dark	dark	orange	0.03—0.04 each
Vitamin A alcohol .	10	16	dark	yellow-green	—	0.2/0.03
Vitamin A acetate .	45	72	dark	yellow	—	0.2/0.03
Vitamin A palmitate	72	84	dark	yellow	—	0.2/0.03
Vitamins D_2 and D_3 .	15	22	dark	—	—	0.5
α-Tocopherol. . . .	32	53	dark	—	—	10
α-Tocopheryl acetate	40	71	dark	—	—	10
Vitamin K_1	61	79	dark	dark	yellow	0.5/1/20

at the same time; this was shown, for example, by Gänshirt and Malzacher [*15*] for the water-soluble vitamins (p. 235), and by Blattná and Davídek for the fat-soluble vitamins [*3*, *7*]. (See, for example, the list set out in Table 25, extracted from the original papers.)

With the aid of these techniques fat-soluble vitamins in multi-vitamin preparations may be detected qualitatively and, to some extent, estimated quantitatively.

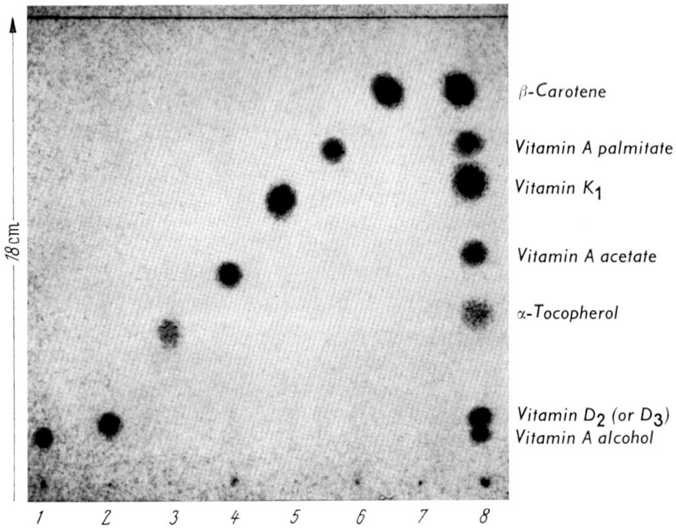

Fig. 102. The separation of fat-soluble vitamins on Silica Gel G (containing a fluorescent substance) with cyclohexane-ether (80+20) in ultraviolet light of short wavelength (time of run, about 60 minutes). *1* = 20 μg all-*trans*-vitamin A alcohol; *2* = 20 μg vitamin D_2; *3* = 30 μg d,l-α-tocopherol; *4* = 20 μg all-*trans*-vitamin A acetate; *5* = 30 μg vitamin K_1; *6* = 20 μg all-*trans*-vitamin A palmitate; *7* = 20 μg all-*trans*-β-carotene; *8* = substances and quantities of *1*—*7*

2. Carotenoids (provitamins A)

All other carotenoids can be derived from the three basic types A, B and C in this group of natural substances by dehydration, cyclization, aromatization, the introduction of functional groups or by oxidative degradation.

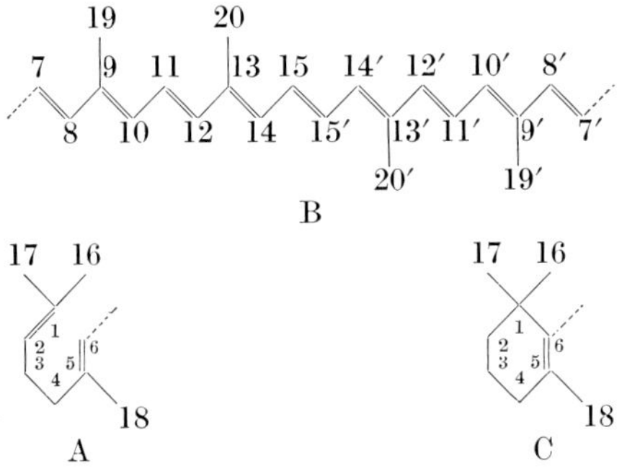

A + B + A = Lycopene
A + B + C = γ-Carotene
C + B + C = β-Carotene

The carotenoids, which are widely distributed throughout the plant and animal kingdoms, are yellow to deep-red pigments belonging to the terpene group. Almost a hundred of these compounds are now known. The carotenoids may be grouped into the hydrocarbons, i.e. the carotenes proper, and their oxygen-containing derivatives (alcohols, aldehydes, carboxylic acids and epoxides). *Cis-trans* isomerism is commonly observed. Many carotenoids must be regarded as provitamins A (for example, β-carotene, cryptoxanthin, torularhodin, echinenone and apo-carotenal etc.) since they are transformed in the human and animal body into vitamin A.

Thin-layer chromatography may be applied to the identification and determination of this group with excellent results. Although synthetic carotenoid materials are easily determined by this method, it has proved advisable in the case of natural extracts to remove the accompanying substances first, using column chromatography for example, and to investigate the individual fractions separately.

As carotenoids undergo rapid change on the dry layers, chromatographic investigations must be carried out immediately after the application. As a safeguard, the application can be conducted under carbon dioxide, and, to avoid degradation, only freshly prepared carotenoid solutions should be used [*53*].

a) Conditions of separation and results

Mottier [*38*] isolated the carotenoids from carrots on loose alumina layers with the aid of benzene. Stahl [*52*] demonstrated the usefulness of the "gradual development technique" by the chromatography of four carotenoids on Silica Gel G, using solvents of different eluting power. Bixin and canthaxanthin could be separated from β-carotene and from lycopene with chloroform (first step) and then with hexane (second step). According to Seher [*49*], β-carotene migrates farther on activated Silica Gel G than the various tocopherols ($Rf \times 100 \cong 80$) when chloroform is used as solvent.

Recently, Montag (*37a*) described the use of thin-layer chromatography to demonstrate artificial and natural lipid stains such as bixin, norbixin, carotene, capsanthin and crocin in foodstuffs, using three different solvents on Silica Gel G layers prepared by the standard method.

Demole [*9*, *10*] used silica gel layers bonded with rice starch for the separation of various carotenoids and developed with a mixture of *n*-hexane and ether (30 + 70). Isler and co-workers [*28*] used calcium hydroxide-Silica Gel G (6 + 1). Petroleum ether-benzene (98 + 2) mixtures could be used for the separation of carotenoid hydrocarbons; carotenoid ketones could be separated with benzene, and hydroxylated carotenoids with benzene-methanol (49 + 1).

The carotenoids also show different rates of migration on activated Silica Gel G layers with petroleum ether (90°–110° C)-benzene mixtures (50 + 50) (see Table 27).

Stahl, Bolliger and Lehnert [*53*] found that carotenoid mixtures cannot be separated on a single adsorbent, and with a single solvent. Using

Table 27. *Approximate Values for Rf × 100 for Various Carotenoids*

System	1	2	3	4	5	6	7	8	9	10
Carotenoid-hydrocarbons										
α-Carotene	—	88	—	—	—	84	75	↑	↑	↑
β-Carotene	96	84	—	—	82	69	75			
7,7′-Dihydro-β-carotene	—	82	—	—	70	55	75			
15,15′-Dehydro-β-carotene	—	74	—	—	—	—	—			
γ-Carotene	—	40—50	—	—	—	§[1]	65	97		
Lycopene	—	10—20	—	—	74	§	56			
3,4,-Dehydrolycopene.	—	—	—	—	—	0	51		100	100
3,4,3′4′-Bisdehydrolycopene	—	—	—	—	—	0	0	↓ 0		
Torulin	—	—	—	—	—	—	—	—	(98)	
Carotenoids containing oxygen										
β-Apo-12′-carotenal	—	—	—	—	—	↑	↑	70		
β-Apo-8′-carotenal .	—	—	—	—	15			64		
β-Apo-8′-lycopenal .	—	—	—	—	—			58		
β-Apo-10′-carotenal	—	—	—	—	—			53		
Lycopenal	—	—	—	—	—			43		
Crocetin dialdehyde	—	—	—	—	—			7,5		
Torularhodin methyl ester	—	—	—	—	—			83	↓ 94	↓ 100
Methylbixin . . .	—	—	—	—	—	0	0—7	13	81	97
Canthaxanthin . .	38	—	43	—	0			↑	63	90
Cryptoxanthin . .	—	—	34	74	—				54	75
Xanthophyll . . .	—	—	—	55	—				↑	35 }
Antheraxanthin . .	—	—	—	—	—					32 }
Zeaxanthin	17	—	—	57	—			0		24
Violaxanthin . . .	—	—	—	—	—				10—11	21
Capsanthin	—	—	—	—	—					16
Capsorubin	—	—	—	—	—					13,5
Bixin	51	—	—	—	—					5,5
Azafrine	—	—	—	—	—	↓	↓	↓	↓	2,0
Echinenone. . . .	—	—	82	—	10	—	—	—	—	—
Rhodoxanthin . .	—	—	16	94	—	—	—	—	—	—
β-Apo-8′-carotenoic acid methyl ester	—	—	—	—	26	—	—	—	—	—
Isozeaxanthin. . .	63	—	—	—	—	—	—	—	—	—
Butter-yellow. . .	—	—	—	—	—	35§	32	94	95 }	100
Indophenol	—	—	—	—	—	10	3	58	86 }	100
Sudan red	—	—	—	—	—	0	8	68	90 }	100

Layer (0.25 mm)	*Solvent*
1 = Silica gel (with rice starch)	*n*-Hexane-ether (30 + 70), [*9, 10*]
2 = Calcium hydroxide-Silica Gel G (6+1)	Petroleum ether-benzene (98 + 2) [*28*]
3 = Calcium hydroxide-Silica Gel G (6+1)	Benzene [*28*]
4 = Calcium hydroxide-Silica Gel G (6+1)	Benzene-methanol (98 + 2) [*28*]
5 = Silica Gel G	Petroleum ether (90—110° C)-benzene (50 + 50)
6 = Calcium hydroxide	Hydrocarbon mixture[2]-methylene dichloride (95 + 5)
7 = Silica Gel G	Undecane-methylene dichloride (80 + 20)
8 = Secondary magnesium phosphate	Carbon tetrachloride
9 = Secondary magnesium phosphate	Benzene
10 = Silica Gel G	Methylene dichloride-ethyl acetate (80 + 20)

6—10 = Separation in groups as described by STAHL, BOLLIGER and LEHNERT [*53*].

[1] Sign indicating tailing.

[2] Pentane-hexane-heptane-octane-undecane (60 + 20 + 10 + 6 + 4).

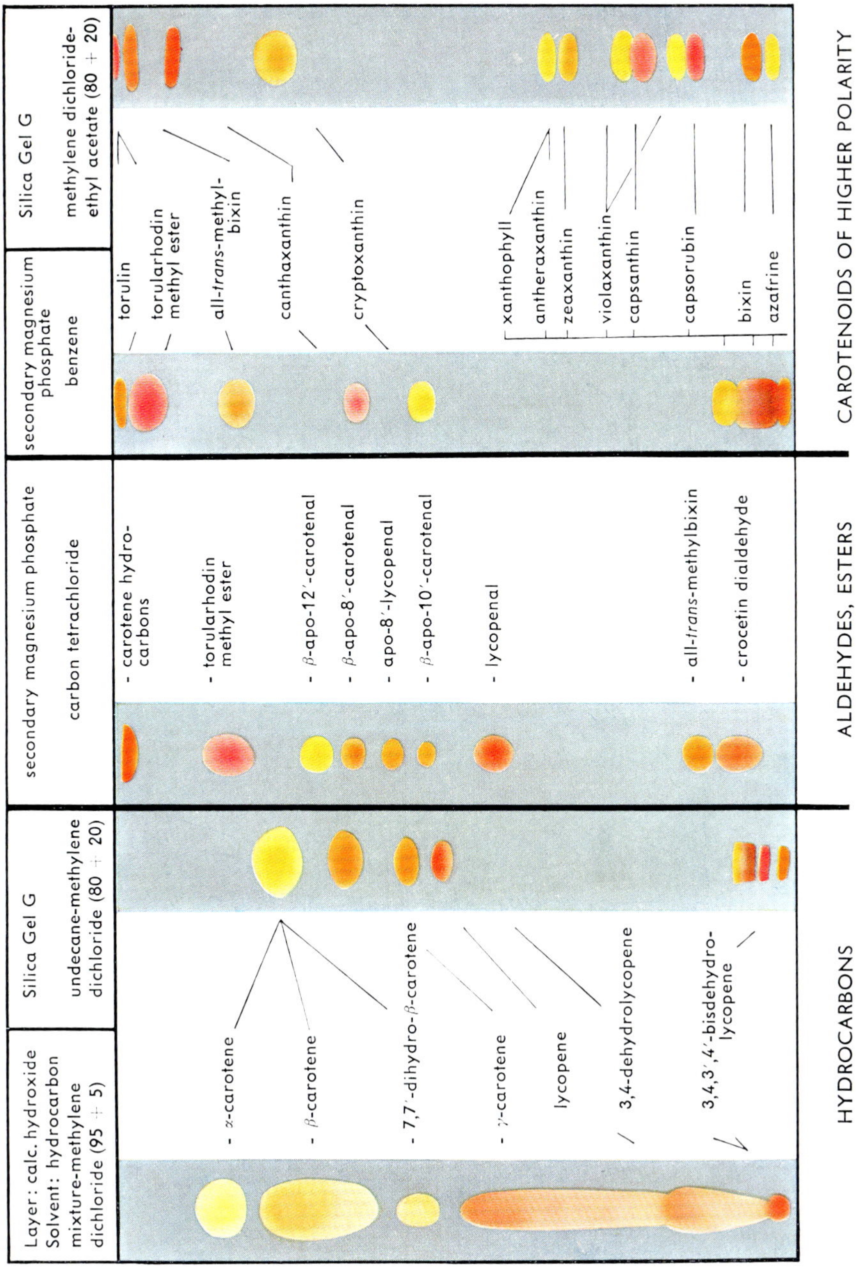

Plate I. Thin-layer chromatogram of mixtures of carotenes and carotenoids

Springer-Verlag, Berlin · Göttingen · Heidelberg · New York
Academic Press Inc., New York · London

calcium hydroxide, magnesium phosphate, and Silica Gel G layers, these authors were able to separate nearly thirty different carotenoid derivatives into individual compounds with suitable solvents (see Table 27 and Table I). To check this method of analysis, plant extracts containing carotenes of unknown composition were investigated chromatographically. These extracts could then be estimated quantitatively, either by the direct or the indirect method. For this purpose, bands were applied on layers 0.5 mm thick (separation distance, 10–12 cm; time of run, about 15 min). It seems desirable to include an even greater number of carotenoids than shown in the scheme mentioned above, and then perhaps to subdivide them further. In all experiments the S-chamber[1] gave better results than the glass vessels previously used. This S-chamber system makes the saturation of the tank unnecessary, and on a plate 40 cm wide, thirty or more mixtures can be applied and compared.

To separate and estimate carotenoid aldehydes in plants, microorganisms and animal tissues, WINTERSTEIN et al. [*67*, *68*, *69*] employed calcium hydroxide-Silica Gel G (80 + 20) layers; a mixture of petroleum ether and benzene (50 + 50) was generally used as the solvent. In certain cases, the content of benzene was modified, and occasionally 1% methanol was added. Plates were frequently used on which the silica gel layer was impregnated with liquid paraffin (the activated plate being immersed in a solution containing 5% liquid paraffin in petroleum ether, and afterwards dried); development was then affected with methanol saturated with liquid paraffin.

The peculiar behaviour of carotenoid aldehydes with regard to adsorption [*67*, *68*] should also be mentioned. The adsorption affinity on unimpregnated layers is a function of constitution rather than of the number of double bonds (as it is with the hydrocarbons). For example, retinene (vitamin A_1 aldehyde) which contains five conjugated double bonds is adsorbed to an appreciably greater extent than β-apo-8′-carotenal which contains nine double bonds. The comparative chromatogram of the synthetic aldehydes (Fig. 103) shows that the C_{40}-, C_{35}-, C_{30}- and C_{25}-aldehydes are adsorbed to a lesser degree than the C_{37}-, C_{32}- and C_{27}-aldehydes which contain one double bond less. It thus appears that in the above case the adsorption affinity depends on whether the α-position of the aldehyde group is substituted by a methyl group, or by a hydrogen atom. On silica gel plates impregnated with liquid paraffin, however, the $Rf \times 100$ values of carotenoid aldehydes increase according to theoretical expectations:

Number of C atoms:	C_{40}	C_{37}	C_{35}	C_{32}	C_{30}	C_{27}	C_{25}	C_{20}
Values found for $Rf \times 100$:	28	34	38	47	49	63	69	85

In addition, THOMMEN [*56*] was able by means of thin-layer chromatography to detect β-apo-8′-carotenal and β-apo-8′-carotenoic acid, which are important colouring agents in egg yolk. The same author [*57*] was able to identify canthaxanthin [*63*] in the red feathers of numerous birds,

[1] Produced by Desaga GmbH., Heidelberg, Germany.

and also in the organs of the flamingo. In the latter case, canthaxanthin was found predominantly in the liver, but it was also present in smaller amounts in the heart and in the kidneys. Using this method Thommen [58] was able recently to detect the presence of citraurin, β-apo-10′-carotenal, β-apo-2′-carotenal and especially β-apo-8′-carotenal, and to determine the latter substance quantitatively in the juice and peel of fresh oranges.

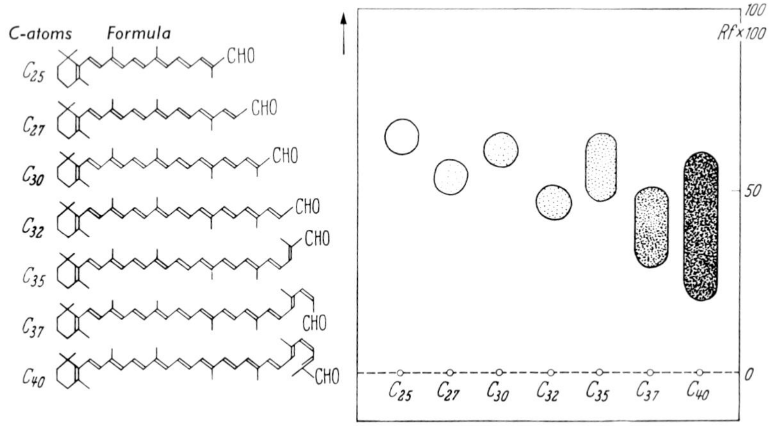

Fig. 103. Chromatogram of the β-apo-carotenal series (C_{25}—C_{40}) on Silica Gel G, developed with petroleum ether-benzene (40+60) [29]

Goodwin [17] used thin-layer chromatography as a technique in investigating the biogenesis of carotenoids in micro-organisms and in bacteria. Davies, Goodwin and Mercer [8] were also able to detect

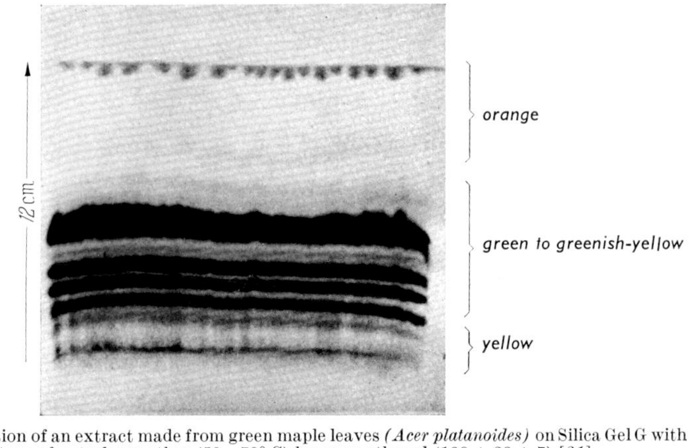

Fig. 104. Separation of an extract made from green maple leaves *(Acer platanoides)* on Silica Gel G with a mixture of petroleum ether (50—70° C)-benzene-ethanol (100 + 20 + 7) [21]

squalene, phytoene and phytofluene in *Rhodospirillum rubrum*, but not lycopersene. Grob et al. [21, 22] and Eichenberger and Grob [13] carried out similar investigations on plants and fungi. The formation and transformation of carotenoids in leaves, blossoms and fruit was followed (see Fig. 104), and evidence that lycopersene occurs naturally in *Neurospora crassa* was obtained.

Thin-layer chromatography may also be applied to differentiate between *cis* and *trans* isomers. Thus, for example, all-*trans*-β-carotene remains at the starting point on active Silica Gel G with methanol, as solvent, while 15,15'-*cis*-β-carotene migrates ($Rf \times 100 \simeq 65$). The formation of elongated spots indicates that isomerisation is taking place during the operation. Secondary magnesium phosphate is suitable as an adsorbent (the plates may be prepared by the method of STAHL); and petroleum ether (30° – 45° C)-ether (95 + 5) mixture can be used as a solvent for the separation of all-*trans*- and 15, 15'-*cis*-β-apo-8'-carotenal (C_{30}) ($Rf \times 100 \simeq 65$ and 88, respectively). Products of decomposition, which may appear due to the prolonged presence of the substance on the dry layer, are adsorbed to a greater degree. All-*trans*-canthaxanthin, 15,15'-*cis*-canthaxanthin, echinenone and β-carotene may be separated on Silica Gel G with methylene chloride-ether (90 + 10) ($Rf \times 100 \simeq 56$, 48, 80 and 90, respectively). Suitable conditions must be established for the separation of every individual mixture of isomers because no general rules exist.

b) Detection and evaluation

Even minute amounts of these pigments can be detected on the plate in daylight because of their intense color; in ultraviolet light (254 mμ and 365 mμ) they appear as dark absorption spots (cf. Table 26). Yellow-orange compounds can be recognised on the chromatogram at levels of 0.1 μg; and as little as 0.01 μg of the intensely red-colored carotenoids can sometimes be detected. The colors can be intensified and stabilised by spraying with a mixture of liquid paraffin and heptane [*53*]. Certain decomposition products will fluoresce in the light of a quartz lamp. Spray reagents of general applicability such as $SbCl_3$ (No. 11 and 12), $SbCl_5$ (No. 13; 20% in chloroform) and concentrated sulphuric acid are less sensitive in detection, but have the advantage of staining the carotenoids in different colors.

To identify carotenoid aldehydes, the chromatogram is sprayed with an alcoholic solution of rhodanine (Reagent No. 130) and subsequently with a concentrated solution of ammonia or with concentrated sodium hydroxide; it is afterwards dried. Of the characteristically deeply colored condensation products, for example, that of retinene, as little as 0.02 to 0.03 μg is still recognisable as orange-red spot. In this way, β-apo-8'-carotenal was detected in the intestinal mucosa, and retinene in eye extracts [*69*].

For quantitative work, the spots obtained may be compared with those from parallel runs using a series of standards. Alternatively, the spots may be scraped off immediately after the separation (when necessary, in an inert atmosphere [*53*]), eluted, and a spectrophotometric analysis then performed.

3. The A vitamins

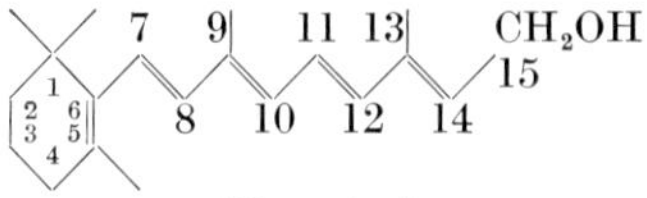

Vitamin A

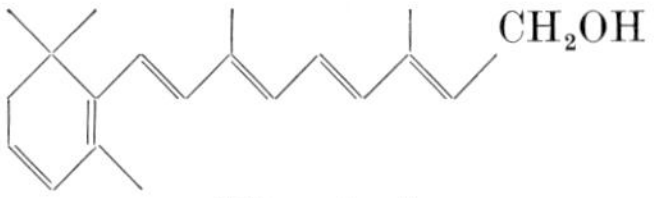

Vitamin A_2

In human and in animal organs vitamin A occurs mainly as the free alcohol, or as aliphatic esters. In plants, vitamin A occurs as a provitamin (the carotenoids) and as vitamin A aldehyde (retinene). The latter was detected by Winterstein et al. [*67*, *68*, *69*] with the aid of thin-layer chromatography and has been discussed in the section on carotenoids. The term "vitamin A" is used for the all-*trans* form or for vitamin A_1, which is the physiologically most active compound of the vitamin A group. Several of the theoretically possible *cis-trans* isomers of vitamin A_1 have now been found in nature, or have been synthesised in the laboratory. 13-*cis*-vitamin A always accompanies vitamin A_1 in cod liver oil for example, where it may reach 35% of the total vitamin A content. Vitamin A_2, which contains an additional double bond in the substituted cyclohexene ring, may also be found in these oils. In thin-layer chromatography, the vitamin A compounds must first be isolated from the natural materials, drugs or foodstuffs, and in some cases the extracts must undergo preliminary purification.

a) Conditions of separation and results

Vitamin A alcohol, acetate, palmitate and other derivatives can be rapidly separated using various systems[1] (compare Table 28). These

Table 28. *Approximate Values for Rf × 100 for various Vitamin A Compounds* (standard conditions; time of run, about 60 min; length of run, about 18 cm)

Vitamin A compounds	Alumina (Fluka or Merck)	Silica Gel G (Merck)		Silica Gel G + *sec.* calcium phosphate 1+1	Silica Gel G impregnated with liquid paraffin
	Benzene	Cyclohexane-ethyl acetate (90+10)	Cyclohexane-ether (80 + 20)	Cyclohexane-ether (85 + 15)	Acetone-water (90 + 10)
All-*trans*-vitamin A alcohol	16	11	10	10	97
All-*trans*-vitamin A acetate	72	46	45	45	79
All-*trans*-vitamin A palmitate	84	69	72	66	10
Anhydrovitamin A	—	73	75	75	—
13-*cis*-Vitamin A acetate	—	—	—	50	—
Retro-Vitamin A acetate	—	—	—	45	—
13-*cis*-Vitamin A palmitate	—	—	—	64—66	—

[1] Recently, Davídek et al. [*7*] carried out the chromatographic investigation of vitamin A alcohol, acetate and palmitate, in addition to some other fat-soluble vitamins, using different solvents on loose alumina layers. According to Mangold et al. [*35*], vitamin A palmitate can be separated from the lipid components on Silica Gel G, using petroleum ether-ether-acetic acid mixture.

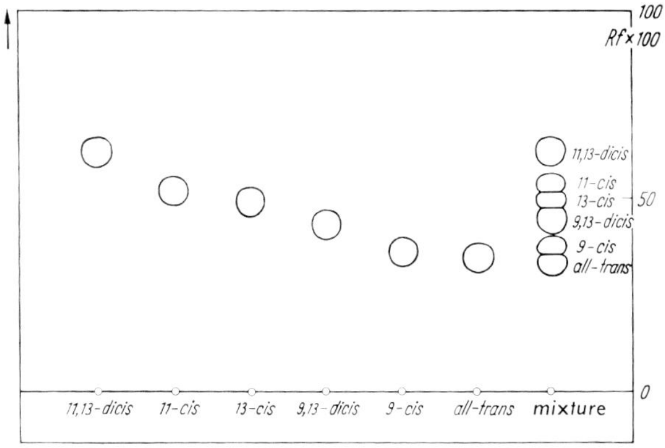

Fig. 105. Separation of six isomeric vitamin A alcohols on Silica Gel G with a mixture of petroleum ether (30—45° C) and methyl heptenone (11 + 2) [*42*]

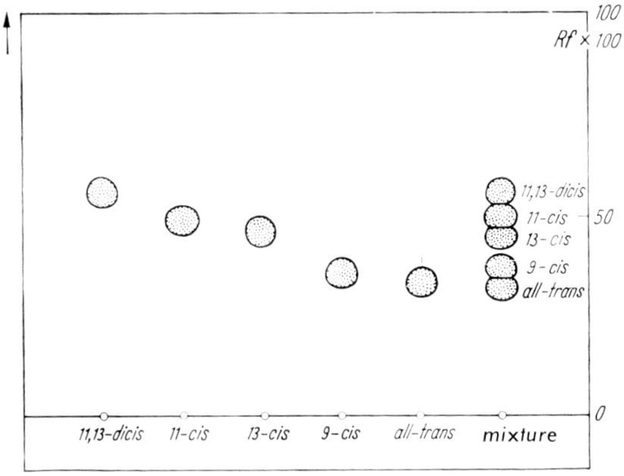

Fig. 106. Separation of five isomeric vitamin A_2 alcohols on Silica Gel G with a mixture of petroleum ether (30—45° C) and methyl heptenone (11 + 3) [*42*]

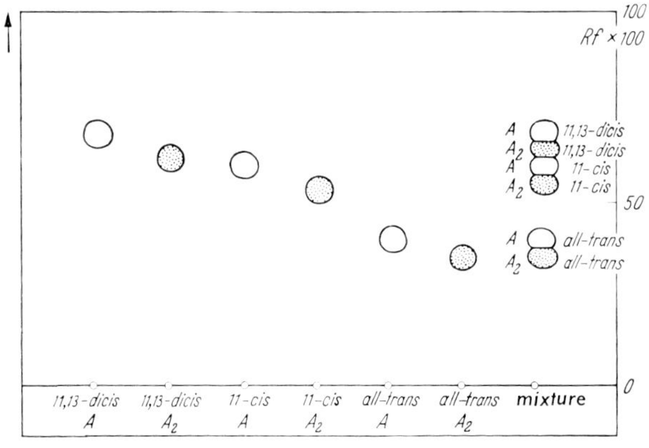

Fig. 107. Separation of hindered and of all-*trans* isomers of vitamin A and vitamin A_2 alcohols on Silica Gel G with a mixture of petroleum ether (30—45° C) and methyl heptenone (11 + 2) [*42*]

compounds are highly sensitive and unstable on dry layers. In addition to anhydrovitamin A, several other degradation products form very rapidly, and are more strongly adsorbed. It is therefore important that the plate should be put into the separation tank and developed immediately after the application of the vitamin A solution. The TLC method has been found useful in the investigation of natural and synthetic concentrates, and for the estimation of the vitamins A in drugs, animal feeds and foods.

Metzger [in *42*] recently succeeded in achieving excellent separation of synthetically produced vitamin A isomers. Portions of 0.4 μg were applied to activated Silica Gel G layers [*51*], and were developed for sixty minutes with a mixture of petroleum ether (30°–45° C) and methylheptenone (11 + 2) or (11 + 3) (see Fig. 105–107). This method will probably prove to be particularly useful in the investigation of natural vitamin A and A_2 concentrates, as well as for the study of visual purple in various animals. Isomers possessing a sterically-hindered *cis*-double bond (11-*cis*-and 11,13-di-*cis*) show an appreciably higher rate of migration than the unhindered *cis* forms (compare Fig. 105 and Fig. 106). Vitamin A_2 compounds migrate at a slower rate than the corresponding vitamin A isomers (compare Fig. 107). Only the separation of the all-*trans* from the 9-*cis* compounds, and of the 11-*cis* from the 13-*cis* compounds, remains incomplete.

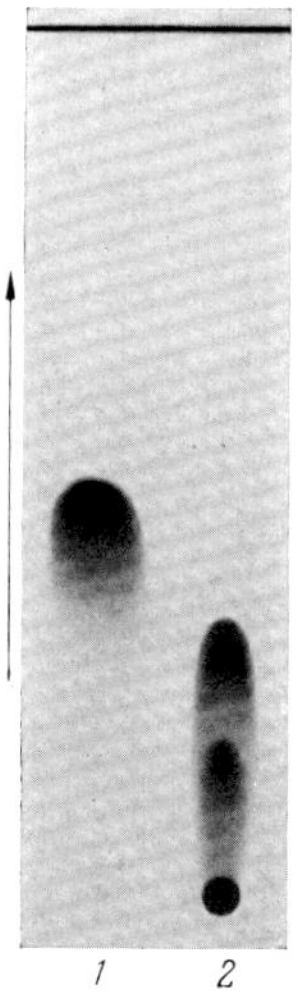

Fig. 108. Chromatogram of $SbCl_3$ and $SbCl_5$ (see text). *1* = $SbCl_3$; *2* = $SbCl_5$

b) Detection and evaluation

Chromatograms can best be evaluated by inspection in ultraviolet light. Numerous vitamin A derivatives show a yellow-green fluorescence at 365 mμ (the smallest detectable quantity is 0.03 μg) but anhydrovitamin A, vitamin A aldehyde and vitamin A_2 derivatives, all appear as brownish spots. With adsorbents containing fluorescent additive in 254 mμ light as little as 0.2 μg of these substances are visible as dark absorption spots. Vitamin A gives a grey-blue color with phosphomolybdic acid (Reagent No. 120 b); a blue color with antimony trichloride (this is the reagent of the Carr-Price reaction; No. 11, 12); and, initially, a deep blue color with concentrated sulphuric acid, the smallest detectable amount being 0.1–0.3 μg. Only the phosphomolybdic acid complex is stable. The intensely yellow vitamin A aldehyde (retinene) produces only a drab brown color with antimony trichloride on Silica Gel G. A more specific reaction can be achieved with rhodanine (Reagent No. 130) [*68*, *69*]. A semi-quantitative estimation can be achieved by comparison with standards. The reliable quantitative estimation of vitamin A compounds has so far proved impossible because of the strong tendency of these compounds to undergo decomposition on active adsorbents.

In addition to vitamin A, other vitamins and substances such as oils, antioxidants or synergists can be identified on thin-layer chromatograms with universal reagents, a fact which has proved to be very useful in the investigation of products containing vitamin A.

BRÜGGEMANN, KRAUSS and TIEWS [*5*, *59*] assumed that the blue color with vitamin A is not due to antimony trichloride but rather to the pentachloride which is present in trace amounts. This point was clarified in our own experiments by the following method: chloroform solutions containing 20% of either pure $SbCl_3$ or $SbCl_5$ were applied to activated Silica Gel G, and were developed with ether. The plates were then sprayed with an 0.5% solution of vitamin A palmitate in chloroform. Both antimony compounds produced blue spots, $SbCl_3$ produced a single spot ($Rf \times 100 \simeq 46$), while $SbCl_5$ formed a zone containing three points of concentration ($Rf \times 100 \simeq 0, 15, 31$) (see Fig. 108). These results appear to prove conclusively that both $SbCl_3$ and $SbCl_5$ react with vitamin A to produce a blue color.

4. The D vitamins

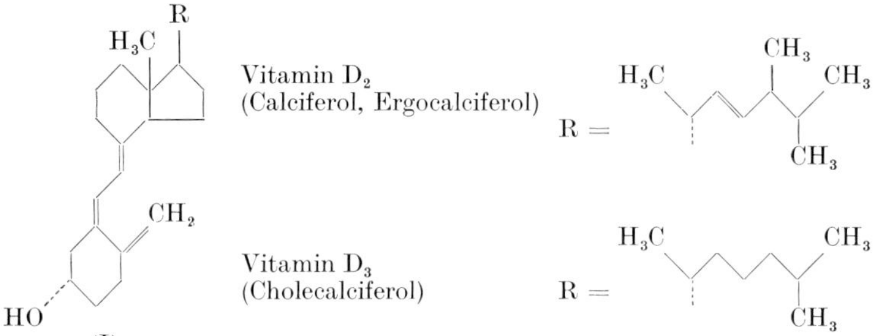

(I)

The D vitamins are sterols which can be produced by irradiation from precursors that are widely distributed throughout the plant and animal kingdoms. Vitamin D_2 and D_3, biologically the most active members, are very similar in chemical structure; they occur in small amounts in both human and animal organs. Oils from the liver and other viscera of various fish contain relatively large quantities of vitamin D_3. Both vitamins are very sensitive substances. The detection of these vitamins by physicochemical methods is chiefly hindered by accompanying substances. Numerous methods of detection and of separation by paper chromatography have been evolved [*23*, *65*], but the methods so far described have proved to be unsatisfactory because of incomplete separation, the instability of the vitamins on the adsorbents used and the labor required. With thin-layer chromatography, we are now able to assess the value of the analytical methods previously used, and to elucidate the behavior of the D vitamins under the influence of various factors.

a) Conditions of separation and results

A valuable study on the thin-layer chromatography of vitamin D has been published by JANECKE and MAAS-GOEBELS [*30*]. The work was carried out on Silica Gel G layers with the solvents hexane, hexane-ethyl acetate (90 + 10) and chloroform (see Table 29). A complete separation of the vitamins D_2 and D_3 from other sterols and degradation products can be achieved by this method; carotenoids migrate with the solvent front. A thin-layer chromatographic investigation of the reaction which takes place during the chemical estimation of vitamin D, has led to some interesting results. During the alkaline saponification and chromatographic examinations on activated adsorbents such as Floridin earth, Superfiltrol and (alkaline) alumina, various transformation products are formed; these could not be identified with certainty, however. The choice of adsorbents used for the purification is critical, and vigorous saponification [*60*] is not advisable.

The vitamins D_2 and D_3 can be separated on impregnated Silica Gel G (plate immersed in a solution of 5% liquid paraffin in petroleum ether, and dried) with an acetone-water mixture (80 + 20) (see Table 29). Silica gel layers of greater thickness (e.g. 0.4–2.0 mm) prepared by the standard method using the adjustable applicator, and employing cyclohexane-ether (50 + 50) as the solvent, have produced good results. Accompanying substances, degradation products, and the other vitamins can easily be separated by this method, which has the additional advantage that larger amounts of the extracts (0.1–1.0 ml) can be applied. All vitamin D : vitamin A ester mixtures used in practice can be separated. If the extract is, for example, applied in bands, quantities below 400 I.U. of vitamin D can be identified and estimated when mixed with 100,000 I.U. or more of vitamin A. This method is suitable for checking the purity of, as well as for the detection and estimation of, the vitamins D.

Table 29. *Approximate Values for Rf × 100 for the Vitamins D, and some Steroids in various Systems* (Layers: standard method; length of run, 12—16 cm)

Substances	Silical Gel G				Alumina	Silica Gel G impregnated with liquid paraffin
	Hexane-ethyl acetate (90 + 10) [*30*]	Chloroform [*30*]	Cyclohexane-ether (50 + 50)	Cyclohexane-ethyl acetate (90 + 10)	Benzene	Acetone-water (80 + 20)
Vitamin D_2	13	43	40	15	22	57
Vitamin D_3	13	43	40	15	22	54
Previtamin D_2	21	53	50	23	30	54
Previtamin D_3	21	53	50	23	30	50
Ergosterol (provitamin D_2) . .	9	33	30	—	—	60
7-Dehydrocholesterol (provitamin D_3) . .	—	33	—	—	—	—
Cholesterol	9	35	30	—	—	45

Thermal isomerization of D vitamins, first appreciated by Velluz et al. [*61*], can easily be followed on thin-layer chromatograms. Thus, on gently warming a solution of the D vitamins in benzene or chloroform in the dark and under nitrogen, about 15% is transformed at 60° C into previtamins D; the rest remains unchanged (see Fig. 109). These previtamins are in turn re-transformed, under the same conditions, into the vitamins D. The equilibrium ratio of vitamin D: previtamin D, and the rate of conversion, depend on the temperature. Thus, the transformation of vitamin D_3 to previtamin D_3 at 20° C in ethanol is much slower, and the equilibrium mixture contains only 7% of previtamin D_3 [*24*]. Vitamin D_2 behaves in the same way. In crystals, the elongated molecular shape corresponding to formula I (p. 223) was found [*6*]; to explain the thermal isomerization in solution an equilibrium mixture of I with the s-*cis* form (II) is assumed [*26*].

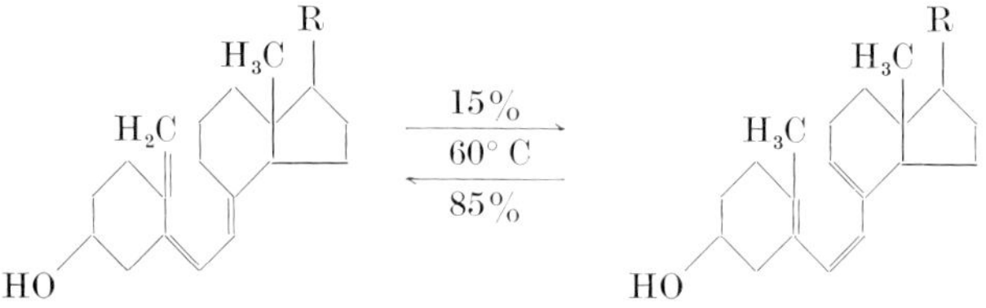

Vitamin D_2 and D_3 (II) Previtamin D_2 and D_3

Isomerization of the vitamins D_2 and D_3 to the previtamins D_2 and D_3 (accompanied by slight decomposition), may also be observed during alkaline saponification as well as during the storage of vitamin D preparations.

Both vitamins and previtamins D are unstable on dry silica gel layers. If the compound is applied to the active layer, and left for even one minute before developing, small amounts of degradation products, which are more strongly adsorbed, can be found. During development with the solvent, however, no further changes occur. It is therefore important that the plate is placed in the separating tank *immediately* after the application of the vitamin solution, developed, and the vitamin or previtamin D zones scraped off and eluted immediately after their localization. In this way, it was possible to reproduce quantitative results, colorimetrically and spectrophotometrically, in more than twentyfour determinations.

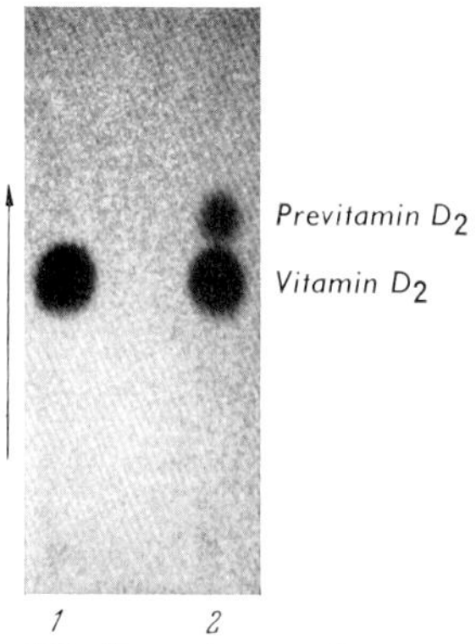

Fig. 109. Chromatogram of pure vitamin D_2 before and after heating in chloroform solution; photograph taken in ultraviolet light (254 mμ). *1* = 400 μg vitamin D_2, crystalline, untreated; *2* = 400 μg vitamin D_2, crystalline, after heating at 60° C for five hours in chloroform solution, in the absence of light and air

The biological value of the previtamins D is considerably lower (by some 40%) than that of the corresponding vitamins [*50*] although, (because of transformations) the absolute activity can be only approximately estimated. According to HANEWALD et al. [*24*], pure previtamin D_3 has only 35% of the antirachitic activity of pure vitamin D_3. This result has been discussed in connection with the kinetic equilibrium previtamin $D_3 \leftrightharpoons$ vitamin D_3, and the activity observed could be deduced from the quantities of vitamin D_3 formed in the test animal.

In our own experiments the previtamin D_2 showed, after purification by thin-layer chromatography, about 56% of the activity of pure vitamin D_2.

In the extraction of the vitamins D, saponification must be avoided, and any warming of the solution during the course of the analysis made as short and as gentle as possible. The active principle can best be extracted from mixtures with chloroform; or, after liberation using aqueous ammonia, with petroleum ether (30°–45° C). This extract can then be concentrated in a rotary evaporator, under vacuum, at 40° C. In this way, less than two per cent of the vitamins present become isomerized.

If vitamin D solutions are exposed to light and air, other compounds develop in addition to these products of isomerization. Crystalline vitamin D is much more stable when pure, and no chemical changes can be detected in it by thin-layer chromatography after twenty four hours exposure to light and air, even at 45° C. When stored in a cool, dark place, and when kept under an inert atmosphere, pure vitamin D may remain stable for several years.

b) Detection and evaluation

JANECKE and MAAS-GOEBELS [*30*] stained vitamin D spots by spraying the chromatogram with a phosphotungstic acid solution (reagent No. 121), and then heating for twenty minutes at 70° C. The vitamins D_2 and D_3 formed light brown stains, the smallest detectable quantity being 0.1–0.2 μg while other sterols stained in different colors. The $SbCl_3$ reagent (No. 11) has also been recommended for the detection. In our experience, the semi-quantitative estimation of small amounts of Vitamin D by visual comparison with a series of standard spots (0.1–0.2 μg) which are only just visible, is difficult. Fat components,

Table 30. *Detection of Vitamin D with the Aid of Spraying Reagents on Silica Gel G Layers*

Indication with			Color of vitamin D_2- and D_3-spots, and determination in		Approximate limits of detection* (Thickness of Layer 0.25 mm)	
reagent	Type of spray	Processing and evaluation after	Long-wave ultraviolet light (fluorescence)	Daylight	μg	I.U.
$SbCl_3$ (reagent No. 11)	strong	5 min/120°	orange*	grey-blue	0.025	1
$SbCl_5$ (reagent No. 13,	weak	1 min	—	orange red*	0.3	12
in chloroform) . . .		5 min/120°	orange*	—	0.3	12
Concentrated sulphuric acid	weak	3 min	—	D_2 brown* D_3 green*	30	1200
	moderate	5 min/120°	orange*	grey-blue*	0.3	12
Phosphotungstic acid (reagent No. 121). .	moderate	20 min/ 70°	—	grey-brown*	0.2	8
Phosphomolybdic acid (reagent No. 120b) .	moderate	1 min	—	grey-blue*	0.3	12
Trichloroacetic acid (reagent No. 143, 1%)	moderate	5 min/120°	bluish*	—	0.1—0.2	4—8
Trifluoroacetic acid (reagent No. 144). .	moderate	5 min/120°	bluish*	—	01.—0.2	4—8

and degradation products of vitamin A, migrate over a similar distance, react similarly with this spray reagent, and, consequently interfere with the results.

The D vitamins can be identified and recognized by inspection of the fluorescent plate in ultraviolet light of short wavelength (0.5 μg is still visible as a dark absorption spot); and by spraying with various reagents (see Table 30). The vitamins D_2 and D_3 can be distinguished from each other with the aid of concentrated sulphuric acid — provided at least 30 μg are present — as brown and green spots, respectively. The most sensitive detection can be obtained with $SbCl_3$, followed by heating the plate to 120° C (0.025 μg; see Table 30 and Fig. 110). The previtamins D behave similarly to the corresponding vitamin D isomers; but other sterols give different stains with different reagents.

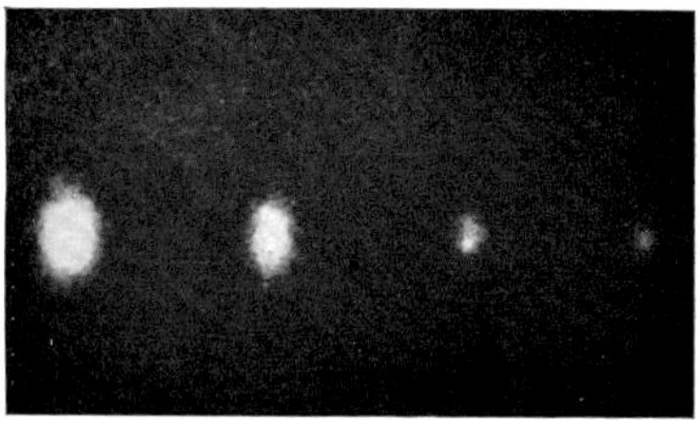

Fig. 110. Limit of detection of vitamin D on the plate after spraying with $SbCl_3$ and heating; viewed in long-wave ultraviolet light

Using the methods described above, vitamin D can be identified in many preparations, and can then be eluted and determined colorimetrically or by spectrofluorimetry [*31*].

c) Determination of vitamin D in various preparations

After numerous experiments, we were able to develop a relatively simple method for the estimation of vitamin D in oily concentrates and various other forms, but this method will be discussed here only briefly[1]. The method is specific, sensitive and easily reproducible. Vitamin D is practically unaffected during the process and a duplicate analysis can be carried out in three to six hours, depending on the method by which the extract is produced.

Depending on the composition of the samples, vitamin D can be extracted and concentrated diretly or following liberation with water and ammonia, and/or after freezing out the fats, Quantities of 0.1—1.0 ml of the extract in question are applied in bands on Silica Gel G layers of 0.4—2.0 mm in thickness, and are developed with cyclohexane-ether (50 + 50). After localization, the vitamin D band is scraped off directly into a chromatographic column, eluted with chloroform, and concentrated in a rotating evaporator under vacuum at 40° C. The residue containing vitamin D is dissolved in sufficient chloroform, mixed, and determined colorimetrically by the well-known method [*60*] using an $SbCl_3$ reagent[2] and acetic anhydride as color inhibitor. It may also be possible to make the estimation spectrofluorometrically [*31*].

The preliminary treatment of the samples for the preparation of an extract suitable for chromatography must be checked in each special

[1] The method will be described in detail in another publication.

[2] 2% acetic anhydride was added to the reagent for greater stability.

case. A quantity of extract used for chromatography which contains at least 800 I.U. (= 20 μg) of vitamin D is sufficient for repeated colorimetric measurements within the optimal range of the colorimeter. Chloroform is used as the solvent for the color reaction. The $SbCl_3$-acetyl chloride reagent (3 ml) is added to the solution containing the vitamin D (2 ml, containing 200—400 I.U.) in matched $^3/_4$-inch cells. It is then mixed, and the complex formed is measured at 500 mμ in the "Spectronic 20"[1] at maximum intensity, 15—45 sec after the addition; a blank solution is treated and read simultaneously[2]. The result is based upon calculation from freshly prepared vitamin D_2 or D_3 standard solutions read directly.

Errors arising from degradation products of vitamin A, which fail to separate and which interfere with the color reaction, can be allowed for and eliminated by the addition of the color inhibitor. Fat components which may still be present react only after one minute and are not measured in our method.

Using this procedure, we analyzed oily and "dry" concentrates containing vitamin D alone, or vitamin D and vitamin A together. Multivitamin tablets and solutions, chocolate and weight-reducing preparations

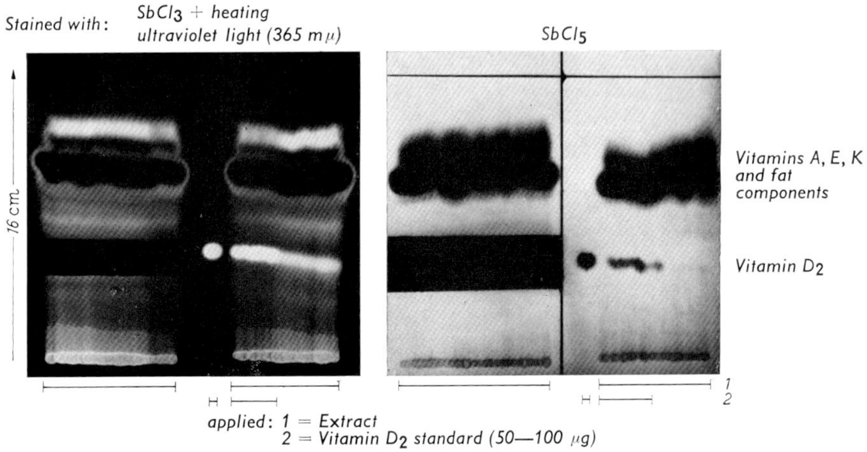

Fig. 111. Chromatograms of multivitamin tablets A 118945, amounts of 0.5 ml extract each (about 800 I.U. vitamin D_2) applied; photograph taken after the scraping off (left) and identification of the vitamin D strip. Pure vitamin D is added in the guide chromatogram (each at right), alone and together with the extract

(containing 100 I.U. D_2/25 g), were also analyzed, and the results obtained were entirely consistent with the comparable biological values. During the separation of an extract solution containing about 800 I.U. of vitamin D, a relative standard deviation of 2.8% can be expected (see example).

[1] Supplied by Bausch and Lomb, Rochester, N. Y., U.S.A.

[2] Smaller quantities of vitamin D can be estimated spectrophotometrically in 1 cm cells, using a modified color reaction (2 ml vitamin D solution + 2 ml reagent).

Analytical example: Multi-vitamin tablets A 118945

Claim:		150 I.U. D_2/tablet
Found by bioassay:		152 I.U. D_2/tablet
Found upon extraction, TLC and colorimetry:	166/164/164/156/ 161/170/166/168/ 155/164/166/168/ 160	164 I.U. D_2/tablet

5. Tocopherols (vitamin E)

The constitution of the known naturally-occurring tocopherols [*1*] may be shown as:

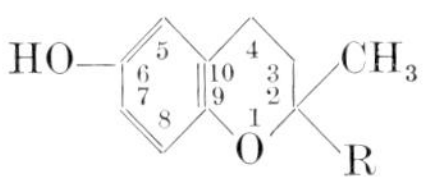

$$R = -(CH_2-CH_2-CH_2-\overset{\overset{\large CH_3}{|}}{C}H)_3-CH_3 \qquad \text{Tocol}$$

$$R = -(CH_2-CH_2-CH=\overset{\overset{\large CH_3}{|}}{C})_3-CH_3 \qquad \text{Tocotrienol}$$

Tocopherol	*Chemical name*
α	5,7,8-Trimethyltocol
β	5,8-Dimethyltocol
γ	7,8-Dimethyltocol
ζ_2	5,7-Dimethyltocol
δ	8-Methyltocol
η	7-Methyltocol
ζ_1	5,7,8-Trimethyltocotrienol
ε	5,8-Dimethyltocotrienol
	[5-Methyltocol, synthetic]

Tocopherols occur widely in nature. They are optically active derivatives of 6-chromanol; they differ in the number and the position of the methyl groups, and – as shown only recently [*19, 20, 37*] – in their terpene side-chains. The naturally-occuring tocopherols are sterically homogeneous; synthetic preparations, however, consist of more or less complex diastereoisomeric mixtures depending on the methods of preparation.

The tocopherols fulfil a multitude of biological functions; α-tocopherol, which occurs most frequently in nature, exhibits the highest activity.

Since the stability of the tocopherols can be improved by esterifying the phenolic-OH group, esters such as the acetate or the succinate are frequently used for practical applications.

The estimation of the tocopherols is difficult because the extraction of this easily oxidized material must be completed without loss and because accompanying substances interfere with the colorimetric determination [*12*]. Both paper chromatography and thin-layer chromatography have been found useful for the elimination of accompanying substances and artifacts [*4*].

a) Conditions of separation and results

SEHER [*47*], when estimating antioxidants, had separated α-tocopherol in one and two dimensions on activated Silica Gel G layers, using toluene, benzene and chloroform as solvents. The same author [*48, 49*] was also able to separate the various tocopherols on Merck Silica Gel G with chloroform, and on FLUKA alumina with the aid of benzene (see Table 31). The application of this technique to concentrated preparations of vitamin E yielded not only a highly satisfactory separation of α-, β- plus γ- and δ-tocopherol, but also separated, simultaneously, the various accompanying substances in the unsaponifiable residues of oils and fats, or in molecular distillates. After spraying, the α-tocopherol was estimated by measuring the size of the spot.

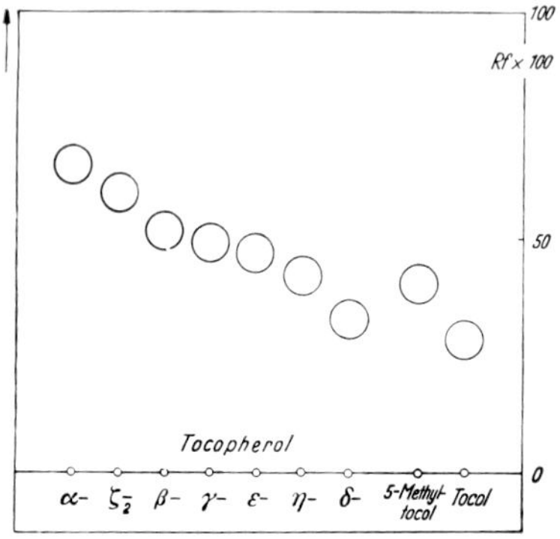

Fig. 112. Separation of tocopherols and related compounds (50 μg of each) on Silica Gel G with benzene-methanol (98 + 2); height of ascent, 18 cm; time of run, 50 minutes (compare Table 31)

The tocopherols also migrate at a different rate on Silica Gel G when a benzene-methanol mixture (98 + 2) is used, and on secondary magnesium phosphate with petroleum ether (30 – 45° C)-ether (85 + 15) as solvent (see Table 31 and Fig. 112). These methods are also suitable for the analysis of natural concentrates and extracts, and for the investi-

Table 31. *Approximate Values of Rf × 100 for Various Tocopherols* (Time of run, about 45 min; length of run of the solvent front, 12—18 cm)

Layer: standard method:	Silica Gel G	Alumina	Silica Gel G	*sec.* Magnesium phosphate	Silica Gel G
Solvent:	Chloroform [*48, 49*]	Benzene [*48, 49*]	Benzene-methanol (98 + 2)	Petroleum ether (30—45° C)-ether (85 + 15)	Cyclohexane-ether (80 + 20)
Tocol	19	—	28	—	—
α-Tocopherol	58	56	65	86	32
β-Tocopherol	35	34	51	76	30
γ-Tocopherol	37	31	48	71	26
ζ_2-Tocopherol.	49	48	59	82	29
δ-Tocopherol	23	21	32	61	21
η-Tocopherol[1].	32	28	41	—	23
ε-Tocopherol[1]	32	30	46	—	26
5-Methyltocol	27	24	40	—	26
α-Tocopheryl acetate .	—	71	—	—	40
α-Tocopheryl succinate	—	—		—	30—40 (tail)
α-Tocopheryl phosphate	—	—	—	—	0

[1] These two tocopherols were put at our disposal by courtesy of Dr. J. GREEN, Vitamins Ltd., Tadworth, England.

gation of oxidation and decomposition products. An optimal separation of critical components such as β- and γ-tocopherol can be achieved by means of the wedged-tip technique. α-Tocopherol and its esters can be readily analyzed on silica gel plates with cyclohexane-ether (80 + 20), and according to WALDI [*66*], the same is true of the solvent mixture cyclohexane-chloroform-glacial acetic acid (60 + 30 + 10). On thicker layers (0.4–2.0 mm), up to 10–20 mg can be applied, and the spots which develop are still compact.

It should be possible to isolate the various tocopherols by partition chromatography on impregnated layers, or by TLC of their nitroso derivatives in one or two dimensions [*36*].

b) Detection and evaluation

SEHER [*47, 48, 49*] used a 1% solution of 2,6-dichloroquinone-chloroimide in ethanol for the staining of α-tocopherol and α-tocopheryl acetate to yellow-brown and pink, respectively. β-Tocopherol produces brown spots, and γ-tocopherol blue spots with ceric sulphate (Reagent No. 25). After heating for ten minutes at 120° C, α-tocopherol gives a pink color with the same spraying solution, and exhibits pronounced bright fluorescence under ultraviolet light. With phosphomolybdic acid (Reagent No. 120b), the free tocopherols give grey-blue spots, even in the cold, the α-compound responding most readily. The esters react less readily, and only after prolonged heating at 110° C. Subsequent treatment of the plates with ammonia vapor two or three minutes after the spraying

Table 32. *Colors given by the Tocopherols after moderate Spraying of the dried Plates with Antimony Pentachloride Reagent. Assay after about 3 minutes* (Limit of detection, about 2 μg; optimal differentiation of color with 20—50 μg)

Tocopherols (d- and d,1-forms, respectively)	Staining reaction with $SbCl_5$ on		
	Silica Gel G	Alumina	*sec.* Magnesium phosphate
α	orange-red	orange-pink	orange-pink
β	light brown	brown	yellow-brown
γ	green	green	lilac
ζ_2	orange-yellow	cream	orange
δ	red-brown	violet	violet
η	russet	lilac	violet
ε	brown	olive-brown	brown
5-Methyltocol	lilac-grey	lilac-grey	—
Tocol	lilac-grey	—	—

clears the background, and all the compounds stain to a uniform grey-blue (the limit of detection being to 1 μg; or after heating, to 0.5 μg). SEHER [*48, 49*] ran a series of amounts (10, 20, 40 and 60 μg) of pure α-tocopherol simultaneously and, after visualizing the spots, obtained a semi-quantitative estimation by comparing the sizes planimetrically.

The red color complex which develops after treatment with iron-dipyridyl reagent[1] can be used for quantitative analyses. 0.5 μg of α-tocopherol becomes visible, but the other tocopherols react at different rates and to different extents. The red color is extracted with water and determined photometrically. The $SbCl_5$ reagent (No. 13, in chloroform) gives a variety of colors with different tocopherols, which may be used to differentiate between them (Table 32), because this reagent responds to the position and number of the methyl groups in the chroman ring, and to the nature of the side-chain. These color complexes are unstable.

As little as 10 μg of tocopherol can be recognized as a dark spot on silica gel layers in short-wave ultraviolet light using the luminescent substance "ZS-Super". Just as in paper chromatography [*62*], sodium fluorescein (0.02%) can be added to the silica gel to produce tocopherol spots which appear, in ultraviolet light, violet against a fluorescent background (the least detectable quantity is about 0.2 μg) or red in daylight (the least detectable quantity is 2 μg).

For quantitative estimation, the spot is first localized (preferably under ultraviolet light) and eluted; the color complex which develops with iron-dipyridyl is measured colorimetrically in the manner described for the corresponding paper chromatographic technique [*62*]. The α-tocopherol content in various preparations and in some natural concentrates was estimated by this method. Smaller amounts of this vitamin (as low as 0.01 μg/ml) can also be detected by spectrofluorometry [*11*].

6. K vitamins and ubiquinones

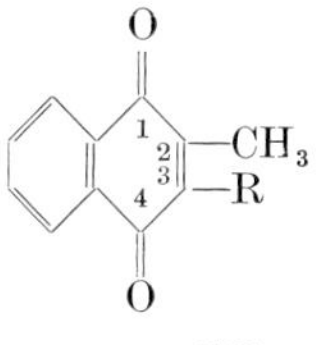

$$R = -CH_2-CH=\overset{\displaystyle CH_3}{\overset{|}{C}}-(CH_2-CH_2-CH_2-\overset{\displaystyle CH_3}{\overset{|}{C}H})_3-CH_3 \qquad \text{Vitamin } K_1$$

$$R = -CH_2-CH=\overset{\displaystyle CH_3}{\overset{|}{C}}-(CH_2-CH_2-CH=\overset{\displaystyle CH_3}{\overset{|}{C}})_6-CH_3 \qquad \text{Vitamin } K_{2(35)}{}^{2}$$

$$R = -H \qquad \text{Vitamin } K_3 \text{ (Menadione)}$$

The vitamins K_1 and K_2 occur widely in nature. Various K_2 vitamins having different numbers of isoprene units in the side-chain have now been described. Synthetic vitamin K_3 is transformed in the human and

[1] The iron-dipyridyl reagent consists of a mixture of a 0.5% solution of α,α′-dipyridyl in ethanol, and a 0.2% solution of ferric chloride in ethanol (which should be stored in the dark); prepared, in the proportion of 1 + 1, shortly before use. The chromatogram must be heavily sprayed.

[2] The number of C-atoms in the side-chain is quoted in parentheses.

animal intestine and provided with a polyisoprenoid side-chain, in order to achieve full vitamin K activity. For determination, the sensitive K vitamins must be worked up very carefully and isolated. Column and paper chromatography [*23*, *65*] have been used for this separation; thin-layer chromatography, which is less time-consuming, has also been applied recently.

a) Conditions of separation and results

The K vitamins can be investigated chromatographically by various methods (compare also Table 26 and Fig. 102). Vitamin K_3 ($Rf \times 100 \simeq 38$) can be separated from the vitamins K_1 and $K_{2(35)}$ ($Rf \times 100 \simeq 61$) on Silica Gel G with cyclohexane-ether (80 + 20). This procedure has been useful in identifying the K vitamins in various preparations. Like the other fat-soluble vitamins, these compounds are relatively unstable on the active layer and must be processed accordingly.

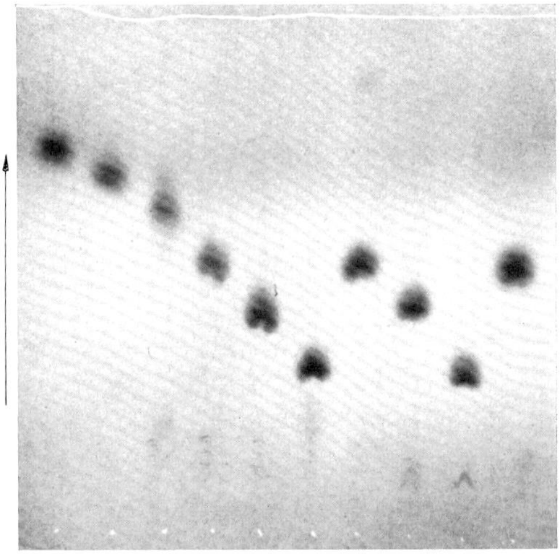

Fig. 113. Separation of vitamins K_1 and K_2 and the related benzoquinones (see text) [*44*]. Time of development, 1 hour; plate sprayed with sulphuric acid after heating to 130° C; 20 μg of each substance applied. *1* ubiquinone (30): x = 5, *2* ubiquinone (35): x = 6, *3* ubiquinone (40): x = 7, *4* ubiquinone (45): x = 8, *5* ubiquinone (50): x = 9, *6* plastoquinone (45), *7* vitamin $K_{2(30)}$, *8* vitamin $K_{2(35)}$, *9* vitamin $K_{2(45)}$, *10* vitamin K_1

GLOOR [*16*] investigated chromatographically the vitamins K_1 and K_2 having side-chains of varying length, and related benzoquinones; he used the solvent system acetone-water (95 + 5) on impregnated Silica Gel G. The plates were immersed in a solution of 5% liquid paraffin in petroleum ether, and dried (see Fig. 113 and Table 33). During this process, the homologous series are separated in a regular manner, while the vitamins K_1, $K_{2(30)}$ and ubiquinone (45) migrate to the same level. Using this method, vitamin $K_{2(45)}$ was detected in *Mycobacterium tuberculosis;* and 90% of ubiquinone (45), together with ubiquinone (50) and ubiquinone (40) were detected in homogenates of rat livers. Thin-layer chromatography has also been used by BILLETER [*2*] to study the transformation of the vitamins $K_{2(30)}$ and $K_{2(10)}$ into $K_{2(20)}$ in the organs of birds and mammals.

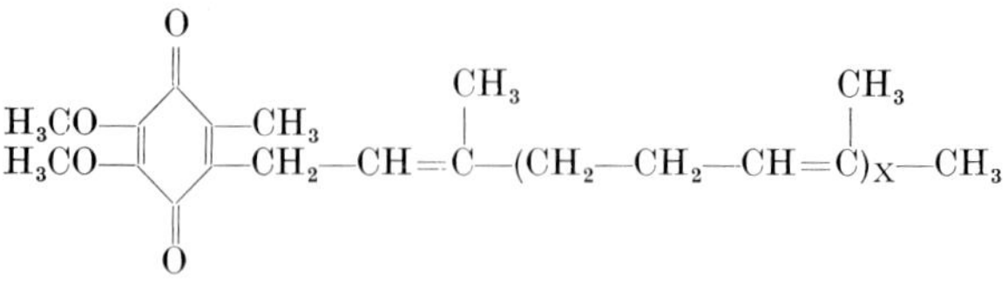

Ubiquinones: x = 0—9

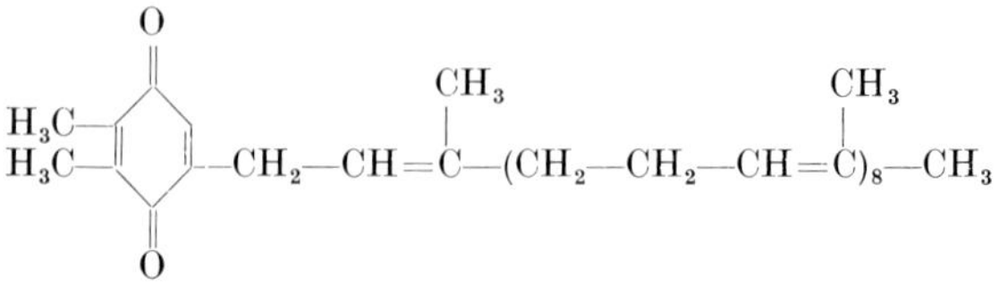

Plastoquinone (45)

Ubiquinones (5)—(50) were separated on mixed layers (Fig. 114). The separation of various ubiquinones has also been reported by Wagner et al. [64], who used a combined method to detect ubiquinones in

Table 33. *Approximate Values for Rf × 100 for the Ubiquinones in various Systems* (Length of run, 10—16 cm)

Ubiquinones	Silica Gel G impregnated with liquid paraffin		Silica Gel G-Kieselguhr G (1+1), impregnated with liquid paraffin
	Acetone-water (95 + 5) [*16*]	Aetone-water saturated with liquid paraffin (90+10) [*64*]	Methanol saturated with liquid paraffin-isopropanol (90+10) [*53*]
(5)	—	—	90
(10)	—	—	85
(15)	—	—	78
(20)	—	—	68
(25)	—	—	60
(30)	74	71	44
(35)	69	63	35
(40)	63	55	25
(45)	52	42	14
(50)	43	31	8

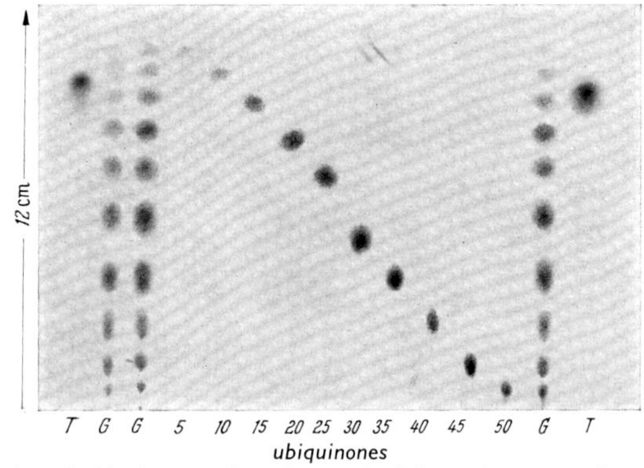

Fig. 114. Separation of ubiquinones (5)—(50) on mixed layers impregnated with liquid paraffin (see Table 33) [*53*]. *T* = DESAGA test mixture. *G* = mixture of ubiquinones. 2 μg of each substance were applied and rendered visible with phosphomolybdic acid reagent (No. 120b)

biological material (for example, ubiquinone(50) in the bovine myocardium) by first separating the lipids from the ubiquinones ($Rf \times 100 \simeq 86$) on Silica Gel G with benzene-chloroform (50 + 50), and then separately analyzing the latter.

b) Detection and evaluation

In ultraviolet light the vitamins K are detectable as dark absorption spots on layers impregnated with a fluorescent indicator, the minimum detectable quantity being 0.5 μg. After exposure for ten minutes to a quartz lamp situated a short distance away, the spots show a yellow fluorescence (the smallest detectable quantity being 0.3 μg of vitamin K_1). By spraying the hot plate with concentrated sulphuric acid, the K vitamins and the ubiquinones can be stained to a violet color (about 3 μg) [*16*]. Phosphomolybdic acid (Reagent No. 120b) is just as unspecific, but is more sensitive. The vitamins K may also be identified using sodium diethyldithiocarbamate [*27*], or xanthane hydride [*46*] or 2,4-dinitrophenylhydrazine [*45*]. The ubiquinones are visible on the impregnated layers in ultraviolet light as fluorescent brown spots. They can also be recognized with antimony trichloride (Reagent No. 11, in chloroform), after subsequent heating for ten minutes at 110° C, as grey-violet spots; or with rhodamine (Reagent No. 129) in ultraviolet light as fluorescent violet spots [*64*]. The comparative method is suitable for the semi-quantitative estimation of the vitamins K and the ubiquinones.

IV. Thin-layer chromatography of water-soluble vitamins[1]

Thin-layer chromatography has proved to be a valuable analytical technique in this field. The method, which is distinguished by its simplicity and by the short time required for the development, is now widely used, although only relatively few publications have appeared [*15, 39, 40*]. Water-soluble vitamins can be separated with the aid of the procedures recommended, but the *Rf*-values may be somewhat different.

A knowledge of the solubility of the active components (which may vary widely) is of great importance in the various stages of extraction, application and separation. The fact that the water-soluble vitamins are much more stable than the fat-soluble compounds on dry silica gel layers, considerably facilitates the application and quantitative determination.

1. Mixed water-soluble vitamins

Various vitamins of the B group, and vitamin C, were separated by Gänshirt and Malzacher [*15*] by the one-dimensional method, on layers prepared by Stahl's method for Silica Gel G containing 2% of the fluorescent indicator "ZS-Super". The solvent mixture, glacial acetic acid-

[1] For some general remarks on the vitamins, and techniques, see p. 210.

acetone-methanol-benzene (5 + 5 + 20 + 70) produced the best results (see Table 34). The method has been developed for combinations involving 1—10 μg of the B vitamins and 5—30 μg of vitamin C, since such combinations occur in pharmaceutical preparations. This method can be used generally, provided 50—100 μg of the mixture is applied. To achieve the complete detection of all the components, two starting spots

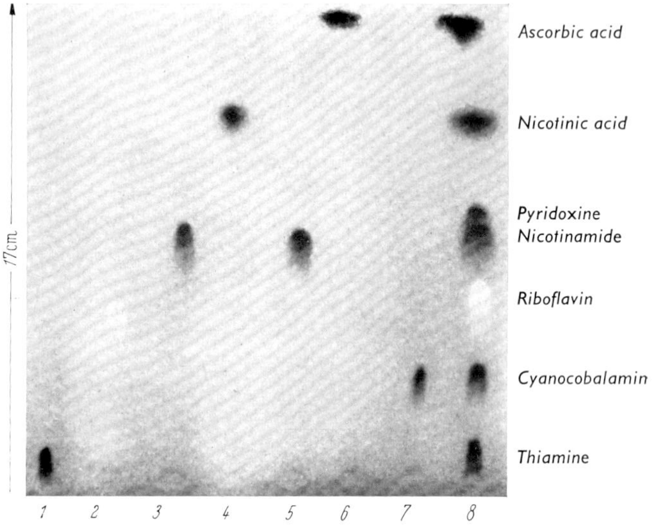

Fig. 115. One-dimensional chromatogram of water-soluble vitamins after separation with water on silica gel mixed with fluorescent indicator (time of run, about 40 minutes; photograph taken in short-wave ultraviolet light). *1* = 30 μg thiamine-HCl, *2* = 3 μg riboflavin, *3* = 30 μg pyridoxine-HCl, *4* = 30 μg nicotinic acid, *5* = 30 μg nicotinamide, *6* = 30 μg l-ascorbic acid, *7* = 10 μg cyanocobalamin, *8* = substances and quantities of *1*—*7*

from every sample must be applied to the plate. For detection, the developed and dried chromatogram (length of run, 19 cm) is inspected under long-wave ultraviolet light. Riboflavin can be recognized as a yellow, fluorescing spot against a dark background. In short-wave ultraviolet light, vitamin B_1, vitamin C and nicotinamide are visible against the bright fluorescent layer as dark absorption spots, riboflavin as a yellowish spot, and vitamin B_6 as a dark blue one. For the identification of biotin, one chromatogram is covered and the other is sprayed with iodoplatinate (reagent No. 76). Biotin becomes visible as a white spot against a pink background. Simultaneously, vitamin B_1 shows a grey, nicotinamide a pale yellow color, and vitamin C a yellow color. To detect vitamin B_6 and pantothenic acid, the lower part of the second, previously-covered chromatogram, is sprayed with dichloroquinone-chloroimide (reagent No. 41; 0.1%) and exposed to ammonia vapor, vitamin B_6 staining to a blue color. The chromatogram is subsequently heated at 160° C for half an hour, and the upper part of the strip is then sprayed with ninhydrin (reagent No. 108; 0.5% in ethanol). After further brief heating at 160° C, pantothenic acid appears as a violet spot.

These mixtures can be partially separated on a similar silica gel layer if water is used as the developing solvent (see Fig. 115); water migrates at approximately the same rate as the solvent mentioned above (see Fig. 116) and, furthermore, it permits immediate development after the application of aqueous extracts to the plate. The comparable results are

Table 34. *Approximate Values for Rf × 100, and the direct Detection of Water-Soluble Vitamins in different Types of Light, on Silica Gel Layers, in the Presence of Fluorescent Substances*

Vitamins	Approximate values for *Rf* × 100 with		Visible in			Approximate limits of detection in μg
	Glacial acetic acid-acetone methanol-benzene (5 + 5 + 20 + 70) [*15*]	Water	Ultraviolet light (254 mμ)	Ultraviolet light (365 mμ)	Day-light	
B_1 (thiamine-HCl and -HNO_3) . .	0	5	violet	—	—	2
B_2 (riboflavin) . .	35	38—45	yellow	yellow	yellow	0.1/0.01/0.3
B_6 (pyridoxine-HCl)	15	52	dark-blue	dark-blue	—	3/10
Nicotinic acid. . .	75	78	dark	—	—	3
Nicotinamide . . .	65	49	dark	—	—	3
Pantothenic acid (Na and Ca salt)	57	89	—	—	—	—
B_{12} (cyanocobalamin)	0	22	dark	violet	red	1/0.5/0.3
Folic acid	0	0	dark	dark-blue	yellow	2/2/2
Biotin	80	70	—	—	—	—
C (l-ascorbic acid) .	30	96	dark-blue	—	—	3

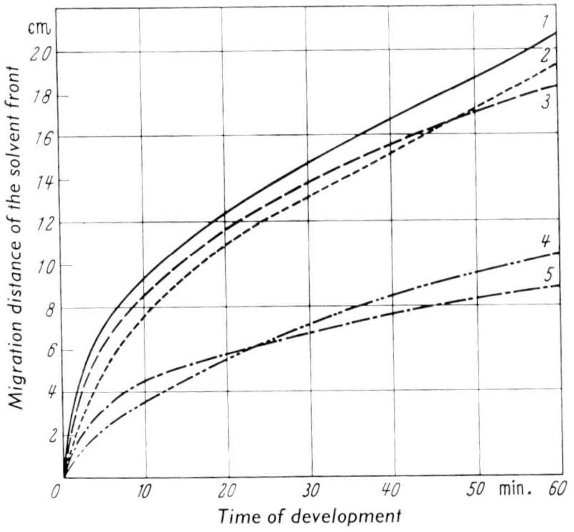

Fig. 116. Migration rates of various solvents on silica gel layers (0.25 mm thick, at 20—23° C); solvents: *1* = water, *2* = cyclohexane-ether (80 + 20), *3* = glacial acetic acid-acetone-methanol-benzene (5 + 5 + 20 + 70) [*15*], *4* = pyridine-glacial acetic acid-water (10 + 1 + 40) *5* = *n*-propanol — 10% ammonia (95 + 5) [*39*]

listed in Table 34, which also shows the direct identification of water-soluble vitamins in light of different wavelengths, and the approximate limits of the identification on the unsprayed chromatogram. As mentioned above, the compounds may be detected with the aid of spraying reagents, or after chlorination with o-tolidine-potassium iodide solution (reagent No. 32). With the latter reagent, most vitamins give colors at very low concentrations: B_1, B_6, pantothenic acid and biotin, a blue color; vitamin B_2, olive-green; nicotinic acid, nicotinamide and folic acid, a grey; and vitamin B_{12}, a violet color. Vitamin C does not react, and solvents containing ammonia or sodium carbonate interfere with the identification. It should be emphasized that large amounts of vitamin B_2 and vitamin B_6 tend to produce tailing spots.

Vitamin B_1 and folic acid, which are very strongly adsorbed on the silica gel layer, may be moved from the starting point by subsequent chromatographic analysis in a direction at right angles to the original, using a suitable solvent.

Both procedures thus permit a rapid, simultaneous identification of the compounds concerned in pharmaceutical preparations and probably also in natural extracts. Some of the vitamins can be estimated visually, after the simultaneous development of a series of standards, as mentioned in other examples.

Various possibilities relating to the separation and selective estimation of individual water-soluble vitamins are discussed separately later.

2. Thiamine (vitamin B_1)

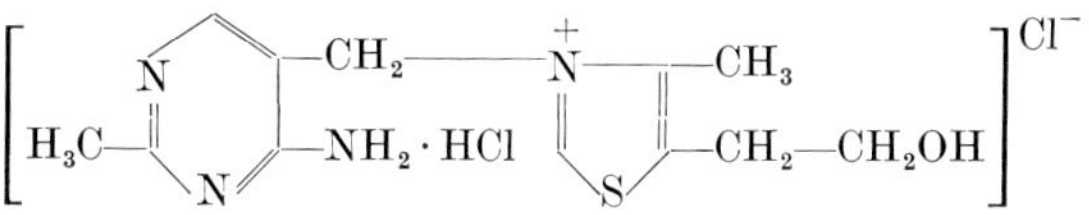

Thiamine hydrochloride

Thiamine occurs in natural products in its free form, as the mono-, di- and tri-phosphoric esters, as thiamine disulphide, and sometimes in other sulphur-containing compounds. In pharmaceutical preparations it is used as the hydrochloride, mononitrate and diphosphoric ester (cocarboxylase); it may be extracted directly with water from most samples. The processing should take place within the p_H range of 4 to 6 since the sensitive thiamine compounds are not destroyed in this range.

a) Conditions of separation and results

Although thiamine migrates only very slowly on silica gel layers with solvents such as glacial acetic acid-acetone-methanol-benzene (5 + 5 + + 20 + 70) [*15*], and water, thiamine hydrochloride and thiamine mononitrate ($Rf \times 100 \simeq 56$, compact spot) can be separated from cocarboxylase ($Rf \times 100 \simeq 33$, zone formation) with the aid of pyridine-glacial acetic acid-water (10 + 1 + 40) at p_H 5.8. Similar results were obtained by STROHECKER [*55*] on the same adsorbent, with 20% urea solution-85% formic acid—n-propanol (1 + 2 + 3).

The satisfactory separation of even larger amounts of all thiamine compounds in the presence of various additional substances can be expected from chromatography on ion-exchange layers [*43*], and from thin-layer electrophoresis, a technique which has also led to satisfactory results with phenols, phenol carboxylic acids [*41*] and amino acids [*25*].

b) Detection and evaluation

Thiamine, a substance which fluoresces violet in short-wave ultraviolet light, is visible in amounts of at least 2 μg. The detection can be rendered more sensitive and more specific by spraying the plate with freshly prepared potassium ferricyanide (Reagent No. 81), when as little as 0.03 μg of oxidized thiamine (thiochrome) can be seen as a bright blue fluorescence under the quartz lamp (365 mμ). Thiamine can also be colored grey with iodoplatinate (Reagent No. 76) or brown on pyridine-containing layers, and with other Reagents [*65*].

A quantitative estimation of thiamine can be made by running a series of thiamine standards and making a visual comparison on the plate after oxidizing to thiochrome, but can probably also be made after elution of the thiochrome with *iso*-butanol.

3. Riboflavin (vitamin B_2)

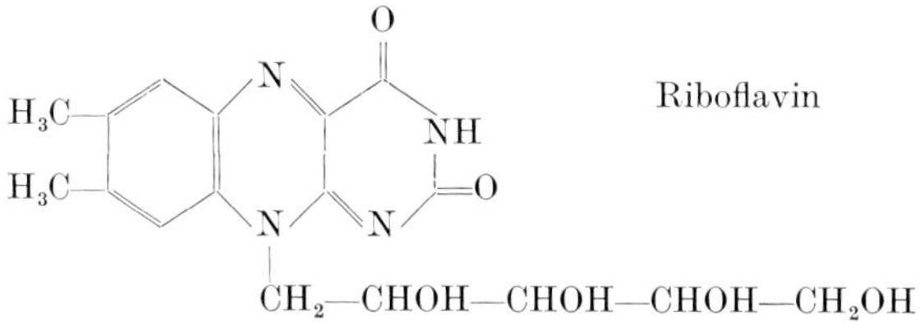

Riboflavin

Riboflavin is sometimes found in the free form but more often as the riboflavin 5′-phosphate ester (flavin mononucleotide), and also as flavin adenine dinucleotide, a coenzyme of d-amino acid-oxidase. Riboflavin is very sparingly soluble in water, but it may be extracted for quantitative analysis with pyridine-glacial acetic acid-water (10 + 1 + 40).

a) Conditions of separation and results

On silica gel layers the approximate $Rf \times 100$-values found for riboflavin are 82, and for riboflavin 5′-phosphate sodium, 45, with the solvent system pyridine-glacial acetic acid-water (10 + 1 + 40). Both compounds can also be separated – in small amounts – with water ($Rf \times 100 \simeq 40$ and 28, respectively) and with a mixture of glacial acetic acid-acetone-methanol-benzene (5 + 5 + 20 + 70) [*15*] ($Rf \times 100 \simeq 35$ and 0, respectively); and can be distinguished from other by-products and degradation products. Ion-exchange chromatography [*43*] and thin-layer electrophoresis [*25*, *41*] are probably also suitable for this separation.

b) Detection and evaluation

Flavins, which are yellow, produce an intense yellowish fluorescence under the quartz lamp. In ultraviolet light at 365 mμ as little as 0.01 μg is still detectable on the plate; at 254 mμ the limit is 0.1 μg; and, in daylight, 0.3 μg of riboflavin can be seen. Visual comparison enables a semi-quantitative estimation to be made, while after scraping off and elution of the spot, colorimetric or fluorimetric determinations appear to be promising.

4. The vitamin B_6 group

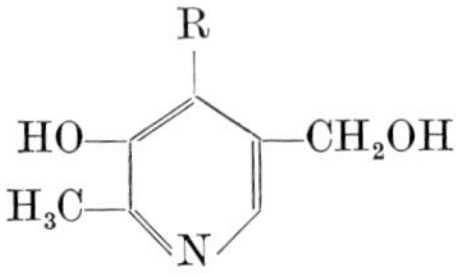

Pyridoxine, R = $—CH_2OH$
Pyridoxal, R = —CHO
Pyridoxamine, R = $—CH_2NH_2$

Vitamin B_6 occurs in all plant and animal cells in three forms: pyridoxine, pyridoxal and pyridoxamine. In the living body, they are usually bound to protein as 5-phosphoric esters, and they are easily transformed one into the other.

a) Conditions of separation and results

An excellent separation of these three derivatives of vitamin B_6 has been described by Nürnberg [*40*], using the gradual development technique in saturated jars recommended by Stahl [*52*]. The first development is carried out on an air-dried silica gel layer with pure acetone as solvent (length of run, 14 cm); then, after evaporation of the acetone with gentle heating, the plate is developed in the same dimension, with the solvent system acetone-dioxane-25% ammonia (45 + 45 + 10) (length of run, 14 cm). The development takes one and a half to two hours, and the spots containing 2—3 μg substance each, can be adequately estimated after the application of a color forming reagent.

After being dissolved in hot methanol, pyridoxine-HCl and pyridoxamine-2 HCl can be pipetted directly onto the plate, but the pyridoxal, which is also soluble and occurs as an "cyclic hemiacetal", is unstable and exhibits a tendency to form the acetal and to produce two or three spots on the chromatogram. If a similar methanol-solution is heated for an hour on a water bath, pyridoxal is transformed quantitatively into pyridoxal methyl acetal, which can then be conveniently separated from the other derivatives of vitamin B_6.

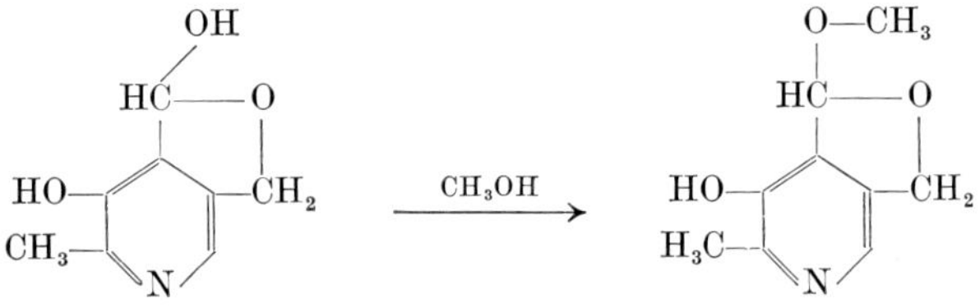

Pyridoxal "cyclic hemiacetal" Pyridoxal methyl acetal

The unchanged pyridoxal can be detected only after the application of an aqueous solution. This is not recommended because this technique of application is more difficult, and the spots are less intense. A satisfactory separation can be achieved on activated silica gel layers, using water or 0.2% ammonia as eluants, the latter yielding more compact spots; pyridoxine-HCl, pyridoxal 5′-phosphate, pyridoxamine-2 HCl and pyridoxal ethyl acetal-HCl (2–3 μg each) all quickly migrate over different distances (see Table 35).

Table 35. *Approximate Values for Rf × 100 for the B_6 Group on Silica Gel G*

Derivatives of vitamin B_6	Solvent		
	Acetone-dioxan-25% ammonia (45 + 45 + 10) [*40*]	0.2% Ammonia	Water
Pyridoxine-HCl	30	53	52
Pyridoxamine-2 HCl . . .	60	16	19
Pyridoxal 5′-phosphate .	—	75	64
Pyridoxal (possibly isomers)	~60 ~66	—	—
Pyridoxal methyl acetal .	45	—	—
Pyridoxal ethyl acetal-HCl	—	44	36

b) Detection and evaluation

Dark blue primary fluorescences are visible under ultraviolet light on the dried layer of silica gel containing a fluorescent indicator. At 254 mμ, as little as 3 μg of pyridoxine and pyridoxamine can be identified; and at 365 mμ, as little as 10 μg. After spraying with 0.4% solution of 2,6-dibromoquinone chloroimide in methanol (Reagent No. 40), or with a 0.1% solution of 2.6-dichloroquinone chloroimide in ethanol (Reagent No. 41), followed by treatment with ammonia vapor, all of the vitamin B_6 compounds stain a bluish color, the smallest detectable quantity being 0.1 μg. The weakest color is produced by pyridoxal (minimum quantity, 0.5–1 μg). Other staining reagents can also be used [*23*, *65*]. The relative contents of the various B_6 components can be estimated by using a series of different amounts of the pure substances treated in the same way.

c) The estimation of vitamin B_6 in multi-vitamin preparations

After extraction with water from pharmaceutical preparations, these compounds can be subjected to chromatographic examination, with water or 0.2% ammonia as a solvent and so estimated. The following procedure is recommended for pyridoxal by Nürnberg [*40*]: pyridoxal methyl acetal is produced by heating the material under examination (containing 10 μg pyridoxal-HCl) for one hour, in absolute methanol; the latter is then removed under reduced pressure, and the residue is dissolved in 50 ml of water. 5.0 ml of this solution is mixed with 5 ml of a p_H 1 buffer and a 2% aqueous solution of sodium tetraphenyl borate added; the mixture is left to stand overnight. The precipitated crystalline mass of the pyridoxal methyl acetal-tetraphenyl borate is filtered off under suction, dissolved in 50 ml of acetone, and applied to the air-dried silica gel layer together with pure pyridoxal treated in a similar manner (using a series of dilutions). The chromatogram is developed with chloroform-ethanol (99 + 1), and estimated after spraying ($Rf \times 100 \simeq 40-50$).

5. Nicotinic acid and nicotinamide

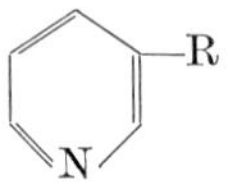

Nicotinic acid, R = —COOH
Nicotinamide, R = —$CONH_2$

Both of these compounds occur widely in nature. The free acid has the same biological activity as the amide; in the body it is transformed into the amide which is present in a bound form as a component of important enzymes.

a) Conditions of separation and results

A simple method has been reported by Nürnberg (*39*) for the separation of nicotinic acid and nicotinamide, found in various pharmaceutical multivitamin preparations. Nürnberg worked with air-dried Silica Gel G layers without tank saturation, and used a freshly prepared mixture of *n*-propanol and 10 per cent ammonia (95 + 5) as solvent. In this procedure, the solvent front ascends in about sixty minutes to a height of eight centimetres, the approximate $Rf \times 100$ values for nicotinic acid and nicotinamide being 35, and 60—70 respectively. After spraying, both pyridine compounds can be identified with certainty, and estimated.

Both vitamins can also be separated and estimated on activated silica gel layers with solvents yielding a slightly higher rate of migration (see Table 36 and Fig. 115.)

Table 36. *Approximate Values for Rf × 100 for Nicotinic Acid and Nicotinamide*

Layer:	Silica gel layer, dried in air	Activated silica gel layer	
Solvent:	n-Propanol-10% ammonia (95+5) [*39*]	Glacial acetic acid-acetone-methanol-benzene (5+5+20+70) [*15*]	Water
Nicotinic acid	35	75	78
Nicotinamide	60—70	65	49

b) Detection and evaluation

As little as 3 μg of both of these substances can be localized as dark absorption spots in short-wave ultraviolet light. According to Nürnberg [*39*], specific and sensitive detection can be achieved when pyridine vapors are excluded, by spraying the air-dried plate with *p*-aminobenzoic acid solution, and subsequent treatment with cyanogen chloride vapor (Reagent No. 88). The maximum color develops after about six minutes. Nicotinamide appears as an orange-red, and nicotinic acid as a red spot. As little as 0.1 μg of the latter substance is still detectable in layers of normal thickness ($\sim$ 0.25 mm).

Semi-quantitative estimation can be made visually by comparison with a series of different concentrations of the pure substance [*39*]. Estimation is possible also by scraping off the untreated spots, eluting, and estimating with cyanogen chloride or cyanogen bromide [*54*].

6. Pantothenic acid and panthenol

$$HOH_2C\text{—}C(CH_3)_2\text{—}CHOH\text{—}CO\text{—}NH\text{—}CH_2\text{—}CH_2\text{—}R$$

Panthothenic acid, $R = \text{—}C(=O)\text{—}OH$
Panthenol, $R = \text{—}CH_2OH$

Pantothenic acid occurs widely in the plant and animal kingdoms, in both the free state and in bound forms such as the co-enzyme A. The alcohol corresponding to pantothenic acid (panthenol) is more easily absorbed by the animal body; it is transformed into pantothenic acid.

a) Conditions of separation and results

Using the mixture recommended by GÄNSHIRT and MALZACHER [*15*] consisting of glacial acetic acid-acetone-methanol-benzene (5 + 5 + 20 + 70), both pantothenic acid (or its sodium or calcium salt) and panthenol, can be identified chromatographically on silica gel layers ($Rf \times 100$ values, for both substances, $\simeq 57$). If water is used as the solvent, pantothenic acid and panthenol migrate over different distances ($Rf \times 100 \simeq 89$ and 67, respectively) and are separated from the degradation products, β-alanine and amino propanol. Absolute ethanol is also a suitable solvent ($Rf \times 100 \simeq 40$ and 65, respectively).

b) Detection and evaluation

Pantothenic acid and panthenol can be decomposed on the plates by heating to 160° C for thirty minutes. After spraying with 0.5% solution of ninhydrin in ethanol (Reagent No. 108) and renewed brief heating to 160° C [*15*], the liberated β-alanine produces violet spots, and aminopropanol wine-red spots. Complete decomposition can also be achieved by treating with 10% trichloroacetic acid and drying for ten minutes at 110° C. The smallest detectable quantity for both substances is 2 μg.

Using these methods, pantothenic acid and its salts, panthenol together with also their degradation products, can be detected rapidly and specifically in various pharmaceutical and cosmetic preparations. Comparison with a series of standards of the corresponding pure substances permits the estimation of the vitamins and of the decomposition products.

The ninhydrin color complex can also be estimated colorimetrically after scraping off the spots, eluting them with 96% ethanol and stabilizing them with a solution of copper nitrate.

7. Cyanocobalamin (vitamin B_{12})

Cyanocobalamin, or vitamin B_{12}, belongs to the group of cobalamins; it occurs widely in nature, but in relatively small quantities when compared with the other B vitamins.

a) Conditions of separation and results

If the preparations are chromatographed on silica gel using the solvent recommended by GÄNSHIRT and MALZACHER [*15*] cyanocobalamin remains at the point of application (see Table 34); it will, however, migrate if water is used ($Rf \times 100 \simeq 22$).

b) Detection and evaluation

For the estimation of relatively large amounts of cyanocobalamin the chromatogram can be examined in light of various wavelengths. As little as 0.3 μg can still be seen in daylight because of its red color; at a wavelength of 365 mμ, 0.5 μg can be recognized, and, at a wavelength of 254 mμ, 1 μg, as a dark absorption spot against the fluorescent background. After chlorination and spraying with o-tolidine-potassium iodide (Reagent No. 32) as little as 0.2 μg can be made visible as a violet spot. For the detection of traces, bioautography should be employed.

8. Folic acid

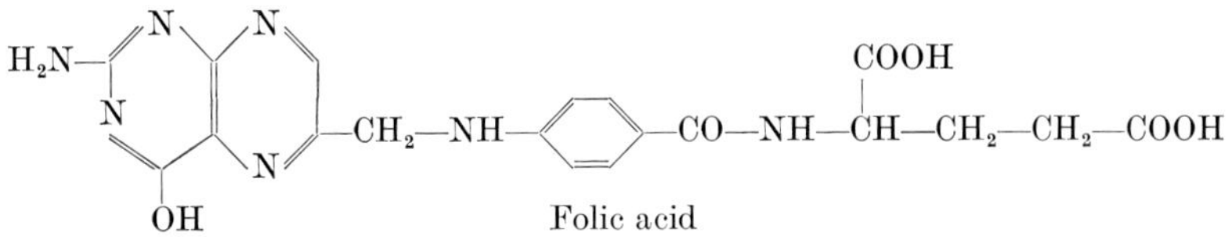

Folic acid

Folic acid (pteroylglutamic acid) occurs in minute amounts everywhere in nature both in the free form and in the form of derivatives in which the glutamic acid radical is linked, as in the peptides, to the pteroyl structure of folic acid. It is liberated from these derivatives in animal tissues by an enzyme. Folic acid is only sparingly soluble in water and organic solvents, but it may be dissolved in alkaline hydroxides or carbonates with the formation of salts.

a) Conditions of separation and results

Neutral solvents are practically incapable of transporting folic acid on silica gel layers (cf. Table 34). This may, however, be achieved with a 1% or a 10% solution of ammonia ($Rf \times 100 \simeq 95$ and 85, respectively). When examining the vitamin, possible impurities are separated in the form of bluish fluorescent spots which are visible in long wave-length ultraviolet light.

b) Detection and evaluation

The yellow folic acid can be identified in daylight at a level of 2 μg; a similar quantity can be seen as a dark absorption spot in ultraviolet light, and, after chlorination, 0.5 μg can be made visible as a grey spot by spraying with o-tolidine-potassium iodide (Reagent No. 32). This staining reaction is, however, impaired by developing the plates with ammonia; a more suitable solvent is therefore highly desirable. The spots can also be stained using other spraying reagents [*65*]. The more sensitive bioautographic methods are also suitable for estimations.

9. Biotin

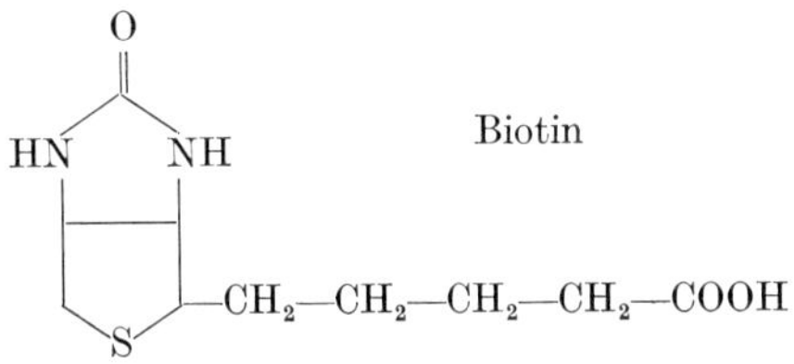

Biotin

This vitamin is probably present in traces in all living cells, both free and in the form of co-enzymes. It is sparingly soluble in water and organic solvents, but it dissolves readily in dilute alkali.

a) Conditions of separation and results

As shown in to Table 34, biotin can be separated from the other water-soluble vitamins on Silica Gel G with glacial acetic acid-acetone-methanol-benzene (5 + 5 + 20 + 70) [*15*] or with water, the $Rf \times 100$ values being 80 and 70, respectively.

b) Detection and evaluation

After spraying with iodoplatinate (Reagent No. 76), as little as 5–10 μg of biotin appear as a white spots against a pink background, and, after preliminary chlorination, even 0.2 μg of biotin can be stained to a blue color with o-tolidine-potassium iodide (Reagent No. 32). Smaller quantities can be detected and estimated by microbiological methods.

10. Ascorbic acid (vitamin C)

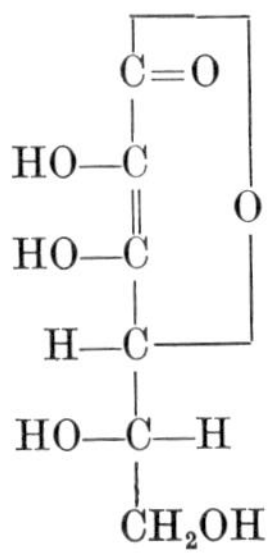

l(+)-Ascorbic acid

l(+)-Ascorbic acid occurs widely in all types of cells. It can also occur in plants combined with proteins in the form of "ascorbigen" from which it can be liberated by mild hydrolysis. For pharmaceutical purposes, it is also used in the form of its salts and esters. Owing to its strong reducing power, ascorbic acid is unstable and is readily oxidized even in air. By dehydrogenation it is transformed reversibly into dehydro-ascorbic acid. Its extraction from tissues should therefore be carried out in an inert atmosphere.

a) Conditions of separations and results

Ascorbic acid and its derivatives can be partially separated on activated silica gel layers using various solvents. The approximate values of $Rf \times 100$ found are shown in Table 37. STROHECKER [55] prepared the dinitrophenylhydrazone of ascorbic acid, and investigated this substance chromatographically with chloroform-ethyl acetate (50 + 50) on a Silica Gel G layer dried in air ($Rf \times 100 \cong 25$).

Table 37. *Values of Rf × 100 for Ascorbic Acid Derivatives on Silica Gel G*

	Solvent		
	Glacial acetic acid-acetone-methanol-benzene (5 + 5 + 20 + 70) [*15*]	Water	Ethanol
l-Ascorbic acid (or its Na or Ca salt)	30	96	22
Isoascorbic acid	35	96	29
Ascorbyl palmitate	85	0	55

b) Detection and estimation

As little as 3 μg can be detected in ultraviolet light of short wavelength. After brief heating to 120° C, these spots show a bluish fluorescence at a wavelength of 365 mμ. The limits of detection are all of the same order, whether using iodoplatinate (Reagent No. 76), which produces yellowish spots on a pink background; phosphomolybdic acid (Reagent No. 120b), which produces a greyish-blue color; ammonium molybdate-citrate buffer (Reagent No. 3), which produces a blue color; or alkaline silver nitrate solution (Reagent No. 137), which produces greyish-black spots but is less sensitive. Cacotheline and titanium dichloride, which yield, respectively, violet and orange colors, have also been suggested as specific spraying reagents [*34*].

All of these methods have proved to be useful for the rapid and selective identification of ascorbic acid in association with other vitamins. In all probability, quantitative estimations can be made in a manner similar to that used in paper chromatography [*34*]: after localization (directly or by means of a reference band) the spot is scraped off, eluted and estimated volumetrically or photometrically. Using this method STROHECKER [*55*] estimated the ascorbic acid content in food products containing soluble carbohydrates by scraping off the brick-red spot of ascorbic acid-dinitrophenylhydrazone, eluting it with 85% sulphuric acid, filtering and determining it photometrically.

The skilful assistance of Mr. A. KÖNIG is gratefully acknowledged as well as the experimental help of Messrs. M. PRÉTÔT and W. BÜRKI.

Bibliography to Chapter C: Vitamins

[*1*] BACHARACH, A. L., and J. GREEN: Chem. & Ind. (London) **1961**, 2135.
[*2*] BILLETER, M., u. C. MARTIUS: Biochem. Z. **334**, 304 (1961).
[*3*] BLATTNÁ, J., u. J. DAVÍDEK: Experientia **17**, 474 (1961).
[*4*] BOOTH, V. H.: Analyt. Chem. **33**, 1224 (1961).
[*5*] BRÜGGEMANN, J., W. KRAUSS u. J. TIEWS: Chem. Ber. **85**, 315 (1952).
[*6*] CROWFOOT, C., and J. D. DUNITZ: Nature (Lond.) **162**, 608 (1948).
[*7*] DAVÍDEK, J., u. J. BLATTNÁ: J. Chromatog. **7**, 204 (1962).
[*8*] DAVIES, B. H., T. W. GOODWIN and E. I. MERCER: Biochem. J. **81**, 40P (1961).
[*9*] DEMOLE, E.: J. Chromatog. **1**, 24 (1958).
[*10*] — Chromatog. Rev. **1**, 1 (1959).
[*11*] DUGGAN, D. E.: Arch. Biochem. Biophys. **84**, 116 (1959).
[*12*] EDWIN, E. E., A. T. DIPLOCK, J. BUNYAN and J. GREEN: Biochem. J. **75**, 450 (1960).
[*13*] EICHENBERGER, W., u. E. C. GROB: Helv. Chim. Acta **45**, 974 (1962).
[*14*] FONTELL, K., R. T. HOLMAN and G. LAMBERTSEN: J. Lipid Res. **1**, 391 (1960).
[*15*] GÄNSHIRT, H., u. A. MALZACHER: Naturwiss. **47**, 279 (1960).
[*16*] GLOOR, U.: Private communication, F. HOFFMANN — La Roche & Co. AG, Basel.
[*17*] GOODWIN, T. W.: Private communication, University of Wales, Aberystwyth Great Britain.
[*18*] GREEN, J.: J. Sci. Food Agric. **9**, 801 (1958).
[*19*] —, D. MCHALE, S. MARCINKIEWICZ, P. MAMALIS and P. R. WATT: J. Chem. Soc. **1959**, 3362.
[*20*] —, P. MAMALIS, S. MARCINKIEWICZ and D. MCHALE: Chem. & Ind. (London) **1960**, 73.
[*21*] GROB, E. C., W. EICHENBERGER u. R. P. PFLUGSHAUPT: Chimia (Switz) **15**, 565 (1961).
[*22*] —, u. A. BOSCHETTI: Chimia (Switz) **16**, 15 (1962).
[*23*] HAIS, I. M., u. K. MACEK: Handbuch der Papierchromatographie, I und II, Jena: VEB Gustav Fischer Verlag 1958 und 1960.
[*24*] HANEWALD, K. H., M. P. RAPPOLD and J. R. ROBORGH: Rec. trav. chim. **80**, 1003 (1961).
[*25*] HONEGGER, C. G.: Helv. Chim. Acta **44**, 173 (1961).
[*26*] INHOFFEN, H. H.: Angew. Chem. **72**, 875 (1960).
[*27*] IRREVERRE, F., and M. X. SULLIVAN: Science **94**, 497 (1941).
[*28*] ISLER, O., et al.: Private communication, F. HOFFMANN — La Roche & Co. AG., Basel.
[*29*] —, R. RÜEGG u. P. SCHUDEL: Chimia (Switz) **15**, 208 (1961).
[*30*] JANECKE, H., u. I. MAAS-GOEBELS: Z. anal. Chem. **178**, 161 (1960).
[*31*] JONES, S. W.: Chem. Eng. News **38**, Nr. 39, 57 (1960).
[*32*] LAGONI, H., u. A. WORTMANN: Milchwissenschaft **10**, 360 (1955).
[*33*] — — Milchwissenschaft **11**, 206 (1956).
[*34*] LINSKENS, H. F.: Papierchromatographie in der Botanik, 2. Aufl., Berlin-Göttingen-Heidelberg: Springer-Verlag 1959.
[*35*] MANGOLD, H. K., and D. C. MALINS: J. Am. Oil Chem. Soc. **37**, 383 (1960).

[36] MARCINKIEWICZ, S., and J. GREEN: Analyst **84**, 304 (1959).
[37] MCHALE, D., J. GREEN and S. MARCINKIEWICZ: Chem. & Ind. (London) **1961**, 555.
[37a] MONTAG, A.: Z. Lebensmitt.-Untersuch. **116**, 413 (1962).
[38] MOTTIER, M.: Mitt. Gebiete Lebensm. u. Hyg. **43**, 118 (1952).
[39] NÜRNBERG, E.: Dtsch. Apoth. Ztg. **101**, 142 (1961).
[40] — Dtsch. Apoth. Ztg. **101**, 268 (1961).
[41] PASTUSKA, G., u. H. TRINKS: Chemiker Ztg. **85**, 535 (1961).
[42] PLANTA, C. v., U. SCHWIETER, L. CHOPARD-DIT-JEAN, R. RÜEGG, M. KOFLER u. O. ISLER: Helv. Chim. Acta **45**, 517 (1962).
[43] RANDERATH, K.: Angew. Chem. **73**, 436 (1961).
[44] RÜEGG, R., u. O. ISLER: Planta Med. **9**, 386 (1962).
[45] SATHE, V., J. B. DAVE and C. V. RAMAKRISHNAN: Nature (Lond.) **177**, 276 (1956).
[46] SCHILLING, K., and H. DAM: Acta Chem. Scand. **12**, 347 (1958).
[47] SEHER, A.: Fette, Seifen, Anstrichmittel **61**, 345 (1959).
[48] — Nahrung **4**, 466 (1960).
[49] — Mikrochim. Acta **1961**, 308.
[50] SHAW, H. W. C., J. P. JEFFERIES and T. E. HOLT: Analyst **82**, 2 (1957).
[51] STAHL, E.: Chemiker Ztg **82**, 323 (1958).
[52] — Arch. Pharm. **292/64**, 411 (1959).
[53] —, H. R. BOLLIGER u. L. LEHNERT: Paper on thin-layer chromatography, Symposium über Carotine und Carotinoide der Dtsch. Ges. f. Ernährung in Mainz (Oktober 1961).
[54] STROHECKER, R.: Vitaminanalysen, E. Merck AG, Darmstadt (1960).
[55] — Private communication, E. Merck AG, Darmstadt.
[56] THOMMEN, H.: Chimia (Switz) **15**, 433 (1961).
[57] — Private communication, F. HOFFMANN — La Roche & Co. AG., Basel.
[58] — Naturwiss. **69**, 517 (1962).
[59] TIEWS, J.: Vitamine u. Hormone 8, 185, 297 (1959).
[60] e. g. United States Pharmacopeia XVI, 910 (1960).
[61] VELLUZ, L., G. AMIARD et A. PETIT: Bull. soc. chim. France **1949**, 501.
[62] Vitamin E-Panel: Analyst **84**, 356 (1959).
[63] VÖLKER, O.: Naturwiss. **48**, 58 (1961).
[64] WAGNER, H., L. HÖRHAMMER u. B. DENGLER: J. Chromatog. **7**, 211 (1962).
[65] WALDI, D.: Chromatographie, E. Merck AG, Darmstadt (1959).
[66] — Private communication, E. Merck AG, Darmstadt.
[67] WINTERSTEIN, A., A. STUDER u. R. RÜEGG: Chem. Ber. **93**, 2951 (1960).
[68] —, u. B. HEGEDÜS: Hoppe-Seyler's Z. physiol. Chem. **321**, 97 (1960).
[69] — — Chimia (Switz) **14**, 18 (1960).

Some recent noteworthy articles

CIMA, L., i R. MANTOVAN: Farmaco, Prat. Ed. **17**, 473 (1962): Cyanocobalamin and hydroxycobalamin.
DAVID, S., et H. HIRSHFELD: Bull. Soc. Chim. France **1963**, 1011: Thiamine and thiamine phosphates.
EGGER, K.: Planta **58**, 664 (1962): Carotenoids and chlorophyll.
NORMANN, A. W., and H. F. DELUCA: Anal. Chem. **35**, 1247 (1963): Vitamin D and related compounds.
SMITH, E. C., and D. E. METZLER: J. Am. Chem. Soc. **85**, 3285 (1963): Riboflavin and its degradation products.
STOWE, H. D.: Arch. Biochem. Biophys. **103**, 42 (1963): Separation of β- and γ-tocopherols.
STROHECKER, R., u. H. M. HENNING: Vitamin-Bestimmungen, Verlag Chemie, Weinheim (1963).
WAGNER, H., u. B. DENGLER: Biochem. Z. **336**, 380 (1962): Quantitative determination of ubiquinones.
WEBER, F., u. O. WISS: Helv. Physiol. et Pharmacol. Acta **21**, 131 (1963): Metabolism of vitamin E in rats.

D. Steroids

Sterols; Pregnane-, Androstane-, and Estrane-Compounds; Bile Acids and Cardiac Glycosides

By

D. WALDI

I. General introduction

1. Structures and nomenclature

The basic skeleton of the compounds described in this chapter is represented by a perhydro-cyclopentenephenanthrene ring system.

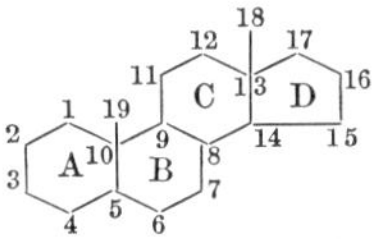

The basic skeleton of steroids

It consists of three six-membered carbon rings, A, B, C, and one five-carbon ring, D. The carbon atoms are numbered with arabic numerals in the sequence shown above. An angular methyl group is usually found on C_{10} and C_{13}. Functional groups ($-OH$, $=O$) of the therapeutically important steroids favor the positions C_3, C_{11}, and C_{17}. A side-chain is often attached at C_{17}.

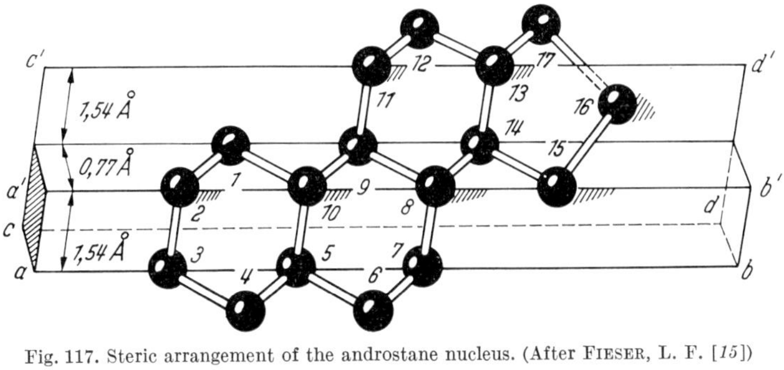

Fig. 117. Steric arrangement of the androstane nucleus. (After FIESER, L. F. [*15*])

The linkage of the ABCD makes possible a number of stereoisomeric compounds [*15*, *23*, *40*, *42*].

By convention, the methyl group (C_{19}) attached to C_{10} is always shown as projecting anteriorly from the plane of the page, i.e., it occupies the β-position. Subsequent investigations by PRELOG et al. [*56, 57*] have proved that the absolute configuration does indeed correspond to that postulated at an earlier period. If the hydrogen on C_5 is also orientated in the β-position (symbolized by a solid line), this indicates a *cis*-configuration of rings A and B. If the hydrogen atom on C_5 is orientated in the α-position (symbolized by a dotted line), this indicates the *trans*-configuration of rings A and B. In practically all steroids, rings BC and DC are linked in the *trans*-configuration, with the exception of the cardenolides and bufadienolides (rings DC = *cis*).

The side-chain on C_{17} of natural steroids usually occupies the β-position, unless otherwise shown. The hydrogen on C_{14} occupies the α-position.

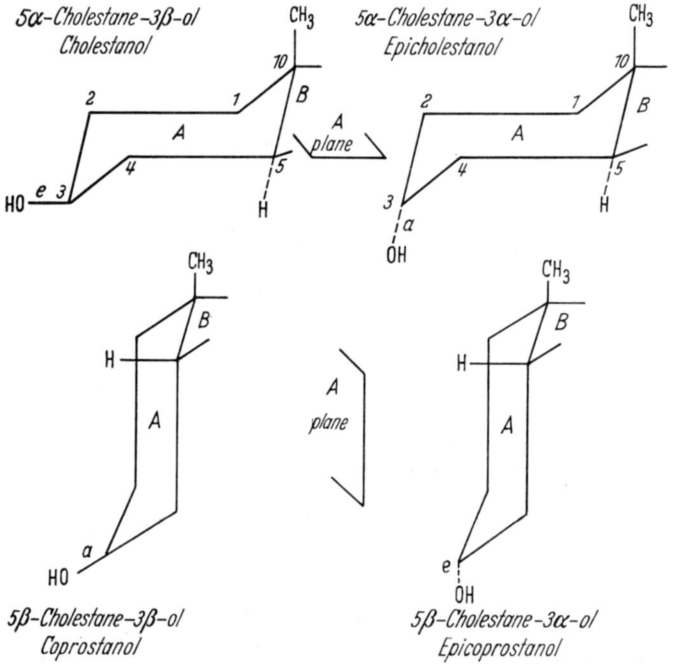

Fig. 118. The four possible forms of cholestanol

The carbon atoms in the six-membered carbon ring of cyclohexane do not lie in one plane, as in benzene, but slightly transposed against each other. This gives rise to configurations known either as the "chair" or "boat"-forms. Steroids, in general, only show the chair-form, as illustrated schematically, in Fig. 118.

BARTON and MORRISON [7] showed that the "configuration" of a substituent is of special significance, depending on whether it lies in the plane of the ring or at right angles to it. Thus, an OH-group on C_3 may lie in the plane of the ring (= equatorial, e) or at right angles to it (= axial, a). Functional groups lying in the plane of the ring (e) not only have a higher polarity but also react more easily (e.g., easier esterification and saponification).

Configuration (*cis* or *trans*-linkage) and conformation thus combine so as to give, in the case of, e.g., cholestanol, four possible compounds which differ in polarity, and may, therefore, be separated chromatographically.

According to ČERNY, JOSKA and LÁBLER [*12*], migrates fastest, on alumina layers with benzene ether (7 + 3), 5 α-cholestane-3 α-ol is followed by 5β-cholestane-3α-ol, then 5β-cholestane-3β-ol and finally, 5α-cholestane-3β-ol having the strongest polarity and migrating least.

2. Chromatographic behavior [*1, 11, 53, 70*]

The hydrocarbon skeleton is responsible for the weak polarity of the steroids. However, polarity depends on the number of functional groups, or on the C/O-Index (ratio of number of C-atoms to O-atoms). Steroids which have long hydrocarbon side-chains, or the cholesteryl fatty acid esters, are especially lipophilic and migrate very rapidly; they are, therefore, developed with hydrophobic solvents. The polarity of functional groups increases in the order: $-OCH_3$, $-COOR$, $-C=O$, $-CHO$, $-OH$, $-COOH$. Steroids having several OH- or COOH-groups are already hydrophilic, especially those which are esterified on C_3 with sugars or sugar acids, which are known for their high polarity and water-solubility. Steroids are excreted from the body in this form or as sulphate esters [*70*, p. 414].

After enzymatic cleavage or acid hydrolysis, steroids can be extracted from aqueous solutions with organic solvents (see p. 255) and can then be separated on thin layers by adsorption chromatography.

NEHER [*53* p. 150] has dealt extensively with secondary reactions which may occur with various adsorbents, especially with activated basic alumina.

In general, because of the short running time, such secondary reactions do not occur on silica gel-layers. In special cases, OH-groups, in the vicinity of another functional group, are protected by acetylation if there is danger of rearrangement. In the majority of cases, it is not important to biologists which metabolities are found. The question, whether, in a given experiment, a qualitative or quantitative change has occurred, is of particular interest.

We shall discuss in this chapter the following substances, in their order of increasing polarity:

Sterols and their esters	C_{27}–C_{29}-Steroids
Progesterone and adrenal cortical hormones = corticosteroids.	C_{21}-Steroids
Androsterone, testosterone and their derivatives	C_{19}-Steroids
Estrogens	C_{18}-Steroids
Bile Acids.	C_{24}-Steroids

A section on cardiac glycosides will follow.

II. The sterols

Sterols were formerly classified into zoo- and phyto-sterols according to their origin in nature. The following formal distinction can be made [*54*]:

Monohydroxy-sterols: C_{27}-sterols, e.g., cholesterol, C_{28}-sterols, e.g., ergosterol, C_{29}-sterols, e.g., stigmasterol;

Di- and tri-hydroxy-sterols, e.g., cholestanetriol.;

The basic skeleton is constituted by cholestane:

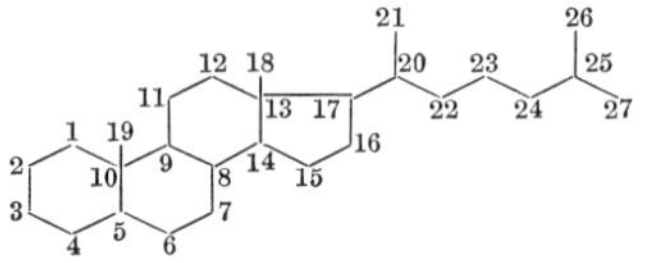

Cholestane

As can be seen from Fig. 118, the C_3-OH substituted cholestane can occur in four steric forms, which can also be separated chromatographically (pp. 261—263).

1. Separation, detection and h*Rf*-values

Various sterols have been separated with own solvents and those described in the literature. The *Rf* × 100-values (h*Rf*) obtained under standard conditions are shown in Table 38. Solvents were classified in order of decreasing polarity. This decrease does not always coincide with

Table 38. *Separation of Sterols on Silica Gel G Layers under Standard Conditions*

Compound	Fluorescent color Reagent no. 123	*Rf* × 100-values with solvents I	II	III	IV	V	VI	VII	VIII	IX	X
Cholesteryl acetate. . .	pink- to rust-red	100	90	79	86	70	71	69	70	61	45
Epidihydrocholesterol (Epicholestanol)	yellow	78	60	43	45	34	31	26	25	23	5
Coprosterol . . . . . . (Coprostanol)	yellow	72	56	49	38	31	26	25	22	18	2
Cholesterol	pink- to rust-red	65	48	35	25	22	21	19	15	13	0
β-Sitosterol	pink- to rust-red	63	45	35	25	20	20	12	13	11	2
Dehydrocholesterol . .	blue-violet	—	45	31	25	—	—	—	—	—	—
Dihydrocholesterol. . . (Cholestanol)	pale-violet	58	45	36	23	17	19	15	13	11	2

	Solvents used (Table 38)	ε
I	Chloroform-acetone (90 + 10) [*89*]	6.75
II	Benzene-ethyl acetate (60 + 20) [*13*]	3.5
III	Cyclohexane-ethyl acetate (70 + 30) [*5*]	3.3
IV	Chloroform [*73*]	5.0
V	Benzene-ethyl acetate (90 + 10) [*13*]	5.0
VI	Toluene-ethyl acetate (90 + 10) [*13*]	2.7
VII	Cyclohexane-ethyl acetate (85 + 15) [*5*].	2.65
VIII	Cyclohexane-chloroform (50 + 50) [*89*]	3.6
IX	Benzene-chloroform (40 + 60) [*90*]	4.0
X	Cyclohexane-chloroform (70 + 30) [*89*]	3.0

the decrease in the calculated dielectric constant ε. The *Rf*-values decrease in the same ratio, so that the order in which substances may be eluted with these solvents remains unchanged. The *Rf*-values do not always agree with those cited in the literature. This is certainly due to differences in experimentation, e.g., the activity of the thin-layers.

Remarks on the *Rf*-values of Table 38

Table 38 shows that *Rf*-values decrease with the increase in hydrophobic properties of the solvents. It will be seen that all solvents conform to this rule, with the exception of solvents III and IV. The *hRf*-values ($Rf \times 100$) cited in the literature are, in part, much higher than those we observed. Thus, VAN DAM obtained $hRf = 12$ for cholesterol with solvent II, with V $hRf = 39$ [*22*], with VI $hRf = 40$ [*21*]. Since the relation is approximatively correct, it must be assumed that the layers we used were much more active. We predried the layers in air for 2 hr and activated them for about 1 hr at 110—120° C. Also, we used saturation of solvent-vapor according to the standard conditions. (Developing chamber lined with filter paper!) It is, therefore, recommended to run cholesterol and cholesterol acetate as standards, to which the *Rf*-values cited in the literature may then be referred.

The *hRf*-value for coprosterol of 16 [*49*] with solvent III, found by WYSS et al. [5], is very low. No comparison can be drawn, since his paper makes no reference to *hRf*-numbers for cholesterol.

The last four compounds of Table 38 cannot be directly separated with the usual adsorbent layers and solvents, but they can be differentiated by their various color reactions. This provides a means of identification (for staining see p. 263).

BRIESKORN et al. [*10*] have detected β-sitosterol in the leaves of *Salvia triloba* and *Pirus malus* by TLC. In addition to location with antimony trichloride (reagent no. 11), infra-red spectra were also used for identification. HÖRHAMMER, WAGNER and LAY [*25*] detected β-sitosterol and oleanolic acid in ginseng roots by TLC.

Table 38a. *Further hRf-Values of Sterols*

Compound	$Rf \times 100$	Solvent	Method	Author
Cholestanone	86	VII	partially modified standard conditions	[*5*]
Coprostanone	71	VII	partially modified standard conditions	[*5*]
4-Cholestenone	63	VII	partially modified standard conditions	[*5*]
5-Cholestene-3 β-7 α-diol	21	II	standard conditions	[*13*]
5-Cholestene-3 β-25-diol	41	II	standard conditions	[*13*]
26-Nor-5-cholestene 3 β-ol-25-one	63	II	standard conditions	[*13*]
3 β-Acetoxy-26-nor-5-cholestene-24-one	74	V	standard conditions	[*13*]
3 β-Acetoxy-5-cholestene-7-one	93	V	standard conditions	[*13*]

2. Cholesterol and cholesteryl esters

Large quantities of these substances are present in both human and animal organs and body fluids. The total amount of cholesterol in the human body is about 200—250 g.

Cholesterol is a relatively stable compound. Hydration of the C_{5-6} double bond yields coprosterol, which is found, e.g., in small quantities in faeces.

Cholesterol can be esterified at the OH-group of C_3 with stearic, oleic, palmitic, linoleic, arachidonic, and other higher fatty acids.

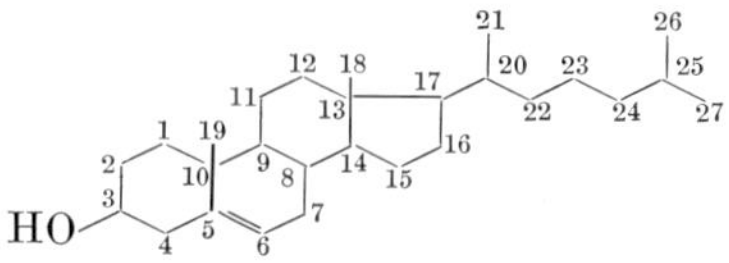

Fig. 119. Cholesterol = 5-cholestene-3 β-ol

a) Ratio cholesterol: cholesteryl esters

In organs and body fluids, this ratio is of importance for the diagnosis of certain diseases. It may be determined by various methods. It is generally assumed that in serum its value is 1:3.5, i.e., 70–75% of cholesterol is present in the form of esters. Since about 75% of cholesterol is plasma protein-bound, some difficulties arise in the quantitative determination of cholesterol and cholesteryl esters.

Only about 20% of these esters are split off during heating for 30 min with dilute HCl.

As can be seen from Fig. 120, during acid hydrolysis, decomposition products, such as linoleic acid ester or 3-cholesteryl chloride, may arise in various quantities.

The choice of an appropriate procedure for extraction is of decisive importance for such quantitative determinations, as is a knowledge of the optimal conditions for saponification. TLC is a valuable aid in establishing suitable methods.

The methods of extraction described in the literature give different values for cholesteryl esters, as may be easily determined by TLC. Furthermore, Table 39 shows that photometric determinations (total cholesterol plus ester-cholesterol) give different readings, also.

Table 39. *Comparison of Methods for the Determination of Cholesterol and its Esters*

	Method: ZÖLLNER [*93*] Extraction, TLC		SCHMIDT VON ELMENDORFF [*71*] Extraction, TLC		SEARCY and BERQUIST [*63*] Photometry		Extraction after acid hydrolysis, TLC
	C*	D*	C*	D*	C*	D*	C*
Cholesterol . .	48	82	48	92	52	70	62
Cholesteryl ester	278	375	213	350	219	295	168
Total cholesterol	206	298	169	297	178	240	159
Chol.: Tot.cholest	1:4.3	1:3.6	1:3.5	1:3.2	1:3.4	1:3.4	1:2.6
Chol.: Chol. ester	1:5.7	1:4.5	1:4.4	1:3.5	1:4.2	1:4.2	1:2.7
Ester quotient	76.5	72.5	71.5	69.0	70.5	70.5	61

* Serum of dogs C and D, respectively; values are given in mg/100 g serum.

The values of Table 39 were obtained as follows: 2 ml of dog serum each, were extracted, following a procedure described by ZÖLLNER and KIRSCH [*93*], with chloroform-methanol, and chloroform was added until the evaporated extract had a volume of 5 ml. Three samples (2 ml each) of the same serum were extracted following the method of SCHMIDT VON ELMENDORFF [*71*] with chloroform-ethanol-ether (3 + 1 + 1), the evaporated extract was also brought to 5 ml by adding chloroform. Total cholesterol and ester-cholesterol were determined in three additional samples using SEARCY and BERQUIST [*63*] method.

For hydrolysis, 2 ml serum were heated by refluxing with 30 ml of 10%-HCl for 25 min. After cooling, extraction was carried out three times with 50 ml ethyl acetate. The extract was dried over sodium sulphate, evaporated and the residue dissolved in 5 ml chloroform.

For *comparative determination* of cholesterol by TLC, 20 μl of extract were placed on a silica gel plate, together with increasing quantities of a 0.01% standard solution of cholesterol. The solvent used was chloroform.

To determine the cholesteryl esters, 10 μl of extract and increasing quantities of a 0.1% cholesteryl ester standard-solution were applied to the layer, with benzene-chloroform (95 + 5) or cyclohexane-chloroform (50 + 50) as solvents. Detection was carried out by treatment with phosphoric acid (reagent No. 123) and subsequent treatment with phosphomolybdic acid (reagent No. 120c).

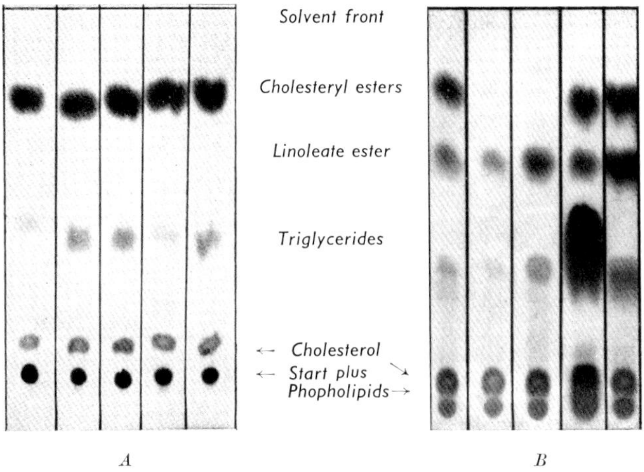

Fig. 120. Serum lipids extracted *(A)* directly, and *(B)* after acid hydrolysis, separated on Silica Gel G layers with cyclohexane-chloroform (50:50). Detection with reagent No. 123, followed by reagent No. 120c (E. Merck AG.)

Fig. 120*A* shows the separation of serum lipids after extraction by SCHMIDT VON ELMENDORFF's method [*71*]. According to HABERMANN, BANDTLOW and KRUSCHE [*19*], polar lipids which remain at the starting point may be further separated into diglycerides, cephalins, lecithins, plasmalogens, sphingomyelins, and lecithins. The higher readings for cholesteryl esters extracted by the ZÖLLNER-KIRSCH [*93*] method have been confirmed by other determinations. Attempts to obtain comparable values using VAN DAM's procedure [*13*, *13a*], in which 3–10 μl dog serum are directly applied on silica gel layers, were not successful.

b) Fractionation of cholesteryl ester fraction

The separation of the serum cholesteryl ester fraction by TLC represents a further progress. WEICKER [*90*] was the first to describe such a separation with carbon tetrachloride on silica gel layers. From our own experiments, it appears that the separation into five different components takes place more easily if the carbon tetrachloride contains 4–5% chloroform (Fig. 121). The chamber must be well saturated with solvent vapors.

c) Quantitative evaluation of cholesterol separation

ZÖLLNER et al. [*95*, *96*] not only separated cholesteryl esters by TLC but also tried to identify and evaluate, quantitatively, the various spots. In an earlier publication [*94*], these authors state that a quantity corresponding to 0.04 μl serum was applied to Silica Gel G layers in strips about 1 cm wide, and developed three times successively with carbon tetrachloride. The separated zones were scraped off, and their

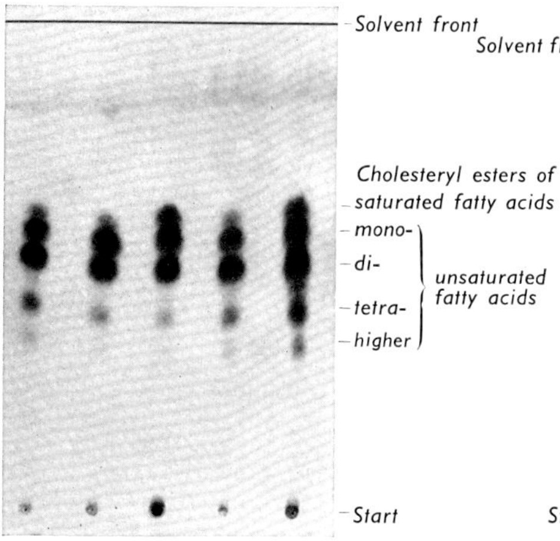

Fig. 121

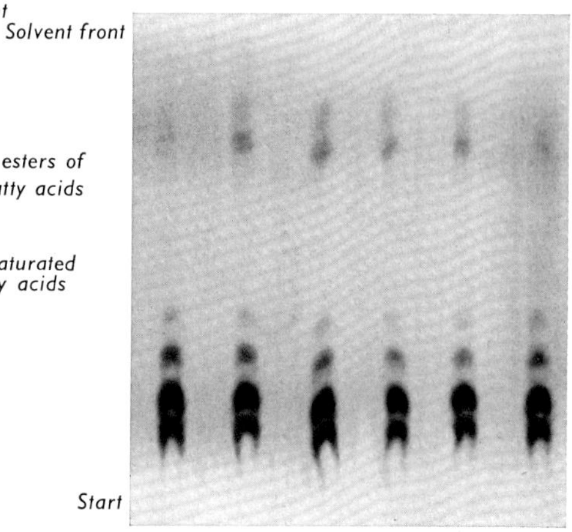

Fig. 122

Fig. 121. Cholesteryl esters from human aortae, isolated by columnchromatography, separated on Silica Gel G layers with CCl_4-chloroform (96:4). Detection with reagent No. 123, followed by spraying with reagent No. 120c

Fig. 122. Cholesteryl esters separated on paraffin-impregnated Silica Gel G with methyl ethyl ketone-acetonitrile (70:30), according to the phase-reversal method described by KAUFMANN et al. [*38*] Detection with reagent No. 123, followed by spraying with reagent No. 120c

cholesterol content was determined according to ZAK et al. [*92*]. A more recent publication [*96*] states that the compounds are spotted on narrower and thinner layers. The solvent was petroleum hydrocarbon (B.P. 50–70° C) containing 1% isopropyl ether. The chromatograms, stained with antimony chloride, are photometrically assayed directly with a device for the evaluation of electrophoresis strips. The readings are shown in column 1 of Table 40, where they are compared with other procedures.

Table 40. *Comparison of Cholesterol Components obtained from Normal Human Serum, using various Methods*

Serum-Cholesteryl Ester of:	TLC according to ZÖLLNER [94] %	[96] %	Column chromat. acc. to AHRENS et al. [2] %	Gas chromat. acc. to KLEIN and JANSSEN [39] %
Stearic acid, Palmitic acid	9	14.1	10.1	6.3
Oleic acid, Palmitoleic acid	25	22.9	18.1	22.2
Linoleic acid	56	48.5	62	60
Arachidonic acid	8.5	12.9	6	10.5
Docosapentaenoic acid	1.5	1.6	1	—

d) Two-dimensional TLC of cholesteryl esters

KAUFMANN, MAKUS and DEICKE [*38*] have described two-dimensional separations of test mixtures. Separation was carried out on Silica Gel G layers in direction 1 by adsorption chromatography, followed

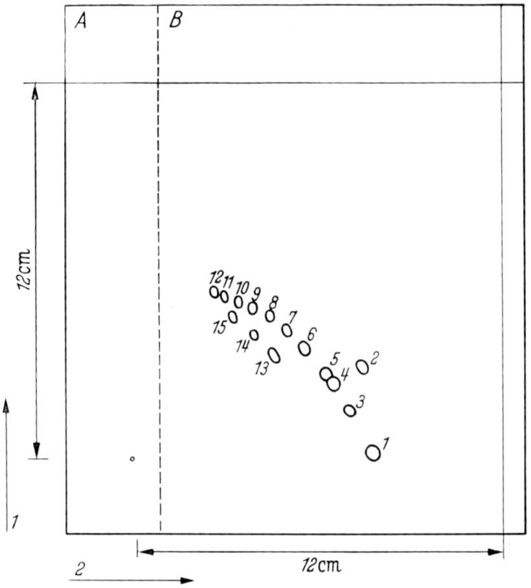

Fig. 123. Scheme of a two-dimensional TLC of cholesteryl esters, according to KAUFMANN et al. [*38*]. For details, see text

by reversed-phase partition chromatography in direction 2. It is to be hoped that this separation can be extended to mixtures of natural cholesteryl esters and thus offer a possibility for further differentiation.

The mixture of cholesteryl esters was developed with tetralin-hexane (25 + 75) followed by complete removal of the solvent in a vacuum oven (5 Torr) at 110° C for 45 min. The plates were put to cool in a desiccator containing P_2O_5. Its right-hand portion B (see Fig. 123) was then placed cautiously into petroleum hydrocarbon-paraffin (95 + 5)[1]. The petroleum ether was removed in 5 min by blowing with cold air. Direction 2 was developed with methyl ethyl ketone-acetonitrile (70 + 30).

The solvent should be 80%-saturated with paraffin oil. 70 ml ketone and 30 ml acetonitrile are mixed together; 80 ml of this mixture is saturated with paraffin oil and after decanting of the excess of paraffin oil, the remaining 20 ml of the mixture are added.

Table 41. *Rf × 100-Values of various Cholesteryl Esters according to* KAUFMANN [*38*]

No.	Cholesteryl ester	Solvents I	II	III	No.	Cholesteryl ester	Solvents I	II	III
1	Cholesterol	2	2	64	9	Laurate	43	58	33
2	Formate	—	30	60	10	Myristate	45	59	29
3	Acetate	10	16	58	11	Palmitate	48	62	25
4	Propionate	25	25	55	12	Stearate	50	64	23
5	Butyrate	20	30	52	13	Linolenate	26	42	39
6	Caproate	30	40	47	14	Linoleate	32	48	33
7	Caprylate	36	47	42	15	Oleate	40	55	28
8	Capriate	40	54	38					

I: Tetralin-Hexane (25 + 75).
II: Tetralin-Hexane (50 + 50).
III: Methyl ethyl ketone-acetonitrile (70 + 30) 80%-saturated with paraffin oil.

KAUFMANN [*38*] states that cholesteryl esters can be detected with phosphomolybdic acid reagent (Reagent No. 120a). Other reagents mentioned are: antimony trichloride (No. 12), Rhodamine B (No. 129), phosphotungstic acid (No. 121), phosphoric acid (No. 123), vanillin-sulphuric acid (No. 151) and antimony pentachloride (No. 13).

e) Determination of acid in a cholesteryl ester (alkaline saponification)

For conclusive identification of the acid component of cholesteryl esters alkaline saponification is carried out. The fatty acids are isolated and separated.

The extract, from 1 g organ or 1 ml serum, is obtained either according to BLOOR (p. 144), SCHMIDT VON ELMENDORFF [*71*] or ZÖLLNER and KIRSCH [*93*]. It is evaporated to dryness in an atmosphere of nitrogen. 1 ml of 2N-methanolic caustic potash solution is added and the whole, again in a N_2-atmosphere, is heated for 2 hr, in a water-bath. The alkaline mixture is then acidified with 2 ml of 2N-sulphuric acid, and the fatty acids are extracted with three 2 ml portions of petroleum hydrocarbon (B.P. 60—80° C). The combined extracts are dried with a little anhydrous sodium sulphate, the filtrate is slowly concentrated by evaporation, and a certain volume of chloroform added to the residue.

The isolated fatty acids were then separated according to the procedures described in the chapter on lipids (pp. 157—160).

MAHADEVAN et al. [*43*] deal with TLC of cholesteryl arachidonate and other cholesteryl esters.

For the sake of completeness, we shall mention the relevant paper-chromatographic publications of LABARRÈRE et al. [*41*], MICHALEC [*48, 49, 50*], KAUFMANN

[1] Liquid Paraffin for I.R. spectroscopy, E. Merck A.G., Darmstadt, Germany. Petroleum ether Ph. Helv. V. (B.P. 40—60° C).

et al. [28—37], BOLLIGER et al. [9], ZOTTI et al. [97], MARTIN [45], PEEREBOOM et al. [55] and SWARTWOUT et al. [80]. It remains to be seen whether layers of cellulose powder are more successful for separation than paper. Regarding separation by column chromatography, the reader is referred to SCHÖN and GEY [72].

III. The C_{21}-, C_{19}- and C_{18}-Steroids

Notes on structural formulas with nucleus of C_{21}-Steroids

Pregnane compounds (progesterones) and corticosteroids

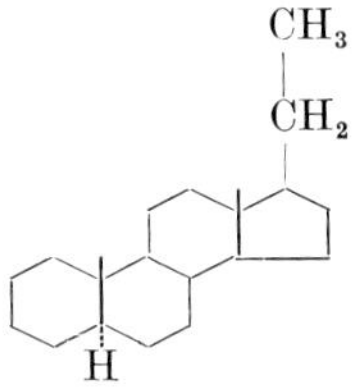

5α-Pregnane = *trans*-form = Allopregnane

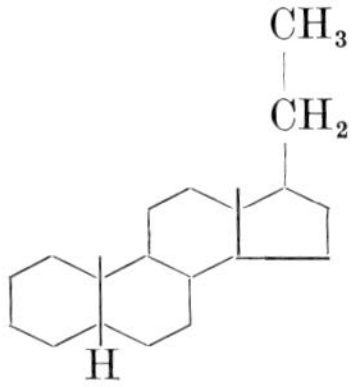

5β-Pregnane = *cis*-form

With nucleus of C_{19}-Steroids

Androstane compounds:

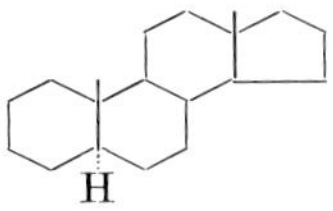

5α-Androstane

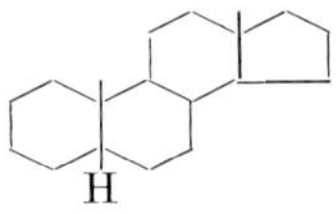

5β-Androstane = Etiocholane = Testane

With nucleus of C_{18}-Steroids

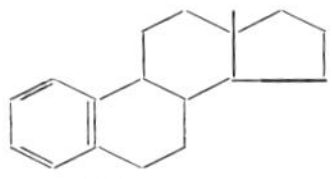

Estrane

Examples:

C_{21}-Steroids from Table 42. (The numbers refer to the number column of Table 42)

Nr. 27

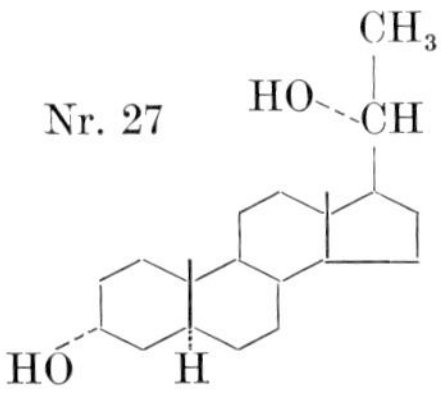

Allopregnane-3α,20α-diol

Nr. 28

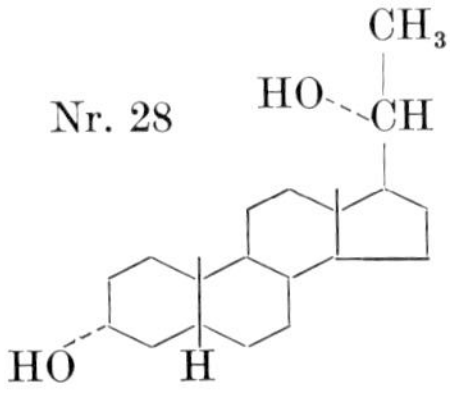

Pregnane-3α,20α-diol

Nr. 3

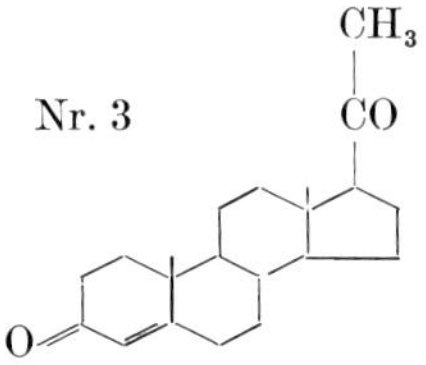

Progesterone

Nr. 12

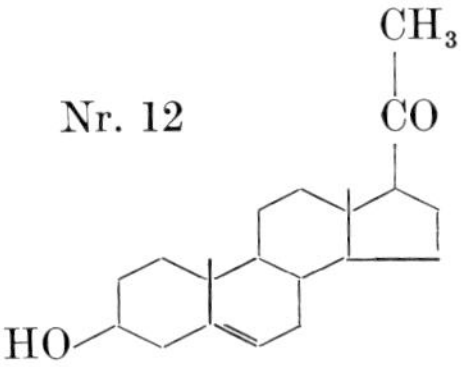

Pregnenolone

Corticosteroids

Nr. 22

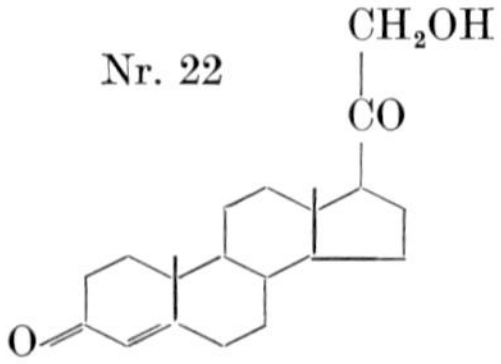

Desoxy-corticosterone (DOC = Cortexon)

Nr. 34

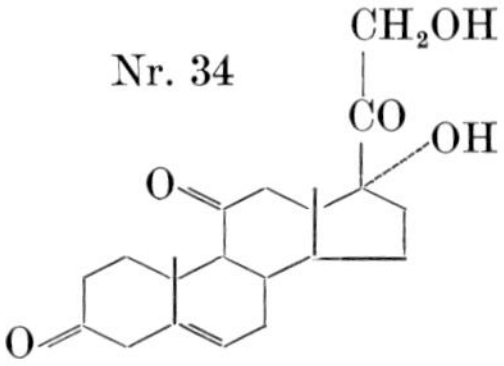

Cortisone

Nr. 35

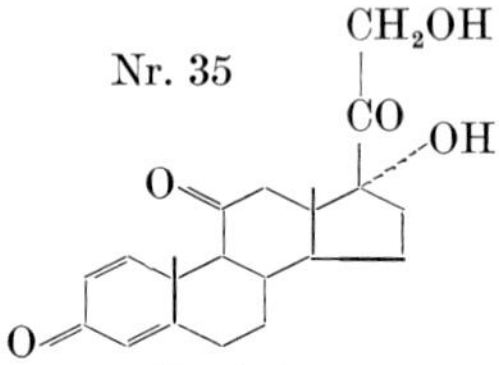

Prednisone

Nr. 41

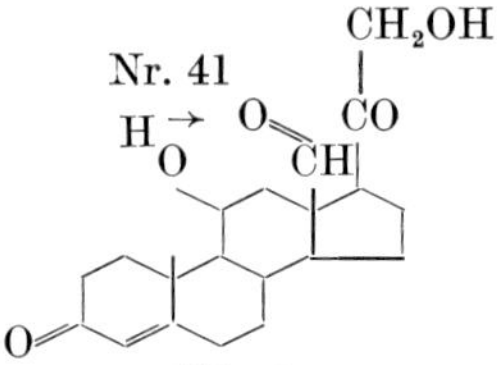

Aldosterone

Nr. 38

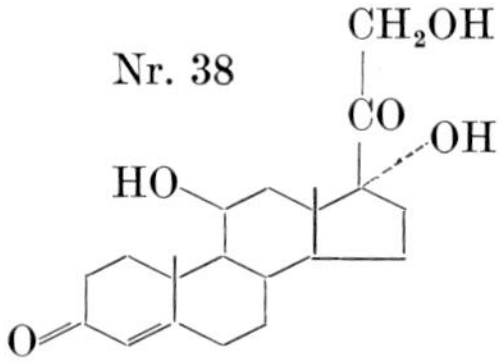

Cortisol, Hydrocortisone

Nr. 39

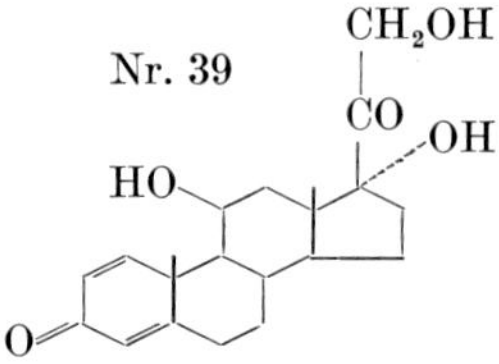

Prednisolone

C_{19}-Steroids = Androstane compounds

Nr. 2

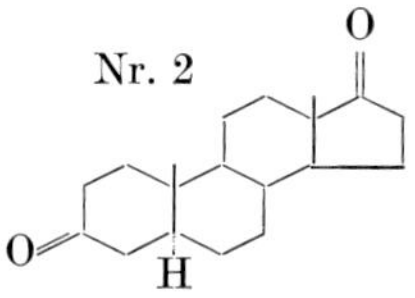

Androstane-3,17 dione

Nr. 4

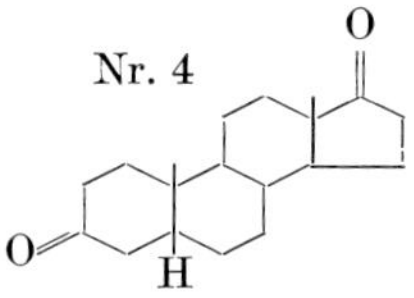

Etiocholane-3,17-dione

Nr. 26

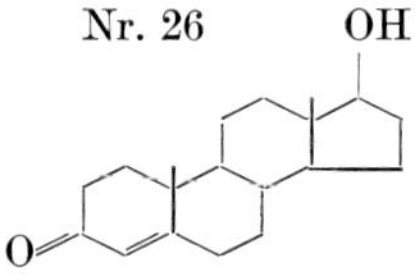

Testosterone

Nr. 25

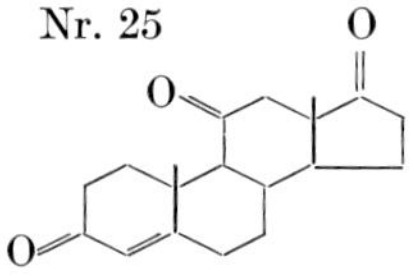

Androsterone

C_{18}-Steroids = Estrogens

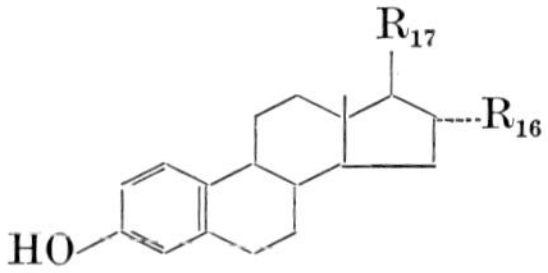

No. 6 = Estrone R_{17} = O, R_{16} = H
No. 21 = Estradiol R_{17} = OH, R_{16} = H
No. 37 = Estriol R_{17} = OH, R_{16} = OH

Compound showing estrogenic activity

Nr. 20

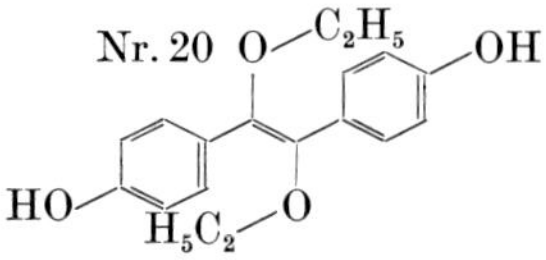

Diethylstilbestrol, Estromon

1. Separation of steroids on Silica Gel G layers (evaluation of Table 42 and comparison with the *Rf*-values of PC)

Mention has already been made on p. 251 of the behavior of steroids during paper partition chromatography with various systems, as well as of adsorption chromatography on various adsorbents in columns. In this section we shall discuss those regularities which are observed in TLC and which are shown in Table 42.

Steroids behave on Silica Gel G layers as in other chromatographic procedures, i.e., compounds of low polarity migrate most rapidly, as might be expected.

Diketones have the highest *Rf*-values (Table 42, No. 2, 3 and 4). The sequence is maintained in solvents I–VI. The *trans*-form (No. 2, androstane-3,17-dione) is markedly more hydrophobic than the *cis*-form (No. 4, ethiocholane-3,17-dione). On filter paper, progesterone (No. 3) migrates more rapidly with the Bush A-system ($hRf = 77$) than the two diones No. 2 and No. 4 which occupy the same level at $hRf = 70$. Table 42 shows clearly that 4-androstene-3,17-dione (No. 9) is more hydrophilic than No. 2 and 4; in the Bush-system, No. 9 has *hRf* 54.

4-Androstene-3,17-dione (No. 9) has a conjugated double bond. Such compounds also show lower *Rf*-values on filter paper than a steroid with isolated double bonds (No. 3). Both on paper and on Silica Gel G layers, a separation of saturated steroids and of steroids with only a single isolated double bond cannot be achieved. For instance, 5-androstene-3α-ol-17-one (No. 15) and androstane-3α-ol-17-one (No. 17) have the same migration rate with practically all solvents. The equatorial OH-group of etiocholane-3α-ol-17-one (No. 23) is strongly polar, and this compound thus remains behind the others.

While the equatorial OH-group on C_3 in etiocholane-3α-ol-17-one (No. 23) confers high polarity on the molecule compared with etiocholane-3β-ol-17-one (No. 11) with an axial OH-group, steroids no. 15 and No. 16 are hardly separable. In PC, No. 15 and 16 have the same *hRf*, i.e. 42. Androsterone (No. 17), however, migrates further with the Bush A-system ($hRf = 57$).

With the exception of cholesterol (No. 8), which migrates further on filter paper with the Bush solvents, estrone (No. 6) is the most hydrophobic steroid with a *single* OH-group. A phenolic OH-group is not as hydrophilic as an alcoholic group.

Estradiol (No. 21) has two OH-groups, one phenolic, one alcoholic, and it is interesting to note that estromone (diethylstilboestrol) (No. 20) behaves chromatographically rather like estradiol.

The most rapidly migrating of the steroids with three oxygen atoms, shown in Table 42, is pregnane-3,20-dione-21-ol (No. 14), followed by 17α-hydroxyprogesterone (No. 18). The difference between No. 14 and 18 is due to the presence of the conjugated double bond.

The next of the three-oxygen steroids shown in Table 42 is desoxy-corticosterone, No. 22. It is more hydrophilic than pregnane-3,20-dione-21-ol

No.	Steroids (Esters are printed in italics)	No. of O-Atoms	*Rf* × 100 with solvent I	II	III	IV	V	VI	VII	VIII
1	*Testosterone propionate*	—	31	65	72	75	83	86	<80	<80
2	Androstane-3,17-dione	2	30	62	72	72	80	82	<80	<80
3	Progesterone	2	28	59	71	68	77	80	<80	<80
4	Etiocholane-3,17-dione	2	17	59	69	70	79	75	<80	<80
5	*1-Dehydro-testosterone propionate*	—	18	58	68	70	75	73	<80	<80
6	Estrone	2	13	56	58	63	72	65	<80	<80
7	*6-Chlor-6-dehydro-17 α-hydroxyprogesterone acetate*	—	20	55	70	75	75	75	<80	<80
8	Cholesterol	1	25	52	69	67	—	73	<80	<80
9	4-Androstene-3,17-dione	2	18	50	69	65	70	70	<80	<80
10	Dihydro-testosterone	2	19	48	68	67	67	68	<80	<80
11	Etiocholane-3 β-ol-17-one	2	16	48	55	53	66	68	<80	<80
12	Pregnenolone	2	17	41	52	58	66	66	<80	<80
13	4-Pregnene-3 α-ol-20-one	2	18	40	49	60	65	67	<80	<80
14	Pregnane-3,20-dione-21-ol	3	18	37	53	59	62	66	<80	<80
15	5-Androstene-3 α-ol-17-one (Dehydroandrosterone)	2	16	37	48	54	55	63	<80	<80
16	Dehydro-epiandrosterone	2	13	36	48	53	55	56	<80	<80
17	Androstane-3 α-ol-17-one Androsterone	2	14	34	46	52	53	65	<80	<80
18	17 α-Hydroxyprogesterone	3	11	31	47	60	54	64	<80	<80
19	17 α-Methyltestosterone	2	12	30	46	50	54	69	<80	<80
20	Estromone (diethyl stilb-oestrol, no steroid)	2	4	28	45	45	54	31	<80	<80
21	Estradiol	2	5	32	35	47	45	56	<80	<80
22	11-Desoxy-corticosterone	3	15	28	45	52	55	70	<80	<80
23	Etiocholane-3 α-ol-17-one	2	9	27	38	45	50	60	<80	<80
24	6-Dehydro-17 α-hydroxy-progesterone	3	12	27	42	59	59	66	<80	<80
25	4-Androstene-3,11,17-trione	3	19	26	60	56	58	78	<80	<80
26	Testosterone	2	15	26	45	48	50	66	<80	<80
27	Allopregnane-3 α,20 α-diol	2	10	22	39	44	46	62	71	<80
28	Pregnane-3 α,20 α-diol	2	5	12	24	34	35	61	65	75
29	*Cortisone acetate*	—	—	10	30	44	58	60	55	68
30	*Prednisone acetate*	—	—	8	29	42	54	56	55	70
31	16-Methylene-1-dehydro-11 β, 17α-dihydroxy-progesterone	4	—	5	18	34	45	42	43	35
32	*Hydrocortisone acetate*	—	—	4	20	36	45	55	44	38
33	*Prednisolone acetate*	—	—	—	15	30	38	49	36	29
34	Cortisone	5	—	—	10	20	23	31	20	15
35	Prednisone	5	—	—	8	17	18	18	16	12
36	*Dexamethasone*	5	—	—	6	11	8	16	10	10
37	Estriol	3	—	—	4	9	15	17	13	19
38	Hydrocortisone	5	—	—	3	6	8	23	8	11
39	Prednisolone	5	—	—	—	4	5	14	8	8
40	16-Methylene prednisolone	5	—	—	—	3	12	17	8	11
41	Aldosterone	5	—	—	—	6	25	12	20	22

Remarks on Table 42. Active Silica Gel G layers prepared according to standard conditions. The *hRf*-values are mean values obtained from several separations.

Solvents: I = chloroform; II = chloroform-ethyl acetate (80 + 20); III = chloroform-acetone (90 + 10); IV = chloroform-acetone (80 + 20); V = cyclohexane-chloroform-glacial acetic acid (70 + 20 + 10); VI = methylene chloride-acetone

No.	Detection: Phosphoric acid, Reagent No. 123 UV	DL	Acetic-anhydride-H_2SO_4 Reag. No. 63 UV	DL	Antimony trichloride Reag. No. 12 UV	DL	Vanillin-H_2SO_4 Reagent No. 151 DL
1	sgrg	grgu	lor	dgubr	lpr	hbl	dgr
2	fv	gbr	hblgr	gugr	lv	hblgr	org
3	fblv	sg	lg	hg	hgrbl	sgbr	orbr
4	brg	gbr	dg	ggr	sv	—	dbr
5	rbr	hr	lbrv	orr	lgbr	rsa	rv
6	lor	r	lgor	zr	sfv	rsa	v
7	dgrbl	dgr	dgrbl	dblbr	lggr	grbl	pv
8	lzr	gu	lorr	rsar	lvbl	vbl	pv
9	dgugr	grbl	lr	dbrr	fv	—	brr
10	dgrbl	dgrbl	rr	dgugr	sfgr	—	bei
11	fggr	sgbr	dgugr	grbr	fv	hbl	dgrbl
12	lv	gbr	lv	v	dgubl	gubl	v
13	v	hvgr	fv	brgu	lbrr	gubl	dv
14	bei	sg	ggr	ggr	fv	—	gbr
15	lv	brv	pgor	vr	rsa	p	rv
16	lv	hv	pgor	vr	rsar	p	rv
17	gbr	hbei	dgrgu	rbr	lrv	grgu	v
18	dbrg	bei	lvbl	bl	lvr	ggr	zr
19	fgbr	gbr	lgbr	br	lpr	p	brr
20	dbl	dbrv	brv	rv	p	dvbr	bl
21	lg	gor	lgr	hr	lgbr	lhr	gr
22	dgrgu	dgubr	lzr	dgubl	dbr	brr	rv
23	sgrgu	shgu	dbr	fbrr	fv	hbl	gr
24	lggr	gbr	hgrbr	gugr	lrbr	gu	zr
25	sgrgu	shgu	fgrg	hggu	sfv	—	bei
26	dgrbl	dgrbl	lgor	gubr	lgbr	bl	brr
27	frsa	sfr	sfg	sbrr	fv	hgubl	hgrbl
28	fgugr	shgu	lggr	hbrgu	fggr	hgu	p
29	v	gbr	fv	hbr	fgr	—	brr
30	dbr	br	dbrr	brv	fv	—	brr
31	rbr	rsar	rbr	rsar	rv	vbl	rv
32	gbr	grgu	lgrg	ggu	fg	gbr	brr
33	grgu	br	dbrr	brv	br	vbr	grgu
34	fv	hg	sg	hgr	sgrbl	—	hbrr
35	dbr	dbei	dbr	vbr	hbr	hbei	hbr
36	dbr	br	dbr	dblgr	hbr	hbei	hgr
37	lor	rbr	lvr	zr	sfv	hgr	dv
38	gbr	gbr	dbr	hbr	hbr	—	hbrr
39	gbr	br	dbr	br	lgbr	hbr	dgr
40	brr	rbr	dbr	rbr	br	brv	rv
41	lgr	gbr	dbr	br	fgr	—	gr

(80 + 20); VII = chloroform-glacial acetic acid (90 + 10); VIII = methylene chloride-glacial acetic acid (90 + 10).

Colors: DL = daylight; d = dark; h = bright; l = brilliant; f = pale; s = weak; bl = blue; bei = beige; br = brown; g = yellow; gr = green; gu = grey; r = red; rr = rust-red; rsa = pink; zr = brick-red; or = orange; p = purple; v = violet.

(No. 14), but is not unequivocally more hydrophilic than 17α-hydroxy-progesterone (No. 18).

A clearly assured sequence cannot be established with solvent II for steroids No. 22, 23 and 25, all having three oxygen atoms. If other solvents are considered, the following sequence is found: No. 25 (3-keto-groups), No. 22 and, finally, No. 24.

With the Bush A-system, androsterone (No. 25) on paper has a still lower *hRf*-value (= 11) than testosterone (*hRf* = 19). Methyltestosterone (No. 19) has a higher value than testosterone (No. 26), (Table 42) just as in PC. However, with steroids No. 14, 18, 22, and 24, hydrogen bonding is possible, so that the corresponding OH-groups do not confer such highly hydrophilic properties on their molecules. In the Bush B-system[1], desoxy-corticosterone (No. 22, *hRf* = 36) has, on filter paper, a much lower value than pregnanediol (No. 28, *hRf* = 50). Allopregnanediol (No. 27) and pregnanediol (No. 28) separate well on Silica Gel G layers. The "early pregnancy test", described in Chapter G, is based on the detection of these two steroids in urine, by TLC.

Steroids with five oxygen atoms, e.g. cortisones and cortisols, are much more hydrophilic, and accordingly migrate more slowly. These compounds, which have a moderately strong polarity generally can be well separated with solvents V–VII.

Aldosterone (No. 41) is altered after a few hours in chloroform solution. With solvents containing acetic acid, No. 41 shows especially high *Rf*-values, exceeding even those for cortisone (No. 34). With PC, steroid No. 41 does not migrate as far on B_5, C and E_2B[2] as No. 34. With B_5, however, it exceeds cortisol (No. 38); with C it equals No. 38, and with E_2B it lags behind.

On Silica Gel G layers, estriol (No. 37) takes an intermediary position between cortisone (No. 34) and cortisol (No. 38).

2. Survey of the literature concerning TLC of steroids

In some cases, corticosteroids, but also steroids of weaker polarity, have been separated with non-polar solvents using the multiple development technique [*58*, *89*].

Instead of taking pure steroids, which are either too polar or too sensitive to oxidation, chromatography may be carried out on their more lipophilic esters, e.g. acetates, formates, benzoates, etc.

To produce the acetates, about 20 μg of sterols are dissolved in pyridine to which one drop of acetic anhydride is added. After leaving overnight, excess solvent is removed by gently heating in vacuo.

The reader is referred to the papers of SMITH [*70*] and HANČ [*23*] concerning the preparation of other derivatives as well as the isolation of steroids from biological material.

[1] Designation for the solvents of the Bush system in PC.
[2] Designation for solvents given by BUSH [*11*].

Remarks on Table 43.

Solvent: I = Cyclohexane – ethyl acetate, (85 + 15); II = n-hexane – ethyl acetate, (25 + 75); III = cyclohexane – ethyl acetate, (30 + 70); IV = cyclohexane – ethyl acetate, (75 + 35); V = cyclohexane – ethyl acetate, (67 + 33); VI = cyclohexane – ethyl acetate, (50 + 50); VII = cyclohexane – ethyl acetate, (40 + 60).

No.	Compound	$Rf \times 100$	Solvent	Method	Author
1	14-Etiene 3 β-ol-17 β-carboxylic acid methyl ester-3-acetate	47	I	Modified standard conditions	
2	8,14-Etiene-3 β-ol-17 β carboxylic acid methyl ester-3-acetate	47	I	Modified standard conditions	
3	Etiane-3 β-ol-17 β-carboxylic acid methyl ester-3-acetate	47	I	Modified standard conditions	
4	Etiane-3 β,14 β-diol-17 β-carboxylic acid methyl ester-3-acetate	22	I	Modified standard conditions	
5	14,15 α-Etiane-3 β-ol-17 β-carboxylic acid methyl ester-3-acetate	23	I	Modified standard conditions	
6	8,14 α-Etiane-3 β-ol-17 β-carboxylic acid methyl ester-3-acetate	18	I	Modified standard conditions	
7	Etiane-3 β-15 α-diol-17 β-carboxylic acid methyl ester-3-acetate	16	I	Modified standard conditions	
8	Etiane-3 β, 12 β-diol-17 β-carboxylic acid methyl ester	54 8	II III	Modified standard conditions	
9	Etiane-3 β, 12 β-diol-17 β-carboxylic acid methyl ester-3-acetate	40	III	Modified standard conditions	[5]
10	Etiane-3 β, 12 β-diol-17 β-carboxylic acid methyl ester-3,12-diacetate	52	III	Modified standard conditions	
11	Etiane-3 β, 12 β, 14 β-triol-carboxylic acid methyl ester-3,12-diacetate	42	III	Modified standard conditions	
12	Etiane-3 β, 14 β-diol-12-one-17 β-carboxylic acid methyl ester-3-acetate	41	IV	Modified standard conditions	
13	Etiane-3 β-ol-12-one-17 β-carboxylic acid methyl ester-3-acetate	48	III	Modified standard conditions	
14	Etiane-3 β-diol-11-one-17 β-carboxylic acid methyl ester-3,12-diacetate	32	III	Modified standard conditions	
15	Pregnane-3 α-ol-11,20-dione-3-acetate	43	III	Modified standard conditions	
16	Etiane-3 α-ol-11-one-17 β-carboxylic acid methyl ester-3-acetate	58	III	Modified standard conditions	
17	Etiane-3 α-11 β-diol-17 β-carboxylic acid methyl ester-3-acetate	52	III	Modified standard conditions	
18	5-Androstene-3 β-ol-17-one	28	V	Loose layers	
19	5-Androstene-3 β-ol-17-one-3-acetate	59	V	Loose layers	
20	17 α-Ethinyl-5-androstene-3 β-17 β-diol	35	V	Loose layers	
21	17 α-Ethinyl-5-androstene-3 β-17 β-diol-3-acetate	55	V	Loose layers	
22	17 α-Ethinyl-5-androstene-3 β,17 β-diol-3,17-diacetate	75	V	Loose layers	
23	17 α-Dibromoethinyl-5-androstene-5 β-brom-3 β, 6 β, 17 β-triol-3,17-diacetate	27	V	Loose layers	
24	5,6 α-Oxidoandrostane-3 β-ol-17-one-3-acetate	34 66	V VI	Loose layers	
25	5,6 β-Oxidoandrostane-3 β-ol-17-one-3-acetate	38 59	V VI	Loose layers	
26	5,6 α-Oxido-17 α-ething-androstane-3 β, 17 β-diol-3-acetate	80	VI	Loose layers	[3, 4]
27	5,6 β-Oxido-17 α-ethinyl-androstane-3 β, 17 β-diol-3-acetate	69	VI	Loose layers	
28	Cortisone	78	V	Loose layers	
29	17 α-Ethinyl-4-androstene-17 β-ol-3-one-17-acetate	51 86	V VII	Loose layers	
30	17 α-Dibromoacetyl-4-androstene-17 β-ol-3-one-17-acetate	39 75	V VII	Loose layers	
31	17 α-Acetyl-4-androstene-17 β-ol-one-17-acetate	35 68	V VII	Loose layers	
32	4-Androstene-3,17-dione	34 65	V VII	Loose layers	

Recently, a number of publications dealing with TLC-separation of steroids have appeared. In order to reach a correct comparison, it is pointed out that the *Rf*-values only serve as a guide and that the methods used differ, in part, from those described in this book. They, therefore, show only a rough agreement with the *Rf*-values shown in Table 42.

BARBIER et al. [*5*] who chromatographed steroids in addition to various sterols used a hydrous mass for spreading (1 + 3) so that much thinner layers are produced. Solvents and *Rf*-values are shown in Table 43. Length of run was 14 cm. For detection, the latter authors used Reagents No. 12, 52 and 120.

The author has observed with other classes of compounds that *Rf*-values and the sequences of individual compounds differ greatly on thin, loose layers as compared with more retentive adhering thin layers (see also p. 2). In view of the many advantages presented by adhering layers — even those without gypsum[1] — the shorter running time, which is the advantageous feature of loose layers, loses its attraction. The first results of steroid separations on loose layers, published by AKHREM, and KUZNETSOVA [*3, 4*], are shown in Table 43. Reagents used were concentrated sulphuric acid and Reagents No. 12, 52 and 120.

ČERNÝ, JOSKA and LÁBLER [*12*], also HEŘMAŇECK, SCHWARZ and ČEKAN [*24*] used loose layers of alumina.

ČERNÝ et al. [*12*] developed sterols, steroids (pregnane and androstane compounds) and genins with the solvents: hexane-benzene (70 + 30), benzene, benzene-ether (70 + 30), ether, ether-ethanol (98 + 2 or 96 + 4). According to the increasing polarity of these solvents, the weakly polar sterols were developed with the first-named solvent, and the latter were used for the polar steroids. For detection, they used concentrated sulphuric acid as well as Dragendorff's reagent (No. 60) and morin (No. 90e).

HEŘMAŇECK et al. [24] used mixtures of petroleum ether-benzene and benzene-ethanol of varying composition. Steroids were detected with antimony trichloride (No. 12).

STRUCK [*79*] reported the separation and quantitative determination of estrone, 17β-estradiol and 16α, 17β-estradiol. The plates were developed according to the standard method. The solvent used was benzene-ethanol (90 + 10). Estrone *Rf* 0.51, estradiol *Rf* 0.30 and estriol *Rf* 0.11 were determined spectrophotometrically at 242 mμ after extracting the spots with 0.5N-NaOH (80% alcoholic). BARBIER et al. [*6*] also dealt with TLC of estrogens.

BEIJLEVELD [*8*] used TLC to control the purity of steroids and for the examination of steroid preparations. His 7 photostats attest the excellence of the separations achieved. MATIS et al. [*46*] demonstrated that corticosteroids can be separated also on calcium sulphate (gypsum) layers. DANNENBERG and NEUMANN [*14*] used TLC in their studies on the dienol-benzene conversion of steroids. JEGER and his group [*26a*] also used TLC to investigate conversions. WEHRLI [*90a*] used TLC, as well as PC, for purposes of identification.

[1] Silica Gel H and Alumina H manufactured by E. Merck A.G. Darmstadt, Germany.

It is obvious that TLC can be utilized in the steroid field with most satisfactory results.

3. Detection of separated steroids with spray reagents

Some of the steroid compounds may be recognized merely by their color reactions. However, the experiences gained in this field have not yet reached the stage where one can deduce the existence of a specific structure from the formation of some color, as is the case, for example, with cardiac glycosides detected by the chloramine-trichloroacetic acid reaction (see p. 276).

Sufficient indications exist to favor the view that silica gel layers are particularly well suited for specific and selective color reactions. SIBLIKOVA and HAIS [*64*] have investigated color reactions of 85 different steroids on filter paper. WACHSMUTH and VAN KOECKHOVEN [*88*] also investigated color reactions of steroids. To detect steroids on inorganic thin layers, the most suitable reagents are: phosphoric acid (No. 123), acetic anhydride-sulphuric acid (No. 63), antimony trichloride (No. 12) and vanillin-sulphuric acid (No. 151).

Own experiments have shown that 15%-phosphoric acid, which was originally recommended by NEHER and WETTSTEIN [*52*, *53*] for PC, is particularly specific for the detection of all steroids. We find that in TLC, spraying with a 30–40% o-phosphoric acid is advantageous. As a rule, spraying continues until the plate is uniformly saturated. Before the reaction, lipid zones appear already as white "grease spots." After heating for 7–15 min at 110–120° C, the steroids fluoresce strongly in UV-light (365 mμ). Initially, the colors are weak but intensify later on. It is, therefore, necessary to observe the intensification of the colors during heating. The plate should be examined after about 8 min in daylight and under a U.V. lamp followed by heating for another 7 min at 110° C. It is not advantageous to heat for a longer period.

Cholesterol and its esters, which under UV-light appear a pink-red after initial heating, later turn a rust-red color.

Estrogens, under UV-light, first exhibit a brilliant yellow, then orange and, finally, orange-red. Pregnenolone and dehydroandrosterone show a brilliant violet. Testosterone and dehydrotestosterone appear dark blue-grey in daylight. The color appears first as bright-blue, turning to dark green-blue after further heating, finally becoming dark blue-grey. If heating is too prolonged, spots turn grey-black. A number of compounds such as pregnanediol, allopregnanediol and progesterone show only pale fluorescence in UV-light. On the other hand, aldosterone shows as a brilliant green.

All those colors persist for a short time only. The plates sprayed with phosphoric or sulphuric acids adsorb water from the atmosphere, so that the colors gradually start to run. For this reason, the plate should be immediately covered with a second glass plate which prolongs the life of the colors. Color photographs are indicated for purposes of documentation.

Steroids which do not appear distinctly after reacting with phosphoric acid (7 min, 100° C) can be located as dark blue spots by slight spraying with freshly prepared 1.5%-alcoholic phosphomolybdic acid (Reagent No. 120c) followed by heating for 3 min. Although coloration may already be achieved with the phosphomolybdic acid reagent, it is preferable to pre-treat the sample with phosphoric acid.

A further "specific" reagent for steroids is provided by the modified LIEBERMANN-BURCHARD reagent (No. 63). After copious spraying, the chromatogram is heated to 110° C for 15–20 min. During heating, the colors should be examined in daylight and under long-wave UV-light.

Anitimony trichloride (Reagent No. 12) does not produc any characteristic color differences with steroids. Following spraying and heating for 15–20 min, spots appear which exhibit non-specific fluorescent colors in UV-light.

Vanillin-sulphuric acid (Reagent No. 151) is particularly well suited for checking identity. The reagent must be freshly prepared and must not be sprayed too heavily on the plate. After heating to 110° C for 5–10 min, colored steroid spots are located. However, they do not fluoresce in UV-light. Here too, one can prolong the life of the colors by covering the thin layers with a second glass plate.

4. Special applications of TLC in the steroid field

a) Investigation of the stability of estrone derivatives in tablets

FOKKENS and POLDERMANN [*16*] have investigated estrene-derivatives. The stability of various 17 β-hydroxy-4-estrene-compounds, substituted in the 17 α-position, was investigated. Significant differences in the degree of stability may be compensated for by the addition of antioxidants.

Development proceeded according to the standard method using n-heptane-acetone (50 + 25). Steroids were located as fluorescent spots after spraying with antimony trichloride (Reagent No. 12). Their *Rf* × 100-values were:

17α-Ethinyl-4-estrene-17β-ol	48
17α-Methyl-4-estrene-17β-ol	50
17α-Ethyl-4-estrene-17β-ol	57
17α-Propyl-4-estrene-17β-ol	57
17α-Allyl-4-estrene-17β-ol	59

b) Rapid analysis for the control of enzymatic steroid conversions

Enzymatic syntheses of steroids may be checked by TLC, e.g., when introducing hydroxyl-groups in the steroid skeleton with cultures of a suitable microorganism. Secondary reactions, as well as completeness of the reaction, may thus be recognized. METZ [*47*] described routine surveillance of culture solutions. During production, 200 ml of a 0.05% suspension were extracted with chloroform, and the extracts analyzed by TLC. For pilot experiments in the laboratory, samples of only 2 ml suspension will provide sufficient quantities of steroids for identification. In these experiments, the extract is put on a narrow strip of filter paper from whose tip it is transferred to the silica gel layer.

Metz separated steroids on Silica Gel G by developing with chloroform-acetone or chloroform-ethyl acetate mixtures. He suggested the following reagents for detection: phosphoric acid (No. 123), vanillin-phosphoric acid (No. 149), anisaldehyde-sulphuric acid (No. 9), antimony trichloride (No. 12). For quantitative determination against a standard, staining is carried out with perchloric acid (2% aqueous solution) by heating to 150° C for 10 min after spraying. With reducing steroids, 2,3,5-triphenyltetrazolium chloride (Reagent No. 145) is used.

c) Detection of steroids in human urine

In chapter G of this book, an early-pregnancy test is described, which is based on the detection of pregnane-3α,20α-diol and allo-pregnane-3α, 20α-diol in urine.

This test can also be positive if gestagenes (progesterone derivatives) are administered to the subject, or if the kidneys are stimulated to excessive activity by overexertion (stress). In such cases, additional steroid excretions will be found on the thin-layer chromatogram, e.g., pregnanetriol. In normal pregnancy, such steroids hardly ever show up. The appearance of larger quantities of pregnanetriol in the urine indicates the existence of an adrenogenital syndrome. Stárka et al. [*75*] demonstrated this on loose alumina layers. Reisert et al. [*61*] separated, on 50 cm long Silica Gel G layers, the 17-ketosteroids from urine. A chromatogram given by the authors [*51*] shows the decreased excretion of 17-ketosteroid in case of Cushing's syndrome.

80% of the steroids excreted in the urine are in the form of glucuronides, the remainder being sulphates (estrogene), or unbound. The acid hydrolysis required presents advantages compared with an enzymatic hydrolysis, since it is faster. However, it may produce changes in more sensitive steroids. Special precautions are needed to isolate the corticosteroids, since oxidative changes easily occur in the side-chain at C-17. The side-chain can be split off to give androgens (C-17 ketosteroids). Gestagenes (progesterone-derivatives) are, as a rule, slightly more stable, as shown by the early-pregnancy test. It is preferable to hydrolyze under the mildest conditions possible, and in an atmosphere of nitrogen.

IV. Bile acids

Bile acids behave, during chromatography, according to their polarity, rather like corticosteroids. If a bile acid has a peptide-like linkage with an amino acid, as is the case with taurocholic acid or glycocholic acid, solvents containing highly hydrophilic components, like methanol or water, must be used.

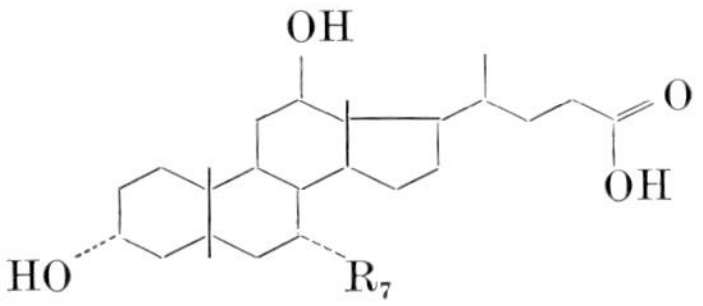

Cholic acid, R_7 = OH
Desoxycholic acid, R_7 = H

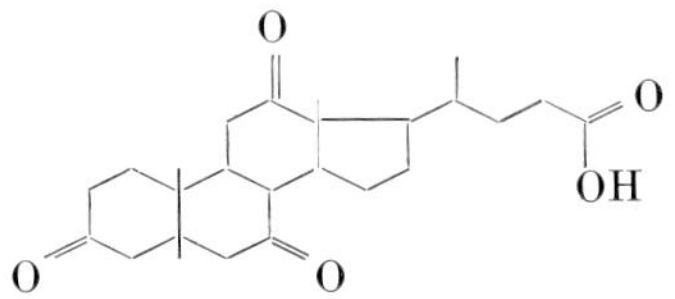

Dehydrocholic acid

GÄNSHIRT, KOSS and MORIANZ [*17*] were the first to separate various bile acids on Silica Gel G layers with the solvents toluene-glacial acetic acid-water (Table 44a, Nos. IX and X) and butanol-glacial acetic acid-water (XI) (Fig. 125 and 126). The spots were evaluated, quantitatively, by photometry after elution.

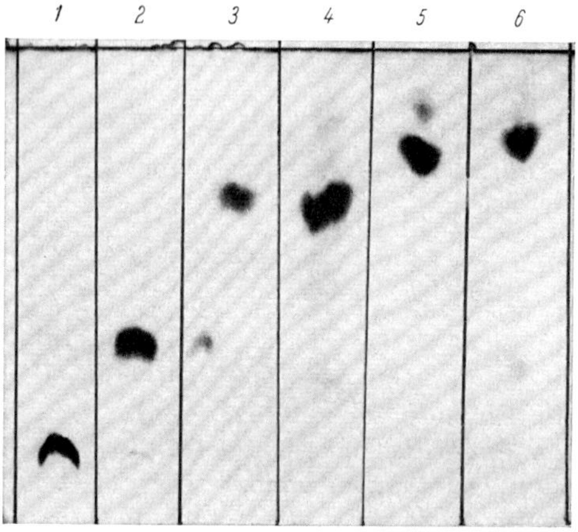

Fig. 124. Separation of some free bile acids: cholic acid (*1*), desoxycholic acid (*2*), dehydrocholic acid (*3*), 3,12-diacetoxy-cholanic acid (*4*), 3α-hydroxy-(7,8)-11,12-choladienic acid (*5*), 3α-hydroxy-(8,14)-11,12-choladienic acid (*6*) on Silica Gel G with chloroform-glacial acetic acid (90 : 10). Detection: phosphoric acid (Reagent No. 123) followed by phosphomolybdic acid (Reagent No. 120c) (E. Merck AG.)

Table 44a. *Separation of Bile Acids on Silica Gel Layers with various Solvents*

Bile acids	*Rf* × 100-values with solvents										
	I	II	III	IV	V	VI	VII	VIII	IX	X	XI
1. Cholic acid	0	0	0	0	0	12	13	55	21	53	73
2. Desoxycholic acid	12	16	13	25	25	54	70	85	38	65	81
3. Dehydrocholic acid	14	21	70	65	65	78	90	97	37	65	70
4. 3,12-Diacetoxycholanic acid	19	25	—	66	—	—	—	—	20	—	75
5. 3α-Hydroxy-(7,8)-11,12-choladienic acid	28	33	—	78	—	—	—	—	38	—	77
6. 3α-Hydroxy-(8,14)-11,12-choladienic acid	22	33	—	75	—	—	—	—	39	—	67

Solvents

I Cyclohexane-chloroform-glacial acetic acid (70 + 20 + 10)
II Toluene-chloroform-glacial acetic acid (70 + 20 + 10)
III Methylene chloride-glacial acetic acid (90 + 10)
IV Chloroform-glacial acetic acid (90 + 10)
V Methylene chloride-ethyl acetate-glacial acetic acid (70 + 20 + 10)
VI Chloroform-acetone-glacial acetic acid (70 + 20 + 10)
VII Methylene chloride-acetone-glacial acetic acid (70 + 20 + 10)
VIII Cyclohexane-chloroform-methanol-glacial acetic acid (20 + 60 + 10 + 1)0
IX Upper phase of: Toluene-glacial acetic acid-water (50 + 50 + 10) [*17*]
X Toluene-glacial acetic acid-water (50 + 50 + 5) [*17*]
XI Butanol-glacial acetic acid-water (100 + 10 + 10) [*17*]

Thin-layer chromatograms of bile acids according to GÄNSHIRT *et al.* [17]

Nos. *1* and *12* are mixtures of the following bile acids: *2* cholic acid, *3* desoxycholic acid, *4* chenodesoxycholic acid, *5* lithocholic acid, *6* glycocholic acid, *7* glycochenodesoxycholic acid, *9* taurocholic acid, *10* taurodesoxycholic acid, *11* taurochenodesoxycholic acid

Fig. 125. Separation of free bile acids and separation of the conjugated bile acids, developed on Silica Gel G layer with the upper phase of the system: toluene-glacial acetic acid-water (50 + 50 + 10). Detection: Reagent No. 120a

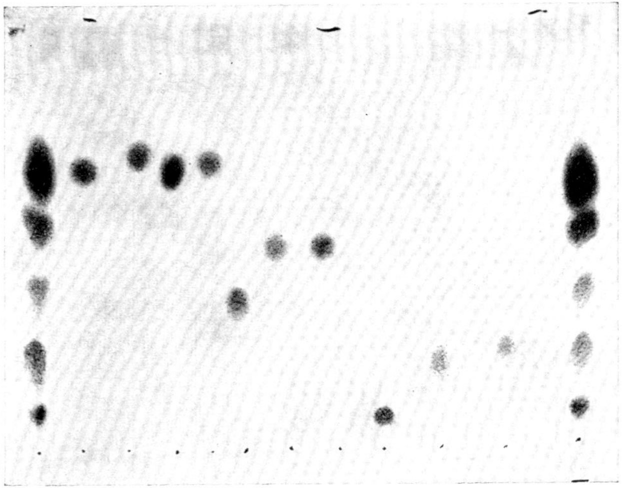

Fig. 126. Separation of conjugated bile acids and separation of free bile acids on Silica Gel G layer, developed with butanol-glacial acetic acid-water (100 + 10 + 10). Detection: as for Fig. 125

Fig. 124 shows that cholic acid, dehydrocholic acid and desoxycholic acid separate readily on Silica Gel G layers, using chloroform-glacial acetic acid (90 + 10) as solvent. Detection of conjugated bile acids is best achieved with vanillin-sulphuric acid (No. 151). The Liebermann-Burchard reagent (No. 63) shows spots in daylight which fluoresce with various colors in UV-light. Dehydrocholic acid reacts neither with Reagent No. 63 nor with phosphoric acid (No. 123); but it can be located by subsequent spraying with phosphomolybdic acid (No. 120c) followed by heating. Color reactions with antimony trichloride (reagent No. 12) are less specific.

While this book was in the press, there appeared a publication by Hofmann [*26*] which also deals with the TLC of bile acids.

Sjoevall [*65—69*] reported on the PC and determination of bile acids. Hamilton and Dieckert [*20*] separated bile acids, also steroids [*21*] and saponins [*22*] on glass-fiber paper.

Table 44b. *Color Reactions of Bile Acids*

Bile acids	Detection					
	Vanillin-H_2SO_4 (No. 151)	Acetic anhydr. H_2SO_4 (No. 63)		Phosphoric acid (No. 123)		Antimony trichloride (No. 12)
	DL	DL	UV	DL	UV	UV
1. Cholic acid.	blgu	g	lg	g	lg	fv
2. Desoxycholic acid. . .	blgu	g	lgrg	g	lg	—
3. Dehydrocholic acid . .	br	—	—	—	—	—
4. 3,12-Diacetoxycholanic acid.	dgu	ggr	dgr	—	—	—
5. 3α-Hydroxy-(7,8)-11,12-choladienic acid . . .	v	gbr	dbr	—	—	fv
6. 3α-Hydroxy-(8,14)-11,12-choladienic acid . .	v	gbr	dbr	—	—	fv

For key to color-symbols, see Table 42, p. 263.

V. Cardiac glycosides

A number of herbs contain poisonous glycosidic mixtures which act on the heart. These cardiac glycosides are used in the therapy of circulatory decompensation. Cardiac glycosides are characterized by a steroid skeleton and most possess a five-membered lactone ring at C_{17}. The OH-group at C_3 is linked by an ether group with one or more sugars (cf. Table 45).

The best known medicinal herbs containing such substances are: *Digitalis purpurea* L. and D. *Lanata Ehrh*; *Strophantus Kombé* Oliv. and *Str. gratus* (Wall et Hook), *Franch* and *Urginea maritima* (L). Baker (= *Scilla maritima* L.). Extracts from *Convallaria majalis L.*, *Adonis vernalis* L., *Nerium oleander* L., and others are also used. The glycoside content of these herbs, and, hence, also their activity depends on a number of factors, e.g., environment, age of plants, time of harvest and, especially, the

Table 45

Arrangement of Digitalis Glycosides according to their Polarities

D = Digitoxose *Ac* = Acetyl *G* = β-Glucose *OH* = Hydroxyl group

Increasing polarity →

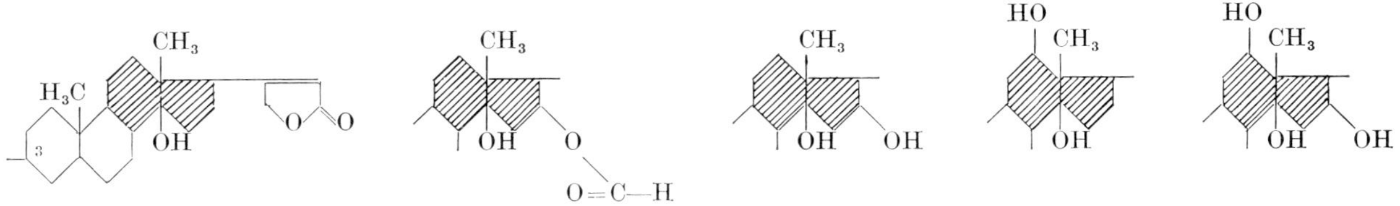

Series:	A	E	B	C	D
Linked to C_3:					
HO- Aglycones	1 Digitoxigenin	2 Gitaloxigenin	5 Gitoxigenin	6 Digoxigenin	9 Diginatigenin
D-D-D-O- Secondary glycosides	3 Digitoxin	4 Gitaloxin	7 Gitoxin	8 Digoxin	12 Diginatin
Ac | *G-D-D-D-O* Primary glycosides of *Digitalis lanata*	10 Lanatoside A	11 Lanatoside E	13 Lanatoside B	15 Lanatoside C	18 Lanatoside D
G-D-D-D-O- Primary glycosides of *Digitalis purpurea*	14 Desacetyl-Lanatoside A (= Purpureaglycoside A)	16 [Desacetyl-Lanatoside E]	17 Desacetyl-Lanatoside B (= Purpureaglycoside B)	19 [Desacetyl-Lanatoside C]	20 [Desacetyl-Lanatoside D]

Increasing polarity →

mode of drying. Recently, specific "chemical races" have been discovered. Since individual glycosides differ strongly in their action, and, since a mixture of glycosides is present in the herbs, it is absolutely essential to determine their degree of activity. At present, this is still preferably done by biological testing. The preparations are standardized to a specific biological activity (e.g., a frog-unit). Attempts have been made for some time to replace these rather laborious biological test methods with more precise physicochemical procedures. Chromatography offers this possibility, including a quantitative evaluation of mixtures of cardiac glycosides.

STOLL et al. [*78*] were able to separate chromatographically on silica gel columns these substances, because of their differences in polarity. Polarity, as well as water-solubility, increases with the number of OH-groups present in the basic structure and also with the number of glycosidically-linked sugars (Table 45).

ZAFFARONI [*91*] was the first to separate steroid mixtures on filter paper impregnated with formamide or ethylene glycol. REICHSTEIN, SCHINDLER et al. [*59, 60*] used this method for the chromatography of glycosides of *Digitalis* and *Strophantus*, and found many new compounds.

KAISER [*27*] carried out a systematic investigation of the PC of digitalis glycosides. Table 45 was compiled from his results. As will be seen from the following publications, PC is also a valuable aid for the investigation of mixtures of cardiac glycosides by TLC.

TSCHESCHE et al. [*81–87*] were the first to make use of TLC to separate and identify various steroid glycosides.

STAHL and KALTENBACH [*74*] showed that the most important cardiac glycosides can be separated on Silica Gel G layers with the solvent methylene chloride-methanol-formamide (80 + 19 + 1). A mixture of 14 different compounds was well separated in a length of run of 10 cm (Fig. 127). It is necessary, though, that the quantity applied for each individual glycoside should be very small, (i.e., approximately 1 μg) and that there is a satisfactory tank saturated with solvent vapors. (TS, p. 18). The authors prefer trichloracetic acid-chloramine (Reagent No. 31) for detection of the separated compounds.

STEINEGGER and VAN DER WALT [*77*] found that the cardiac glycosides of the white and red sea-onion (*Urginea maritima var.*) may be separated on Silica Gel G with the upper phase of the mixture ethyl acetate-pyridine-water (50 + 10 + 40), and showed this in an illustration. STEIDLE [*76*] described the UV-spectrometric evaluation of cardiac glycosides separated by TLC, using *Scilla*-glycosides as an example. While this chapter was in the press, there appeared a remarkable paper by SJÖHOLM [*72a*] on the TLC of 28 digitalis glycosides. Two-dimensional TLC was carried out with the solvents ethyl acetate-methanol-water (80 + 5 + 5) and chloroform-pyridine (60 + 10) on Silica Gel G layers. One-dimensional chromatograms of glucose-containing digitalis glycosides were developed with the solvent mixture methyl ethyl ketone-chloroform-formamide (50 + 20 + 10).

GÖRLICH [*18*] reported on TLC of steroid glycosides from *Nerium oleander* L. and their subsequent colorimetric determination. To judge from the chromatograms reproduced in his paper, the following two solvents produce a very good separation of *Oleander* extracts on Silica Gel G layers: I. ethyl acetate-chloroform (90 + 10), and II. methyl ethyl ketone-toluene-methanol-glacial acetic acid-water (80 + 10 + 5 + 2 + 6). In a two-dimensional chromatogram (direction of migration 1 = solvent II, direction of migration 2 = solvent I), a neat separation of 14 steroid glycosides was achieved.

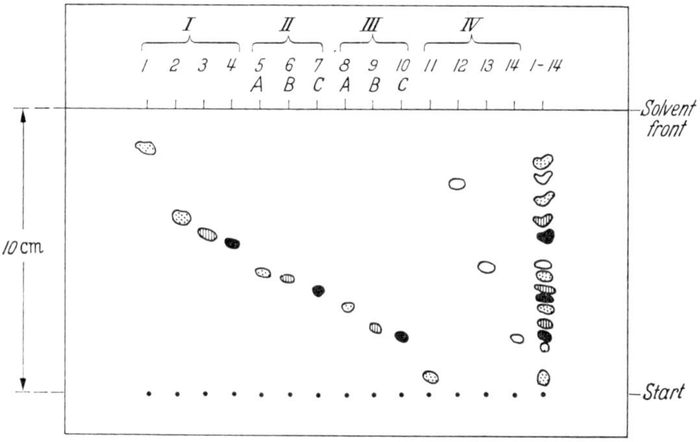

Fig. 127. Separation of cardiac glycosides according to STAHL et al. [*74*] *I* sec. Glycosides; *II* Digilanides; *III* Desacetyl digilanide; *IV* Digitaloides *1*—*14* (*1* acetyldigitoxin, *2* digitoxin, *3* gitoxin, *4* digoxin, *5* digilanide A, *6* digilanide B, *7* digilanide C, *8* desacetyl digilanide A, *9* desacetyl digilanide B, *10* desacetyl digilanide C, *11* k-strophantoside, *12* cymarine, *13* proscillaridine A, *14* scillaren 1—14 = Mixture. Separated on Silica Gel G with methylene chloride-methanol-formamide (80:19:1). Detection: Reagent No. 31. Colors: under UV-light (365 mμ): bright-yellow = □ brown-yellow = ▒ bright-blue = ▥ violet-blue = ■

For the quantitative determination of cardiac glycosides separated by TLC, it is convenient to apply larger amounts. Our own experiments [*89*] have shown that multiple development is most suitable. Standard conditions were observed throughout on Silica Gel G layers and the solvent mixtures consisted of components of varying polarities.

When separating glycosides with but slight differences in polarity, it has been found useful to choose a solvent of low polarity, which is then used repeatedly (multiple development).

By using the solvent mixture cyclohexane-acetone-glacial acetic acid (49 + 49 + 2) on Silica Gel G for multiple development, the following results are obtained:

During the *first* development, the aglycones of the digitalis glycosides reach about the middle of the 10 cm long path; the secondary glycosides are below this point, and separate also. Approximate values ($Rf \times 100$) obtained were: digitoxigenin = 68, gitoxigenin = 51, digoxigenin = 48, digitoxin = 42.

The *second* development brings the secondary glycosides within the *hRf-range* of 40—70 whereas the primary glycosides remain under 20.

During the *third* and *fourth* development, there occurs good separation of primary glycosides, which are found in the same sequence as in PC.

With chloroform-methanol as solvent (90 + 10), triple development gives the following values: digitoxin = 73, gitoxin = 55, digoxin = 47.

Apart from cardiac glycosides, a number of other steroid glycosides and aglycones was also investigated by TLC. For instance, SANDER [*62*] used the standard method and separated diosgenin ($hRf = 60$) from gitogenin ($hRf = 20$) with cyclohexane-ethyl acetate (70 + 30). Vanillin-phosphoric acid-reagent (No. 149) was used for location. The reader is referred to pp. 201, 202 for details on the separation of triterpenoids.

Chloramine-trichloroacetic acid reagent (No. 31) is stated by KAISER [*27*] to be especially well suited for the detection of cardiac glycosides. The antimony trichloride reagent (No. 11 or 12) is less sensitive. Good results are also obtained with phosphoric acid reagent (No. 123) and with sulphuric acid-acetic anhydride-reagent (No. 63). Experience has shown that the various digitalis glycoside series may be differentiated by their various fluorescent colors, as shown in the table below. These colors differ, however, from those described for PC.

Digitalis glycosides	Fluorescent color with chloramine-trichloro-acetic acid in UV-light	Color after reaction with phosphoric acid (cf. Fig. 127 Reagent No. 123) in day light
Row A	Brown-yellow	Blue-black
Row B	Light blue	Brown-violet
Row C	Violet-blue	Grey-violet

SJÖHOLM [*72a*] prefers the following mixture to locate digitalis glycosides on thin-layer chromatograms: 0.5 ml anisaldehyde, 5 ml 70%-perchloroacetic acid, 10 ml acetone and 40 ml water. After spraying, the plates are heated for 4–5 min to 75–80° C and the colors are noted in daylight and under UV-light (365 mμ). It is important to note changes in color and fluorescence during the first hour. Sensitivity in daylight is 0.1 μg, and in UV-light 0.02 μg.

Bibliography to Chapter D. Steroids

[*1*] ABELSON, D., and R. BROOKS: Paper Chromatography of Steroids. London: Cambridge Univ. Press 1960.
[*2*] AHRENS, E. H. jr., W. INSULL jr., J. HIRSCH, W. STOFFEL, M. L. PETERSON, J. W. FARQUHAR, T. MILLER and H. J. THOMASSON: Lancet **1959 I**, 115.
[*3*] AKHREM, A. A., i A. I. KUZNETSOVA: Med. Prom. SSSR **15**, 57 (1961).
[*4*] — — Doklady Akad. Nauk. SSSR. **138**, 591 (1961).
[*5*] BARBIER, M., H. JÄGER, H. TOBIAS u. E. WYSS: Helv. Chim. Acta **42**, 2440 (1959).
[*6*] —, i S. I. ZAV'YALOV: Izvest. Akad. Nauk. SSSR. Otdel. Khim. Nauk (7), **1960**, 1309.
[*7*] BARTON, D. H. R., G. A. MORRISON u. L. ZECHMEISTER: Fortschr. Chem. org. Naturstoffe **19**, 165 (1961).
[*8*] BEIJLEVELD, W. M.: Pharm. Weekblad **97**, 190 (1962).
[*9*] BOLLIGER, H. R., M. L. QUAIFE and R. P. GEYER: Anal. Chem. **31**, 950 (1959).
[*10*] BRIESKORN, C. H., H. KLINGER u. W. POLONIUS: Arch. Pharm. **294**, 389 (1961).
[*11*] BUSH, I. E.: The Chromatography of Steroids. Oxford 1961.
[*12*] ČERNÝ, V., J. JOSKA a. L. LÁBLER: Collection Czechosl. Chem. Commun. **26**, 1658 (1961).
[*13*] VAN DAM, M. J. D. et al.: J. Chromatog. **4**, 26 (1960).
[*13a*] — Bull. soc. chim. Belges **70**, 122—124 (1961).
[*14*] DANNENBERG, H., u. H.-G. NEUMANN: Liebigs Ann. Chem. **646**, 148 (1961).

[15] Fieser, L. F., u. M. Fieser: Steroids. Reinhold Publishing Corp. N. Y. 1959.
[16] Fokkens, J., and J. Poldermann: Pharm. Weekblad **96**, 657 (1961).
[17] Gänshirt, H., F. W. Koss u. K. Morianz: Arzneimittel-Forsch. **10**, 943 (1960).
[18] Görlich, B.: Planta Med. **9**, 442 (1961).
[19] Habermann, E., G. Bandtlow u. B. Krusche: Klin. Wschr. **39**, 816 (1961).
[20] Hamilton, J. G., and J. W. Dieckert: Arch. Biochem. Biophys. **82**, 203 (1959).
[21] — — Arch. Biochem. Biophys. **82**, 212 (1959).
[22] — — Arch. Biochem. Biophys. **82**, 220 (1959).
[23] Hanč, O.: Hormone, 2. Auflage. Jena: VEB Fischer Verlag 1959.
[24] Heřmánek, S., V. Schwarz a. Z. Čekan: Collection Czechoslov. Chem. Commun. **26**, 1669 (1961).
[25] Hörhammer, L., H. Wagner u. B. Lay: Pharm. Ztg. **106**, 1307 (1961).
[26] Hofmann, A. F.: J. Lipid Res. **3**, 127 (1962).
[26a] Immer, H., M. Lj. Mihailovic, K. Schaffner, D. Arigoni u. O. Jeger: Helv. Chim. Acta **45**, 753 (1962).
[27] Kaiser, F.: Chem. Ber. **88**, 556 (1955).
[28] Kaufmann, H. P., u. W. H. Nitsch: Fette, Seifen, Anstrichmittel **56**, 154 (1954).
[29] — — et al.: Fette, Seifen, Anstrichmittel **57**, 473 (1955).
[30] — — Fette, Seifen, Anstrichmittel **58**, 234, 492, 985 (1956).
[31] —, u. E. Mohr: Fette, Seifen, Anstrichmittel **60**, 165 (1958).
[32] —, u. H. Schnurbusch: Fette, Seifen, Anstrichmittel **60**, 1046 (1958).
[33] —, u. D. K. Chowdhury: Chem. Ber. **91**, 2117 (1958).
[34] —, u. H. Schnurbusch: Fette, Seifen, Anstrichmittel **61**, 523 (1959).
[35] — — Fette, Seifen, Anstrichmittel **62**, 1 (1960).
[36] —, u. Z. Makus: Fette, Seifen, Anstrichmittel **62**, 153 (1960).
[37] —, u. G. Schmidt: Fette, Seifen, Anstrichmittel **62**, 164 (1960).
[38] —, Z. Makus u. F. Deicke: Fette, Seifen, Anstrichmittel **63**, 235 (1961).
[39] Klein, P. D., u. E. T. Janssen: J. biol. Chem. **234**, 1417 (1959).
[40] Klyne, W.: The Chemistry of the Steroids; Methuen & Co. Ltd., London-New York: John Wiley & Sons Inc. 1957.
[41] Labarrère, J. A., J. R. Chipault and W. O. Lundberg: Anal. Chem. **30**, 1466 (1958).
[42] Lettré-Inhoffen-Tschesche: Über Sterine, Gallensäuren und verwandte Naturstoffe, Bd. I und II. Stuttgart: Ferd. Enke Verlag 1959.
[43] Mahadevan, V., and W. O. Lundberg: Lipid Res. **3**, 106 (1962).
[44] Mangold, H. K.: private communication.
[45] Martin, A. P.: Biochem. Biophys. **25**, 408 (1959).
[46] Matis, J., O. Adamec and M. Galvánek: Nature **194**, 477 (1962).
[47] Metz, H.: Naturwiss. **48**, 569 (1961).
[48] Michalec, Č.: Naturwiss. **42**, 509 (1955).
[49] — V. Jirgl u. J. Podzimek: Experientia **13**, 242 (1957).
[50] —, and J. Strasek: J. Chromatog. **4**, 254—257 (1960).
[51] Mottier, M., and M. Potterat: Anal. Chim. Acta **13**, 46 (1955).
[52] Neher, R., u. A. Wettstein: Helv. Chim. Acta **34**, 2278 (1951).
[53] — Chromatog. **1**, 122, 205 (1958).
[54] — Chromatog. **1**, 231 (1958).
[55] Peereboom, J. W. C., J. B. Roos and H. W. Beekes: J. Chromatog. **5**, 500 (1961).
[56] Prelog, V.: Helv. Chim Acta **36**, 308 (1953).
[57] —, u. H. Meier: Helv. Chim. Acta **36**, 320, 325 (1953).
[58] Prusíková, M.: Experientia **15**, 460 (1959).
[59] Reichstein, T., u. O. Schindler: Helv. Chim. Acta **34**, 108 (1951).
[60] — E. Schenker u. A. Hunger: Helv. Chim. Acta **37**, 680 (1954).
[61] Reisert, P., et al.: Experientia **19**, 84 (1963).
[62] Sander, H.: Naturwiss. **48**, 303 (1961).
[63] Searcy, R. L., and L. M. Bergquist: Clin. Chim. Acta **5**, 192 (1960).
[64] Siblíková, O., a. I. M. Hais: Českoslov. Farmac. **7**, 1—13 (1958).
[65] Sjoevall, J.: Acta Chem. Scand. **6**, 1552 (1952); **13**, 711 (1959).
[66] — a. G. A. D. Haslewood: Biochem. J. **57**, 126 (1954).

[67] SJOEVALL, J.: Ark. Kemi 8, 317 (1955).
[68] — Clin. Chim. Acta **4**, 652, 793 (1959).
[69] — Clin. Chim. Acta **5**, 33 (1960).
[70] SMITH, I.: Chromatographic and Electrophoretic Techniques, Volume I. pp. 355—362 and 409—460. W. HEINEMANN, Medical Books Ltd., London 1960.
[71] SCHMIDT v. ELMENDORFF, H.: Z. ges. exp. Med. **130**, 142—145 (1958).
[72] SCHÖN, H., u. F. GEY: Hoppe Seyler's Z. physiol. Chem. **303**, 81 (1956).
[72a] SJÖHOLM, I.: Svensk Farm. Tidskr. **66**, 321 (1962).
[73] STAHL, E.: Pharmaz. Rdsch. **1**, Nr. 2, 1 (1959).
[74] —, and U. KALTENBACH: J. Chromatog. **5**, 458 (1961).
[75] STÁRKA, L., and J. MALÍKOVÁ: J. Endocrinol. **22**, 215 (1961).
[76] STEIDLE, W.: Planta Med. **9**, 435 (1961).
[77] STEINEGGER, E., u. J. H. VAN DER WALT: Pharm. Acta Helv. **36**, 599 (1961).
[78] STOLL, A., E. ANGLIKER, F. BARFUSS, W. KUSSMAUL u. J. RENZ: Helv. Chim. Acta **34**, 1460 (1951).
[79] STRUCK, H.: Mikrochim. Acta **1961**, 634.
[80] SWARTWOUT, R. R., J. W. DIECKERT, O. N. MILLER and J. G. HAMILTON: J. Lipid Res. **1**, 281 (1960).
[81] TSCHESCHE, R., W. FREYTAG u. G. SNATZKE: Chem. Ber. **92**, 3053 (1959).
[82] —, u. G. SNATZKE: Ann. Chem. **636**, 105 (1960).
[83] —, u. A. K. S. GUPTA: Chem. Ber. **93**, 1903 (1960).
[84] — F. LAMPERT and G. SNATZKE: J. Chromatog. **5**, 217 (1961).
[85] — W. HAMMERSCHMIDT u. G. SNATZKE: Ann. Chem. **642**, 199 (1961).
[86] — H. SCHWARZ u. G. SNATZKE: Chem. Ber. **94**, 1699 (1961).
[87] —, G. WULF: Chem. Ber. **94**, 2019 (1961).
[88] WACHSMUTH, H., u. L. VAN KOECKHOVEN: Anal. Chim. Acta **22**, 41 (1960).
[89] WALDI, D.: unpublished.
[90] WEICKER, H.: Klin. Wschr. **37**, 763 (1959).
[90a] WEHRLI, W.: Helv. Chim. Acta **45**, 1206 (1962).
[91] ZAFFARONI, A., R. B. BURTON and H. E. KEUTMANN: J. Biol. Chem. **177**, 109 (1949).
[92] ZAK, B., R. C. DIEKENMAN, E. G. WHITE, H. BURNETT and P. J. CHERNEY: Am. J. Clin. Pathol. **24**, 1307 (1954).
[93] ZÖLLNER, N., u. K. KIRSCH: Z. ges. exp. Med. **134**, 10 (1960).
[94] — — u. G. AMIN: Verhandl. deut. Ges. inn. Med. **66**, 677 (1960).
[95] —, u. G. WOLFRAM: Klin. Wschr. **39**, 817 (1961).
[96] — — u. G. AMIN: Klin. Wschr. **40**, 273 (1962).
[97] DE ZOTTI, G., P. CAPELLA u. G. JACINI: Fette, Seifen, Anstrichmittel **61**, 1114 (1959).

Some recent noteworthy articles

AVIGAN, J., DEWITT S. GOODMAN and D. STEINBERG: J. Lipid. Res. **4**, 100 (1963): Sterols and steroids.
BENNETT, R. D., and E. HEFTMANN: J. Chromatog. **9**, 348 (1962): Corticosteroids.
DIAMANTSTEIN, T., and K. LÖRCHER: Z. anal. Chem. **191**, 429 (1962): Estrogens.
DYER, W. G. et al.: Steroids **1**, 271 (1963): C(19)-steroids.
FAUCONNET, L., and M. WALDESBÜHL: Pharm. Acta Helv. **38**, 423 (1963): Cardiac glycosides.
HAMILTON, J. G.: Arch. Biochem. Biophys. **101**, 7 (1963): Bile acids.
JOHNSON, D. F., R. D. BENNETT and E. HEFTMANN: Science **140**, 198 (1963): Sterines in higher plants.
KELLER, M., and A. HAUSER: Gynaecologia **150**, 275 (1960): Estrogens.
KRITCHEVSKY, D., D. S. MARTAK and G. H. ROTHBLAT: Anal. Biochem. **5**, 388 (1963): Bile acids.
LISBOA, B. P., and E. DICZFALUSY: Acta Endocrinol. **40**, 60 (1962): Estrogens.
PEEREBOOM, J. W. C., and H. W. BEEKES: J. Chromatog. **9**, 316 (1962): Steroids.
STAIB, W., u. K. DÖNGES: J. Chromatog. 8, 25 (1962): Purifying of urinary steroids for IR spectroscopy.
STAUDINGER, J. H., and O. NISHIKAZE: Klin. Wschr. **40**, 1014 (1962): Aldosterones in urine.

E. Organic Bases

I. Alkaloids

By

D. WALDI

1. Introduction

Most alkaloids are medium-strong bases occurring in plants. They are markedly lipophilic in the free base form and thus they can usually be extracted from an alkaline aqueous phase with ether, chloroform, etc. The alkaloids may be separated by paper chromatography as salts using acidic solvent systems. Separation of the free bases has only been possible since the introduction of formamide impregnated paper. It has been shown that the many separations described for formamide impregnated paper can be carried out also by chromatography on Silica Gel G or Alumina G layers. The Alumina G layers are basic, and neutral solvents, such as chloroform, are usually used. Silica Gel G is weakly acidic and alkaloids are retained more strongly according to their basicity and dissociation constants. This can easily be changed, however, by preparing basic Silica Gel G layers (p. 37), or by adding a strong base, e.g. diethylamine, to the developing solvent.

If the alkaloids do not migrate with chloroform, a more strongly polar solvent should be added, for example, 5–20% of methanol. If the alkaloids travel near the solvent front in chloroform, the less polar cyclohexane should be added.

Besides purely inorganic adsorption layers, cellulose powder can be used [*41*].

The rapid identification of one or more alkaloids in the analysis of pharmaceuticals is of interest, especially in cases of intoxications.

A systematic paper chromatographic procedure has been worked out for the analysis of alkaloids [*45*]. It can now be adapted to thin-layer chromatography with further simplification and a considerable saving of time [*46*].

Silica Gel G is preferred and impregnation of the thin layer is not necessary. To allow for variations in adsorption, which affects the *Rf*-values, a well characterized commercial alkaloid or alternatively, the dye Rhodamine B, is also chromatographed. The values obtained for the unknown constituents are correlated with those of the standard (R_{St}-values).

2. Procedure for systematic analysis of alkaloids

Silica Gel G layers, 250 μ, are used which had been prepared by the standard procedure (pp. 7—9). The solution to be analyzed should contain between 0.05 and 5% of alkaloids. It is not necessary to convert salts into the corresponding free bases, since the solvent systems used are strongly basic.

a) Allocation to groups and determination of optimum amounts for application

In a *preliminary experiment*, increasing amounts of alkaloid solution, e.g. 1 × 5 mm³ to 10 × 5 mm³, are applied at the starting points, and for comparison, 5mm³ of a 0.5% alcoholic solution of Rhodamine B, or better, of a 1% strychnine or reserpine solution is spotted also. The chromatogram is developed using solvent system III, i.e. a mixture of cyclohexane-chloroform-diethylamine (50 + 40 + 10). After drying in air, it is examined in UV-light (365 mμ) and sprayed with iodoplatinate solution (Reagent No. 76). In contrast to the Dragendorff reagent (No. 60 or 60a), the alkaloids appear after a short time with this reagent, some of them in various colors (Table 46). Characteristic color changes are often observed, they can furnish further information for identification.

The reference substances, Rhodamine B and reserpine, have *hRf*-values of 20 in solvent system III. Alkaloids with *hRf*-values greater than 30 are assigned as group II, and those with a lower *hRf*-value as group I. Each group is chromatographed separately with appropriate solvents. It is most suitable to apply amounts of about 50 μg.

b) Group I, alkaloids with *h*R*f*-values from 0—30 (Table 46)

Apart from the color reaction, the *hRf*-values obtained on Silica Gel G with solvent systems I (chloroform-acetone-diethylamine, 50 + 40 + 10) and II (chloroform-diethylamine, 90 + 10) are also useful for identification. It is advisable to chromatograph Rhodamine B, or better, the alkaloids suggested for the preliminary experiments, as standards. Then, the unknown alkaloid can be determined using Table 46. In doubtful cases, the substance is also chromatographed on Alumina G and/or alkaline Silica Gel G layers with solvent systems VI and VIII. *hRf*-Values, the composition of the solvent system and the dye used for comparison, are given in Table 46.

c) Group II, alkaloids with *h*R*f*-values over 30

If the approximate values obtained in the preliminary experiment are over 30, the substance is chromatographed on normal Silica Gel G layers with solvent systems IV: cyclohexane-diethylamine (90 + 10) or V: benzene-ethyl acetate-diethylamine (70 + 20 + 10), and, if necessary, again with III, as in the preliminary experiment. The *hRf*-values are to be found in Table 46 or Fig. 128. The suspected alkaloid or alkaloids are chromatographed as standards alongside the sample. In every case, the substance under investigation should be applied at several points. Then, it is possible to use various specific spray-reagents such as diazotized sulphanilic acid. For spraying, the other streaks are covered by 2 glass plates, so that several reactions can be performed on one plate. In every case the fluorescence colors (365 mμ) should be evaluated before and after spraying (Table 46).

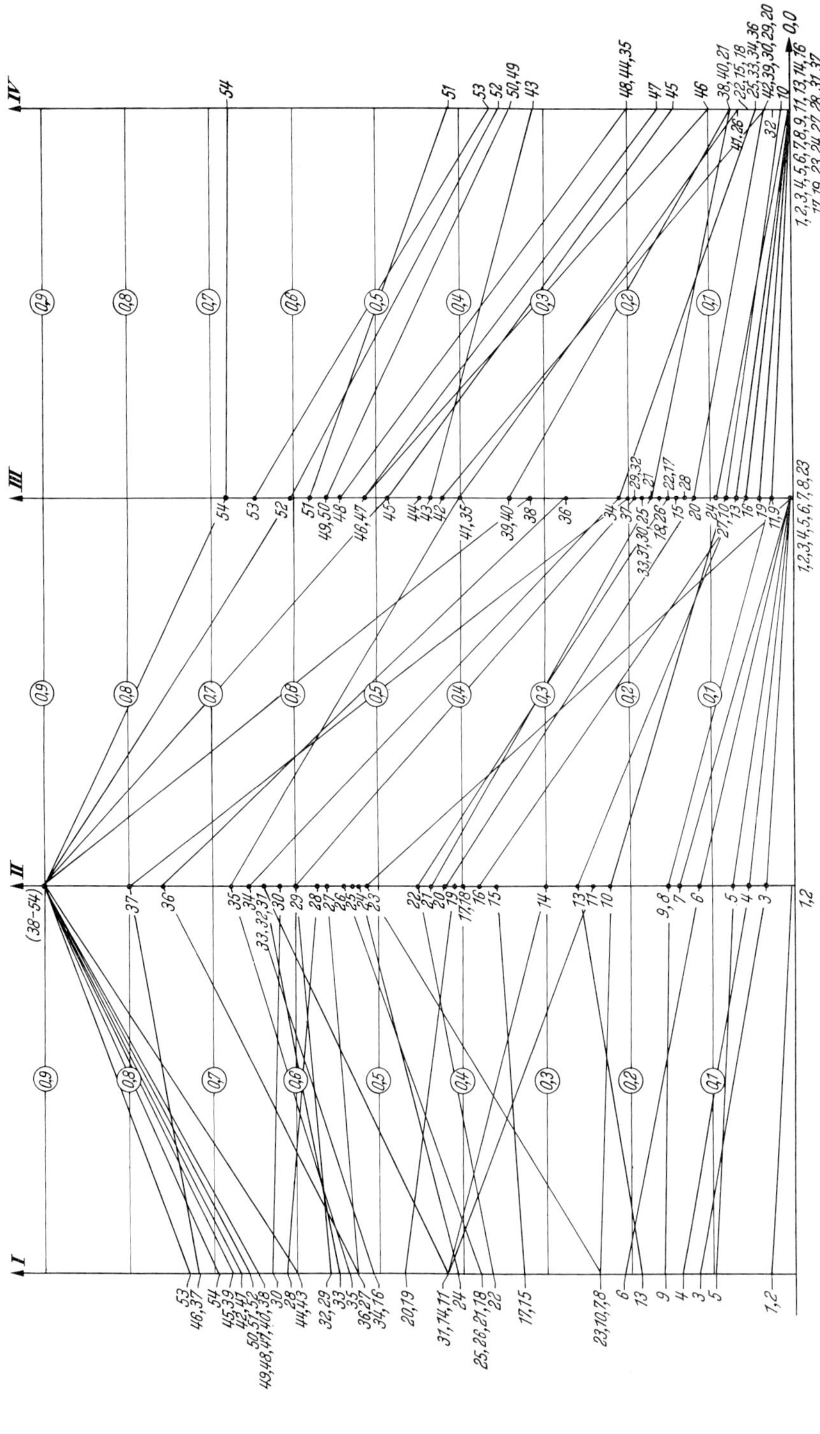

Fig. 128. *Rf*-values of alkaloids 1—54 (groups I and II) on Silica Gel G layers using solvents I—IV (see p. 284)

Table 46. *Alkaloids of Group I,*

Coating Material[1]		S	S	S	S	S	A	A	S
Pretreatment			none	none	none	none	none	none	0.1 N NaOH
No.	Solvent[2]	I	II	III	IV	V	VI	VII	VIII
1	Narceine	3	0	0	0	0	0	0	0
2	Cupreine	3	0	0	0	0	0	0	46
3	Sarpagine	12	4	0	0	0	0	0	0
4	Ergometrine	14	6	0	0	2	3	0	64
5	Morphine	10	8	0	0	3	3	0	34
6	Dihydroergotamine	21	12	0	0	3	7	0	61
7	Serpentine	24	15	0	0	4	0	0	0
8	Ergotamine	24	16	0	0	3	10	5	59
9	Boldine	16	16	3	0	5	24	6	58
10	Dihydromorphinone	24	23	8	1	11	5	8	16
11	Ergometrinine	42	25	3	0	8	12	10	62
12	Ephedrine	—	—	—	—	—	—	—	—
13	Quinine	19	26	7	0	17	9	18	43
14	Dihydroergocristine	42	30	3	0	7	15	7	69
15	Hordenine	33	36	14	5	28	0	15	35
16	Ergocristine	51	38	14	5	13	46	15	70
17	Quinidine	33	40	15	0	25	12	18	50
18	Atropine	38	40	16	5	12	0	10	17
19	Colchicine	47	41	4	0	4	11	0	57
20	Ajmaline	47	42	12	3	30	6	13	56
21	Cinchonine	38	44	17	7	27	0	22	40
22	Homatropine	37	45	15	5	23	4	24	15
23	Ergotaminine	24	51	0	0	14	42	15	68
24	Pilocarpine	41	52	9	0	13	32	25	55
25	Codeine	38	53	16	4	26	12	27	35
26	Dihydrocodeine	38	54	18	6	28	10	30	25
27	Serpentinine	53	56	8	0	10	0	3	12
28	Ergocristinine	61	57	13	0	20	0	27	70
29	Scopolamine	56	60	19	3	34	30	0	52
30	Yohimbine	63	62	18	3	37	33	15	60
31	Brucine	42	63	18	0	19	50	54	12
32	Cephaeline	56	63	19	2	23	25	17	37
33	Rauwolscine	55	63	18	4	36	36	15	68
34	Dihydrocodeinone	51	65	21	4	30	48	43	18
35	Apoatropine	54	67	40!	20	26	15	40	16
36	Strychnine	53	76	28	5	38	57	60	22
37	Reserpine	72	80	20	0	46	63	35	69
									Alkaloids of
38	Physostygmine	65	> 90	32	4	44	59	50	46
39	Aconitine	68	> 90	35	3	49	36	60	65
40	Bulbocapnine	65	> 90	35	7	54	78	70	48
41	Emetine	67	> 90	40	6	45	38	58	50
42	Papaverine	67	> 90	42	3	47	85	84	70
43	Cotarnine	60	> 90	43	31	45	0	25	0
44	Scopoline	60	> 90	44	20	44	46	50	37
45	Lobeline	68	> 90	48	14	48	55	60	55
46	Narcotine	72	> 90	51	10	57	81	79	72
47	Thebaine	65	> 90	51	16	50	71	76	40
48	Aspidospermine	65	> 90	54	20	49	50	60	65

[1] S = Silica Gel G; A = Alumina G. [2] Solvent see end of table.

hRf-Values and Colors for Identification

No.	Remarks				
	Fluorescence color in UV-light, 365 mμ	Color with iodoplatinate reagent	Shape of spot	No. of spots.	Numerical order on PC
1	—	deep blue	round	—	2
2	brownish-yellow	red-brown	round	—	5
3	—	beige	round	—	1
4	blue-violet	white (on pink)	round	—	3
5	—	deep blue	round	—	4
6	blue-violet	brownish	round	—	13
7	dark brown	red-brown	round	—	7
8	blue-violet	pink	round	—	14
9	violet	beige	round	—	9
10	—	brownish-yellow	round	—	8
11	blue-violet	blue-violet	round	—	11
12	—	light brown	extended!!!	—	43
13	blue	yellow-white	round	—	41
14	blue-violet	brownish	round	—	18
15	—	white (on pink)	round	—	36
16	blue-violet	beige/light brown	round	—	16
17	blue	light yellow	round	1	44
18	—	blue-violet	round	—	37
19	—	light grey	round	—	12
20	bluish	beige	round	2	45
21	—	beige-brown	round	—	33
22	—	blue-violet	round	—	35
23	blue-violet	pink	round	—	15
24	—	light brown	round	—	6
25	—	heather-color	round	—	30
26	blue	blue-violet	round	—	32
27	yellow-green	yellow-brown	extended	—	19
28	blue-violet	light brown	round	—	17
29	—	violet	round	—	28
30	green-blue	light-yellow	round	3	31
31	—	violet-brown	round	1	20
32	blue-violet	white (on pink)	round	3	22
33	yellow-green	pale beige	round	1	42
34	—	violet	round	2	25
35	—	blue-violet	round	—	52
36	—	yellow	extended	—	24
37	green-yellow	white (on pink)	round	3	34
Group II					
38	—	pink	round	1	23
39	—	red-brown	round	3	46
40	blue	white (on pink)	round	—	29
41	blue	red-brown	round	3	39
42	yellowish	yellow	round	—	26
43	green-yellow	violet	extended	1	21
44	—	white (on pink)	round	—	27
45	—	red-brown	round	—	47
46	blue	light yellow	round	—	40
47	—	red-brown	round	—	38
48	—	white (on pink)	round	—	49

Table 46 (continued).

Coating Material		S	S	S	S	S	A	A	S
Pretreatment		none	none	none	none	none	none	none	0.1 N NaOH
No.	Solvent	I	II	III	IV	V	VI	VII	VIII
49	Tropacocaine	65	> 90	56	34	45	58	78	35
50	Arecoline	66	> 90	56	34	48	0	0	0
51	Hydrastinine	66	> 90	58	41	50	0	25	0
52	Psicaine new	66	> 90	60	35	53	83	82	59
53	Cocaine	73	> 90	65	36	58	84	77	62
54	Sparteine	70	> 90	68	68	55	0	55	5

No.	Solvents	Standard	$Rf \times 100$
I	Chloroform-acetone-diethylamine (50 + 40 + 10)	Rhodamine B	58
II	Chloroform-diethylamine (90 + 10)	Rhodamine B	49
III	Cyclohexane-chloroform-diethylamine (50 + 40 + 10)	Rhodamine B	20
IV	Cyclohexane-diethylamine (90 + 10)	Butter yellow	45
V	Benzene-ethyl acetate-diethylamine (70 + 20 + 10)	Butter yellow	44
VI	Chloroform	Butter yellow	85
VII	Cyclohexane-chloroform (30 + 70) + 0.05% of diethylamine (3 drops)	Butter yellow	85
VIII	Methanol	Rhodamine B	53

The numbering of the alkaloids 1—37 is based on increasing *hRf*-values in column II; that of the alkaloids 38—54 on column III.

Note: There are only a few alkaloids that are unstable under the strongly basic conditions suggested. Quaternary bases, berberine for example, have several stable forms under such conditions, each with a different solubility so that streaky spots result. For this reason, it is better to replace diethylamine by acetic acid if quaternary bases are chromatographed on Alumina G layers.

Berberine (*hRf* 55). The solvent system cyclohexane-chloroform-acetic acid (45 + 45 + 10) on alumina (Merck) is used.

Capsaicine (*hRf* 57) also has a phenolic OH group. Two spots were obtained on Silica Gel G using the solvent system cyclohexane-chloroform-acetic acid (70 + 20 + 10). Both the second spot (*hRf* 22) as well as the main spot, stain with iodoplatinate (Reagent No. 76).

Ephedrine (*hRf* 49) cannot be chromatographed in a basic solvent system (Table 46). The spot is uniformly round using chloroform-methanol-acetic acid (25 + 65 + 10) and gives a color with 2′,7′-dichlorofluorescein (Reagent No. 42).

The results obtained to date for the thin-layer-chromatographic separation of particular classes of alkaloids are dealt with below. The results are summarized in tables for ease of reference.

3. Classes of alkaloids

a) Opium alkaloids

As early as 1953, BORKE and KIRSCH [*2*] had reported the separation of opium alkaloid mixtures using the "chromatostrip" method [*12*]. They applied layers prepared from a mixture of silicic acid, magnesium oxide, gypsum, and a luminescent inorganic compound, and brought to

Alkaloids of Group II

No.	Remarks: Fluorescence color in UV-light, 365 mμ	Color with iodoplatinate reagent	Shape of spot	No. of spots.	Numerical order on PC
49	—	violet	round	—	50
50	—	white (on pink)	round	—	—
51	steel-blue	blue-violet	extended	1	10
52	—	yellow	round	—	51
53	—	violet	round	—	48
54	—	violet	round	—	54

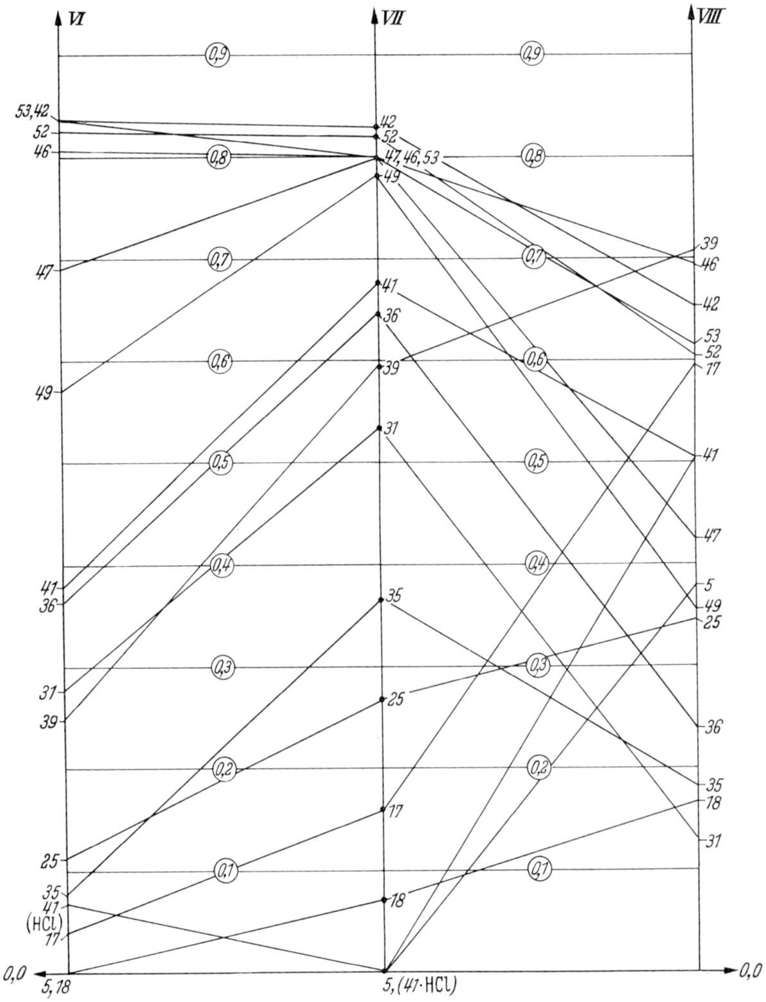

Fig. 129. Comparison of h*Rf*-values of some alkaloids (Nos. Table 46) on Alumina G layers (columns VI and VII) and values obtained on alkaline Silica Gel G layers (column VIII). Solvents VI chloroform, VIII methanol, VII cyclohexane-chloroform (30 : 70) + 0.05% of diethylamine

pH 6.6 with phosphate buffer. Dioxane was used as solvent. STAHL (1956) among others, drew attention to the possibility of separating alkaloid mixtures by TLC. In 1960, MACHATA [*19*] separated various other pharmaceuticals as well as opium alkaloids (see p. 326) using methanol as developing solvent on Silica Gel G layers. With this method, he detected a tincture of opium in the stomach of a suicide. BÄUMLER and RIPPSTEIN [*1*] investigated thoroughly the analysis of the poppy alkaloids by thin-layer chromatography. They worked with Silica Gel G layers and used the hydrophilic solvent system VI (see Table 47). Detection was achieved by spraying with Dragendorff reagent (No. 60 or 60a) or the iodoplatinate reagent (No. 76). They also observed the fluorescence given by some compounds. Later experiments showed that the *hRf*-values given by these authors were excellently reproducible. TEICHERT, MUTSCHLER and ROCHELMEYER [*40–42*] recommended alkaline Silica Gel G layers for the separation of the morphines and concomitant alkaloids; chloroform-ethanol (90 + 10) was used as developing solvent. The latter author pointed out that cellulose layers can be used also. Solvent system X was used with formamide impregnated cellulose and mixture IX for non-impregnated plates (Table 47). If a quantitative photo-

Table 47. *hRf-Values of Opium-Alkaloids on different Layers using Solvents I—XI.* (Adsorbents: S = Silica Gel G; A = Alumina G; C = cellulose powder)

Coating Material	S	S	S	A	S	S	S	S	S	C	S
					(0.1 N NaOH)			(0.5 N KOH)		Formamide	
Solvent	I	II	III	IV	V	VI	VII	VIII	IX	X	XI
Narceine	3	0	0	0	0	23	—	—	—	—	—
Dihydromorphinone (dilaudide)	24	23	8	8	16	28	5	13	27	6	14
Dihydrocodeinone (dicodide)	51	65	21	43	18	29	10	28	34	63	—
Morphine	10	8	0	0	34	40	2	2	27	0	24
Morphine ethyl ether (dionine)	—	—	—	—	—	37	14	37	44	57	—
Dihydrocodeine (paracodine)	38	54	18	30	25	—	6	22	34	—	—
Codeine	38	53	16	27	35	43	12	33	41	37	26
Acetyldihydrocodeinone (acedicone)	—	—	—	—	—	—	24	59	—	90	—
Dihydro-hydroxycodeinone (eucodal)	—	—	—	—	—	—	47	70	79	75	—
Thebaine	65	90	51	76	40	—	—	—	—	—	—
Papaverine	67	90	42	84	70	82	74	78	86	89	74
Cotarnine	60	90	43	25	0	—	—	—	—	—	—
Narcotine	72	90	51	79	72	82	78	81	92	94	69

Solvent systems: I = Chloroform-acetone-diethylamine (50 + 40 + 10) [*46*]; II = Chloroform-diethylamine (90 + 10) [*46*]; III = Cyclohexane-chloroform-diethylamine (50 + 40 + 10) [*46*]; IV = Cyclohexane-chloroform (30 + 70), with 0.05% of diethylamine [*46*]; V = Methanol [*46*]; VI = Methanol-acetone-triethanolamine (50 + 50 + 1.5) [*1*]; VII = Chloroform-ethanol (90 + 10) [*41*]; VIII = Chloroform-ethanol (80 + 20) [*41*]; IX + dimethylformamide-diethylamine-ethanol-ethyl acetate (5 + 2 + 20 + 75) [*41*]; X = Benzene-heptane-chloroform-diethylamine (60 + 50 + 10 + 0.2) [*41*]; XI = Methanol [*19*].

metric determination of the alkaloids should follow the separation, the adsorbent must be purified beforehand. WINKLER and AWE [*47*] chromatographed some alkaloids from *Papaver rhoeas*, e.g. rhoeadine, alongside hydrastinine on Alumina G layers using chloroform-methanol (90 + 10) or benzene-tetrahydrofuran (95 + 5) as solvents.

NEUBAUER and MOTHES [*21*] separated the 10 most important poppy alkaloids on "manually" prepared Silica Gel G layers (10 × 20 cm) with benzene-methanol (80 + 20) as solvent. They give the following approximate *hRf*-values: morphine 11; codeine 21; laudanine 26; cryptopine 34; protopine 38; thebaine 40; laudanosine 42; narcotiline 58; papaverine 63; narcotine 68. They used the method, which was developed also for quantitative determinations, to accelerate the identification and selection of the "chemical species" of various types of papaveraceae.

VAN PINXTEREN and VERLOOP [*25*] chromatographed the alkaloids contained in tinctures of opium on Silica Gel G layers prepared by the standard procedure using carbon tetrachloride-butanol-methanol-6 N aqueous ammonia (40 + 30 + 30 + 2). For TLC of pseudo-morphine (*hRf* 4), morphine (35) and narcotiline (85), they used the same solvent system but with an ammonia content of 3.3.

It is obvious that the separation of the opium alkaloids by TLC is most conveniently carried out on Silica Gel G layers. The choice of the most suitable solvent system depends on the problems of separation.

b) Tropane alkaloids

In his first paper on TLC [*34*], STAHL pointed out that the tropane alkaloids can be separated, and he reproduced a chromatogram developed

Table 48. *hRf-Values of Tropane Alkaloids* (Adsorbents: S = Silica Gel G; A = Alumina G; C = Cellulose powder)

Coating Material	S	S	S	A	S (0.1 N NaOH)	S	S	S (0.5 N KOH)	C (Formamide)
Solvent	I	II	III	IV	V	VI	VII	VIII	IX
Tropine	—	—	—	—	—	—	16	3	13
Atropine	38	40	16	10	17	17	37	36	15
Homatropine	37	45	15	24	15	—	—	—	—
Apoatropine	54	67	40	40	16	—	44	44	74
Belladonnine	—	—	—	—	—	—	26	17	69
Cocaine	73	90	65	77	62	61	—	—	—
Scopolamine	56	60	19	0	52	—	73	83	53
Scopoline	60	90	44	50	37	—	—	—	—
Tropacocaine	65	90	56	78	35	—	—	—	—
Psicaine new	66	90	60	82	59	—	—	—	—

Solvent systems: I = Chloroform-acetone-diethylamine (50 + 40 + 10) [*46*]; II = Chloroform-diethylamine (90 + 10) [*46*]; III = Cyclohexane-chloroform-diethylamine (50 + 40 + 10) [*46*]; IV = Cyclohexane-chloroform (30 + 70), with 0.05% diethylamine [*46*); V = Methanol [*46*]; VI = Methanol-acetone-triethanolamine (50 + 50 + 1.5) [*1*]; VII = Dimethylformamide-diethylamine-ethanol-ethyl acetate (5 + 5 + 30 + 60) [*41*]; VIII = 70% ethanol- 25% ammonia (99 + 1) [*41*]; IX = First run 15 cm with heptane-diethylamine (100 + 0.2), second run 10 cm with benzene-heptane-chloroform-diethylamine (60 + 50 + 10 + 0.2) [*41*].

with butanol saturated with acetic acid. Subsequent work is summarized in Table 48.

Note: Tropic acid can be chromatographed on silica gel with cyclohexane-chloroform-acetic acid (60 + 20 + 20) as solvent system (*hRf* 47). Detection: Cerium-reagent (No. 25), blue-violet fluorescing spot.

c) Indole alkaloids

When separating mixtures of indole alkaloids by thin-layer chromatography, it is best to apply the substance in an inert gas atmosphere, and to exclude UV-light during spotting and developing. A suitable apparatus is described and illustrated on p. 12. There is no danger of isomerization during the short time needed for development. The adsorbents and solvent systems tried to date for *Rauwolfia* alkaloids are summarized in Table 49, and those for the ergot alkaloids in Table 50. Other indole alkaloids are included in Table 52.

Table 49. *hRf-Values of Rauwolfia Alkaloids* (Adsorbents S = Silica Gel G; A = Alumina G; C = cellulose powder)

Coating Material	S	S	S	A	S (0.1 N NaOH)	C (Formamide)
Solvent	I	II	III	IV	V	VI
Sarpagine	12	4	0	0	0	3
Serpentine	24	15	0	0	0	6
Ajmaline	47	42	12	13	56	28
Serpentinine	53	56	8	3	12	—
Yohimbine	63	62	18	15	60	33
Rauwolscine	55	63	18	15	68	—
Rescinnamine	—	—	—	—	—	51
Reserpine	72	80	25	35	69	59
Reserpinine	—	—	—	—	—	89

Solvent systems: I = Chloroform-acetone-diethylamine (50 + 40 + 10) [*46*]; II = Chloroform-diethylamine (90 = 10) [*46*]; III = Cyclohexane-chloroform-diethylamine (50 + 40 + 10) [*46*]; IV = Cyclohexane-chloroform (30 + 70 + 0.05% of diethylamine [*46*]; V = Methanol [*46*]; VI = Heptane-methyl ethyl ketone (50 + 50), in ammonia atmosphere [*41*].

A. HOFMANN [*7*] separated the indole alkaloids of the Mexican magic drug, *Ololiuqui*, using chloroform-methanol (95 + 5) on alumina, or chloroform-methanol (70 + 30) on Silica Gel G layers. He obtained the series, in order of increasing *Rf*-values, chanoclavine, Substance E, d-lysergic acid amide, elymoclavine, lysergol, d-iso-lysergic acid amide, Substance F. WINKLER [*48*] was able to separate a number of the principal alkaloid constituents and their degradation products from *Voacanga africana* STAPF. KUMPF and SCHMIDT [*13*] separated *Pleiocarpa* alkaloids on a Silica Gel G layer with the solvent, chloroform-methanol (96 + 4). They used a saturated solution of ceric (IV) sulphate in 60% sulphuric acid for visualization. Rochelmeyer's group investigated the analysis of the ergot alkaloids in detail [*11, 11a, 40*].

The "simple" indole derivatives are dealt with in the following section (pp. 292—301).

Many indole alkaloids are easily detected in UV-light; some fluoresce, whereas others are converted to fluorescent products by UV-irradiation. In addition, the perchloric acid-ferric chloride reagent is frequently used for visual detection of *Rauwolfia* alkaloids. Sarpagine can be identified by spraying with conc. nitric acid. A modified van Urk reagent (No. 48) is preferred for ergot alkaloids. p-Dimethylamino cinnamaldehyde or vanillin are used instead of the usual p-dimethylamino benzaldehyde. This can be combined to advantage with subsequent oxidation with *aqua regia* vapor as described on p. 298. The limit of sensitivity is about 0.50 μg. To detect alkaloids on formamide-impregnated cellulose layers, the formamide must first be removed by heating in a vacuum drying oven.

SCHLEMMER and LINK [*31*] as well as ULLMANN and KASSALITZKY [*43*] report a quantitative spectrophotometric determination (see p. 46) of *Rauwolfia* alkaloids separated by thin-layer chromatography. The latter

Table 50. *hRf-Values of Ergot Alkaloids*
(Adsorbents: S = Silica Gel G; C = cellulose powder)

Coating Material	S	S	C (Formamide)	Indicator	
Solvent [1]	I	II	IV	Fluorescence in UV-light	VAN URK reagent (No. 50)
Dihydroergotamine	0	11	—	—[2]	↑
Dihydroergocristine	0	14	—	—[2]	
Ergometrine	3	17	—	↑	
Ergometrine	8	30	—		
Ergotamine	13	51	6		
Ergosine	13	51	11	bright blue	blue color
Ergocristine	28	69	67	fluorescence	reaction
Ergocornine	28	69	73		
Ergocryptine	28	69	83		
Ergotaminine	34	75	50		
Ergosinine	34	75	65		
Ergocristinine	45	81	90		
Ergocorninine	45	81	93		
Ergocryptinine	45	81	97	↓	
Coating Material Solvent [1]	S; I	S; II	S; III		
Chanoclavine	0	2	4	—[2]	
Elymoclavine	2	13	15	—[2]	↓
Peniclavine	4	17	20	+	green
Isopeniclavine	8	30	35	+	green
Agroclavine	15	38	45	—[2]	blue
Setoclavine	18	42	50	+	green
Isosetoclavine	20	46	60	+	green

[1] Solvent systems: I Chloroform-ethanol (95 + 5) [*11*], cf. Fig. 13, p. 15
II Chloroform-ethanol (90 + 10) [*11*]
III Chloroform-ethanol (80 + 20) [*11*]
IV Developed twice: 1. heptane-benzene-chloroform (25 + 30 + 15)
2. heptane-benzene (25 + 30) [*40*]

[2] Yellow fluorescence after intensive UV-irradiation.

authors show that the separation and the quantitative determination of reserpine and rescinnamine are possible also in the presence of emulsifiers such as polyethylene oxide derivatives.

Quantitative determination of Ergot alkaloids after separation by thin-layer chromatography

(a) The visual method, by estimation of the size of spots, is only semiquantitative. For this method, very small amounts must be applied (0.1—30 μg of alkaloid). For a detailed procedure see pp. 47, 48.

(b) Photometric method after elution of the alkaloid zones [*11a*]. The alkaloid mixture (0.1—0.3 mg of the alkaloids) is applied as a band (see p. 13), and marked quickly under UV-light after development. Then the zones (at least 100 μg of alkaloid) are scraped off quantitatively, using a spatula shaped like a coal shovel with sides bent up and a knife-edge at the end, and placed in a centrifuge tube. 2.00 ml of methanol-water-acetic acid (45 + 45 + 10) is added to each tube to extract the alkaloid, and the tube allowed to stand for 8—10 min under occasional stirring with a glass rod. Then 4.00 ml of the modified van Urk reagent is added (125 mg of p-dimethylamino benzaldehyde dissolved in a cooled mixture of 65 ml of AR conc. H_2SO_4, 35 ml of distilled water and 0.05 ml of a 10% ferric chloride solution added; stable for 1 week). After repeated stirring, the silica gel is centrifuged off for 7 min at 3500—4000 r.p.m. The optical density of the clear blue reaction solution can be measured against a blank at 550 mμ. If the values are too high, one volume of extraction mixture is diluted with two volumes of reagent.

d) Ipecacuanha alkaloids [*37*]

The ipecacuanha alkaloids can be separated on Silica Gel G (standard method) using the solvent chloroform-methanol (85 + 15) (Fig. 130).

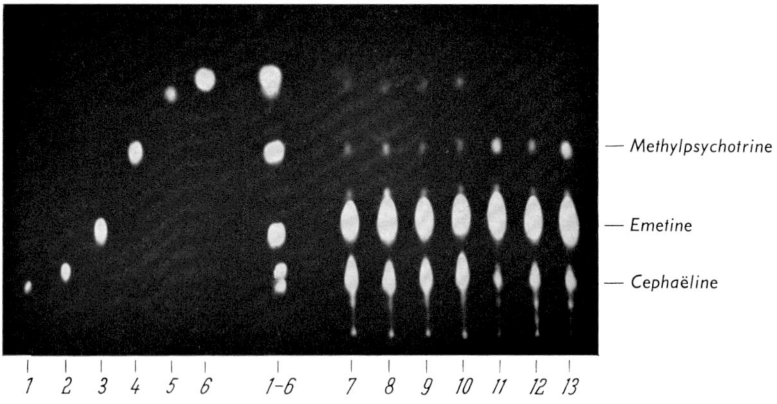

Fig. 130. Thin-layer chromatogram of Ipecacuanha alkaloids (~0.1 μg.) photographed by UV-light (365 mμ) after the iodine reaction [*37*]. Further details in text. *1* Cephaeline, *2* Psychotrine, *3* Emetine, *4* O-Methylpsychotrine, *5* Protoemetine, *6* Emetamine, *7—10* Extracts of drugs from root of *Uragoga acuminata* KARSTEN, *11—13* from the officinal *Uragoga ipecacuanha Ball* (Rio-drug)

This is best done in a "saturated" tank; development over a 10 cm path requires about 30 min. The most sensitive method of detection uses about 10 ml of a 0.5% solution of iodine in chloroform as spray-reagent, followed by warming to 60° C for 15 min. The alkaloids can be recognized by their intensive fluorescence in UV-light (365 mμ) (Table 51).

Table 51. *Thin-Layer Chromatography of Ipecacuanha Alkaloids on Silica Gel G*

Alkaloid	hRf-value	Color after iodine reaction: Daylight	Color after iodine reaction: UV-light	Limit of detection (μg)
Cephaeline	13	light brown	light blue	0.006
Psychotrine	16	light brown	yellow, blue border	0.001
2-Dehydro-isoemetine	18	(light beige)	turquoise	0.002
2-Dehydroemetine	21	(light beige)	turquoise	0.002
Emetine	28	yellow	yellow-bluish	0.002
o-Methylpsychotrine	47	yellow	yellow	0.0008
Protoemetine	63	pale beige	blue	0.01
Emetamine	67	light beige	yellow	0.003

Stahl gives the following directions for separating Rio- and Cartagena-Ipecacuanha drugs:

100 mg of finely powdered drug are treated with 1 drop of conc. ammonia in a small test-tube, 5.0 ml of chloroform added, and mixed vigorously with a glass rod. After 3—4 hr, the slurry is filtered and some 5 mm^3 used for TLC. 5 mm^3 of a 0.01% solution of emetine and cephaeline (1 + 1) are applied alongside for comparison. The zone sizes of this standard solution correspond to the Cartagena drug extract; the Cephaeline zone is considerably smaller with a Rio drug extract (Fig. 130).

e) Alkaloids of various classes

Table 52. *Various Classes of Alkaloids*

Coating Material	S	S	S	A	S (0,1 N NaOH)	S (0,5 N KOH)	C (Formamide)	S	S
Solvent	I	II	III	IV	V	VI	VII	VIII	IX
Brucine	42	63	18	54	12	—	—	17	—
Strychnine	53	76	28	60	22	—	—	19	—
Physostygmine	65	90	32	50	46	—	—	—	—
Aspidospermine[1]	65	90	54	60	65	—	—	—	—
Lobelanidine	—	—	—	—	—	—	24	—	—
Lobeline	68	90	48	60	55	—	69	—	—
Lobelanine	—	—	—	—	—	—	92	—	—
Sparteine	70	90	68	55	5	1	—	—	—
Coniine	—	—	—	—	—	9	—	—	—
Anabasine	—	—	—	—	—	16	—	—	—
Nicotine	—	—	—	—	—	44	—	57	—
Arecoline	—	—	—	—	—	50	—	—	—
Quinine	19	26	7	18	43	—	—	56	36
Quinidine	33	40	15	18	50	—	—	—	42
Cinchonine	38	44	17	22	40	—	—	—	31
Cinchonidine	—	—	—	—	—	—	—	—	23

Solvent systems: I = Chloroform-acetone-diethylamine (50 + 40 + 10) [*46*]; II = Chloroform-diethylamine (90 + 10) [*46*]; III = Cyclohexane-chloroform-diethylamine (50 + 40 + 10) [*46*]; IV = Cyclohexane-chloroform (30 + 70), with 0.05% of diethylamine [*46*]; V = Methanol [*46*]; VI = Chloroform-ethanol (90 + 10) [*41*]; VII = Benzene-heptane-diethylamine (10 + 60 + 0.2) [*41*]; VIII = Methanol-acetone-triethanolamine (50 + 50 + 1.5) [*1*]; IX = Chloroform-ethanol (90 + 10].

[1] Lehner and Schmutz [*16*] report the separation of other *Aspidosperma* alkaloids by TLC on Silica Gel G layers using methylcellosolve as solvent.

Table 52 gives conditions for some *Lobelia* and other indole alkaloids; also for steam-volatile alkaloids and the most important constituents of *Cinchona* bark.

MÜLLER and HONERLAGEN [*20*] describe a fluorometric estimation of quinine-quinidine mixtures separated by TLC. In the preparation of the layer, they used acetone instead of water.

Meanwhile, TLC has been used most successfully for purification, identification and testing for purity of alkaloids [*3, 10, 16, 22, 22a, 28, 29, 30*]. See also Arch. Pharm. **295**, 524 and 605 (1962).

II. "Simple" Indole Derivatives

By

EGON STAHL

1. Introduction

The indole group of compounds is conveniently divided into the so-called "simple" indole derivatives (no additional rings, see Table 53) and the indole alkaloids, often of complicated structure (p. 288), and indole dyes. The boundaries between the groups are naturally continuous.

A number of simple indole derivatives play important roles [*39*] in physiological processes.

Serotonin can be regarded — with acetyl choline and adrenaline — as the third physiological substance active in the human nervous system. The greater part of the serotonin (which is probably formed from tryptophane by way of 5-hydroxy-tryptophane) occurs in the organism in an inactive protein-bound form. This bond can be cleaved by a number of compounds, for example, by the indole alkaloid reserpine, and especially by lysergic acid diethyl amide. The free serotonin is then broken down by amine-oxidase and excreted in the urine as 5-hydroxy-indolyl-3-acetic acid [*15*].

KÖGL and co-workers [*14*] isolated a highly active growth substance from urine, when searching for plant auxin, which they called heteroauxin. The discovery of this indolyl-3-acetic acid has raised the question whether this and other "simple" indole derivatives act as growth regulators [*9, 17, 33, 39*].

The detection of simple indole derivatives is also of importance in work aiming to clarify the biogenesis of the indole alkaloids.

One important reason for the many lacunae that exist in this field of research, in spite of intensive work, is the very small concentrations of these substances in plant and animal organisms. A further complicating factor in their isolation is the often poor stability of the indole derivatives concerned.

Paper chromatography [*18*] and paper electrophoresis [*4*] afforded a valuable enrichment of the methods for identification of indole derivatives. Thin-layer chromatography on standardized thin layers of Silica Gel G was a further advance in analytical technique. The decomposition of labile substances is minimized by the short running-time (15–35 min). Moreover, separations are distinctly better than in previous methods, and they are less adversely affected by accompanying compounds than on paper. Most important, however, was the finding that with TLC it is possible to reach the limits of (visual) detection in the nonagram region (1 ng = 10^{-3} μg).

2. Preparation of material for analysis

The solution to be analyzed is chromatographed in a preliminary experiment in which it is spotted in increasing amounts (e.g., 1, 10, 50 mm³).

If the concentration of indole derivatives is too small, the solution is concentrated under nitrogen, at room temperature, in a rotary evaporator. In any case, however, it is best to effect a preliminary separation of the material to be analyzed, into hydrophilic, lipophilic, and acidic, basic and neutral fractions. Most of the simple indole derivatives under discussion can be extracted stepwise in their undissociated form from appropriately buffered aqueous media by ethyl acetate (or, if necessary, ether or chloroform). Tryptophane, hydroxytryptophane, some urinary melanogens and decomposition products of ascorbigen remain in the aqueous phase.

A generally applicable procedure is not given here because it must be adapted to particular problems and materials. Here, the stages of the fractionation procedure would be controlled by thin-layer chromatography.

3. Coating materials and solvents

Best results are obtained using Silica Gel G layers prepared by the standard procedure (pp. 7—9). Alumina G layers can only be used successfully for neutral or basic substances. The first solvent systems to find favor was the mixture of isopropanol-25% aqueous ammonia-water (85 + 5 + 15) which is also much used in paper chromatography. Running time with this solvent is up to 2 hr.

The following solvent mixtures have proved the best of many tried for the substances given in Table 53.

A (alkaline system): methyl acetate-isopropanol-25% ammonia (45 + 35 + 20) [*36*].

S (acidic system): chloroform-96% acetic acid (95 + 5) [*36*].

For more highly polar metabolism products of tryptophane, the chloroform in mixture S is partly replaced by methanol [*5*].

Sp (acidic, polar) chloroform-methanol-glacial acetic acid (75 + 20 + 5) [*5*].

The mixture should be freshly prepared each day from pure solvents and no more than 10 chromatograms (20 × 20 cm) should be run before refilling the chamber. The plates are developed by the ascending technique in a "saturated" chamber.

Chromatography should be done in a partially darkened room. The usual 10 cm run can be shortened by half for particular problems. By this means, the limit of detection is lowered and the running time

Table 53. *hRf-Values, Color Reactions (on 1 μg of substance), and*

No.	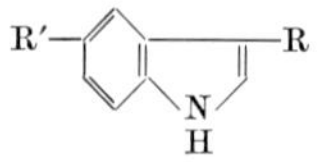	R	R′
I	Indole	—H	—H
II	Skatole	$—CH_3$	—H
III	3-Hydroxymethyl indole	$—CH_2OH$	—H
IV	Indole-3-aldehyde	—CHO	—H
V	Indole-3-acetaldehyde	$—CH_2 \cdot CHO$	—H
VI	Indole-3-acetic acid	$—CH_2 \cdot CO_2H$	—H
VII	β-Indolyl-3-propionic acid	$—[CH_2]_2 \cdot CO_2H$	—H
VIII	β-Indolyl-3-butyric acid	$—[CH_2)_3 \cdot CO_2H$	—H
IX	β-Indolyl-3-acrylic acid	$—CH{:}CH \cdot CO_2H$	—H
X	[5-Hydroxyindolyl-3]-acetic acid .	$—CH_2 \cdot CO_2H$	—OH
XI	Indolyl-3-acetonitrile	$—CH_2 \cdot C{:}N$	—H
XII	Tryptamine	$—[CH_2]_2 \cdot NH_2$	—H
XIII	Serotonin	$—[CH_2]_2 \cdot NH_2$	—OH
XIV	Gramine	$—CH_2 \cdot N(CH_3)_2$	—H
XV	DL-Tryptophane.	$—CH_2 \cdot CH(NH_2) \cdot CO_2H$	—H
XVI	DL-5-Methyl-tryptophane.	$—CH_2 \cdot CH(NH_2) \cdot CH_2O$	$—CH_3$
XVII	DL-5-Hydroxy-tryptophane	$—CH_2 \cdot CH(NH_2) \cdot CO_2H$	—OH
XVIII	Indolyl-3-acetamide	$—CH_2 \cdot CO \cdot NH_2$	—H
XIX	Isatin		
XX	Anthranilic acid.		
XXI	Ergot alkaloid* (Ergotamine, ergocristine, ergometrine)		

* Intense blue fluorescence in UV-light (365 mμ).

Limits of Detection (length of run 10 cm) of "Simple" Indole Derivatives [36]

Rf-values × 100 in solvent systems A[1]	S[1]	Color with p-dimethylaminobenzaldehyde Reagent, No. 48	Limit of detection (in μg)	Color with formaldehyde Reagent, No. 64	Fluorescence in UV (365 mμ), with formaldehyde reagent	Limit of detection (in μg)
84	73	dark red to violet	0.1	pale green	green	0.01
87	78	blue	0.1	yellow	yellow	0.01
84	45	pink	0.1	yellow-pink	yellow	0.1
81	20	pink	1	orange with 2,4-dinitrophenyl hydrazine reagent		
86	46	reddish brown	1	yellow with 2,4-dinitrophenyl hydrazine reagent		
31	28	blue, tinge of violet	0.05	yellow	yellow with green border	0.005
38	34	blue	0.1	yellow	yellow	0.01
40	38	blue	0.1	yellow	yellow	0.01
33	29	pink	0.5	pale beige	green	0.1
19	4	blue to violet	0.05	pale yellow to beige	deep violet	0.05
85	46	grey	0.1	pale beige	yellow	0.1
77	0	blue-green	0.1	yellow	yellow with blue border	0.01
65	0	grey	0.1	yellow	brown	0.01
77	0	(yellow to beige)	(1)	grey-blue	green	0.5
23	0	blue-green	0.05	yellow	yellow with blue border	0.005
28	0	blue	0.05	turquoise	yellow green	0.005
14	0	blue-grey	0.05	pale yellow to beige	yellow	0.005
83	12	blue	0.1	pale yellow to beige	yellow	0,01
75	27	remains orange	1	—	—	—
33	40	becomes intensive yellow	0.1	pale beige	blue, turning brown	0.01
83	4	blue-violet	—	blue	green	

[1] On Silica Gel G layers, 250 μ, with tank saturation.

is reduced. Results obtained under the given conditions are summarized in Table 53. The *hRf* values should be regarded as approximate only. They are increased by higher humidity and when other substances are running in the same region of the chromatogram (= displacement process).

DIAMANTSTEIN and EHRHART [5] separated a series of tryptophane metabolites on Silica Gel G layers, about 300 μ thick, using systems A and Sp. They worked with "saturated" chambers and applied the substances 2 cm from the edge of the plates. The separation distances were 12 and 15 cm. Table 54 shows the *hRf*-values, the fluorescence, and the colors obtained with Ehrlich's reagent (van Urk's color reaction).

Since tryptophane, indole and anthranilic acid were also chromatographed by these workers [5], a comparison can be made with the *Rf*-values obtained by of STAHL and KALDEWEY as given in Table 53.

Table 54. *hRf-Values and Detection of 10 Metabolic Products of Tryptophane* [5]

Substance	*Rf* × 100 in solvent		Detection	
	A	Sp	Fluorescence	Color with *p*-dimethylaminobenzaldehyde reagent
Tryptophane	25	7	—	violet
Indole	90	98	blue	violet
Indicane	61	14	brown	brown
DL-Kynurenine	32	11	green-blue	yellow-brown
3-Hydroxykynurenine	16	6	yellow-green	orange
Kynurenic acid	45	18	green after 12 hr	—
Xanthurenic acid	45	26	grey	—
Anthranilic acid	45	91	light blue	yellow
3-Hydroxyanthranilic acid	31	75	light blue	yellow
3-Amino-3-hydroxyacetophenone	88	93	blue	yellow-brown

4. Two-dimensional development

In all cases with unknown mixtures of natural products one-dimensional chromatography with solvents A and S should always be followed by a two-dimensional development. Thus, Table 53 and Fig. 132 show that not all substances are separated by either the basic solvent (Fig. 131, St. 1) or the acidic solvent (Fig. 131, St. 2). Therefore, one combines the effects of the two systems, and in this way, 15 and more simple indole derivatives can be detected.

The mixture to be analyzed is applied at starting-point St. I. A mixture of indole derivatives of known composition is spotted at St. 1 and 2. In the analysis of "indole auxins", it is convenient to develop first in direction 1 with the basic system. Then the ammonia is removed from the layer by placing the plate in a desiccator over conc. H_2SO_4, and evacuation to 10–15 mm Hg. The desiccator is refilled several times with nitrogen. Then the plate is left *in vacuo* for 60 min.

Subsequently, it is developed in direction 2 using the acidic solvent system.

Fig. 131 shows that, under the conditions given above, substances are separated into groups according to their polarity:

Acids (4, 5, 6, 7, 8), for example, lie in a definite region. Weakly polar compounds (14, 15) migrate the farthest in both solvents. The more strongly basic amines (9, 10) and the amide tested lie together in one

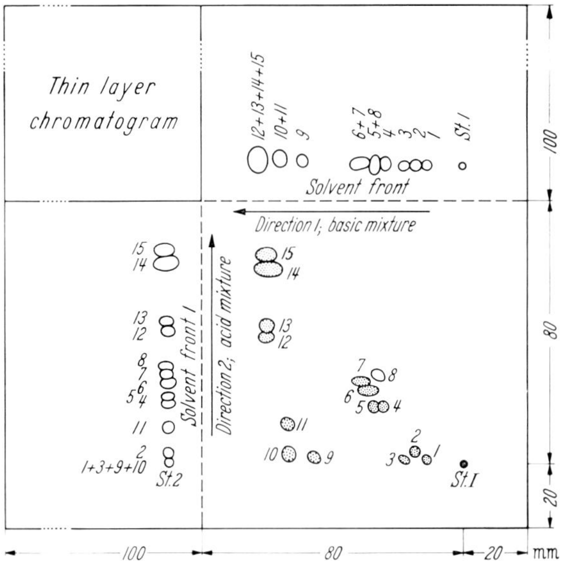

Fig. 131. Two-dimensional thin-layer chromatogram of 14 simple indole derivatives and anthranilic acid. The mixture applied at the starting point (St. I, 1 and 2) contains 0.5 μg. of each substance. The success of the separation with the basic solvent A and the acidic solvent S can be seen from the comparison chromatograms run from starting points St. 1 and 2, respectively. *1* = 5-hydroxytryptophane; *2* = (5-hydroxy-indolyl-3)-acetic acid; *3* = tryptophane; *4* = indolyl-3-acetic acid; *5* = (-indolyl-3-)acrylic acid; *6* = β-(indolyl-3-) propionin acid; *7* = γ-(indolyl-3-) butyric acid; *8* = anthranilic acid; *9* = serotonin; *10* = tryptamine; *11* = indolyl-3-acetamide; *12* = 3-hydroxymethylindole; *13* = indolyl-3-acetonitrile; *14* = indole; *15* = skatole [*36*]

group. Acids (1, 2, 3) containing a further amino or hydroxyl group in the molecule are most strongly adsorbed and have the lowest *Rf*-values.

Diamantstein and Ehrhart [5] separated the 10 tryptophane metabolites (Table 54) by two-dimensional thin-layer chromatography. They used the chloroform-methanol-glacial acetic acid mixture (Sp) in direction 1 (15 cm), dried for 60 min at room temperature and then chromatographed in direction 2 (12 cm) with the basic solvent A. The approximate position of the compounds can be found from the values given in the Table 54.

5. Detection

Simple indole derivatives give color reactions with numerous reagents. It is not unusual to find colored and/or fluorescent zones appearing after chromatography with an acidic solvent, if the plate is simply allowed to stand untreated.

a) Chemical methods of detection

The most specific of the many detection reactions described in the literature is the van Urk color reaction (No. 48), and the most sensitive is the Prochazka fluorescence reaction (No. 64). To obtain optimum color formation in the van Urk reaction, the conditions prescribed (see No. 48) must be followed exactly. The subsequent exposure to *aqua regia* vapors should be done very carefully: traces of vapor are sufficient to intensify the color, but an excess turns the whole layer yellow.

PÖHM [*26*] suggests that a condensation reaction of p-dimethylaminobenzaldehyde takes place, at the 2-position of the indole derivative thus leading to the blue to red Rosindole dyes.

The Prochazka reaction [*27*] using formaldehyde-hydrochloric acid (No. 64) gives only yellow reaction products, which fluoresce strongly in long-wave UV light (Fig. 132).

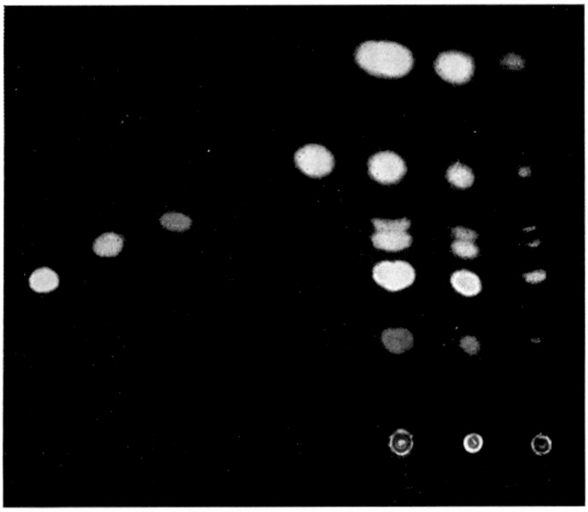

Fig. 132. Thin-layer chromatogram of indole derivatives after visual detection with Prochazka reagent; photographed in UV-light. Silica Gel G, solvent S [*36*]

For indole derivatives substituted with a CH_2R group in the 3-position, the limit of detection (length of run 10 cm.) lies between 0.005 to 0.01 μg (= 5–10 nonagrams). Those containing a dimethylamino-, aldehyde or alcohol-group, show distinctly less fluorescence in this reaction.

According to JEPSON and STEVENS [*8*], a very sensitive reagent for the detection of serotonin and tryptamine is a 0.25% solution of ninhydrin in acetone containing 10% of acetic acid. A blue-green fluorescence is obtained on heating for 2–3 min at 90–100° C. As little as 0.02 μg of serotonine can be detected on paper chromatograms in this way.

Chloroform phase

	UV-fluorescence before	UV-fluorescence after	Color reaction after
	spraying with Reag. No. 48		
q:	red +	red +	yellowish-greenish
p:	white*	white*	—
o:	greenish	—	—
n:	blue +	blue	—
m:	weak	white* or blue	—
b:	white*	—	—
a:	—	—	blue-violet
c:	white* or greenish	white* or greenish	orange
f:	yellow +	yellow	yellow or*
g:	—	—	pink
i:	—	—	pale blue
k:	white*	white*	yellowish
l:	white*	white*	

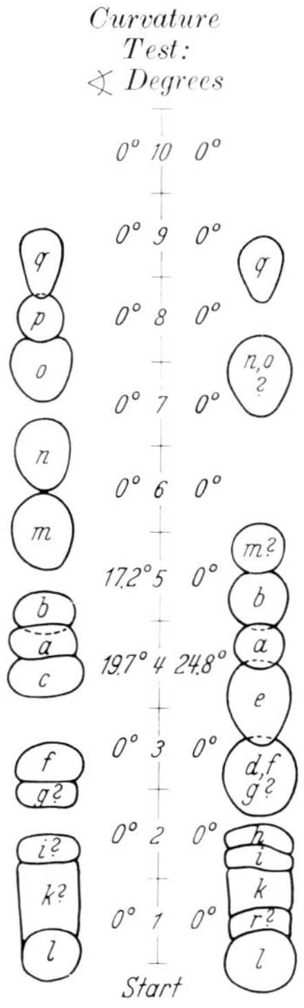

Crude extracts

Color after	UV-fluorescence before	UV-fluorescence after
spraying with Reag. No. 48		
yellowish-greenish	red +	red +
—	blue	blue
—	blue or white*	weak
—	—	white* or yel.
blue-violet	—	—
—	dark	dark
pink*	blue or light	yellow
red	white* (bracketed: red, pale blue)	white* to greenish (bracketed: red to blue)
pale blue		
yellowish	greenish	
blue	blue *	
brown	yellow +	yellow +

Test

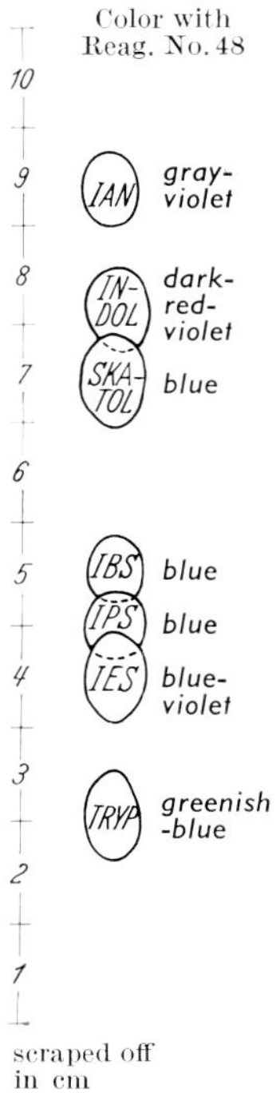

* = weak, + = intense, small letters in the fiirst column = chromatogram zones

Fig. 133. Visual and biological evaluation of thin-layer chromatograms of extracts from young fruit-stems of *Fritilliara meleagris*. (Silica Gel G layer, solvent: isopropanol — 25% ammonia-water (85 + 5 + 15). IES = indole-3-acetic acid; IPS = β-indolyl-3-propionic acid; IBS = β-indolyl-3-butyric acid; IAN = indolyl-3-acetonitrile; TRYP = tryptophane (KALDEWEY [9])

During investigation of plant and animal extracts, one should not omit examining the chromatogram under short and long-wave UV-light before spraying, and to mark fluorescing zones (Fig. 133).

b) Biological test of the growth regulators activity

Avena coleoptiles are generally used for testing of the plant regulators activity. An agar block containing the substance is placed unilaterally on the cut surface of the decapitated coleoptile and the curvature is taken as a relative measure for the activity of the substance ("Coleptile curvature test").

In the "straight growth" (coleoptile section) test one measures the increase in length of the coleoptile cylinders when placed in the solution to be tested. Söding [*33*] and Linser and Kiermayer [*17*], amongst others, give detailed instructions. Kaldewey [*9*] examined the procedure critically in relation to the evaluation of thin-layer chromatograms, and showed that the growth-regulating activity of the scraped-off silica gel zones can be tested efficiently by both methods. He gives the following details:

After chromatography, the area to be tested is marked by drawing a thick line down both sides of the migration route with a 3—4 mm broad spatula, using the marking template. The silica gel scraped off is blown away. The migration route, now about 10 mm broad, is further divided by thin crosslines, using a needle, into 10 sections (*Rf* 0—0.1; 0.1—0.2 etc.). The the sections are scraped off one by one, starting from the origin. When the silica gel from one section has been scraped into a pile, it is transferred to celluloid foil by lifting up one end of the plate and tapping gently. For the curvature test, the silica gel from each section is scattered over a block of 6—12 small agar blocks. One drop of water is added and the blocks allowed to stand for 30 minutes in a darkened humid chamber. One small agar block is then placed unilaterally on the cut surface of the coleoptile. Blank experiments with the silica gel gave no curvature; control experiments with indole-3-acetic acid zones of known concentration gave the expected degrees of curvature.

In the straight growth test, the silica gel can be transferred directly from the celluloid foil to a buffer solution in the small trough containing the inverted coleoptile sections. The mixture under investigation should be applied several times on each chromatogram alongside a known test solution. This makes it possible to visualize substances in one of the migration routes in the usual way after the parallel area had been scraped off for the biological test. Fig. 133 shows very clearly the results of the test worked out by Kaldewey. The latter author investigated both crude extracts and the chloroform-soluble fraction from young fruit-stems of *Fritillaria meleagris* by this procedure. Both chromatograms show that the extracts contain numerous compounds[1]. Their color before and after spraying is given in the columns on either side of the chromatogram. The result of the curvature-test (in degrees) is shown between them. In both cases, section 4 of the chromatogram shows the highest growth hormone activity. The test-chromatogram is shown on the right. The results mentioned here only briefly call for greatest caution in evaluating crude plant extracts [*9*] (see also p. 144).

[1] Only a few zones could be made visible in paper-chromatographic examinations undertaken for the same reasons.

6. Further applications

GMELIN and VIRTANEN [*6*] used thin-layer chromatography to separate the breakdown products of glucobrassicine, i.e., the first "bound auxin". They obtained the following *hRf*-values using Silica Gel G layers with chloroform + 1% of ethanol as solvent: Indole 66, skatole 70, 3.3′-di-indolylmethane 55, indolyl-(3)-acetonitrile 42, indole(3)-aldehyde 14.5, 3-hydroxymethyl indole 9.5. Sulphur, which was also found as a breakdown product (*hRf* 75), was detected as a dark brown spot after spraying the plate with ammoniacal silver nitrate solution and short heating.

DIAMANTSTEIN and EHRHART [*5*] used the method for the detection of tryptophan metabolites occurring in the urine, of patients with chronic myelosis. They applied 50–80 mm^3 of the urine sample and chromatographed ascending with the acidic solvent system (Sp. see p. 293). A preliminary concentration of the urine is not necessary. Displacements of *Rf*-values by the natural organic and inorganic constituents of the urine did not occur.

III. Amines

Procedures for the paper-chromatographic separation of amines have been reviewed recently by STEIN VON KAMIENSKI [*38*]. This treatise also includes methods for the isolation of amines from plant material and procedures for the preparation of 2,4-dinitro-α-naphthol compounds and picrolonates of amines. These derivatives exhibit typical crystal structures, most can be isolated and purified easily by sublimation (see also p. 174, Fig. 89).

1. Thin-layer chromatography of amines

TEICHERT, MUTSCHLER and ROCHELMEYER [*40*] described the fractionation of amines on layers of Silica Gel G or cellulose: 1 to 10 μg. of the amine hydrochlorides are applied to the layers as solutions in 70% ethanol. The h*Rf*-values and experimental details are given in Table 55 (see also p. 160, Fig. 82).

Layers of buffered silica gel are prepared by slurrying Silica Gel G with either a mixture of equal volumes of 0.2 M *prim.* potassium phosphate and 0.2 M *sec.* sodium phosphate solutions, or with 0.15 M sodium acetate solution. Primary and secondary amines are identified more clearly in mixtures, by chromatographing them as 3,5-dinitrobenzamides (DNB) on Silica Gel G layers using chloroform-ethanol (96%, 99 + 1) as the developing solvent. The following h*Rf*-values are obtained [*40*]: methylamine-DNB, 14; dimethylamine-DNB, 47; ethylamine-DNB, 22; diethylamine-DNB, 68; n-propylamine-DNB, 38; *iso*-butylamine-DNB, 42; *iso*-amylamine-DNB, 50.

Preparation of 3.5-Dinitrobenzamides (DNB):

50 mg. of an amine hydrochloride or 25 mg. of a free amine is dissolved in 5 ml. of water. This solution is placed in a separatory funnel and 15 ml. diethyl

ether, 0,25 ml. pyridin and a solution of 250 mg. 3,5-dinitrobenzoyl chloride in 1 ml. benzene are added. Then, 5.5 g. potassium carbonate is added while under agitation and cooling. After 20 min., the aqueous layer is drawn off and the ether phase is washed twice with 5 ml. 1% sulphuric acid and then with water. The ether solution of the DNB derivatives is dried over anhydrous sodium sulphate and filtered. After evaporation of the ether, the 3,5-dinitrobenzamide is recrystallized from 50% aqueous ethanol. 5 to 25 μg. of a 1% solution of a DNB derivative in diethyl ether or chloroform is applied to the chromatogram.

Table 55. *hRf-Values of Amines on different Layers and with different Solvents [40]*

Amines	Silica Gel G layer			Buffered Silica Gel G layer		
	I[1]	II	III	IV	V	VI
Methylamine	—	11	10	10	7	12
Ethylamine	—	15	13	20	14	19
n-Propylamine	—	30	19	30	23	31
i-Amylamine	—	49	39	45	44	58
Cadaverine	3	6	2	1	1	—
Putrescine	3	4	2	1	1	—
Ethanolamine	30	18	11	10	10	7
Histamine	41	26	2	3	3	2
Tyramine	56	44	38	55	40	28
Phenylethylamine	66	56	37	55	42	—
Benzyle amine	70	49	36	50	42	—
Tryptamine (s. p. 294)	60	54	43	90	45	—

Solvents: I, Ethanol (96%)—25% aqueous ammonia (80 + 20); II, Phenol-water (80 + 30); III, Upper phase of the mixture n-butanol-acetic acid-water (40 + 10 + 50); IV, 70% Ethanol on a phosphate-buffered layer; V, Solvent III on acetate-buffered layer; VI, Upper phase of the mixture amyl alcohol-acetic acid-water (40 + 10 + 50).

2. Thin-layer electrophoresis of amines

PASTUSKA and TRINKS [23] fractionated amines by thin-layer electrophoresis in the apparatus described on p. 24. This instrument has been equipped recently with a cooling device (see also Honeggers' apparatus, Fig. 26). PASTUSKA and TRINKS worked with layers of Silica Gel G impregnated with boric acid and found the following M_G-values: methyl amine (1.00), ethyl amine (0,69), n-butyl amine (0.82), ethanol amine (1.09), ethylene diamine (0.32), and triethanol amine (0.51). These M_G-values are related to methyl amine-$M_G = 1.00$. The experimental conditions are described here in detail:

The slurry of Silica Gel G is prepared with a 3% borax solution instead of water, and simply poured onto the glass plates, 8/20 cm. The layers are dried for 20 to 25 min. and 5—10 μg. of the various substances are applied as salt solutions at a distance of about 7 cm. from the edge, pointing to the anode. A mixture of 80 ml. ethanol, 30 ml. dist. water and 1 g. anhydrous sodium acetate, adjusted to pH 12 with 40% sodium hydroxide, serves as the electrolyte. This solution has to be renewed for each experiment. The samples are run for 2 hr. at a field intensity of 10 V/cm.

Reference is made here to the combination of thin-layer ionophoresis (electrophoresis) with thin-layer chromatography, as described by HONEGGER [*7a*], pp. 447, 448. Fig. 177 demonstrates an application of this two-dimensional technique for the fractionation of a complex mixture of amines and amino acids. The experimental conditions are given on p. 448.

3. Detection

Suitable spray reagents are solutions of ninhydrin (Reag. No. 108), Fast Blue-Salt B (Reag. No. 61), Sodium nitroprusside (Reag. No. 112), Nitric acid (Reag. No. 132), and vanillin (Reag. No. 147). 3.5-Dinitrobenzamides may be visualized with a solution of iodine in chloroform (Reag. No. 73) or with α-naphthylamine (Reag. No. 98). They appear as violet spots upon irradiating the chromatograms for 15 min. using a UV lamp without a filter.

IV. Organic bases in tars

Thin-layer chromatography of tar constituents[1] is of value for following synthesizing procedures. The possibility of fractionating coal tars and charcoal tars (see Fig. 136) has been recognized early [*34a*, *35*]. PETROWITZ [*24*, *24a*] chromatographed some tar bases, such as the isomeric methyl quinolines, on "manually" prepared layers, 750 μ thick, of Silica Gel G. SCHORN [*32*] investigated, in the present author's laboratory, the possibilities of separating and detecting basic constituents of tars on standard Silica Gel G layers. The h*Rf*-values he determined are presented in Table 56.

The compounds listed in group I are best resolved by TLC on layers of Silica Gel G with solvent system I. Quinoline and pyridin derivatives are chromatographed on Silica Gel G layers which had been prepared with 0.1 M sodium acetate solution, instead of water. The acidic solvent II yields the most favorable fractionation of quinoline derivatives, whereas, solvent III is suitable for resolving pyridin derivatives.

It is important to apply only very small amounts of the substances, i.e., 0.5 to 3 μl. of 0.1% solutions of the bases. Highly active layers, must be avoided, good results are obtained with air-dried layers.

Detection. A solution of iodine in chloroform (Reag. No. 73) can be used as a non-specific spray reagent for detecting the bases listed in Table 56. The various substances appear as brown spots which soon fade. Antimony pentachloride solution (Reag. No. 13) is very suitable for detecting the bases of group I (Table 56); acridine reacts yellow, carbazol green, anilin pink, p-xylidine light purple, diphenyl amine blue and β-naphthylamine appears gray. The bases of groups II and III exhibit orange to red colors after having been sprayed with modified Dragendorff reagent (Reag. No. 60a).

[1] Gesellschaft für Teerverwertung m.b.H., Duisburg-Meiderich, Germany.

Table 56. *hRf-Values of 18 Tar Bases*

Layer	Silica Gel G[1]	Silica Gel G[1] + 0,1 M Na-acetate		Group
Solvent	I	II	III	
Aniline	38	—	—	
p-Xylidine	44	—	—	
β-Naphthylamine	44	72	86	I
Acridine	50 (45)[2]	38	89	
Carbazole	58 (97)	85	92	
Diphenylamine	69	87	91	
2-Methylquinoline (Quinaldine)	40 (27)	18	74	
4-Methylquinoline (Lepidine)	33 (22)	25	74	II
Isoquinoline	30 (22)	35	74	
Quinoline	35 (30)	39	79	
2,4,6-Trimethylpyridine (2,4,6-Collidine)	24	7	23	
2,6-Dimethylpyridine (2,6-Lutidine)	28	6	36	
2,4-Dimethylpyridine (2,4-Lutidine)	22	8	38	III
2,5-Dimethylpyridine (2,5-Lutidine)	26	10	48	
4-Methylpyridine (γ-Picoline)	21	10	49	
Pyridine	22	13	55	
3-Methylpyridine (β-Picoline)	24	15	59	
2-Ethylpyridine	30	12	62	
Time for 10 cm run	25 min	35—40 min	30—35 min	

[1] Charge No. 200473.
[2] Values given by PETROWITZ [*24*] are in ( ).

Solvent: I, Benzene-methanol (95 + 5); II, Ethyl acetate-methanol-formic acid (80 + 10 + 10); III, Ethyl acetate-methanol-acetic acid (75 + 20 + 5).

Bibliography to Chapter E. Organic Bases

[*1*] BÄUMLER, J., u. S. RIPPSTEIN: Pharm. Acta Helv. **36**, 382 (1961).
[*2*] BORKE, M. L., and E. R. KIRCH: J. Am. Pharm. Assoc., Sci. Edit. **42**, 627 (1953).
[*3*] CHOCHIN, J., u. J. W. DALY: Experientia **18**, 294 (1962).
[*4*] DENFFER, D. v., M. BEHRENS u. A. FISCHER: Naturwiss. **39**, 258 (1952).
[*5*] DIAMANTSTEIN, T. u. H. EHRHART: Hoppe-Seyler's Z. physiol. Chem. **326**, 131 (1961).
[*6*] GMELIN, R., u. A. J. VIRTANEN: Ann. Acad. Sci. Fennicae Ser. A, II. Chem. 107 (1961).
[*7*] HOFMANN, A.: Planta Med. **9**, 354 (1961).
[*7a*] HONEGGER, C. G.: Helv. Chim. Acta **44**, 173 (1961).
[*8*] JEPSON, J. B., and B. I. STEVENS: Nature (Lond.) **172**, 772 (1953).
[*9*] KALDEWEY, H.: Habilitationsarbeit Saarbrücken 1961.
[*10*] KHANNA, K. L., A. E. SCHWARTING, A. ROTHER and J. M. BOBBITT: Lloydia **24**, 179 (1961).
[*11*] KLAVEHN, M., u. H. ROCHELMEYER: Deut. Apotheker-Ztg. **101**, 477 (1961).
[*11a*] —, — u. I. SEYFRIED: Deut. Apotheker-Ztg. **101**, 75 (1961).

[12] KIRCHNER, J. G., J. M. MILLER and G. J. KELLER: Anal. Chem. **23**, 420 (1951).
[13] KUMP, CH., u. H. SCHMID: Helv. Chim. Acta **45**, 1090 (1962).
[14] KÖGL, F., A. I. HAAGEN-SMIT u. H. ERXLEBEN: Hoppe-Seyler's Z. physiol. Chem. **228**, 90 (1934).
[14a] —, u. D. G. F. R. KOSTERMANS: Hoppe-Seyler's Z. physiol. Chem. **235**, 301 (1935).
[15] LEWIS, G. P.: "5-Hydroxy-tryptamine". London, New York: Pergamon Press 1958; s. a. V. ERSPAMER in Fortschritte der Arzneimittelforschung, Vol. 3. Basel/Stuttgart: Birkhäuser 1961.
[16] LEHNER, H., u. J. SCHMUTZ: Helv. Chim. Acta **44**, 444 (1961).
[17] LINSER, H., u. O. KIERMAYER: Methoden zur Bestimmung pflanzlicher Wuchsstoffe. Wien: Springer 1957.
[18] LINSKENS, H. F.: Papierchromatographie in der Botanik, 2nd ed Berlin-Göttingen-Heidelberg: Springer 1959.
[19] MACHATA, G.: Mikrochim. Acta **1960**, 79.
[20] MÜLLER, K. H., u. H. HONERLAGEN: Arch. Pharm. **293/30**, 202 (1960).
[21] NEUBAUER, D., u. K. MOTHES: Planta Med. **9**, 466 (1961).
[22] PAILER, M., u. W. G. KUMP: Arch. Pharm. **293**, 646 (1960).
[22a] —, u. R. LIBISELLER: Monatsh. Chem. **93**, 403, 511 (1962).
[23] PASTUSKA, G., u. H. TRINKS: Chem. Ztg. **86**, 135 (1962).
[24] PETROWITZ, H. J.: Materialprüfung **2**, 309 (1960).
[24a] — Chemiker-Ztg. **85**, 143 (1961), s. a. Erdöl u. Kohle **14**, 923 (1961).
[25] PINXTEREN, J. A. C., u. M. E. VAN VERLOOP: Pharm. Weekblad **97**, 1 (1962).
[26] PÖHM, M.: Arch. Pharm. **286**, 509 (1953).
[27] PROCHÁZKA, Z., in: J. M. HAIS u. K. MAČEK: Handbuch der Papierchromatographie. Jena: VEB Fischer 1958.
[28] RISTIĆ, S., u. A. THOMAS: Arch. Pharm. **295**, 524 (1962).
[29] ROBLES, M. A., u. R. WIENTJES: Pharm. Weekblad **96**, 379 (1961).
[30] ROTHER, A., J. M. BOBBITT u. A. E. SCHWARTING: Chem. & Ind. (London) **1962**, 654.
[31] SCHLEMMER, F., u. E. LINK: Pharm. Ztg. **104**, 1349 (1959).
[32] SCHORN, P. J.: unpublished.
[33] SÖDING, H.: Wuchsstofflehre. Stuttgart: G. Thieme-Verlag 1952.
[34] STAHL, E.: Pharmazie **11**, 633 (1956).
[34a] — Parf. u. Kosm. **39**, 564 (1958).
[35] — Arch. Pharm. **292**, 411 (1959).
[36] —, u. H. KALDEWEY: Hoppe-Seyler's Z. physiol. Chem. **323**, 182 (1961).
[37] — In K. PAECH and M. V. TRACEY, Modern Method of Plants Analysis. Vol. V. Berlin-Göttingen-Heidelberg: Springer-Verlag 1962.
[38] STEIN VON KAMIENSKI, E.: In H. F. LINSKENS, Papierchromatographie in der Botanik. Berlin-Göttingen-Heidelberg: Springer-Verlag 1959.
[39] STOWE, B. B.: Progress in the Chemistry of Organic Narural Products, Vol. 17, p. 249. Wien: Springer 1959.
[40] TEICHERT, K., E. MUTSCHLER u. H. ROCHELMEYER: Deut. Apotheker Ztg. **100**, 283 (1960).
[41] — — — Deut. Apotheker-Ztg **100**, 477 (1960).
[42] — — — Z. anal. Chem. **181**, 325 (1961).
[43] ULLMANN, E., u. H. KASSALITZKY: Arch. Pharm. **295**, 37 (1962).
[44] VAN URK, H. W.: Pharm. Weekblad **66**, 473 (1929).
[45] WALDI, D.: Arch. Pharm. **292**, 206 (1959).
[46] —, K. SCHNACKERZ and F. MUNTER: J. Chromatog. **6**, 61 (1961).
[47] WINKLER, W., u. W. AWE: Arch. Pharm. **294**, 301 (1961).
[48] — Naturwiss. **48**, 694 (1961).

Some recent noteworthy articles

AGURELL, A., and E. RAMSTAD: Lloydia **25**, 67 (1962): PC and TLC of *Clavine* alkaloids.
CONE, N. J., R. MILLER and N. NEUSS: J. Pharmac. Sci. **52**, 688 (1963): *Vinca* alkaloids.

Döpke, W.: Arch. Pharm. **295**, 605 (1962): TLC of some alkaloids groups.
Farnsworth, N. R., and K. L. Euler: Lloydia **25**, 186 (1962): Alkaloid screening of extracts.
Gröger, D., and D. Erge: Pharmazie **18**, 346 (1963): Ergot alkaloids.
Halmekoski, J.: Suomen Kem. **36** B, 58 (1963): Adrenaline derivatives.
Heacock, R. A., and M. E. Mahon: Can. J. Biochem. Physiol. **41**, 487 (1963): Hydroxyskatoles.
Kutacek, M.: Biologia Plantarium (Praha) **4**, 226 (1962): Gibberelins.
Macmillan, J., and P. J. Soter: Nature **197**, 790 (1963): Gibberellins.
Moll, F.: Arch. Pharm. **296**, 205 (1963): Hemlock alkaloids, piperidine bases.
Sandberg, F., and K. H. Michel: Lloydia **26**, 78 (1963): Amaryllidaceae alkaloids.
Seiler, N., G. Werner and M. Wiechmann: Naturwiss. **50**, 643 (1963): Quantitative estimation of fluorescing indole derivatives.
Sembdner, C., R. Gross and K. Schreiber: Experientia **18**, 584 (1962): Gibberellins.
Schmid, E., L. Zicha, J. Krautheim and J. Blumbreg: Med. exp. **7**, 8 (1962): Biogenic amines and metabolites.

F. Pharmaceutical Products

By

H. Gänshirt

Thin-layer chromatography is increasingly being used to identify pharmaceutical mixtures and also to quantitatively determine their composition. Only small amounts of material and low expenditure of time and funds are required. Thin-layer chromatography is not limited by the choice of spray-reagents as necessitated in paper-chromatography by the nature of the carrier material.

The method may be used for analytical control as well as toxicological investigations. In routine control analyses an unequivocal separation of the known contents is demanded, using all the space available on the plate. In toxicological investigations the aim is chiefly to assay qualitatively a number of medicaments and poisons. These substances are identified, after separation, by their *Rf*-values and with specific spray reagents. A number of methods may be used for both types of investigations.

To give a better overall picture, drugs are classified according to their therapeutical effects. Methods for their separation will be described in Section I of this chapter. Procedures for analytical control are given in Section II and procedures for toxicological work in Section III. Stability tests of drugs are becoming increasingly important, and TLC is of particular service in this respect. Such applications are extensively discussed in Section IV. TLC can also be successively used in recognizing drugs and to detect adulteration (p. 375).

Thin-layer chromatography of vitamins, alkaloids, steroids and other natural products which are also of therapeutical significance is treated in several chapters of this book.

I. Groups of therapeutical materials

1. Analgesics, antipyretics and antirheumatics

The behavior in TLC of drugs listed in Table 57 and Fig. 134 has been investigated in detail [*14*]. Silica Gel G layers are made by the standard method (pp. 7—9) and 2% of luminescent material ZS-Super (Messrs. Riedel de Haen) are added. After separation, all the substances may be recognized in UV-light of 254 mμ as dark absorbing spots. If plates are used without a fluorescent material, they may be sprayed with indicators such as fluorescein-sodium-, morin-, Rhodamine-B-solutions etc. (Reag. No. 90). A series of bases react with iodoplatinate (Reag. No. 76). After reacting with chlorine, a thorough spraying with benzidine-potassium iodide solution will also make all the substances visible

Table 57. h*Rf-Values of various Analgesics, Antipyretics and Antirheumatics in some Solvents* [*14*]

		Rf-values × 100 in the solvent:			
		Methyl ethyl ketone	Cyclohexane-Acetone (40 + 50)	Cyclohexane-Chloroform-Acetic acid (40+50+10)	Cyclohexane-Chloroform-Pyridine (20 + 60+5)
1	Sodium, Noramidopyrine-methansulphonate (Novalgin)	0	—	0	0
2	Cinchophen (Atophan)	3	0	24	—
3	Acetylsalicylic acid (Aspirin)	11	13	42	10
4	Phenazone (Antipyrin)	19	32	15	30
5	Amidopyrin (Pyramidon)	30	58	4	45
6	Phenacetin	57	64	24	37
7	Ethenzamidum (o-ethoxy-benzoic acid amide)	58	—	40	50
8	Paracetamol (N-acetyl-p-aminophenol)	60	—	0	10
9	Phenylbutazone	65	—	57	—
10	Salicylamide	69	70	27	21

with various colors (Reag. No. 32). This method of detection is exceptionally dependent on the time of reaction with chlorine, which should be about 10 sec. Noramidopyrine-methanesulphonate (Novalgin) remains at the origin with the solvents described here because of its high polarity.

The analgesic Carisoprodol (N-isopropyl-2-methyl-2-n-propyl-1,3-propanediol dicarbamate) can be chromatographed on Silica Gel G layers with the solvent cyclohexane-chloroform-acetic acid (4 + 5 + 1). Iodoplatinate (Reag. No. 76) has been of value as a spray-reagent.

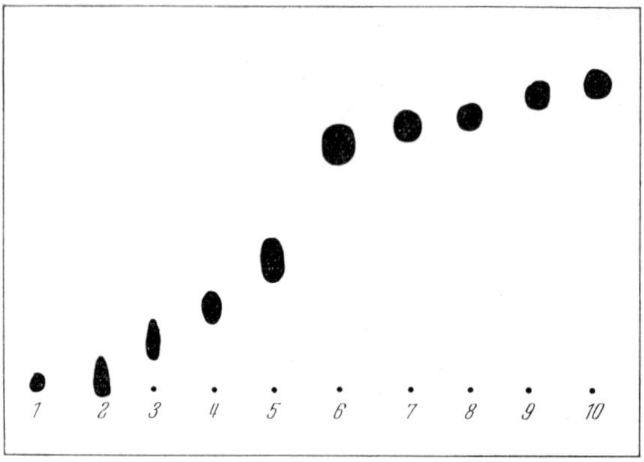

Fig. 134. Thin-layer chromatogram of various analgesics, antipyretics and antirheumatics. Methyl ethyl ketone, Silica Gel G layers [*14*]. Explanation of figures in Tab. 57

2. Analeptics

a) Purines

Theobromine, theophylline and caffeine may be separated on buffered Silica Gel layers (pH 6.8) with the solvent chloroform-96% ethanol (9.1) [*38*] (Table 58). By spraying with alcoholic iodine-potassium solution and then with a mixture of 96% ethanol and 25% hydrochloric acid in equal parts, caffeine and theophylline are colored red, theobromine grey and phenazone brown. The method is sensitive down to 1 μg.

Table 58. h*Rf-Values of Purines*

Purine	Chloroform 90 96% Ethanol 10 pH 6.8 buffered silica gel [*38*]	Ethyl acetate 90 Methanol 10 12 N HCL 0.25 [*1*]	Ethyl acetate 80 Methanol 10 Acetic acid 10 [*1*]	Chloroform 95 Methanol 5 [*1*]
Theobromine .	22	25	36	26
Theophylline .	37	41	50	23
Caffeine	57	36	41	51

Preparation of buffered silica gel layers: 25 g Silica Gel G are well ground with 50 ml of a mixture of equal parts of 0.2 M *prim.* potassium phosphate and 0.2 M *sec.* sodium phosphate solution and 5 plates are spread, using the Stahl-applicator. After preliminary drying in hot air, the plates are dried in a drying-cupboard at 110° C.

BAEHLER [*1*] separated the three purines on Silica Gel G layers with various solvents. After separation, the purines were detected by sublimation onto a cooled glass plate laid over the chromatogram (sensitivity

1–5 μg.). Barbituric acids were detected in the same way. Uric acid, xanthin, hypoxanthin and 6-mercaptopurine, as well as caffeine, theophylline and xanthin, have been separated by other authors using TLC [*7*].

b) Coramin, Pentetrazol, Micoren, Benegride, Camphor, Lobeline

These substances are completely different in chemical structure. After chromatography with methyl ethyl ketone and an acidic solvent-mixture on Silica Gel G containing a luminescent material (p. 54), they show the properties given in Table 59 [*14*]. As Fig. 135 shows, the spot

Table 59. h*Rf*-*Values and Detection of various Analeptics* [*14*]

No.	Analeptic	*Rf* × 100 in the solvent: Methyl ethyl ketone	Cyclohexane Chloroform-Acetic acid (40 + 50 + + 10)	Detection of active hydrogen (Reag. No. 32)	UV 254 mμ	DRAGENDORFF (Reag. No. 60)
1	Micoren Spot A	51	{A₁ 52	—	+	—
			A₂ 60	—	+	—
	Spot B	5	2	—	—	orange
	(N-crotonyl-ethylamino-butyric acid-dimethyl amide and N-crotonyl propylamino-butyric acid-dimethylamide)					
2	Lobeline (Lobeline hydrochloride)	20	5	—	+	orange
3	Nikethamid (Coramin)	33	22	blue	+	orange
4	Pentetrazol (Cardiazol). . .	55	?	—	—	—
5	Benegride (Eucraton)	65	55	'dark blue	—	—
6	Camphor	73	85	—	—	orange

Plate: Silica Gel G with 2% luminescent material ZS-Super (RIEDEL DE HAEN)
Amount applied: 20—50μg of each substance (in the form of extracts or dilution of the corresponding preparation)
Separation distance: 10 cm; (+) positive, (—) negative reaction

of the polar base lobelin is rather drawn-out in adsorption chromatography if methyl ethyl ketone is used as the developing solvent. In the acidic solvent, cyclohexane-chloroform-acetic acid, Micoren is split into 3 spots. The two adjacent spots, which cannot be distinguished with methyl ethyl ketone as solvent, will probably associate with the two amides contained in Micoren. These spots are not stained with Dragendorff-reagent, in contrast to a third spot, near the origin, which could not be identified. Camphor and Pentetrazol cannot be detected on layers containing luminescent material. Even with the spray reagents used hitherto, the sensitivity of detection is so poor that amounts under 30 μg cannot be unambiguously identified. Detection on fluorescent layers is more sensitive, when followed by spraying with Rhodamin B- and sodium carbonate solutions (Reag. No. 101 B), as described on page 362.

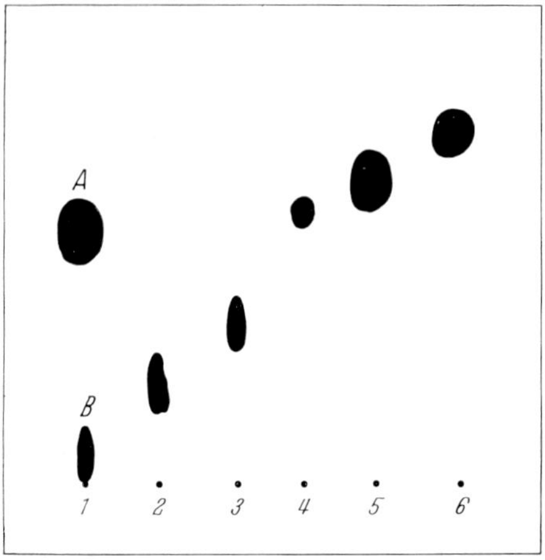

Fig. 135. Thin-layer chromatogram of various analeptics. Methyl ethyl ketone on Silica Gel G layers [*14*]. Explanation of figures in Tab. 59

3. Thin-layer chromatography of various anti-histaminics of the phenothiazine series and of psychologically active products

BÄUMLER and RIPPSTEIN [*4*] fractionated phenothiazine and psychologically active drugs with methanol-acetone-triethanolamine (50 + 50 + 1.5) on Silica Gel G layers (Table 60). Identification could be effected by applying several reagents.

The tranquilizer Meprobamate (2-methyl-2n-propyl-1.3-propanediol dicarbamate) was isolated from urine and investigated by thin-layer chromatography [*10*]. The layers were made with silica gel and rice starch as a binding agent. Cyclohexane-ethanol (85 + 15) was used as the solvent. Identification was achieved using a spray of conc. sulphuric acid. After extraction from the silica gel the substance was also quantitatively determined by reaction with a 0.2% solution of hydroquinone in conc. sulphuric acid. Exact estimates of the reproducibility of this method are not available. Meprobamate may also be chromatographed on silica gel layers prepared by a standard method (pp. 7—9) using chloroform-diethylamine (90 + 10) or cyclohexane-acetone-diethylamine (70 + 20 + 10). Identification is possible under UV-light after spraying with 0.2% 2′7′-dichlorfluorescein solution (No. 42). The neuroleptic Perphenazine may be chromatographed by the standard method on Silica Gel G with chloroform-diethylamine (90 + 10) and detected by spraying with Rhodamin B (No. 129) or iodoplatinate (No. 76).

Table 60. h*Rf-Values and Color Reactions of various Anti-histaminics of the Phenothiazin Series and Psychologically Active Preparations* [*4*]

	Rf-value × 100	Colors of spots with reagents 1, 2, 3		
Acepromazine . . .	33—35	1 orange	2 red	3 red brown
Promazine	37—39	1 blue	2 orange	3 violet
Imipramine. . . .	44—46	1 light blue	2 orange	3 white
Thioradizine . . .	47—49	1 light blue	2 red	3 red brown
Chlorpromazine . .	49—50	1 light blue	2 orange	3 red brown
Diethazine	53—55	1 blue	2 orange	3 violet
Promethazine. . .	55—57	1 blue	2 orange	3 violet
Chlorprothixenum .	55—57	1 luminous blue	2 orange	3 yellowish
Profenamine Hydrochloride .	57—59	1 blue	2 orange	3 violet
Levomepromazine .	62—64	1 blue	2 orange	3 violet
Chlordiazepoxide .	84—86	1 brownish	2 orange	3 yellow
Phenothiazine . .	89—91	1 dark	2 red violet	3 blue

Plate: Silica Gel G, coarser particles removed with the fine meshed sieve of Pharmacop. Helv., otherwise prepared by standard methods (pp. 7—9).
Separation distance: 10 cm.
Identification: UV-light. 2 Dragendorff reagent. 3 Palladium chloride solution (0.5% indilute hydrochloric acid).

4. Bacteriostatic and bactericidal substances

a) Pharmaceutical phenols, tars and tar-oil components

Ways of separating phenols occurring in essential oils are discussed on pp. 197—200, (see Table 21), and TLC of tar-bases is described on pp. 303—304, (see Table 56).

The chromatographic behavior of many phenols, phenol carboxylic acids, phenol aldehydes, dimethyl phenols and binuclear phenols was investigated by Pastuska and Petrowitz using manually prepared Silica Gel G layers, 0.5 mm thick. The results of this work were summarized in a review. Particular attention was paid to the connection between the structures of phenols and their *Rf*-values (comp. with p. 199) [*31, 32*]. For phenols and phenol carboxylic acids the solvents listed in Table 61 were used. In the same table the h*Rf*-values are given of some of the pharmaceutically more interesting materials of this class. Phenolaldehydes, dimethyl phenols and the corresponding methyl ethers were chromatographed with the solvents benzene and benzene methanol (45 + 5). The behaviour of some binuclear phenols with the solvents shown in Table 62 was also investigated.

Phenols and phenol carboxylic acids were separated by Pastuska and Trinks by thin-layer electrophoresis [*30*]. In the course of investigations on the products arising from methylation of β-resorcylic acid, many other phenolic substances were obtained and identified by TLC [*22*]. According to Waldi [*42*], separation of phenol, o-, m- and p-cresols is also possible on Silica Gel G or Alumina G layers, prepared by the standard method, if the volatilization of these substances is prevented by appropriate measures.

In an investigation of steam-volatile alcohols and phenols in foodstuffs, these substances were fractionated after convertion to the corresponding dinitrobenzoates [9] (pp. 341, 342). The method is well suited for the detection of small amounts of phenols or higher alcohols in aqueous-alcoholic solutions, but not for the separation of homologous series of alcohols or phenols. For instance, thymol was detected in a solution of 5 mg thymol in 1 ml ethanol and 1000 ml water.

Table 61. h*Rf-Values of a few Phenols in two Solvents* [31, 32]

Phenols	*Rf* × 100 in the solvent:	
	Dioxane-Benzene-Acetic acid (25 + 90 + 4	Methanol-Benzene-Acetic acid (8 + 45 + 4)
Pyrogallol . . .	32	45
Resorcinol . . .	56	52
Phenol	76	60
Guaiacol	83	72

Hexylresorcinol and Hexachlorophene have been separated on Silica Gel G layers with methyl isobutyl ketone as the developing solvent [14]. If a luminescent material was incorporated in the layers (p. 54), the substances could be recognized as absorbing spots in UV-light (254 mμ). The two phenols were made visible also by spraying with diazonium salt solutions (Reag. No. 37 or 61), Hexachlorophene *Rf* × 100 = 50; hexylresorcinol *Rf* × 100 = 90. Dichlorophenol and Hexachlorophene could be

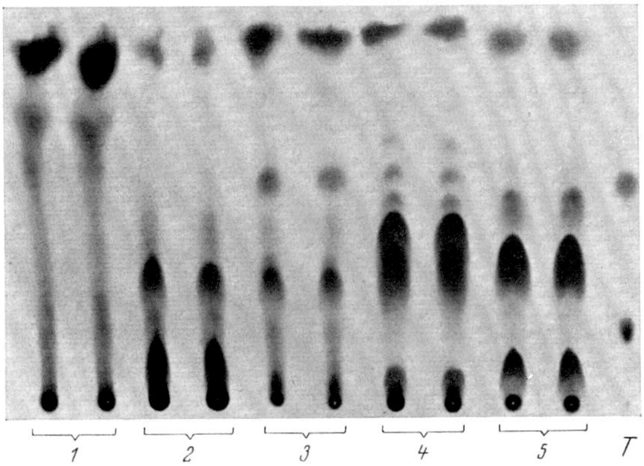

Fig. 136. Separation of medicinally-used tars (STAHL). *1* Coal tar, *2* Wood tar DCB, *3* Juniper tar, *4* Birch tar, *5* Beech tar oil, and *T* test mixture. Butter-yellow upper spot, Sudan Red G, lower spot

separated with acetic acid-saturated n-heptane on layers of silicic acid containing starch as a binding agent [6]. Identification was possible with ferric-chloride/potassium ferricyanide reagent, prepared by dissolving 50 mg potassium ferricyanide and 1000 mg ferric chloride in 10 ml water. This solution should always be prepared fresh. Quantitative analysis of the mixture was possible by UV-spectrophotometry after the spots being scraped off and extracted (see pp. 45—47).

Table 62. h*Rf-Values and Detection of Tar-Oil Components* [*33*]

Tar Components	*Rf*-values × 100 in the solvents			Detection
	Chloroform	Benzene	Benzene-Methanol (95:5)	
Isoquinoline	—	7	22	Dragendorff: orange
β-Naphthol	16	21	38	diaz. Benzidine: red-violet
α-Naphthol	23	28	46	diaz. Benzidine: blue violet
Quinoline	30	9	30	Dragendorff: orange
Acridine	36	8	45	Dragendorff: orange
8-Methyl quinoline	59	—	—	Dragendorff: orange

On 17.5 × 4.5 cm strips, 4 g silica gel was mixed with water to a thin paste and transferred manually as evenly as possible. The evenness of the layer was examined with test substances. For separation, 2000 μg tar-oil was applied 2.5 cm from the lower edge.

Coal-tar, wood-tar DAB 6, juniper-tar, birch-tar and beech-tar oil were chromatographed (Fig. 136) by Stahl [*35, 37*] on Silica Gel G layers prepared by the standard method. Benzene was used as solvent without saturation of the tanks, and good discrimination was obtainable by this method.

Tars were applied as 10% solutions in benzene or chloroform. The spray reagent was antimony pentachloride (No. 13). The plate was heated to 110° C for 10 min. to develop the colors.

Tar-oils were investigated by Petrowitz [*33*]. Some of their contents are of pharmaceutical interest. Their chromatographic behavior is summarized in Table 62. As the plates were prepared "manually" the *Rf*-values given deviate from those obtained by the standard method.

b) Thin-layer chromatography of sulphonamides

Sulphonamides can be separated by paper-chromatography with acidic or basic solvents. The *Rf*-values and the shape of the spots are strongly dependent on the base – or acid – concentration of the atmosphere in the tank [*18*]. Attempts to separate sulphonamides by TLC were, therefore, particularly interesting. The sulphonamides described below were investigated using the solvents listed in Table 63 [*14*]. The resulting *Rf*-values were tabulated. As Fig. 137 shows, it was possible to separate, using only one solvent, a number of sulphonamide mixtures taken arbitrarily from the sulphonamides listed. As the detection limit of the p-dimethylaminobenzaldehyde spray is low, only small amounts (under 10 μg) are required and thus, very good separations are achieved.

With the exception of sulphisomidine, the sulphonamides were extracted with acetone from commercially prepared tablets. A sulphisomidine solution was prepared by diluting Aristamide ampouls with ethanol. Volumes containing about

Table 63. h*Rf*-*Values of Sulphonamides in various Solvents* [*14*]

No.	*Sulphonamide*	Solvent: *Rf*-values × 100				
		Chloroform Methanol (80 + 15)	Cyclohexane-Acetone-Acetic acid (40 + 50 + + 10)	Methyl ethylketone Acetic acid (75 + 5)	Acetone-Methanol-Diethylamin (90+10 +10)	Methyl ethylketone pyridine (75 + 5)
1	Sulphadimidine (sulphanilyl amino-4.4-dimethylpyrimidine)	66	56	76	41	63
2	Sulphafurazole (5-sulphanilyl-amino-3-4-dimethyl isoazole)	47	55	79	30	67
3	Sulphapyridine (2-sulphanilyl-amino-pyridine)	55	54	71	39	63
4	Sulphaquanidine (sulphanilylguanidine)	13	28	53	33	43
5	Sulphisomidine (4-sulphanilyl-amino-2.6-dimethylpyrinidine)	45	33	46	32	39
6	Sulphacetamide (sulphanilylacetamide)	37	49	72	21	59
7	Sulphathiazole (2-sulphanilyl-amino-thiazole)	46	38	64	32	49
8	Sulphanilamide (p-aminobenzol-sulphonamide)	33	43	68	69	67
9	Sulphathiourea (sulphanilylthiourea)	20	56	72	49	65

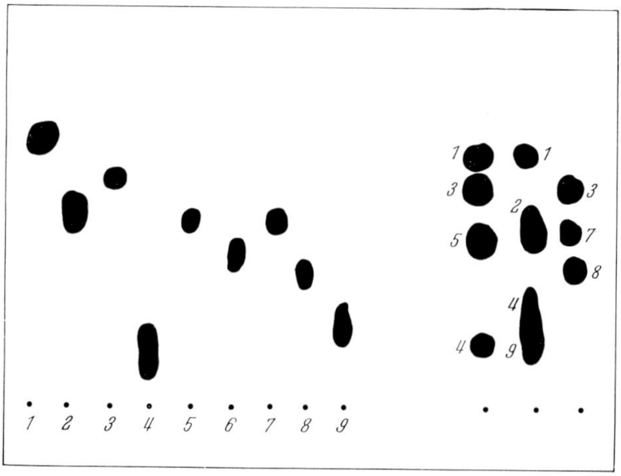

Fig. 137. Sulphonamide in chloroform-methanol (80 : 15) [*14*]. Explanation of figures in Tab. 63

10 μg of the appropriate sulphonamide were applied. The silica gel layers were prepared by the standard method (pp. 7—9). Luminescent material was added to the adsorbent slurry. Migration-distance 10 cm. After separation, the sulphonamides could be seen as absorbing spots in UV-ligth (254 mμ). In addition, they could be stained yellow with p-dimethylaminobenzaldehyde (Reagent No. 48).

In a review on TLC by WOLLISH et al. [*43*], a further solvent system for the separation of sulphonamides is presented, chloroform-heptane-ethanol (1 + 1 + 1). The sulphonamides separated by WOLLISH and his coworkers and their *Rf*-values are listed in Table 43.

Table 64. *Separation of some Sulphonamides with the Solvent Chloroform-Heptane-Ethanol* (1 + 1 + 1) [*43*]

Sulphonamide	*Rf* × 100
N^1-Acetyl-(3,4-dimethyl-5-isoxazyl) sulfanilamide (Acetylgantrisin)	90
N^1-(3,4-Dimethyl-5-isotazyl) sulfanilamide (Sulfafurazol)	70
2,4-Dimethoxy-6-sulphanilamido-1,3-diazine (Sulfadimethoxine)	40
p-Aminobenzenesulphonic acid, (Sulphanylic acid)	2

Silica Gel G layers prepared by the standard method (pp. 7—9).
Amount applied: 1 μg in 0.01 ml acetone.
Detection: p-dimethylaminobenzaldehyde (Reag. No. 48).

The sulphonamide Palladin (5-methylsulphadiazin) may be chromatographed in cyclohexane-methanol-diethylamine (60 + 30 + 10) on silica gel or alumina layers [*42*].

c) Antibiotics

α) Penicillins

Various penicillins can be separated on Silica Gel G under the conditions given in Table 65 [*11*]. The penicillins are visualized after chromatography with iodine azide reagent. Double spots appear for several penicillin derivatives (see Table 65). Although the formation of these spots was discussed in the original work, using procain-penicillin as an example, the reaction involved was not explained. As a complete separation of all the penicillins investigated is not possible, NUSSBAUMER [*28*] tried to characterize frequently used penicillin derivatives by chromatography of the break-down products after acid hydrolysis. Even by this method discrimination is only possible in certain cases. NICOLAUS et al. [*25*] separated 6-aminopenicillin on Silica Gel G layers from the basic penicillins G, V, and dimethoxypenicillin with the acidic solvent, butanol-water-acetic acid (40 + 40 + 1). They evaluated the chromatogram

Table 65. h*Rf-Values of various Penicillins* [*11*]

Pen		Acetone-Methanol (50 + 80)		Isopropanol-Methanol (30 + 70)	
		white spot	yellow spot	withe spot	yellow spot
Benzyl penicillin	Pen G-K	16		22 48	
Phenoxymethyl penicillinic acid, potassium salt . .	Pen V-K	68		64	
Phenoxymethyl penicillinic acid	Pen V-Acid	68	54	74	24
α-Phenoxyethyl penicillin potassium salt	Me-Pen-V-K	72		68	
p-Methyl phenoxymethyl penicillin	Pen P-K	62		58	
Phenylmercaptomethyl penicillin	Pen PM-K	4 58		8 54	
Procaine-penicillin . . .	Proc. Pen.	62	52	60	42
N-N^I-Dibenzylethylene-diamine-di-penicillin G.	DBED Pen	58	52	56	24
N-N′-Dibenzyl ethylene-diamine dipenicillin V .	DBED V	64	52	64	24
Diethylaminoethanolester hydroiodide of penicillin V	Pulmo 500	72		4 60	
Cl-Benzylpyrrolidyl-methyl-benzimidazol penicillin G.	Megacillin	64	91	60 100	84

Plate: Silica Gel G applied by the standard method (pp. 7—9).
Separation distance: 12 cm in "saturated tank"
Detection: iodine-azide solution (Reag. No. 75).
Limit of detection: 1—2 μg.

microbiologically as described for the tetracyclins. As 6-aminopenicillin shows practically no microbiological activity, it was converted to benzylpenicillin after separation. The limit of detection of this method is about 0.1 μg.

β) Miscellaneous classes

Thin-layer chromatography has been successfully applied to the separation of various rifomycins [*35*]. In the rifomycin series the following transformations are known:

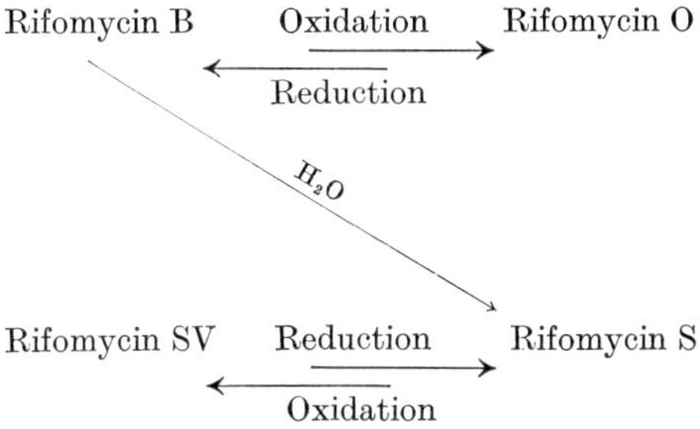

(Rifomycin B changes to rifomycin O if treated with oxidizing agents and the latter can be reduced back to rifomycin B. Both rifomycin B and rifomycin O are converted in aqueous solution to a substance of high antibiotic activity, rifomycin S, which can be changed to rifomycin SV by mild reduction with ascorbic acid.)

In an experiment to separate the four substances on paper, rifomycin O was reduced to rifomycin B and rifomycin S to rifomycin SV due to the weak reducing properties of the chromatographic paper. These substances could be detected by their UV-spectra after elution of the spots. By thin-layer chromatography on Silica Gel G, rifomycin B could be separated from rifomycin O, rifomycin SV from rifomycin S, and rifomycin B from rifomycin SV. The most favorable solvent proved to be acetone. Owing to their characteristic coloring, the rifomycins can be seen on a chromatogram containing more than 10 μg. More sensitive is the microbiological detection (0.1–0.5 μg) as described for the TLC of tetracyclins. If water-free solvents are used, it is necessary to expose the plates to steam before evaluation. The diffusion of the antibiotics through the agar is thereby facilitated.

TLC has also been applied in the elucidation of the structures of cytostatically effective antibiotics of the myrothecium species, i.e. the *verrucarines* and *roridines*. The same method was used for isolating 2-(p-chloroanilino)-5-(p-chlorophenyl)-3-(isopropylimino)-3.5-dihydrophenazine, from by-products of its synthesis [*34*].

γ) Tetracyclines

Several tetracyclines have been investigated in many solvents using various adsorption layers [*25*].

On Silica Gel G the following have been separated using 10% citric acid or 10% tartaric acid solution:

1. Desoxytetracyclin from tetracyclin, oxytetracyclin, chlorotetracyclin and dimethyltetracyclin.
2. Oxytetracyclin from tetracyclin, chlorotetracyclin, dimethyltetracyclin and desoxytetracyclin.

With 10% citric acid solution saturated with butanol, oxytetracyclin and dimethyltetracyclin could be separated from tetracyclin, chlorotetracyclin and desoxytetracyclin.

Tetracyclin and chlorotetracyclin could be separated from the corresponding anhydro-derivatives on Kieselguhr G layers. The chromatograms may be evaluated chemically or microbiologically. For chemical detection, the chromatograms are sprayed with hydrochloric acid and heated for a few minutes to 50° C; the tetracycline can be recognized as yellow spots. Desoxytetracyclin can be stained by coupling with a diazonium salt. Sensitivity is about $0.1-1\,\mu g$. The sensitivity may be markedly increased by microbiological evaluation using an agar culture as described below. The latter is poured onto the chromatogram, triphenyltetrazole added and then inoculated with *Bacterium subtilis*. This method, however, requires considerable time and effort.

Preparation of silica gel and kieselguhr layers

25 g Silica Gel G or Kieselguhr G are well shaken for 30 sec with 50 ml water in a 200 ml flask. From this suspension the chromatographic plates are prepared with the Desaga applicator. 10 mins. is allowed for setting of the plate and then 30 min. activation at 110° C. The plates are kept in a desiccator over silica gel.

Carrying out the microbiological evaluation of the chromatograms

The thin-layer plates are laid in the specially prepared Plexiglas container described in the original reference. 50 ml sterile (adjusted to p_H 5.9 if *Sarcina lutea* be used or pH 7 for inoculation with *Bacterium subtilis* and *Staphylococcus aureus*) is put in a 500 ml flask at 50° C and 0.7 ml of a 5% solution of triphenyltetrazole in 50% methanol is added.

For the analysis of rifomycins this solution is inoculated with 0.3 ml of a 30—35% suspension of *Sarcina lutea*, for analysis of tetracyclins and penicillins with 0.5 ml of a 35% suspension of *Staphylococcus aureus*. The mixture is well stirred after 30 sec. and finally decanted with the greatest care. It is then left for 15 mins. at room temperature in a sterile environment until the agar has hardened. To exclude air, which would prevent the reduction of the tetrazolium compound to formazan, 25 ml of an agar protective layer (15 g agar, distilled water 1000 g, sterilized) is poured at 50° C over the inoculated culture medium. This layer is cooled for 15 mins. and the treated plates left for 1 hour at 0° C. During this period, the antibiotic diffuses through the agar before the culture of microorganisms develops. The treated plates are then kept at 37° C for at least 16 hours. The separated antibiotics will be seen as restricted yellow zones on a red-brown background.

5. Hypnotics

The barbiturates listed in Table 67 have been fractionated in numerous solvents [*42*]. A mixture of Silica Gel G and Alumina G (1 + 1) was found to be a particularly good adsorbent (see p. 31). It was not possible to separate all the barbiturates in one dimension as in paper-chromatography. This would be of partricular interest in toxicological investigations. However, the results derived by using isopropanol-cyclohexane-25%-ammonia (65 + 25 + 10) are in many cases useful in the control analysis of preparations containing combinations of barbiturates. If inorganic luminescent materials (Reag. No. 90d or h) are added to the adsorbent, the barbiturates can be seen as absorbing spots in UV-light. Specific spray reagents can also be applied. The behavior of numerous barbiturates and other hypnotics has been described by authors [*4*] who used chloroform-acetone (90 + 10) on Silica Gel G layers (see Table 67) see also section III, p. 326.

The separation shown in Fig. 37, p. 43 of three known bromine-containing soporifics, Acetylcarbromal, Carbromal and Bromisoval, were achieved on Silica Gel G layers.

Method: About 30 μg of a single substance, dissolved in an appropriate solvent, are applied to silica gel layers containing luminescent material (p. 54). To remove pyridine, the plate is dried after separation, for 15 mins. at 150° C. The soporifics are then seen as absorbing spots in UV-light (254 mμ), they are particularly sharp after treatment with chlorine and spraying with toluidine-potassium iodide solution (Reag. No. 32).

Table 66. h*Rf-Values of Barbiturates* [*42*]

Barbiturate	*Rf* × 100
Barbital	43
Allobarbital	44
Cyclobarbital	45
Methylphenobarbital	51
Aprobarbital	52
Phenobarbital	55
Amobarbital	58
Buthalital-Natrium	67
Methitural	68
Methaqualone-hydrochlorid (not barbiturate)	90

Solvent: Cyclohexane-isopropanol-25% ammonia solution (25 + 65 + 10).
Adsorption-layer: Silica Gel G + Alumina G (1 + 1).

For chromatography of several barbiturates and "non-barbiturates" FRAHM et al. [*13*] recommend TLC on Silica Gel G with isopropanol-25% aqueous ammonia-chloroform (45 + 10 + 45). As a spray-reagent, 1% mercurous nitrate solution is used. *Rf*-values were not reported.

Table 67. *TLC of Barbiturates and other Hypnotics* [*4*]

Barbiturate	*Rf* × 100	Color of spot with Hg [*1*] nitrate
Esanin[1]	26—28	light grey
Methyprylon	34—36	white
Thalidomide	44—46	white (yellow in UV-light)
Barbital	48—50	black
Phenobarbital	53—55	white
Cyclobarbital	54—56	white
Heptabarbital	54—56	white
Amobarbital	55—57	black
Aprobarbital	56—58	black
Dihydroprylone	57—59	black
Butalbital	59—61	black
Pentobarbital	62—64	black
Allobarbital	62—64	black
Hydantal	67—69	white
Secobarbital	69—71	black
Gluthethimide	77—79	black
Hexobarbital	84—86	black

Solvent: Chloroform-acetone (90 + 10).
Adsorbent: Silica Gel G, coarser particles removed with fine meshed sieve of Pharm. Helv. otherwise prepared by standard method (pp. 7—9).
Spray-reagents: 1. After spraying with fluorescein (Reag. No. 90c) all the substances may be seen as adsorbing spots under UV-light (254 mμ). 2. Hg(1)-nitrate solution (Reag. No. 127).

[1] The preparation Esanin contains 3 active substances of which only one was identified.

6. Local anaesthetics

To date, only preliminary investigations have been undertaken [*14*]. It was found that local anaesthetics gave very well defined spots on strongly alkaline Silica Gel G layers (prepared with 0.5 N-NaOH instead of water) with the neutral solvents given in Table 68. Benzocaine could be separated from Procaine and Tetracaine. If the plates are prepared with the addition of 2% luminiscent material (p. 54), the substances may be recognized after separation as dark spots in UV-light (254 mμ). For staining, dimethylaminobenzaldehyde – or Dragendorff – reagents (Nos. 48, 60a) are suitable; see p. 327.

Table 68. h*Rf*-*Values of some Local Anaesthetics* [*14*] *on Alkaline Layers*

Local anaesthetic	*Rf* × 100 in the solvent[1]:	
	Chloroform-Methanol (80 + 10)	Cyclohexane-Chloroform-Methanol (50 + 30 + 10)
Procaine (Novocaine)	60	32
Tetracaine (Pantocaine)	65	32
Benzocaine (Anaesthesin)	74	44

[1] *Plate:* 25 g Silica Gel G and 0.5 g luminescent material were mixed with 50 ml 0.5 N sodium hydroxide and 5 plates prepared from this slurry (p. 37). The layers were dried for 2 hours at 120° C.

7. Thyreostatics

2-Thiouracil, 3.5-diiodo-L-tyrosine, DL-thyroxin and 2-mercapto-benzimidazole have been chromatographed with the solvents listed in Table 69 [*14*]. The substances could be distinguished by comparing their

Table 69. h*Rf*-*Values of some Thyreostatics in various Solvents*

Thyreostatics	*Rf*-values × 100 in the solvent:		
	I Butanol-Methanol-5N NH_3 (60 + 20 + 20)	II Chloroform-Acetone-Butanol-Methanol-5 N NH_3 (20 + 12 + 20 + + 20 + 18)	III Cyclohexane-Acetone-Pyridine (40 + 50 + 10)
3,5-Diiodo-L-tyrosine (Dan. Pharm. 48, dextrorotatory in N-HCl)	37	41	2
2-Thiouracil	46	50	53
Triiodothyronine	50[1]	(—)	(—)
DL-Thyroxine (3,5,3′,5′-tetraiodo-DL-thyronine	51	51	4
2-Mercaptobenzimidazole	72	88	70
Diiodothyronine	75[1]	(—)	(—)

Plate: Silica Gel G with [*14*] and without [*42*] added luminescent material.
[1] Values according to WALDI [*42*].
(—) Not evaluated.

behavior in two solvents (I and I or II and III). The best spot pattern was obtained with solvent (II). Methods of detection are given in Table 70.

Table 70. *Detection of some Thyreostatics* [*14*]

	UV 254 mμ Reag.No.90d	DRAGENDORFF Reag. No. 60a	Rhodamine B Reag. No. 129	Iodine platinate Reag. No. 76
2-Thiouracil	+	—	—	+ (white)
2-Mercaptobenzimidazole	+	+ (orange)	+ (blue)	+ (white)
3,5-Diiodo-L-thyrosine	+	—	—	—
DL-Thyroxine	+	—	—	—

2-Thiouracil, 3.5 diiodo-thyroxine and DL-thyroxine were dissolved in ammonia, 2-mercaptobenzimidazole in methanol. Volumes containing about 50 μg substance were applied.

The behavior of triiodo-thyronine and diiodo-thyronine in solvent (I) is also known [*42*]. These substances were detected with diazotized sulphanilic acid (Reag. No. 37).

8. Sympathomimetic drugs

Adrenalin and noradrenalin can be chromatographed with aqueous ethanol (70%). Sodium bisulphite is added to eliminate possible oxidative changes during chromatography. After spraying with iodine solution (Reag. No. 73) as little as 0.005 μg of both substances may be recognized in UV-light.

Method: 25 g silica gel is slurried with 50 ml Sörensen buffer solution (pH 6.8) with the addition of 500 mg sodium bisulphite, and spread. After predrying with a heat-ventilator, the plates are dried in a drying-cupboard at 110° C.

After acetylation both substances can also be separated on Silica Gel G layers with a neutral solvent [*42*] (p. 342).

II. Commercial preparations

Thin-layer chromatographic procedures for the clinical laboratory have been described by LIEBICH [*21*]. He emphasized not so much complete separation of the given substances as the rapid identification of common preparations and mixtures. Several chromatograms have been prepared with a number of simple solvents and evaluated with appropriate sprayreagents. For instance, descriptions were given of the detection of acetylsalicylic acid, Ethenzamidum, ethyl p-aminobenzoate, Phenazone, acetylcholine hydrochloride, Amidopyrine, caffeine, codeine phosphate, quinine sulphate, ephedrine, sodium Barbital, narcotine hydrochloride, Phenobarbital and salicylamide. Silica Gel G layers prepared by the standard method (pp. 7—9) were used for the separations.

For the analysis of medicaments of known composition a clear separation of all the components to be determined is required. A number of preliminary experiments are necessary for complicated mixtures, for instance nose-drops, as shown in Table 71, or vitamins in a compound vitamin preparation as listed in the same table. Each trial chromatogram is simply copied on transparent paper as described on p. 43; it usually becomes possible to find suitable conditions for separation by comparing copies of chromatograms made from a fairly large number of chromatograms. An important preliminary is the choice of a suitable solvent for separating the active substances from the base. If possible, a neutral solvent or solvent-mixture of low boiling point should be used to avoid changes in the materials to be separated as they are extracted or transferred to the thin-layer plate. Auxiliary substances also extracted can interfere with the chromatographic separation. This should also be remembered when choosing a solvent. NUSSBAUMER [*29*] discussed this point in detail using the example of a number of penicillin preparations.

With regard to the *amount of substance* to be separated, the following may be noted: as the formulation of the preparation is known, enough material must be applied to ensure that the least easily-recognized component is identifiable with certainty after separation. The amounts of the other components of the preparation are thus fixed. If the ratio of the amounts is less than 1:10, it is advisable to apply several spots of varying concentrations. The volume of solvent, in which the active substances should be dissolved, should be about 0.002–0.010 ml.

Table 71 shows the separation of various pharmaceutical products, carried out in accordance with the foregoing remarks. While the compositions of the *solvents* should be rigorously adhered to, the instruction for identification should be treated only as a guide.

The following notes on the nature of pharmaceutical preparations and methods of extraction might be of value:

For G: Ilvico-juice "Merck"[1]

Composition:

Codeine phosphate	10	mg
Ilvin (1-2′-pyridyl-1-(p-bromophenyl)-3-dimethylaminopropane)	3	mg
N-Methylephedrine hydrochloride	7.5	mg
Cebion (ascorbic acid)	50	mg
Phenazone	150	mg
Sodium salicylicate in 5 ml juice	50	mg

For E: This combination occurred in the form of suppositories with Witepsol H as foundation. To extract the active substance, the suppositories are melted

[1] In this method, determination of the codeine phosphate, ilvin and N-methylephedrine contents should be done after alkaline treatment. Thin-layer chromatography is carried out on silica gel layers and a comparison made with test substances on the same plate (pp. 47—48). A portion of the Phenazone accompanies these into the organic phase during distribution and is qualitatively detectable alongside the components to be quantitatively estimated. The quantitative determination of Phenazone as well as of the remaining components, ascorbic acid and sodium salicylate, may be done by direct analysis or colorimetry.

Table 71. *TLC-Separation of Combinations of Medical Preparations*

	Combination of medical preparations	$Rf \times 100$	Identification	Coating material	Solvent		Literature and — as far as is known — the form of medicine from which the active ingredients may be isolated
A	Coffeine Benzoic acid	25 77	UV 254 mμ and Reag. 30 UV 254 mμ and Ferric chloride peroxide Reag. 118	Silica Gel G, Merck + 2% Luminescent material ZS-Super "Riedel de Haen"	Chloroform Cyclohexane Acetic acid	80 20 10	[*14*] ampoules coffeine-sodium-benzoate ampouls may be chromatographed in the same way
B	β-Diphenylhydramine hydrochloride 8-Chlorotheophyllin	40 80	UV 254 mμ or Dragendorff reagent, No. 60a	Silica Gel G Merck, + 2% ZS-Super "Riedel de Haen"	n-Butanol Acetic acid Water	60 15 15	[*14*] Tablets
C	Nicotinic acid-3-butoxy-ethyl ester Nonylic acid vanillylamide	64 93	Bromocyanide, Reag. No. 88 "Genuine Blue B," Reag. No. 61	Silica Gel G, Merck	Acetic acid Butanol	10 90	[*15*] Liniment
D	Coffeine Ilvine Propylphenazone	30 55 75	Rhodamine B, Reag. No. 129 or iodoplatinate, Reag. No. 76	Silica Gel G, Merck	Chloroform Cyclohexane Diethylamine	60 30 10	[*42*]
E	Theophylline Papaverine Phenobarbital	17 50 78	UV 254 mμ	Silica Gel G, Merck + 2% ZS-Super "Riedel de Haen"	Benzene Ethanol Acetic acid	70 12 5	[*15*] Suppositories

Table 71. (Continued)

	Combination of medical preparations	$Rf\times 100$	Identification	Coating material	Solvent		Literature and — as far as is known the form of medicine from which the active ingredients may be isolated
F	Paracetamol	23	Ferric-chloride potassium ferricyanide, see p. 312	Silica Gel G, Merck	Diethylamine	5	[*14*] Suppositories
					Methyl ethyl ketone	85	
	Coffeine	39	Reag. No. 30				
	Amidopyrin	57	Ferric-chloride potassium ferricyanide, see p. 312				
G	Bromopheniramine maleate	29	Iodine solution, Reag. No. 73 and iodoplatinate, Reag. No. 76, sprayed in alternation	Silica Gel G, Merck	Methanol	30	[*26*]
					Acetate buffer, pH 4.6	70	
	Codeine	40					
	N-Methyl ephedrine	51					
	Phenazone	71					
H	Chloramphenicol	5	Rhodamine B, Reag. No. 129 or iodoplatinate, Reag. No. 76	Silica Gel G, Merck	Diethylamine	10	[*42*]
	Hydrocortisone	15			Chloroform	90	
	Naphazolin nitrate	35					
	Ilvin	60					
I	Coffeine	16	Reag. 30	Silica Gel G, Merck + 2% ZS-Super "Riedel de Haen"	Methanol	1	[*14, 15*] Tablets (As little as 2 μg free salicylic acid can be recognized)
	Phenacetin	45	Ferric-chloride/potassium ferricyanide, see p. 312		Acetic acid	18	
					Ether	60	
					Benzene	120	
	Salicylamide	57	Ferric-chloride, Reag. No. 62				
	Acetyl salicylic acid	75					
	Salicylic acid	85					

K	Acetyl salicyclic acid	9	Lum. mat. and fluorescence-indicators, Reag. No. 90 or Rhodamine B, Reag. No. 129	Alumina G, Merck	Chloroform		[*42*]
	Ephedrine	11					
	Codeine	36					
	Phenazone	46					
	Coffeine	61					
L	Bromisoval	15	UV_1 254 mμ	Alumina G, Merck + 2% SZ-Super "Riedel de Haen"	Chloroform		[*14*] Tablets (dividing distance 12 cm)
	Acetcarbromal	28					
	Persedon	44					
	Phenacetin	74					
	Amidopyrine	83					
M	Vitamin B_1	0	Iodoplatinate, Reag. No. 76, or "Genuine blue" B, Reag. No. 61	Silica Gel G, Merck	Acetic acid Acetone Methanol Benzene	5 5 20 70	[*16*] Tablets see also p. 235
	Vitamin B_6	15	Dibromoquinonechlorimide, Reag. No. 40				
	Vitamin C	30	Iodoplatinate, Reag. No. 76				
	Vitamin B_2	35	UV 366 mμ, yellow fluorescence				
	Calcium pantothenate	57	After thermal degradation: Ninhydrine, Reag. No. 108				
	Nicotinamide	65	Bromocyanide, Reag. No. 88				
	Biotin	90	Iodoplatinate, Reag. No. 76				

by heating in a mixture of water and methanol (20 + 80) and the uvula-mass is frozen out in a refrigerator and centrifuged.

For C: The active substances occurred in a liniment. An appropriate amount of the liniment was mixed to a paste with methanol and transferred to the plate.

For M: The distance of migration of the solvent must be about 19 cm, if a sharp separation is to be obtained. To be sure of unambiguous identification of all the components, two chromatograms should be prepared. Chromatography is carried out in the dark. For further details of the method and other means of separating water-soluble vitamins, see pp. 235—247.

III. Use of TLC in toxicological investigations

Machata [23] has described the separation of various medicinal products for toxicological and clinical purposes. Alkaloids must be present as free bases and acidic poisons as free acids. The "extraction" method of Stas-Otto was used before chromatography. Silica Gel G layers were prepared with the help of an applicator illustrated in the original article. The two solvents used and the substances investigated are listed in Table 72.

Table 72. h*Rf-Values for the Separation of various Medical Preparations in Toxicological Investigations* [23]

Substances	Approx. *Rf* × 100 in methanol	*Substances*	Approx. *Rf* × 100 in chloroform-ether (85 + 15)
Hydromorphon hydrochloride	14	Papaverine	73
		Salicylic acid	9
Metamphetamin hydrochloride	16	Theophylline	20
		Phenazone	30
Methadon	18	Amidopyrine	36
Morphine	23	Coffeine	38
Codeine	26	Theobromine	40
Dolantin, Pethidine	36	Phenobarbital	42
Phenmetrazin hydrochloride	38	Phenacetin	45
		Barbital	48
Nicotin	50	Cyclobarbital	48
Phenazone	67	Aprobarbital	50
Amidopyrine	67	Allobarbital	54
Noscapine	69	Carbromal	67

Method: 30 g of Silica Gel G was stirred into 60 ml water and spread onto glass plates with the Stahl applicator. The layer-thickness was wet, 0.27, dry 0.25 mm. The plates were activated by drying at 100° C for 30 min. After applying the spots, the basic materials were separated with methanol, the acidic extracts extracted by the Stas-Otto procedure with chloroform-ether (85 + 15). For chromatographing complicated mixtures, a separation distance of 14 cm was sufficient.

The spray reagents used were: 1. Dragendorff-reagent No. 60a; 2. acetic acid-iodine-potassium iodide solution corresponding to No. 73 (for alkaloids); 3. acetic-acid-potassium permanganate solution No. 86 (for reducing substances); 4. fluorescein-sodium solution No. 90c (to detect UV-absorbing substances); 5. acetic acid-iron chloride solution, corresponding to No. 62 (for pyrazalones); 6. Zwicker's reagent for barbiturates, No. 87.

As an example, the detection of opium alkaloids in stomach contents is given (see also Alkaloids, p. 286). By the same method, and using chloroform-ether (85 + 15) as solvent, MACHATA and KISSER [*24*] fractionated hydantoin derivatives (Phenylethyl hydantoin h*Rf* 15, diphenyl hydantoin *Rf* 25, and N-methylphenyl ethyl hydantoin h*Rf* 45).

BÄUMLER and RIPPSTEIN [*4*] described the TLC of alkaloids and basic medicaments (Table 73), opium alkaloids and synthetic substitutes (Table 74). In both cases, the solvents were methanol-acetone-triethanolamine (50 + 50 + 1.5).

(Tartaric acid and alkaline extracts were prepared from the various specimen with ether. If morphine derivatives were suspected, ammonia was added after the acidic ether-precipitation and drawn off after having been refluxed for about 30 min. with chloroform. The remainder, after evaporation of the ether or chloroform, was dissolved in 1 ml methanol or methylene chloride and spots (1/1000, 1/100, 1/20) applied to the starting points.)

Table 73. *TLC of Alkaloids and Basic Medical Preparations* [*4*]

Substances	*Rf* × 100	Detection and color		
		UV-light	DRAGENDORFF (Reag. No. 60a)	$FeCl_3$-solution Reag. No. 62
Atropine	15—19	—	violet	
Brucine	16—18	—	orange	
Strychnine	18—20	dark	orange	
Quinine	55—57	luminous light blue	red	
Methylphenidare hydrochloride	55—57	—	red	
Nicotine	56—58	dark	red-violet	
Novocaine	58—60	dark	orange	
Cocaine	60—62	dark	orange	
Procaine	60—62	—	red	
Oxybuprocaine	60—62	blue	orange	
Coffeine	68—70	—	—	
Phenazone	70—72	dark	orange	orange
Amidopyrine	74—76	dark	violet	red-violet
Propylphenazone	83—85	dark	red	pale yellow
Reserpine	85—87	luminous green	yellow	

Plate: Silica Gel G. Coarse particles removed with fine-meshed sieve of Pharm. Helv., otherwise prepared by standard method.
Separation distances: 10 cm.
Solvent: Methanol-acetone-triethanolamine (50 + 50 + 1.5).

BÄUMLER also described various examples of the practical application of TLC. For instance, in a case of suicide, a hypnotic (Glutethimide) was found in the stomach fluid after precipitation with sulphuric acid-ether. In the urine of a drug addict, TLC showed Butalbital in the acidic ether extract and Amidopyrine in the alkaline ether extract, i.e., two components of the preparation Optalidon. In testing cachets for opium, an alcoholic extract was compared by thin-layer chromatography using an alcoholic extract of *Pulvis opii* as well as papaverin, narcotine, codeine, thebaine, morphine and ephedrine. The content of the cachet could be recognized as *Pulvis opii et ipecacuanhae* with ephedrine. TLC has also been used by other authors to detect narcotics in the urine and in tablets [*41*].

Table 74. *TLC of Opium Alkaloids and Synthetic Substitutes* [*4*]

Substances	*Rf* × 100	Detection and Color	
		UV-light	DRAGENDORFF (Reag. No. 60a)
Narceine	22—24	blue	violet
Dextromethorphane	22—24	—	red
Laevorphanol	27—29	—	red
Hydromorphone hydrochloride	27—29	dark	yellow
Hydrocodone	28—30	—	orange
Thebacone	29—33	yellowish	orange
Codetyline	36—38	—	red
Morphine	39—41	dark	orange
Thebaine	40—42	dark	red
Codeine	42—44	—	red
Methadone (Polamidon)	47—49	—	red
Ketomebidone	55—57	dark	orange
Pethidine	55—57	—	red
Noceapine	81—83	luminous blue	orange
Papaverine	81—83	luminous yellow	orange
Dextromoramide	86—88	—	orange

Plate: Silica Gel G. Coarse particles removed with fine-meshed sieve of Pharm. Helv., otherwise prepared by standard method.
Separation distance: 10 cm.
Solvent: Methanol-acetone-triethanolamine (50 + 50 + 1.5).

VIDIC [*39*] separated Dextromoramide (2,2-diphenyl-3-methyl-4 morpholino-butyryl-pyrrolidine) by various paper-chromatographic methods from other narcotics and in addition, he used TLC on silica gel (see Table 75).

Table 75. *TLC of Dextromoramide and some Narcotics* [*39*]

Substances	*Rf* × 100
Dextrometorphane	26
Methadone	42
Normethadone	59
Dextromoramide	85

Plate: Silica Gel G prepared by the standard method (pp. 7—9).
Solvent: 0.1 N-Methanolic ammonia solution.

VIDIC and SCHÜTTE [*40*] used TLC to complement a paper chromatographic analysis worked out for over 70 toxicologically important basic poisons and medicinal products. Effective pre-separation is possible by an extraction and re-extraction method. The final analysis may be carried out on very small samples. Recently, COCHIN and DALY [*7a*] recommended silica gel layers and 6 solvents for rapid identification of 18 various analgetics and epileptic drugs in urine.

TLC of insecticides: see p. 359.

IV. Stability tests for pharmaceutical substances by thin-layer chromatography

In the development of a drug, a knowledge of its stability is of decisive importance. To formulate the carrier materials with the desired properties, it is important to know the behavior of the effective substance at various pH-values and also its susceptibility to light, etc. The substances are dissolved in suitable solvents and exposed to the various conditions. After a definite time, specimens are removed which may then be quickly examined by TLC, both qualitatively and quantitatively, using the correct experimental procedure (pp. 44—57). In this way, it is possible to assess the kinetics of a reaction involving any particular change. As other factors are involved in the final form of the preparation, however, which make it more difficult to predict how long it will remain stable, it is also necessary to test the preparation at various temperatures encountered during storage. The medicament itself, or an extract therefrom, may be chromatographed and compared with the results from the test on the original substance.

1. Stability tests for preparations containing nicotinic acid esters

Pyridine derivatives may be easily detected with chloro- or bromocyanide which form glutaconaldehyde, which can be visualized by condensation with primary aromatic amines to form colored Schiff bases.

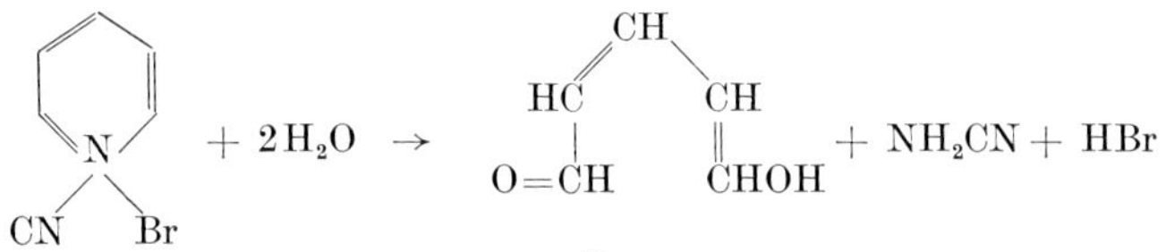

I.

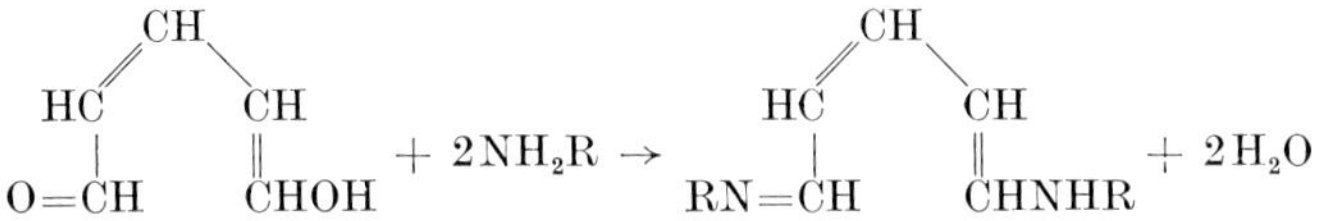

II.

This color reaction is used in combination with thin-layer chromatography to determine the stability of nicotinic acid esters and nicotinic acid amides in pharmaceutical preparations. For example, the stability of nicotinic acid-3-butoxy ethylester in ointments and liniments has been investigated. In the unchanged ester, there is some free nicotinic acid produced by saponification and this is separated by thin-layer chromatography. Both are then detected by the color-reaction described above [*15*].

Method: The thin-layer plates are prepared in the usual manner (pp. 7—9). Ointments are made into a paste with acetone, and liniments with methanol, filtered, and a sufficient quantity of extract applied to a Silica Gel G layer in amounts of about 100 μg of substance.

Solvent: n-Butanol-acetic acid (10 + 1). Separation distance 10 cm. The plates are exposed to bromocyanide vapor and sprayed with 0.05% benzidine solution in 2 N acetic acid. Nicotinic acid-3-butoxy ethylester stains red, nicotinic acid, brown.

2. Stability tests of pharmaceutical preparations containing phenol esters

An interesting example of a stability test of esters is that of preparations containing (4,4′diacetoxy diphenyl)-(pyridyl-2)-methane (Bisacodyl), in which the half-ester and the deacetylated compound cound be separated from the non-saponified substance by TLC.

Method: From acetone or ethanol extracts of the preparations, an aliquot corresponding to about 100 μg Bisacodyl is transferred to a thin-layer plate containing luminescent material (p. 54).

Solvent: Xylene-methylethylketone (1 + 1). Solvent front migration: 10 cm. Evaluation in UV-light (254 mμ). Partially saponified Bisacodyl shows 3 spots, whereas Bisacodyl itself and the completely de-acetylated compound each yield a single spot. As a control, a spot of the extract is sprayed with concentrated ammonia and the plate quickly heated to 100° C. This causes saponification and the test for stability may be carried out using this as reference standard instead of the partially or completely de-acetylated compound.

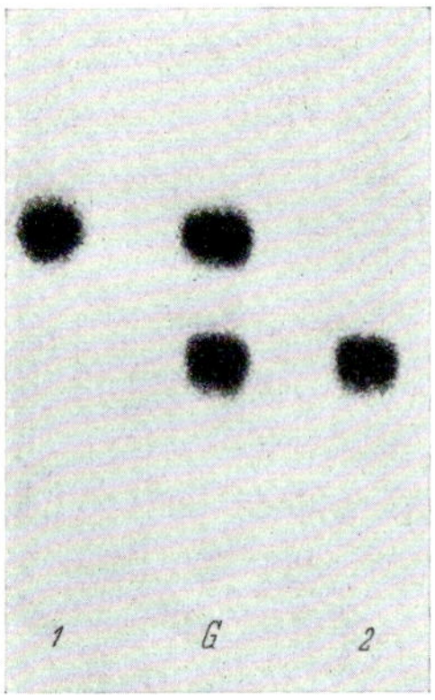

Fig. 138. Separation of hydrocortisone-alcohol and its acetate [*14*] in ethyl acetate-xylene-methanol (90 + 5 + 5) on a Silica Gel G layer, 500 μ. *1* Hydrocortisone acetate, *2* Hydrocortisone alcohol, *G* mixture

3. Preparation containing steroid esters

Using the solvent mixture, ethyl acetate-xylene-methanol (90 + 5 + 5), corticosteroid alcohols can be separated from their 21-esters (Fig. 138). The separation works better with Silica Gel G layers 500 μ thick. If the alcohol arising from saponification exceeds 10%, quantitative estimation may be carried out in UV (p. 51) or after coloring with tetrazolium compounds. This procedure is of particular interest as all other known methods have been time-consuming and open to errors. Cope [*8*], for instance, has discussed the difficulties arising from paper-chromatographic separation and subsequent colorimetric determination. The method of visual comparison (pp. 47—48) give very good results if the alcohol arising from saponification is under 10%.

Method: Extract the medicament to be tested with ethanol. For coating 5 plates (20 × 20 cm) of about 500 μ layer thickness, mix 50 g Silica Gel G with 1 g luminous material ZS-Super, Riedel de Haen. Stir evenly into 104 ml water and prepare a 500 μ thick adsorption layer using Stahl's adjustable thin-layer applicator (p. 7). When the silica gel layers are well set, dry for 1 hr. at 120° C and keep for at least 4 hours over KOH before use. The solvent, ethyl acetate-xylene-methanol (90 + 5 + 5), is used for developing in a "saturated chamber" to a height of 14 cm. The spots may be seen in UV-light and quantitatively determined (p. 51).

Separation is also possible with other solvents [*5*].

4. Stability test for Δ^4-17β-hydroxyestrone derivatives

Preparations in tablet form containing steroids with a basic formula I (4-estrone-17 β-ol), in which R represents an alkyl radical (allyl-, propyl-, ethyl-, or ethinyl) have been investigated by thin-layer chromatography on Silica Gel G after various storage conditions. The plates were prepared by the standard method [*12*].

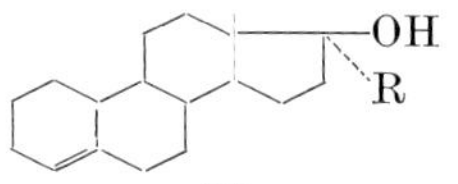

(I)

Heptane-acetone (90 + 10) was used as solvent. It can be plainly seen from the photographs of the chromatograms, which accompany the report, that after unfavorable storage conditions, numerous degradation products occur. The course of degradation is markedly dependent on the 17-alkyl group and can be influenced by addition of antioxidants.

5. Stability test for preparations containing nicotinamide

The stability of nicotinamide in various pharmaceutical preparations may be tested in a manner similar to that used for nicotinic acid esters (Combionta sweets, Nicobion ampoules, Multibionta drops, Hormo-Gerobioni capsules, and Polybion ampoules) [*27*].

The limit of detection of nicotinic acid and nicotinamide is established after TLC separation of test mixtures. In 10 μg of nicotinamide + 0.1 μg free nicotinic acid, the free acid is still detectable.

Method: An amount of extract containing 5—10 μg nicotinamide is applied to a Silica Gel G layer.

Solvent: n-Propanol-aqueous ammonia, 10%, (95 + 5). Detection: Reagent No. 88. Orange-red spot appears, h*Rf* 60—70 (nicotinamide), red spot at h*Rf* 53 (nicotinic acid).

6. Stability test for the neuroleptic Perphenazine

Alcoholic or water solutions of this neuroleptic are sensitive to light. After a long period of exposure to light, the Perphenazine solution contains Perphenazine-sulphoxide and an unknown substance as breakdown products [*26*].

Adsorption layer: Silica Gel G layers are prepared by the standard method (pp. 7—9). Amount applied: about 20 μg of Decentan. During development, the substance is protected against degradation by shielding the chromatography jar with paper.

Solvent: n-Butanol-acid-water (4 + 1 + 1). Separation distance: 14 cm. After chromatography, the preparation is sprayed with p-dimethylaminobenzaldehyde-sulphuric acid containing $FeCl_3$. In the middle of the chromatogram, unaltered Perphenazine may be seen, below it a degradation product not identified further, and near the starting point, Perphenazine-sulphoxide.

7. Test for 1-N-methyl-piperidyl-(4′)-pyrazolones

The sensitivity to oxidation of 1-N-methyl-piperidyl-(4′)-3-phenyl-4-hydroxy-4-benzyl-pyrazol-5-one (I) has been investigated by JUCKER and

LINDEMANN [*19*]. After leaving the solution of (*1*) in 75% ethanol for a long time, or after allowing hydrogen peroxide to act on an alkaline solution of (*1*), a substance may be isolated on layers of aluminium oxide with a mixture of ammoniacal isopropanol-dioxane (50 + 100) as the solvent. From the chemical behavior, the IR-spectra, and by syntheses, it appeares that there is neither N-oxidation nor ring opening. There occurs only hydroxylation in position 4, keeping the original ring structure the same.

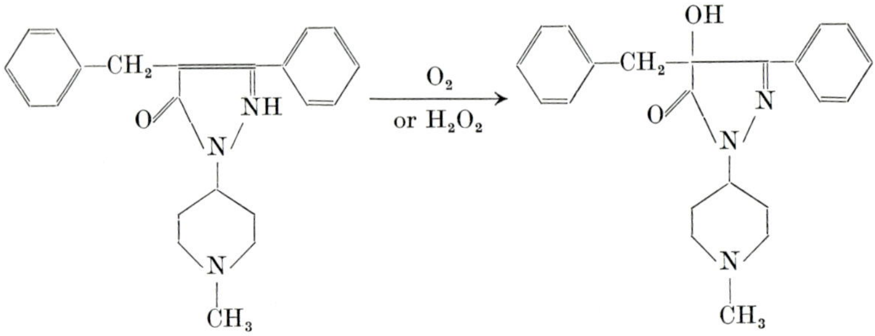

(I) (II)

Substance (II), isolated from the alcoholic solution as well as from the alkaline solution, and also the synthetic compound, exhibit the same thin-layer chromatographic behavior.

8. Breakdown of Chlorodiazepoxide in acidic medium

BÄUMLER and RIPPSTEIN [*3*] could isolate a yellow substance by allowing aqueous hydrochloric acid to act on Chlorodiazepoxide (I). This substance was identified, on the evidence of physico-chemical data, as 2-amino-5-chlorobenzophenone (II).

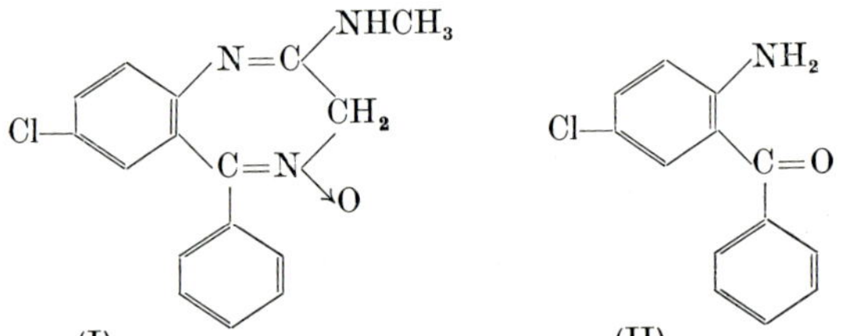

(I) (II)

By TLC on Silica Gel G layers, the following *hRf*-values were obtained:

	Rf × 100 in solvent:	
Methanol-Acetone-Triethanolamine	(10 + 10 + 0.3)	Benzene
Chlorodiazepoxide	85	0
Breakdown product	80	50

The breakdown product reacts neither with Dragendorff-reagent nor with m-dinitrobenzene solution or ferric chloride solution, whereas Chlorodiazepoxide is colored by Dragendorff-reagent (No. 60a). Very small amounts may be recognized after diazotation and coupling with β-naphthol solution. Acid degradation may be used to detect small amounts of Chlorodiazepoxide in urine or blood after acidification and alkaline extraction with ether.

9. Stability of a preparation containing ergot alkaloids (*Guttae secalis* "Stada")

It has been possible to follow, quantitatively, the course of isomerization of alkaloids, during several weeks, with the help of thin-layer chromatography [*20*].

Ergometrin → Ergometrinin
Ergotamine → Ergotaminin
Ergocristin → Ergocristinin

The alkaloid bases were liberated by addition of bicarbonate and precipitated with methylene chloride. After centrifugation, a known volume was applied as bands to a Silica Gel G layer. The alkaloids, made visible by the effect of the solvent on account of their self-luminescence in UV-light, were scraped off, eluted, and evaluated colorimetrically after adding the reagent (see pp. 46 and 290).

Bibliography to Chapter F. Pharmaceutical Products

[*1*] Baehler, Br.: Helv. Chim. Acta **45**, 309 (1962).
[*2*] Bäumler, J.: Praxis **50**, 841 (1961).
[*3*] —, u. S. Rippstein: Helv. Chim. Acta **44**, 2208 (1961).
[*4*] — — Pharm. Acta Helv. **36**, 382 (1961).
[*5*] Beijleveld, W. M.: Pharm. Weekbl. **97**, 190 (1962).
[*6*] Bravo, R. O., u. F. A. Hernández: J. Chromatog. **7**, 60 (1962).
[*7*] Cerri, O., u. G. Maffi: Boll. chim. farm. **100**, 940 (1961).
[*7a*] Cochin, J., u. J. W. Daly: Experientia (Basel) **18**, 294 (1962).
[*8*] Cope, C. L.: Quantitative Paper Chromatography of Steroides, p. 18. Cambridge: University Press 1960.
[*9*] Dhont, J. H., and C. de Rooy: Analyst **86**, 527 (1961).
[*10*] Fiori, A., and M. Marigo: Nature (London) **182**, 943 (1958).
[*11*] Fischer, R., u. H. Lautner: Arch. Pharm. **294/66**, 1 (1961).
[*12*] Fokkens, J., and J. Poldermann: Pharm. Weekbl. **96**, 657 (1961).
[*13*] Frahm, M., A. Gottesleben u. K. Soehring: Arzneimittel-Forsch. **11**, 1008 (1961).
[*14*] Gänshirt, H.: Unpublished.
[*15*] —, u. F. Malzacher: Arch. Pharm. **293/65**, 925 (1960).
[*16*] — — Naturwissenschaften **47**, 279 (1960).
[*17*] Härri, E., W. Loeffler, H. P. Sigg, H. Stähelin, Ch. Stoll, Ch. Tamm u. D. Wiesinger: Helv. Chim. Acta **45**, 839 (1962).
[*18*] Hais, J. M., u. K. Macek: Handbuch der Papierchromatographie I, S. 614. Jena: VEB Gustav Fischer Verlag 1958.
[*19*] Jucker, E., u. A. Lindemann: Helv. Chim. Acta **44**, 1249 (1961).

[20] Klavehn, M., H. Rochelmeyer u. J. Seyfried: Deut. Apotheker-Ztg. **101**, 75 (1961).
[21] Liebich, H.: Deut. Apotheker-Ztg. **99**, 1246 (1959) and **100**, 393 (1960).
[22] Lyman, R. L., A. L. Livingston, E. M. Bickoff and A. N. Booth: J. org. Chem. **23**, 756 (1958).
[23] Machata, G.: Mikrochim. Acta **1960**, 79.
[24] —, u. W. Kisser: Arch. Toxicol. **19**, 327 (1962).
[25] Nicolaus, B. J. R., C. Coronelli u. A. Binaghi: Il Pharmaco, Ed. Pr. **16**, 349 (1961).
[26] Nürnberg, E.: Arch. Pharm. **292/64**, 610 (1959).
[27] — Deut. Apotheker-Ztg. **101**, 142 (1961).
[28] Nussbaumer, P. A.: Pharm. Acta Helv. **37**, 65 (1962).
[29] — Pharm. Acta Helv. **37**, 161 (1952).
[30] Pastuska, G., u. H. Trinks: Chemiker-Ztg. **85**, 535 (1951).
[31] — Z. anal. Chem. **179**, 427 (1961).
[32] —, u. H.-J. Petrowitz: Chemiker-Ztg. **86**, 311 (1962).
[33] Petrowitz, H.-J.: Materialprüfung **2**, 309 (1960).
[34] Ragazzi, E.: Boll. Chim. Farm. **100**, 402 (1961).
[35] Sensi, P., C. Coronelli u. B. J. R. Nicolaus: J. Chromatog. **5**, 519 (1961).
[36] Stahl, E.: Pharm. Rundschau **1**, Nr. 2, 1 (1959).
[37] — Parfümerie u. Kosmetik **39**, 564 (1958).
[38] Teichert, K., E. Mutschler u. H. Rochelmeyer: Deut. Apotheker-Ztg. **100**, 283 (1960).
[39] Vidic, E.: Arch. Toxikol. **19**, 254 (1961).
[40] —, u. J. Schütte: Arch. Pharm. **295**, 342 (1962).
[41] Völksen, W.: Krankenhaus-Apotheker **11**, 5 (1961), Beilage zur Deut. Apotheker-Ztg.
[42] Waldi, D.: Unpublished.
[43] Wollish, E. G., M. Schmall and M. Hawrylyshyn: Anal. Chem. **33**, 1138 (1961).

Some recent noteworthy articles

Cochin, J., and J. W. Daly: J. Pharmacol. Exp. Therap. **139**, 154 (1963): Barbiturates and some nonbarbiturate hypnotics.

Gänshirt, H.: Arch. Pharm. **296**, 73 (1963): Separation and detection of some commonly used drugs.

Halmekoski, J.: Suomen Kem. B **36**, 58 (1963): TLC of sympathomimetica of the adrenaline-type on layers impregnated with complex forming reagents.

—, and H. Hannikainen: Suomen Kem. B **36**, 24 (1963): Structure and R_M-values of homologous phenols.

Karpitschka, N.: Mikrochim. Acta **1963**, 157: Sulfonamides.

Klein, S., and B. T. Khoe: J. Pharm. Sci. **51**, 966 (1962); **52**, 404 (1963): Sulfonamides.

Lehmann, J., and V. Karamustafaoglu: Scand. J. Clin. & Lab. Invest. **14**, 554 (1962): Barbiturates.

Mellinger, J., and C. E. Keeler: J. Pharm. Sci. **51**, 1169 (1962): Phenothiazine drugs.

Reisch, J., H. Bornfleth and J. Rheinbay: Pharm. Ztg. **107**, 920 (1962); **108**, 1182 and 1183 (1963): Sulfonamides and other drugs.

Šaršunová, M.: Pharmazie 18, 748 (1963): TLC of local anaesthetics an loose layers.

Schulze, W.: Diplomarbeit. Technische Hochschule Braunschweig 1962: Antihistamines and phenothiazine drugs.

Thoma, F.: Tuberkulosearzt **16**, 362 (1962): Stability of p-aminosalicylic acid solutions.

G. Thin-Layer Chromatography in Clinical Diagnosis and Pharmacology

By

D. WALDI

I. Introduction

Simple and reliable micromethods for the detection of metabolic changes are very much needed in the medical research laboratory. These methods should be rapid and capable of being carried out in series. Shortly after the first successful separation of amino acid mixtures by paper chromatography, this technique was used extensively in medical laboratories for analyzing such substances. Electrophoresis (ionophoresis), on various layers[1], was subsequently introduced as a valuable diagnostic procedure for separating mixtures of ionic compounds. In particular, it was rapidly adopted as a diagnostic method for analyzing proteins. The mixtures are applied as bands; after separation, they can be evaluated quantitatively with the aid of various self-recording photoelectric instruments.

For the rapid separation of lipids, present in small quantities in all body fluids and in larger amounts in organ extracts, no generally applicable separation technique has been available. TLC has proved invaluable since both lipids and medically important hydrophilic substances such as amino acids, nucleic acids and sugars may be separated more readily and quickly than with other methods.

Several chapters in this volume contain examples of the separation of substances which are of significance in medical diagnosis. The following summary indicates possible applications in the fields of clinical diagnosis and pharmacology. The prerequisite for this, however, is a careful study of the best methods of concentration (e.g. see cholesterol and cholesteryl esters, p. 255) and both qualitative and quantitative comparison with a reasonable number of normal and pathological cases.

Material studied	Possible determinations for
Urine	Fats, pp. 137—181.
Blood (serum).	Vitamins, pp. 210—247.
Faeces	Sterols, steroids, bile acids, pp. 249—276.
	"simple indole derivatives pp. 292—301.
Other body fluids (secretions).	Porphyrin derivatives [*1*]
Organ extracts	Amino acids, proteins, pp. 391—435.
	Nucleic acids, nucleotides, pp. 440—458.
(Microorganisms)	Sugars, pp. 461—468.

This chapter will henceforward be dealing with special TLC methods and standard conditions applicable to diagnosis which are not described

[1] Electrophoresis on Silica Gel G and Kieselguhr G layers provides quicker and better differentiation of single bands than acetate foil [*3*] (see pp. 24—25).

in detail in other chapters. TLC can be used also for studying the metabolism and elimination products of various drugs in the human organism and for the detection of metabolic abnormalities which may be induced by them. References to pertinent work are made here. In view of the very recent application of TLC, this edition can only contain a limited number of recommendations and suggestions.

II. Excretion products in urine

1. Metabolism of steroids in humans (steroid hormones)

a) Observations on the female cycle

In the female monthly cycle, various steroid hormones controlled by the hypophysis are mobilized during the different phases. Fig. 139 illustrates the relationships as known to-day.

Maximum progesterone secretion occurs during menstruation roughly on the 21st day. Progesterone will, at the same time, be metabolized and excreted in the urine as pregnanediol glucuronate and allopregnanediol glucuronate. The day of maximum progesterone formation, its level, and the amount of pregnanediol secreted in urine vary with the individual and are determined by various biological factors. Gynecologists are primarily interested in disturbances which can be recognized by variations in the excretion of pregnanediol. With TLC, pregnanediol and other steroids can be quickly and easily assayed quantitatively during the cycle [*6*].

The test is carried out similarly to the pregnancy test described below. If, during the determination of the menstrual cycle, steroid secretion is checked every four days, a chromatogram like that reproduced in Fig. 140 will be obtained.

Under certain conditions, the determination of pregnanediol can give a useful indication of conception-free days [*4*]. It should be remembered that, while estrogen-gestagen mixtures are prescribed in tablet form as an aid in diagnosing pregnancy, various steroids are also used as contraceptives. In these cases, normal balance in steroid content is disturbed, as can be easily confirmed by TLC. After administration of estrogenic substances and during the treatment with cortisone, we found values for pregnanediol and allopregnanediol similar to those observed in pregnancy; a number of further steroids such as pregnanetriol are also secreted in greater quantity. These substances can be easily differentiated. It should be pointed out, however, that tests of this kind are in their early stages and a great deal remains to be done.

b) Early pregnancy test[1]

Of the many biological and chemical methods of pregnancy detection published over the years, only a few have proved successful in practice.

[1] The chemicals and instruments needed for the menstrual cycle and pregnancy tests are produced by C. DESAGA, Heidelberg, Germany. Details of the method are published by this company in a brochure.

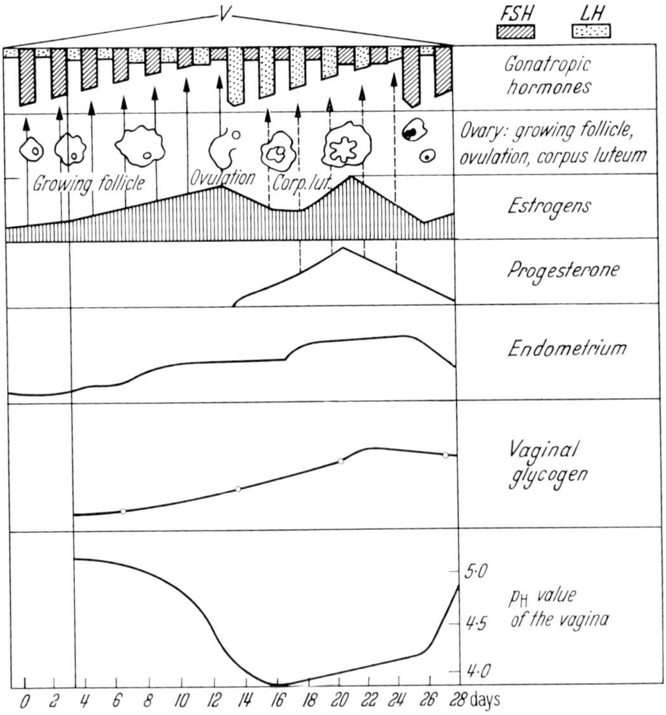

Fig. 139. Physiological changes during the menstrual cycle (H. E. NIEBURGS)

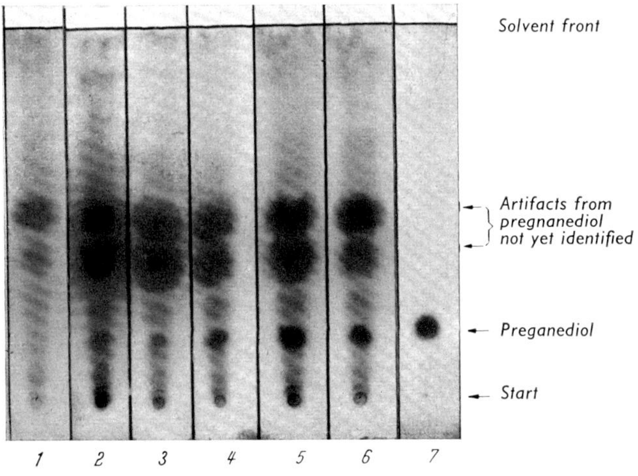

Fig. 140. Detection of abnormally high pregnanediol secretion during the menstrual cycle. Every four days an extract was produced and applied as follows: *1* = 4th day; *2* = 8th day; *3* = 13th day; *4* = 16th day; *5* = 21st day; *6* = 24th day; *7* = pregnanediol standard solution. Adsorbent: Silica Gel G. Solvent: Chloroform-acetone (90+10). Time of run: 30 min. Detection: phosphoric acid reagent No. 123, plus spraying with phosphomolybdic acid reagent No. 129c

The generally-known biological detection techniques of Aschheim-Zondek with mice, Sulmann and Black, and Kuppermann and Greenblatt with rabbits, and Galli-Mainini with toads are all based on the detection of fairly large quantities of gonadotropic hormone. These biological processes require animals whose maintenance is reasonably expensive, and, in addition, the test takes from 2 to 100 hr and the sensitivity can vary considerably. A certain percentage of error has always to be allowed for. The new rapid immunologic test, with 93–95% of certainty, does not detect the present hormones and will be positive only 21 days after conception.

In the early pregnancy test described below, the pregnane-3a,20α-diol and allo-pregnane-3a,20α-diol, which are present in urine in increased quantity during pregnancy, are detected by TLC.

A few days after conception (5–8 days), the pregnanediol value in urine is clearly above the maximum value for the menstruation phase; values rise even further, whereas, usually they fall away again at the end of the cycle. Providing that the experimental details are observed, this test is highly reliable. Accuracy of prediction of 99–100% has been confirmed by tests on about 800 samples.

Method [*5, 6*]: α) Hydrolysis and extraction from urine

25 ml urine is sufficient for a routine examination, although 50 or 100 ml are recommended for initial tests. 100 ml filtered urine is mixed in a 500 ml round-bottom flask with 10 ml concentrated hydrochloric acid (or correspondingly less with smaller test quantities) and boiled for 20 min under reflux in a fume cupboard after adding boiling chips. When cooled, the hydrolyzate[1] is extracted three times with 80–100 ml cyclohexane in a separatory funnel of requisite size. The combined cyclohexane extracts are extracted twice with 100 ml 10–15% sodium hydroxide solution and washed with water. The sodium hydroxide solution and washing water are discarded. The cyclohexane extract thus purified is finally evaporated to dryness in an evaporating dish on a water bath. A few ml of chloroform are added and concentrated to 0.3 ml in a tube of approx. 3 ml capacity (correspondingly less with smaller quantities).

β) TLC of extracts

40 mm^3 (4 × 10 mm^3) of the extracts and corresponding standard solutions are applied to Silica Gel G layers prepared by the standard method. A mixture of chloroform-acetone (90 + 10) is used as developing solvent. The chamber is saturated (p. 15). Length of run is 10 cm, and duration of run approx. 25 min.

The method can be carried out by applying the following samples to the TLC plates:

[1] Hydrolysis with strong acids will destroy approx. 75% of the original pregnanediol. Attempts to find less drastic hydrolysis conditions are in progress. If successful, it will be possible to use smaller samples of urine (*6*).

1. 1 mm^3 pregnanediol standard solution[1] = 1 μg (5′ Agla syringe)
2. 30 mm^3 extract of urine to be tested (corresponding to 10 ml urine)
3. 40 mm^3 urine extract from a woman known to be pregnant.
4. 50 mm^3 extract of urine to be tested.
5. 2 mm^3 pregnanediol standard solution = 2 μg (10′)
6. 40 mm^3 extract of urine to be tested.
7. 3 mm^3 pregnanediol standard solution = 3 μg (15′)
8. 40 mm^3 urine extract from a woman known to be pregnant.
9. 40 mm^3 normal urine extract from a non-pregnant woman.

In routine examinations, 20 different extracts of 40 mm^3, each, can be applied at the same time, as well as 2 and 3 mm^3 of the standard solution, twice.

γ) Recognition and evaluation

The chief steroids are immediately recognizable as fat-like spots when sprayed with phosphoric acid (Reagent No. 123). Spraying must be applied uniformly until the layer is transparent. Pregnanediol and allopregnanediol are situated in the lower part of the chromatogram. For *hRf* values, see Table 42, No. 27 and 28, III, p. 262. Allopregnanediol gives a clear ivory-colored spot, though, hitherto, this spot has only been found in menstruation, when hormone treatment is involved, and in pathological cases (inflammatory adnexa).

After spraying with phosphoric acid, the plates are dried by heating at approx. 110° C. After 10 min a length-wise half of the chromatogram may be sprayed with phosphomolybdic acid (Reagent No. 120) and heated to the same temperature for another 2–3 min. This reaction is so sensitive that the pregnanediol is visible in almost every extract. Evaluation of the test depends essentially on the amount of pregnanediol separated, and it is important that any allopregnanediol present should also be detected using phosphoric acid as indicator. The extracts not sprayed with phosphomolybdic acid reagent exhibit the pale grey-green fluorescent color typical of pregnanediol in long-wave UV-light, when heated further at 100° C (for a total of 25–30 min). This occurs only if this steroid is present in sufficient quantity as a result of pregnancy or hormone treatment. Spraying a plate with vanillin-sulphuric acid (Reagent No. 151) and heating for 7–10 min at 110° C will make it possible to evaluate the pregnanediol content equally well, if a standard solution is present. The colors on the chromatogram can be preserved longer if it is covered with another glass plate (see colors of steroids, p. 267).

c) Further methods for detection of steroids (Adrenogenital syndrome, Cushing syndrome)

The appearance of rather large quantities of pregnanetriol and 17-ketosteroids in conjunction with the adrenogenital syndrome is mentioned on p. 269 in the chapter on steroids. Increases in androgens are found where one or more enzymes are missing (Block). Cushing syndrome, caused by the hyperactivity of the suprarenal cortex, leads to increased formation of corticosteroids.

[1] A 0.1% standard solution (E. Merck) is included in the group of reagents for cycle tests. See footnote to p. 336.

2. Metabolites of drugs

The study of the metabolism of drugs in the healthy and diseased organism is of the greatest importance. TLC has been of considerable help in the detection of metabolites. Two examples will illustrate this:

a) WAGNER [8] has reported tests made by BICKEL and VUILLEUMIER on the determination of the metabolites of Medomine (see formula in Fig. 141). After administration of labelled Medomine, the total activity of the compound can be detected in the urine. Paper radiograms of urine extracts show four radioactive zones, i.e., four metabolites. Metabolite I is separated on Silica Gel G layers into two equal zones (Ia = *hRf* 85 and Ib = *hRf* 73) with a solvent consisting of chloroform-glacial acetic acid (98.5 + 1.5). Rather large amounts of extract are then chromatographed on a column (silica gel, approx. 100 mesh, ether + 2% glacial acetic acid) and the fractions tested by TLC for homogeneity. This analysis shows that metabolite Ia is 5-ethyl-5-(3′-oxo-Δ^1-cycloheptenyl)-barbituric acid (Fig. 141). In metabolite Ib, the keto-group is reduced to alcohol.

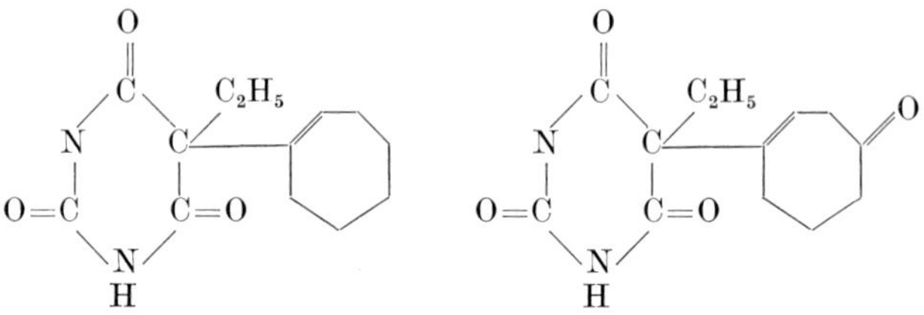

Medomine Metabolite Ia

Fig. 141. Structure of Medomine and one metabolite

b) If dihydrogeraniol labeled with C^{14} is fed to rats, four metabolites are found on a thin-layer chromatogram of their urine after acid hydrolysis and extraction with ether. Two of these compounds are dicarboxylic acids (see formulae).

Dihydrogeraniol:

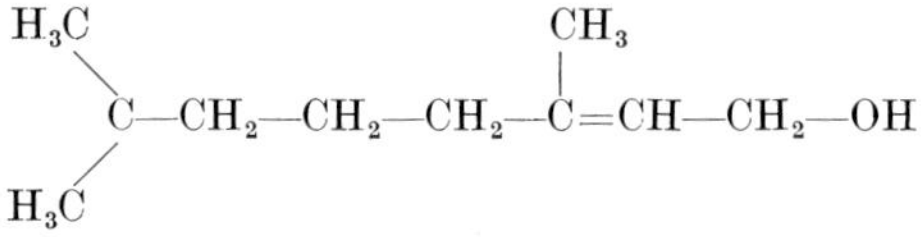

Dihydrohildebrandt acid:

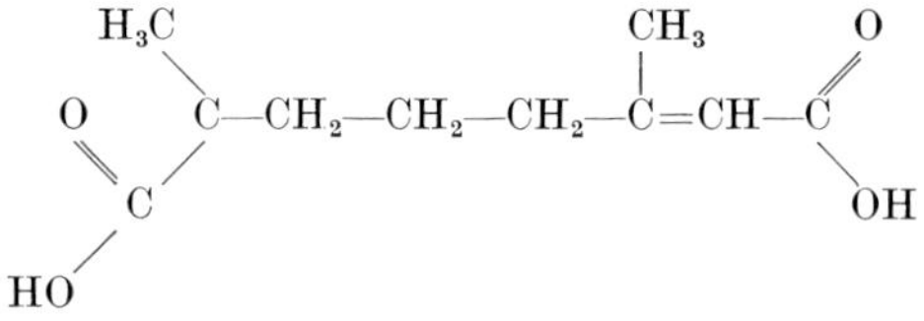

Hildebrandt acid:

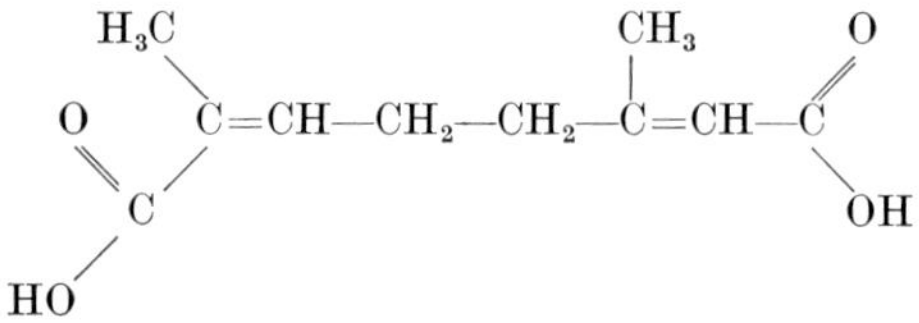

Fig. 142. Dihydrogeraniol and two metabolites found in urine

III. Lipids in faeces and in faecaliths

In a study by WILLIAMS, SHARMA, MORRIS and HOLMAN [*9*], the lipid composition of faeces is compared with that of faecaliths from the caecum.

Chloroform-methanol (2 + 1) was used for extraction and the lipid extract was developed on Silica Gel G layers with the solvent petroleum hydrocarbon-ether-glacial acetic acid (80 + 20 + 1). 2′,7′-Dichlorofluorescein was used for visualization under ultraviolet light. The components were extracted and determined quantitatively by weighing.

Table 76. *Lipid Composition of Faeces and Faecaliths* [*9*]

h*Rf*	Substances	% in faeces	% in faecaliths
80—85	Cholesteryl esters and others	5—10	10—25
65—75	Triglycerides	10—15	2— 5
45—50	Free fatty acids	12—20	25—36
30—35	Unknown substances	22—32	24—29
26	Cholesterol and others	10—15	4—10
12—22	Unknown substances	10—15	3— 8
0—10	Phospholipids	8—10	5—10

Faecaliths were found to contain considerably more cholesteryl esters and free fatty acids than faeces. The constituent fatty acids of lipid classes were analyzed by gas liquid chromatography.

IV. Investigation of substances in blood (serum)

Determination of alcohol in blood

It was pointed out on p. 196 that very small amounts of alcohols can be separated and detected by TLC in the form of their 3,5-dinitrobenzoates. It is obvious that this method can also be used for detecting alcohol in blood.

Method

a) Extraction and esterification

1.0 ml blood is mixed with 10 ml physiological saline (0.9%), and the alcohol extracted 5—6 times with 10 ml alcohol-free ether in a 50 ml separatory funnel (see p. 356). The ether extract is dried over anhydrous sodium sulphate, filtered, washed with ether, and then mixed with 500 mg 3,5-dinitrobenzoyl chloride. Esterification is completed after boiling for 30 min under reflux. Excess acid chloride is removed by adding approx. 30 ml water to the cooled ether solution and adjusting the pH to 9—10 with 5—10% sodium hydroxide. After shaking in a separatory funnel, the ether phase is separated with the DNB ester. The DNB esters left in the aqueous phase are extracted 3—4 times with alcohol-free benzene. The ether and benzene residues are combined, dried over anhydrous sodium sulphate, and filtered. This filtrate is concentrated to approx. 5 ml, and transferred, quantitatively, to a 10 ml measuring flask, which is filled to the mark with benzene.

b) Preparation of standard solution

1.0 ml of exactly 0.1% aqueous ethanol solution is treated in the manner described above. 1 mm^3 of this standard solution corresponds to 0.1 μg alcohol.

c) Application of the DNB ester solution under investigation and comparison of spot sizes

50 mm^3 of the DNB ester solution is applied to starting points 1, 3, 5, 7, etc., on a 20 $\times$ 20 cm silica gel plate, and 1–10 mm^3 of the standard solution (0.1–1.0 μg alcohol = 0.2–2.0‰ blood alcohol) on the even-numbered starting points. Carbon tetrachloride-cyclohexane-ethyl acetate (75 + 10 + 15), is used as developing solvent in a saturated tank. Rhodamine B solution (Reagent No. 129) is used to render the DNB esters visible. The alcohol content can now be quite accurately determined by visual comparison of spot sizes. If 1 ml blood contains 0.8‰ alcohol for instance, the spot size — under the conditions given — would correspond to an application of 0.4 μg alcohol (4 mm^3 comparison solution).

Chapters A, p. 154, and D, p. 255, should be referred to for further details of detection of lipids in serum by TLC.

V. Organ extracts

Determination of adrenaline and noradrenaline in suprarenals

The quantitative detection of adrenaline and noradrenaline in body fluids and organs is of significance. These hydrophilic compounds are extremely sensitive to oxygen. They oxidize to colored adrenochromes through the action of atmospheric oxygen. Methods of chemical determination of values of commercially available adrenaline solutions are discussed by Dibbern and Picher [*2*].

The corresponding acetyl compounds are much more stable. These derivatives, which are relatively simple to produce, can be separated on Silica Gel G layers using chloroform-methanol (90 + 10) as solvent. Vanillin-sulphuric acid (Reagent No. 151) is sprayed on for detecting these substances. The triacetates give the following *hRf* values and colors:

Adrenaline ~ 56 (pink), noradrenaline ~ 43 (grey) [*7*].

The isolation and acetylation of adrenaline and noradrenaline in the suprarenals of a cat may be taken as an example:

After dissection, the organs are placed in a 0.1% sodium pyrosulphate solution, and worked up as quickly as possible. The tissue is triturated to a completely homogeneous state, using part of the solution in a mortar together with approx. 10 g sand and 0.5 g sodium pyrosulphate. 0.5 ml 6 N hydrochloric acid and 10 ml water are then added and homogenization continues for 5—7 min. The solution is then filtered from the sand and washed twice with water. The filtrate, which is only slightly opaque, is mixed with a few drops of a starch solution, and a solution of 3 g potassium iodide and 2 g iodine in 45 ml water is added drop-wise until a permanent blue coloration is obtained. Excess iodine is destroyed by adding 0.1 N sodium thiosulphate solution till colorless.

The filtrate is neutralized by gradually adding sodium bicarbonate (pH paper) and, subsequently, 2 g sodium bicarbonate and 1 ml acetic anhydride. The solution is shaken in a separation funnel for 7—10 min, and the carbon dioxide thus formed released from time to time. After standing for 5 min, 15 ml methylene chloride is used for extraction, which is repeated four times. The methylene chloride extracts are combined, dried over 5 g freshly dried sodium sulphate, filtered, and again washed with methylene chloride. The filtrate is concentrated in an evaporating dish on a water bath, and the residue reduced to a volume of 0.3 ml in a 3 ml glass tube. Depending on the adrenaline content, either 25 or 50 mm³ of the solution is applied (Fig. 143). The adrenaline and noradrenaline content can be determined quantitatively by comparing with standards [7].

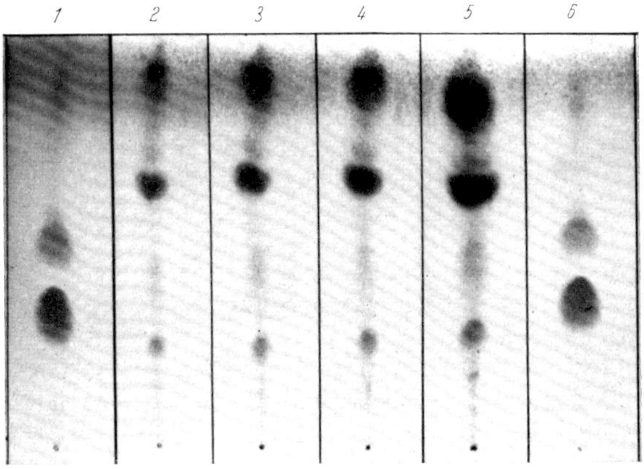

Fig. 143. Detection of adrenaline and noradrenaline taken from the suprarenals of a cat. Adsorbent: Silica Gel G. Solvent: Chloroform-methanol (90 + 10). Detection: Reagent No. 151. Acetylated products applied: *1* and *6* = mixtures of pure adrenaline (uppermost spot) and noradrenaline (lower spot); *2*—*5* = increasing amounts of organ extract

Details of extraction and TLC analysis of lipids obtained from organs are described on pp. 144—145.

Bibliography to Chapter G. TLC in Clinical Diagnosis and Pharmacology

[*1*] Demole, E.: J. Chromatog. **1**, 24 (1958).
[*2*] Dibbern, H. W., u. H. Picher: Arzneimittel-Forsch. **11**, 317 (1961).
[*3*] Martin, H.: Private communication.
[*4*] Reimann-Hunziker, R., u. W. Wild: Münch. med. Wschr. **103**, 1264 (1961).
[*5*] Waldi, D., F. Munter u. E. Wolpert: Med. exp. (Basel) **3**, 45 (1960).
[*6*] — Klin. Wschr. **40**, 827 (1962).
[*7*] — Arch. Pharm. **295/32**, 125 (1962).
[*8*] Wagner, H.: Mitt. Gebiete Lebensm. u. Hyg. **51**, 416 (1960).
[*9*] Williams, J. A., A. Sharma, L. J. Morris and R. Holman: Proc. Soc. Exptl. Biol. Med. **105**, 192 (1960).

Some recent noteworthy articles

Funck, F. W., and L. Zicha: Med. Exp. (Basel) **7**, 1 (1962): Metabolites from prednisolone and derivatives.
Jatzkewitz, H., and K. Sandhoff: Biochim. Biophys. Acta **70**, 354 (1963): Tay-Sachs disease.

Pataki, G., and M. Keller: Z. klin. Chem. (Berlin) **1**, 157 (1963): Estimation of amino acids in blood.
Schmid, E., L. Zicha, J. Krautheim and J. Blumberg: Med. Exp. (Basel) **7**, 8 (1962): Biogenic amines and their metabolites.
Waldi, D.: Ärztl. Labor. **9**. 221 (1963): Progesterone in serum.
Walz, D., A. R. Fahmy, G. Pataki, A. Niederwieser and M. Brenner: Experientia **19**, 213 (1963): Amino acids in urine.
Zöllner, N.: Z. klin. Chem. (Berlin) **1**, 18 (1963): Plasma lipids.
—, and G. Wolfram: Klin. Wschr. **40**, 1101 (1962): TLC of plasma lipids.

H. Synthetic Organic Materials

(Industrial intermediates)

By

H. Gänshirt, D. Waldi and Egon Stahl

In many branches of industry, synthetic organic products are in frequent use. Some of these materials are only added in small quantities to other products for special purposes. Detection of those compounds and their changes are of interest.

Particularly important are compounds with which the human being must come into contact, more or less by necessity, and which were found to be dangerous to health.

For clarification, this chapter will be divided into I, synthetic dyestuffs, II, additives occuring particularly in food and articles of personal use and III, other additives and synthetic products.

I. Synthetic dyestuffs[1]

For several years now it has been possible to separate mixtures of dyestuffs on alumina spread in a thin layer on glass plates. Mottier et al. [*29* and *29a*] have thoroughly applied this method. They have used aqueous alcohol as solvent. Lagoni and Wortmann [*21*] have described a circular technique with loose layers for detecting food dyes.

Our experiments have shown that, in addition to the known disadvantages of loosely spread layers, the given solvents [*54*] cannot be used without further specification. In the following section, the separation methods have been investigated with a larger number of dyestuffs. Besides Silica Gel G and Alumina G, "Alusil"[2] and MN-cellulose powder have proved their value in TLC.

1. Fat-soluble dyestuffs

Stahl [*50*] has already shown that fat-soluble dyes may be separated on standard Silica Gel G layers, and he gave a black and white photo-

[1] Possibilities of TLC for separation of mixtures of *natural* pigments are described in the chapters, Carotenoids, p. 216, colored Table 1, and anthocyans pp. 379—380; reference is made to a recent paper by Montag [*28*].

[2] Proprietary name for a mixture of Alumina G and Silica Gel G (1 + 1).

graph of such a thin-layer chromatogram. The colored illustration between pp. 346 and 347 demonstrates the possibilities of separation of mixtures of fat-soluble pigments [*22*]. Already after a migration of 5–6 cm, numerous fat-soluble dyestuffs can be characterized clearly. For identification, however, it is necessary to adhere to the standard conditions (p. 27) and also to chromatograph simultaneously the Desaga test mixture of fat-soluble dyes. In Table 77 below, the h*Rf*-values and specific indices of a series of such dyestuffs are given [*54*].

The dependence of the sharpness of separation on the application technique is shown by Fig. 11 on page 13.

Table 77. h*Rf*-*Values of Fat-Soluble Dyes on Silica Gel C with Benzene as Solvent*

Dyestuffs	*Schultz No.*	*Col. Ind. No.*	*Rf* × 100 Main spot	Minor spot(s)	Color of main spot
Sudan Orange	*31*	*11920*	4	—	yellow orange
Sudan III	*532*	*26100*	12	30; 41*	carmine red
Sudan R	*149*	*12150*	13	32	carmine red
Sudan Blue G	—	*61525*	20	12; 35	blue
Sudan Black B . . .	—	*26150*	28	0; 4; 16*; 63*	steel blue
Sudan Orange RR . .	*92*	*12140*	29	14	bright red
Butter Yellow	*28*	*11020*	40	—	orange yellow
Sudan Violet BR . .	—	*61705*	53	—	violet blue

* faintly visible.

Schorn and Stahl [*45*] determined the h*Rf*-values of a number of other dyes. Amounts of 0.2 μg were applied and developed on Silica Gel G layers with benzene under the standard conditions (CS). For further identification, the plate was sprayed with concentrated hydrochloric acid and the colors shown in Table 78 were noted.

Table 78. h*Rf*-*Values of some Azobenzene Derivatives on Silica Gel G Layer (Charge No. 200473) Solvent: benzene*

Dyestuffs	*Rf* × 100	After spraying with conc. hydrochloric acid
p-Hydroxyazobenzene, *trans*-	12	orange-yellow
p-Aminoazobenzene, *trans*-	19	orange
Benzolazonaphthol, *trans*-	44	orange
p-Dimethylaminoazobenzene, *trans*-	55	red
p-Dimethylaminoazobenzene, *cis*-	39	red
p-Methoxyazobenzene, *trans*-	64	orange-yellow
p-Methoxyazobenzene, *cis*-	12	orange-yellow
Azobenzene, *trans*-	72	yellow
Azobenzene, *cis*-	30	yellow
Guajazulene } Test mixture	76	
Sudan Red G } Test mixture	22	
Indophenol } Test mixture	8	

Montag [28] reported the detection of a number of synthetic and natural fat-soluble dyes in food with the help of TLC. He included exact working descriptions for the identification of carotene and bixin in margarine and cheese, and of paprika and curcuma-pigments in sausage.

Schetty and Kuster [41] succeeded in separating the isomeric 1:2 chromium and cobalt-complexes of the o-hydroxy-o'-carboxyazo-series with methanol on alumina layers. TLC was used, with equally good effect, to detect the isomers of 1:2 metal complexes of azo and azomethine dyes [42].

More recently, Davidek [6a] has described the separation, already mentioned in the introduction, of fat-soluble food colors on loose alumina layers.

Special applications: dyestuffs in gasoline-fuel

The manufacturers of the various brands of gasoline are obliged to add dyestuffs to their products for recognition purposes. According to Häusser [16] and Machata [23], these dyes may be detected by TLC from distillation residues of gasolines. Häusser [16] prefers an intermediate purification. 5—10 ml of the gasoline sample is concentrated to one-third and chromatographed with petroleum ether in an Allihn tube filled with active aluminum oxide. The dye zones are eluted with acetone, filtered and evaporated to dryness. The enriched dyestuffs are dissolved in a few drops of acetone and applied to a Silica Gel G layer. Benzene is used as the developing solvent. BP and NORD-ÖL excluded, the gasolines used in Germany may be distinguished (ARAL, ARALIN, BP Super, DEA, DEA-Super, ESSO, ESSO Extra, FANAL, FANAL-Super, NORD-ÖL Special, SHELL, SHELL-Super). The *Rf*-values are given in Häusser's report. The method is of value in investigations of fuel thefts.

In addition, the less volatile components of gasoline may be detected in filtered UV-light.

2. Water-soluble dyestuffs

Water or alcohol-soluble dyes can be separated with more polar acidic or basic solvents. These dyes often consist of several components and thus, one or two smaller spots appear, in addition to the main fraction.

a) Indicator dyes

Table 79. h*Rf*-*Values of Indicator Dyes on Alusil-Layers* [54]

Dyestuffs	*Rf* × 100	Color of spots
Chlorophenol Red	27 (78)[1]	violet (yellow)[1]
Bromocresol Purple	40 (0)	violet (yellow-brown)
Cresol Red	42 (60 and 77)	orange-yellow (bright red and yellow)
m-Cresol Purple	43 (71)	orange-yellow (yellow)
Bromophenol Blue	48 (39)	blue-violet (red-violet)
Bromochlorophenol Blue	48	blue-violet
Bromothymol Blue	65	green-brown
Benzyl Orange	67	yellow
Methyl Orange	67	yellow-orange
Thymol Blue	74	orange-yellow
Phenolphthalein	83	colorless, red with alkali
p-Ethoxychrysoidine	84	orange

[1] The h*Rf*-values and colors of accompanying dyes are given in ( ).

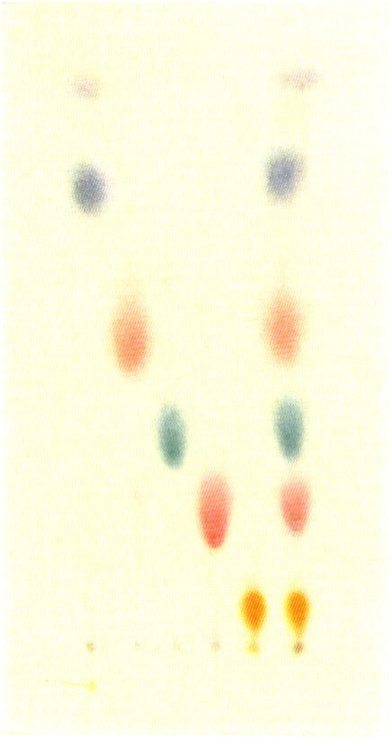

5 4 3 2 1 G

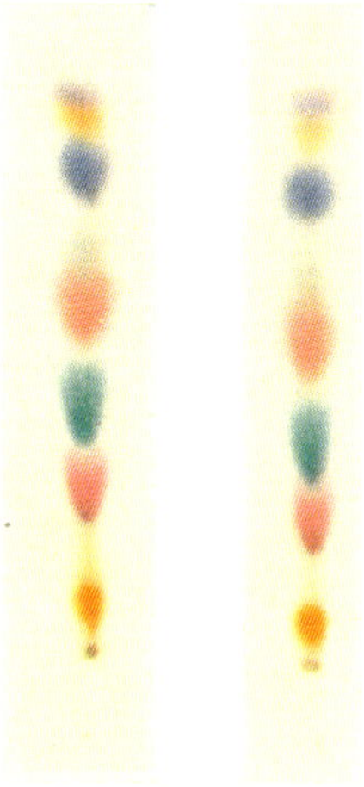

G 1—5 G

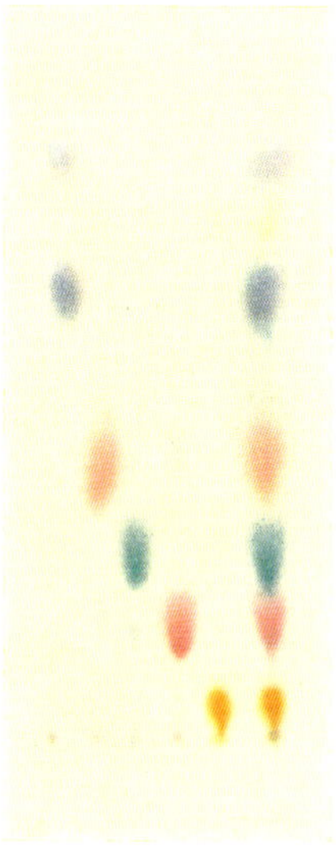

5 4 3 2 1 G

G 5 4 4+5 3 1 1+3 2 2+4 G

Thin layer chromatograms of Sudan Dyes in original size [22].

In ascending order: *1* Sudan Orange G; *2* Sudan Red R; *3* Sudan Blue G; *4* Sudan Orange RR; *5* Sudan Violet BR. Silica gel G (standard procedure). Solvent: Benzene; length of run: 5—6 cm (!). Preservation of the upper chromatograms: After spraying with Neatan, the layer is covered with a thin adhesive tape, and carefully stripped off (see p. 44). The lower chromatogram, after spraying with Neatan, is directly taken off the plate without tape.

Springer-Verlag, Berlin · Göttingen · Heidelberg · New York
Academic Press Inc., New York · London

In order to identify and test the purity of indicator dyes their h*Rf*-values were determined on Alusil layers (p. 31) under standard conditions (pp. 7—9). Ethyl acetate-methanol-5N ammonia solution (60 + 30 + 10) (Table 79) was used as developing solvent. Migration time was 30 min. 2 μl of a 25% solution of the dyes in methanol were applied.

PASTUSKA and TRINKS [*31*] described thin-layer electrophoresis of indicator dyes.

b) Dyes for microscopy

In Table 80, the h*Rf*-values of dyes frequently used in histology, bacteriology and biology are given [*54*]. With Silica Gel G layers, the solvent, chloroform-acetone-isopropanol-sulphurous acid (5—6% SO_2) (30 + 40 + 20 + 10) was used. 2 μl of a 25% solution of the dyes in methanol were applied.

Table 80. h*Rf*-*Values of Dyes used in Microscopy* [*54*]

Dyestuffs	*Schultz No.*	*Col. Index No.*	*Rf* × 100	Color
Acridine Orange	*902*	*46005*	41	yellow
Alkali Blue	*811*	*42765*	16 and 34	blue (blue)[1]
Brilliant Green.	*760*	*42040*	59 (0)[1]	green
Brilliantcresyl Blue	*992*	*51010*	21 and 52	green (green)
Erichromazurol S	*841*	*43825*	39	dark violet
Chromazurol S				
Gentian Violet	*783*	*42535*	43 and 48	red (violet)
Methyl Violet 2 B				
Crystal Violet	*78*	*42555*	43	violet
Light Green, yellowish . . .	*765*	*42095*	11 (0)	green
Malachite Green	*754*	*42000*	35 (0)	green
Metanil Yellow	*169*	*13065*	39	yellow
Methylene Blue B	*1038*	*52015*	9	blue-green
Methylene Green	*1040*	*52020*	*18*	green
Victoria Blue	*822*	*44945*	51	blue

[1] The h*Rf*-values of the minor spots and their colors are given in ( ).

According to SCHORN and STAHL [*45*], the dyes listed in Table 81 are best separated with the mixture n-propanol-formic acid (80 + 20) on Silica Gel G layers. The developing time is 70—90 min in a saturated chamber for a separation distance of 10 cm. The dyestuffs (Table 81) with the index[1] fluoresce intensively in UV-light (365 mμ) and, therefore, very small amounts may be detected.

Table 81. h*Rf*-*Values of a few Fluorescent Dyes* [*45*]

Dyestuffs	h*Rf*	Dyestuffs	h*Rf*
Methyl Green	0	Nile Blue	49
Diazine Green	12	Fuchsin	55
Pyronin G[1]	20	Rhodamine B[1]	62
Acridine Orange G[1]. .	40	Fluorescin[1]	74

[1] Fluorescein dyes.

c) Ink and staining dyes

TLC-separation of ink dyes has been carried out by WOLLENWEBER [*56*]. He used layers of MN-Cellulose Powder 300 G and as the developing solvent, the upper phase of the mixture n-butanol-acetic acid-water (50 + 10 + 40).

The same author [*56*] has described TLC of rather water-soluble dyes, e.g., anthraquinone derivatives. He used layers of acetylated cellulose-powder (MN 300 G/AC., see p. 32); a freshly prepared mixture of ethyl acetate-tetrahydrofurane-water (6 + 35 + 47) was filled into the tank about two hours before insertion of the plate. The following h*Rf*-values were obtained: 1,4-dihydroxyanthraquinone = 6; 4-amino-l-hydroxyanthraquinone = 11; and 1,4-diaminoathraquinone = 21. (Migration time: 60 min.)

d) Cellulose and wool dyes

MECKEL et al. [*24*] have succeeded in separating cellulose and wool dyestuffs on alkaline Silica Gel G and Alumina G layers, as suggested by STAHL [*51*]. To prepare the slurry for spreading, the adsorbent was mixed with 2.5% sodium carbonate solution instead of water. A freshly prepared mixture of butyl acetate-pyridine-water (50 + 45 + 25) was used as the developing solvent. The separations are sharper on alkaline Silica Gel G than on Alumina G layers. The detection limit lies at a factor of about 10 lower than that of the paper-chromatographic procedure used hitherto. The dyestuffs investigated (e.g., Sirius, Benzamine, Palatine dyes) generally consist of several components of various colors.

e) Synthetic food colors

Paper-chromatographic methods have been used chiefly for characterizing the synthetic food colors suggested by the Dyestuff Commission of the Deutsche Forschungsgemeinschaft (DFG). The 8th communication of this commission lists a number of solvents for developing paper chromatograms. Samples of the dyes and colored copies of the chromatograms, as well as their spectra, are also given[1].

WOLLENWEBER [*56*] and the present authors [*54*] have found that the dyestuffs given in Table 82 can be clearly distinguished with a *single solvent* on layers of MN-Cellulose Powder 300 G. The solvents tested,

1. n-propanol-ethyl acetate-water (60 + 10 + 30) [*54*] and
2. 2.5% aqueous sodium citrate solution-25% ammonia solution (80 + 20) [*56*]

yield different patterns of separation. They may, therefore, be used with good effect on two-dimensional thin-layer chromatograms. The migration-time of solvent I is 90 min.

---

[1] Communication No. 8 of the Dyestuff-Commission of DFG, 2nd Ed., published by Verlag Franz Steiner, Wiesbaden, Germany, 1957.

Table 82. h*Rf*-*Values of Food-Dyes on MN-Cellulose Layers* [*56*]

DFG-8th Comm.	Commercial name	*Schultz No.*	*Col. Ind. No.*	DFG-6th Comm.	*Rf* × 100	
					Solvent I	Solvent II
Yellow 1	Genuine Yellow (also Extra) Acid or Fast Yellow Acidic Yellow Jaune Solide	*172*	*13 015*	*23*	53 (50)¹	58 (—)¹
Yellow 2	Tartrazine Jaune Tartarique Hydrazine Yellow 0 F.D. & Co. Yellow No.5	*737*	*19 140*	*64*	25 (—)	72 (—)
Yellow 3	Quinoline Yellow (also Water-Soluble or Extra) Quinoline Yellow	*918*	*47 005*	*97*	39 (27)	12 (20)
Yellow 4	Chrysoine 5 Tropeoline 0 Resorcine Yellow	*186*	*14 270*	*26*	88 (—)	34 (—)
Yellow 5	Yellow 27 175 N	—	—	*30*	31 (—)	29 (—)
Orange 1	Orange GGN or GGL	—	*15 980*	*32*	56 (69)	45 (—)
Orange 2	Yellow Orange S Sunset Yellow FCF F.D. & Co. Yellow No.6	—	*15 985*	*29*	55 (69)	42 (24)
Red 1	Azorubin	*208*	*14 720*	*38*	64 (75)	12 (—)
Red 2	Genuine Red E	*210*	*16 045*	*39*	57 (71)	20 (—)
Red 3	Amaranth S Naphtha Red S	*212*	*16 185*	*40*	27 (—)	31 (—)
Red 4	Brilliant Ponceau 4 RC Cochineal Red A	*213*	*16 255*	*41*	33 (—)	55 (—)
Red 5	Ponceau 6 R Scarlet 6 R	*215*	*16 290*	*42*	16 § (—)	76 (—)
Red 6	Scharlach GN	—	—	*34*	64 (73)	90 (94)
Blue 1	Indanthrene Blue RS	*1228*	*69 800*	*104*	0	0
Blue 2	Indigotin I or Ia Indigo Carmine F.D. & Co. Blue No. 2	*1309*	*73 015*	*105*	26 (—)	19 (7)
Black 1	Brilliant Black BN	—	*28 440*	*58*	20 § (—)	10 (—)

¹ h*Rf*-values of minor spots given in ( ); § = drawn out spots (tail-formation).

II. Substances in food and household articles

1. Antioxidants and preservatives

To chromatograph antioxidants or preservatives, it is mandatory to isolate them first. For instance, for the detection of antioxidants in animal or plant fats, the fat is dissolved in cyclohexane or petroleum ether and extracted with 75% ethanol. The larger part of the antioxidants enter the alcoholic phase, which may then be chromatographed.

A solution of 1 g fat in 1—2 ml cyclohexane or petroleum ether (b.p. 40—60° C) is mixed for 2 min. with 1 ml ethanol (75%). After separation of the two phases, the alcoholic layer is removed and 0.1 to 0.2 ml of this solution is chromatographed.

A few antioxidants, e.g., butyl hydroxyanisol, are not extracted by this method: this compound may be isolated by steam distillation (see p. 187).

If spots are detected after the chromatographic separation which might indicate the presence of antioxidants or preservatives, it is necessary to confirm their identity by simultaneous chromatography of corresponding test substances, or by the use of an evaluation template as described by SEHER [*46*]. Components of essential oils and taste-correctants may simulate antioxidant additives (cf. Fig. 144).

Semi-solid or fluid samples should be kept in glass vessels in order to avoid contact with rubber or synthetic materials as the "analytical" sample can easily extract plasticizers and antioxidants from the latter.

SEHER [*46*] has developed a method for finding synthetic antioxidants in food fats and other food products. The majority of the substances shown in Fig. 144 may be distinguished by two-dimensional chromatography. Identification of unknown materials can be accomplished with a stencil (Fig. 144). The factors influencing the chromatographic patterns have been thoroughly investigated, and the working conditions described below should be followed exactly.

Method:

Silica Gel G layers of 250 μ thickness are prepared with the Desaga applicator and dried at 120° C. Chloroform is first allowed to rise to a height of 12 cm, to ensure that a moisture content within defined limits is obtained. Then, the plates are air-dried and put into an evacuated desiccator for 20 min. Three spots of the material to be analyzed and 2 spots of the 3-colored test mixture (prepared as on p. 27) are applied, as shown in Fig. 144. By using this method of application, separations by the 1st solvent (chloroform) and the 2nd (benzene) are recorded and a two-dimensional chromatograph is obtained. The test dyes are chromatographed simultaneously to ensure the greatest accuracy. The separation distance for the 1st solvent (chloroform) is 10 cm; the plate is then air dried, rotated through 90°, and put into the second solvent (benzene). The separation-distance is also 10 cm. *Detection:* The air-dried plate is sprayed evenly with a 20% solution of phosphomolybdic acid (Reag. No. 120a) until it is a homogenous yellow color. After 1—2 min the blue spots of the more strongly reducing antioxidants appear. Then the plate is treated with ammonia vapor, making the background a pure white, whereas the substances become clearly visible as dark blue or partly violet or green-tinged spots. Compounds of less reductive capacity may be recognized after a final heating for 10 min at 120° C.

As Fig. 144 shows, Ionol and butylhydroxytoluene are not separated. The unlabelled spot below Ionol derives from an impurity in the diphenyl-p-phenylenediamine. The components remaining at the starting point, gallates, nordihydroguaiaretic acid and components of guaiac resin can be separated on a second two-dimensional chromatogram, although there are certain difficulties.

Applications: Several commercial preparations containing antioxidant mixtures were investigated with the above method. In the course of these investigations, it was found that the butyl-hydroxyanisoles (BHA) on the market were not uniform. Moreover, the butyl-hydroxyanisoles were found to be isomers that were inseparable by the procedure given, namely, 2-*tert.* butyl-4-hydroxyanisole and 3-*tert.*-butyl-4-hydroxy-anisole, the impurities 2.5-di-*tert.* butyl-4-hydroxy-anisole, hydro-quinone-monomethyl ether and 4-*tert.* butoxyanisole. The major compound was also quantitatively estimated by simultaneous chromatography of known amounts of pure 3-*tert.* butyl-4-hydroxyanisole. Photostats of these chromatograms were evaluated planimetrically (see pp. 48—49).

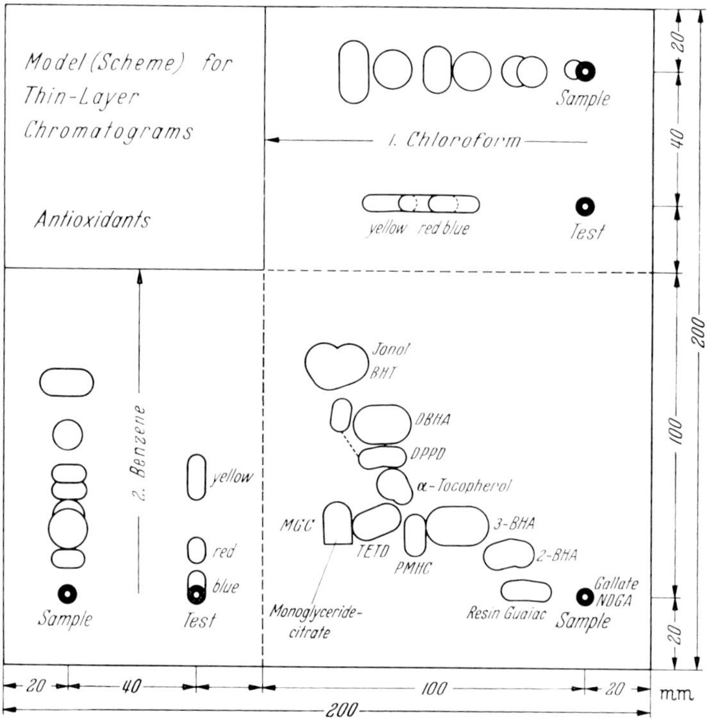

Fig. 144. Schemes for evaluating thin-layer chromatograms of antioxidants, acc. to SEHER [*46*]. Measurements in mm. BHT = butyl-hydroxytoluene; DBHA = dibutyl-hydroxyanisole; DPPD = diphenyl-p-phenylenediamine; TETD = tetraethyl-thiuramdisulphide; PMHC = pentamethyl hydroxychromane; BHA = butyl-hydroxyanisole; MGC = monoglyceride citrate; NDGA = nordihydroguaiaretic acid

As BHA and the impurities show the same color reactions and similar spectral behavior. Thin-layer chromatographic analysis appears to be the only reliable method for determining the purity of BHA.

In the course of further investigations, SEHER [*47*] separated, by a different method, the polyhydroxy compounds that remain at the start in the above procedure. Adsorption layers were used consisting of 95.5% Silica Gel G and 4.5% oxalic acid. Gallates and other antioxidants were separated by MEYER [*25*] on silica gel-kieselguhr layers with mixtures of hexane-acetic acid (Fig. 145). The plates were developed with the same solvent twice. 5% Phosphomolybdic acid was used as spray reagent. Fat soluble dyes, and other substances soluble in the alcohol used to extract the fat, often make the one-dimensional separation of antioxidants difficult. This is particularly true for gallates. In such cases, good results may be obtained by two-dimensional chromatography with the solvents chloroform and hexane-acetic acid.

DAVIDEK and POKORNY [*6*] investigated the behavior of nordihydroguaiaretic acid, propyl gallate, *tert.* butyl-hydroxyanisole and ascorbyl palmitate on loose polyamide layers. The plates were laid nearly horizontally in the jars. A good solvent was found to be methanol-acetone-water (60 + 20 + 20) or (60 + 10 + 30). To test the method, synthetic mixtures of fat and the above-mentioned antioxidants were extracted, and the antioxidants detected qualitatively. Although only four antioxidants were separated, the authors consider this method superior to the procedure of SEHER [*46*] because of the simplicity of the preparation of the layers and the better separating capacity of polyamide as compared with silica gel. However, the method is not to be recommended because these thin-layers are very sensitive, give poor separation and can be sprayed only in a moist state.

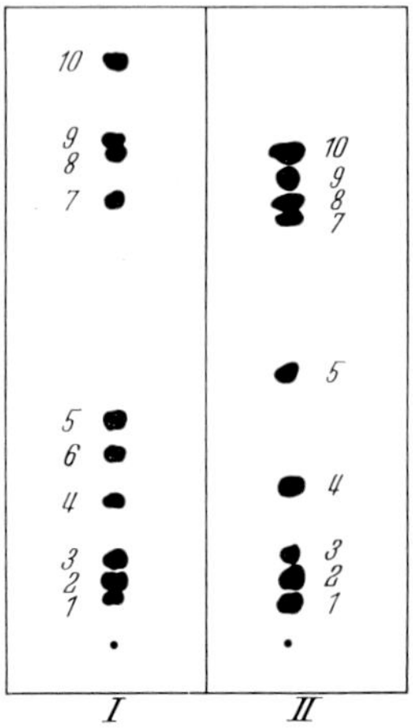

Fig. 145. Thin-layer chromatograms of antioxidants [*25*]. *I* Plates: Silica gel-kieselgur (25 + 5); solvent: hexane-acetic acid (40 + 10). *II* Layers: Silica gel-kieselgur (20 + 10); solvent: hexane-acetic acid (60 + 10). *1* Nordihydroguaiaretic acid. *2* Propyl gallate. *3* Butyl gallate. *4* Octyl gallate. *5* Dodecyl gallate. *6*[1] Vanillin. *7* Butyl-hydroxyanisole. *8*[1] Eugenol. *9*[1] Thymol. *10* Butyl-hydroxytoluene

[1] *6*, *8* and *9* were chromatographed along with the others as these taste-correctors or components of essential oils can easily simulate the presence of added antioxidants.

SEHER [*48*] has further described thin-layer chromatographic separation of tocopherols. It was not only possible to resolve the various tocopherols, but also to separate them from related materials accompanying the tocopherols in natural extracts. Quantitative evaluation of the chromatogram was possible if known amounts of pure substances were applied alongside the mixture to be analyzed. Details may be found in the chapter on vitamins, pp. 210—248.

Detection of inhibitors in insulating oils [*37*, *38*]

TLC was used as an auxiliary for better evaluation of the insulating oils used in high-tension transformers. Insulating oils consisting of petroleum fractions can be broken down by TLC into paraffins, naphthenes and aromatics. More important, however, is the possibility of detection of inhibitors which serve to delay the natural ageing processes of such oils. A number of commercially obtainable inhibitors have been chromatographed with the solvents benzene and hexane on silica gel layers prepared in the standard manner (cf. Tab. 83). If stability tests are applied to the insulating oils, it is possible to follow the behavior of the inhibitors without preliminary separation from the insulating oil. (Table 83).

Table 38. *Rf-Values and Color Reactions of some Commercial Inhibitors* [*38*]

Chemical name of Inhibitor	*Rf values* × 100		Color reactions with antimony pentachloride at 120° C (Reag. No. 13)
	Benzene	Hexane	
2,6-Di-*tert*-butyl-4-methylphenol	81	36	violet
2,6-Di-*tert*-butylphenol	79	33	yellow
4,4′Bis(2,6-di-*tert*-butylphenol)	79	11	canary-yellow
4,4′-Methylene-bis(2,6-di-*tert*-butyl-phenol)	78	10	red
Dodecyl-o-cresol	71	7	dark yellow
Ortho-*tert*-butylphenol	65	—	yellow ochre
N-phenyl-2-naphthylamine	63	—	brown
Dodecyl-phenol	56	—	brown-red
Diphenyl-picryl-hydrazyl	55	—	bright yellow
4,4′Thio-bis(6-*tert*-butyl-o-cresol)	47	—	yellow
N,N′-Diphenyl-p-phenylenediamine	46	—	red
4,4′-Methylene-bis(6-*tert*-butyl-o-cresol)	46	—	wine-red
2,6-Di-*tert*-butyl-l-methoxy-p-cresol	41	—	red
2-*Tert*-butyl-4-hydroxyanisol	28	—	yellow-brown
2,4,6-Tri-*tert*-butylphenol	13	—	brown
2,6-Di-*tert*-l-dimethyl-amino-p-cresol	0	—	yellow
4,4′-Isopropylidene-diphenol	0	—	brownish

Preservatives

The p-hydroxybenzoic acid esters used as preservatives have been separated on Silica Gel G layers by Gänshirt and Morianz [*13*]. These authors used Silica Gel G to which 2% of the luminescent material ZS-Super (Riedel de Haen) had been added. A mixture of pentane-acetic acid (88 + 12) was employed as the developing solvent. After separation, the preservatives could be recognized under short-wave UV-light (254 mμ). Separation of the four substances shown in Fig. 46 could only be achieved with rigid adherence to the conditions described below under "Method". On the other hand, it is easy to separate esters whose alcohol components differ by two C-atoms, provided they are not branched. For instance, mixtures of esters methyl and propyl were separated by TLC and, after separation, the esters were eluted and quantitatively determined by UV-spectroscopy. Sources of error entering into this procedure have been investigated in detail (see p. 54).

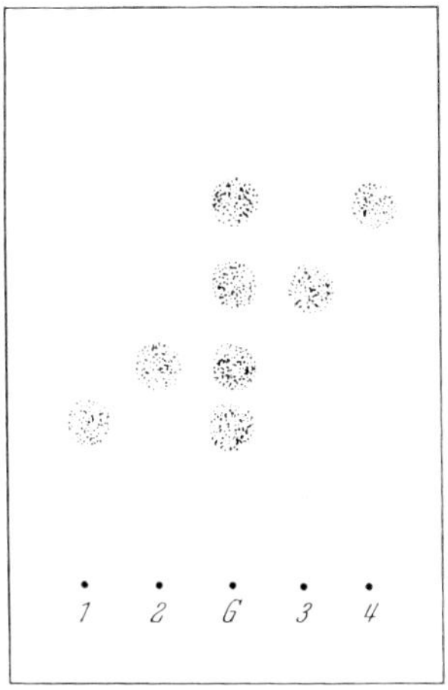

Fig. 146. Separation of p-hydroxybenzoic acid esters [*13*]. *1* Methyl ester. *2* Ethyl ester. *3* n-Propyl ester. *4* n-Butyl ester. *G* Mixture

Method:

To prepare five 20 × 20 cm thin-layer plates, 25 g Silica Gel G is mixed with 0.5 g of the luminescent material, ZS-Super[1] (Riedel de Haen), and 45 ml

[1] Riedel-de Haen, Seelze, Germany.

water. The slurry resulting from thorough stirring is immediately applied to the plates by means of a thin-layer applicator. When the silica gel layers are set, they are dried in an oven for 2 hrs. at 160°C and finally kept in an evacuated desiccator over potassium hydroxide. These rigorous drying conditions are necessary to prevent demixing of the solvent, pentane-acetic acid (80 + 12), on the plates. For the same reasons, it is essential to use anhydrous acetic acid. The separation distance is 13 cm. The solvent is removed after development by warming for an hour at 60° C.

2. Plasticizers

In preparing plastic products, organic fluids of high boiling point are used as plasticizers. Some of these substances are highly toxic and should, therefore, never be employed when manufacturing packing material which will come into contact with foodstuffs or medicinal goods. Peereboom [*32*] has succeeded in separating the plasticizers used in the U.S.A. in making food packages. For better recognition, the Silica Gel G layer was supplemented with 0.005% of the water-soluble fluorescence-indicator Ultraphor WT[1]. The plasticizers themselves or the corresponding extraction residue of the packing material were dissolved in ether and 10 mm³ of approximately 5% solutions were applied to the plates.

Table 84. R_{St}-*Values of Plasticizers in various Solvents, referred to Dibutyl Sebacate* = 100 [*32*]

	Isooctane-ethyl acetate (90 + 10)	Benzene-ethyl acetate (45 + 5)	Dibutyl ether-hexane (80 + 20)
Triacetin (Glycerol triacetate)	18	34	17
Ethylphthalyl ethyl glycolate	22	66	30
Acetyl triethyl citrate	26	51	29
Triphenylphosphate	33	80	50
Tricresyl phosphate	42	86	69
Butylphthalyl butyl glycolate	43	90	65
2-Ethyl hexyl diphenyl phosphate	46	77	58
Diethyl phthalate	51	79	60
Acetyl tributyl citrate	53	85	70
Di-n-butyl phtalate	74	103	84
Diisobutyl adipate	83	86	85
Dibutyl sebacate	100	100	100
Dinonyl phthalate	101	118	114
Di-2-ethyl hexyl phosphate	114	116	115
Butyl stearate	161	123	128
Paraflex G 2 (epoxide of natural glyceride)	Several spots	Several spots	Several spots

For developing, the solvents tested in Table 84 were used. After separation, a few of the plasticizers could be detected by their fluorescence under UV-light (366 mμ), and the phthalate derivations as dark absorbing spots. The plasticizers could be identified also by various color reactions (Reag. Nos. 120b, 151 and others). The R_{St}-values with re-

[1] BASF, Ludwigshafen am Rhein, Germany.

ference to dibutylsebacate = 100, are given in Table 84. Three critical pairs of substances: 5–6, 5–7 and 7–9 (see Table 84) could not be separated in the solvents given, but identification by color reactions was possible. Paraflex, prepared from a natural product, is of complex composition and gives a large number of spots on the chromatogram.

Braun [3] has chromatographed many plasticizers using the solvent methylene chloride on Silica Gel G layers prepared by the standard method. The possibilities of separating compounds of similar structure and of various classes of substances were noted (Table 85). 1–5% Solutions of the various plasticizers in benzene or ether were prepared and volumes containing about 100 μg of substance were applied. To test thin plastic sheets for plasticizers, these were extracted with methylene chloride. Antimony pentachloride (Reag. No. 13) appeared to be the most generally suitable spray reagent. Most of the plasticizers could be recognized as brown spots on a light background after spraying and heating to 120° C; a few could be seen only under UV-light. In addition, the phthalic acid esters could be stained with resorcinol solution and the phosphoric acid esters with a diazonium reagent (Reag. No. 37).

Table 85. h*Rf*-*Values of Plasticizers on Silica Gel G with Methylene Chloride as Solvent* [3]

Phthalic acid esters		*Phosphoric acid esters*	
Dimethyl phthalate	38	Trichlorethyl phosphate	13
Dibutyl phthalate	52	Trioctyl phosphate	24
Dihexyl phthalate	57	Diphenyloctyl phosphate	38
Dioctyl phthalate	59	Triphenyl phosphate	42
Dinonyl phthalate	60	Diphenylcresyl phosphate	43
Didecyl phthalate	63	Tricresyl phosphate	49
Di-(2-ethylhexyl)-phthalate	58		
Benzylbutyl phthalate	53	*Citric acid esters*	
Dimethoxy ethyl phthalate	10	Triethyl citrate	15
Dicyclohexyl phthalate	53	Tributyl citrate	20
Dimethylcyclohexyl phthalate	48	Acetyltriethyl citrate	19
Diisodecyl phthalate	57	Acetyltributyl citrate	32
Diisononyl phthalate	56	Acetyltri-2-ethyl hexyl citrate	52
Adipic acid esters		*Various plasticizers*	
Adipic acid dioctyl ester	49	Acetylricinoleic acid methyl ester	33
Adipic acid-(2-ethylhexyl)-ester	50	Acetylricinoleic acid butyl ester	40
Adipic acid dinonyl ester	50	Benzenesulphonic acid butyl amide	38
Adipic acid benzyloctyl ester	50		
Adipic acid polyester	0	Benzenesulphonic acid N-methyl amide	23
Sebacic acid esters		Thiodibutyric acid di-(2-ethylhexyl) ester	52
Sebacic acid dibutyl ester	35		
Sebacic acid dioctyl ester	47	p-Oxybenzoic acid-2-ethyl-hexyl ester	16
Sebacic acid-(ethylhexyl)-ester	47		
Sebacic acid polyester	0	Glyceryl triacetate	18

Note: On saponification of ester softeners, the constituent acids are obtained along with the alcohols. Both components may be identified chromatographically, as will be described in the following section.

3. Alcohols and acids

a) C_1 to C_4-alcohols

The volatile, simple alcohols can be chromatographically separated in the form of suitable derivatives and can then be detected in trace amounts. The following procedure has been worked out by Waldi [*54*]:

Preparation of 3,5-dinitrobenzoic acid ester (DNB-Esters)

Alcohols occurring in aqueous solution are first enriched by extracting with alcohol-free ether[1]. The ether extract is dried over freshly dehydrated sodium sulphate and then mixed with 3,5-dinitrobenzoyl chloride and boiled for 30 min under reflux. To remove the unused acid chloride, water is then added and the solution is made alkaline (pH 9–10) with 5–10% sodium hydroxide. The ether phase containing the DNB-esters is then separated in a separating funnel and the aqueous phase extracted

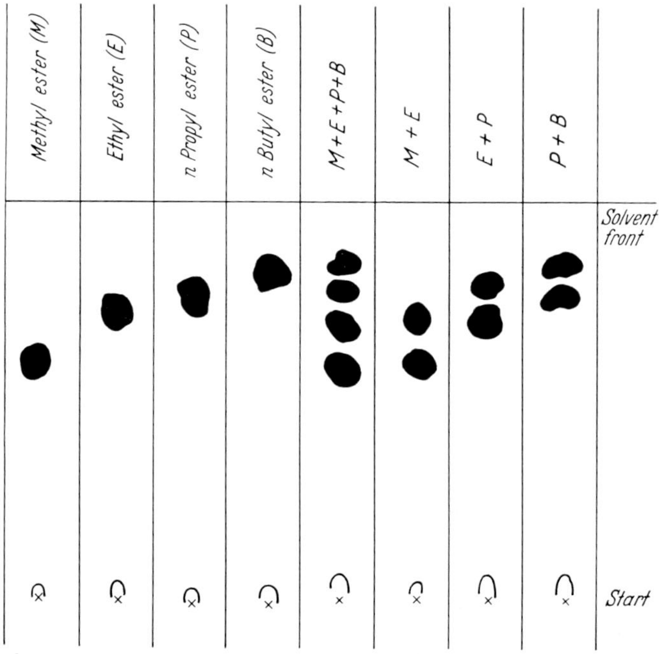

Fig. 147. DNB-esters of the lower alcohols on a Silica Gel G layers developed with cyclohexane-carbon tetrachloride-ethyl acetate (10+75+15). Detection: Rhodamine B, Reag. No. 129 [*54*]

3 or 4 times with alcohol-free benzene[1]. The combined ether and benzene extracts, dried over sodium sulphate, are evaporated to dryness and the residue dissolved in a small, exactly measured amount of benzene. This DNB-ester solution may then be chromatographed.

[1] Commercial ether or benzene are boiled for about 60 min. under reflux with an excess of 3,5 dinitrobenzoyl chloride and then distilled.

Before analyzing unknown mixtures, it is necessary to test the above procedure with 1 ml of a 0.1% aqueous solution of the alcohols in question. In this case, the DNB esters are dissolved in 10 ml benzene and 5 mm^3 (0.5 μg) alcohol simultaneously chromatographed for comparison.

Adsorption layers and solvents. The DNB-ester solutions to be examined are applied alongside a standard, and a good separation is obtained with the solvent cyclohexane-carbon tetrachloride-ethyl acetate (10 + 75 + 15) (Migration distance, 10 cm, saturated atmosphere) on Silica Gel G layers prepared by the standard method. Fig. 147 shows the results of this procedure.

The unspecific Rhodamine B-reagent (No. 129) has proved very useful as an indicator. Besides inspection of the chromatogram by day light, one can also recognize the DNB-esters in UV-light as dark spots on the red-fluorescing plates.

This method was used, e.g., for the determination of alcohol in blood as described on p. 342.

b) Glycerol and glycols

The following alcohols may be separated on air-dried Silica Gel G layers with the solvent chloroform-acetone-5N-ammonia solution (10 + 80 + 10), (migration-distance, 10 cm, saturated atmosphere): glycerol (h*Rf* 35), ethylene glycol (h*Rf* 70) and 1,2-propylene glycol (h*Rf* 85). Benzidine-periodate solution (No. 18) is used as indicator [*54*].

Prey et al. [*36*] separated glycerol (h*Rf* 38) from glycol (h*Rf* 45) on normal Silica Gel G layers with butanol-water (90 + 10). Mannitol and sorbitol were not separated but remained in the starting point (h*Rf* 5). The separation of the two first named compounds can be improved by chromatographing on layers impregnated with 0.1 N boric acid. Dichromate-sulphuric acid was used as spray reagent. It yielded white spots on a yellow-brown background.

Hromatka and Aue [*17*] confirm a linear relationship between the logarithm of the *Rf*-values and the number of C-atoms in diols. On Silica Gel G layers, abs. ethanol or dioxane were used as solvents for ethylene glycol, 1,3-propylene glycol, 1,6-hexanediol, 1,7-heptanediol, 1,9-nonanediol, 1,10-decanediol and 1,13-tridecanediol.

c) Organic acids

Mixtures of carboxylic acids may be separated on Silica Gel G layers with basic or acidic polar solvents. The following solvent mixtures have been used so far:

I. Methanol-5N-ammonia solution (80 + 20) [*54*].
II. Ethanol (96%)-water-ammonia solution 25% (100 + 12 + 16) [*4*].
III. Benzene-methanol-glacial acetic acid (90 + 16 + 8) [*34*].
IV. Benzene-dioxane-glacial acetic acid (90 + 25 + 4) [*34*].
V. Diisopropyl ether-formic acid-water (90 + 7 + 3) saturated with polyethylene glycol, M. Wt. 1000, on Kieselguhr G-polyethylene glycol layers [*19*].

Experimental details are given in the work of BRAUN and GEENEN [*4*]: the acids were dissolved as ammonium salts, 2% in methanol-water (1 + 1) and 2 mm³ (40 μg) of these solutions were applied. The salts were chromatographed on Silica Gel G layers, with solvent II (running time 120 min.), migration-distance of 10 cm, in plain chambers. Only the acids listed in column III of Table 86 were developed in a saturated atmosphere, the conditions otherwise being the same, (migration time 110 min.). The *Rf*-values given by PETROWITZ and PASTUSKA [*34*] were obtained on manually prepared thicker silica gel layers and they should, therefore, only be considered as indicative.

The procedure worked out by KNAPPE and PETERI [*19*] appears to be particularly useful:

30 g Kieselgur G "Merck" and 0.05 g sodium diethyl dithiocarbaminate were carefully stirred in a mortar into a mixture of 45 ml dist. water and 15 g polyethylene glycol, M.Wt. 1000. The slurry was applied to glass plates (20 × 20 cm) by the standard method (pp. 7—9) and dried for 30 min at 100° C. After development (length of run 12 cm), the plates were heated for 10 min at 100° C. After cooling, they were sprayed with a solution of 0.04 g Bromocresol Purple in 100 ml 50% ethanol, adjusted to pH 10 with sodium hydroxide; yellow spots on a blue background were obtained.

Table 86 shows that the h*Rf*-values of the dicarboxylic acids increase both for basic and acidic solvents as chain length increases. The *cis-trans*-isomers, maleic acid (h*Rf* 7) and fumaric acid (h*Rf* 23) can be separated with solvent III, but it was not possible to separate the fatty acids C_1-C_{10}; C_{12}, C_{14}, C_{18} and C_{20} with the basic mixture (see also pp. 167—174).

Table 86. h*Rf*-*Values of Carboxylic Acids on Silica Gel G Layers with various Solvents* (see p. 357)

Acids	Solvents II[1]	III	IV	V[3]	Acids	Solvents II[2]	I[2]
Oxalic acid	5	0	0	14	Citric acid	5	15
Malonic acid	14	13	5	21	Tartaric acid	8	—
Succinic acid	30	28	23	28	Phthalic acid	26	—
Glutaric acid	39	35	28	36	Tricarbalylic acid	—	35
Adipic acid	43	42	34	43	Terephthalic acid	73	—
Pimelic acid	53	47	36	55	Benzoic acid	76	—
Suberic acid	54	50	40	67	p-Toluylic acid	76	—
Azelaic acid	56	53	43	82	Methyl succinic acid	—	80
Sebacic acid	67	55	47	92	o-Phosphoric acid	0	—

[1] Standard method without chamber saturation.
[2] Standard method with chamber saturation.
[3] Kieselgur G-polyethylene glycol-layers, standard method [*19*].

PREY and co-workers [*36*] separated formic, acetic, lactic and pyruvic acids with pyridine-petroleum hydrocarbon (25 + 50) or with ethanol-ammonia-water (80 + 4 + 16) on Silica Gel G layers. Oxalic acid stays at the starting point with both solvents. The di-sulphuric acid ester of dihydro indanthroazine was recommended for visualizing the spots on the plates [*44a*].

For the n-monocarboxylic acids (C_4 to C_{10}) chromatographed by HROMATKA and AUE [*17*] on Silica Gel G with abs. ether, there was a linearity between the log *Rf*-values and the number of C-atoms. There was, however, a different gradient for the even-numbered acids (C_4—C_{10}) than for the odd-numbered (C_5—C_{11}).

For detection of acids or their salts developed with acidic or basic solvents, Bromocresol Green (Reag. No. 22) is recommended. The chromatograms developed with acidic solvents, however, should first be heated for 60 min. at 120° C to remove the acetic acid. After spraying, the acids are recognized as blue spots on a yellow background. For equal amounts of substance, the intensity of the color and size of the spots decreases with increasing chain-length. The detection limit lies between 0.8 and 8 μg, depending on the acid.

4. Insecticides

A large number of institutions, e.g., the Plant Protection Authorities and the Plant Protection Section of the World Health Organization, are concerned with the analytical detection of plant protection substances which may be genuinely toxic. In food control, the recognition of toxic insecticides is of considerable importance and it is of particular interest to know how long a time needs to elapse for sprays applied to fruit and vegetables to lose their toxic properties. The application of large amounts of this kind of product occasionally leads to severe cases of poisoning; the toxicologist should, therefore, be acquainted with means for the detection of insecticides.

The chemical analytical methods used so far are usually nonspecific. Column chromatographic procedures have been successfully employed [*2, 35, 40*]. Although these separations require large quantities, this method has been used for enriching insecticide residues. Müller, Ernst and Schoch [*30*] described paper-chromatography and thin-layer-chromatography as an additional means of identification.

Amounts of 15—20 μg can be detected using paper chromatography for the analysis of insecticides [*9, 11, 27, 39, 55* et al.]. The sensitivity of detection may be increased by combining it with a biological test [*26, 52*]. As the insecticides are mainly strongly lipophilic, impregnation of the paper before development is essential.

Table 87. h*Rf*-*Values of Insecticides on Silica Gel G with Hexane-Acetone* (80 + 20)

Insecticides	*Rf* × 100 according to [*1*]	*Rf* × 100 according to [*54*]
Diazinon	76—82	—
Parathion (E 605)	65—68	62—64
Metasystox	62—64	—
Malathion	52—54	50—54
Chlorthion	43—45	40—44
Fac	20—26	—
Rogor	4—7	8

Fischer and Klingelhöller [*12*] first reported the separation and identification of thiophosphoric acid esters from biological materials by TLC on Silica Gel G layers. These authors fractionated the extracts with the solvent: methanol-methylene chloride-ammonia (10 p.c.) (20 + 80 + 3). The substances were chromatographed at 30–31° C in saturated atmosphere (pp. 14—17). A mixture of 3% sodium azide with 0.1 N aqueous iodine solution (1 + 1) was used as spray reagent (Table 90). Bäumler and Rippstein [*1*] separated thiophosphoric acid esters on Silica G with n-hexane-acetone (80 + 20) (Table 87). Detection was carried out with a weakly acidic 0.5% palladium chloride solution.

The same authors [*1*] chromatographed chlorinated hydrocarbons on more highly activated alumina layers with n-hexane. The layers were heated at 200–220° C for 2 hrs (Table 88). We [*54*] obtained good separations of chlorinated hydrocarbons on Silica Gel G with cyclohexane-chloroform (80 + 20) or petroleum hydrocarbon-carbon tetrachloride (50 + 50) as solvents (Table 88) (Detection see p. 362).

Table 88. h*Rf*-*Values of Chlorinated Hydrocarbons*

Layer:	Alumina highly active [*1*]	Alumina	Silica Gel G[1]	
Solvent:	I	I	II	III
Aldrin	78—82	Front	60—62	60—63
DDT	59—61	83 (94)	52—56	50—53
Perthane	48—50	—	—	—
γ-HCH	39—40	35	27—28	24—26
Dieldrin	17—19	28	15—16	12—14
Methoxychlorine	10—12	—	8—10	5— 8

Solvents: I = n-hexane; II = cyclohexane-chloroform (80 + 20); III = petroleum hydrocarbon-carbon tetrachloride (50 + 50).

[1] Authors' results [*54*].

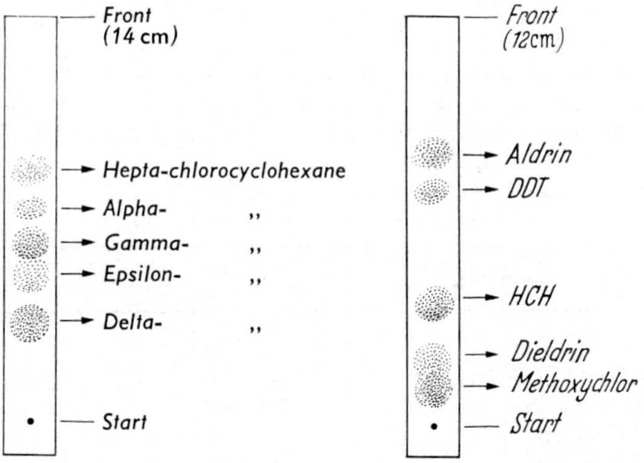

Fig. 148. *Left:* Separation of isomeric hexachlorocyclohexanes on Silica Gel G with solvent I of Table 89. Detection: Treatment of the fluorescein-plates with Rhodamine B and sodium carbonate. *Right:* Separation of various chlorinated hydrocarbons on Silica Gel G with solvent II of Table 89. Detection: as above

In the course of the preparation of hexachlorocyclohexane, mixtures of isomers are obtained. As the individual isomers are very different in their insecticidal efficacy a separation procedure is highly important to the analyst. As Table 89 and Fig. 148 show, it is possible to separate the isomers on Silica Gel G layers (standard method) with a migration-distance of 14 cm. [*54*].

Petrowitz [*33*] also studied the separation of some chlorinated hydrocarbons. He used silica gel layers manually prepared and developed with simple solvents such as n-hexane, cyclohexane or petroleum hydrocarbon (b.p. 50–70°).

Table 89. h*Rf*-*Values*[1] *of Isomeric Hexachlorocyclohexanes on Silica Gel G* [*54*]

HCH-isomers	Solvents	
	I	II
Hepta-	56—60	54—56
Alpha-	40—43	38—40
Gamma-	33—36	37—39
Epsilon-	31—33	37—39
Beta-	25—28	28—30
Delta-	14—17	14—16

Solvents: I = cyclohexane-chloroform (80 + 20); II = petroleum hydrocarbon-carbon tetrachloride (50 + 50).

[1] On less active layers, rather higher values are obtained; separation, however, is little affected.

Deters [*7*] has briefly reported that pentachlorophenol migrates well on acidic Silica Gel G layers prepared with 0.05 N oxalic acid instead of water. On developing with chloroform, h*Rf*-values of about 50 are obtained in saturated atmosphere.

Table 90. h*Rf*-*Values of 9 Thiophosphoric acid Esters and a few Cleavage Products* [*12*]

Substance	Silica Gel G layer	
	I[1]	II
E 605	42	84
p-Nitrophenol	81	90
Diazinon	45	86
Potasan	42 and 56	88
Coumarin residue	80	92
Phenkapton	40 and 68	98—100
Systox	43 and 58	86
Chlorthion	31	82
Cl-Nitrophenol	67	90
Metasystox	30, 40 and 48	87
Melathion	32, 44, 58 and 66	99—100
Thiometon	34, 28 and 58	90

[1] Solvent I = methylene chloride-methanol-ammonia solution (80 + 20 + 3); II = methylene chloride-methanol-ammonia solution, 10 per cent. (65 + 35 + 5).

Detection of insecticides: Although the thin-layer chromatographic separation offered no difficulties, it was not easy to find suitable color-reactions for detecting the substances on the plates.

Quite successful is the use of the method given by BÄUMLER and RIPPSTEIN [*1*]: The plate is sprayed with a solution of 0.5 g NN-dimethyl-p-phenylenediamine-hydrochloride in sodium ethylate solution (1 g. Na in 100 ml ethanol). The plate is moistened with water and held for about 1 min. under a UV-lamp without filter. BÄUMLER and RIPPSTEIN worked with alumina layers. On Silica Gel G this reaction is somewhat less sensitive.

In studies with fluorescent reagents, we [*54*] found a very sensitive method which permits detection of as little as 0.02 μg hexachlorocyclohexane. This method has proved of value for a whole series of other classes of substances. One starts with layers impregnated with fluorescein-sodium (Reag. 90f). If such plates are left exposed to air for 1–2 hrs. after development, chlorinated hydrocarbons may be recognized as rose-colored spots, in amounts of over 5 μg, or by a quenching of a blue-violet fluorescence under UV-light. This coloration may also be obtained immediately after development if one brings the plate carefully into contact with steam (from a boiling water-bath). Color intensity is maximal only for a definite degree of moisture content. The necessary moistening may, however, be repeated without damage.

On spraying with Rhodamine B (Reag. No. 129), some of the chlorinated hydrocarbons may be seen under UV-light as rose-colored fluorescent spots. When a 10% sodium carbonate solution is uniformly sprayed, the remaining compounds also exhibit a yellow fluorescence in UV-light.

We improved this method of coloring by covering the plate sprayed with Rhodamine B with a wet filter paper previously dipped in 10% sodium carbonate solution. The paper must be drawn across from one side so that no air bubbles form beneath it. Then a second filter paper, also soaked in sodium carbonate, is laid on top and finally a third dry filter paper pressed down with a glass plate. After 15 min., the glass plate is removed and after drying the papers (about 2 hrs.) they are removed. Red spots which strongly fluoresce in UV-light are obtained from the insecticides. On the filter papers taken off, the same spots fluoresce yellow in UV-light. Paper and plates may be kept for years for demonstration purposes. The papers are improved if the 0.04% aqueous sodium fluorescein solution used in preparing the fluorescein plates contains 0.5% sodium chloride.

Chlorobenzenes must first be nitrated. They then react with Rhodamine B or fluorescein in the manner described.

PETROWITZ [*33*] uses the following procedure for identifying DDT and HCH-isomers: ethylamine is sprayed on the plates, which are then heated for 20 min. at 100° C. Another spray of a mixture of 0.1 N-silver nitrate solution and nitric acid, D 1.4 (10 + 1) is applied and the layer is exposed to UV- or sun light.

To detect the *thiophosphoric acid esters* shown in Table 90, FISCHER and KLINGELHÖLLER [*12*] recommended spraying with conc. iodine azide solution (3% NaN_3 in 0.1 N iodine solution). The white spots should be labelled immediately (limit of detection 1–5 μg). Nitrophenol is made visible by exposing to ammonia vapor.

Pyrethrin and synergists

Besides synthetic products, plant insecticides such as pyrethrin mixtures[1] and rotenone are finding increased application. In contrast to many synthetic substances, these natural products are not harmful to warm-blooded animals. Moreover, their application has not yet caused any resistance. In a number of commercial preparations one finds, in addition to synthetic insecticides, those derived from plants, and also synergists (e.g., piperonyl butoxide, bucarbolate). These synergists markedly increase the insecticidal effect of pyrethrins. The natural pyrethrins show pronounced differences in their effectiveness and sensitivity to light-catalyzed autoxidation [*51a*].

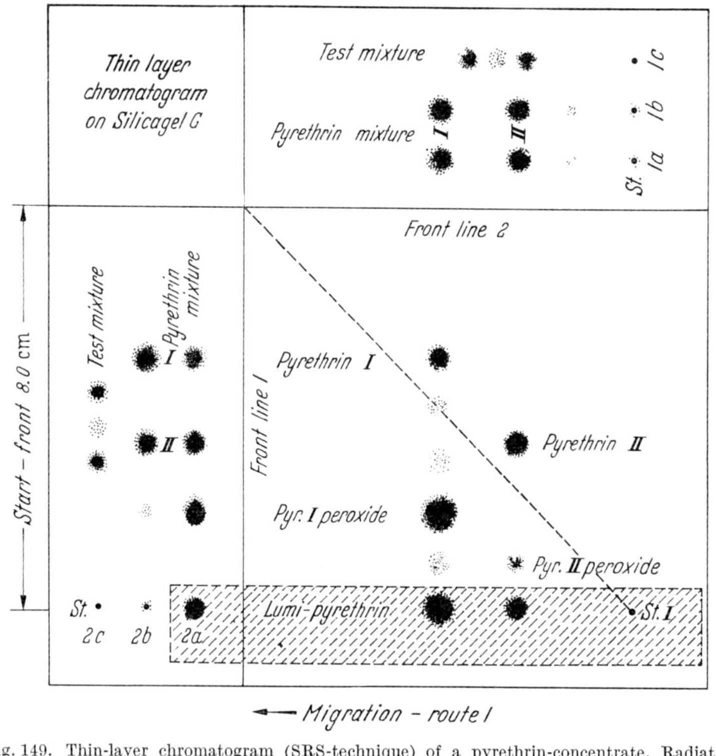

Fig. 149. Thin-layer chromatogram (SRS-technique) of a pyrethrin-concentrate. Radiation (the shadowed region) followed the separation along direction 1. All the spots which showed up with antimony trichloride, 2,4-dinitrophenylhydrazine, or potassium iodide/acetic acid/starch, are shown as blackdots. (For further details, see text)

In studying pyrethrin inactivation by radiation effects, the best results have been obtained by two-dimensional TLC [*51a*]. A few crude pyrethrin concentrates dissolved in benzene (5%) were applied to a Silica Gel G layer and pyrethrin I and pyrethrin II were separated. UV

[1] Spickett [*49*] has already shown that pyrethrin I and II may be separated on silica gel-gypsum layers.

or sun light were allowed to act on these substances (Fig. 149, dotted region) and the same solvent was then used to develop in the second dimension. The highly polar oxidation products, appearing after irradiation of the chromatogram, lie below the respective pyrethrins. The sizes of the zones allow conclusions to be drawn about the speed of decomposition. As intermediate products, pyrethrin peroxides were found which are no longer effective as insecticides, and also, as end-products, the similarly ineffective lumi-pyrethrins.

For general applications of this "SRS-technique", the reader is referred to p. 36.

Solvents and layers

Silica Gel G layers prepared by the standard method were used for separating pyrethrins and synergists. For developing in saturated atmosphere, the following chloroform-isoeluotropic solvents are suitable:

1. Benzene-methyl ethyl ketone (90 + 10),
2. Benzene-ethyl acetate (85 + 15),
3. Carbon tetrachloride-ethyl acetate (80 + 20),
4. Hexane-methyl ethyl ketone (80 + 20) and
5. Hexane-ethyl acetate (75 + 25).

The h*Rf*-values given in Table 91 are obtained with the last named solvent.

Table 91. h*Rf*-*Values and Reactions* [*51a*]

Substances	*Rf* × 100 (mean) Direction 1	Direction 2	$SbCl_3$ ($SbCl_5$)*	P_2O_5 24 MoO_3	2,4-dinitroph.	KJ-AcOH-starch
Pyrethrin I	50	62	+	+	+	—
Pyrethrin II	30	41	+	+	+	—
Isopyrethrin I	57	72	+	+	+	—
Isopyrethrin II	35	49	+	+	+	—
Pyrethrin I peroxide	—	23	(+)	(+)	+	+
Pyrethrin II peroxide	—	11	(+)	(+)	+	+
Isopyrethrin I peroxide	—	37 (23)	(+)	(+)	+	+
Isopyrethrin II peroxide	—	17	(+)	(+)	+	+
Lumipyrethrin	—	0	+	+	+	+ (+)
Allethrin	48	—	+*	+		
Piperonyl butoxide[1]	35	—	+	+		
Bucarpolate	23	—	+*	+		
S 421	67	—	—	—	Chlorine detection	
Butter Yellow	42	53	red	red	red	red
Indophenol	35	45	yellowish	yellowish	yellowish	yellowish
Sudan Red G	27	36	red	red	red	red

[1] Better separation from the pyrethrins is sometimes achieved with other chloroform-isoeluotropic mixtures.

Detection

For the visualization of the compounds listed in Table 91, the following reagents give good results:

1. Antimony trichloride (Reag. No. 11) heating for 10 min. at 110° C. pyrethrin I and II are recognizable as grey-green spots; isopyrethrin shows a grey brown. In UV-light (365 mμ), isopyrethrins show a blue fluorescence, pyrethrins a brown-yellow fluorescence. The pyrethrin-synergist piperonyl butoxide stains violet. The limit of detection for pyrethrins is around 2—3 μg.

2. Antimony pentachloride (Reag. No. 13) reacts in the cold, but it gives less color distinction. It has been used particularly for visualizing allethrin and bucarpolate (heating for 10 min. + 120° C). Brown spots are obtained.

3. Phosphomolybdic acid (Reag. No. 120b). Pyrethrins and synergists appear blue on a yellow background after heating the chromatogram (unspecific). Limit of detection for pyrethrin, about 1 μg.

4. 2.4-Dinitrophenylhydrazine (Reag. No. 52a). Pyrethrins and oxidation products react with a yellow color which intensifies towards orange.

5. The potassium-iodide-starch reaction (Reag. No. 85) has been used for the detection of pyrethrin peroxides.

6. Detection of chlorine in octachloro dipropylether (synergist S 421, manufactured by BASF) and in analogous compounds: spray the chromatogram with alcoholic 2N-potassium hydroxide and heat for 20 mins. at 120° C. Then spray on 1% silver nitrate solution in nitric acid (30%). On exposure to UV or sunlight, a grey-violet spot appears. Limit of detection, 2—3 μg.

Of particular interest is a *biological* method for the detection of insecticidal constituents of chromatographically separated mixtures. STAHL [*51a*] applied the highly sensitive *Aedes*-larva-test of BRUCHFIELD and HARTZELL [*5*], and also the less recommendable *Drosophila*-test.

Method: If the position of the insecticides on the chromatogram is unknown, the various zones are scraped off one by one in the manner described for the growth hormones test (p. 300). The scrapings are distributed in vessels containing the experimental organisms. For the *Aedes*-test, 10 *Aedes aegyptici*-larvae (age 5—8 days, length 2—4 mm) are each put into a hollow polished glass dish (30 mm ∅ capacity 0.4 ml), filled with 0.2 ml rain-water. Then, the sample to be examined, still adsorbed onto the silica gel, is shaken in and a cover-slip put on. The movements of the larvae are observed for 12 hrs. For the very effective pyrethrins I and II (amounts about 1—10 μg), the larvae show immobility after only 10 min. and severe convulsions; exitus within 12 hours with light falling from one direction only; the unharmed larvae show photophobic behavior.

5. Artificial flavors and sweeteners

a) Artificial sweeteners (saccharin, dulcin) [*54*]

To detect saccharin and/or dulcin, their acidified aqueous solution is extracted with ethyl acetate. The ethyl acetate extract contains the saccharin. The acidic, aqueous phase is then made alkaline with sodium hydroxide and dulcin is similarly extracted with ethyl acetate. The two extracts thus obtained may be chromatographed on Silica Gel G by the standard method. With the solvent chloroform-acetic acid (90 + 10), the h*Rf*-value of saccharin is about 30 and that of dulcin about 50. For detection, Rhodamine B (Reag. No. 129) is first sprayed on, followed by spraying with silver nitrate solution, Reag. No. 137.

b) Flavors

Experimental details for isolating and separating flavors frequently encountered in food-stuffs and cosmetic preparations will be found in the chapter on terpene derivatives etc. (pp. 186—210).

III. Other additives and synthetic products

1. Polyphenyl mixtures (organic cooling agents for reactors)

Mixtures of polyphenyls have attracted attention for cooling reactors. At the high working temperatures of about 400° C, complex mixtures may arise as a result of radiation and/or temperature effects. GEISS and SCHLITT [*14*] reported on the possibilities of thin-layer chromatographic analysis of mixtures of polyphenyls of the type: $C_6H_5\text{-}(C_6H_4)_n\text{-}C_6H_5$, $n \geq 0$.

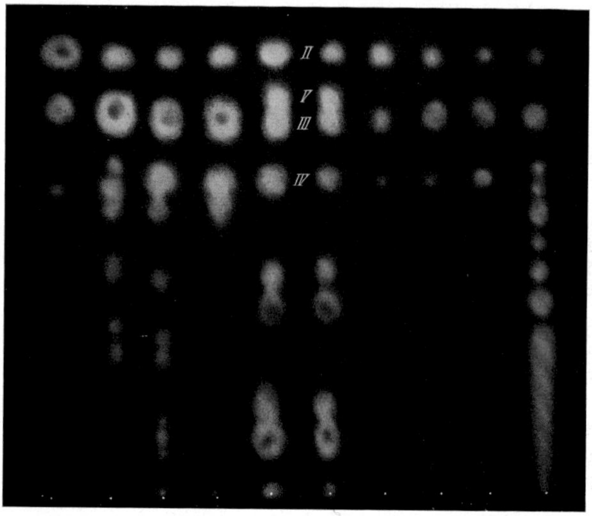

Fig. 150. UV-photograph of a thin-layer chromatogram of technical polyphenyl mixtures. The amount of substance applied was only 0.2 or 2 μg. *II* = o-terphenyl, *III* = *m*-terphenyl, *IV* = p-terphenyl and *V* = 2′,2″-quaterphenyl (GEISS and SCHLITT) [*14*]

These authors followed the standard conditions and they have shown that the activity of the adsorbent is of exceptional importance for the success of the separation. For instance, a separation of o-, m-, and p-terphenyl could be obtained only with a mixture of Aluminium Oxide G "Merck" and the Aluminum Oxide for TLC manufactured by Fluka. The layers were activated for 2 hrs. at 110° C, cooled, and kept in a desiccator over phosphorus pentoxide. In a number of cases, the samples were spotted onto plates warmed to 40° C. 0.1 μg of the pure substance and 0.2–2 μg of technical mixtures were applied. o-Terphenyl was chosen as a reference substance; the R_{St}-values relative to it are given in Table 92. Heptane was used as the solvent, but in a few cases, n-pentane was used. The length of run was 15 cm in the S-chamber system (pp. 18—19).

Spraying with a 0.3% ceric sulphate solution in conc. nitric acid was shown to be a highly sensitive detector. The polyphenyls occur in various colors even in daylight; in long-wave ultra-violet light (366 mμ) they show intensive fluorescence (Fig. 150) and amounts as small as 0.05 μg may be recognized easily.

Table 92. *R_{St}-Values and Detection of Polyphenyls on Alumina Layers [14]*

No.	Polyphenyls	R_{St}-values	Ceric sulphate-HNO_3 Daylight	UV (366 mμ)[1]
I	Biphenyl	1.10	—	—
II	o-Terphenyl (= St.)	1.00	light yellow	=
III	m-Terphenyl	0.92	red	=
IV	p-Terphenyl	0.85	brown	=
V	2′,2″-Quaterphenyl	0.88	light brown	=
VI	3′,3″-Quaterphenyl	0.69	violet	=
VII	4′,4″-Quaterphenyl	0.45	—	white
VIII	4′,3″,3‴-Quinquaphenyl	0.36	blue	=
IX	4′,4″,4‴-Quinquaphenyl	0.06	—	white
X	4′,3″,2‴,4‴-Hexaphenyl	0.29	green-yellow	=
XI	Dixenylmethane	0.56	yellow	=
XII	Triphenylene	0.78	brown-yellow	=

[1] Fluorescent colors are the same as those seen in daylight.

2. Ferrocene derivatives

To follow the reaction during the synthesis of ferrocene derivatives, Schlögl et al. [*43, 44*] used TLC on Silica Gel G. Numerous ferrocenylcarbonyl compounds, ferrocenyl carbinols, esters and ethers, and N-containing derivatives of ferrocene could be separated with benzene and benzene-ethanol mixtures as solvents. The carbonyl compounds, particularly interesting for preparation, were developed at various times in I benzene, II benzene-ethanol (30 + 1) and III benzene-ethanol (30 + 2) and the *Rf*-values were determined.

Detection: The color of ferrocene derivatives makes it unnecessary to look for a special staining process, in most cases. Alkyl ferrocenes, and derivatives without a chromophoric group at the ferrocene nucleus, are colored yellow, monoacyl ferrocenes exhibit an orange and diacyl compounds a red color. Detection of less intensely colored ferrocene derivatives can succeed if they are transformed into more deeply colored derivatives of the ferrocinium ion, e.g., with the help of an oxidizing reagent.

The *Rf*-values of about 50 ferrocenes are graphically presented in the original publication [*43*]. The *Rf*-values increase with increasing ethanol concentration in the solvent and with increasing "hydrocarbon characteristics" of the substances investigated. 1,1′-Disubstituted products migrate more slowly than the corresponding monosubstituted products. As a consequence of this, it is possible, for example, to decide quickly whether acylferrocenes prepared by the Friedel-Crafts reaction are mono- or di-substituted.

Alkyl ferrocenes, like ferrocene itself, migrate near the solvent front with the solvents mentioned. They can be chromatographed with n-hexane as solvent [*44*]. As the acylferrocenes remain at the starting point with hexane, they can be easily separated from the alkyl ferrocenes. Thus, it is possible, for instance, to follow the reduction of acyl ferrocenes to alkyl ferrocenes.

3. Detection of organo-tin stabilizers

Organic tin compounds are used especially for stabilizing polyvinyl chloride. Such substances have been investigated chromatographically on silica gel layers by TÜRLER and HÖGL [*53*]. Dibutyl tin dilaurate, dichloride, dioleate, dimaleate and tributyl tin chloride have been chromatographed with n-butanol-acetic acid (60 + 1) saturated with water. The dibutyl compounds migrated, independently of the acid residue, to equal distances, whereas the toxic tributyl compounds, which occur as impurities in the dibutyl compounds, were clearly separated. With the same technique, dibutyl tin salts could be separated from dioctyl tin salts. Dibutyl, dioctyl, dibenzyl and tribenzyl tin salts have been segregated in groups using the solvent water-n-butanol-ethanol-acetic acid (40 + 20 + 20 + 2). The separated substances were detected by spraying with a solution of dithizone in chloroform (10 mg/100 ml), upon which both dialkyl- and trialkyl tin compounds were colored. Spraying with diphenylcarbazone solution (Reag. No. 57) only stains the dialkyl tin salts.

Note: As the layer had been prepared by spraying on a kieselgel-suspension, an even thickness could not be achieved. These experiments should be repeated under standard conditions in order to establish more reproducible *Rf*-values.

4. Cyclourethane, Phosphineoxide etc.

1. Cyclodiurethanes are dissolved in cyclohexanone and 70 μg substance is applied to Silica Gel G layers prepared by the standard method [*18*]. Benzene – ethyl acetate – cyclohexanone (50 + 10 + 20) is used as solvent. After removal of the solvent (2 hrs at 120° C) the substances are detected by spraying with a saturated solution of chlorine in carbon tetrachloride. The chlorine held on the layer is removed by leaving the plate for an hour. Then, aqueous potassium iodide-starch solution is sprayed on. The detection limit for oligo-urethane is about 0.1 μg.

2. ERTEL and HORNER [*10*] have investigated separation and detection of phosphine oxides and phenylbenzyl sulphur compounds on Silica Gel G layers. As solvents, they used acetone, mixtures of acetone-benzene and chloroform. Chromic-sulphuric acid (Reag. No. 34) and potassium permanganate-sulphuric acid (Reag. No. 86a) were recommended as spray reagents.

5. Nitramine explosives

The explosive Hexogen (hexahydro-1,3,5-trinitro-s-triazine) is obtained by nitration of hexamethylenetetramine. To follow the course of the reaction and to find by-products arising during the synthesis, HARTHON [*15*] worked out a thin-layer chromatographic method for the nitramines concerned, as shown in Table 93.

The plates were prepared by the standard method with Silica Gel G and about 30 μg of the individual substances were applied as methanolic solutions. As a solvent, petroleum ether (b.p. 40–60°)-acetone (20 + 12) was used. After separation, the nitramines could be recognized as dark spots after spraying with fluorescent indicator solution (Reag. No. 90a–i). Very small amounts (0.5 μg) can be detected as blue-violet spots on a faint brown background if they are sprayed with a 1% solution of diphenylamine in ethanol and the plates are illuminated with a 125 W mercury pressure lamp.

Table 93. h*Rf-Values of Nitramine Explosives on Silica Gel G*
Solvent: Petroleum hydrocarbon-acetone (20 + 12) [*15*]

Nitramine explosives	*Rf*-values × 100
1-Acetyl octahydro-3,5,7-trinitro-1,3,5,7-s-tetrazine	16
Octahydro-1,3,5,7-tetranitro-s-tetrazine (Octogen)	48
3,7-Dinitro-1,3,5,7-tetrazabicyclo-(3,5,1)-nonane	62
Hexahydro-1,3,5-trinitro-s-triazine (Hexogen)	71
2,4,6-Trinitro-2,4,6-triazaheptane-1,7-diolacetate	93
2,4,6,8-Tetranitro-2,4,6,8-tetrazanonane-1,9-diolacetate	93

6. Photochemicals

In investigating the numerous chemicals required in preparing and developing color films, thin-layer chromatography has been used extensively and there is a large amount of experimental material available, much as yet unpublished. A paper by EGGERS [*8*] gives 7 color photographs of thin-layer chromatograms, without reporting the separation and detection conditions. It appears that mixtures of o-, m-, and p-aminophenols and N-methyl-p-aminophenols can be separated, as well as a series of naphthylamine (1)-mono-and di-sulphonic acids. Suggestions for the choice selection of appropriate solvents for the TLC of phenols may be found on p. 312; for the sulphonic acids, solvents of acidic character should probably be used and a good choice is possibly provided by the very highly polar solvents listed on p. 376.

For visualization, the partly known procedures of color-film "development" may be used in addition to reagents for the detection of phenols and amines.

Bibliography to Chapter H. Synthetic Organic Materials

[*1*] BÄUMLER, J., u. S. RIPPSTEIN: Helv. Chim. Acta **44**, 1162 (1961).
[*2*] BECKMANN, H. F.: Anal. Chem. **26**, 922 (1954), and **20**, 610 (1948).
[*3*] BRAUN, D.: Kunststoffe **52**, 2 (1962).
[*4*] —, u. H. GEENEN: J. Chromatog. **7**, 56 (1962).
[*5*] BRUCHFIELD u. HARZELL: J. Econ. Entomol. **48**, 210 (1955), ref. [*51*].
[*6*] DAVIDEK, J., u. J. POKORNÝ: Z. Lebensm.-Untersuch. u. Forsch. **115**, 113 (1961).
[*6a*] — — u. G. JANICEK: Z. Lebensm.-Unters. u. Forsch. **116**, 13 (1961).
[*7*] DETERS, R.: Chem. Ztg. **86**, 388 (1962).

[8] Eggers, J.: Phot. u. Wiss. **10**, 40 (1961).
[9] Eichenberger, J., u. L. Gay: Mitt. Gebiete Lebensm. u. Hyg. **51**, 423 (1961).
[10] Ertel, H., u. L. Horner: J. Chromatog. **7**, 268 (1962).
[11] Fischer, R., and N. Otterbeck: Sci. Pharm. **27**, 1 (1959).
[12] —, u. W. Klingelhöller: Arch. Toxikol. **19**, 119 (1961).
[13] Gänshirt, H., u. K. Morianz: Arch. Pharm. **293**, 1065 (1960).
[14] Geiss, F., u. H. Schlitt: Publ. of EURATOM, AUR I-1-Chemistry, Nov. 1961.
[15] Harthon, J. G. L.: Acta Chem. Scand. **15**, 1401 (1961).
[16] Häusser, H.: Arch. Kriminol. **125**, 72 (1960).
[17] Hromatka, O., u. W. A. Aue: Mh. Chem. **93**, 503 (1962).
[18] Kern, W., K. J. Rauterkus u. W. Weber: Makromol. Chem. **43**, 98 (1961).
[19] Knappe, E., u. D. Peteri: Z. anal. Chem. **188**, 184 u. 352 (1962).
[20] Kröller, E.: Fette, Seifen, Anstrichmittel **64**, 85 (1962).
[21] Lagoni, H., u. A. Wortmann: Milchwissenschaft **10**, 360 (1955); **11**, 206 (1956).
[22] Lichtenberger, W.: Private communication and Z. anal. Chem. **185**, 111 (1962).
[23] Machata, G.: Arch. Toxikol. **127**, 1 (1961).
[24] Meckel, L., H. Milster u. U. Krause: Textil-Praxis **16**, 1032 (1961).
[25] Meyer, H.: Deut. Lebensm. Rundschau **57**, 170 (1961).
[26] Mills, P. A.: J. Assoc. Offic. Agr. Chemists **42**, 734 (1959).
[27] Mitchell, L. C.: J. Assoc. Offic. Agr. Chemists **36**, 1183 (1953); **40**, 294 (1957); **41**, 781 (1958).
[28] Montag, A.: Z. Lebensm.-Untersuch. u. Forsch. **116**, 413 (1962).
[29] Mottier, M.: Mitt. Gebiete Lebensm. u. Hyg. **47**, 372 (1956).
[29a] —, u. M. Potterat: Analyt. Chim. Acta **13**, 46 (1955).
[30] Müller, R., G. Ernst u. H. Schoch: Mitt. Gebiete Lebensm. u. Hyg. **48**, 152 (1957).
[31] Pastuska, G., u. H. Trinks: Chem. Ztg. **86**, 135 (1962).
[32] Peereboom, J. W. C.: J. Chromatog. **4**, 323 (1960).
[33] Petrowitz, H. J.: Chem. Ztg. **85**, 867 (1961).
[34] —, u. G. Pastuska: J. Chromatog. **7**, 128 (1962).
[35] Press, J. M.: Bull. World Health Organization **20**, 153 (1959).
[36] Prey, V., H. Berbalk u. M. Kausz- Mikrochim. Acta **1962**, 449.
[37] Rey, E., u. L. Erhart: Bull. schweiz. elektrotech. Ver. **52**, 401 (1961).
[38] — Elektrotechn. Z. Ausg. B **13**, 299 (1961).
[39] San Antonio, J. P.: J. Assoc. Offic. Agr. Chemists **43**, 721 (1960).
[40] Sándi, E.: Z. anal. Chem. **167**, 241 (1959).
[41] Schetty, G., u. W. Kuster: Helv. Chim. Acta **44**, 2193 (1961).
[42] — Helv. Chim. Acta **45**, 809, 1095 (1962).
[43] Schlögl, K., H. Pelousek u. A. Mohar: Mh. Chem. **92**, 533 (1961).
[44] — A. Mohar u. H. Pelousek: Mh. Chem. **92**, 921 (1961).
[44a] — Naturwissenschaften **46**, 447 (1959).
[45] Schorn, P. J., u. E. Stahl: Unpublished.
[46] Seher, A.: Fette, Seifen, Anstrichmittel **61**, 345 (1959).
[47] — Nahrung **4**, 275 u. 466 (1960).
[48] — Mikrochim. Acta **1961**, 308.
[49] Spickett, R. G.: Chem. & Ind. (London) **1957**, 561.
[50] Stahl, E.: Chem. Ztg. **82**, 323 (1958).
[51] — Arch. Pharm. **292**, 411 (1959).
[51a] — Arch. Pharm. **293**, 531 (1960).
[52] Strache, F., u. J. Indinger: Deut. Lebensm. Rundschau **57**, 197 (1961).
[53] Türler, M., u. O. Högl: Mitt. Gebiete Lebensm. u. Hyg. **52**, 123 (1961).
[54] Waldi, D.: Unpublished.
[55] Winteringham, F. P., A. Harrisson and R. G. Bridges: Nature (London) **177**, 86 (1956).
[56] Wollenweber, P.: J. Chromatog. **7**, 557 (1962).

Some recent noteworthy articles

Badger, G. M., J. K. Donnelly and T. M. Spotswood: J. Chromatog. **10**, 397 (1963). TLC of polycyclic aromatic hydrocarbons on partially acetylated cellulose.

Beroza, M.: Agric. Food Chem. **11**, 51 (1963). 3,4-Methylenedioxyphenyl synergists.
Bürger, K.: Z. analyt. Chem. **192**, 280 (1963). Organic tin compounds.
Curtis, R. F., and G. T. Phillips: J. Chromatog. **9**, 366 (1962). Thiophene derivatives.
Deters, R.: Holz als Roh- u. Werkstoff **21**, 362 (1963). Analysis of wood preservatives.
Druding, L. F.: J. Chem. Educ. **40**, 536 (1963). TLC of ink pigments.
Haub, H. G., and H. Kämmerer: J. Chromatog. **11**, 487 (1963). Polynuclear compounds (novolacs).
Heide, R. F. v. d., and O. Wouters: Z. Lebensm.-Untersuch. u. -Forsch. **117**, 129 (1962). Antioxydants in polyethylenes.
Korte, F., and J. Vogel: J. Chromatog. **9**, 381 (1962). Lactones, lactames, and thio-lactones.
Kucharczyk, N., J. Fohl and J. Vymetal: J. Chromatog. **11**, 55 (1963). Aromatic hydrocarbons and heterocyclic compounds.
Pastuska, G., and H.-J. Petrowitz: J. Chromatog. **10**, 517 (1963). *cis-trans* Isomeric carboxylic acids.
Ronkainen, P.: J. Chromatog. **11**, 228 (1963). Keto acids.
Salo, T., K. Salminen and K. Fiskari: Z. Lebensm.-Unters. u. -Forsch. **117**, 369 (1962). Pesticide residues.
— — Suomen Kem. B **35**, 146 (1962). Synthetic food colors.
Schneider, H., and J. Hofstetter: Dtsch. Apoth. Ztg. **103**, 1423 (1963). Synthetic dyestuffs in drugs.
Stahl, E., and J. Pfeifle: Z. analyt. Chem. **200** (1964). Organic iodine compounds.
Walker, K. C., and M. Beroza: J. Assoc. Offic. Agr. Chemists. **46**, 250 (1963). Insecticides.

I. Hydrophilic Constituents of Plants,

especially of Medicinal Plants

By

Egon Stahl and P. J. Schorn

Quite a large number of plant constituents with more or less pronounced lipophilic characteristics have already been dealt with in preceding chapters. There remain, however, groups of hydrophilic substances which cannot be included therein, or in subsequent chapters. Many of the compounds mentioned here are characterized by the presence of phenolic hydroxyl groups, either free, methylated, or etherified with sugars. Medicinal plant extracts obtained with diluted ethanol generally contain mixtures of these compounds. Methods for the analysis of such extracts are needed. It appears that, under suitably chosen conditions, many plant extracts can be satisfactorily characterized by TLC.

I. Natural α- and γ-pyrone derivatives

Pyrone derivatives are widespread in the vegetable kingdom. The most important α-pyrone derivatives are the coumarins, the isocoumarins and the 6,7- and 7,8-furocoumarins.

Their formulae indicate that these substances are unsaturated aromatic lactones. Nearly all natural coumarin derivatives have, at least in the 7-position, a free or esterified OH-group.

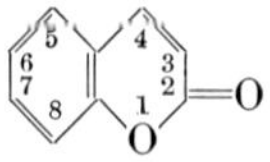
Coumarins

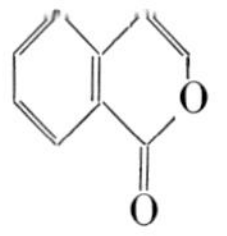
Isocoumarins

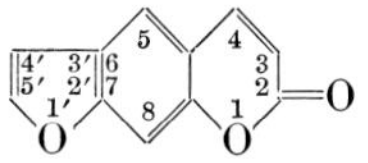
6,7-Furocoumarins

Along with the methyl esters (esters of vinylogous-homologous carboxylic acids), the corresponding hydrophilic glycosides are found. Numerous compounds of this group display a more or less marked fluorescence. Their many derivatives can often be sublimed directly out of the plant material. They are similar to the γ-pyrone derivatives both as far as polarity and their behavior during chromatography are concerned. The relationships between structure and *Rf*-values (p. 378) for the γ-pyrone derivatives apply to α-pyrones as well.

In analyses of plant material, γ-pyrone derivatives are much more commonly encountered. This large number of compounds may be classified into the group of natural chromones and into flavonoids.

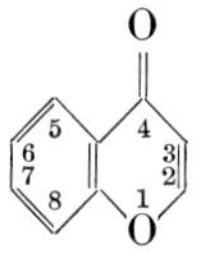
Chromones
(Benzo-γ-pyrone)

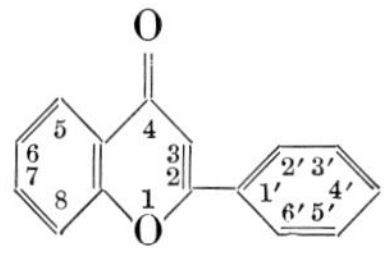
Flavones
(2-Phenylbenzo-γ-pyrone)
(light to dark-yellow)

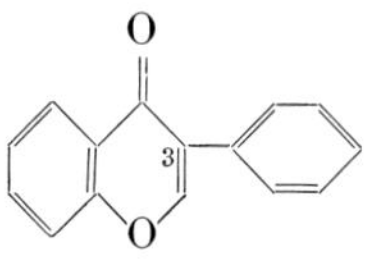
Isoflavones
(3-Phenylbenzo-γ-pyrone)
(colorless)

Flavones occur in differing states of oxidation and can be categorized as follows:

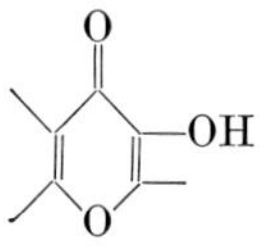
Flavonols
(light- to dark-yellow)

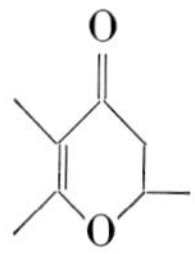
Flavanones
(colorless)

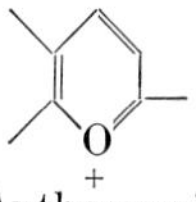
Anthocyanidin
(red, blue, violet)

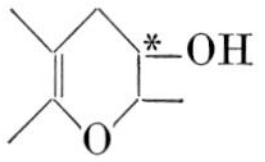
Catechin
(colorless)

A further subdivision can be made based on the number and position of substituents, which are mainly OH-groups. Etherification with sugars most commonly occurs at C_3, and rarely at C_5, C_7 or other positions. MEIER and FÜRST [*31*] recently isolated a 5,3′-dihydroxy-3,6,7,8,4′,5′-hexamethoxy-flavone which they named digicitrin. Further information has been provided by VENKATARAMAN [*55*] who published on flavonoids, FREUDENBERG and WEINGES [*13*] on catechins, and SCHMID [*48*] on chromones. Analytical techniques, including methods of isolation, have been summarized by GEISSMANN [*14*]. DEAN [*9*] and others have described the naturally occurring coumarins. Good examples of isolation are to be found in the numerous publications on this subject by SPÄTH [*49*] as well as by BAERHEIM SVENDSEN [*2*]. Paper chromatographic identification is dealt with by REPPEL [*44*].

1. Concentration from plant material

The parent substances shown in the above formulae are, of course, lipophilic. As they normally contain OH-groups, however, which can be linked with sugars, their derivatives can be hydrophilic and thus a general scheme for a method of concentration cannot be suggested.

a) Anthocyanins (anthocyanidin + sugar) can often be extracted using a cold homogenate of the crushed material with 1% aqueous or methanolic hydrochloric acid. After acid hydrolysis, 15 min. at 100° C., the anthocyanidins can be extracted with amyl alcohol. Further details are given by BATE-SMITH [*3*] and others.

b) Flavonoids and coumarin derivatives, including numerous unwanted impurities, can be obtained by exhaustive extraction using methanol. Glycosides, on the other hand, can be extracted from the plant material with boiling water. In nearly every case, the methanol extracts are carefully and substantially concentrated to a high degree, then diluted with a little water and, quite often, extracted with ethyl acetate-methanol (95 + 5). The di- and triglycosides remain in the aqueous phase, while the monoglycosides and aglycones pass into the ester phase. Further purification can be achieved by precipitation reactions with neutral and/or basic lead acetate. In recent years, unwanted substances have been separated out and satisfactorily, crystallizing fractions have been obtained using polyamide column chromatography [*12, 15, 23, 35*].

2. Coating materials and solvents

As with other groups of substances, separation on a thin, fine-grained layer is much better than can be achieved by column chromatography. It seems helpful, however, to provide a short resumé of earlier experiences because the adsorbents and solvents that have previously proved successful in column chromatography can also be used in TLC.

a) Column chromatography of flavonoids

It was realized early that alumina is not suitable for chromatography of flavonoids because of the formation of stable complexes. Calcium sulphate, which has been used in column chromatography of anthocyanidins [*26*], would seem more suitable. Successful separation of catechins [*6, 7*] and synthetic anthocyanidins [*49a*] can be achieved using moist silica gel as stationary phase and ether as mobile phase. In several papers, "Magnesol", a magnesium trisilicate hydrate[1], is recommended [*24, 42*]. Flavonoid mixtures can be separated in the magnesol columns with water-saturated ethyl acetate. Ion-exchange columns have also been used to purify crude extracts [*14*]. Separation in columns of cellulose powder has proved satisfactory only with simple mixtures. Better results for micropreparative pre-separations have been obtained

---

[1] A magnesium silicate for use in TLC is being manufactured by Messrs. Woelm, Eschwege, Germany, 15 g. of this product is applied to the plates as a slurry with 45 ml. water.

using filter cardboard instead of the normal filter paper. During recent years, the qualities of polyamide powder (Perlon, Ultramide, Silon, etc.) as a sorbent have been recognized.

Polyamide has been used to separate mixtures of low-molecular tanning substances, flavones, chalkones, flavanones, polyhydroxyphenols, quinones, and DNP-amino acids. Pertinent literature references are to be found in [*12*]. These investigations indicate that polyamide chromatography is suitable for the separation of phenols and their derivatives. The amide groups in the polyamide are responsible for their retentive powers due to the formation of hydrogen bondings with phenolic OH-groups. As a result, the phenols are retained according to the number of OH-groups but o-diphenols are retained like a monophenol. A substance adsorbed in this manner may be eluted by replacement with a suitable solvent.

HÖRHAMMER [*23*] described a series of solvents arranged in order of increasing eluent effect: water → ethanol → methanol → acetone → diluted alkaline solutions → formamide → dimethylformamide.

b) Thin-layer chromatography

Although there exists considerable information, summarized by HÄNSEL [*20*], PROCHÁZKA [*43*], and others, about the applicability of paper chromatography for the fractionation of coumarins [*2*, *44*] and flavonoids, separations on fine-grained cellulose layers have not yet been reported. Results obtained by our group [*51*] indicate, however, that silica gel layers give better resolutions.

α) Silica Gel G layers

In studies of biogenesis of isoflavones, GRISEBACH [*16*, *17*] has had good success in separating radioactively labelled flavonoids on Silica Gel G layers. Mixtures of benzene and ethanol, e.g., 92 + 8, were used as solvents. BILLEK [*5*] used TLC of coumarin and its precursors isolated from woodruff; benzene served as the developing solvent.

10 mg. Methanol extract was separated on a Silica Gel G layer with benzene; the coumarin zone was scraped off and extracted with chloroform. The residue was sublimed. A further separation by paper chromatography was then carried out [*5*].
Those techniques were used also by WEYGAND and co-workers [*62*, *63*] in similar studies.

TLC proved useful in detecting and characterizing digicitrin, a substance recently isolated from *Digitalis purpurea* L. by MEIER and FÜRST [*31*]. The following h*Rf* values were obtained with Silica Gel G layers using benzene-ethyl acetate (75 + 25) as solvent:

5,3′-dihydroxy-3,6,7,8,4′,5′-hexamethoxy-flavone (digicitrin)	38—40
3,5,6,7,8,3′,4′,5′-octamethoxy-flavone	24—28
5-hydroxy-3,6,7,8,3′,4′,5′-heptamethoxy-flavone	64—66
5,3′-dibenzyloxy-3,6,7,8,4′,5′-hexamethoxy-flavone	68—71
2-hydroxy- ω,3,4,5,6-pentamethoxy-acetophenone	44—47

PARIS [*36*–*38*] also reports very satisfactory separation of flavonoid mixtures on silicic acid-starch layers. On layers 1 mm. thick, ethyl acetate-chloroform or ethyl acetate-methanol (e.g. 95 + 5) or a water-saturated mixture of hexane-isopentanol-acetic acid were used as solvents.

TLC is of great value in the characterization of natural drugs and the easy recognition of adulterants. Silica Gel G layers have, in fact, been used by Hörhammer, Wagner and Lay [22] to detect the very common adulteration of *Radix pimpinellae* by the roots of acanthus (*Heracleum spondylium* L.). Differentiation is quite simple due to the widely different content of coumarin derivatives.

1 g. of the fine-ground root is extracted for 6—8 hrs. with frequent shaking using 10 ml. of petroleum ether. The filtrate is then concentrated to 1 ml. and applied in a volume of 10—15 mm³ on a Silica Gel G layer 250 μ thick. Chloroform is used as the developing solvent. In a "saturated" chamber, the length of run is 12 cm. The chromatograms are first examined in long-wave UV light, then sprayed with 10—20 ml. antimony pentachloride solution (Reagent No. 13) and heated to 110° C. for 2—3 min. Differences are summarized in Table 94.

Table 94. *Differentiation of Radix pimpinellae from Radix heraclei by TLC* [22]

Pimpinella root			*Heracleum* root		
h*Rf*	$SbCl_5$ daylight	UV light unsprayed	h*Rf*	UV light unsprayed	$SbCl_5$ daylight
8	—	dark	8	dark	—
15	light brown	—	15	—	light brown
20	light brown	—	20	—	light brown
25	—	—	—	dark	—
—	—	Pimpinellin	40	brown	blue
—	—	Spondin	43	bright blue	—
45	brownish				—
		Isopim-pinellin	50	brown	yellowish green
		Isobergap-ten	55	dark blue	yellowish green
60	—	dark	60	dark	—
70	brown	—	70	—	brown
90	brown	—	90	—	brown

In order to distinguish alcoholic drug extracts, such as tinctures, fluid extracts, etc., more accurately, Stahl and Schorn [51] investigated more than 50 pyrone derivatives and also aromatic hydroxyacids, tannins, anthracene derivatives and lichenous substances on Silica Gel G layers. The considerable variations in polarity of these types of compounds made it necessary to use a range of solvents with differing elution effects. With the exception of the solvent used for lichenous substances we find the following to be a generally applicable solvent combination:

Toluene Benzene Chloroform ↓	+	Ethyl acetate or ethyl formate (25—50 parts)	+	Formic acid 1—10%
lipophilic part		*fundamental component*		*acid part*

Solvents of this type appear to have a similar wide applicability for silica gel layers as the Partridge mixture [41] has in paper chromatography. This is evident from behavior of the three-color test mixture in such solvents (Fig. 151).

The test mixture was originally applied only for determining the activities of silica gel and alumina layers when using benzene and similar eluants.

Table 95 defines the established conditions for separation of the different classes of compounds. Further variation of the basic type of solvent may be necessary to solve special problems of separation.

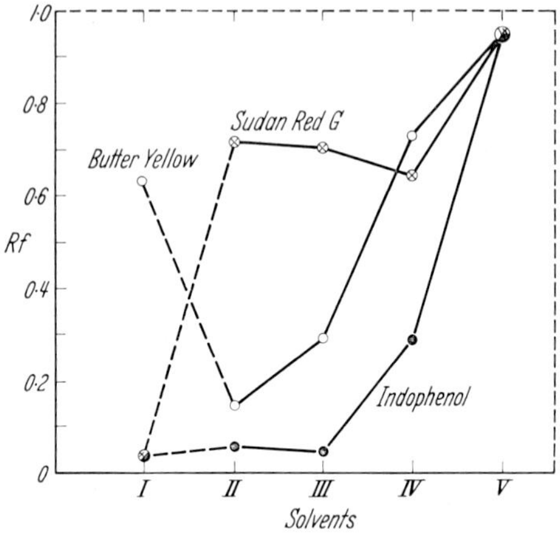

Fig. 151. Variation in the *Rf*-values of the test mixture influenced by different solvent-mixtures on Silica Gel G layers. (Composition of the solvents, see Tab. 95)

Fig. 152 summarizes results obtained with solvent III. The legend gives information about *Rf*-values and the possibilities of detection.

The photograph of a thin-layer chromatogram under UV-light (Fig. 153) indicates that useful results can be obtained also with alcoholic drug extracts in this way. With solvent III, the glycosides remain at the starting point. Separation of flavone glycosides and anthocyanins is possible with solvent V. The following h*Rf* values have been obtained on a Silica Gel G layer under standard conditions: *Robinin 16, rutin 28, naringin 41, hyperoside 47, apigenin-7-glucoside 55, quercitrin 63.*

Table 95. *Solvents for TLC of Hydrophilic Plant Substances* [*51*]

No.	Solvent	Mixture	Layer	Substance group
I	Benzene-chloroform	50+50	Silica Gel G, 0.5 N-oxalic acid buffered	Lichen substances
II	Chloroform-ethyl acetate-formic acid	50+40+10	Silica Gel G	Tannin-containing plant extracts
III	Toluene-ethyl formate-formic acid	50+40+10	Silica Gel G, 0.3 M-sodium acetate buffered	α- and γ-Pyrones, hydroxy carboxylic acids, hydroquinone derivatives
IV	Benzene-ethyl formate-formic acid	75+24+1	Silica Gel G	Anthracene derivatives
V	Ethyl acetate-methyl-ethyl ketone-formic acid-water	50+30+10+10	Silica Gel G	Flavone glycosides, anthocyanins

As there were no pure anthocyanins[1] available for purposes of comparison, the suitability of solvent V was tested with extracts of blossoms

[1] During printing, a paper by BIRKOFER and co-workers [*5a*] dealing with TLC of anthocyanins on polyacrylonitril-perlon layers (see p. 33) has appeared. These authors discuss also TLC of hydroxy-cinnamic acids and sugars.

containing anthocyanin (1 g. powdered blossoms was extracted cold with 10 ml. of methanol containing 0.01% hydrochloric acid).

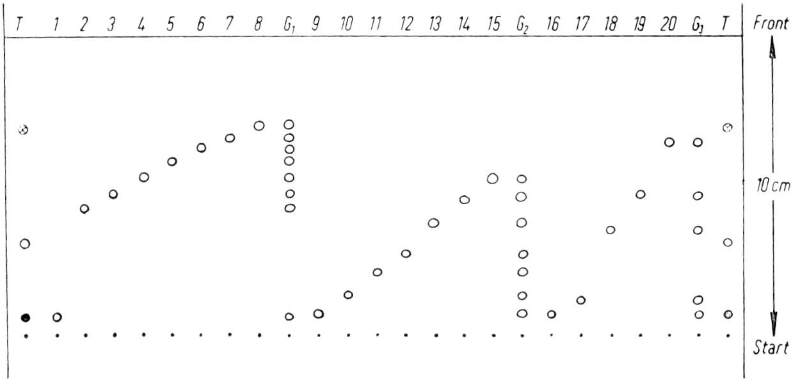

Fig. 152. Scheme showing a thin-layer chromatogram of coumarin derivatives (*1—8*), flavonoids (*9—15*) and hydroquinone derivatives (*16—20*). 0.1 µg. each were applied. The fluorescent colors of the coumarins are given in ( ) brackets, of flavenoids after spraying with Reagent No. 55 in [ ] brackets. The hydroquinones are made visible by means of Millons reagent.

T = test mixture: Butter Yellow (○) + Sudan Red G (×) + Indophenol (●)

1 Esculin, h*Rf* 6 (blue)
2 Esculetin, h*Rf* 42 (blue)
3 Scopoletin, h*Rf* 49 (blue)
4 4-Methyl-umbelliferone, h*Rf* 52 (blue)
5 Xanthotoxin, h*Rf* 56 (yellow)
6 Bergapten, h*Rf* 61 (yellow)
7 Imperatorin, h*Rf* 64 (yellow)
8 Athamantin, h*Rf* 68 (blue)

G_1 Coumarin mixture (*1—8*)

9 Robinetin, h*Rf* 8 [orange]
10 Morin, h*Rf* 12 (yellow-green)
11 Taxifolin, h*Rf* 22 [orange]
12 Quercetin, h*Rf* 28 [orenge]
13 Kaempherol, h*Rf* 40 [yellow-grey]
14 Naringenin, h*Rf* 46 [brown-grey]
15 Chrysin, h*Rf* 52 [orange]

G_2 Flavone mixture (*9—15*)

16 Arbutin, h*Rf* 5
17 Methylarbutin, h*Rf* 12
18 Hydroquinone, h*Rf* 35
19 Hydroquinonemonomethyl-ether, h*Rf* 46
20 Hydroquinonedimethyl-ether h*Rf* 65

G_3 Mixture (*16—20*)

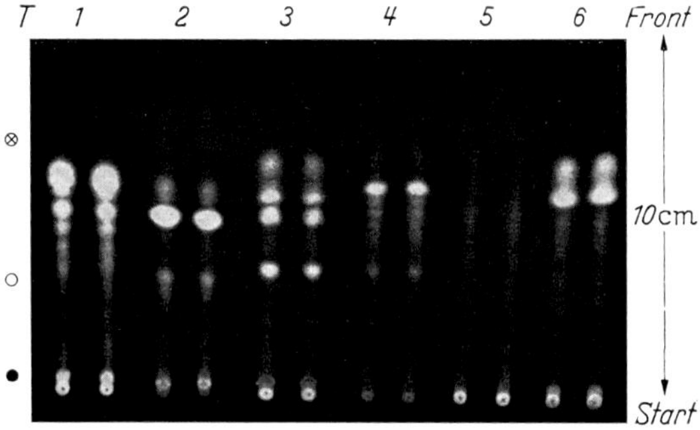

Fig. 153. Distinction between alcoholic extracts from umbelliferous drugs. The untreated thin-layer chromatogram was photographed in long-wave UV-light.

1 Fruits of *Ammi majus* L.
2 Fruits of *Ammi visnaga* L.

h*Rf* values and fluorescence of main components of these two fruits:

Khellin (brown) h*Rf* 47
Visnagin (yellow-grey) h*Rf* 45
Khellol (yellow-grey) h*Rf* 30
Khellol-glycoside (blue) h*Rf* 0

3 *Radix Angelicae* (Tincture Erg. B. 6)
4 *Rhizoma Imperatoriae* (Erg. B. 6)
5 *Radix Levistici* (DAB. 6)
6 *Radix Pimpinellae* (Tincture DAB. 6)

In the chromatogram of *Flor. paeonial*, five anthocyanin spots were obtained (h*Rf* = 17, 25, 39, 48, 62), and four red spots, each from Flor. *Malvae arboreae* (h*Rf* = 8, 13, 30, 44) and Flor. *Malvae silvestris* (h*Rf* = 8, 15, 22, 43); in the chromatogram of Flor. *Cyani*, three spots were observed (h*Rf* = 8, 14, 27) [*53*].

Relation between Rf-values and chemical structure. Recent results indicate that, with the above acidic solvents, on Silica Gel G layers, the same rules of thumb are valid as those formulated by BATE-SMITH and WESTALL [*4*] for paper chromatography of flavonoids. The solvents, n-butanol-glacial acetic acid-water (40 + 10 + 50) and m-cresol-glacial acetic acid-water (50 + 2 + 48), are normally used for separating flavonoids on paper, and the following relationships refer to these systems:

1. The *Rf*-value decreases with a rise in the number of free OH-groups. C_{15}-compounds with the same number of OH-groups show almost the same *Rf*-values (see also 5).
2. Methylation of an OH-group causes a rise in the *Rf*-value of $^1/_3$—$^1/_2$ less than the elimination of the OH-group.
3. Acetylation usually increases the *Rf*-value considerably.
4. Glycoside formation causes a reduction of the *Rf*-value; the effect with glucose coincides, to some extent, with the introduction of an OH-group. Differences between *Rf*-values of a monoside and a bioside are much less than between a flavonoid linked by two different OH-groups with one sugar each.
5. Ortho- or vicinal-positioning of substituents can lead to an exception to the rules as an unexpected rise of *Rf*-values is often found in such cases due to internal hydrogen bonding.

β) Polyamide layers

The success of column chromatography led us to try fine-grained polyamide layers [*50*]. Our experience indicates, however, that data relating to separation and spot size are better compared with data obtained by PC, and they have not been published for this reason. Recently, DAVIDEK and DAVÍDKOVÁ [*8*] described the preparation of loose polyamide layers[1] and the separation of an elder (Sambucus) extract with 80% methanol into three zones. In the same year, EGGER [*11*] published a detailed investigation of the prospects of differentiating flavonol-glycosides on polyamide layers.

Preparation of polyamide layers [*10*]: Polyamide powder PP 15/SP (BASF, Ludwigshafen, Germany) is suspended in ethyl acetate, and this suspension is applied to a glass plate and spread out as a layer, between 200 and 500 μ thick. After drying, 5—30 μg. of sample is applied per starting point. Chamber saturation (p. 15) was employed for development.

As the layers do not adhere well to the glass plates, it is advisable to moisten the starting point area by dipping the plate into the solvent up to this area. With lengths of run of 20 cm., the developing time is 2—3 hrs.

A polyamide powder for thin-layer chromatography has recently been introduced commercially by Messrs. Woelm, Eschwege, Germany, with the following directions for use:

For each 5 glass plates, 20 × 20 cm., 5 g polyamide powder should be mixed with 45 ml. chloroform-methanol (2 + 3) in an Erlenmeyer flask and this suspension spread on the plates using the applicator set at 300 μ.

---

[1] "Silon" from North Bohemian Chemical Works, Rudnik, Lovosice, Czechoslovakia.

In his investigation of the behavior of 14 glycosides of the aglycones, kaempferol, quercetin, and myricetin, EGGER [*11*] used the following *solvents*: 1. = ethanol-water (60 + 40), 2. = water-ethanol-acetylacetone (40 + 20 + 10) and 3. = water-ethanol-methylethylketone-acetylacetone (65 + 15 + 15 + 5). The 3-monosides separated satisfactorily from the 3-biosides and these, in turn, from the 3,7-glycosides. Thus, the rates of migration were dependent, to a large extent, on the glycosidic pattern

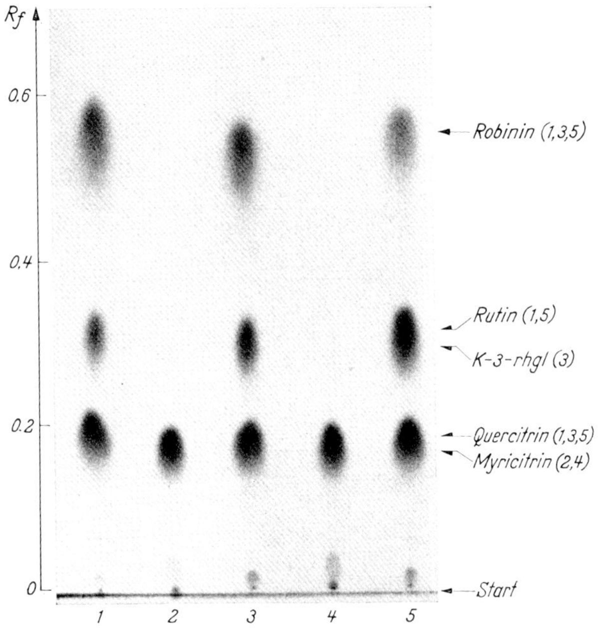

Fig. 154. Separation of flavonol-glycoside types on a polyamide layer using solvent 3 (see text). The untreated thin-layer chromatogram was photographed in UV-light (EGGER)

and not on the aglycone. With solvents 1 and 2, h*Rf*-values for flavonol-3-monosides was 22, for 3-biosides, between 32 and 39, and for 3,7-diglycosides, between 57 and 72. Fig. 154 illustrates separation of a glycoside mixture into the three groups mentioned.

3. Visualization

Some pyrone derivatives, such as anthocyanins, can be recognized on a chromatogram by their color. It is essential before applying spray reagents, to examine the chromatogram in long- and short-wave UV-light and to mark fluorescent as well as dark absorbing zones by such means as pricking round with a needle. Treatment by a base-solution is normally carried out first by means of ammonia vapor only, and later with a *more strongly* alkaline solution. In both cases, colors are noted by daylight and UV-light.

A rough distinction can be made from the colors shown in Table 96. Data on finer color differentiation are available in the papers of HÄNSEL [*20*] and GEISSMANN [*14*].

The numerous spray reagents described in the literature are rather nonspecific. The following basic reagent types may be used: a) alkali (ammonia vapor and Reagent No. 82); b) metal salts (complex formation, Reagent Nos. 2, 20, 62); c) inorganic or organic boron compounds (complex formation, Reagent No. 55); d) acids intensify fluorescence (Reagents Nos. 142 and 143), with added aldehydes color reactions occur (Reagent No. 150); e) diazonium salts react with phenols (Reagents Nos. 37 A and 61). Most suitable spray reagents are stable diazonium salts [*34*]. Reagent No. 40 is also used for detecting phenols.

Table 96. *Color of Pyrone Derivatives before and after Alkali Treatment in Daylight and UV Light* [*14, 20*]

Type of compounds	Untreated		After alkali treatment	
	daylight	UV 365 mμ	daylight	UV 365 mμ
Flavones	colorless to yellow	brown, blue spots	yellow	yellow
Flavonols	light to dark yellow	yellow, yellow-green	yellow	yellow
Flavonol-3-glycosides	colorless-yellowish	brown, dark	yellow	yellow, yellow-green
Flavanones	colorless	colorless (254 mμ, dark)	colorless	yellow
Catechins	colorless	colorless (254 mμ, dark)	colorless, becoming partially dark	dark
Anthocyanins	red (acid)	brown, dark	blue	dark
Coumarins	colorless	colorless, blue, yellow or dark	colorless, yellow	colorless, yellow-green, blue or dark

MEIER and FÜRST [*31*] detect digicitrin and its derivatives with a 0.5% aqueous potassium permanganate solution which gives yellow zones on a violet background. According to REZNIK and EGGER [*45*], Benedicts reagent is very satisfactory for detecting phenolic o-dihydroxy groups in coumarins, flavonoids and cinnammic acid derivatives. Non-fluorescent natural coumarin can be recognized by lightly spraying with N-sodium hydroxide. The sodium salt of coumaric acid shows up intensively yellowish-green in UV light. After spraying with N-hydrochloric acid, the various substances may be extracted from the thin layer in their original lactone form.

The peroxide-ferric-chloride reaction (reagent No. 118) is also suitable for making coumarins visible.

Visualization by treatment with alkali, and with the diphenyl-boric acid-β-amino-ethyl ester (Reagent No. 55); "Naturstoffreagens A", Messrs. Roth, Karlsruhe, Germany) as described by NEU [*33*], can be highly recommended.

II. Lichen components

Lichen components have received a great deal of attention because of the antibiotic qualities occurring in a number of lichens and their extracts. These properties arise, in particular, from the d-usninic acid present in the *Usnea*, *Evernia*, *Letharia*, and *Parmelia* species. Approximately 70 further compounds have been isolated from lichens which can be classified as depsides, depsidones, dibenzofuranes and lactone-carboxylic acids. The quantitative and qualitative variations in content are helpful for the taxonomic identification of some lichens [*21*]. Paper chromatography [*21*, *57*] has often been used for separating these compounds. For purposes of identification, it is advisable to chromatograph the hydrolysis products of lichen acids as well as the naturally occurring acids themselves.

Approx. 50 mg. finely powdered dried material is repeatedly extracted with 0.5 ml. boiling benzene, and then several times with 0.5 ml. boiling acetone. The two filtered extracts are combined and concentrated to 0.5 ml.; 10—100 mm^3 of this concentrate is applied. For acid hydrolysis, the dry residue is mixed with 1—2 drops of concentrated sulphuric acid, diluted with 2 ml. water after 5—10 min., and then the hydrolysis products extracted with ether [*57*]. This ether solution is chromatographed.

On acid Silica Gel G layers, the main components of commercially available lichen extracts are rapidly detected by comparison with pure reference substances[1]. A 0.5 N-oxalic acid solution has been used instead of water to prepare the silica gel suspension. Benzene-chloroform (50 + 50) is used as solvent. Examination of sprays employed to date [*57*] indicated that the most satisfactory results can be obtained with anisaldehyde-sulphuric acid (Reagent No. 9). The most commonly occurring lichen acids show the following h*Rf* values and color reactions: Vulpinic acid, h*Rf* 80 (yellow), usninic acid, h*Rf* 65 (violet), evernic acid, h*Rf* 11 (red), and orcin produced by hydrolysis, h*Rf* 3 (red).

III. Phloroglucinol butanones

(*Filix*-phloroglucides)

The components of male fern rhizome (*Dryopteris filix-mas* (L.) Schott) which are effective against tape-worm (taenia) have been identified as phloroglucinol derivatives.

Zwimpfer and Büchi [*64*] give details of current developments in chemical research and of separation of these compounds by paper chromatography. The *Filix*-phloroglucides shown in Table 97 differ considerably in their taenicide effects.

These substances can be concentrated by extracting the powdered drug with ether. The ether extract is then mixed thoroughly with magnesium oxide and the water-soluble "crude Mg-filicines" extracted with water. By adding acetic acid, a mixture called "crude filicines" is obtained [*1*]. A 1% solution of this is used for TLC and a quantity of 2—4 mm^3 is applied to a plate.

---

[1] E. Brunner, Steinmauer, Zürich, Switzerland.

Details of separation techniques were published at roughly the same time by two research groups. VON SCHANTZ and co-workers [47] used Silica Gel G layers prepared with 0.1 M citric acid-0.2 M di-sodium-hydrogen-phosphate (McIlvaine buffer, pH 6), and a petroleum ether-chloroform-ethanol mixture as solvent. Length of run and conditions of saturation of chamber are not given; the chromatogram was removed from the trough after 60 mins. In the work of my group [52] Silica Gel G layers buffered with a 0.3 M sodium acetate solution were used and developed twice, for the sake of better separation, with ethyl acetate as solvent (chamber saturation; 10 cm. length of run). In addition, the "reversed-phase technique" was employed. Silica Gel G layers were impregnated with paraffin-petroleum ether (5 + 95) (p. 37), and methanol-formic acid-water (75 + 10 + 15) was used as solvent. Results are shown in Table 97.

Table 97. h*Rf*-*Values of Phloroglucin Derivatives Isolated from Dryopterisfern Species*

Phloroglucinol butanones	Formula	Silica Gel G buffered I	Silica Gel G buffered II	"Reversed phase" III	Color with Fast Blue Salt B
Filicinic acid (1)*	$C_8H_{10}O_3$	—	0	100	red-violet
Desaspidinol (1)	$C_{11}H_{13}O_4$	—	75	87	orange-red
Methylphlorobutyrophenone** (1)	$C_{11}H_{14}O_4$	—	55	83	red-violet
Butyryl filicinic acid (1)	$C_{12}H_{16}O_4$	—	5	81	red-orange
Aspidinol (1)	$C_{12}H_{16}O_4$	41	73	78	red-violet
Phloropyron (2)	$C_{21}H_{26}O_7$	—	50	72	orange
Flavaspidic acid (2)	$C_{24}H_{30}O_8$	7	9	70	orange-red
Desaspidin (2)	$C_{24}H_{30}O_8$	82	12	70	orange
Aspidin (2)	$C_{25}H_{32}O_8$	85	33	11	yellow
Albaspidin (2)	$C_{25}H_{44}O_{12}$	87	26	11	orange-red
Filixic acid (3)	$C_{36}H_{44}O_{12}$	90	16	3	orange-red

* Number of rings in molecule in ( ); ** = DF_x [52].

I = acid-buffered Silica Gel G (see text); solvent: petroleum ether-chloroform-anhydrous ethanol (47.5 + 47.5 + 5) [47].

II = alkaline-buffered Silica Gel G; solvent: ethyl acetate, double development with chamber saturation [52].

III = Silica Gel G impregnated with paraffin; solvent: methanol-formic acid-water (75 + 10 + 15) [52].

Comparison of h*Rf*-values from weakly acid layers (I) with those from weakly basic layers (II) shows the significant effects of buffering. With the latter, variations in acidity in the *Filix* phloroglucides are clearly evident [64]. The "reversed-phase technique" shows a relation between the number of phloroglucinol rings and the h*Rf*-value, which is of definite interest. With one ring, h*Rf*-values range between 78 and 100, with two rings, the value is either 70 (number of C atoms 24) or 10, where there is an additional methyl group.

Visualization: Reaction with stable Fast Blue Salt B (Reagent No. 61) gives the colors shown in Table 97. Information about colors obtained from reaction with other diazonium salts is to be found in a publication by von Schantz [*47*]. The latter author uses as a spray also: 1% ferric chloride solution + 1% potassium ferricyanide solution (1 + 1), adding 10 drops of concentrated nitric acid to 10 ml. of this mixture. Deep blue spots are produced by this method, and a blue reaction has also been obtained using our favorite reagent, the Folin-Ciocalteau reagent (No. 122). In another study, von Schantz describes the possibility of evaluating thin-layer chromatograms quantitatively. The scraped-off zones are extracted and Fast Blue Salt solution is added to the eluted filicins. The dye solution obtained is used for quantitative photometric estimation [*47*].

Application: Using TLC, a new *Filix* compound (DF_x) was discovered in a relatively short time and identified as methyl phlorobutyrophenone [*52*, *53*]. Quantitative evaluation has given the first exact data on indication of the composition of crude filicins obtained in various ways [*47*].

IV. Anthracene derivatives

1,8-Dihydroxy-anthracene derivatives are responsible for the purgative effects of a number of natural drugs. They may be present in the form of anthraquinone, anthrone and anthranol, and additional OH-groups can be linked with sugars [*48*, *56*]. The significant differences in their

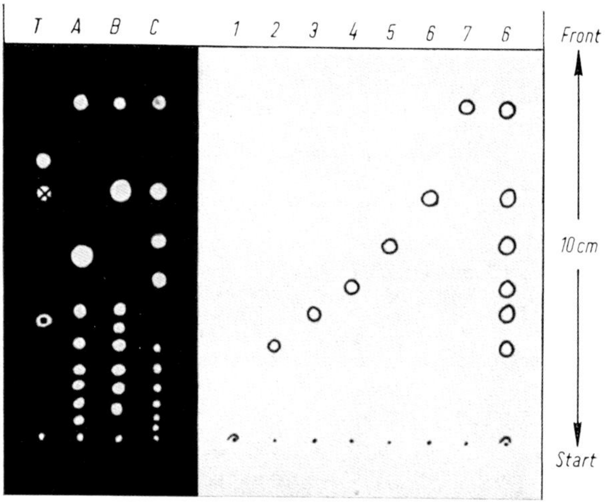

Fig. 155. Separation of anthracene derivatives and of corresponding plant extracts (solvent IV, Silica Gel G layer). *1* Aloin, h*Rf* 0 (orange-brown); *2* Sennidin A, h*Rf* 24 (yellow-brown); *3* Emodinanthrone, h*Rf* 34 (yellow-brown); *4* 1.8-dimethoxy-anthraquinone-carboxylic acid-(3)-methyl ester (Sandoz), h*Rf* 38 (yellow-yellowish red); *5* Rhein, h*Rf* 49 (pink-orange); *6* Emodin, h*Rf* 63 (pink-orange); *7* Alizarin, h*Rf* 83 (pink-orange); *A* Tinct. Aloes (DAB. 6); *B* Extr. *Frangulae fluidum* (DAB. 6); *C* Tinct. *Rhei vinosa* (DAB. 6). Fluorescent colors in long-wave UV-light after spraying with alcoholic- potassium hydroxide are shown in brackets.

purgative effects make it essential to determine the individual compounds in natural mixtures. A number of compounds of this type can be separated on Silica Gel G layers. Fig. 155 shows a thin-layer chromatogram developed with benzene-ethyl formate-formic acid (75 + 24 + 1).

Various drug extracts chromatographed alongside indicate that numerous unknown compounds are present. A 1% alcoholic potassium hydroxide solution is used for visualization. The anthracene derivatives fluoresce mainly yellow, orange, or red in long-wave UV-light.

TEICHERT and co-workers [*54*] describe the advantage of quantitative spectrophotometric determination of Aloin by TLC of the extract. Details of this procedure are not given, however. JANIAK and BÖHMERT [*25*] also used TLC for quantitative estimation of Aloin. Benzene-glacial acetic acid (60 + 30) was used to separate Sennidin A + B (h*Rf* 65) from Rhein (h*Rf* 72) on Silica Gel G layers. This method was helpful for controlling methods for the photometric estimation of Sennosid.

V. Phenolic carboxylic acids and their derivatives

Some of the phenolic carboxylic acids listed in Table 98 are freely present in plant material, while others are normally obtained by alkali fusion of lignins, tannins, coumarins, flavonoids, lignanes, lichen acids, etc.

LYMAN and co-workers [*30*] investigated in 1957 the possibilities of separating 26 phenol derivatives on silicic acid layers (chromatostrip technique). The majority of the *Rf* values given by these authors are reproduced in Table 98 (cols. A, to A_3). KRATZL and PUSCHMANN [*29*] used Silica Gel G layers for micro preparation of radioactively labelled degredation products of lignin.

They were able to separate vanilloxylacetyl (h*Rf* 6), syringic acid (24), p-hydroxybenzoic acid (31), vanillin (58), p-hydroxybenzoylacetyl, p-hydroxybenzaldehyde (h*Rf* of both 67), etc., using water-saturated isoamyl ether-n-butanol (75 + 25) as solvent. Water-saturated n-butyl ether-glacial acetic acid (90 + 9)

Layers and solvents for Table 98.

A_1: Silicic acid-starch layer ~ 0.5 mm.; ether-Skellysolve B* (70 + 30) [*30*]
A_2: Silicic acid-starch layer ~ 0.5 mm.; ethyl acetate-Skellysolve B* (75 + 25) [*30*]
A_3: Silicic acid-starch layer ~ 0.5 mm.; acetone-Skellysolve B* (25 + 75) [*30*]
B_1: Silica Gel G layer ~ 0.5 mm.; benzene-dioxane-glacial acetic acid (90 + 25 + 4) [*39*]
B_2: Silica Gel G layer ~ 0.5 mm.; benzene-methanol-glacial acetic acid (90 + 16 + 8) [*39*]
C_0: non-impregnated Silica Gel G layer (standard conditions); n-butyl ether (water saturated)-glacial acetic acid (90 + 9) [*19*]
C_1: Silica Gel G layer impregnated with 0.01 M Na_2MoO_4; solvent as for C_0 [*19*]
C_2: Silica Gel G layer impregnated with 0.01 M Na_2WO_4; solvent as for C_0 [*19*]
C_3: Silica Gel G layer impregnated with 0.01 M borax; solvent as for C_0 [*19*]
D_0: non-impregnated Silica Gel G layer; ethyl acetate-isopropanol-water (65 + 24 + 11) [*19*]
D_1: Silica Gel G layer impregnated with 0.01 M Na_2MoO_4; solvent as for D_0 [*19*]
D_2: Silica Gel G layer impregnated with 0.01 M Na_2WO_4; solvent as for D_0 [*19*]
D_3: Silica Gel G layer impregnated with 0.01 M borax; solvent as for D_0 [*19*]

* Skellysolve B = hexane fraction manufactured by Skelly Oil Company, Kansas City, Missouri, USA.

Table 98. h*Rf*-*Values of Phenolic Carboxylic Acids under differing Conditions* [*19*, *30*, *39*]

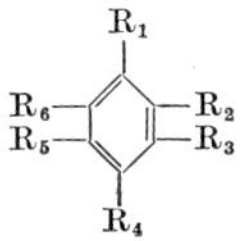	Substituents					h*Rf*-values for layer and solvent												
	R_1	R_2	R_3	R_4	R_5	A_1	A_2	A_3	B_1	B_2	C_0	C_1	C_2	C_3	D_0	D_1	D_2	D_3
Salicylic acid	—COOH	—OH	—	—	—	72	88	48	64									
m-Hydroxybenzoic acid	—COOH	—	—OH	—	—				49	51								
p-Hydroxybenzoic acid	—COOH	—	—	—OH	—				55	60								
Anisic acid	—COOH	—	—	$—OCH_3$	—	63	92	39	84	69								
p-Hydroxyphenyl-acetic acid.	$—CH_2—COOH$	—	—	—OH	—	35	72	14										
p-Coumaric acid . . .	—CH:CH—COOH	—	—	—OH	—				49	52								
Protocatechuic acid. .	—COOH	—	—OH	—OH	—	32	55	10	32	39	38	10	24	10	81	71	15	4
Homocatechuic acid .	$—CH_2—COOH$	—	—OH	—OH	—						23	6	13	5	55	66	13	3
Hydrocaffeic acid . .	$—CH_2CH_2—COOH$	—	—OH	—OH	—						27	14	19	9	57	80	23	3
Caffeic acid	—CH:CH—COOH	—	—OH	—OH	—				24	43	31	14	22	8	65	85	31	2
Vanillin acid.	—COOH	—	$—OCH_3$	—OH	—	46	76	29	54	61	45	44	39	34	82	93	38	53
Guaiacyl-acetic acid .	$—CH_2—COOH$	—	$—OCH_3$	—OH	—						29	22	24	10	59	69	19	21
Guaiacyl-propionic acid	$—CH_2CH_2—COOH$	—	$—OCH_3$	—OH	—						34	29	32	25	56	84	29	40
Ferula acid	—CH:CH—COOH	—	$—OCH_3$	—OH	—	42	76	16	50	58	35	34	36	28	63	87	37	50
Isoferula acid	—CH:CH—COOH	—	—OH	$—OCH_3$	—				43		30	25	31	25	60	86	33	45
Veratric acid	—COOH	—	$—OCH_3$	$—OCH_3$	—				73	70								
Gentisic acid.	—COOH	—OH	—	—	—OH	35	65	17	30	40								
Homogentisic acid . .	$—CH_2—COOH$	—OH	—	—	—OH	67	85	43										
α-Resorcyclic acid . .	—COOH	—	—OH	—	—OH	21	61	73										
β-Resorcyclic acid . .	—COOH	—OH	—	—OH	—	57	85	19	54	52								
γ-Resorcyclic acid . .	—COOH	—OH	—	—	R_6—OH	10	15	10										
Gallic acid.	—COOH	—	—OH	—OH	—OH				18	23								
Syringic acid	—COOH	—	$—OCH_3$	—OH	$—OCH_3$	33	57	16	48	60								
Eudesmic acid. . . .	—COOH	—	$—OCH_3$	$—OCH_3$	$—OCH_3$	43	75	31										

was used for TLC of the mixture syringic acid, vanillin acid, and p-hydroxybenzoic acid. Ferulic acid (h*Rf* 41) was separated from sinapic acid (27) with water-saturated n-butylether-formic acid (99 + 1).

PASTUSKA [*39*] has made a detailed study of TLC of phenols and phenolic carboxylic acids on Silica Gel G layers. The values given in Table 98, cols. B_1 and B_2, have been further supplemented by a private communication of the author. A further study [*40*] contains information about thin-layer electrophoresis of phenols and phenolic carboxylic acids on buffered Silica Gel G and Kieselgur G layers. The M_g values of 17 phenolic carboxylic acids are referred to m-hydroxybenzoic acid. Experimental details are to be taken from the original paper. To resolve complex mixtures, the combination of adsorption-TLC in direction 1 and thin-layer electrophoresis in direction 2, suggested by HONEGGER, should be very helpful. Very similar phenolic carboxylic acids may be separated by impregnating the Silica Gel G layer with chelate-forming salts. HALMEKOSKI [*19*] investigated the effect of Silica Gel G layers impregnated with tungstate, molybdate and borax, using five different solvents (Table 98, col. C and D). The influence of chelate formation is shown; *Rf*-values in cols. C and D with index 0 may be compared with values for 1, 2 and 3.

Impregnation was carried out by mixing the dry Silica Gel G with a 0.01 M sodium molybdate, sodium tungstate or borax solution instead of water, and spreading and drying under standard conditions.

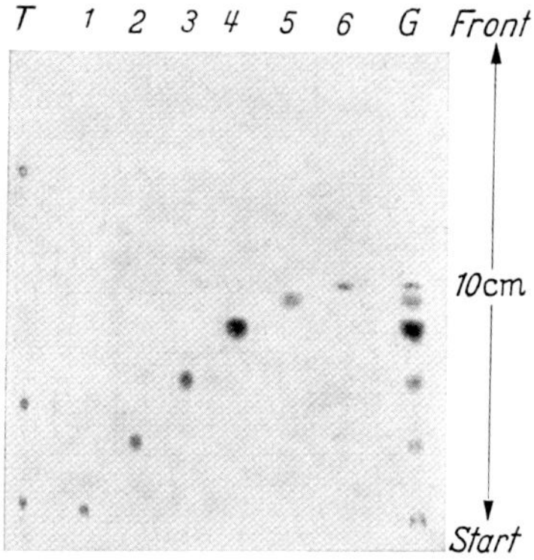

Fig. 156. Aromatic hydroxy carboxylic acids after separation and visualization with phosphomolybdic acid (solvent III; chamber saturation; Silica Gel G layer). *1* Chlorogenic acid, h*Rf* 7; *2* m-digallic acid, h*Rf* 27; *3* gallic acid, h*Rf* 39; *4* caffeic acid, h*Rf* 47; *5* gentisic acid, h*Rf* 51; *6* ferulic acid, h*Rf* 56; *G* mixture (*1*—*6*)

Phenol carboxylic acid mixtures occurring in plant material have been chromatographed on Silica Gel G layers with the solvent toluene-ethyl formate-formic acid (50 + 40 + 10). This solvent mixture has also proved satisfactory for separating coumarin and flavone compounds. The separation of various tannins and allied products on silica gel layers is described in the same study [*51*]. Chloroform-ethyl acetate-formic acid (50 + 40 + 10) were used in this case, and the following h*Rf*-values were obtained under standard conditions (chamber saturation): epigallocatechin 14, epicatechin 23, m-digallic acid 26, gallic acid 37, pyrogallol 49.

Visualization: Most phenol derivatives exhibit a more or less intense self adsorption in ultra-violet light. If Silica Gel G layers containing an inorganic fluorescent agent[1] are used (e.g., LYMAN and co-workers [*30*]), the phenols show up as dark spots on a fluorescent back-ground in short-

[1] A silica gel with fluorescent additive (254 mμ) has recently been introduced commercially by Messrs. E. Merck A.G., Darmstadt, Germany. For further details, see p. 33.

wave UV light (254 mμ). The zones are marked, and then a diazonium salt solution can be sprayed on to develop the color. The Fast Blue Salts in normal commercial use (Reagent No. 61) have proved very suitable for this purpose. PASTUSKA [*39*] summarized the colors occurring with different coupling reactions. He prefers a diazotized benzidine reagent. Phosphomolybdic acid solutions (Reagent No. 120a) are more sensitive, with quantities of up to 0.1 μg showing clearly as blue spots (Fig. 156). The Folin-Denis reagent [*19*] and the more sensitive Folin-Ciocalteau reagent (No. 122) have been used in the same way. The acidic properties of phenol carboxylic acids can be detected using Bromocresol Green (Reagent No. 22) etc., though when employing acidic solvents, acid must first be removed from the layer.

VI. Bitter principles and saponins

1. Bitter principles

Bitter tasting oxygen-containing organic natural substances free of nitrogen and sulphur, which cannot be included in any of the known categories of compounds, are classified as bitter principles. Many valuable details about these compounds are summarized by KORTE [*27, 28*]. A "quantitative determination" of a bitter principle can be made, according to WASICKY [*58*], by determining the "bitterness limit", i.e., the degree of dilution by water at which the solution is still bitter to the taste. The bitter power of a compound can thus be expressed numerically. Quassia wood, for example, still tastes bitter in a dilution of 1:45.000, whereas the alkaloid brucin has a factor of approx. 1:4.500.000.

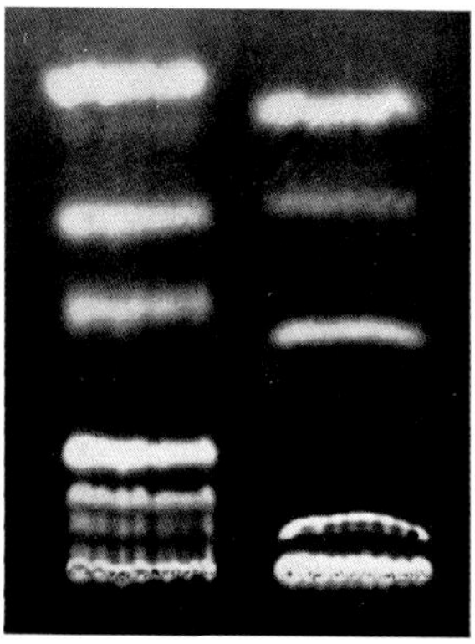

Fig. 157. Thin-layer chromatogram of two different bitterwoods; unsprayed, photographed in long-wave UV-light. Right: Jamaican quassia; left: Brazilian bitterwood [*61*]. For details, see text

Alcohol extracts (1 + 10) from two different bitterwoods were compared on Silica Gel G layers with chloroform-methanol (90 + 10) as solvent in a saturated chamber (Fig. 157). The position of the bitter principles in the chromatogram can be determined by scraping off the fluorescent and other zones, a millimeter at a time, and testing their bitterness [*18*].

The scrapings are suspended separately in 0.5 ml. pure ethanol, and a series of dilutions are produced with water. In this way, it has been established that, contrary to previous belief, Jamaica quassia (*Picrasma excelsa* Planchon) contains a range of bitter principles, and that these compounds are not identical with those contained in the Brazilian wood *Aeschrion creneta* Vel. [*61*].

TLC should be of great value to the brewing industry[1] in separation and quantitative evaluation of different bitter principles in hops. Preliminary experiments [*53*] indicate that the above-mentioned conditions for separation can be adopted. In an earlier study by RIGBY and BETHUNE [*46*], TLC was used for investigating essential oils in hops.

2. Saponins

Natural substances which foam like soap in aqueous solutions are known as saponins. Chemically, they belong to the category of triterpene glycosides [*59*] or steroid glycosides [*60*]. Their property of haemolyzing blood, i.e., erythrocytes, is particularly worthy of note. The different methods of quantitative determination of this haemolyzing effect (= haemolytic index) are described by GSTIRNER [*18*], who also gives precise experimental details of Wasicky's bitterness factor.

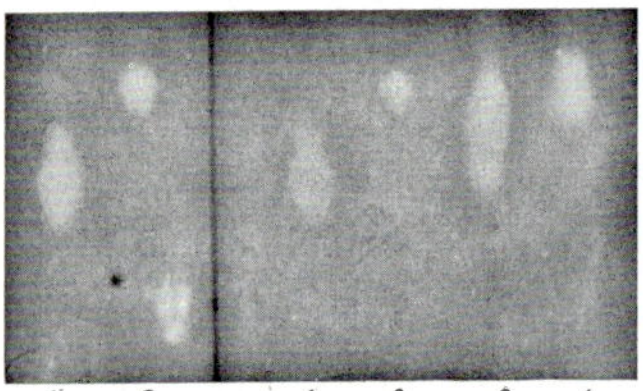

Fig. 158. Segment of a thin-layer chromatogram covered with a blood gelatine film. The haemolyzing saponins show as light spots. The "Merck" saponin (1) consists of two components, and is not identical to the Brasilian saponin extract (2), the quillaja- (3), or the sarsaparilla-saponin (4)

As in plants, these substances are normally present as mixtures. Separation and detection of individual saponins by chromatography is important.

Separation and detection of saponins [*61*]: "Merck" saponin and several drug extracts containing saponin have been chromatographed on Silica Gel G layers under standard conditions with isopropanol-water-formic acid (70 + 24 + 6). In order to detect haemolyzing compounds on the chromatogram, a blood gelatine suspension was poured on to the plate and the formation of haemolytic zones observed. In contrast to the normally opaque red blood gelatine layer, the saponin zones are transparent and almost colorless as a result of the action of the saponin.

This effect is even clearer in the original than on the section of a thin-layer chromatogram shown in Fig. 158.

Preparation and application of blood gelatine solutions: a) 100 ml. 0.9% sodium chloride solution is added to 4.5 g. gelatine powder, and after standing 30 mins. at room temperature, the mixture is heated, with stirring, to about 80° C. in a water bath.

b) After cooling the gelatine solution (a) to 45° C., 6 ml defibrinated cow's blood (see below) is stirred in, and this blood gelatine suspension immediately poured on to the chromatogram as a thin film. Spilling of the blood gelatine is prevented by sticking adhesive tape about 1 cm wide round the edge of the plate to form a kind of trough. After application the plate is left to cool in a horizontal position, preferably on a cooling block, until the film has set. After one hour or less, the red blood gelatine film will be transparent (varnish color) wherever saponins occur on the

---

[1] See also Y. KUROIWA and H. HASHIMOTO: J. Inst. Brew. **67**, 347 and 352 (1961).

chromatograms, while the rest of the film will remain a non-transparent red (opaque color) (Fig. 158).

Note: If acidic or basic solvents are used, they must be completely removed from the adsorption layer before the blood gelatine is applied.

Defibrinated cow's blood: Approx. 300 ml. blood from a freshly slaughtered cow is collected in a 1 litre widenecked flask, and stirred strongly with a wooden stick until the fibrin agglutinates. The gelatinous residue is removed by "filtration" through several sheets of muslin. This defibrinated blood can be kept only for 1—2 days at 3—4° C. As soon as the opaque "solution" becomes transparent due to haemolysis, it can no longer be used.

Bibliography to Chapter I. Hydrophilic Constituents of Plants

[*1*] Ackermann, M., u. M. Mühlemann: Pharm. Acta Helv. **21**, 157 (1946).
[*2*] Baerheim Svendsen, A.: Zur Chemie norwegischer Umbelliferen. Diss. Oslo 1954.
[*3*] Bate-Smith, E. C.: Partition Chromatography. Biochem. Soc. Symposia (Cambridge) **3**, 68 (1951).
[*4*] —, and R. G. Westall: Biochim. et Biophys. Acta **4**, 427 (1950).
[*5*] Billek, G.: Private communication.
[*5a*] Birkofer, L., Ch. Kaiser, H.-A. Meyer-Stoll u. F. Suppan: Z. Naturforschg. **17, b** 352 (1962)
[*6*] Bradfield, A. E., M. Penney and W. B. Wright: J. Chem. Soc. **1947**, 32.
[*7*] — — J. Chem. Soc. **1948**, 2249.
[*8*] Davídek, J., u. E. Davídková: Pharmazie **16**, 352 (1961).
[*9*] Dean, F. M. in L. Zechmeister: Fortschritte der Chemie organischer Naturstoffe, Bd. 9. Wien: Springer-Verlag 1952.
[*10*] Egger, K.: Private communication.
[*11*] — Z. anal. Chem. **182**, 161 (1961).
[*12*] Enders, H.: Z. anal. Chem. **181**, 331 (1961).
[*13*] Freudenberg, K., u. Kl. Weinges in L. Zechmeister: Fortschritte der Chemie organischer Naturstoffe, Bd. 16. Wien: Springer-Verlag 1958.
[*14*] Geissmann, T. A., in K. Paech u. M. V. Tracey: Moderne Methoden der Pflanzenanalyse, Bd. III. Berlin-Göttingen-Heidelberg: Springer-Verlag 1955.
[*15*] Grassmann, W. and co-workers, see [*12*].
[*16*] Grisebach, H.: Z. Naturforsch. **14**b, 802 (1959).
[*17*] —, u. L. Patschke: Chem. Ber. **93**, 2326 (1960).
[*18*] Gstirner, F.: Prüfung und Verarbeitung von Arzneidrogen, Bd. 1. Berlin-Göttingen-Heidelberg: Springer-Verlag 1955.
[*19*] Halmekoski, J.: Suomen Kemistilehti **35**, 39 (1962).
[*20*] Hänsel, R. in H. F. Linskens: Papierchromatographie in der Botanik. Berlin-Göttingen-Heidelberg: Springer-Verlag 1959.
[*21*] Hess, D.: Planta **52**, 65 (1958).
[*22*] Hörhammer, L., H. Wagner u. B. Lay: Pharmazie **15**, 645 (1960).
[*23*] —, —, u. W. Leeb: Naturwissenschaften **44**, 513 (1957), for further publications see [*12*].
[*24*] Ice, C. H., and S. H. Wender: Anal. Chem. **24**, 1616 (1952); J. Am. Chem. Soc. **75**, 50 (1953).
[*25*] Janiak, B., u. H. Böhmert: Arzneimittel-Forsch. **12**, 431 (1962).
[*26*] Karrer, P., u. F. M. Strong: Helv. Chim. Acta **19**, 25 (1936).
[*27*] Korte, F.: Arch. Pharm. **286**, 257, 295 (1953).
[*28*] — H. Barkemeyer u. I. Korte in L. Zechmeister: Fortschritte der Chemie organischer Naturstoffe, Bd. 17. Wien: Springer-Verlag 1959.
[*29*] Kratzl, K., u. G. Puschmann: Holzforschung **14**, 1 (1960).
[*30*] Lyman, R. L., A. L. Livingston, E. M. Bickoff and A. N. Booth: J. Org. Chem. **23**, 756 (1958).
[*31*] Meier, W., u. A. Fürst: Helv. Chim. Acta **45**, 232 (1962).
[*32*] Müller, K. H., B. Christ u. G. Kühn: Arch. Pharm. **295**, 41 (1962).

[*33*] Neu, R.: Naturwissenschaften **44**, 181 (1957).
[*34*] — Mikrochim. Acta **1957**, 196.
[*35*] — Nature (London) **182**, 660 (1958).
[*36*] Paris, R.: Pharm. Acta Helv. **36**, 176 (1961).
[*37*] — (Lecture) Journées Internationales d'Etude des Méthodes de Séparation Immédiate et de Chromatographie, 13—15 Juin 1961, Paris.
[*38*] — Prod. pharm. **15**, 347 (1960); cf. [*36*].
[*39*] Pastuska, G.: Z. anal. Chem. **179**, 355 (1961).
[*40*] —, u. H. Trinks: Chem. Ztg. **85**, 535 (1961).
[*41*] Partridge, S. M.: Biochem. J. **42**, 238 (1948).
[*42*] Pearl, I. A., and E. E. Dickey: J. Am. Chem. Soc. **73**, 863 (1951); **74**, 614 (1952).
[*43*] Procházka, Ž. in I. M. Hais u. K. Macek: Handbuch der Papierchromatographie, Bd. I. Jena: VEB Fischer Verlag 1958.
[*44*] Reppel, L.: Pharmazie **12**, 654 (1957).
[*45*] Reznik, H., u. K. Egger: Z. anal. Chem. **183**, 196 (1961).
[*46*] Rigby, F. L., and J. L. Bethune: Proc. Am. Soc. Brewing Chem. **1955**, 174.
[*47*] Schantz, M. v.: Planta Med. **10**, 22 u. 98 (1962).
[*48*] Schmid, H., in L. Zechmeister: Fortschritte der Chemie organischer Naturstoffe, Bd. 11. Wien: Springer-Verlag 1954.
[*49*] Späth, E., and co-workers, see Dean [*9*].
[*49a*] —, and D. H. Rosenblatt: Anal. Chem. **22**, 1321 (1950); cf. [*14*].
[*50*] Stahl, E.: Lecture, GDCH-Tagung, Fachgruppe Analyt. Chem., München, 27. Oktober 1960, cf.: Z. anal. Chem. **181**, 305 (1961).
[*51*] —, u. P. J. Schorn: Z. physiol. Chem., Hoppe-Seyler's **325**, 263 (1961).
[*52*] — — Naturwissenschaften **49**, 14 (1962).
[*53*] — — Unpublished.
[*54*] Teichert, K., E. Mutschler u. H. Rochelmeyer: Z. anal. Chem. **181**, 325 (1961).
[*55*] Venkataraman, K., in L. Zechmeister: Fortschritte der Chemie organischer Naturstoffe, Bd. 17. Wien: Springer-Verlag 1959.
[*56*] *Lectures*, Tagung Dtsch. Ges. Arzneipflanzenforsch. in: Pianta Med. **7**, 336 bis 449 (1959).
[*57*] Wachtmeister, C. A., in H. F. Linskens: Papierchromatographie in der Botanik, 2. Aufl. Berlin-Göttingen-Heidelberg: Springer-Verlag 1959.
[*58*] Wasicky, R.: Leitfaden für die pharmakognostischen Untersuchungen in Unterricht und Praxis. Leipzig und Wien: Deuticke-Verlag 1936.
[*59*] Steiner, M., u. H. Holtzem: In Moderne Methoden der Pflanzenanalyse, Bd. III, S. 67. Berlin-Göttingen-Heidelberg: Springer-Verlag 1955.
[*60*] Stoll, A., u. E. Jucker: In Moderne Methoden der Pflanzenanalyse, Bd. III, S. 176. Berlin-Göttingen-Heidelberg: Springer-Verlag 1955.
[*61*] Weigert, E., P. J. Schorn and E. Stahl: Unpublished.
[*62*] Weygand, F., u. H. Wendt: Z. Naturforsch. **14**b, 421 (1959).
[*63*] —, H. Simon, H. G. Floss u. U. Mothes: Z. Naturforsch. **15**b, 765 (1960).
[*64*] Zwimpfer, G., u. J. Büchi: Pharm. Acta Helv. **37**, 224 (1962).

Some recent noteworthy articles

Birkofer, L., C. Kaiser, W. Koch, M. Donike and D. Wolf: Z. Naturforsch. **18**b, 631 (1963), Acetylated anthocyanins.
Böhme, H., and L. Kreutzig: Dtsch. Apoth. Ztg. **103**, 505 (1963), PC and TLC of Aloe-drugs.
Fikenscher, L. H., and R. Hegnauer: Planta medica **11**, 348 (1963), Constituents of male fern.
Freudenberg, K., and K. Weinges: Ann. Chem. **668**, 91 (1963), Tanning substances.
Gerritsma, K. W., and M. C. B. van Rheede van Oudtshoorn, Pharmac. Weekbl. **97**, 765 (1962), Aloe-drugs.
Hess, D., and Ch. Meyer: Z. Naturfosch. **17**b, 853 (1962), TLC of anthocyanins.
Hörhammer, L., H. Wagner and B. Salfner: Arzneimittel-Forschung **13**, 33 (1963), Flavones in *Matricaria chamomilla.*

HÖRHAMMER, L., H. WAGNER and G. BITTNER: Arzneimittel-Forschung **13**, 537 (1963), South African Aloe-drugs.
NYBOM, N.: Fruchtsaft-Industrie 8, 205 (1963), TLC of anthocyanins in fruit juices.
PARIS, R. R., and M. PARIS: Bull. Soc. Chim. France **1963**, 1597, TLC of anthocyanins.
SIEPER, H., R. LONGO and F. KORTE: Arch. Pharmaz. **296**, 403 (1963), TLC of polyhydroxy anthraquinones.
STAHL, E., and P. J. SCHORN, Sci. Pharmac. **31**, 157 (1963), Constituents of male fern.
SCHANTZ, M. VON: Planta medica **10**, 22, 98 (1962), TLC of phloroglucinols in male fern and their quant. estimation.
STEINEGGER, E., and J. GEBISTORF, Pharmac. Acta Helv. **37**, 343 (1962), Flavones in Solanum species.
STEINEGGER, E., and J. GEBISTORF, Sci. Pharmac. **31**, 298 (1963), Characterization of Tilia drugs by TLC.

J. Amino Acids and Derivatives

By

M. BRENNER, A. NIEDERWIESER and G. PATAKI

I. Introduction

Free amino acids and peptides are markedly hydrophilic compounds which dissolve only slightly in non-aqueous solvents. This should be borne in mind when sampling and preparing materials for thin-layer chromatography as well as when selecting the proper solvent. A few data on solubilities of some amino acids in various solvents are summarized in Table 99.

Table 99. *Solubility of some Amino Acids in different Solvents*
Moles per liter at 25° C[1]

	Glycine	DL-Norleucine	L-Aspartic acid
Water	2.886	0.0866	0.0375
20% Ethanol	1.343	0.0516	0.0149
80% Ethanol	0.0278	0.0130	0.00070
Ethanol	0.00039	0.00104	0.0000116
Methanol	0.00426	0.00854	
n-Butanol	0.0000959	0.000336	
Acetone	0.0000305	0.0000793	

[1] E. J. COHN and J. T. EDSALL [*1*].

Furthermore, amino acids are amphoteric. They also exhibit a pronounced capacity for binding metals. It therefore makes a difference whether they are free, protonated (hydrochlorides), deprotonated (alkali salts) or complexed with heavy metal ions when applied to the chromatogram. It is also of importance whether they are chromatographed in an acid, neutral or basic medium, and at what pH the color reactions developing the spots are carried out. Particular attention should be paid to the effect of water on amino acid salts. Depending on concentration, an amino acid hydrochloride decomposes in water more or less completely into non protonated amino acid and HCl. This decomposition is complete at high dilution. As all chromatographic techniques involve dilution, chromatography of a specified ionic form requires buffering of both the thin-layer and the solvent. Attention may also be drawn to chemical changes which, under certain conditions, may occur more rapidly and in a more specific way on chromatograms than known from test tube experiments. The difference between the ninhydrin reactions performed on paper or in solution, respectively, constitutes a well known example.

The problem of amino acid separation was the starting point for the development of paper chromatography. The mechanism of separation was at first thought to be based on partition of the amino acids between water bound to cellulose by a swelling process and a "mobile" phase immiscible with water, e.g., water saturated phenol. This attitude has changed somewhat since it was discovered that the mobile phase must not necessarily be immiscible with water. The liquid involved in the swelling of cellulose obviously differs with regard to mobility and solvent capacity from the liquid which is merely surrounding the cellulose in such a way as to simulate a two phase system; liquid-liquid partition may, in some cases, be accompanied or even replaced by adsorption on the cellulose surface.

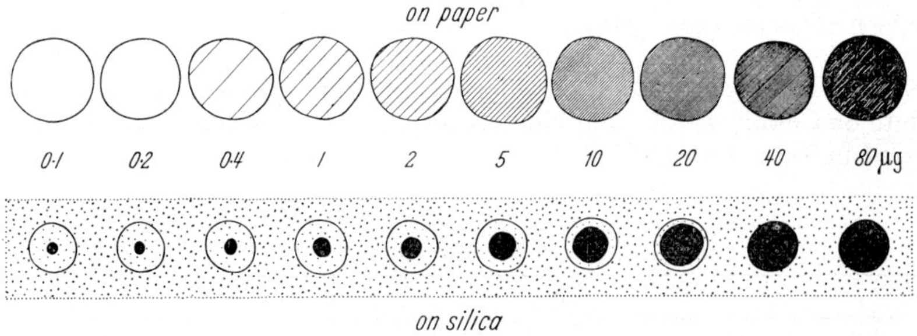

Fig. 159. Spreading on Silica Gel G (layer thickness 0.25 mm., air dried) and Whatman paper No. 1 of DNP-serine spotted in acetone

Each spot was formed upon application under identical conditions of 10 μl. of a solution containing the amount of DNP-serine indicated in the figure. At concentrations less than 70 μg./10 μl., the material spotted on silica covers but a fraction of the area wetted by the acetone, while the material spotted on paper covers the total wet area

Thin-layer chromatography on silica gel, developed especially for separating lipophilic substances, at first seemed less suitable in the case of hydrophilic substances. However, silica gel, like cellulose, contains a considerable amount of water, depending on its state of hydration. Its outstanding suitability for chromatography of amino acids is, therefore, not too astonishing. There is also a basic similarity to paper chromatography in practice, and, in fact, a great amount of experience published on paper chromatography may be closely followed when applying thin-layer chromatography on silica to free amino acids. Therefore, as far as preparation of material, solvents and detection reactions are concerned, the use of the procedures developed for paper chromatography[1] is suggested.

However, thin-layer chromatography on Silica Gel G has two decisive advantages over paper chromatography:

1. Less spreading of substance zones, i.e.,
   a) when applying solutions at the start (see Fig. 159)
   b) during chromatography (there is slower diffusion, see Fig. 160 and the paragraph dealing with differences between thin-layer and paper chromatography, p. 126).
2. Saving in time.

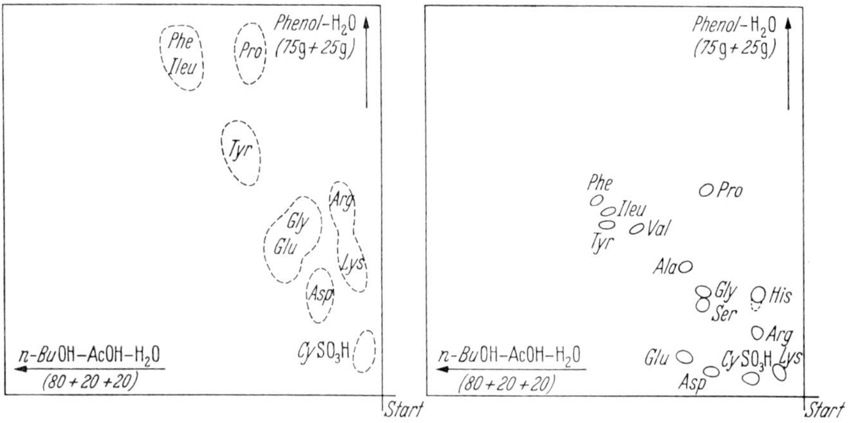

Fig. 160. Comparison of thin-layer and paper chromatography. Two-dimensional chromatograms, original size × 0.57

Left: 10 amino acids on Whatman paper No. 1, time required about 1 (BuOH-AcOH-H_2O) and $2^1/_2$ hrs. (Phenol-H_2O)
Right: 14 amino acids on Silica Gel G, time required about $1^1/_2$ (BuOH-AcOH-H_2O) and 2 hrs. (Phenol-H_2O)
Spotting: 1 µl. of an aqueous solution containing 1 µg. of each amino acid. Ascending technique: detection by the ninhydrin reagent as modified by MOFFAT and LYTLE [*3*] (cf. Reagent No. 109)

II. General technique

1. Method for layer preparation

The preparation of layers is, in a general way, described on pp. 7—9. While that section recommends activation of the layer by heating before application of the material to be chromatographed, we consider such a pre-treatment inadvisable, *at least* as far as chromatography of compounds discussed in this chapter is concerned:

[1] I. M. HAIS and K. MACEK, in [*2*].

When the plate is heated to a little over 100° C., the gypsum in the Silica Gel G ceases to properly fulfill its function as a binder and the layer becomes soft, almost powdery. Layers "activated" at high temperatures (e.g., 2 hrs. at 140° C. according to Cherbuliez et al. [*4*] or even 4 hrs. at 140° C. according to Nicolaus [*5*]) become increasingly sensitive to atmospheric moisture[1], and it is not surprising to observe, with such layers, quite considerable variations of *Rf* values.

Therefore, we coat our plates and then leave them **overnight for drying in the open air.** The separations described below (amino acids, peptides, DNP amino acids, PTH amino acids) were all obtained on air dried layers [*7*—*9*]. Narrow plates (50 × 200 mm.) are suitable only for preliminary tests, since layers on such plates are uneven as a rule.

When using certain solvents, a further pre-treatment of the layers may be necessary. This is referred to in the section on chromatography of DNP amino acids (p. 422).

Preliminary tests were carried out with "Alox Fluka for use in thin-layer chromatography"[2] (a suspension of 20 g. Alox in 60 ml. H_2O was spread on a plate in the usual manner and the layer was dried as described above). The results were unsatisfactory because of the slow flow rate of the mobile phase, a mixture of n-butanol-glacial acetic acid-water (80 + 20 + 20). In fact, this solvent requires four hrs. for a 10 cm. run. The amino acid separation, however, was equivalent to the separation achieved on silica gel.

Mutschler and Rochelmeyer [*10*], in an endeavor to meet the requirement for a well defined stationary phase, apply 50 ml. of a mixture of equal proportions of 0.2 M KH_2PO_4 and 0.2 M Na_2HPO_4 to 25 g. of silica gel rather than water alone. Then they dry, without further activation, for only 30 min. at 110° C. Our experience, so far, suggests that this type of buffering is not essential. In fact, it may have quite a disturbing effect on any attempts towards recovery of the chromatographed material.

Mottier [*11*] has developed a variation of TLC. He activates "Merck Alumina for Chromatography" (No. 1097) for 45 min. at 300°—500° C. and applies it to the plate in the dry state without using a binder. Furthermore, he does not chromatograph the free amino acids, but rather their sodium salts. Generally speaking, his approach is quite different from that followed by Stahl.

2. Sample application and running of chromatograms [*7*, *8*, *12*]

a) Application of samples: Samples (optimal amount 0.5—2 μg. in 0.5 μl. of liquid) are applied on a line exactly 15 mm. from the lower edge of the plate. The distance of the outermost spots from the lateral edges should be at least 15 mm. and the interval between spots at least 8 mm. Wait a few minutes before placing the plate into the chromatography tank to allow for complete evaporation of the liquid from the layer; special care should be taken if high boiling or acidic liquids were used for dissolving the sample.

b) Solvent: Only freshly prepared mixtures of components of greatest possible purity should be used.

c) Ascending technique: 100—120 ml. solvent are placed into the tank which should be completely lined with filter paper and shaken thoroughly (saturation of the vapor phase, see p. 15) before the plates (200 × 200 mm.) are inserted. The tank should close tightly and must not be opened during chromatography.

d) Horizontal technique [*13*]: A description is given in the general section (pp. 22—24). Saturation of the gas phase occurs automatically.

e) Temperature: Working at any given temperature is less important than keeping that temperature constant. With viscous solvents, e.g., phenol/water, a high temperature (50—60° C.) may help to speed up the solvent flow.

---

[1] cf. R. E. Kirk and D. F. Othmer [*6*].

[2] Fluka AG., Buchs SG., Switzerland.

f) Length of run: 10 cm. is normally adequate. If longer or shorter distances are chosen, this should be noted when stating *Rf* values, except in the case of chromatograms with single component solvents (see pp. 106—114). If the length of the run is limited by a dividing line across the layer, the plate should be removed from the tank as soon as the solvent reaches this line (see pp. 114—123).

III. Amino acids

1. Preparation of solution to be spotted

Amino acids should be as free from impurities as possible. If controls for comparison of *Rf* values etc. are required, it is advisable to use a commercially available set of pure amino acids[1] and to prepare solutions of 1 mg. per 1 ml. of each substance to be tested for. The best solvent is a mixture of water with about 10%, by volume, of n-propanol. Such solutions can be kept in the refrigerator or even at room temperature for 2–4 weeks. Relatively insoluble amino acids, e.g., tyrosine and cystine, require 0.1 N-hydrochloric acid as a solvent. The amount normally spotted is 0.5 or 1 μl. corresponding to 0.5 or 1 μg. of each amino acid. When using *standard solutions* in dilute hydrochloric acid, it is advisable to dissolve and spot the unknown material in the same solvent. Before running the chromatogram, the plate should be aerated for 15–20 min. in order to remove excess hydrochloric acid. In paper chromatography, hydrochloric acid present is often buffered by exposure of the paper to ammonia vapor. With silica, however, care should be taken as ammonia adheres quite strongly and a supposedly neutral chromatography system may turn out to be basic. Amino acids obtained by evaporation from acidic hydrolyzates of proteins or peptides (see below) mostly exist in the hydrochloride form. An appropriately diluted solution in water corresponds approximately to a solution of the standard amino acids in 0.1 N hydrochloric acid. Amino acids in animal and vegetable extracts, as well as in body fluids such as urine, serum, etc., should be freed from contaminating foreign material before spotting. This end is best achieved if these amino acids are chromatographed as their DNP derivatives.

The experimental details given in paragraphs 2 and 3 below are based on procedures developed for paper chromatography. A quantitative recovery of the amino acids originally present in a condition suitable for chromatography is an exception rather than a rule.

2. Hydrolysis of proteins and peptides

a) Acid hydrolysis: If hydrolysis is carried out without precaution, considerable losses, particularly of tyrosine, may occur. Special care should be taken if carbohydrates arepresent (darkening of the hydrolyzate due to humin formation). Acid hydrolysis d estroys tryptophane in particular (for an exception, see below under

[1] Fluka, Buchs SG., Switzerland; E. Merck, AG., Darmstadt, Germany; Cyclo Chemical Corp., Los Angeles, Calif.; California Corp. for Biochemical Research, Los Angeles, Calif.; Mann Research Laboratories Inc., New York; Nutritional Biochemicals Corp., Cleveland, Ohio, U.S.A.

"strongly acidic ion exchangers") and, to a lesser extent, serine and threonine (the original content in hydroxy amino acid may be determined quantitatively by comparing hydrolyzates at 24 and 72 hrs.). Reproducible results may be obtained by a hydrolysis method used in this laboratory as a first step in quantitative amino acid analysis[1].

6 N-Hydrochloric acid: The substance is mixed with a 200—500 fold excess of 6 N hydrochloric acid (distilled two or three times) in a thick glass tube designed for evacuation and sealing. (The excess acid relates to the amount of protein present and should be chosen particularly high if the material contains carbohydrate.) The liquid is then frozen by immersing the tube in a mixture of acetone and solid CO_2. After replacement of the air by pure nitrogen, the tube is evacuated to approx. 1 mm. and subsequently sealed. During this operation, the tube is kept in a wooden beaker filled with solid CO_2 and equipped with a handle. Hydrolysis is carried out for 18—72 hrs. in a thermostat at 110° C. ± 1°. After hydrolysis, the hydrochloric acid is removed by slow evaporation in a vacuum desiccator containing NaOH.

Strongly acidic ion exchangers: A recently described method for acid hydrolysis allows a quantitative recovery of tryptophane and lysergic acid. As a further advantage, this method avoids formation of humin. According to M. Pöhm [*15*], 0.05—0.2 g. of the peptide, 1 g. of Amberlite IR-112 (H) per millival amide-nitrogen and 3—10 ml. of 80 % ethyl alcohol are heated in a nitrogen atmosphere in a sealed tube to 90—95° C. for 6—10 hrs. After cooling, the amino acids are eluted from the resin by a 10 % ammonium hydroxide solution.

Earlier experiments with Dowex-50 in 0.05 N-hydrochloric acid [*16*] have shown that free aspartic acid, serine, and threonine are liberated very rapidly. Valine and isoleucine, however, are liberated more slowly than in 6 N-hydrochloric acid. Peptide bonds involving cystin and cysteic acid are said to be very resistant to this type of hydrolysis. About 25 % of the glutamic acid present is converted to pyrrolidone carboxylic acid. With dilute ammonia solutions, basic amino acids are only partially removed from the ion exchange resins.

Further details about hydrolysis methods, with particular reference to the use of formic acid, acetic acid, stannous chloride, sulphuric acid and hydriodic acid, are listed in the books by Hais and Macek[2] and by Linskens[3].

b) Alkaline hydrolysis: In a sealed tube, 5—10 mg. of the material, 65 mg. of $Ba(OH)_2 \cdot 8\,H_2O$ and 1 ml. of water are kept for 24 hrs. at a temperature of 125 to 130° C. The cooled reaction mixture is adjusted to pH 6 with 2 N H_2SO_4, heated to boiling and centrifuged to separate $BaSO_4$. The latter is washed with water. The supernatant and the washings are combined, evaporated to dryness and the residue is dissolved in 0.5—1 ml. of water or 0.1 N HCl. Cystine and β-hydroxy amino acids are partially destroyed during this procedure, while tryptophane is retained.

3. Free amino acids in biological material

Aqueous extracts of animal or plant tissues, press juices, homogenates and body fluids, generally contain – in addition to free amino acids – peptides[4], proteins, carbohydrates, urea, salts and emulsified lipids.

a) Separation of amino acids from proteins and polysaccharides: Solvent precipitation: Proteins and polysaccharides may be precipitated by addition of organic solvents. Alcohol [*18*] and acetone are used mostly as they are easily removed after the process. Even without removal, they do not impair the subsequent chromatographic analysis of the filtrate. Awapara [*19*] mixes aqueous solutions with 4—8 parts of alcohol, centrifuges after several hours and washes the residue with 80 % alcohol. According to Hais and Macek[5], 1 ml. of plasma is evaporated to dryness in

[1] Chromatography on ion exchange resins, described by Stein and Moore [*14*].
[2] Hais and Macek (p. 415, 482 in [*2*]).
[3] Linskens (p. 149 in [*17*]).
[4] Removal of smaller peptides from amino acids is normally impossible. See section on peptides, p. 409.
[5] Hais, I. M., and K. Macek (p. 780 in [*2*]).

a vacuum desiccator containing H_2SO_4. The residue is left for 2 hrs. with 8 ml. of acetone containing 1% of conc. aqueous HCl. Undissolved material is then centrifuged off, washed two or three times by resuspending in acetone-HCl and centrifugation, and discarded. The combined supernatants are evaporated at 37° C. in a stream of dry air. The residue is dissolved in 0.5 ml. of water, the resulting solution is extracted three times with an equal quantity of ether and again evaporated in a desiccator containing H_2SO_4. For spotting, the residue is dissolved in 20—100 ml. of water.

Ion exchange resins: On highly cross-linked ion exchange resins of the strong base or the strong acid type, proteins are — unlike amino acids — only slightly retained. On a column, a molecular sieve mechanism makes them behave similar to carbohydrates and inorganic cations or anions, respectively. See paragraph on desalting procedures.

Gel filtration: A new method for the separation of high molecular from low molecular weight compounds is gel filtration with Sephadex [*20*][1,2,3], a cross-linked insoluble but highly swelling dextran. Large molecules are excluded, but smaller molecules or inorganic salts diffuse without hindrance into the gel. If an aqueous solution of a mixture of such substances is passed through a Sephadex column, the high molecular weight portion will be found in the first fractions. Smaller molecules are retained for some time and can be eluted quantitatively with more water or diluted salt solution. The main advantage of this "dialysis" process is the time it saves; in fact, only 60 min. are required for a separation on a small scale.

b) Destruction of urea: With a trace of urease, urea in urine is converted within 24 hrs. into CO_2 and NH_3. These gases are removed upon subsequent partial evaporation of the water present [*25*].

c) Desalting: One may use either electrodialysis, as described by CONSDEN, GORDON and MARTIN[4] [*26*], or else demineralization by a suitable ion exchange resin[5] [*27—29*].

Electrolytic desalting partially converts arginine into ornithine, but 10—30% of histidine, lysine, methionine, proline and tyrosine are lost [*30*]. Occasionally, *desalting by means of an ion exchange resin also results in losses:* arginine and, to some extent, lysine are not retained by strongly basic resins; but, on the other hand, they are incompletely eluted from acid resins.

In our laboratory [*31*], the following ion exchange process, based on a method described by DRÈZE et al. [*32, 33*], has proved successful for removing both *salts* and *soluble carbohydrates.*

Depending on the problem under investigation, a 1 × 2 cm. bed of Dowex-50 or of Dowex-2 is prepared in a 1 × 10 cm. glass tube and operated in the following manner:

α) *Desalting of basic amino acids or tryptophane:* Dowex-50, 8% DVB, 200 to 400 mesh (H^+-form) is washed with 10 ml. N HCl followed by water until the eluate is neutral. 1—4 ml. of the test solution (the pH of which is adjusted to < 6 with HCl in order to decompose carbonates) are placed on the column. The rate of flow is set to 1 ml. per 3 min. The resin is then washed with 20 ml. 0.5 N HCl. The amino acids are then eluted with 4 N HCl, the first 3 ml. of the eluate being discarded. The following 5 ml. are collected and repeatedly evaporated to dryness in vacuo, each time adding a few ml. of water. The residue is dissolved in water and chromatographed.

β) *Desalting of neutral and acid amino acids:* Dowex-2, 10% DVB, 200—400 mesh, (OH^--form) is washed with 20 ml. of 2 N NaOH (CO_2 free) followed by boiled (CO_2 free) water until the eluate is neutral. 1—4 ml. of the test solution is placed on the column and the rate of flow is set to 1 ml. per 3 min. The resin is then washed with 20 ml. of water followed by 10 ml. of N acetic acid which changes the color of the resin from brown to light yellow. When the acid front reaches the bottom of the

[1] General methodology and desalting: [*21*].
[2] Gel filtration of proteins, peptides and amino acids: [*22, 23*].
[3] Peptide separation: [*24*].
[4] cf. also LINSKENS (p. 32, in [*17*]).
[5] cf. HAIS, I. M., and K. MACEK (p. 415, in [*2*]).

resin bed, one starts to collect 5 ml. of the eluate. Repeated evaporation, as described in α, yields the amino acids ready to be dissolved and chromatographed.

d) Removal of lipids [*19*]: After the removal of proteins by precipitation [*19*] and centrifugation, the clear supernatant is thoroughly extracted with 3 parts of chloroform. The aqueous (upper) layer should contain the major proportion of the amino acids originally present.

e) Examples: In a recently published paper on the detection of amino acids in serum [*34*], an ion exchange procedure for the isolation of the amino acid fraction was checked with regard to individual amino acid recovery. 1.5 ml. portions of three samples, i.e.,

1) fresh serum,
2) the same quantity of serum with known amounts of amino acids added, and
3) an amino acid mixture similar to the amino acids present in serum, dissolved in water,

were separately passed through a 5 × 0.4 cm. bed of Zeocarb 225 (H). Inorganic anions, proteins, sugars and other uncharged substances were eliminated by washing with 50 ml. of water. The amino acids then were eluted with 20 ml. of 10% NH_3 and the eluate was evaporated to dryness in vacuo. The residues, consisting of neutral amino acids and of salts of acidic amino acids with ammonia or with basic amino acids, were quantitatively analyzed.

Table 100. *Recovery of Amino Acids after Filtration through a Column of an Acid Ion Exchanger* (Zeocarb 225) (Cook and Luscombe [*34*])

Amino acid (abbreviations cf. Table 103, p. 400)	Recovery of amino acids added to serum %	Recovery from a synthetic amino acid mixture %
Glu (HN_2) . . .	113	—
Cit	102	—
Thr	74	93
Pro	76	85
Met	—	41
Phe	84	107
Ser	91	110
Val	90	93
Leu	—	88
Ileu	79	—
Ala	101	122
Gly	—	124
Orn	—	77
Arg	82	53
Average	89	83

Table 101. *Literature on the Extraction of Amino Acids from Biological Material*

Material	Authors
Potatoes	Dent et al. [*35*]
Rice	Parihar [*36*]
Leaf juice (potato plants) . . .	Reindel and Bienenfeld [*37*]
Biological material	Biserte et al. [*38*]
Mushrooms and micro-organisms	Close [*39*]
Ascites and pleurapunctates . .	Knauff et al. [*40*]

Table 100 is based on the study mentioned and gives a good idea of the extent of the amino acid losses which have to be taken into account.

The authors explicitly attribute the poor recovery of certain amino acids to shortcomings of the ion exchange process described.

Table 101 may serve as a guide to some additional applications of the extraction methods mentioned in this paragraph.

4. Solvents and their efficiency for amino acid separation

One phase systems: In view of the extremely limited solubility of amino acids in organic solvents (see Table 99), systems suitable for chromatography must generally contain water. Solvents such as methanol, ethanol or acetone may be used as markedly polar organic components. Often a fairly satisfactory separation can be achieved with such systems. They give relatively diffuse zones, however, and tend to cause tailing. This tendency may be checked by adding a few per cent of glacial acetic acid (see solvent G below), and by reducing the amounts spotted to about 2 μg. per amino acid. Smaller zones result also when methanol or ethanol are replaced by propanol, butanol and the like, but this advantage is somewhat offset by reduced rate of flow, largely due to increased viscosity. Zones generally spread less (i.e., separation is better) the higher the solvent viscosity. Phenol, as a solvent, provides a good example for this effect. If viscous alcohols are replaced by nonpolar fluids of lower viscosity, a solubilizer such as methanol, pyridine or glacial acetic acid will help to restore miscibility with water. With systems of this latter type, separations may be very satisfactory and rapid.

Two phase systems: There are two disadvantages. First, preparation takes time since phase separation is usually slow. Second, there is an

Table 102. *Solvents[1] for TLC of Amino Acids on Silica Gel G*

Symbol	Components	Grams	Milliliters
A	96 % Ethanol-water	63 + 37	70 + 30
B	n-Propanol-water	64 + 36	70 + 30
C	n-Butanol-glacial acetic acid-water	60 + 20 + 20	80 + 20 + 20
D	Phenol-water[2]	75 + 25	—
E	n-Propanol-34 % ammonia solution	67 + 33	70 + 30
F	96% Ethanol-34 % ammonia solution	77 + 23	70 + 30
G	Methyl ethylketone-pyridine-water-glacial acetic acid		70 + 15 + 15 + 2
H	Chloroform-methanol-17 % ammonia solution		40 + 40 + 20

[1] E. Mutschler and H. Rochelmeyer [*10*] use, on buffered layers (p. 394), solvents consisting of 70 % ethanol, 96 % ethanol-25 % ammonia (80 + 20) or 96 % ethanol-25% ammonia-water (70 + 10 + 20).

E. Nürnberg [*41*] runs two-dimensional chromatograms in n-propanol-water (50 + 50) and phenol-water (100 + 40).

According to Jost, Rudinger and Šorm [*42*], phenol-water (90 + 30) is suitable for separating tyrosine, N-methyl-tyrosine and O-methyl-tyrosine.

[2] 75 g. of melted phenol are mixed with 25 ml. of water, and NaCN (about 20 mg.) or another antioxidant is added. The mixture should be cooled to room temperature before use.

extreme sensitivity even to slight variations in temperature. For chromatography, the organic layer is generally used as the mobile phase. The water layer may serve to help equilibration of the vapor phase in the

Table 103. *Rf-Values*[1] *of the Main Amino Acids in Solvents A—F of Table 102* (ascending technique, solvent migration 10 cm.), and $R_{Leucine}$-values[2] in solvent G of Table 102 (horizontal technique[3], 4 hours). Amount spotted about 0.5 μg.

Amino acid	Abbreviation	*Rf* × 100 in solvent						$R_{Leucine}$[2]
		A	B	C	D	E	F	G
Alanine	Ala	47	37	22	29	39	40	51
β-Alanine	β-Ala	33	26	22	30	30	29	35
α-Amino-i-butyric acid	α-AiB			27				61
α-Amino-n-butyric acid	α-AnB			27				59
β-Amino-i-butyric acid	β-AiB			25				48
γ-Amino-n-butyric acid	γ-AnB			27				45
ε-Amino-n-caproic acid	ε-ACo			34				68
α-Amino-n-caprylic acid	α-ACy	66	65	59	69	58	60	143
Arginine	Arg	04	02	06	19	10	06	19
Aspartic acid	Asp	55	33	17	06	09	07	18
Cysteic acid	$CySO_3H$	69	50	10	04	17	21	53
Cystine	$(Cys)_2$	39	32	09	12	27	22	18
Dibromotyrosine	DBT			60				168
Glutamic acid	Glu	63	35	24	10	14	15	32
Glycine	Gly	43	32	18	24	29	34	45
Histidine	His	33	20	05	32	38	42	18
Homocystine	$(Hcys)_2$			17				28
Hydroxyproline	Hypro	44	34	16	38	28	31	50
Isoleucine	Ileu	60	53	43	49	52	58	92
Leucine	Leu	61	55	44	48	53	58	100
Lysine	Lys	03	02	03	09	18	11	11
Methionine	Met	59	51	35	49	51	60	92
Methioninesulphone	$Met.O_2$							66
1-Methyl-histidine	Mehis							09
Norleucine	Nleu	61	57	45	52	53	59	102
Norvaline	Nval	56	50	36	42	49	57	77
Ornithine	Orn			04				14
Phenylalanine	Phe	63	58	43	55	54	60	109
β-Phenylserine	β-Φ-Ser			41				108
Proline	Pro	35	26	14	50	37	30	40
Sarcosine	Sar	31	22	12	37	34	31	32
Serine	Ser	48	35	18	20	27	31	47
Taurine	Tau			18				79
Threonine	Thr	50	37	20	26	37	40	51
Tryptophane	Try	65	62	47	63	55	58	122
Tyrosine	Tyr	65	57	41	47	42	51	107
Valine	Val	55	45	32	40	48	56	72
Asparagine	$Asp(NH_2)$			14				43
Glutamine	$Glu(NH_2)$			15				47
Glycinamide	$Gly(NH_2)$			16				62

[1] Each figure given is the arithmetic mean of the results of 4 (solvents A, B, D, E, F) or of 9 separate measurements (solvent C). Factors affecting *Rf*-values are discussed by M. BRENNER, A. NIEDERWIESER, G. PATAKI and A. R. FAHMY [*12*].

[2] $R_{Leucine}$ = *Rf*-value in reference to leucine. Each figure given is the arithmetic mean of the results of 9 separate measurements.

[3] M. BRENNER and A. NIEDERWIESER [*13*], see p. 22.

tank but often it is discarded. From a practical point of view, one may replace the saturated organic phase by an almost identical solvent mixture which is not quite saturated, however, with regard to its water content. It is found, for instance, that the *Rf*-values in phenol-water change only slightly if water saturated phenol, which, on a weight basis, contains about 71% phenol, is replaced by 80% phenol. In fact, 75% phenol is a preferred solvent (Table 102).

In Table 102, several solvents suitable for separating amino acids are listed. They are partly known from their use in paper chromatography. Table 103 contains the corresponding *Rf*-values.

Consulting the diagrams in Fig. 161 facilitates the selection of the solvent most suitable for a specific purpose. It is interesting to note that, in neutral solvents such as ethanol- or n-propanol-water mixtures, the acidic amino acids travel much faster than lysine and arginine which, indeed, show very small *Rf*-values (Table 103). The difference might be due to cation exchange. In fact, AHRLAND et al. [*43*] reported that the titration curves of silica gel and of a slightly acid ion exchange resin are similar (cf. chapter on inorganic TLC). Furthermore, it is seen that a hydroxyl group in the molecule does not necessarily reduce the *Rf* value. Depending on its interaction with the solvent, the reverse may be true (cf. Ser/Gly, Thr/Ala, Hypro/Pro and Tyr/Phe in solvents A and E).

Separation efficiency is by far best with chloroform-methanol-17% ammonium hydroxide (40 + 40 + 20), n-butanol-glacial acetic acid-water (80 + 20 + 20) and phenol-water (75 g. + 25 g.).

The combination of solvents C and D (Fig. 160 [*7*]) and, even more so, a combination of solvents H and D (Fig. 162 [*9*]) is suitable for two-dimensional chromatography. This latter combination will separate all protein amino acids in addition to β-alanine and γ-amino-n-butyric acid, except leucine and isoleucine. Detection of these isomers becomes possible with another technique, i.e., continuous TLC [*13*], using solvent G (Fig. 163).

In one-dimensional thin-layer chromatography of amino acids, the *Rf* values so far observed usually do not exceed a limit of about 0.7. With all solvents examined, only about $^2/_3$ of the distance between start and front is used in separation. Increasing the water content of the solvent raises small *Rf*-values more than high ones and thus reduces the separation effectively achieved.

Supplementary notes: It should be remembered that the *Rf*-value of a pure substance is not a constant, but is influenced by a number of factors [*12*]. Fig. 164, for instance, reveals a relation between *Rf*-value and the quantity of substance applied.

Further, there is a certain dependence on the properties of other materials present. Acidic amino acids chromatographed in a mixture with other amino acids in a n-propanol-water solvent (70 + 30) travel faster than is known from chromatograms of the pure compounds.

In the case of DNP amino acid mixtures, *Rf*-values depend somewhat on the ratio of the components. Dinitrophenol has a particularly marked influence, as was already noticed in paper chromatography [*45*]. In the

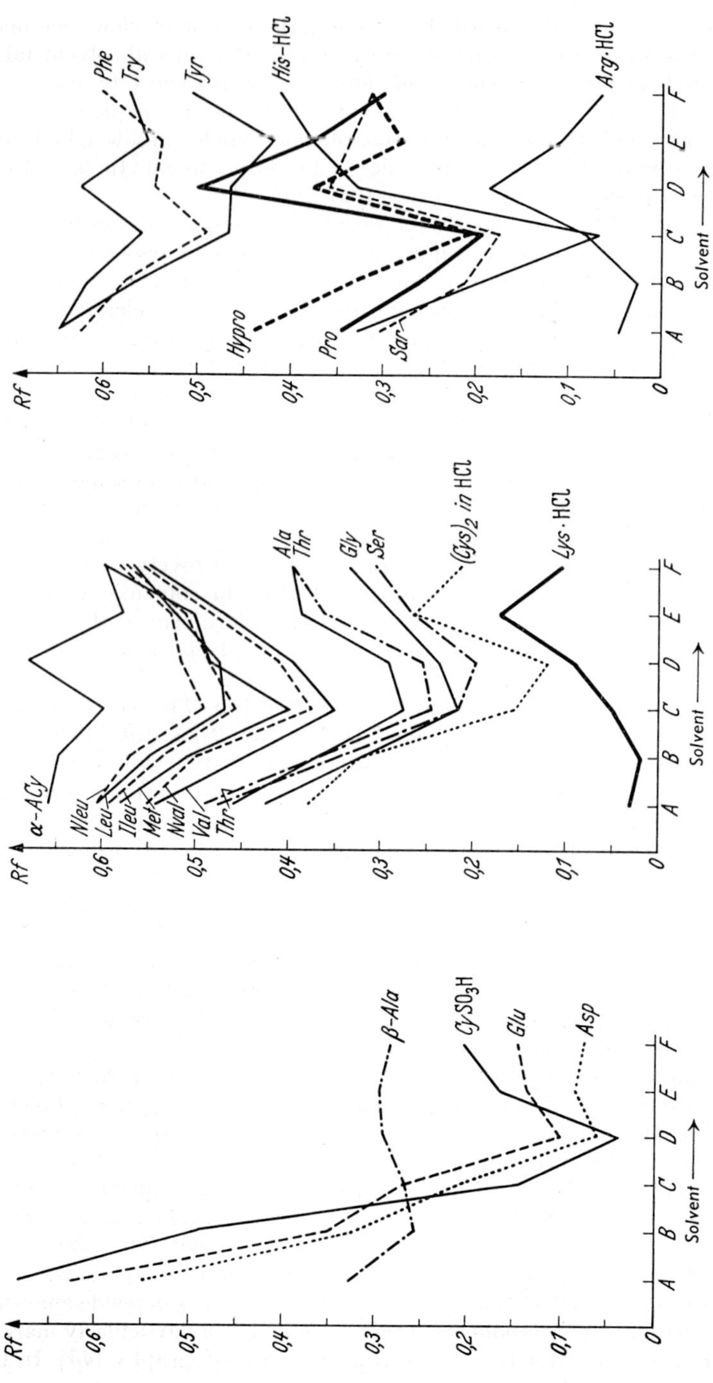

Fig. 161. Graph of *Rf*-values of the main amino acids in solvents A—F of Table 102 [7]. Note the differences in response to changes in the solvent. Abbreviations are explained in Table 103

second run of two-dimensional chromatograms, *Rf*-values are not so much affected by the relative proportions of components as by previous treatment of the layer itself, i.e., chromatography in the first dimension

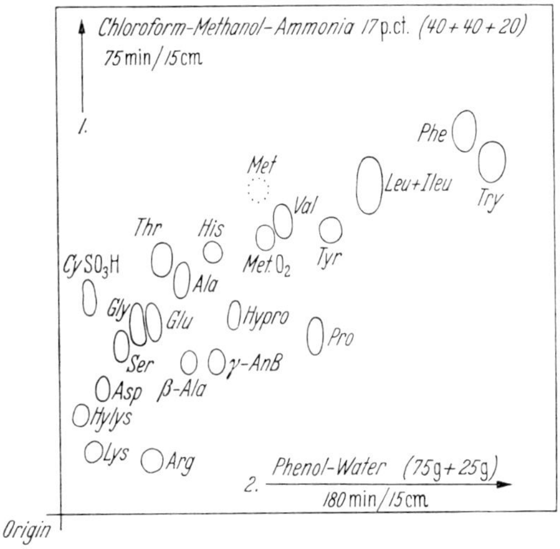

Fig. 162

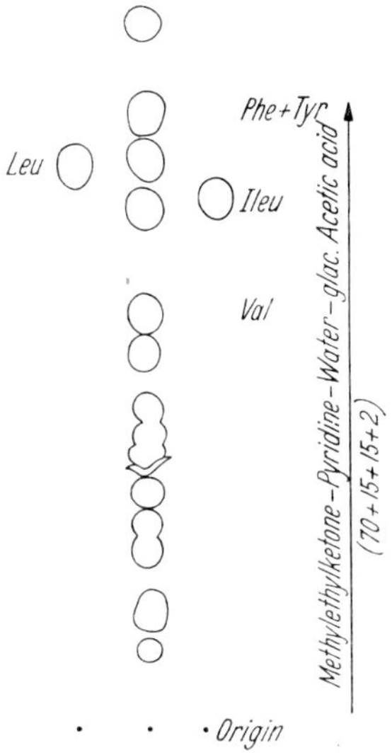

Fig. 163

Fig. 162. Two-dimensional separation of a performic acid oxidized [*44*] mixture of 20 protein amino acids + β-alanine + γ-amino-n-butyric acid (γ-AnB), ascending technique

0.5 μg. of each amino acid are spotted in a total volume of 0.5 μl. of 0.1 N HCl. After the run in the first dimension, the layer is dried by placing the plate for 20 min. into a well ventilated hood. The methionine spot indicated by a dotted circle is not observed after performic acid oxidation. Detection by means of ninhydrin (Reagent No. 108)

Fig. 163. One-dimensional separation in a continuous flow chamber ([*13*], pp. 22—24) to detect leucine and isoleucine in the presence of 18 protein amino acids + β-alanine + γ-amino-n-butyric acid

Quantity applied: 0.5 μg. of each amino acid is spotted in a total volume of 0.5 μl. of 0.1 N-HCl. Chromatogram run for $4^1/_2$ hrs. Detection by means of ninhydrin (Reagent No. 108). Performic acid oxidation [**44**] is essential if methionine is present. Identification of leucine and isoleucine is unambiguous if, for comparison, a sample of each of these amino acids is chromatographed in a parallel experiment on the same plate

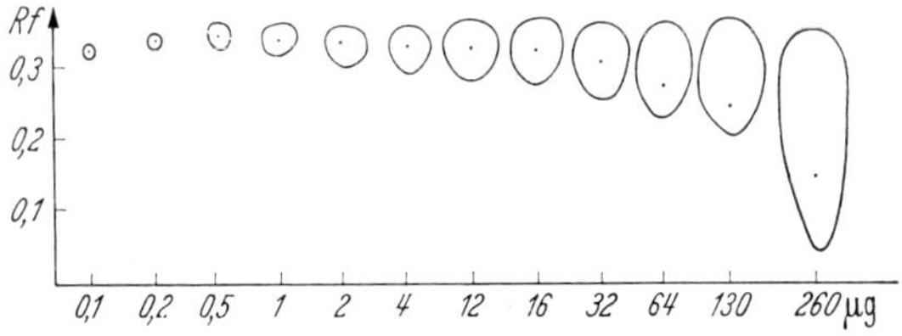

Fig. 164. Relation between *Rf*-value and amount spotted in the case of chemically pure glycine

Location of highest glycine concentration is marked by a dot. Amounts indicated were spotted each time in a volume of 1 μl. Ascending technique with n-propanol-water (70 + 30) as the solvent and intermediate drying. As such "pre-treatment" can be and generally is standardized, *Rf*-values for the second dimension are normally quite constant. In two-dimensional chromatography, mixtures therefore display characteristic spot patterns which are hardly obscured by variations of *Rf*-values. For this reason, it is advisable not to chromatograph an unknown sample on its own but together with a standard mixture containing just enough of each substance for spots to be still visible after two-dimensional separation. Compounds present in the unknown sample are thus quickly and satisfactorily identified by an increase of the intensity of the corresponding spots.

Note on intermediate drying of two-dimensional chromatograms

Intermediate drying presents no problem where highly volatile solvents are used. The plate should be left in a well ventilated hood for about 15 min., whereupon it can be immediately submitted to the second dimension run. When sensitive substances are present, there is some danger of destruction, especially if the plate must be heated in order to evaporate less volatile solvents such as phenol. Another alternative is to leave the plate in an air current over night, risking oxidation. It would indeed be useful if small and inexpensive vacuum equipment[1] were designed for rapid drying of chromatograms. For the time being, a non-volatile solvent is preferably used in the second dimension only. When used in the first dimension, considerable amounts may be left after drying and thus chromatography in the second dimension may be markedly affected (see "indirectly" obtained *Rf*-values of DNP amino acids, Table 109 and p. 422).

There is, of course, another possibility: leave the plate in an air current for 10 min., heat in a drying cabinet to 60° C. for 15 min. and finally cool for 15 min. in a draught.

If the chromatogram, after intermediate drying, has to be kept for some time before being processed further, the layer should be covered with a glass plate. Protection from daylight is advisable.

5. Detection of amino acids on the chromatogram

At present, there is no reagent known responding exclusively to amino acids. However, a few unspecific reagents responding to several classes of nitrogenous compounds are available and some others have been shown to be specific for one or a few amino acids. The generally adapted procedures are described in the following paragraphs.

a) Ninhydrin: Ninhydrin is still the most popular reagent for amino acids. Although the color reaction of triketohydrinden hydrate with amino acids and small peptides has been known as long as 1910 (Ruhemann), there is still doubt about the course of the reaction. The many theories on the subject have been summarized by Caldin [*46*]. Treatment with complex-forming cations (Cu, Cd, Ca) changes the color obtained with ninhydrin to red and increases color fastness considerably. While the normal method (e.g., [*26, 47, 48*]) gives yellow (proline and hydroxyproline) or violet spots (all other α-amino acids), more specific coloration can be achieved by adding bases such as collidine and benzylamine. Though this may facilitate identification of single amino acids, it does diminish the sensitivity of the reaction.

---

[1] Commercially obtainable vacuum drying cabinets with heating facilities are generally too expensive.

Practical aspects of the ninhydrin reaction are given in the appendix on reagents (No. 108). The proposed addition of acetic acid to the reagent helps to establish a slightly acid medium – as required for the ninhydrin reaction – even where basic solvents were used to run the chromatogram. Prolonged heating (110° C.) of the sprayed plate causes the ninhydrin on the background to turn pink. Minimal amounts of amino acids required for color formation on thin-layer chromatograms are listed in Table 104.

Table 104. *Minimal Amounts (μg.) of each of 19 Amino Acids required for Detection on Thin-Layer Chromatograms (Silica Gel G) by the Ninhydrin Reaction [9]*

Amino acids see Table 103 for abbreviations	One-dimensional chromatogram in n-propanol-water (70 + 30)	Two-dimensional chromatogram of an amino acid mixture as shown in Fig. 162
Ala	0.009	0.05
β-Ala	0.01	0.06
Arg	0.01	0.06
Asp	0.1	0.2
$CySO_3H$	0.01	0.1
Glu	0.04	0.4
Gly	0.001	0.006
His	0.05	0.5
Hypro	0.05	0.1
Leu	0.01	0.2
Lys	0.005	0.03
Met	0.01	0.4
Phe	0.05	0.2
Pro	0.1	0.5
Ser	0.008	0.1
Thr	0.05	0.1
Try	0.05	0.5
Tyr	0.03	0.1
Val	0.01	0.2

α) *Stabilization of colors obtained with ninhydrin:* After color development with the ninhydrin reagent, which should not contain any complex-forming buffer such as citrate, KAWERAU and WIELAND [*49*] spray the chromatogram with a copper nitrate solution, (Reagent No. 108). The Cu-complex obtained in this way is, however, only colorfast if the presence of free acids is excluded. Therefore, immediately after spraying, the chromatogram must be exposed to ammonia vapor. Moreover, the chromatogram must be protected from moisture as the Cu-ninhydrin complex dissociates reversibly between pH 7 and 9 and irreversibly above pH 9. As a protective, a collodion film, according to BARROLLIER [*50*], may be used (see. DNP amino acids, documentation, p. 426).

β) *Polychromatic ninhydrin reaction:* The reagent described by MOFFAT and LYTLE [*3*] has proved to be satisfactory (Reagent No. 109). Similar effects on colors have been achieved by WOIWOD [*51*] using collidine and by HARDY et al. [*52*] using cyclohexylamine or dicyclohexylamine.

b) Chlorine/tolidine test (Reagent No. 32): This test, originally developed for detecting substances containing the grouping NH–CO and described in the section on peptides, may be applied to free amino acids as well. However, minimal amounts required for color formation are relatively large.

c) Other reagents. Table 105a summarizes some experience gained from paper chromatography. Suitably modified techniques may well be applied to thin-layer chromatography. While the excellent separation effects provided by TLC render specific reagents less important for the detection of specific amino acids present in mixtures, they may still help in detecting such amino acids when present in intact peptides.

Table 105a. *Predominantly Unspecific Reagents for Detecting Amino Acids*

Type	Reag. No.	Reagent	Authors
	108	Ninhydrin	Caldin [46]
			Tsukamoto and Komori [48]
			Patton and Chism [54]
			Toennies and Kolb [55]
			Consden et al. [47]
			Ruhemann [56—60]
		Ninhydrin + cobaltous chloride	Wiggins and Williams [61]
polychromatic	109	Ninhydrin + copper nitrate + collidine	Moffat and Lytle [3]
		Ninhydrin + collidine	Woiwod [51]
		Ninhydrin + cyclohexylamine or dicyclohexylamine	Hardy et al. [52]
		Ninhydrin + phenol	Hais[1]
		Isatin	Saifer and Oreskes [62]
			Noworytko and Sarnecka-Keller [63]
			Acher et al. [64]
			Grassmann and v. Arnim [65]
polychromatic		Isatin + zinc acetate + pyridine + alloxan	Barrollier et al. [66]
			Saifer and Oreskes [62]
			Rosebeek [67]
		Na-1,2-Naphthoquinone-4-sulphonate	Kofrányi [68]
			Consden [69]
	96	. . . .	Müting [70]
			Giri and Nagabhushanam [71]
polychromatic		Na-1,2-Naphthoquinone-4-sulphonate + zinc acetate + quinoline	Barrollier et al. [66]
		4,5-Diacetyl-cyclohexene-(1)	Riemschneider and Preuss [72]
	32	Chlorine/tolidine[2]	Reindel and Hoppe [73]

[1] see p. 419, 753 in [2].
[2] see p. 411.

Table 105b. *Relatively Specific Reagents for Amino Acids*

Amino acid	Reag. No.	Reagent	Authors
Arginine	97	α-Naphthol (or oxine)/NaOBr (or NaOCl) "Sakaguchi reaction"	Bhattacharya et al. [74]
			Roche et al. [75]
			Jepson and Smith [76]
			Acher and Crocker [77]
			Sakaguchi [78—80]
	113	Na-Nitroprusside/potassium ferricyanide	Roche et al. [75]
		Na-α-Naphtholate/diacetyl	Tuppy [81]

(Contin. of Table 105b)

Amino acid	Reag. No.	Reagent	Authors
Cysteine	111	Sodium nitroprusside	Toennies and Kolb [55] Winegard et al. [82]
	cf. 76	Iodoplatinate	Toennies and Kolb [55] Winegard et al. [82]
	75	Iodine-azide	Kirby-Berry et al. [83] Awe et al. [84] Chargaff et al. [85]
	145	Tetrazolium salts	Burton et al. [86]
		p-Aminodimethylaniline/potassium ferricyanide	Toyada [87]
Cystine	111	Sodium nitroprusside/NaCN	Winegard et al. [82]
	cf. 76	Iodoplatinate	Winegard et al. [82]
	75	Iodine-azide	Kirby-Berry et al. [83] Awe et al. [84] Chargaff et al. [85]
		p-Aminodimethylaniline/potassium ferricyanide	Toyada [87]
Glycine		o-Phthalaldehyde	Patton and Foreman [88]
Histidine	37	Diazotized sulphanilic acid or sulfanilamide ("Pauly reagent")	Frank and Petersen [89] Bray et al. [90] Kirby-Berry et al. [83] Pauly [91, 92]
	61	"Echtblausalz B" (diazo-reagent)	
		Diazotized p-chloroaniline	Edlbacher [93, 94]
		Diazotized b-bromoaniline	
		Diazotized p-anisidine	Sanger and Tuppy [95]
		Bromine	Kirby-Berry et al. [83]
		o-Phthalaldehyde	Patton and Foreman [88]
Hydroxy-proline		Isatin/Ehrlich reagent	Jepson and Smith [76]
		Ninhydrin modification	Clarkson [96]
		Periodate/acetylacetone + ammonium acetate	Schwartz [97]
		Ninhydrin	
Lysine	147	Vanillin	
Methionine	cf. 76	Iodoplatinate	Winegard [82]
	75	Iodine-azide	Kirby-Berry et al. [83] Awe et al. [84] Chargaff et al. [85]
		Potassium permanganate	Dalgliesh [98]

(Contin. of Table 105b)

Amino acid	Reag. No.	Reagent	Authors
Ornithine	147	Vanillin	
Proline		Isatin	Acher et al. [64]
		Ninhydrin	
Serine		Periodate/acetylacetone + ammonium acetate	Schwartz [97]
	106	Periodate/Nessler reagent	Consden [69], Consden et al. [99]
		Periodate/KI + starch	Metzenberg and Mitchell [100]
		1,2-Dinitrobenzene-enediol-reaction	Fearon and Boggust [101]
Threonine		Periodate/Na-nitroprusside + piperidine	Schwartz [97]
		Periodate/Na-nitroprusside + piperazine ("Rimini reagent")	Edward and Waldron [102]
	106	Periodate/Nessler reagent	Consden [69], Consden et al. [99]
		Periodate/KI + starch	Metzenberg and Mitchell [100]
		1,2-Dinitrobenzene-enediol-reaction	Fearon and Boggust [101]
Tryptophane	48	p-Dimethylaminobenzaldehyde + HCl (Ehrlich reagent)	Smith [103] Dalgliesh [104] Pachéco [105]
	154	Cinnamaldehyde + HCl	Wieland and Bauer [106] Jerchel and Müller [107]
		$NaNO_2$ + HCl	Fischer [108]
	64	Formaldehyde reagent	Procházka [109]
		Modified Salkowski reagent	Linser et al. [110]
	37	Diazotized sulphanilic acid	Erspamer [111]
	38	Diazotized p-nitroaniline	Erspamer [111]
		Diazotized benzidine	Clerk-Bory et al. [112]
		Diazotized ethyl-α-naphthylamine	Dalgliesh [104] Ekman [113]
		o-Phthalaldehyde	Patton and Foreman [88]
Tyrosine		α-Nitroso-β-naphthol/HNO_3	Acher and Crocker [77] Gerngross et al. [114]
	37	Diazotized sulphanilic acid (Pauly reagent)	Frank and Petersen [89] Bray et al. [90] Kirby-Berry et al. [83]
	61	"Echtblausalz B" (diazo reagent)	
	122	Phosphomolybdic tungstic acid (Folin-Ciocalteu reagent)	Kudzin et al. [115] Folin and Ciocalteu [116]
		Folin-Denis reagent	Kudzin et al. [115] Folin and Denis [117]
		Millon reagent	Durant [118] Millon [119]

The fluorescence test sometimes used in paper chromatography depends, according to OPIENSKA-BLAUTH et al. [*53*], on a reaction of the amino acids with aldehyde groups formed from carbohydrates in the paper by heat and is, therefore, not applicable to chromatograms run on silica gel.

IV. Peptides

Peptides, like amino acids, are generally hydrophilic. Thin-layer chromatographic techniques developed for amino acids are therefore, in principle, equally applicable to peptides. There are, however, limits to this analogy. The number, the nature and the sequence of the amino acid residues have a bearing on solubility and adsorption behavior of peptides. Accordingly, conditions of chromatography may have to be adapted, or other methods of separation may have to be employed. Peptides with masked functional groups such as intermediates in peptide synthesis, are, as a rule, less hydrophilic than those without protective groups.

Peptides exist along with amino acids in biological material[1], usually however in small amounts and often in a conjugated form (phosphopeptides, peptidyl nucleotides, glucopeptides, lipopeptides, peptide-protein complexes). Smaller free peptides behave much like amino acids during extraction. At times, their separation from the amino acids may be very difficult.

There may be no other way than to run a two-dimensional chromatogram, elute each of the separated substances and characterize it by further chromatography, and eventual hydrolysis. A combination of paper-chromatography and paper-electrophoresis often has been successfully used to separate such mixtures (ANFINSEN and others [*122*]). According to HONEGGER [*123*], thin-layer chromatography may be combined with thin-layer electrophoresis (pp. 433—435).

Column chromatography on ion-exchange materials, on a cellulose[2] or dextran-basis[3], offers additional possibilities, especially with regard to a prefractionation. DEAE-cellulose, for instance, contains diethylaminoethyl groups. It does not adsorb neutral and basic amino acids but does adsorb neutral peptides, which are eluted with water or water saturated with carbon dioxide [*125*].

Gel-filtration with Sephadex[4] separates substances which differ in molecular weight [*21, 24, 126, 127*]. In its effect, it corresponds more or less to dialysis, but it works faster. LINDNER et al. [*128*] extracted a dry preparation from the posterior lobe of pituitary glands (pig, oxytocin and vasopressin activity 2—3 units/mg.)

---

[1] See, e.g., [*120, 121*].

[2] DEAE-cellulose, ECTEOLA-cellulose, carboxymethyl-cellulose, phosphoryl-cellulose [*124*].

[3] DEAE-Sephadex, AB. Pharmacia, Uppsala, Sweden.

[4] Sephadex is a trade name for cross-linked dextran. In the dry state, it represents a fine grained powder. In water, it swells markedly thereby forming a semi-transparent gel which is packed into a chromatography cloumn like alumina. Its polar character is almost exclusively due to its hydroxyl groups. With increasing number of cross-links, the porosity of the material decreases, and at a certain degree of cross-linking, molecules of a certain size can no longer penetrate the Sephadex particles. With a given grade of Sephadex, smaller molecules permeate to a greater extent and, therefore, migrate more slowly through the column than larger ones. Sephadex may be obtained from AB Pharmacia, Uppsala, Sweden; U.S. representative: Pharmacia Fine chemicals, Inc., 501 Fifth Ave., New York 17, N. Y.

with a pyridine acetate buffer, neutralized, filtered through a column of Sephadex G-25, eluted with the same buffer and obtained two ninhydrin positive fractions. The first one contained oxytocin and vasopressin as peptide-protein-complexes. In the second fraction, there was only low molecular inactive material. Treatment with M formic acid at 70° C. for 10 min. brought about dissociation of the complexes and on refiltration through Sephadex G-25 followed by elution with formic acid, there appeared, in a slowly moving band, vasopressin and oxytocin with an activity of about 100 units/mg. each.

Separation of amino acids from peptides or of salts from peptides on Sephadex is less effective. The situation here is complicated by the fact that migration rates depend more on experimental conditions than on molecular size [*22*].

Finally, fractionation by counter current distribution may be mentioned. This technique is very powerful, indeed, but its application requires relatively complicated apparatus[1].

Mixtures of peptides result from partial hydrolysis of proteins. Separation of the individual components and establishment of their amino acid sequence form the basis for uncovering the chemical structure of proteins. For information on degradation methods yielding peptides, the reader is referred to SANGER [*130*] (partial hydrolysis), CRAIG et al.[2] (oxidative fission of disulfide bridges), H. ZUBER [*131*] (enzymatic hydrolysis), and SJÖQUIST [*132, 133*] (Edman degradation of peptides).

Synthetic peptides are becoming increasingly available as new synthetic methods are being developed. There are, for instance, bradykinin and its analogues [*134–136*], oxytocin analogues [*137*], polymyxin [*138, 139*] and corticotropically active polypeptides [*139a*]. In all these fields, paper chromatography was, up to now, the general analytical procedure. Presently, however, thin-layer chromatography seems to be taking over its role[3]. Tables 106 and 107 give some examples of thin-layer chromatography of peptides.

For chromatographing protected longer chain peptides (up to about 10 amino acid residues), GUTTMANN [*143*] recommends silica gel or alumina layers with dimethylformamide or aqueous acetic acid (5–10% H_2O) as solvents. Spots were detected by the chlorine-iodine reaction, and *Rf*-values between 0.3 and 0.9 were observed. The reproducibility was not very satisfactory; the method may, however, be applied with good results as a test for purity. SCHELLENBERG [*144*] works with chloroform-acetone (90 + 10) and (80 + 20), cyclohexane-ethyl acetate (50 + 50) as well as with chloroform-methanol (90 + 10).

Separation efficiency, in the case of complex peptides of similar structure, is limited (VOGLER [*145*]). In the course of structural investigations on polymyxin B_1, four isomeric cyclodecapeptides, with structures tentatively attributed to the natural product by W. HAUSMANN [*147*] and by BISERTE and DAUTREVAUX [*148*], were synthesized by VOGLER et al. [*146*]. These cyclic oligopeptides, designated by the symbols 8γ, 8α, 7γ, 7α, differ partly from each other in the number of ring amino acids and in the way of linking the side chains to the ring.

[1] CRAIG, L., and D. CRAIG (p. 290 in [*129*]).
[2] CRAIG, L. C., W. M. KÖNIGSBERG and T. P. KING (p. 70 in [*124*]).
[3] see [*134—139, 139a*].

Table 106. h*Rf*-*Values*[1] *of Pairs of Isomeric Dipeptides on Silica Gel G in Two Solvents*[2] [*140*]. For abbreviation see [*142*]

Dipeptide pair	I	II
H · Ala-Gly · OH/H · Gly-Ala · OH	15—16	15—14
H · Gly-Hypro-OH/H · Hypro-Gly · OH	08 / 13	12—13
H · Gly-Leu · OH/H · Leu-Gly · OH	37—35	27 / 33
H · Ala-Leu · OH/H · Leu-Ala · OH	45 / 37	30—31
H · Gly-Phe · OH/H · Phe-Gly · OH	36—38	32 / 37
H · Gly-Pro · OH/H · Pro-Gly · OH	08—09	08—07
H · Gly-Ser · OH/H · Ser-Gly · OH	12—14	12—14
H · Phe-Ala · OH/H · Ala-Phe · OH	48—45	42—40
H · Gly-Val · OH/H · Val-Gly · OH	32 / 27	22—24

I: n-Butanol-glacial acetic acid-water (80 + 20 + 20);
II: n-Propanol-water (70 + 30).

[1] h*Rf*-Values joined by a dash are seen to be different when both compounds are chromatographed separately on the same plate. Separation of a mixture does not occur as long as the normal TLC procedure (without continuous development) is used.

[2] Other suitable solvents are ethanol-34% ammonia solution (70 + 30) and n-propanol-34% ammonia solution (70 + 30).

The behavior of these cyclic oligopeptides in thin-layer chromatography has been thoroughly investigated [*145*]:

Ascending technique

In the following solvents, no *Rf*-differences in the *Rf*-region, 0.5—0.9, were detected between 7γ, 7α and polymyxin B_1:

n-butanol-pyridine-glacial acetic acid-water (v/v)

30	20	6	24
15	5	8	12
15	3	10	12
30	3	23	24

n-butanol-pyridine-glacial acetic acid-water-ethyl acetate (v/v)

5	1	2	2	2

n-butanol-glacial acetic acid-water (v/v) (upper phase)

4	1	5

isopropanol-pyridine-glacial acetic acid-water (v/v)

10	5	4	4
4	8	1	1

ethanol-pyridine-glacial acetic acid-water-ethyl acetate (v/v)

5	1	2	2	2

Horizontal technique, continuous solvent flow:

Continuous solvent flow-chromatograms [*13*] in n-butanol-pyridine-glacial acetic acid-water (30 + 20 + 6 + 6) produced, after 8—10 hours, (spots at upper end of the plate) no significant *Rf*-differences. In ethyl acetate-pyridine-glacial acetic acid-water (50 + 10 + 10 + 10), the three substances covered, in 14 hours, exactly the same distances (20 mm.) and all gave sharply edged, round spots.

For the detection of peptides, the methods developed in paper chromatography may be adapted. The popular ninhydrin reaction is not always sufficiently sensitive with higher peptides; it fails completely with cyclic peptides unless there are side chains containing free amino groups.

Table 107. *TLC of Peptides and Protected Peptides on Silica Gel G in the Solvents A, B and* [1, 2, 3] *according to* Riniker [*141*]. For abbreviations, see Kappeler and Schwyzer [*142*]

	$Rf \times 100$	
	A	B
H · Pro · OH	10	11
Z · Pro · OH	39	81
H · Pro · OtBu	81	34
Z · Pro · OtBu	86	87
Z · Val-Lys (BOC) · OH	57	87
Z · Val-Lys · OH	26	49
Z · Val-Tyr · OCH_3	86	82
Z · Val-Tyr · OH	48	78
Z · Val-Tyr-Pro · OtBu	84	81
Z · Val-Tyr-Pro · OH	45	75
Z · Val-Tyr-Pro · OtBu	73	58
H · Val-Tyr-Pro · OH	30	46
H · Val-Tyr-Val-His-Pro-Phe · OCH_3	76	22
H · Val-Tyr-Val-His-Pro-Phe · OH	47	18
H · Val-Tyr-Val-His-Pro-Phe · OH	51	24
H · Asp-Arg-Val-Tyr-Val-His-Pro-Phe · OH (=Hypertensin II)	21	03
H · Asp-Arg-Val-Tyr-Val-His-Pro-Phe · OH	26	05
H · Asp(NH_2)-Arg-Val-Tyr-Val-His-Pro-Phe · OH	20	02
H · Asp(NH_2)-Arg-Val-Tyr-Val-His-Pro-Phe · NH_2	19	03
H · Asp(NH_2)-Arg-Val-Tyr-Val-His-Pro-Phe · OCH_3	21	05
c-[Val-Orn-Leu-Phe-Pro-Phe-Phe-Asp(NH_2)-Glu(NH_2)-Tyr] (=Tyrocidin A)	32	40
(Val-Orn-Leu-phe-Pro)$_2$ (=Gramicidin)	45	36
(Val-Lys-Leu-phe-Pro)$_2$ (=Lysin-Gramicidin)	38	31
H · Arg-Pro-Pro-Gly-Phe-Ser-Pro-Phe-Arg · OH (=Bradykinin)	12	01
Z · Glu(OtBu)-His-Phe-Arg(NO_2)-Try-Gly · OH	53	53
Z · Glu-His-Phe-Arg(NO_2)-Try-Gly · OH	24	42
BOC · Ser-Tyr-Ser-Met-Glu(OtBu)-His-Phe-Arg-Try-Gly · OH	38	28
H · Ser-Tyr-Ser-Met-Glu-His-Phe-Arg-Try-Gly · OH	18	
H · Lys(BOC)-Pro-Val-Gly-Lys(BOC)-Lys(BOC)-Arg-Arg-Pro-Val-Lys(BOC)-Val-Tyr-Pro · OtBu	32	22
Z · Glu(OtBu)-His-Phe-Arg(NO_2)-Try-Gly-Lys(BOC)-Pro-Val-Gly-Lys(BOC)-Lys(BOC)-Arg-Arg-Pro-Val-Lys(BOC)-Val-Tyr-Pro · · OtBu	39	43
Z · Val-Tyr-Val-His-Pro-Phe · OCH_3	77[1]	81[2]
H · Val-Tyr-Val-His-Pro-Phe · OCH_3	59[1]	66[2]

A: sec-Butanol-3 % ammonia solution (100 + 44);
B: n-Butanol-glacial acetic acid-water (100 + 10 + about 30); upper phase.

[1] Dioxane-water (90 + 10).
[2] Methanol.
[3] Water, methanol, acetone, dioxane and dimethyl formamide, either alone or mixed with each other, are generally less useful for the separation of protected higher peptides because of tailing and formation of diffuse spots.

Generally more useful, and also more sensitive, (minimal amount required about 0.1 μg.) is the chlorine/tolidine reagent of Reindel and Hoppe [*73*] as modified in the following manner [*149, 8*]:

About equal volumes (20—50 ml. of each) of 1.5 % $KMnO_4$-solution and 10 % HCl are placed into a suitable sized photo-development tank with a grating of glass rods 2—3 cm. above the bottom. The plate, with the chromatographed substances to be detected, is placed on the grating, and the tank is covered with a glass plate. After 15—20 min., the chromatogram is removed from the tank and aerated in a well ventilated hood for 2—3 min. Before spraying, the smell of chlorine should have completely disappeared. The spray-reagent (Reagent No. 32) should be applied with greatest care. Thus, even poorly separated substances may be seen at least for a moment as separate spots. The washing process with 2% acetic acid, commonly used with paper chromatograms, is omitted. If a chlorine tank is available, gaseous chlorine is passed through water in a wash bottle and then introduced into a Desaga tank or a similar vessel. The plates to be chlorinated are left in this atmosphere for 5—10 min.

If iodine is used for halogenation, the peptides may be detected by spraying with a starch-solution (Reagent No. 72).

A combination of the ninhydrin and the chlorination-techniques is particularly useful. The plates are first sprayed with ninhydrin and heated and the sites of the spots are localized. In the second step, the chlorine/tolidine reaction is performed; the pre-treatment with ninhydrin does not affect its result. Spots appearing only on halogenation are thus easily distinguished from ninhydrin positive compounds.

Schellenberg [*144*] reports a variation of the well known fluorescence test with morin; however, amounts of material required are rather large (2 μg.). — The fluorescence test of Sanger and Tuppy [*95*] would seem, in the absence of cellulose, to be rather uncertain. Unlike paper chromatograms, thin-layer chromatograms on silica may, of course, be subjected to conditions resulting in the destruction of organic materials by charring (Reagents No. 34, 134).

For elution [*150*], the corresponding area of the layer is carefully removed and suspended in an appropriate solvent. In the case of Silica Gel G layers, water should not be used as the gypsum dissolves and may interfere with later operations. n-Butanol and glacial acetic acid have been found to be satisfactory. Filtration of the suspension yields a solution from which the eluted substance is easily recovered.

V. N-(2,4-dinitrophenyl)-amino acids and 3-phenyl-2-thiohydantoins

Dinitrophenylamino acids (DNP-amino acids) and phenylthiohydantoins (PTH-amino acids) are obtained when reaction products of proteins or peptides with dinitrofluorobenzene [*151—154*] or phenyl mustards [*155*] are properly degraded. Their separation from reaction mixtures and their identification are of considerable practical importance because they constitute essential steps in the process of sequential analysis of peptide structures. Numerous authors have worked on this problem[1].

Substitution in the NH_2 group, the carboxyl group, or both, changes the amino acids into acidic, basic or neutral compounds, i.e., the zwitterionic character is lost. However, depending on the nature of the substituent, the derivatives still are more or less polar. This must be borne in mind when the solvent is selected. Substitution products still containing a free carboxyl and a free amino group, e.g., mono-acyl derivatives

[1] See [*8*] for a review of the literature. An excellent summary was written by Biserte et al. [*156*].

of basic amino acids, monoesters of acidic amino acids or ethers of hydroxy-amino acids, behave similarly to free amino acids in chromatography.

A. Dinitrophenyl amino acids

The formation of the dinitrophenyl amino acids (DNP-amino acids), as mentioned above, is illustrated by the following formulae.

Of course, free amino acids may also react with dinitrofluorobenzene (DNFB). Such a conversion may be useful if an amino acid mixture is not directly amenable to chromatography, e.g., because of salts or other foreign material present. Unlike free amino acids, the DNP-amino acids are often easily brought in a form suitable for chromatography. Besides, the natural color of these derivatives facilitates quantitative analysis by colorimetry. There is no concern about color yield as, for instance, with the ninhydrin reaction.

DNP-amino acids are light-sensitive. They should be kept in the dark and be exposed to daylight only for short periods.

SANGER-*Scheme of End Group Determination*

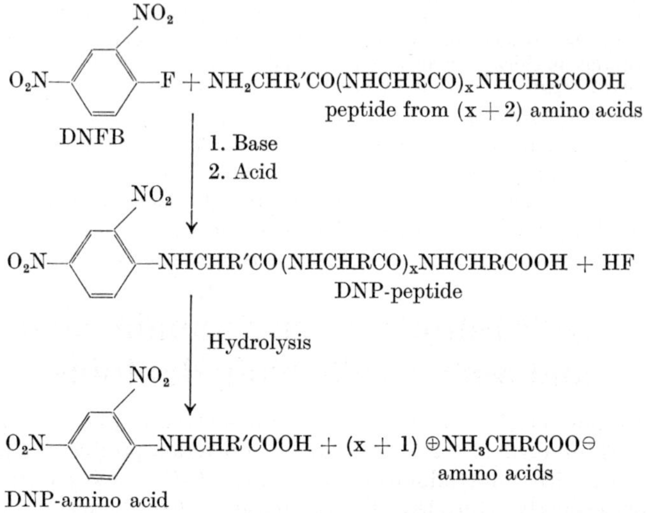

1. Dinitrophenylation

The dinitrophenylation procedures at hand differ with regard to reaction conditions and need for apparatus. An outstanding review has been written by BISERTE et al. [*156*]. We must restrict ourselves to a brief summary:

a) Amino acids

α) Preparation of DNP-amino acids[1]: According to LEVY and CHUNG [*157*] 5 mmoles of the amino acid and 1 g. of anhydrous Na_2CO_3 in 20 ml. water are mixed

[1] A collection is commercially available from Mann Research Laboratories Inc., 136 Liberty Street, New York, U.S.A.

with 5 mmoles of 2,4-dinitrofluorobenzene (DNFB) in 9-times its weight of acetone (for amino acids with two DNFB-reactive groups twice and for histidine $2^1/_2$-times this amount of DNFB is required). The suspension is well shaken for 30—90 min. at 40° C. in the dark. Most DNFB dissolves during this time, and the amount remaining is then extracted with ether. The water solution is carefully acidified by adding 1—2 ml. of conc. hydrochloric acid. The DNP amino acid precipitates as an oil or a solid and is isolated by ether extraction or filtration. Oily derivatives usually crystallize upon evaporation of their solution in ether (DNP-glutamic acid presents particular difficulties). Recrystallization is done by dissolving in benzene, adding some ethanol, and diluting the hot solution with petroleum ether. The more polar DNP-amino acids are best recrystallized from aqueous methanol, and those insoluble in ether may be purified by dissolving in dilute hydrochloric acid and adding a base (e.g., pyridine) for precipitation.

Regarding the preparation of DNP-cysteic acid and the various monoderivatives of cysteine, cystine, histidine, lysine, ornithine and tyrosine, the reader is referred to the review of BISERTE [*156*]; the solubility of these products in water introduces a few particular problems.

Melting-point data may be found in papers by PORTER and SANGER [*154*], LEVY and CHUNG [*157*] and DU VIGNEAUD et al. [*158*].

β) Quantitative dinitrophenylation of a mixture of amino acids (e.g., a protein hydrolyzate): According to WALLENFELS [*159*], the dried hydrolyzate of 2—5 mg. of air-dried performic acid oxidized protein is dissolved with rigorous stirring (magnetic stirrer) in 2 ml. of CO_2-free water at room temperature. An aliquot (1.2 ml.) is transferred into a small reaction vessel equipped with a magnetic stirrer and diluted with 1.8 ml. of CO_2-free water and 0.1 ml. of 3.1 N-KCl. The temperature is raised to 40.0° C. ± 0.1° (thermostat). With an auto-titrator, 0.2 N NaOH is added to a pH of 8.9 under vigorous stirring. About 0.1 ml. of 2,4-dinitrofluorobenzene (p.A. "Merck", No. 2966) is added in the dark and the pH is held at 8.9 for 100 min. by means of the auto-titrator. A recorder connected to the auto-titrator helps to follow the course of the reaction. Dinitrophenylation is practically complete after 50 min.; afterwards, the experiment indicates the hydrolysis rate of dinitrofluorobenzene (formation of dinitrophenol). In order to isolate the DNP-amino acids, the excess of 2,4-dinitrofluorobenzene is removed by extracting twice with 5 ml. of peroxide-free ether [*160, 161*], and the aqueous phase is acidified with 0.5 ml. of 6 N-hydrochloric acid. The ether soluble DNP-amino acids are then extracted with five 4 ml. portions of peroxide-free ether; the extracts are combined and made up with ether to exactly 25 ml. For chromatography, 1 ml. of this solution is removed, reduced to a small volume and quantitatively spotted by means of a capillary pipette.

The aqueous phase still contains the "acid soluble" DNP-amino acids. Dissolved ether is removed *in vacuo*, and the volume is made up with CO_2-free water to exactly 10 ml. 0.5 ml. of this solution is evaporated *in vacuo* to dryness, and the residue is extracted with the least possible volume of acidified acetone (2 ml. of 6 N hydrochloric acid made up to 25 ml. with acetone). The extract is filtered and quantitatively spotted for chromatography.

b) Peptides

α) Dinitrophenylation according to LOCKHARD and ABRAHAM[1]. 50—150 μg. of the peptide is dissolved in 0.1 ml. of a 1.5% aqueous solution of trimethyl ammonium carbonate (pH 9.3), and 0.2 ml. of a 5% alcoholic solution of dinitrofluorobenzene is added. The mixture is left in the dark for $2^1/_2$ hours. The ethanol is then evaporated *in vacuo*, an additional quantity (0.24 ml.) of trimethyl ammonium carbonate solution and 1 ml. of ether are added, the mixture is thoroughly shaken, phases are separated by centrifugation, the upper phase is discarded and the aqueous layer is evaporated to dryness *in vacuo*.

---

[1] Micro method, see [*162*].

β) Total hydrolysis of a DNP-peptide: The residue of the aqueous layer (see α) is dissolved in 0.1 ml. of 6 N hydrochloric acid. The solution is heated for 9 hrs. at 105° C. in a sealed tube under nitrogen and the hydrolyzate is then diluted with two volumes of water. Ether-soluble DNP-amino acids are obtained from this solution by extracting three times with an equal volume of ether or ethyl acetate respectively (di-DNP-histidine).

c) Polypeptides and proteins

α) Dinitrophenylation: LEVY and LI [*163*] dissolve at 40° C. at least 0.2 μmols of the material in 3 ml. of 0.05 N aqueous KCl, and adjust the pH to 8 using 0.05 N potassium hydroxide and an auto-titrator. About 0.1 ml. of dinitrofluorobenzene is added and the solution is thoroughly stirred in the dark at a constant pH and at constant temperature. The end of the reaction is indicated when consumption of alkali ceases. The solution is then extracted three times with ether and the aqueous phase is acidified in order to precipitate the dinitrophenyl derivatives. The precipitate is centrifuged off, washed with water, acetone, and ether, and dried in a desiccator over P_2O_5.

β) Partial hydrolysis of a DNP-protein: Hydrochloric acid or enzymes may be used in the way described for underived proteins. Hydrolysis is, however, significantly slower. The products of partial hydrolysis of a DNP-protein are analytically very interesting. During electrophoretic or chromatographic separation, dinitrophenylated peptides and amino acids, originating from N-terminal parts of the polypeptide chains of the protein and from DNFB-reactive amino acid side chains, are easily distinguished because of their colors. While separation of DNP-peptides from free peptides and amino acids is not complicated (talcum pretreated with hydrochloric acid adsorbs only DNP-compounds, paragraph δ), extraction from acidic aqueous solutions by means of organic solvents often does not satisfactorily differentiate between N-terminal α-DNP-peptides and non-terminal DNP-peptides. The latter ought to remain in the aqueous phase because of their free NH_2 group, but this rule may be misleading [*164*].

γ) Total hydrolysis of a DNP-protein: The DNP-protein is hydrolyzed in a sealed tube for 16 hrs. at 105° C. with 10 times its amount of 5.7 N twice distilled hydrochloric acid. N-Terminal amino acids of the original protein now appear as DNP-amino acids. However, DNP-glycine [*165, 166*] and DNP-proline [*167*] are partially destroyed, particularly if tryptophane is present [*168*]. Di-DNP-tyrosine partly loses its O-DNP-group [*166*]; DNP-cystine is destroyed so much that even its formation should be prevented by previous oxidation of the protein with performic acid [*44*].

δ) Separation of DNP-amino acids from a total hydrolyzate: The hydrolyzate is diluted until it is about 1-normal with respect to hydrochloric acid. This solution is extracted five times with peroxide-free ether [*160, 161*] and, if histidine is present, five times with ethyl acetate; the extracts are washed three times with 0.1 N hydrochloric acid. All the extracts are combined (fraction A: ether soluble DNP-amino acids and dinitrophenol) and the aqueous phase is combined with the washings (fraction B: free amino acids and acid soluble dinitrophenyl derivatives such as DNP-arginine, DNP-cysteic acid, mono-DNP-derivatives of cysteine, cystine, histidine, lysine, ornithine and tyrosine; if the extraction was only done with ether, a part of the di-DNP-histidine may also be found in this fraction).

Fraction A: If much dinitrophenol is present, its removal is recommended. This can be done in two ways:

Sublimation: When heated in a high vacuum to 70°—80° C. in an apparatus designed by MILLS [*169*], dinitrophenol largely evaporates. Some DNP-methionine and — as observed in our laboratory — also some di-DNP-cystine may be lost. The ethyl acetate solution is evaporated in such a way as to deposit its solid contents in the form of a thin film adhering to the wall of the container. At times, it may be necessary to interrupt the sublimation process in order to recast the film, from, e.g., acetone.

Adsorption (Ion-exchange): [*170*]. A quantity of DNP-amino acids equivalent to a few μ-moles is dissolved in 0.5 ml. of methanol and placed on a column of anionotropic aluminum oxide[1] (1 × 10 cm.). After rinsing the column with 0.5 ml. of methanol, the dinitrophenol is quantitatively eluted with 2% acetic acid. The DNP-amino acids are desorbed by first passing a small amount of 0.1 N NaOH (to avoid CO_2-evolution and consequent breaking of the column) and then a 1% solution of $NaHCO_3$. From the eluate, after addition of a small excess of HCl, the DNP-amino acids are isolated by extraction with ether and evaporation of the extract.

The residue left after removal of the dinitrophenol is dissolved in about 1 ml. of acetone per 10 mg. of the original protein and about 1 μl. amounts of this solution are spotted for chromatography.

Fraction B: In spite of the free amino acids and salts present, the acid soluble DNP-amino acids may often be directly chromatographed. The solution is repeatedly evaporated to dryness, each time some water being added. For chromatography, the residue is dissolved in about 1 ml. of 0.5 N hydrochloric acid or of glacial acetic acid, per 10 mg. of original protein, and about 1 μl. quantities of these solutions are spotted. Before running the chromatogram, completely evaporate the volatile acid from the layer!

In order to remove the free amino acids, the residue obtained above may be dissolved in 2 ml. of N hydrochloric acid and passed through a column (diameter 2.5 cm.) of a mixture of 20 g. Hyflo-Super-Cel and 50 g. talc (pre-treated with 0.01 N and N hydrochloric acid) [*171*]. Unlike the free amino acids, all the DNP-amino acids (apart from DNP-cysteic acid) are adsorbed. The column is washed with 100 ml. of N hydrochloric acid and finally, the DNP-amino acids are eluted with alcohol-HCl (alcohol: N hydrochloric acid = 40 + 10), or preferably with ammoniacal alcohol (alcohol: 0.3% ammonia solution = 40 + 10). The effluent is evaporated to dryness and the residue, which now is less contaminated by foreign material, is chromatographed as indicated above.

2. Solvents and efficiency of separation

Solvents used in PC for separating free amino acids are, as a rule, equally efficient in TLC. With the DNP-amino acids, this analogy holds only to a very limited degree. PC of DNP-amino acids has often been found to be a delicate matter since there is a strong tendency for tailing and because the *Rf*-values are highly dependent on the quantities applied to the chromatogram and on the presence of other DNP-amino acids in the sample. Solvents recommended include the "toluene system" of Biserte and Osteux [*45*], n-butyl alcohol-0.1% ammonia solution [*172*], and 1.5 M phosphate buffer [*173*]. On silica layers, the DNP-amino acids do not sufficiently move in the "toluene-system." DNP-leucine has an *Rf*-value of 0.25 only. Moreover, there is a strange distortion of the spots, so-called "beards" being produced[2]. If the layer is inactivated before use, by exposure for at least one night, to vapors of the aqueous phase of the toluene-system, the *Rf*-value of DNP-leucine increases to 0.66, and excellent separations, in spite of the "beards" mentioned, are obtained on two-dimensional chromatograms (see below). In Braunitzer's solvent [*172*], streaks are produced. The phosphate buffer [*173*] is completely useless on silica layers: streak formation and considerable spot spreading due to diffusion obstruct separation; buffer-

[1] Alumina "Merck" is treated for 10 min. with an excess of N hydrochloric acid and then washed free of acid by decanting with water.

[2] The term "beard" was introduced by Hais and Macek, see p. 147 in [*2*].

ing has no effect and adding a few per cent of glacial acetic acid – often a remarkably successful trick – makes little difference in this case. In addition to the modified "toluene-system," the solvents listed in the following paragraphs, a) and b), have proved to be useful. For the preparation of these solvents, reagents of a defined standard should be used.

Reagents for solvent preparation

Ammonia, 0.8 N:	25% Ammonia "Merck," diluted with distilled water.
Ammonia, 34%:	Commercial quality.
tert.-Amyl alcohol:	Fractionate *tert.*-amylalcohol pract., "Fluka[1]," using a short column and collect the fraction boiling at 100.5°—102.0° C.
Ethylene chlorohydrin:	2-Chloroethanol puriss. "Fluka."
Benzene:	Extract commercial benzene 3 times with $^1/_{10}$ its volume of conc. sulphuric acid, wash with water, 2 N Na_2CO_3 solution and again with water, dry over calcium chloride and distill, using a short column.
Benzyl alcohol:	Shake with a saturated solution of sodium bisulphite, wash with 2 N Na_2CO_3 solution, dry over sodium sulphate and distill *in vacuo* in a nitrogen atmosphere using a short column (benzaldehyde present in benzyl alcohol exerts a considerable influence on *Rf*-values.
n-Butanol:	n-Butanol for chromatography, "Merck."
Chloroform:	Distill twice using a short column or pass through an Al_2O_3-column[2] and use immediately. (Alcohol-free chloroform, on standing, soon develops some phosgene.)
Acetic acid: Methanol: n-Propanol:	Distill commercially available products using short columns.
Pyridine:	Reflux over barium oxide for 24 hrs. and distill using a short column.
Toluene:	As for benzene.

a) Solvents for chromatography of acid- and water-soluble DNP-amino acids not extractable by ether

n-Propylalcohol-34% ammonia solution (70 + 30). In this system, DNP-arginine, DNP-cysteic acid, mono-DNP-cystine, α-DNP-histidine, di-DNP-histidine[3], ε-DNP-lysine and O-DNP-tyrosine, each present alone or in commonly occurring mixtures may be identified by ascending chromatography. A run requires about two hours. Table 108 gives the *Rf*-values and indicates further means for spot differentiation. Although separation of DNP-arginine and ε-DNP-lysine is incomplete, both of them can be detected because of the color difference produced in the ninhydrin reaction. DNP-cysteic acid and mono-DNP-cystine constitute a pair never encountered in the same experiment. Note that it is necessary to remove excess acid after applying the sample solution to the layer. To this end, the plate is heated in a current of air to about 60° C. for 10 min., and then allowed to cool for about 15 min.

---

[1] Fluka, Buchs, SG., Switzerland.

[2] G. Wohlleben, Angew. Chem. *68*, 752 (1956).

[3] im-DNP histidine is also a member of the group of acid soluble DNP derivatives. However, Zahn and Pfannmüller [*174*] report that it is not detectable in hydrolyzates of corresponding DNP-peptides because of its instability. The present investigation was, therefore, restricted to α-DNP- and di-DNP-histidine.

Table 108. *Identification of Acid and Water Soluble DNP-Amino Acids**, by TLC on Silica Gel G in the system n-propanol-34% ammonia (70 + 30). Ascending technique, solvent migration 10 cm., amounts spotted 0.5—1 μg.

	Rf × 100**	Color	UV-Absorption (360 mμ)	Color with ninhydrin
Mono-DNP-$(Cys)_2$	29	yellow	+	brown
DNP-$CySO_3H$	29	yellow	+	yellow
α-DNP-Arg	43	yellow	+	yellow
ε-DNP-Lys	44	yellow	+	brown
O-DNP-Tyr	49	colorless	+ ***	violet
α-DNP-His	57	yellow	+	yellow
Di-DNP-His	65	yellow	+	yellow

* For abbreviations of amino acids, see Table 103.
** Average of the results of 6 separate measurements.
*** see "Documentation." p. 426.

n-Butanol-34% ammonia solution (80 + 20): DNP-arginine, DNP-citrulline, DNP-cysteic acid, di-DNP-histidine and DNP-taurine are water soluble DNP-derivatives and are formed as a result of dinitrophenylation of the urinary amino acids in the presence of excess dinitrofluorobenzene. Fig. 65 gives an indication of the separation obtained with a model mixture.

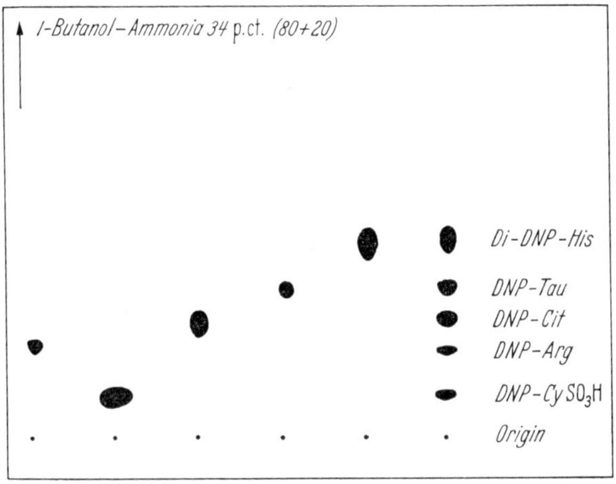

Fig. 165. Model separation of water-soluble DNP-amino acids occurring in dinitrophenylated urine. Ascending technique [*175*]

Amounts spotted 0.5 μg. UV-photocopy see p. 426. For abbreviations of amino acids, see Table 103; Tau = taurine, Cit = citrulline

b) Solvents for chromatography of acid-insoluble DNP-amino acids extractable by ether

Molecules of DNP-amino acids tend, like other carboxylic acids, to associate with each other. Addition of a moderate amount of glacial acetic acid to solvents used for chromatography reduces association and thus seems to abolish one major cause of "tail" formation. This "acetic acid effect" may also be observed in solvents containing pyridine. As a corollary to its influence on tailing, acetic acid quite amazingly increases the "eluting power" of the solvent. However, the acetic acid concentration should not exceed 0.5 to 5% (v/v). A larger proportion of acetic acid diminishes separation efficiency just as, e.g., addition of too much water.

α) Solvents for general separation

No. 1: Toluene-pyridine-ethylenechlorohydrin-0.8 N ammonia solution (100 + 30 + 60 + 60) ("Toluene-system" [*45*]).

The upper layer serves as the solvent for chromatography while the lower is used for pre-treatment of the layer (p. 422). This system causes spots to acquire long "beards." In spite of this disadvantage, separations are excellent. The system is, therefore, recommended as solvent for the first run of two-dimensional chromatograms. For the present, it is indeed considered to be indispensable. Its merits thus compensate for the inconvenience involved by the need for a pre-treatment of the layer. For instance, it separates DNP-phenylalanine from DNP-methionine, DNP-β-alanine from the DNP-leucines, DNP-norleucine from DNP-valine, DNP-α-amino-n-butyric acid from DNP-proline, DNP-alanine from DNP-sarcosine, as well as the group of the di-DNP-amino acids (except di-DNP-cystine) from the group of the smaller DNP-amino acids, and the latter from the DNP-derivatives of the acidic amino acids, which remain at the start.

No. 2: Chloroform-benzyl alcohol-glacial acetic acid (70 + 30 + 3). This system gives symmetrical spots and separates 2,4-dinitrophenol[1] and 2,4-dinitroaniline[1] from all DNP-amino acids.

No. 3: Chloroform-*tert.*-amyl alcohol-glacial acetic acid (70 + 30 + 3). This solvent separates similarly to No. 2. DNP-valine and 2,4-dinitrophenol[1] move closer to DNP-leucine. As a compensating advantage, *tert.*-amyl alcohol is more stable and more volatile than benzyl alcohol.

No. 4: Benzene-pyridine-glacial acetic acid (80 + 20 + 2).

In *continuous* chromatography [*13*] (pp. 22—24), this system is particularly suitable for separating the less polar DNP-amino acids, e.g., the derivatives of the isomeric leucines. 2,4-dinitroaniline[1] migrates fastest ($Rf_{\text{DNP-leucine}} = 1.28$).

No. 5: Chloroform-methanol-glacial acetic acid (95 + 5 + 1).

Di-DNP-tyrosine and di-DNP-lysine, which are not separated by any one of the above systems, move differently when subject to *continuous* chromatography [*13*] (pp. 22—24) in solvent No. 5, provided the layer is air-dried at less than 50% relative humidity; if difficulties are still encountered, try ether-glacial acetic acid (95 + 5) as the solvent.

Table 109 gives *Rf*-values in the solvents mentioned. In Table 110 the time required per run is indicated. In addition, the Table gives a few facts related to separation efficiency.

---

[1] 2,4-dinitrophenol and 2,4-dinitroaniline are by-products of the synthesis and acidic hydrolysis of DNP-peptides.

Separation by two-dimensional procedures. Unambiguous identification of a DNP-derivative out of the relatively large group of acid-insoluble, ether-extractable DNP-amino acids usually requires a two-dimensional

Table 109. *Rf × 100 of Ether-Soluble DNP-Amino Acids*[1] *Observed after One-Dimensional Ascending or Horizontal* [*13*] *TLC on Silica Gel G in Solvents 1—5* Each figure represents average results of 6 observations. Amounts spotted 0.5—1 μg.

	1[2]	2		3		4[3]			5[3]		
	As-cend-ing	Ascending	indi-rect[4]	Ascending	indi-rect[4]	As-cend-ing	Horizontal	indi-rect[4]	As-cend-ing	Horizontal	indi-rect[4]
Solvent migration →	15 cm	10 cm	10 cm	10 cm	10 cm	15 cm	DNP-leu: 10 cm		15 cm	DNP-leu: 10 cm	
DNP-α-AnB	46	72	44	73	42	52	52	55	79	85	75
DNP-α-ACy	79	92	66	83	57	105	108	109	108	101	106
DNP-Ala	34	54	35	60	34	32	33	38	59	66	58
DNP-β-Ala	27	71	57	73	50	89	98	100	99	95	102
DNP-Asp	02	13	08	09	13	06	05	11	07	06	06
DNP-Glu	01	26	17	31	21	12	12	23	12	12	14
DNP-Gly	27	32	22	40	23	17	18	22	31	38	31
DNP-Ileu	64	83	63	81	57	107	107	107	100	101	104
DNP-Leu	66	82	62	80	54	100	100	100	100	100	100
DNP-Nleu	69	82	60	80	52	86	90	88	101	100	98
DNP-Met	55	70	39	69	38	43	43	47	72	81	74
DNP-Met. O_2	17				04	03	03	02	10	10	07
DNP-Phe	67	75	46	74	41	44	46	52	81	86	76
DNP-Pro	29	65	41	67	38	58	59	62	78	84	75
DNP-Sar	23	56	35	57	32	34	35	41	59	65	60
DNP-Ser	15	11	10	11	10	09	10	14	07	08	07
DNP-Thr	20	17	13	15	12	12	14	20	09	11	11
DNP-Try	65	69	38	69	31	23	25	33	54	61	49
DNP-Val	53	79	56	77	51	76	81	85	91	98	86
DNP-Nval	56	77	52	76	48	65	70	75	86	95	89
Di-DNP-$(Cys)_2$	—	03	02	01	01	00	00	02	00	02	02
Di-DNP-His	53	11	09	08	04	05	04	08	12	16	14
Di-DNP-Lys	74	56	35	60	30	12	13	19	66	73	65
Di-DNP-Orn	70	34	23	40	20	06	06	10	39	46	39
Di-DNP-Tyr	76	58	35	60	30	17	16	19	57	65	57
2,4-DNP-OH[5]	41	100	76	83	55	22	21	23	148	102	111
2,4-DNP-NH_2[6]	90	90	84	72	63	115	128	129	131	101	115

[1] For abbreviations of amino acids, see Table 103.
[2] see remarks on the use of the "toluene" system.
[3] *Rf*-values in reference to DNP-leucine.
[4] After "pre-treatment" of the layer by chromatography in the "toluene" system (No. 1) and intermittent drying (see text).
[5] 2,4-DNP-OH = 2,4-dinitrophenol.
[6] 2,4-DNP-NH_2 = 2,4-dinitroaniline.

technique. The run in the first dimension is done in the "toluene system" of BISERTE and OSTEUX [*45*] (solvent No. 1). For the run in the second dimension, one of the solvents Nos. 2–5 may be selected. Fig. 166 shows the separation of a standard mixture containing 0.2 μg. of each DNP-amino acid, using solvents 1 and 2 in combination. No separation results within the leucine group, within the valine group and between di-DNP-lysine and di-DNP-tyrosine.

The separation of the latter compounds may be achieved by a combination of solvents 1 and 5 (Fig. 167), or using No. 5 alone. Combining solvents 1 and 4 (Fig. 168) separates the derivatives of the isomeric leucines and of the isomeric valines.

Table 110. *Notable Separation Effects Observed after One-Dimensional TLC on Silica Gel G in Solvents 1—5, and Time Required*

Solvent No.	Separation of the DNP-amino acids from		Separation of DNP-derivatives from		Separation of the di-DNP-derivatives from	time required
	Dinitro-aniline	Dinitro-phenol	Leu, Ileu, Nleu	Val, Nval	Tyr, Lys	
1	+	—[1]	—	—	—	1 h/15 cm
2	+	+	—	—	—	$1^1/_2$h/10 cm
3	—[5]	—[2]	—	—	—	1 h/15 cm
4[6]	+	—[3]	+	+	—	2—3 h
5[6]	—[4]	—[4]	—	+	+	2—3 h

[1] Spot of 2,4-dinitrophenol is between spots of DNP-val and DNP-ala.
[2] Spot of 2,4-dinitrophenol is near spot of DNP-leu.
[3] Spot of 2,4-dinitrophenol is between spots of DNP-ala and DNP-gly.
[4] Spots of 2,4-dinitrophenol and 2,4-dinitroaniline are just above spot of DNP-leu.
[5] Spot of 2,4-dinitroaniline is between spots of DNP-phe and DNP-met.
[6] Horizontal chromatography [*13*]; DNP-leu migrates about 10 cm. in $2^1/_2$ hrs.

The characteristic spot patterns are hardly distorted by fluctuation in *Rf*-values. For identification, an unknown sample may be chromatographed together with a standard mixture containing just enough of each of the DNP-amino acids in question (0.2 μg.) to remain visible after two-dimensional chromatography[1] (Fig. 166). In most cases, the compound in the sample is unequivocally indicated by the increased intensity of its spot. The reader is also referred to Table 109. Along with *Rf*-values observed on one-dimensional chromatograms, it states values observed in solvents 2–5 after prior chromatography in the "toluene-system" ("indirectly" obtained Rf-values). These *Rf*-values are well reproducible provided the instructions for working with the "toluene-system" and intermittand drying, given in the text below, are carefully followed.

Note on the use of the "toluene-system"

Pre-treatment of the thin-layer (equilibration): A tank lined with filter paper is loaded with the lower phase of the "toluene-system." A thick bent glass rod serves as a grate on the bottom of the tank. Two plates coated with silica layers are placed on the center of the grate with the coated side outwards, and each is inclined to touch a tank wall, with its upper edge. To avoid solution getting from the filter paper to the thin-layer, the latter is divided in two by a heavy pencil stroke parallel to the upper edge of the plate. Leave overnight. The silica gel will take up a good

[1] A solution of 1 mg. of each DNP-amino acid in total volume of 5 ml. of acetone keeps in a refrigerator for at least 4 weeks. For the experiments, 1 μl. should be used.

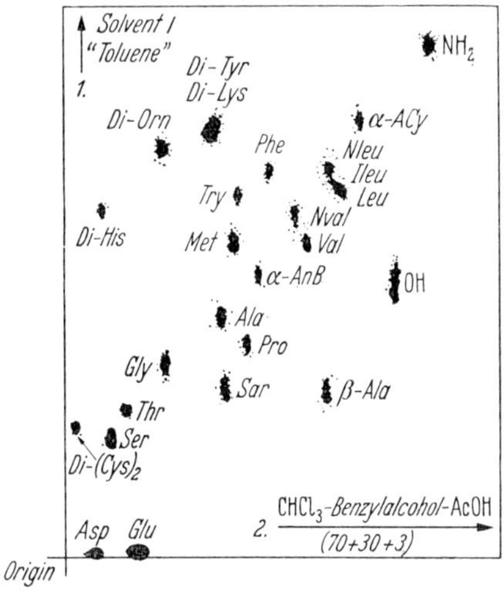

Fig. 166. Two-dimensional chromatogram of a standard mixture containing 0.2 μg. of each DNP-amino acid, using solvents 1 and 2

Ascending technique. UV-photocopy, original size: 12 × 10 cm. [8]. Abbreviations: Amino acid abbreviations (see Table 103) refer to mono-DNP-derivatives; Di = Di-DNP-derivatives; OH = 2,4-dinitrophenol; NH_2 = 2,4-dinitroaniline

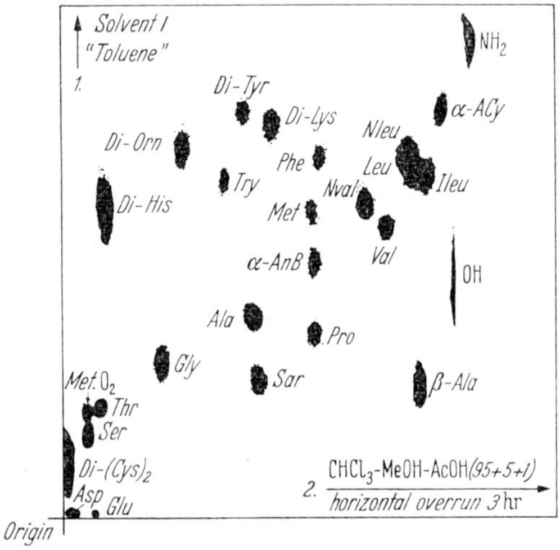

Fig. 167. Two-dimensional chromatogram of a mixture containing 1 μg. of each DNP-amino acid, using solvents 1 (ascending technique) and 5 (continuous horizontal solvent flow) [13]

Abbreviations, see Fig. 166. UV-photocopy, as in Fig. 166, but exposure-time extended. Original size, 13 × 13 cm. [8]

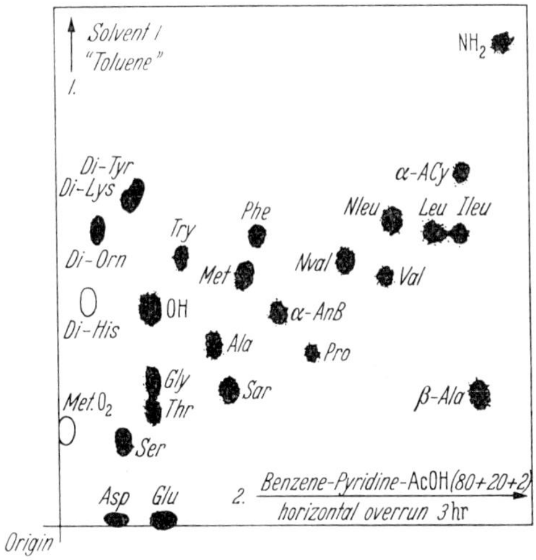

Fig. 168. Two-dimensional chromatogram of a mixture containing 1 μg. of each DNP-amino acid, using solvents 1 (ascending technique) and 4 (continuous horizontal solvent flow [13])

Abbreviations, see Fig. 166. Di-His and Met. O_2 were omitted from this mixture. UV-photocopy, original size, 15 × 14 cm [8]

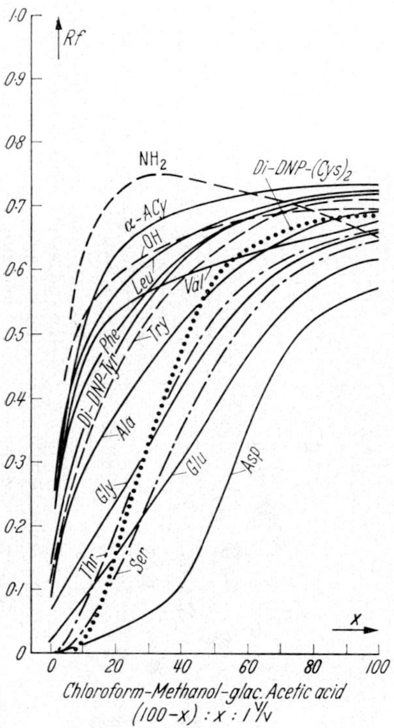

Fig. 169

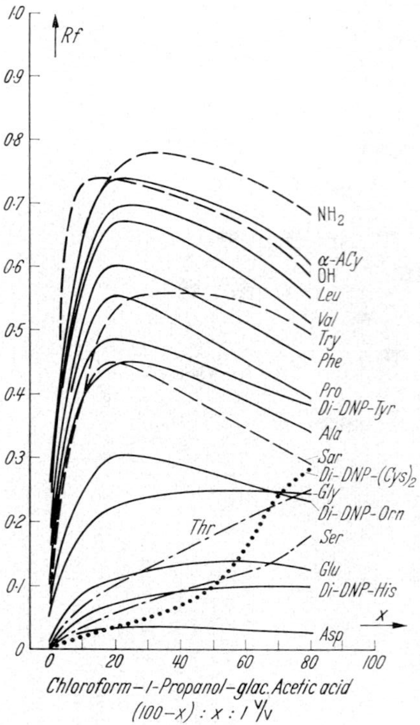

Fig. 170

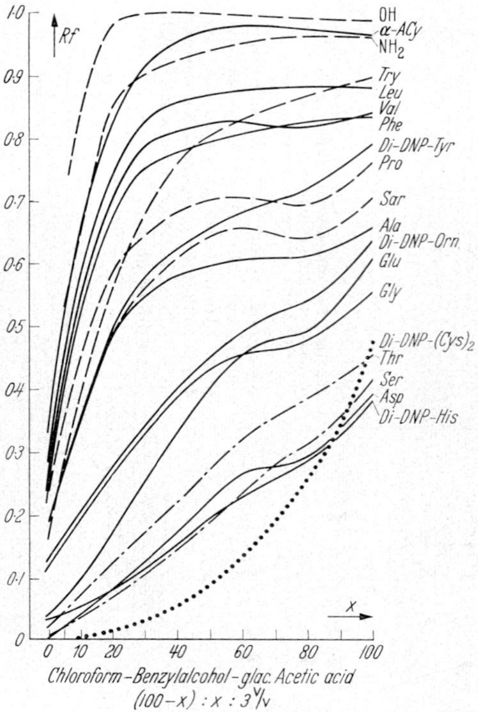

Fig. 171

Fig. 169—171. Ether-soluble DNP-amino acids, 2,4-dinitrophenol and 2,4-dinitroaniline: *Rf*-values as a function of the ratio chloroform: R-OH in solvents composed of chloroform, R-OH, and glacial acetic acid. R = CH_3 (Fig. 169, R = n-C_3H_7 (Fig. 170), R = C_6H_5-CH_2 (Fig. 171)

Ascending technique: solvent migration 10 cm.; abbreviations: see Fig. 166 and Table 103 deal of moisture. This cannot be recognized by its appearance. However, it is demonstrated by the fact that substances spotted on the layer diffuse considerably during such treatment.

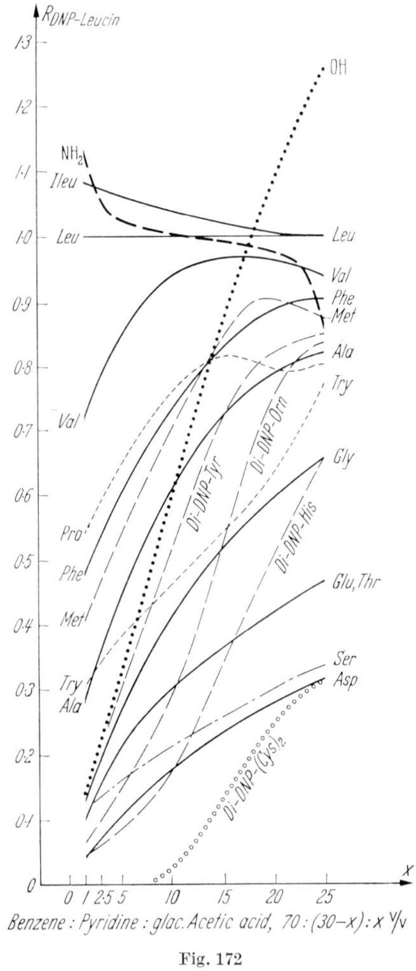

Fig. 172

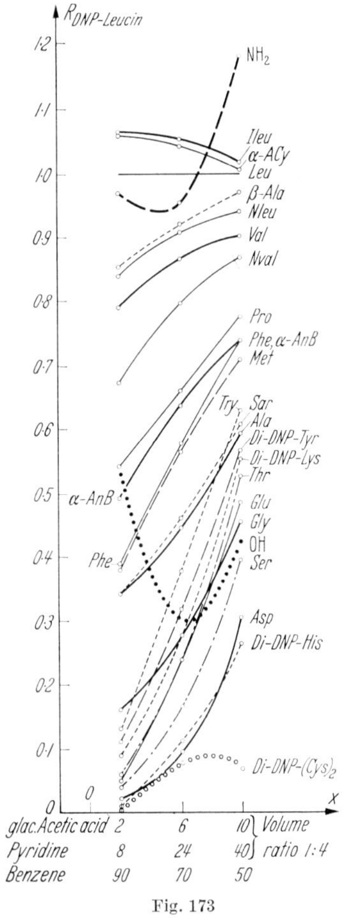

Fig. 173

Fig. 172. Ether-soluble DNP-amino acids, 2,4-dinitrophenol and 2,4-dinitroaniline: $R_{\text{DNP-leucine}}$-values[1] as a function of the ratio pyridine: glacial acetic acid in solvents containing a constant proportion of benzene

Ascending technique: solvent migration, 14 cm. Amounts spotted about 0.5 μg. Abbreviations, see Fig. 166 and Table 103. Di-DNP-lysine and di-DNP-tyrosine invariably migrate together, DNP-β-alanine always exhibits $R_{\text{DNP-leucine}} \approx 0.9$

Fig. 173. Ether-soluble DNP-amino acids, 2,4-dinitrophenol and 2,4-dinitroaniline: $R_{\text{DNP-leucine}}$-values[2] as a function of the ratio benzene: pyridine: glacial acetic acid in solvents characterized by a constant glacial acetic acid: pyridine ratio

Ascending technique: Solvent migration, 14 cm. Amounts spotted, about 0.5 μg. Abbreviations, see Fig. 166 and Table 103

[1, 2] See footnotes 1 and 2, p. 426.

Chromatography on a plate pre-treated in this manner is probably based on partition between two liquid phases. In any case, the effect of the pre-treatment is entirely lost if the plates are left in the open air for 45 min. This fact should not be overlooked when using the plates.

Spotting of samples: If, between pre-treatment and chromatography, a plate is left unprotected for some time in the air and if the logarithms of the *Rf*-values then observed are plotted against the logarithm of the time elapsed, a linear decline already starts after about 2 min. of exposure. For this reason, a pre-treated layer should be covered immediately with a glass plate after removal from the tank, leaving a margin of about 1.7 cm. width for spotting. If spotting is done within no more than 5 min., the interference is negligible. The volume transferred per spot should not exceed 1 μl. as the evaporation rate of the applied solution is less on pre-treated moist layers than on dry ones. After spotting, the covering plate is carefully removed and chromatography is started without delay using 120 ml. of the upper phase of the "toluene-system" (ascending technique).

Intermittent drying: After the first run, the plate is left for 10 min. in a current of air (well ventilated hood), heated in a drying cabinet at 60° C. for 10 min. and cooled in a current of air for 15 min. Extended drying is not recommended because the air might cause partial decomposition of the DNP-amino acids: oxidation of DNP-methionine, for instance, results in formation of a material behaving chromatographically like di-DNP-histidine. The layer must be covered with glass, if immediate chromatography in the second dimension is not possible. Keep in the dark!

β) Solvents for special problems

The graphical representation in Figs. 169–173 will be found useful when looking for solvents with special characteristics.

3. Documentation

Apart from O-DNP-tyrosine, all the DNP-derivatives mentioned are yellow. Amounts of 0.1 μg. (one-dimensional chromatogram) or 0.5 μg. (two-dimensional chromatogram) yield spots which are easily visible if the plate is examined in transmitted daylight. The chromatogram should be evaluated within a few hours as the spots gradually fade. The layer may be preserved by spraying with a mixture of paraffin and ether [*176*]

---

[1] *Rf*-values in reference to DNP-leucine. The *Rf*-value of the standard, DNP-leucine, depends on the acetic acid content, x, of the solvent in the following manner:

x	0	1.5	3	5	10	15	20	25
Rf-value of DNP-leucine	0.03	0.28	0.37	0.46	0.62	0.68	0.76	0.60

In solvents with $x < 5$, general migration is obviously so slow as to exclude effective separation of mixtures by the ordinary ascending technique. Separation is still possible, however, by horizontal chromatography with continuous solvent flow [*13*].

[2] *Rf*-values in reference to DNP-leucine. The *Rf*-values of the standard, DNP-leucine, depends on the proportion of benzene in the solvent in the following manner:

Proportion of benzene in solvent	90	70	50% v/v
Rf-value of DNP-leucine	0.17	0.51	0.70

If the proportion of benzene in the solvent exceeds 70% (v/v), general migration becomes so slow as to exclude separation of mixtures by the ordinary ascending technique. Separation is still possible, however, by horizontal chromatography with continuous solvent flow [*13*].

or preferably, with a dispersion of a plastic[1] [*50, 177*] (pp. 43—44). The resulting film is easily peeled off from the glass plate and may be handled further like a paper chromatogram.

The simplest thing to do is to fix pergamyne paper to the plate using paper clips, then to examine the chromatogram in transmitted daylight and to indicate the spots and their area on the paper, using ink and a soft pen. With a little practice, even very weak pale yellow spots are still visible on the white silica background. One may prefer to observe the plate in transmitted UV-light (light source → layer → glass plate → spectacles → eyes). The DNP-amino acids are seen as dark spots. Small quantities of weakly absorbing materials are not easily detected since the silica gel-gypsum layer itself appears dark. However, upon addition of 0.5 g. of zinc silicate[2] to 25 g. of Silica Gel G, the layer becomes fluorescent and even O-DNP-tyrosine is seen in quantities as small as 0.06 μg.

Particularly convenient documentation is provided by a *UV-photostat*. The sensitive side of Gevaert's "Gevacopy" paper is directly pressed to the layer. UV-light (360 mμ)[3] is allowed to fall for a few seconds through the plate and onto the photographic paper, and then a positive picture is developed in the usual manner. The copies reproduced in Figs. 166–168 were obtained this way. Maximal sensitivity results if exposure is restricted to such an extent as to produce only a grey background of the negative. A copy thus prepared represents a rather balanced picture of the chromatogram. On longer exposure, the background on the negative becomes increasingly black, and the white spots diminish in diameter and ultimately disappear. This phenomenon may be used to break up partly overlapping spots into individual smaller spots.

B. Phenylthiohydantoins

The formation of the phenylthiohydantoins (PTH-amino acids), referred to in the introduction, is illustrated by the following formulae shown on p. 428 [*178, 179*].

This scheme was proposed by Edman [*155*]; later on other laboratories elaborated a generally applicable procedure [*180, 181*].

PTH-Amino acids are also formed, of course, from phenyl mustard oil (phenylisothiocyanate, PITC) and free amino acids [*182*]. Like DNP-amino acids, they are often more readily separated from foreign material than are the free amino acids. Identification of amino acids in mixtures may, therefore, be achieved by converting them into their PTH-derivatives and subsequent chromatography. Quantitative determination may be based on UV absorption [*182*].

---

[1] A material specially selected for conserving thin-layer chromatograms, "Neatan," is supplied by Merck AG, Darmstadt, Germany.

[2] Zinc silicate luminescent material P1, Type 118-2-7, General Electric, Cleveland, Ohio, U.S.A., or Mn-activated zinc silicate supplied by Leuchtstoffwerk GmbH, Heidelberg, Germany.

[3] Uvanalys-Laborgerät 57 US, supplied by W. Bälz & Sohn KG, Heilbronn, Neckar, Germany.

Scheme of the EDMAN-*Degradation:*

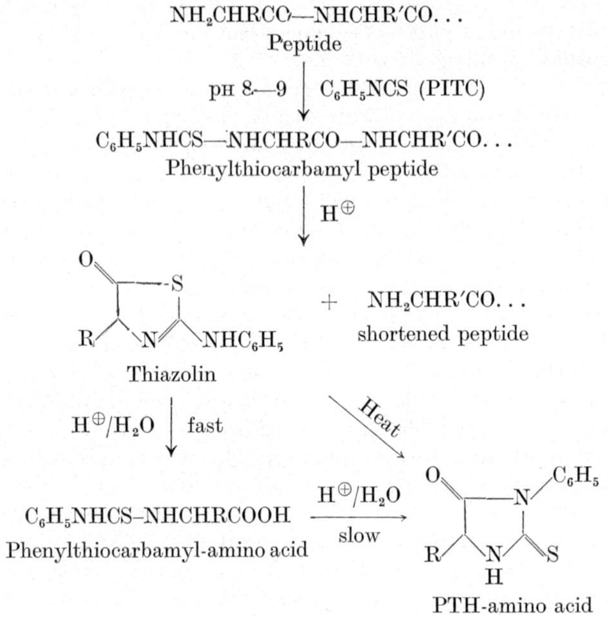

1. Preparation of phenylthiocarbamyl derivatives and transformation into PTH-amino acids

a) Amino acids

SJÖQUIST describes a micro method [*181*]; for preparing larger quantities, SJÖQUIST's method may be suitably modified or EDMAN's instructions [*183*] may be followed.

α) Preparation of the PTH-amino acids[1]**.** 10 mmoles of an amino acid are dissolved in 25 ml. of water and 25 ml. of pyridine, the solution is adjusted to pH ~ 9 (indicator paper) by adding N sodium hydroxide and heated to 40° C. 2.4 ml. of phenyl mustard oil are added, the mixture is well stirred and the pH is held constant by addition of sodium hydroxide. After about 30 min., NaOH consumption ceases. Excess phenyl mustard oil and most of the pyridine are removed by extraction with several portions of benzene. An amount of hydrochloric acid equivalent to the sodium hydroxide used is added. Precipitation of the phenylthiocarbamyl-amino acid may be completed by partial evaporation of the solvent. The precipitate is filtered off. Note: phenylthiocarbamyl-arginine and phenylthiocarbamyl-histidine require, for precipitation, pH 7 and 3.5, respectively.

For transformation into PTH-amino acids, the phenylthiocarbamyl derivatives are refluxed for 2 hrs. in 30 ml. of N hydrochloric acid. The solution is then evaporated to dryness and evaporation is repeated several times after addition of water. The crude PTH-amino acid (yield 80—90%) is recrystallized from a glacial acetic acid-water mixture, or from alcohol or water. Melting points of 18 PTH-amino acids are reported [*181*]. Note: When preparing the tryptophane derivative, reflux in glacial acetic acid instead of hydrochloric acid. In the PTH-derivatives

[1] A collection is obtainable, e.g., from Mann Research Laboratories Inc., 136 Liberty Street, New York 6, N.Y., U.S.A.

of the di-amino-monocarboxylic acids, the amino group not in the α-position is linked to a phenylthiocarbamyl residue; the β-hydroxy and β-mercapto amino acid derivatives definitely tend to eliminate water or H_2S, respectively. All PTH-amino acids must be protected from light. Optically active PTH-amino acids, although quite susceptible to racemization, are reported to be useful in configuration assignment by rotatory dispersion measurements [*184*].

β) Quantitative conversion of an amino acid mixture (e.g., a peptide hydrolyzate) into PTH-amino acids [*182*]**:** 0.5—1 mg. of a peptide or a protein is hydrolyzed in a sealed quartz tube with 0.3 ml. of twice distilled, constant boiling hydrochloric acid (5.7 N) for 22 hrs. at 110° C. in a nitrogen atmosphere. The hydrolyzate is evaporated to dryness *in vacuo* over potassium hydroxide, with repeated addition of water. The residue is mixed with 250 μl. of a triethylamine-acetic acid buffer and 250 μl. of a phenylisothiocyanate solution in acetone (corresponding to 6 μl. of PITC), and the mixture is left to react in a closed tube for $2^1/_2$ hrs. at 25° C. (water bath). The solvent is removed, first for 15 min. *in vacuo*, and then overnight, over phosphorus pentoxide in a high vacuum, and the residue (phenylthiocarbamyl derivatives of amino acids) is dissolved in a mixture of 100 μl. of glass-distilled water and 200 μl. of glacial acetic acid saturated with hydrogen chloride. The resulting solution is kept for 6 hrs. at 25° C. (water bath). The solvent and HCl are then removed over potassium hydroxide as described above. PTH-Amino acids in the residue are identified by chromatography; the residue from a similar experiment, run without amino acids added, will serve as a control. PTH-Cysteine and PTH-cystine are not formed under these conditions. Detection of peptide-bound cystein (or cystine) is possible, however, by a procedure yielding PTH-cysteic acid. This PTH-derivative is formed, at least to some extent, upon PITC-treatment of the hydrolyzate of a performic acid oxidized peptide. SJÖQUIST's oxidation method is similar to one described by HIRS [*44*].

Reagents

Buffer:	2 ml. of 2 N acetic acid (analytical grade) and 1.2 ml. of triethylamine (Eastman Kodak, reflux for 3 hrs. with 5% (w/w) of phthalic anhydride and distill using a short column) are mixed and made up with distilled water to a volume of 25 ml. This solution is added to 25 ml. of acetone (analytical grade, reflux with potassium/permanganate and distill using a short column); pH 10.1.
Phenyl isothiocyanate:	distill *in vacuo*.
HCl-glacial acetic acid:	glacial acetic acid, analytical grade, is saturated at room temperature with gaseous HCl (dried with conc. sulphuric acid) using a quartz vessel.

b) Peptides

Among the many procedures known [*132, 133, 155, 185—188*], the "normal" degradation process of SJÖQUIST et al. [*132—133*] and SJÖQUIST's "special method" [*133*] are outstanding. The latter is reported to guarantee maximal yields of the PTH-amino acids.

When the "special method" is applied, the ordinary by-products, mono- and diphenylthiourea, are formed in particularly large quantities. However, diphenylthiourea (DPTU) does not interfere with the PTH-amino acids on thin-layer chromatograms (Fig. 176). Monophenylthiourea (MPTU), on the other hand, moves in paper- and in thin-layer chromatograms [*189*] very much like PTH-glycine (Fig. 175). Nevertheless, PTH-glycine may be detected by a specific color reaction of MPTU (see p. 432).

The usual procedure for separating a PTH-amino acid from its conjugate peptide involves extraction by ethyl acetate. Unlike other PTH-derivatives, PTH-histidine, PTH-arginine and products arising from the presence of cysteic acid remain in the acidic aqueous solution. Of these, PTH-histidine may be completely extracted from the aqueous phase after removal of most of the volatile acid by careful evaporation and neutralization (pH ≈ 7) of the residual solution, preferably by triethylamine.

With peptides containing cysteic acid, the Edman degradation proceeds well up to the point where cysteic acid becomes N-terminal; afterwards, difficulties may arise[1].

2. Solvents and efficiency of separation

With solvents recommended for paper chromatography of the PTH-amino acids [*182, 190–192*], separations on Silica Gel G are unsatisfactory. Good results were obtained with the solvents listed in Table 111.

Table 111. *Rf × 100 of PTH-Amino Acids and of Mono- and Diphenylthiourea.* For abbreviations, see Table 103 and [1, 2, 3]

Amounts spotted: 0.5 μg. in 0.5 μl. of methanol or acetone. Ascending technique, solvent migration 10 cm. Each figure represents average result of 6 observations.

PTH-Derivatives of	Solvent I	Solvent II	Solvent III	Solvent IV
Ala	16	68	39	11
Arg	00	01	00	00
Asp	00	01	13	00
Asp(NH_2)	00	23	07	00
Glu	01	04	17	00
Glu(NH_2)	01	28	08	00
Gly	10	56	33	05
His	01	29	00	02
Ileu	40	77	57	37
Leu	40	77	60	37
Lys	12	71	34	03
Met	33	75	51	13
Met.O[1]	01	40	12	01
Phe	28	74	50	18
Pro	60	82	65	21
Thr	04	45	15	00
Try	13	62	39	10
Tyr	03	47	21	01
Val	32	74	55	23
MPTH[2]	12	54	31	03
DPTH[3]	43	76	67	22

I : *Chloroform;*
II : *Chloroform-methanol* (90 + 10);
III : *Chloroform-formic acid* (100 + 5);
IV : "*Heptane*"-*system:* n-heptane-ethylene chloride-formic acid-propionic acid (90 + 30 + 21 + 18), use 100 ml. of upper phase.

Specification of chemicals used:

Chloroform: stabilized with 1.5% ethanol (Fluka[4])
Formic acid: water-free (Fluka[4]), chemically pure
Heptane: n-heptane ASTM purum (Fluka[4])
Propionic acid: purum (Fluka[4])
Ethylene chloride: 1,2-dichloroethane, chemically pure (Fluka[4])
Methanol: commercial grade, distill once using a short column.

[1] Methionine sulphoxide
[2] Monophenyl thiourea
[3] Diphenyl thiourea
[4] Fluka AG, Buchs SG, Switzerland

---

[1] Degradation of glutamine peptides: D. G. SMYTH, W. H. STEIN and S. MOORE: J. Biol. Chem., *237*, 1845 (1962).

PTH-Aspartic acid and PTH-glutamic acid are separated by chloroform-methanol-formic acid (70 + 30 + 2) (Fig. 174).

Cherbuliez et al. [*4*] recommend heptane-pyridine-ethylacetate (50 + 30 + 20) for the separation of PTH-glycine, PTH-proline and PTH-leucine.

Rf-values: Table 111 gives *Rf*-values observed when the solvents listed were used. Within wide limits, there is no dependence on the amount of material spotted. For PTH-proline, e.g., this is true in the range between 0.05 and 72 μg. (chloroform-methanol 90 + 10); if the volume spotted is constant (0.5 μl. of a solution in methanol), spot size (after chromatography) is found to be roughly proportional to the logarithm of the amount of substance present.

Separation and identification

Earlier described methods [*8*] are preferably replaced by the following procedure [*193*].

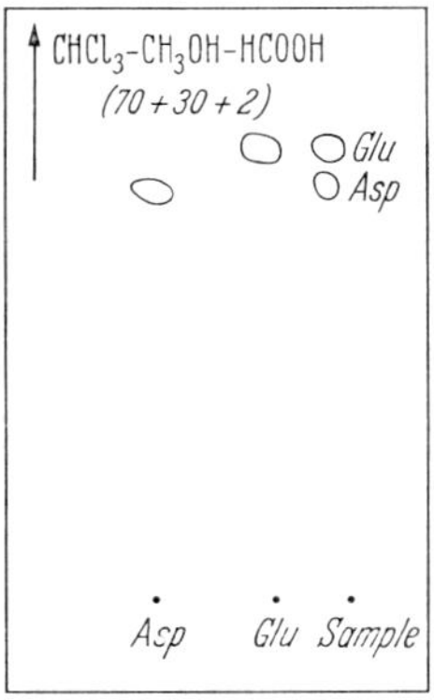

Fig. 174. One-dimensional chromatogram: identification of PTH-aspartic acid and PTH-glutamic acid

Amounts spotted: 0.5 μg. in 0.5 μl. of methanol. Ascending technique. Solvent migration, 11 cm. Detection by the chlorine/tolidine test or by UV-light (270 m μ.)

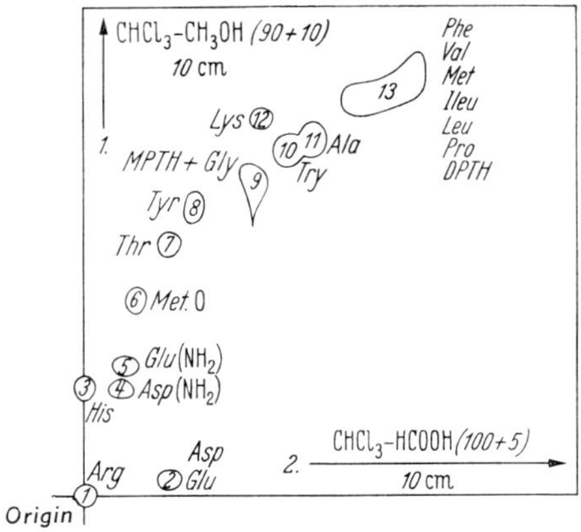

Fig. 175

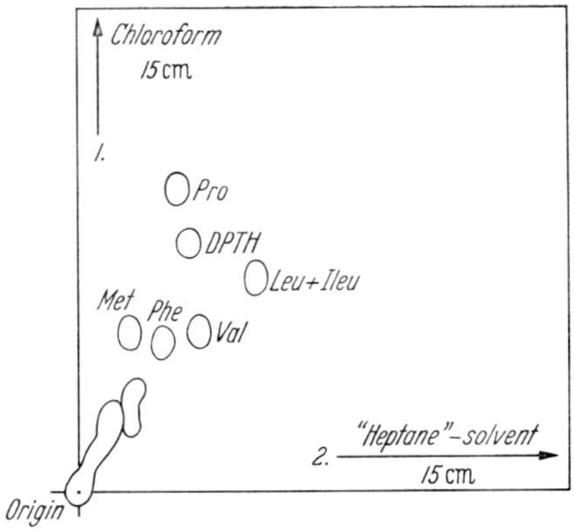

Fig. 176

Fig. 175. Two-dimensional chromatogram of a mixture containing 0.5 μg. of each PTH-amino acid + 0.5 μg. of MPTU + 0.5 μg. of DPTU and spotted in 0.5 μl. of methanol or acetone

Ascending technique. Detection by the chlorine/tolidine test or by UV-light (270 mμ.). Only three out of a total of 13 spots represent more than one substance. For identification of spots 2, 9 and 13, see text, Fig. 174 and 176

Fig. 176. Two-dimensional chromatogram of a mixture containing 0.5 μg. of each PTH-amino acid + 0.5 μg. of MPTU + 0.5 μg. of DPTU and spotted in 0.5 μl. of methanol or acetone

Ascending technique. Detection by UV-light (270 mμ.). Substances composing spot No. 13 of Fig. 175 are now separated, except PTH-leucine and PTH-isoleucine. See text.

Two two-dimensional (Figs. 175 and 176) and one one-dimensional chromatograms (Fig. 174) are run simultaneously. Separation is complete, except for PTH-glycine and monophenylthiourea (MPTU) on one hand, and PTH-leucine and PTH-isoleucine on the other hand.

PTH-glycine may be detected even in the presence of MPTU (Fig. 175) by its color reaction with ammonia (Schramm et al. [*189*]); a slightly modified procedure is described below.

Identification of PTH-leucine and of PTH-isoleucine (Fig. 176) requires the following steps: Removal from the plate of the portion containing the PTH-leucine-isoleucine spot visible in UV-light (see below) – elution with methanol –, evaporation of the eluate to dryness – hydrolysis of the residue (which should amount to at least 1 μg. per PTH-amino acid) for 12 hours with 6 N hydrochloric acid – [*8*], repeated evaporation of the hydrolyzate *in vacuo* with repeated addition of water – dissolution of the residue in water – spotting the solution in successive 1 μl. portions, each time allowing the water to evaporate from the silica layer, and continuous chromatography in methyl ethylketone-pyridine-water-glacial acetic acid (70 + 15 + 15 + 2) (p. 403).

3. Detection of phenylthiohydantoins

The chlorine/tolidine test (Reagent No. 32) described in the chapter on peptides (pp. 412—413) is very useful; the minimal amount required for detection is about 0.05 μg. An alternative is offered by fluorescence quenching already known in paper chromatography [*192*]. Thin layers become fluorescent in UV (270 mμ.) if 0.8% of luminescent zinc silicate (P 1, Type 118-2-7, General Electric, Cleveland, Ohio, U.S.A.; Mn-activated Zn-silicate, Leuchtstoffwerk GmbH, Heidelberg, Germany) is added to Silica Gel G, and quenching is observed in the presence of PTH-amino acids, the minimal amount required being about 0.1 μg.

PTH-Glycine specifically develops a dark red permanent color if the layer is moistened by a water spray and subsequently exposed to gaseous ammonia (an open bottle containing conc. ammonia solution will do); the minimal amount of PTH-glycine required is about 0.08 μg.

Iodine azide (Reagent No. 75b), as suggested by Sjöquist [*191*] and more recently by Cherbuliez [*4*], appears to be far from ideal. On layers of Silica Gel G, we observed white spots on a light blue background; they were barely visible and disappeared rapidly. Grote's reagent [*194*] (Reagent No. 115), although producing a better color, is not recommended because its application is reported to be laborious and time consuming [*190*].

VI. Thin-Layer Ionophoresis and Thin-Layer Ionophoresis Chromatography

By

C. G. HONEGGER

1. Thin-layer ionophoresis

The ionophoretic separation of mixtures on thin inorganic layers – thin-layer ionophoresis – has been described independently by PASTUSKA and TRINKS [*195*] as well as by us [*123*]. Thin-layer chromatography had been worked out by STAHL [*196–200*] to a simple and standardized method which proved so successful in the separation of lipophilic and hydrophilic substances and was superior to paper chromatography. It was of interest to extend this method to thin-layer ionophoresis. Principally, no difficulties were expected from this method as CONSDEN, GORDON and MARTIN [*99*], in 1946, had already used silica gel with good success as support for the ionophoretic separation of amino acids and peptides. Their somewhat complicated technique, found little application in contrast to paper ionophoresis.

As the experimental conditions have already been described in detail on pp. 24—25, we shall discuss here mainly the application, as well as the advantages and disadvantages, of thin-layer ionophoresis on various coating materials.

The main advantages of thin-layer chromatography over paper chromatography are: a) less diffusion leads to smaller spots and thus higher sensitivity is achieved; b) shorter separation time; c) possibility of using corrosive spray reagents for the detection of the separated substances. The latter point is equally valid for thin-layer ionophoresis, whereas point b) shorter separation time, applies at best to low-voltage electrophoresis. The use of adsorption-layers on "loose" glass plates, which are not built into a cooling arrangement, has certain disadvantages. Because of the relatively thick glass plates (3.5 mm.), an effective cooling, necessary for separations at higher voltage (over 500 volts), is difficult. The use of thinner glass plates avoids this disadvantage to a certain extent. Because of the vulnerability of the layer, cooling which, in contrast to PASTUSKA and TRINKS [*195*], we consider necessary at 4–500 volts, is more difficult to achieve with the upper side of the plate.

Comparison between paper, Silica Gel G, Kieselguhr G and Alumina G as coating materials. The relations between migration rate and diffusion have been investigated under almost identical conditions with a known mixture (cystein, mezcalin, cadaverine, methylamine). The ionophoresis was carried out (for 1 hr. in Na-citrate buffer, 0.1 N, pH 3.8, 440 V) on paper (Whatman No. 1), Silica Gel G (Merck), Kieselguhr G (Merck) and Alumina G (Merck). The plates were stained with ninhydrin. An exact comparison of the results is difficult as the original pH (3.8) of the buffer is changed differently upon moistening the various layers: for instance, on Silica Gel G to pH ~ 3, on Kieselguhr G to pH ~ 2.5, on Alumina G to pH ~ 4.5, and on paper to pH ~ 3.5. Furthermore, the degree of moisture on the various layers does not exactly accord. The comparison of the migration rates shows that the values are higher for Kieselguhr G and paper than for Silica Gel G and Alumina G. However, it has to be considered that the values are influenced by adsorption which can be different from layer to layer. From the shape of the spot[1], it can be seen that cadaverine, for instance, shows a higher adsorption (tail formation) on Silica Gel G than on paper. The size of the spots in a low-loading range (0.25–1 μg.) per substance is smallest for Alumina G, followed by Kieselguhr G < Silica Gel G < paper. The evaluation is rendered difficult by the varying sensitivity of the ninhydrin reaction on the different materials as well as by differences in the buffer content which can influence the diffusion.

By comparing the loading capacity of a given substance, it can be established that Silica Gel G and Alumina G show a smaller capacity than Kieselguhr G and paper, but here again, fluctuations in the buffer content and coloring properties can influence the evaluation.

From these facts, the conclusion can be drawn that thin-layer ionophoresis in low voltage ranges – for the present, at least – is superior to paper ionophoresis, whereas, *Kieselguhr G has the optimum properties as coating material*. Therefore, we can agree only to a certain extent, to the advantages of Kieselguhr G and Silica Gel G over paper namely, higher migration rates, less adsorption, etc.; as described by Pastuska and Trinks [*195*].

The *reproducibility* of thin-layer ionophoresis is in no way inferior to the separation on paper. When choosing between the "thin-layer" and "paper" methods of ionophoresis, the advantages and disadvantages have to be considered from case to case. The different separating effects on the various coating materials will play a decisive role.

Thin-layer ionophoresis has, so far, been applied only to the separation of substances of low molecular weight. Pastuska and Trinks [*195*] were particularly concerned with the separation of organic acids and published the *Rf*-values of 36 substances (borate buffer) on Silica Gel G and Kieselguhr G. Our work includes the separation of amines, amino acids [*123*] (Na-citrate buffer pH 3.8), phosphoric acid esters (Na-citrate buffer pH 3.8) and sugars (Na-borate buffer pH 12.5 and 10) on Silica Gel G, Kieselguhr G and Alumina G.

2. Thin-layer ionophoresis chromatography

This two-dimensional separation method in which ionophoresis occurs in one direction and chromatography is applied in the second dimension benefits not only from the thin-layer ionophoretic properties but also from the known advantages of thin-layer chromatography

[1] See [*201*].

It is, therefore, rather superior to the method on paper. Its field of application, as the one of thin-layer ionophoresis, extends especially to mixtures of ionic compounds or substances which can be made to migrate in an electric field. In this sphere, it is superior to two-dimensional chromatography because of the use of two completely different physical properties for the separation. It has been applied to the separation of amino acids and amines [*123*] (Fig. 177).

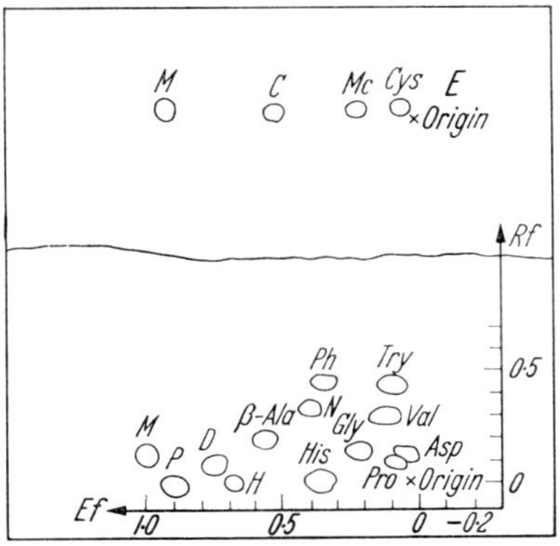

Fig. 177. Thin-layer ionophoresis-chromatogram of a mixture of amines and amino acids on Silica Gel G (plate 20 × 20 cm., 1—8 μg. per substance spotted in a total volume of 2 μl. of water)

Experimental conditions for chromatogram Fig. 177

In the upper part of the plate, a standard mixture was simultaneously separated by ionophoresis as a control and for the evaluation of *Ef*-values[1]. For the abbreviations of the amino acids, see p. 400; Cys = cysteine, C = cadaverine, D = dimethylamine, H = histamine, M = methylamine, Mc = mescalin, N = noradrenaline, Ph = phenyl ethylamine, P = (1.3)-propylenediamine.

1. Ionophoresis: Buffer pH 2: 2 N acetic acid-0.6 N formic acid (1 + 1); degree of moisture 160%; 460 V, 12.6 mA, 1 hr.

2. Chromatography: n-Butanol-glacial acetic acid-water (80 + 20 + 20), ascending technique, $2^1/_2$ hrs.

Detection: Ninhydrin (Reagent No. 108).

[1] Relative migration values calculated by dividing migration distance of a given substance by migration distance of cysteine (*Ef*-value 0) to the one of methylamine (*Ef*-value 1). Substances which migrate to the anode are indicated with negative *Ef*-values.

Bibliography to Chapter J. Amino Acids and Derivatives

[*1*] Cohn, E. J., and J. T. Edsall: "Proteins, Amino Acids and Peptides as Ions and Dipolar Ions". pp. 203 and 201. New York: Reinhold Publishing Corporation, 330 West 42nd St. 1943.
[*2*] Hais, I. M., u. K. Macek: Handbuch der Papierchromatographie. Jena: VEB Gustav Fischer Verlag 1958.
[*3*] Moffat, E. D., and R. I. Lytle: Analyt. Chem. **31**, 926 (1959).
[*4*] Cherbuliez, E., Br. Baehler u. J. Rabinowitz: Helv. Chim. Acta, **43**, 1871 (1960).
[*5*] Nicolaus, B. J. R.: J. Chromatog. **4**, 384 (1960).
[*6*] Kirk, R. E., and D. F. Othmer: Encyclopedia of Chemical Technology. Vol. 2, page 773. New York: Interscience Publishers Inc. 1948.
[*7*] Brenner, M., u. A. Niederwieser: Experientia **16**, 378 (1960).
[*8*] — — u. G. Pataki: Experientia **17**, 145 (1961).
[*9*] Fahmy, A. R., A. Niederwieser, G. Pataki u. M. Brenner: Helv. Chim. Acta **44**, 2022 (1961).

[10] MUTSCHLER, E., u. H. ROCHELMEYER: Arch. Pharmaz. **292**, 449 (1959).
[11] MOTTIER, M.: Mitt. Lebensm. u. Hyg. **49**, 454 (1958).
[12] BRENNER, M., A. NIEDERWIESER, G. PATAKI u. A. R. FAHMY: Experientia **18**, 101 (1962).
[13] BRENNER, M., u. A. NIEDERWIESER: Experientia **17**, 237 (1961).
[14] MOORE, S., D. H. SPACKMAN and W. H. STEIN: Analyt. Chem. **30**, 1185 (1958); **30**, 1190 (1958).
[15] PÖHM, M.: Naturwiss. **48**, 555 (1961).
[16] PAULSON, C., F. E. DEATHERAGE and E. F. ALMY: J. Am. Chem. Soc. **75**, 2039 (1953).
[17] LINSKENS, H. F.: Papierchromatographie in der Botanik, Zweite Aufl. Berlin-Göttingen-Heidelberg: Springer-Verlag 1959.
[18] DE VERDIER, C.-H., and G. ÅGREN: Acta Chem. Scand. **2**, 783 (1948).
[19] AWAPARA, J.: Arch. Biochem. **19**, 172 (1948).
[20] TISELIUS, A.: Experientia **17**, 433 (1961).
[21] FLODIN, P.: J. Chromatog. **5**, 103 (1961).
[22] PORATH, J.: Biochim. et Biophys. Acta **39**, 193 (1960).
[23] — Clin. Chim. Acta **4**, 776 (1959).
[24] STEPANOV, V., D. HANDSCHUH u. F. A. ANDERER: Z. Naturforsch. **16**b, 626 (1961).
[25] BOISSONNAS, R. A., u. S. LO BIANCO: Experientia **8**, 425 (1952).
[26] CONSDEN, R., A. H. GORDON and A. J. P. MARTIN: Biochim J. **41**, 590 (1947).
[27] REDFIELD, R. R.: Biochim. et Biophys. Acta **10**, 344 (1953).
[28] CARSTEN, M. E.: J. Am. Chem. Soc. **74**, 5954 (1952).
[29] PIEZ, K. A., E. B. TOOPER and L. S. FOSDICK: J. Biol. Chem. **194**, 669 (1952).
[30] STEIN, W. H., and S. MOORE: J. Biol. Chem. **190**, 103 (1951).
[31] BUCHNER, H.: Dissertation Universität Basel 1957, S. 84.
[32] DRÈZE, A., et A. DE BOECK: Arch. intern. physiol. et biochem. **60**, 201 (1952).
[33] — S. MOORE and E. J. BIGWOOD: Analyt. chim. Acta **11**, 554 (1954).
[34] COOK, E. R., and M. LUSCOMBE: J. Chromatog. **3**, 75 (1960).
[35] DENT, C. E., W. STEPKA and F. C. STEWARD: Nature (London) **160**, 682 (1947).
[36] PARIHAR, D. B.: Naturwiss. **41**, 502 (1954).
[37] REINDEL, F. u. W. BIENENFELD: Z. physiol. Chem. **305**, 123 (1956).
[38] BISERTE, G., P. BOULANGER et P. PAYSANT: Bull. soc. chim. biol. **40**, 2067 (1958).
[39] CLOSE, R.: Nature (London) **185**, 609 (1960).
[40] KNAUFF, H. A., H. SCHMAIR u. H. ZICKGRAF: Z. physiol. Chem. **318**, 73 (1960).
[41] NÜRNBERG, E.: Arch.Pharm. **292**, 610 (1959).
[42] JOST, K., J. RUDINGER and F. ŠORM: Coll. Czech. Chem. Comm. **26**, 2496 (1961).
[43] AHRLAND, S., I. GRENTHE and B. NORÉN: Acta Chem. Scand. **14**, 1059, 1077 (1960).
[44] HIRS, C. H. W.: J. biol. Chem. **219**, 611 (1956).
[45] BISERTE, G., et R. OSTEUX: Bull. soc. chim. biol. **33**, 50 (1951).
[46] CALDIN, D. J. MC.: Chem. Rev. **60**, 39 (1960).
[47] CONSDEN, R., A. H. GORDON and A. J. P. MARTIN: Biochem. J. **38**, 224 (1944).
[48] TSUKAMOTO, T., and T. KOMORI: Chem. Pharm. Bull. (Tokyo) **8**, 913 (1960).
[49] KAWERAU, E., and TH. WIELAND: Nature (London) **168**, 77 (1951).
[50] BARROLLIER, J.: Naturwiss. **48**, 404 (1961).
[51] WOIWOD, A.: J. Chromatog. **3**, 278 (1960).
[52] HARDY, T. L., D. O. HOLLAND and J. H. C. NAYLER: Analyt. Chem. **27**, 971 (1955).
[53] OPIENSKA-BLAUTH, J., M. SANECKA and M. CHAREZINSKI: J. Chromatog. **3**, 415 (1960).
[54] PATTON, A. R., and P. CHISM: Analyt. Chem. **23**, 1683 (1951).
[55] TOENNIES, G., and J. J. KOLB: Analyt. Chem. **23**, 823 (1951).
[56] RUHEMANN, S.: J. Chem. Soc. **99**, 792, (1911):
[57] — J. Chem. Soc. **99**, 1306 (1911);
[58] — J. Chem. Soc. **99**, 1486 (1911);
[59] — J. Chem. Soc. **97**, 1483 (1910);
[60] — J. Chem. Soc. **97**, 2025 (1910).

[61] WIGGINS, L. F., and J. H. WILLIAMS: Nature (London) **170**, 279 (1952).
[62] SAIFER, A., and I. ORESKES: Analyt. Chem. **28**, 501 (1956).
[63] NOWORYTKO, J., i M. SARNECKA-KELLER: Acta biochim. Polon. **2**, 91 (1955).
[64] ACHER, R., C. FROMAGEOT et M. JUTISZ: Biochim. et Biophys. Acta **5**, 81 (1950).
[65] GRASSMANN, W., u. K. v. ARNIM: Ann. Chem. Liebigs **519**, 192 (1935).
[66] BARROLIER, J., J. HEILMAN u. E. WATZKE: Z. physiol. Chem. **304**, 21 (1956).
[67] ROSEBEEK, S.: Chem. Weekblad **46**, 813 (1950).
[68] KOFRÁNYI, E.: Z. physiol. Chem. **299**, 129 (1955).
[69] CONSDEN, R.: Nature (London) **162**, 359 (1948).
[70] MÜTING, D.: Naturwiss. **39**, 303 (1952).
[71] GIRI, K. V., u. A. NAGABHUSHANAM: Naturwiss. **39**, 548 (1952).
[72] RIEMSCHNEIDER, R., u. K. PREUSS: Mh. Chem. **90**, 924 (1959).
[73] REINDEL, F., u. W. HOPPE: Chem. Ber. **87**, 1103 (1954).
[74] BHATTACHARYA, K. R., J. DATTA and D. K. ROY: Arch. Biochem. **84**, 377 (1959).
[75] ROCHE, J., N. VAN THOAI and J. L. HATT: Biochim. et Biophys. Acta **14**, 71 (1954).
[76] JEPSON, J. B., and I. SMITH: Nature (London) **172**, 1100 (1953).
[77] ACHER, R., and CH. CROCKER: Biochim. et Biophys. Acta **9**, 704 (1952).
[78] SAKAGUCHI, S.: J. Biochem. (Tokyo) **5**, 25 (1925).
[79] — J. Biochem. (Tokyo) **5**, 133 (1925).
[80] — J. Biochem. (Tokyo) **37**, 231 (1950).
[81] TUPPY, H.: Sitzungsberichte (Wien) **162**, 342 (1953).
[82] WINEGARD, H. M., G. TOENNIES and R. J. BLOCK: Science **108**, 506 (1948).
[83] KIRBY-BERRY, H., H. E. SUTTON, L. CAIN and J. S. BERRY: Univ. Texas Pubs. **5109**, 22 (1951).
[84] AWE, W., I. REINECKE u. J. THUM: Naturwiss. **41**, 528 (1954).
[85] CHARGAFF, E., C. LEVINE and C. GREEN: J. Biol. Chem. **175**, 67 (1948).
[86] BURTON, R. B., A. ZAFFARONI and E. H. KEUTMANN: J. Biol. Chem. **188**, 763 (1951).
[87] TOYADA, H.: J. Bull. Chem. Soc. (Japan) **9**, 263 (1954).
[88] PATTON, A. R., and E. M. FOREMAN: Science **109**, 339 (1949).
[89] FRANK, H., u. H. PETERSEN: Z. physiol. Chem. **299**, 1 (1955).
[90] BRAY, H. G., W. V. THORPE and K. WHITE: Biochem. J. **46**, 271 (1950).
[91] PAULY, H.: Z. physiol. Chem. **42**, 517 (1904).
[92] — Z. physiol. Chem. **94**, 288 (1915).
[93] EDLBACHER, S., H. BAUR, H. R. STAEHELIN u. A. ZELLER, Z. physiol. Chem. **270**, 158 (1941).
[94] — Z. physiol. Chem. **157**, 106 (1926).
[95] SANGER, F., and H. TUPPY: Biochem. J. **49**, 463 (1951).
[96] CLARKSON, T. W.: Biochim. et Biophys. Acta **18**, 453 (1955).
[97] SCHWARTZ, D. P.,: Analyt. Chem. **30**, 1855 (1958).
[98] DALGLIESH, C. E.: Nature (London) **166**, 1076 (1950).
[99] CONSDEN, R., A. H. GORDON and A. J. P. MARTIN: Biochem. J. **40**, 33 (1946).
[100] METZENBERG, R. L., and H. K. MITCHELL: J. Am. Chem. Soc. **76**, 4187 (1954).
[101] FEARON, W. R., and W. A. BOGGUST: Analyst **79**, 101 (1954).
[102] EDWARD, J. T., and D. M. WALDRON: J. chem. Soc. **1952**, 3631.
[103] SMITH, I.: Nature (London) **171**, 43 (1953).
[104] DALGLIESH, C. E.: Biochem. J. **52**, 3 (1952).
[105] PACHÉCO, H.: Bull. soc. chim. biol. **33**, 1915 (1951).
[106] WIELAND, TH., u. L. BAUER: Angew. Chem. **63**, 511 (1951).
[107] JERCHEL, D., u. R. MÜLLER: Naturwiss. **38**, 561 (1951).
[108] FISCHER, A.: Planta **43**, 288 (1954).
[109] PROCHÁZKA, Ž.: Chem. Listy **47**, 1643 (1953).
[110] LINSER, H., H. MAYR u. F. MASCHEK: Planta **44**, 103 (1954).
[111] ERSPAMER, V., and G. BORETTI: Arch. int. Pharmacodyn. 88, 296 (1951).
[112] CLERC-BORY, M., H. PACHÉCO et CH. MENTZER: Compt. rend. **238**, 525 (1954).
[113] EKMAN, B.: Acta Chem. Scand. **2**, 383 (1948).
[114] GERNGROSS, O., K. VOSS u. H. HERFELD: Ber. dtsch. chem. Ges. **66**, 435 (1933).

[*115*] KUDZIN, S. F., R. M. DE BAUN and F. F. NORD: J. Am. Chem. Soc. **73**, 4615 (1951).
[*116*] FOLIN, O., and V. CIOCALTEU: J. Biol. Chem. **73**, 627 (1927).
[*117*] — and W. DENIS: J. Biol. Chem. **12**, 239 (1912).
[*118*] DURANT, J. A.: Nature (London) **169**, 1062 (1952).
[*119*] MILLON, E.: Compt. rend. **28**, 40 (1949).
[*120*] BRICAS, E., and CL. FROMAGEOT: Advances in Protein Chem. **8**, 4 (1953).
[*121*] SCHACHTER, M.: Polypeptides which affect smooth muscles and blood vessels. Pergamon Press 1960.
[*122*] KATZ, A. M., W. J. DREYER and C. B. ANFINSEN. J. Biol. Chem. **234**, 2897 (1959).
[*123*] HONEGGER, C. G.: Helv. Chim. Acta **44**, 173 (1961).
[*124*] Biochem. Preparations **8**, 39, 43, 45, 47(1961).
[*125*] YANARI, S.: Biochim. et Biophys. Acta **45**, 595 (1960).
[*126*] PORATH, J., and P. FLODIN: Nature (London) **183**, 1657 (1959).
[*127*] GELOTTE, B. J.: J. Chromatog. **3**, 330 (1960).
[*128*] LINDNER, E. B., A. ELMQUIST and J. PORATH: Nature (London) **184**, 1565 (1959).
[*129*] Weissberger A. Technic of Organic Chem. Vol. III, part I, 2 nd ed. New York: Interscience 1956.
[*130*] SANGER, F.: Advances in Protein Chem. **7**, 1 (1952).
[*131*] ZUBER, H.: Chimia (Switz.) **14**, 405 (1960).
[*132*] SJÖQUIST, J., B. BLOMBÄCK and P. WALLÉN: Arkiv Kemi **16**, 425 (1961).
[*133*] — Arkiv Kemi **14**, 291 (1959).
[*134*] VOGLER, K., R. O. STUDER u. W. LERGIER: Helv. Chim. Acta **44**, 1495 (1961).
[*135*] GUTTMANN, ST., u. R. A. BOISSONNAS: Helv. Chim. Acta **44**, 1713 (1961).
[*136*] — J. PLESS u. R. A. BOISSONNAS: Helv. Chim. Acta **45**, 170 (1962).
[*137*] HUGUENIN, R. L., u. R. A. BOISSONNAS: Helv. Chim. Acta **44**, 213 (1961).
[*138*] VOGLER, K., R. O. STUDER, W. LERGIER u. P. LANZ: Helv. Chim. Acta **43**, 1751 (1960).
[*139*] STUDER, R. O., K. VOGLER u. W. LERGIER: Helv. Chim. Acta **44**, 131 (1961).
[*139a*] SCHWYZER R., u. H. KAPPELER: Helv. Chim. Acta **44**, 1991 (1961).
[*139b*] SCHWYZER R., u. H. DIETRICH: Helv. Chim. Acta **44**, 2003 (1961).
[*140*] BRENNER, M., u. G. PATAKI: Helv. Chim. Acta **44**, 1420 (1961).
[*141*] RINIKER, R., CIBA AG., Basel, Private communication.
[*142*] KAPPELER, H., u. R. SCHWYZER: Helv. Chim. Acta **44**, 1137 (1961).
[*143*] GUTTMANN, ST., SANDOZ AG., Basel, Private communication.
[*144*] SCHELLENBERG, P.: Angew. Chem. **74**, 118 (1962).
[*145*] VOGLER, K., Hoffmann-La Roche AG., Basel, Private communication.
[*146*] VOGLER, K., R. O. STUDER, P. LANZ, W. LERGIER u. E. BÖHNI: Experientia **17**, 223 (1961).
[*147*] HAUSMANN, W.: J. Am. Chem. Soc. **78**, 3663 (1956).
[*148*] BISERTE, G., et M. DAUTREVAUX: Bull. soc. chim. biol. **39**, 795 (1957).
[*149*] BRENNER, M., J. P. ZIMMERMANN, J. WEHRMÜLLER, P. QUITT, A. HARTMANN, W. SCHNEIDER u. U. BEGLINGER: Helv. Chim. Acta **40**, 1497 (1957).
[*150*] NIEDERWIESER, A., and G. PATAKI: Chimia (Switz.) **14**, 378 (1960).
[*151*] SANGER: F.: Biochem. J. **39**, 507 (1945).
[*152*] — Biochem. J. **40**, 261 (1946).
[*153*] — Biochem. J. **45**, 562 (1949).
[*154*] PORTER, R., and F. SANGER: Biochem. J. **42**, 287 (1948).
[*155*] EDMAN, P.: Acta Chem. Scand. **4**, 283 (1950).
[*156*] BISERTE, G., J. W. HOLLEMAN, J. HOLLEMAN-DEHOVE and P. SAUTIÈRE: J. Chromatog. **2**, 225 (1959); **3**, 85 (1960).
[*157*] LEVY, A. L., and D. CHUNG: J. Am. Chem. Soc. **77**, 2899 (1955).
[*158*] DAVOLL, H., R. A. TURNER, J. G. PIERCE and V. DU VIGNEAUD: J. Biol. Chem. **193**, 363 (1951).
[*159*] WALLENFELS, K., u. A. ARENS: Biochem. Z. **332**, 217 (1960).
[*160*] LI, C. H., and L. ASH: J. Biol. Chem. **203**, 419 (1953).
[*161*] TUPPY, H., u. G. BODO: Mh. Chem. **85**, 807 (1954).
[*162*] LOCKHART, I. M., and E. P. ABRAHAM: Biochem. J. **58**, 633 (1954).
[*163*] LEVY, A. L., and C. H. LI: J. Biol. Chem. **213**, 487 (1955).

[164] DESNUELLE, P., and C. FABRE: Biochim. et Biophys. Acta, **18**, 49 (1955).
[165] PORTER, R. R.: Methods in Med. Research **3**, 256 (1951).
[166] MIDDLEBROOK, W. R.: Biochem. J. **59**, 146 (1955).
[167] SCANES, F. S., and B. T. TOZER: Biochem. J. **63**, 282 (1956).
[168] THOMPSON, A.: Nature (London) **168**, 390 (1951).
[169] MILLS, G. L.: Biochem. J. **50**, 707 (1952).
[170] TURBA, F., u. G. GUNDLACH: Biochem. Z. **326**, 322 (1955).
[171] BAILEY, K., and F. R. BETTELHEIM: Biochim. et Biophys. Acta **18**, 495 (1955).
[172] BRAUNITZER, G.: Chem. Ber. **88**, 2025 (1955).
[173] LEVY, A. L.: Nature (London) **174**, 126 (1954).
[174] ZAHN, H., u. H. PFANNMÜLLER: Biochem. Z. **330**, 97 (1958).
[175] FAHMY, A. R.: Unpublished.
[176] HEFENDEHL, F. W.: Planta Medica. **8**, 65 (1960).
[177] LICHTENBERGER, W.: Z. anal. Chem. **185**, 111 (1962).
[178] EDMAN, P.: Nature (London) **177**, 667 (1956).
[179] — Acta Chem. Scand. **10**, 761 (1956).
[180] FRAENKEL-CONRAT, H., and J. I. HARRIS: J. Am. Chem. Soc. **76**, 6058 (1954).
[181] SJÖQUIST, J.: Arkiv Kemi **11**, 129 (1957).
[182] — Biochem. et Biophys. Acta **41**, 20 (1960).
[183] EDMAN, P.: Acta Chem. Scand. **4**, 277 (1950).
[184] DJERASSI, C., K. UNDHEIM, R. C. SHEPPARD, W. G. TERRY and B. SJÖBERG: Acta Chem. Scand. **15**, 903 (1961).
[185] FRAENKEL-CONRAT, H.: Methods of Biochem. Analysis, Vol. 2, page 383. New York: Interscience Publishers 1955.
[186] CHERBULIEZ, E., BR. BAEHLER, M. C. LEBEAU u. A. R. SUSSMANN: Helv. Chim. Acta **43**, 896 (1960).
[187] — A. R. SUSSMANN u. J. RABINOWITZ: Pharmac. Acta Helv. **36**, 131 (1961).
[188] HARRIS, J. I., u. P. ROOS: Biochem. J. **71**, 434 (1959).
[189] SCHRAMM, G., J. W. SCHNEIDER u. A. ANDERER: Z. Naturforsch. **11**b, 120 (1956).
[190] LANDMANN, W. A., M. P. DRAKE and J. DILLAHA: J. Am. Chem. Soc. **75**, 3638 (1953).
[191] SJÖQUIST, J.: Acta Chem. Scand. **7**, 447 (1953).
[192] EDMAN, P., u. J. SJÖQUIST: Acta Chem. Scand. **10**, 1507 (1956).
[193] PATAKI, G.: Dissertation, Universität Basel (1962).
[194] GROTHE, J. W.: J. Biol. Chem. **93**, 25 (1931).
[195] PASTUSKA, G., u. H. TRINKS: Chemiker Ztg. **85**, 535 (1961).
[196] STAHL, E.: Pharmazie **11**, 633 (1956).
[197] — Chemiker Ztg. **82**, 323 (1958).
[198] — Parfümerie und Kosmetik **39**, 564 (1958).
[199] — Arch. Pharm. **292**, 411 (1959).
[200] — Pharm. Rdsch. **1**, Nr. 2, 1 (1959).
[201] WEBER, R.: Helv. Chim. Acta **36**, 424 (1953).

Some recent noteworthy articles

1. Amino acids

a) *Cellulose*

NEHER, R., and A. VON ARX: J. Chromatogr. **12**, 329 (1963).

b) *Silica*

OPIENSKA-BLAUTH, J.: Intern. Symp. on TCL, Elsevier, in press: Urinary amino-acids.

2. Peptides and proteines

a) *Silica*

WIELAND, TH., and U. GEBERT: Analyt. Bioch. **6**, 201 (1963), Peptides, Modification of the Edman Reaction.
PATAKI, G.: Chimia (in press): PTH-Peptides.

b) *Sephadex*

WIELAND, TH. , G. LÜBEN and H. DETERMANN: Exper. **18**, 430 (1962): Enzymes.

3. DNP-Amino acids

WALZ, D., A. R. FAHMY, G. PATAKI, A. NIEDERWIESER and M. BRENNER: Experientia **19**, 212 (1963), DNP-Amino Acids from urine.

KELLER, M., and G. PATAKI: Helv. **46**, 1687 (1963), DNP-Amino acids from sperma.

4. PTH-Amino acids

CHERBULIEZ, E., BR. BAEHLER, J. MARSZALEK, R. H. SUSSMANN and J. RABINOWITZ: Helv. **46**, 2446 (1963), 17, PTH-Amino Acids.

5. 5-Dimethylamino-l-naphthalenesulfonyl-amino acids

DAVID, J. B., T. C. FRENCH and J. M. BUCHANAN: J. Biol. Chem. **238**, 2178 (1963), Synth. and TCL of 5-dimethylamino-l-naphthalene-sulfonyl-amino acids.

6. Iodo-Amino Acids

HOLLINGSWORTH, D. R., M. DILLARD and P. K. BONDY: J. Lab. Clin. Med. **62**, 346 (1963).

K. Nucleic Acids and Nucleotides

By

HELMUT K. MANGOLD

I. Introduction

1. Nucleic acids and their hydrolysis products

The nucleic acids are polymers which form the prosthetic group of nucleoproteins.

There are two types of polynucleotides which are markedly different in their structures: deoxyribonucleic acids (DNA) have molecular weights as high as 10 million, whereas ribonucleic acids (RNA) usually have molecular weights below 300,000.

Deoxyribonucleic acids are more resistant to alkaline hydrolysis than ribonucleic acids. Acid hydrolysis of deoxyribonucleic acids yields purine bases and "thymic acid". Each type of nucleic acid is attacked by different enzyme systems. The action of enzymes isolated from pancreas yields "oligonucleotides".

Both types of nucleic acids are composed of mononucleotides which are built according to the following scheme:

Base-Carbohydrate-o-Phosphoric Acid

Nucleosides are formed from nucleotides by removal of the phosphoric acid group. Deoxynucleosides contain the purine bases guanine or adenine or the pyrimidines cytosine or thymine. Some deoxyribonucleosides contain 5-methyl cytosine and 5-hydroxymethyl cytosine. The ribonucleosides isolated from ribonucleic acids contain guanine and adenine or the pyrimidines cytosine and uracil.

Purines are bound to the sugar phosphates as glycosides at N_9, pyrimidines at N_3; the phosphoric acid moiety is bound to either C_2 or C_3 of the carbohydrate.

The following summary gives the structures of the constituents of the two types of nucleic acids:

Bases

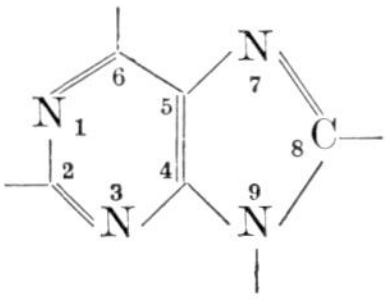

Purines

Adenine (6-Amino-purine)
Guanine (2-Amino-6-hydroxy purine)

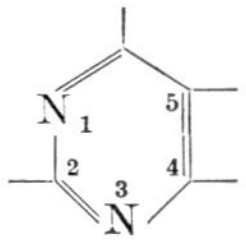

Pyrimidines

Cytosine (2-Hydroxy-6-amino-pyrimidine)
Uracil (2,6-Dihydroxy-pyrimidine)
Thymine (2,6-Dihydroxy-5-methyl-pyrimidine)
Also 5-Methylcytosine and 5-Hydroxy-methyl cytosine.

Carbohydrates

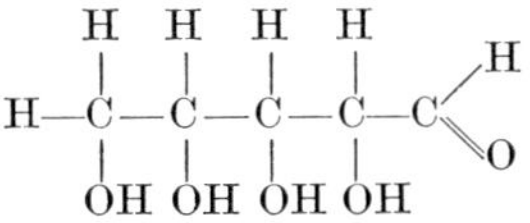

Ribose

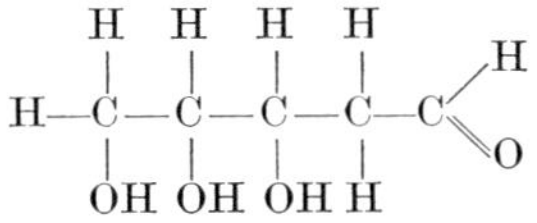

2-Deoxyribose

Examples of a *Nucleoside* and a *Nucleotide*

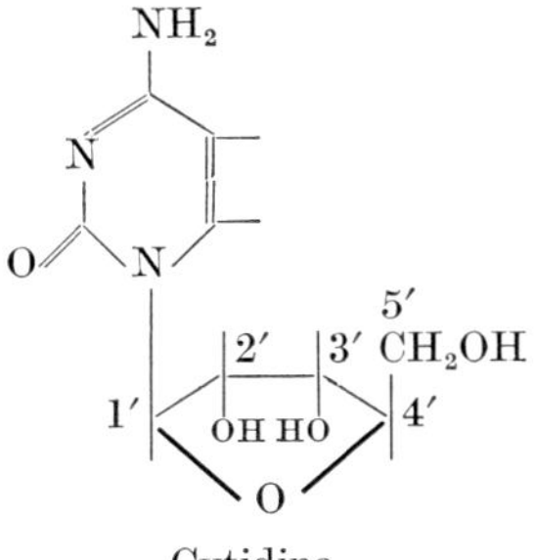

Cytidine

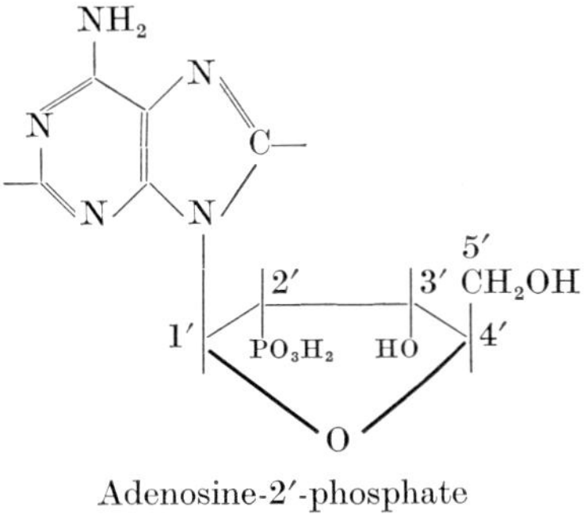

Adenosine-2′-phosphate

2. Nucleotide coenzymes

The nucleotide coenzymes are structurally related to the mono-nucleotides from nucleic acids. They are, however, not nucleic acid constituents. Typical nucleotide coenzymes are, e.g., adenosine triphosphate (ATP) and flavine-adenine-dinucleotide (FAD). Many other coenzymes are phosphoric acid esters of adenosine, guanosine, cytidine or uridine. For example, the following five coenzymes are derived from cytidine diphosphoric acid (CDP); CDP-choline, CDP-colamine, CDP-diglyceride, CDP-glycerol, and CDP-ribitol.

3. Older methods of nucleic acid analysis

Elemental analyses of phosphorus and nitrogen and determination of N/P ratios were for a long time essential procedures of nucleic acid analysis. The individual purines and pyrimidines were identified in nucleic acid hydrolyzates and semi-quantitatively determined by isolation on a small scale. Various color reactions, polarography and microbiological techniques were also utilized for analytical purposes. Today all of these methods are of historical interest only.

4. Newer methods for the separation of nucleic acid hydrolyzates and of nucleotide coenzymes

Great strides have been made during the last 15 years in applying chromatographic techniques to the nucleic acid field. High molecular weight nucleic acids can now be fractionated by chromatography on ion exchange columns or by electrophoresis. Paper chromatography and other partition methods, also ion exchange chromatography on columns or on sheets, permit the efficient fractionation of complex nucleic acid hydrolysates. Quantitative analyses are possible with minute amounts of nucleic acids by determining the amount of bases, nucleosides and nucleotides after their chromatographic separation.

These modern methods of nucleic acid analysis are extensively described in Volume I of the standard monograph of nucleic acid chemistry, which was published by CHARGAFF and DAVIDSON [*14*] in 1955. SOBER and PETERSON [*81*] summarized chromatographic techniques used for fractionating nucleic acids. COHN [*22*] described the separation of nucleic acid derivatives. A most thorough discussion of latest results and many yet unpublished findings were recently also presented by COHN [*23*]. The chemistry of nucleoside polyphosphates is comprehensively dealt with in a handbook edited by BOYER, LARDY and MYRBÄCK [*6*].

5. The U.V. spectra of nucleic acid derivatives[1]

The amount of a purine or pyrimidine in solution is usually determined by U.V. spectrophotometry. The absorption maxima and molar extinction coefficients of these substances, as well as of ribonucleosides and of ribonucleotides, are presented in Table 112.

6. Color reactions

The biuret reaction (see p. 445) and the arginine test of WEBER [*96*] are used for the detection of peptides in nucleic acid preparations. Two

[1] For details see the article by G. H. BEAVEN, E. R. HOLIDAY and E. A. JOHNSON in Vol. I of "The Nucleic Acids," E. CHARGAFF and J.N. DAVIDSON, Editors [*14*]. The following firms distribute useful brochures and tables: Pabst Laboratories, Division of Pabst Brewing Company, 1037 W. Mc Kinley Ave., Milwaukee 5, Wis., U.S.A., California Corporation for Biochemical Research, 3625 Medford St., Los Angeles 63, California, USA., C.F. Boehringer & Soehne G.m.b.H., Mannheim, Germany.

Table 112. *The U.V. Spectra of Nucleic Acid Derivatives*

	Molecular weight	Normality of solution	Absorption Maximum mμ	Millimol Extinction-coefficient
Purines and Pyrimidines				
Adenine	135.13	0.1 HCl	262	13.1
Guanine	151.13	0.1 HCl	249	11.1
Cytosine	111.10	0.1 HCl	276	10.0
Uracil	112.09	Water	260	8.2
Thymine	126.11	0.1 HCl	265	7.95
5-Methylcytosine	125.13	0.1 HCl	283	9.8
5-Hydroxymethyl cytosine	141.14	0.1 HCl	279	9.7
Ribosides				
Adenosine	267.25	Water	259	15.4
Guanosine	283.24	Water	252	13.7
Cytidine	243.22	0.1 HCl	280	13.0
Uridine	244.20	Water	262	10.0
Riboside-2′- and -3′-monophosphates				
Adenosine-2′-monophosphate	347.23	Water	260	15.0
Adenosine-3′-monophosphate				
Guanosine-2′-monophosphate	363.24	0.01 HCl	280	12.9
Guanosine-3′-monophosphate				
Cytidine-2′-monophosphate	323.21	0.01 HCl	278	12.7
Cytidine-3′-monophosphate				
Uridine-2′-monophosphate	324.20	Na salt	260	10.0
Uridine-3′-monophosphate		in water		

other color reactions are applied for distinguishing ribo- and deoxyribonucleic acids and for identifying them in the presence of each other: the DISCHE reaction is specific for deoxyribonucleic acids, and the phloroglucinol test described by v. EULER and HAHN is used for the detection of ribonucleic acids.

a) The Dische-reaction [*29, 30*]

A solution of 50—500 μg of nucleic acid in 1 ml of water is heated to 100° C with 2 ml of a solution of 1 g diphenylamine and 2.75 ml conc. sulfuric acid in 100 ml of glacial acetic acid. The presence of deoxyribonucleic acid is indicated by a blue color (abs. max. 595 mμ). Pyrimidine deoxyribonucleosides and nucleotides are not hydrolyzed under these conditions and do not yield a blue color.

b) The reaction of v. EULER and HAHN [*8, 32, 61*]

The solution containing about 2 mg of ribonucleic acid per ml is heated for 50 min on a water bath with 8 ml of a 0.1% solution of ferric chloride in conc. hydrochloric acid-glacial acetic acid (1 + 6). The reaction mixture is cooled and 25% phloroglucinol in conc. hydrochloric acid-glacial acetic acid-water (1 + 2 + 1) is added. The mixture is heated on a water bath for 4 min. The color reaches maximum intensity after keeping the solution at room temperature for 10 hr (abs. max. 680 mμ). Deoxyribonucleic acid does not react under these conditions.

7. Procedures for the isolation of nucleic acids

Different procedures for the isolation of nucleic acids yield products differing widely in composition and behavior. The presence of nucleolytic enzymes in most plant and animal tissues is one of the reasons for the great variation of nucleic acid preparations. In isolating nucleic acids, it is advisable to work at low temperatures to retard the action of enzymes. Deoxyribonucleases may be inhibited by sodium citrate [*60*]; ribonucleases are inactivated by guanidine hydrochloride [*95*] or dodecyl sulfate [*43*].

Concentrated salt solutions and heating should be avoided. Dilute acids attack nucleic acids, and dilute alkali hydrolyzes ribonucleic acids rapidly.

It is generally assumed that highly viscous nucleic acid solutions correspond most closely to the natural product. Good preparations should be free of proteins, polysaccharides, lipids and inorganic salts. Pure nucleic acids contain about 9.2% of phosphorus and exhibit a U.V. absorption maximum between 257 and 261 mμ at pH 7.

a) Preparation of sodium deoxyribonucleate from calf thymus according to SIGNER and SCHWANDER [*76, 78*]

Nucleoproteins are extracted with fairly concentrated sodium chloride solution. They are then split by saturating the solution with sodium chloride. The sodium deoxyribonucleate is purified by repeated precipitation from its aqueous solution with ethanol.

450 g of fresh calf thymus is cut into small pieces and then further ground in a meat grinder together with pieces of dry ice. Small portions of the brei are homogenized at 0° C with a total of 4 l of ice-cold M-sodium chloride solution containing 0.01 Moles of sodium citrate per liter. The gel-like homogenate is slowly stirred at 0° C for several days until it becomes a reddish, highly viscous liquid. 830 ml-portions of this solution are poured into 4.2 l portions of ice cold water to precipitate nucleoproteins. The precipitate is stirred in 0.01 M-sodium citrate solution and washed repeatedly with 8 l of 1% sodium chloride solution containing 0.01 M-sodium citrate. The nucleoproteins are twice dissolved in portions of 4 l M-sodium chloride-0.01 M-sodium citrate solution and precipitated by pouring these solutions into water.

The nucleoproteins are redissolved by stirring into 5 l of a 10% sodium chloride solution containing sodium citrate. This solution is made up to 6 l with saturated sodium chloride solution, and crystalline sodium chloride is added to saturation. The liquid is stirred for four days at 0° C and then stored in a refrigerator for two weeks. A suspension of 480 g of "Celite 545"[1] in 1.5 l of saturated sodium chloride solution is added, and the mixture is vigorously stirred for 24 hr. Finally, the suspension is filtered at 40—50 mm through a Celite plate, 5—10 mm thick, which is placed between two filter papers.

The filter cake still contains deoxyribonucleate. It is stirred with 1.5 l of saturated sodium chloride for about 24 hr and then filtered. The two filtrates are combined and 10 g of "Hyflo Super Cel"[1] per liter is added. The slurry is stirred for several hours and filtered again through a plate of celite. One volume of the clear filtrate is poured into 1.5 volumes of ethanol to precipitate the nucleate. The fibrous mass is washed several times with 70% ethanol. The precipitate is squeezed and dissolved in 4.5 l of ice-cold water by stirring for several days. This aqueous solution is poured into twice its volume of ethanol to precipitate the nucleate. The salt is washed successively with 80%, 96% and abs. ethanol and finally with diethyl ether and dried *in vacuo* over conc. sulfuric acid. The pure white asbestos-

[1] Johns-Manville Corp., Celite Division, 270 Madison Ave., New York 16, N. Y., U.S.A.

like preparation is stored in a dessicator over sodium chloride solution. It contains less than 0.5% protein.

According to the authors, the yield of sodium deoxyribonucleate is 8 g, i.e. 1.8% of the weight of the starting material.

The procedure described by CHARGAFF and ZAMENHOF [*11*] and by KAY, SIMMONS and DOUNCE [*42*] are also often used for the preparation of deoxyribonucleic acids.

b) Isolation of sodium ribonucleate from animal tissues according to VOLKIN and CARTER [*95*]

In this method deoxyribonucleic acid is first removed in the form of its protein complex. Thereafter, the ribonucleic acid is precipitated from guanidine hydrochloride solution and further purified by treating with chloroform and repeated precipitation with ethanol.

After cutting frozen animal tissues into small pieces, one part of the tissue is suspended in three parts of ice cold 0.15 M-sodium chloride-0.02 M-phosphate buffer of pH 6.8 and homogenized at 2—5° C in a mixer for 6 to 8 min. It is recommended that a few drops of octyl alcohol be added to the suspension to avoid excessive foaming. The homogenate is centrifuged at 3000 × g for half an hour at 5° C. The residue contains practically all the deoxyribonucleic acid in the form of protein complexes, and is discarded.

Crystalline guanidine hydrochloride is added to the supernatant to yield a 2 M-solution. This solution is warmed for half an hour in a water bath at 38° C and then kept at 0° C for one hour. A gel-like precipitate of ribonucleic acid containing small amounts of protein is formed.

The crude product is washed twice with an equal volume of ice-cold 2 M-guanidine hydrochloride solution. The suspension of ribonucleic acid in aqueous guanidine hydrochloride solution is shaken at 40° C with an equal volume of a mixture of octyl alcohol-chloroform (1 + 4) for 30 min. After this time, the suspension is clarified by centrifugation. The upper aqueous phase is twice more extracted with octyl alcohol-chloroform under the conditions stated above. The aqueous guanidine hydrochloride solution is adjusted to pH 4.2—4.5 with glacial acetic acid, and two volumes of ice-cold ethanol are added. The white precipitate is centrifuged off and washed twice with ice-cold 70% ethanol. The ribonucleic acid is dissolved in distilled water, and the solution is carefully adjusted to pH 6.8 by the addition of dilute sodium hydroxide solution. If a precipitate of denatured protein forms, it is removed by centrifugation. The clear aqueous sodium ribonucleate is mixed with sufficient M-sodium chloride solution to make it 0.05 M. Two volumes of ice-cold ethanol are added to precipitate the pure ribonucleate. This product is washed twice with 70% ethanol, then isolated as a pure white powder by freeze-drying. A solution of 20 mg of the pure preparation in 1 ml of water does not give the Dische reaction (see p. 443), nor does it give the biuret reaction (heating the sample with 15% sodium hydroxide solution and 0.10% copper sulfate solution). This indicates that it is not contaminated with deoxyribonucleate or with protein.

The authors state that the yield amounts to 20—30% of the ribonucleic acid present in the starting material.

CHARGAFF and co-workers [*13*], CRESTFIELD, SMITH and ALLEN [*25*] as well as KAY and DOUNCE [*43*] have also published procedures for the isolation of ribonucleic acids, which have been widely accepted.

8. Methods for the hydrolysis of nucleic acids

Hydrolytic cleavage of nucleic acids may lead to the formation of oligonucleotides, mononucleotides, purine and pyrimidine bases, sugar phosphates and free carbohydrates. In addition, secondary products may result, e.g., by deamination of amino purines and amino pyrimidines or their nucleosides and nucleotides.

a) Acid hydrolysis

The N-glycosidic purine-carbohydrate linkage is especially labile to acids, both in ribo- and deoxyribonucleic acids. In contrast, pyrimidine nucleotides and nucleosides resist acid hydrolysis. The phosphoric acid moiety is also more easily removed by the action of mineral acids from purine nucleotides than from pyrimidine nucleotides. If boiled with 0.5 N-, N- or 2 N-hydrochloric acid or sulfuric acid for one to two hours, ribonucleic acids will yield the purines, adenine and guanine, in addition to the pyrimidine nucleotides, cytidylic and uridylic acids [*54, 94*]. Deoxyribonucleic acids also yield adenine and guanine if treated with hydrochloric acid at pH 1.6. However, the high-molecular weight structures of the polynucleotide is, to a great extent, preserved. The purine-free high molecular products, "thymic acids," have molecular weights of around 15,000. They can be further purified by dialysis against hydrochloric acid of pH 1.6 to yield "apurinic acids" [*86*].

Heating of apurinic or deoxyribonucleic acids with methanolic hydrochloric acid at 50° C for three to five hours yields cytosine- and thymine deoxyribo-diphosphoric acids [*53*].

Pyrimidine deoxynucleotides and nucleosides may be cleaved by heating in a sealed tube at 175° C with 98—100% formic acid [*94, 100*]. Hydrolysis with 72% perchloric acid is also useful. Maximum yields of purines and pyrimidines are obtained by heating at 100° C for one hour [*59, 100*].

b) Alkaline hydrolysis

Dilute aqueous solutions of sodium or potassium hydroxides split ribonucleic acids into mononucleotides. Deoxyribonucleic acids are not hydrolyzed under these conditions. Hence, it is possible to free deoxyribonucleic acids of contaminating ribonucleic acids by treating with N-sodium hydroxide at room temperature or at 37° C [*74, 85*].

Nucleosides may be obtained from nucleotides by heating them with conc. aqueous ammonia at 175—180° C for three to four hours or by reacting them with boiling pyridine for several days. However, these reactions as well as similar procedures give poor yields and are useless for analytical studies.

c) Enzymatic hydrolysis

Degradation of polynucleotides by the action of enzymes serves as a valuable tool for studying the structure of nucleic acids.

The enzyme ribonuclease I from pancreas [*41, 51*] hydrolyzes ribonucleic acids to mono-, di-, tri- and tetranucleotides. A "core" of the native ribonucleic acids always remains unattacked. Deoxyribonucleic acid is not attacked, but thymic acid is hydrolyzed to some extent. Crystalline ribonuclease I is commercially available[1].

Deoxyribonuclease I from pancreas [*2, 52*] splits high molecular weight deoxyribonucleic acid into oligonucleotides and small amounts of mononucleotides; large portions of the polynucleotide structure remain unattacked. Ribonucleic acids are not degraded by deoxyribonuclease. Crystalline deoxyribonuclease I is commercially available [1].

Low-molecular weight hydrolysis products or ribo- and deoxyribonucleic acids may be separated from high polymer cores by dialysis.

Deoxyribonucleases from snake venoms hydrolize deoxyribonucleic acids more completely than does deoxyribonuclease I from pancreas. Nucleotidases, i. e. enzymes which dephosphorylate mononucleotides to nucleosides, have also been described. However, little is yet known about these enzymes.

The properties and some applications of various nucleases and nucleotidases are described in several comprehensive articles [*14, 51, 52, 60*].

---

[1] Worthington Biochemical Corp., Freehold, N.J., U.S.A.

II. Thin-layer chromatography of nucleic acid derivatives

1. Chromatography on cellulose and on silica gel

Separations of nucleic acid derivatives on filter paper were first described by VISCHER and CHARGAFF [*93, 94*] and by HOTCHKISS [*40*]. CHARGAFF and co-workers applied paper chromatography to the quantitative analysis of ribonucleic acids [*13, 94*], deoxyribonucleic acids [*12, 15*] and apurinic acids [*88*]. MARKHAM and SMITH [*57, 58*], and also CARTER [*11, 19*] and WYATT [*100, 101*] developed and refined the method further. WYATT [*99, 100*] discovered 5-methyl cytosine in plant and animal deoxyribonucleic acids. WYATT and COHEN [*102*] found by paper chromatography 5-hydroxymethyl cytosine in the deoxyribonucleic acid of a bacteriophage.

Paper chromatographic methods for the analysis of nucleic acids are described in detail in several monographs [*5, 34*]. Attention is drawn to a chapter by WYATT in the first volume of the series "The Nucleic Acids" [*14*].

Partition chromatography became the analytical standard method in the nucleic acid field. Ion exchange chromatography on columns (see p. 451) was widely used for isolating mononucleotides and oligonucleotides on a preparative scale. Procedures for chromatographing on starch [*26, 31*] or on adsorbents [*55, 56*] did not find wide acceptance.

RANDERATH was the first to report separations of purines and pyrimidines of nucleosides and mononucleotides [*68, 69, 70, 71*] and also of nucleotide-polyphosphates and nucleotide coenzymes [*71, 72, 73*]. He found TLC on cellulose and Silica Gel G layers to be superior to paper chromatographic methods in regard to the efficiency of separation [*69, 70*]. Thin-layer chromatography on cellulose yields smaller and more discrete spots than can be obtained on paper under identical conditions [*71*]. Moreover, chromatography of nucleic acid derivatives on thin layers requires only a fraction of the time needed in paper chromatography [*70, 71, 72*].

Experimental conditions

a) The stationary phase

Cellulose powder containing plaster of Paris as a binder ("MN 300 G"[1]) or plain cellulose layers ("MN 300"[1]) may be used. Cellulose powder containing gypsum is applied as an aqueous suspension (s. page 32). According to the manufacturers' specifications, such layers should be dried at 105–110° C. for 10 to 30 min. RANDERATH [*71*] recommends drying "MN 300 G" layers at room temperature overnight. The same author applies plain cellulose powder as a suspension in acetone. The coated

[1] Macherey, Nagel & Co., Inh. Dr. Ing. A. Rademacher, Chemische Filtrierpapierfabrik, Düren, Rhld., Germany. U.S. representative: C. A. Brinkmann and Comp., Inc., Cantiague Rd., Westbury, L. I., N.Y.

plates are dried for 3–5 min in a stream of dry air and are then ready for use.

The preparation of Silica Gel G layers is described on pp. 7—9. Reference is made to the work of TEICHERT, MUTSCHLER and ROCHELMEYER [*89*] on the separation of pharmaceutically prominent methyl purines on Silica Gel G containing buffers (see also p. 308).

b) Solvents

TAMM, SHAPIRO, LIPSHITZ and CHARGAFF [*88*] demonstrated that plain distilled water is a good solvent for the paper chromatographic fractionation of purines and pyrimidines as well as their nucleosides. RANDERATH [*69, 71*], also RANDERATH and STRUCK [*70*], found that water yields good separations on thin layers of cellulose in only 45 min. RANDERATH used water also for fractionating bases and nucleosides on Silica Gel G.

RANDERATH states [*71, 73*] that isomeric purine mononucleotides may be separated on cellulose layers with a solvent system described by MARKHAM and SMITH [*58*], i.e. saturated aqueous ammonium sulfate solution-M-sodium citrate-isopropanol (80 + 18 + 2). This solvent migrates in 90 min to a height of 10 cm (see Fig. 178). The solvent *tert.* amyl alcohol-formic acid-water (30 + 20 + 10), which was first described by MICHELSON [*62*], also yields excellent separations of nucleosides on cellulose layers; RANDERATH [*71*] found that such fractionations require about 2 hr. MICHELSON's solvent is better suited for the resolution of mononucleotides than are the other systems described (Table 114).

RANDERATH and STRUCK [*70*] recommend a mixture of n-butanol-acetone-glacial acetic acid-5% aqueous ammonia-water (9 + 3 + 2 + 2 + 4) as solvent for the fractionation of nucleoside monophosphates, diphosphates and triphosphates on paper or on cellulose layers (Table 114). The same solvents mixed in a ratio (7 + 5 + 3 + 3 + 2) yield a mixture which is suitable for fractionating nucleotides on cellulose layers (Table 114) as well as on layers of Silica Gel G [*71*]. Comparable fractionations require 90 min if cellulose is used as the stationary phase, whereas $2^1/_2$ hr are needed with Silica Gel G. Alkaline solvent mixtures are understandably not suitable for fractionating the strongly acidic nucleotides on silica gel containing gypsum. RANDERATH [*69*] mentions that the addition of 15 g of the chelating agent ethylenediamine tetra-acetic acid per 200 ml of solvent improves the separations.

c) Detection methods

Purines and pyrimidines, as well as their nucleosides and nucleotides, may be seen on cellulose plates in the light of a short wave length U.V. lamp[1] [*70*] (see also [*10, 38*]). It is possible to detect as little as 10^{-3} μ moles of adenine derivatives by this method, but larger amounts are required for detection of cytosine and uracil [*72*].

[1] "Mineralight" (Maximum emmission 250 mμ), Ultraviolet Products, Inc., San Gabriel, California, U.S.A.

Inorganic layers absorb U.V. light to a significant extent and, therefore, this detection method is less sensitive for nucleic acid derivatives on such plates. Adding a fluorescent mineral to the Silica Gel G layers or spraying of the finished chromatograms with fluorescein solution [*98*] (Reagent No. 90C) increases the sensitivity of the detection method [*73*].

It is not possible to apply the "photoprint" process [*57*] for the photographic detection of nucleic acid derivatives on plates [*72*] because the glass is opaque to U.V. The color reactions used in paper chromatography of nucleic acid hydrolysates have not yet been applied to thin-layer chromatograms. Reactions for the detection of pentoses in nucleosides and 5′-nucleotides, as well as those for phosphate esters, should be valuable supplements to the U.V. method. Lead tetra-acetate solution, which is often applied for detecting sugars [*9*], should also be useful for localizing nucleosides and nucleotides on plates. The reaction described by HANES and ISHERWOOD [*37*] should likewise aid in visualizing nucleotides on plates.

Applications and results

Table 113 offers a comparison of thin-layer with paper chromatography for the fractionation of purines and pyrimidines and their nucleosides. These substances were run on filter paper and on thin layers of cellulose, Silica Gel G and ECTEOLA ion exchanger.

Table 113. *Separation of Purines, Pyrimidines and their Nucleosides by Paper-Chromatography and Thin-Layer Chromatography using Water as Solvent* [*69*]

	ECTEOLA-layer[1] *Rf* × 100	Cellulose-layer[2] *Rf* × 100	Paper[3] *Rf* × 100	Silica Gel G-layer[4] *Rf* × 100
Adenine	29	30	38	57
Adenosine	56	53	56	75
Guanine	33	37	38	66
Guanosine	50	58	57	80
Cytosine	—	—	—	—
Cytidine	82	80	77	76
Uracil	73	72	75	78
Uridine	84	81	84	85

[1] Capacity of the ion exchanger 0.26 m Eq. N/g.

[2] "MN 300" of Macherey, Nagel & Co., Düren/Rhld., Germany. U.S. representative: C. A. Brinkmann and Comp., Inc., Cantiague Rd., Westbury, L.I., N.Y.

[3] "Ederol 202" of Binzer, Hatzfeld/Eder, Germany.

[4] "Silica Gel G for Thin-Layer Chromatography acc. to STAHL," E. Merck A.G., Darmstadt, Germany. U.S. representative: C. A. Brinkmann and Comp., Inc., Cantiague Rd., Westbury, L.I., N.Y.

The close similarity between the *Rf* values of purines on paper as well as on cellulose and ECTEOLA layers is striking. This is also true of purine nucleosides. Thin-layer chromatography on Silica Gel G yields higher *Rf* values. The rates of migration of the pyrimidines is approximately identical on the three different types of layers and on paper strips. The same applies to pyrimidine nucleosides.

It appears that the separation of bases from the corresponding nucleosides is better on ECTEOLA and on cellulose layers than on Silica Gel G. In contrast, Silica Gel G is more suitable for fractionating the various purines and pyrimidines from each other. No striking differences are found in the efficiency of separation of nucleosides.

Thin-layer chromatography is quite obviously superior to paper chromatography, both in regard to the sensitivity of detection methods and the sharpness of separation. In addition, TLC can be performed in a fraction of the time required for paper chromatographic separations. It can safely be predicted that TLC will become a useful micropreparative tool in nucleic acid chemistry.

Fig. 178 demonstrates that thin-layer chromatography on cellulose is also more suitable for fractionating nucleotides, including nucleoside polyphosphates, than is paper chromatography.

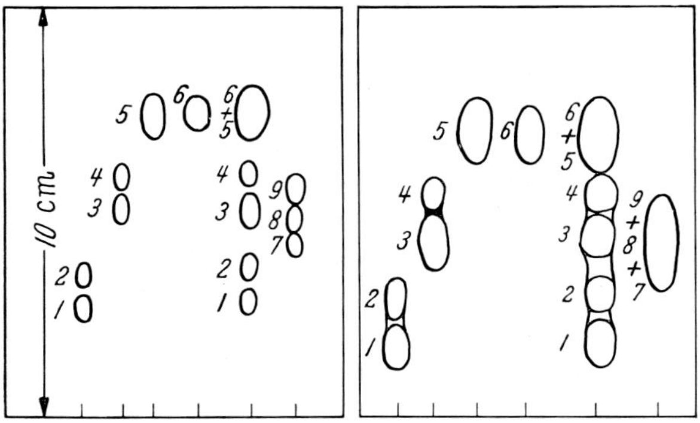

Fig. 178. Comparison of thin-layer chromatography and paper chromatography [*71*]. Nucleotides were fractionated (left) on cellulose layers ("MN 300 G") and (right) on paper (Schleicher & Schüll Nr. 2043 b). The chromatograms were run under identical conditions. Solvent: Saturated aqueous ammonium sulfate solution-M-sodium acetate-isopropanol (80+18+2). Time: $1^1/_2$ hours (TLC), 2 hours (PC). Made visible in UV-light. Amounts: AMP, ADP and ATP 10^{-2} μMoles, each; 2'-, 3/-AMP and GMP 1.5×10^{-2} μMoles; Cytidylic and uridylic acid, 3×10^{-2} μMoles, each.

8 Adenosine-3'-phosphate,
9 Adenosine-2'-phosphate,
10 Guanosine-3'-phosphate,
11 Guanosine-2'-phosphate,
12 Cytidinephosphates,
13 Uridinephosphates,
14 Adenosine-5'-monophosphate,
15 Adenosinediphosphate,
16 Adenosinetriphosphate

RANDERATH [*71*] reports that the weight of cellulose layers per unit area is one-half to one-third that of filter paper, i.e., 4 mg/cm² for cellulose layers, 8—12 mg/cm² for paper. He concludes that the remarkable advantages of TLC on cellulose over paper chromatography can be explained on the basis of difference in the fine structures of the two stationary phases.

The methods described here have, to the author's knowledge, not yet been applied to the analysis of nucleic acid hydrolysates. However, he ventures to predict that they will soon supersede the classical paper chromatographic procedures (see [*73b*]).

Ion exchangers are even more suitable for fractionating nucleotides by thin-layer chromatography than are cellulose or Silica Gel G.

Methods for ion-exchange-TLC of nucleic acid derivatives are dealt with in the following paragraph.

Table 114. *Separation of 5′-Mono-, Di- and Triphosphates on Cellulose*[1]*-Layers* [*69, 70, 71*]

	Solvent 1 $Rf \times 100$	Solvent 2 $Rf \times 100$	Solvent 3 $Rf \times 100$
Adenosine-5′-monophosphate	52	35	38
Adenosine diphosphate	29	17	26
Adenosine triphosphate	16	8	16
Guanosine-5′-monophosphate	—	—	—
Guanosine diphosphate	—	—	—
Guanosine triphosphate	—	—	—
Cytidine-5′-monophosphate	48	30	34
Cytidine diphosphate	27	13	22
Cytidine triphosphate	13	7	13
Uridine-5′-monophosphate	47	30	37
Uridine diphosphate	26	14	25
Uridine triphosphate	13	8	17

[1] "MN 300" of Macherey, Nagel & Co., Düren/Rhld., Germany.
Solvent 1: *tert.* Amyl alcohol-formic acid-water, (3 + 2 + 1); Time: 2 hr.
Solvent 2: n-Butanol-acetone-acetic acid-5% aqueous ammonia-water, (7 + 5 + 3 + 3 + 2). Time: $1^1/_2$ hr.
Solvent 3: n-Butanol-acetone-acetic acid-5% aqueous ammonia-water, (9 + 3 + 2 + 2 + 4). Time: 50—60 min.

2. Separation by ion-exchange chromatography

The hydrolysis products of nucleic acids contain acidic as well as basic functional groups. The net charge of such amphoteric molecules is determined by the pH of their solutions.

Cohn was the first to report (1949–1950) the ion-exchange chromatography of nucleic acid constituents on synthetic exchange resins [*16, 17, 18*]. He pointed out that the fractionation of nucleic acid hydrolysates should be possible both by anion and by cation-exchange chromatography [*23*]. Cohn utilized both kinds of chromatography for the separation of purines and pyrimidines [*16*] and nucleosides and nucleotides [*17, 18, 19, 24, 46*].

The isomeric 2′- and 3′-mononucleotides were first found and isolated by chromatography on columns of "Dowex" ion-exchangers[1]. Several new mononucleotides were discovered by the same technique [*18, 20, 27, 28*]. Smillie [*79*] fractionated the mononucleotides obtained from ribonucleic acids on ion-exchanger paper. He used cellulose paper that had been impregnated with the ion exchanger "Amberlite XE 119"[2]

The sequence of elution of the various degradation products from ion exchange columns should be determined by their net charges. However, experimental observations are not in agreement with this theory. One assumes that this is due to adsorptive properties of the exchange resins [*18, 22*]. Purines are usually more strongly adsorbed than are pyrimidines. Derivatives of uracil, for example, are eluted before guanine derivatives, although one would expect the opposite, according to the net charges of these compounds.

[1] The Dow Chemical Company, Midland, Mich., U.S.A.
[2] Reeve Angel and Co., Clifton, N.J., U.S.A.

A few years ago, PETERSON and SOBER [*67*, *80*] described procedures for the preparation of two types of ion exchangers from cellulose. "DEAE-Cellulose" (Diethylaminoethyl-cellulose) and "ECTEOLA-Cellulose" (obtained by reacting alkaline cellulose with epichlorohydrin and triethanolamine) have since frequently been used for fractionating proteins. BENDICH and co-workers [*3*, *4*], as well BRADLEY and RICH [*7*], also KLOUWEN and WEIFFENBACH [*49*], segregated nucleic acids on the same ion exchange materials. STAEHELIN [*83*] separated nucleoside-monophosphates, -diphosphates and -triphosphates. STAEHELIN, PETERSON and SOBER [*82*] chromatographed di- and trinucleotides from yeast ribonucleic acid on DEAE.

The chromatography of nucleic acids and their hydrolysis products on ion exchangers was discussed in detail by COHN [*14*, *22*, *23*] and also by SOBER and PETERSON [*81*].

RANDERATH is credited with having first described fractionations of low-molecular weight nucleic acid constituents on both ECTEOLA [*68*, *69*, *73*] and on DEAE [*72*, *73*]. He found that TLC on such ion exchangers permits more efficient and rapid separations of mononucleotides than is possible by other methods. Moreover, the detection methods were shown to be more sensitive in TLC than in paper chromatography.

WIELAND and coworkers [*98b*] used layers of DEAE-Sephadex for separating adenosine phosphates.

The chromatographic behavior of various nucleic acid constituents in a given solvent can be predicted with some degree of certainty. This is, according to RANDERATH [*73*], an essential advantage of analytical ion exchange-TLC over partition and adsorption techniques. RANDERATH [*73*] points out that TLC on modified cellulose is also useful for micro-preparative separations.

RANDERATH [*73a*] has recently described the use of polyethyleneimine cellulose for the TLC of nucleotides; WEIMANN and RANDERATH [*96b*] separated oligonucleotides on this ion exchanger. Layers of polyethyleneimine cellulose are particulary suitable for separating mononucleotides and oligonucleotides.

Experimental conditions

a) The ion exchanger

Two firms [1, 2] produce ion exchange powders for TLC. RANDERATH [*73*] used preparations of "Serva Entwicklungslabor." Publications referring to exchangers of the "MN" series are not yet available. The preparation of ECTEOLA and DEAE layers is described on p. 33.

Polyethyleneimine cellulose has recently become commercially available[1]; such preparations may have slightly different properties than self-made batches [*73a*].

---

[1] Serva-Entwicklungslabor, Heidelberg, Römerstraße 118, Germany. U.S. representative: Gallard-Schlesinger, 580 Mineola Ave., Carle Place, L. I., N. Y.

[2] Macherey. Nagel & Co., Inh. Dr.-Ing. A. RADEMACHER, Chemische Filtrierpapierfabrik, Düren, Rhld., Germany. U.S. representative: C. A. Brinkmann and Comp. Inc. Cantiague Rd., Westbury, L.I., N.Y.

KIYASU [*48*] uses layers prepared from mixtures of "Dowex I" or "Dowex 50"[1] with Silica Gel G. These layers do not adhere well to the glass plates. For this reason, a paper strip, 2–3 mm wide, is fastened along the lower edge of the plates to suck up the developing solvents. The suitability of such layers for the fractionation of nucleic acid constituents has not yet been investigated.

b) Solvents

Optimum resolutions of nucleic acid hydrolysates on columns of ion exchange resins are obtained by applying gradient elution, i.e., the ion concentration or the pH of the eluant, or both, are changed stepwise or continuously during chromatography. RANDERATH [*73*] obtained separations on layers of modified cellulose that are equal to column chromatographic fractionations using gradient elution on Dowex resins. The same author assumes that the excellent resolutions obtained by TLC can be explained by a gradient elution effect.

Nucleic acid derivatives are chromatographed in open jars containing a layer 1–1.5 cm high of the developing solvent. The course of chromatography can be observed in the light of a U.V. lamp. The plates are taken out of the jars, dried, and copied on cellophane in U.V. light as soon as the desired resolution is accomplished [*73*].

ECTEOLA may be used as the stationary phase for partition chromatography or as ion exchanger, depending upon the developing solvent.

Distilled water is a suitable solvent for fractionating purines and pyrimidines and also the corresponding ribonucleosides on ECTEOLA layers. This is predominantly a partition process. Satisfactory separations are accomplished within 15 min (see Table 113, p. 449) [*69*].

Mononucleotides and nucleotide polyphosphates can be fractionated by ion exchange-TLC on ECTEOLA. Dilute aqueous sodium chloride solutions serve as solvents (Tables 115 and 117). Satisfactory separations require only 15 min. The *Rf*-values of the nucleotides are a function of the chloride ion concentration of the developing solvent, whereas the rate of migration of the bases and nucleosides is independent of the ion concentration. This is evidenced by Fig. 179 which is taken from a publication of RANDERATH [*69*].

The solvent 0.15 M-sodium chloride is suitable for fractionating nucleoside mono-, di- and triphosphates [*69*] (see Table 117. p. 455), and also for separating purine nucleoside-3′-phosphates from the corresponding pyrimidine nucleotides [*69*] (see Table 115, p. 455). This solvent reaches a height of 8 cm after 15 min.

Especially sharp separations of nucleotides on ECTEOLA result if 0.01–0.07 N-hydrochloric acid is applied as the eluant [*73*] (see Table 117). Generally, 15 min suffice to yield complete separations. Mixtures of 0.1 μ moles of adenosine-diphosphate and adenosine-triphosphate are completely resolved with 0.05–0.07 N-hydrochloric acid in less than 5 min.

---

[1] The Dow Chemical Company, Midland, Mich., U.S.A.

Aqueous hydrochloric acid solutions are also used on DEAE layers [72] (see Table 117, p. 455, and Fig. 181A—C). These solvents do not migrate as fast on DEAE as they do on ECTEOLA; 0.01 N-, 0.2 N- and 0.3 N-hydrochloric acid solutions separate the nucleoside-mono-, -di- and-triphosphates of the same base within 40 min. During this time, the solvent front migrates to a height of 8.5—10 cm. Mixtures of nucleoside triphosphates can be completely resolved with 0.04 N-hydrochloric acid in about $1^1/_2$ hr. The solvent migrates to a height of 12 to 13 cm. A micropreparative separation of a mixture of 2 mg adenosine triphosphate on a 20/20 cm plate requires $1^1/_2$ hr if 0.03 N-hydrochloric acid is used as the developing solvent.

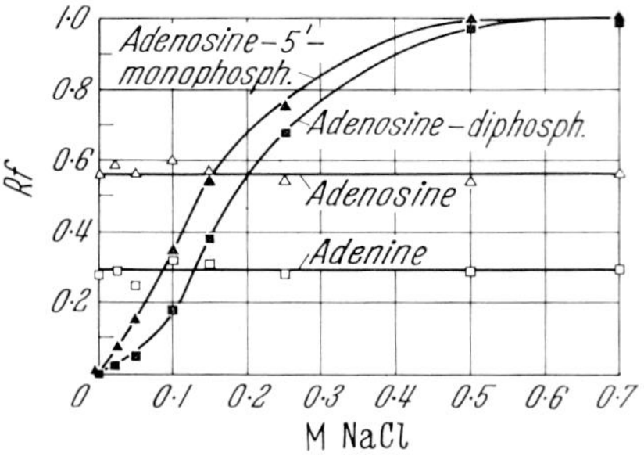

Fig. 179. Separation of nucleic acid derivatives on a thin-layer of ECTEOLA [69]. *Rf* values are plotted as a function of the sodium chloride concentration of the developing solvent. Capacity of the ion exchanger: 0.26 m Eq. N/g. Solvent: 0.1—0.7 M-sodium chloride. Time: 15 minutes (10 cm)

0.3 M to 1 M-Lithium chloride solutions are used for fractionating mononucleotides on polyethyleneimine cellulose. Acidic solvents, such as formic acid-sodium formate buffers of pH 3.4 (0.5—4 M) may be employed as the second solvent in two dimensional TLC [73a]. However, lithium-chloride has to be removed first by washing the layer with abs. methanol. Oligonucleotides are well resolved with 0.7—1.2 M NaCl [96b]. Chromatographing at different temperatures has been recommended for the complete resolution of mixtures of homologous oligonucleotides [96b].

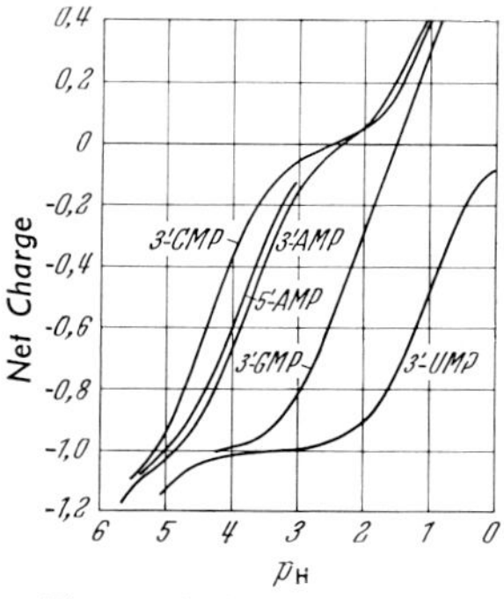

Fig. 180. Relation between net charge and pH [14]

c) Detection methods

ECTEOLA and DEAE layers absorb U.V. light only slightly and, therefore, as little as $5 \times 10^{-4} \mu$ moles of adenine nucleotides can be detected in the light of a U.V. lamp[1] (see also p. 448). Adenine, cytosine and uracil derivatives appear as dark blue fluorescent spots: guanine derivatives exhibit a light blue fluorescence.

Applications and results

Tables 113—117 and Figs. 179—182 demonstrate the kind of separations one can obtain by ion exchange chromatography of nucleic acid derivatives. The greatly different negative charges of nucleoside mono-, di- and triphosphates effect their separation into groups of compounds. Monophosphates migrate faster than diphosphates and these, in turn,

[1] "Mineralight" (Maximum emmission 260 mμ), Ultraviolet Products, Inc., San Gabriel, California, U.S.A.

faster than triphosphates. The negative net charge within each of these groups increases in the sequence: cytosine, adenine, guanine, uracil derivative (Fig. 180).

In accordance with this, cytosine- and adenine-nucleotides have always higher *Rf* values than do the corresponding guanine derivatives and migrate further than the nucleotides of uracil (Tables 116 and 117, Fig. 181).

Table 115. *Separation of Mononucleotides on ECTEOLA Layers [69]*

	Rf × 100
Adenosine-3′-phosphate	48
Guanosine-3′-phosphate	44
Cytidine-3′-phosphate.	71
Uridine-3′-phosphate .	75
High molecular weight ribonucleic acid . .	00

Capacity of the ion exchanger: 0.41 m Eq N/g; Solvent: 0.15 M sodium chloride; Time: 15 min.

Table 116. *Separation of Nucleoside Polyphosphates on DEAE Layers [72, 73]*

	Solvent 1 *Rf* × 100	Solvent 2 *Rf* × 100
Adenosine diphosphate	68	
Guanosine diphosphate	51	
Cytidine diphosphate .	—	
Uridine diphosphate .	25	
Adenosine triphosphate		56
Guanosine triphosphate		41
Cytidine triphosphate.		64
Uridine triphosphate .		18

Capacity of the ion exchanger: 0.7 m Eq N/g; Solvent 1: 0.03 N hydrochloric acid; Time 40 min; Solvent 2: 0.04 N hydrochloric acid; Time: $1^1/_2$ hr.

The efficiencies of ECTEOLA and DEAE layers in regard to the sharpness of separation are about the same, but the spots are especially small and well defined on DEAE [*73*] (Table 117).

Table 117. *Separation of 5′-Mono-, Di- and Triphosphates on Ion Exchange Layers [68, 69, 72]*

	ECTEOLA[1]		DEAE[2]	
	Solvent 1 *Rf* × 100	Solvent 2 *Rf* × 100	Solvent 3 *Rf* × 100	Solvent 4 *Rf* × 100
Adenosine-5′-monophosphate . . .	57	26	45	65
Adenosine diphosphate	36	8	24	48
Adenosine triphosphate	21	—	6	11
Guanosine-5′-monophosphate . . .	55	14	36	60
Guanosine diphosphate	37	3	9	27
Guanosine triphosphate	17	—	5	7
Cytidine-5′-monophosphate	74	31	46	65
Cytidine diphosphate	51	11	31	53
Cytidine triphosphate.	34	—	9	13
Uridine 5′-monophosphate.	80	13	31	49
Uridine diphosphate	63	0	7	15
Uridine triphosphate	44	—	4	4

[1] Capacity of the ion exchanger: 0.41 m Eq. N/g.
Solvent 1: 0.15 M Sodium chloride; Time: 15 min
Solvent 2: 0.01 M Hydrochloric acid; Time: 15 min
Capacity of the ion exchanger: 0.7 m Eq N/g.

[2] Solvent 3: 0.01 N Hydrochloric acid; Time: 40 min
Solvent 4: 0.02 N Hydrochloric acid; Time: 40 min.

Adenine and cytosine nucleotides, which are poorly segrated if hydrochloric acid is used as the eluant, are completely separated with 0.05 to 0.12 M-sodium chloride solutions [*73*]. The spots become somewhat more diffuse with this solvent.

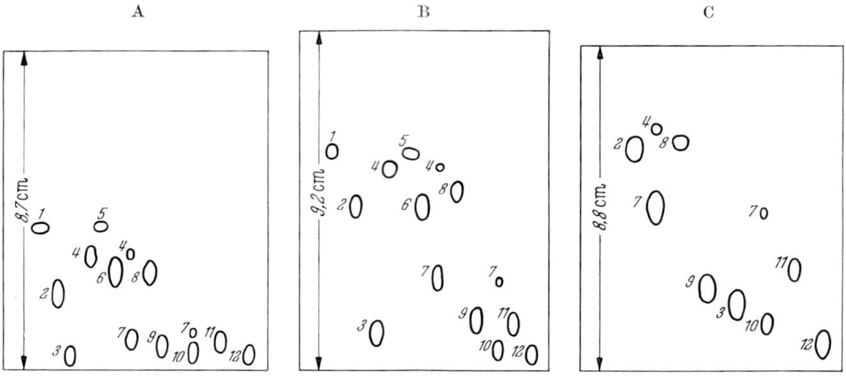

Fig. 181. Thin-layer chromatogram of nucleotides and nucleoside polyphosphates on DEAE [*73*]. Capacity of the ion exchanger: 0.34 m Eq N/g. Solvents: (A) 0.01 N-, (B) 0.02 N-, (C) 0.03 N-hydrochloric acid. Time: (A) and (B) 40 minutes, (C) 100 minutes. *1* Adenosine-5′-monophosphate, *2* Adenosine diphosphate, *3* Adenosine triphosphate, *4* Guanosine-5′-monophosphate, *5* Cytidine-5′-monophosphate, *6* Uridine-5′-monophosphate, *7* Guanosine diphosphate, *8* Cytidine diphosphate, *9* Uridine diphosphate, *10* Guanosine triphosphate, *11* Cytidine triphosphate, *12* Uridine triphosphate

Ribonucleotides and the corresponding deoxyribonucleotides can be resolved on polyethyleneimine cellulose with solutions of lithium chloride in 2–4% aqueous boric acid; the coenzymes DPN, DPNH, TPN, and TPNH may be separated on such layers with 0.3–0.5 M lithium chloride. Sharp fractionations of sugar nucleotides of the UDPG type also are possible with dilute aqueous solutions of lithium chloride [*73a*]

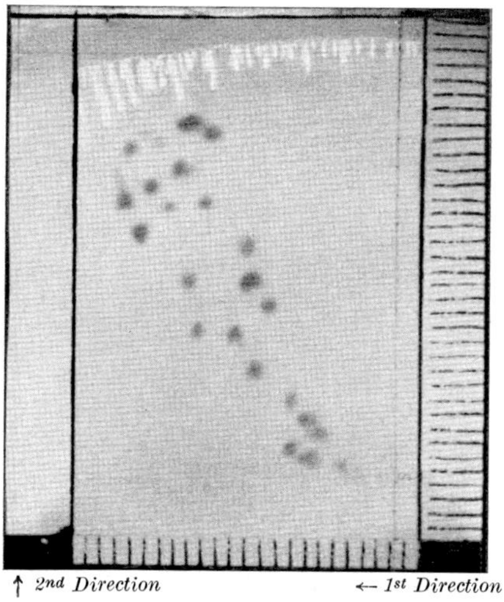

Fig. 182. Two-dimensional thin-layer chromatogram of 22 nucleotides on polyethyleneimine cellulose (courtesy of Dr. K. RANDERATH and Dr. E. RANDERATH)

Thin-layer chromatography is superior to all other analytical techniques used for fractionating nucleotides. Most remarkable is the speed of this method. The high capacity of the ion exchange layer also permits micro-preparative separation.

The methods described above have been used only for separating synthetic mixtures. Applications to the analysis of nucleic acid hydrolysates have not yet been reported.

3. Discussion

The biological properties of nucleic acids are most likely determined by the sequence of their constituents (see, e.g., [*33, 35, 75*]). Hence, it is of great interest to determine the order of mononucleotides in the polynucleotide chain [*15*].

Chromatographic methods have, in recent years, aided in structural studies of nucleic acids. The fractions obtained by chemical or enzymatic hydrolysis were separated and quantitatively determined [*14, 15, 44, 91*]. Chromatography was also used to analyze mononucleotides [*91, 92*] and even oligonucleotides and polynucleotides, synthesized either chemically [*36, 44, 45, 90*] or enzymatically [*50, 65, 66*].

Ion exchange chromatography on Dowex columns served for analytical and preparative fractionations of low-molecular weight mono-, di-, tri- and tetra-nucleotides. Nucleoside polyphosphates and other nucleotide coenzymes are frequently separated on Dowex resins. High molecular weight polynucleotides, including "intact" nucleic acids from plant and animal tissues, are almost exclusively chromatographed on columns of modified cellulose such as ECTEOLA and DEAE. Some pertinent publications were listed on p. 452. Reference is made again to the latest review by Cohn [*23*] on the chromatography of nucleic acids and related substances. This comprehensive discussion contains the theoretical background of ion exchange chromatography and also many yet unpublished results, notably from Cohn's laboratory, also those of Peterson and Sober and of Khorana (see also [*44*]).

Column chromatographic procedures are well worked out, and yield excellent separations. Ion exchange-TLC, however, permits the efficient fractionations of mixtures of low-molecular weight nucleic acid constituents in a shorter time. Thin-layer chromatography on polyethyleneimine cellulose has been found useful for separating oligonucleotides [*73b*]. It should be possible to fractionate apurinic acids as well as genuine ribo- and deoxyribonucleic acids on the same material, or on DEAE and ECTEOLA layers. The same method should be of help for analyzing the carbohydrate moieties of nucleic acids [*84*] (see also p. 462), possibly in the form of their borate complexes (see [*21*]). Other potential applications are methods for the segregation of sugar phosphates (see [*47*]) and o-phosphoric acid and polyphosphoric acids [*77*] (see also p. 482). Reference is made in this connection to publications dealing with the TLC of pteridines [*32*], therapeutically important purines and pyrimidines (see p. 308) and water soluble vitamins (see p. 235). Of great interest is the work of Nürnberg [*64*] on the thin-layer chromatographic analysis of vitamins of the B_6 group and of nicotinic acid amide.

Thin-layer chromatography will facilitate the search for yet unknown purines and pyrimidines (see [*1, 97*]) as well as their nucleosides and nucleotides (see [*20, 27*]) in nucleic acid hydrolysates.

It should be possible to elute the individual nucleic acid constituents from the plates and to determine them quantitatively by U.V. spectrophotometry. Combined applications of TLC and thin-layer electrophoresis [*39*] (see pp. 433—435) also appear promising (see [*14*]).

Bibliography to Chapter K. Nucleic Acids and Nucleotides

[*1*] ADLER, M., B. WEISSMANN and A. B. GUTMAN: J. Biol. Chem. **230**, 717 (1958).
[*2*] ARAKI, T.: Hoppe-Seyler's Z. physiol. Chem. **38**, 84 (1903).
[*3*] BENDICH, A., J. R. FRESCO, H. S. ROSENKRANZ and S. M. BEISER: J. Am. Chem. Soc. **77**, 3671 (1955).
[*4*] — H. B. PAHL, G. C. KORNGOLD, H. S. ROSENKRANZ and J. R. FRESCO: J. Am. Chem. Soc. **80**, 3949 (1958).
[*5*] BOULANGER, P., et J. MONTREUIL: In Chromatographie en chimie organique et biologique, Vol. II, p. 471, sous la direction de E. LEDERER. Paris: Masson et Cie., Editeurs 1960.
[*6*] BOYER, P. D., H. LARDY and K. MYRBÄCK (Editors). The Enzymes, Vol. 2, 2nd Edition. New York-London: Academic Press 1960.
[*7*] BRADLEY, D. F., and A. RICH: J. Am. Chem. Soc. **78**, 5898 (1956).
[*8*] BROWN, A. H.: Arch. Biochem. **11**, 269 (1946).
[*9*] BUCHANAN, J. G., C. A. DEKKER and A. G. LONG: J. Chem. Soc. **1950**, 3162.
[*10*] CARTER, C. E.: J. Am. Chem. Soc. **72**, 1466 (1950).
[*11*] CHARGAFF, E., and S. ZAMENHOF: J. Biol. Chem. **173**, 327 (1948).
[*12*] — E. VISCHER, R. DONIGER, C. GREEN and F. MISANI: J. Biol. Chem. **177**, 405 (1949).
[*13*] — B. MAGASANIK, E. VISCHER, C. GREEN, R. DONIGER and D. ELSDON: J. Biol. Chem. **186**, 51 (1950).
[*14*] — and J. N. DAVIDSON: (Editors) The Nucleic Acids, Chemistry and Biology, Vols. I, II, III. New York-London: Academic Press 1955, 1960.
[*15*] — In Biological Structure and Function, Vol. I, p. 67, T. W. GOODWIN and O. LINDBERG, Editors. London-New York: Academic Press 1961.
[*16*] COHN, W. E.: Science **109**, 377 (1949).
[*17*] — J. Am. Chem. Soc. **71**, 2275 (1949).
[*18*] — J. Am. Chem. Soc. **72**, 1471 (1950).
[*19*] — and C. E. CARTER: J. Am. Chem. Soc. **72**, 4273 (1950).
[*20*] — and E. VOLKIN: Nature (London) **167**, 483 (1951).
[*21*] — and D. G. DOHERTY: J. Am. Chem. Soc. **78**, 2863 (1956).
[*22*] — In Ion Exchangers in Organic and Biochemistry, p. 345, C. CALMON and T. R. E. KRESSMAN, Editors. New York: Interscience Publishers Inc. 1957.
[*23*] — In Chromatography, p. 554, E. HEFTMANN, Editor. New York: Reinhold Publishers 1961.
[*24*] — and F. J. BOLLUM: Biochim. et Biophys. Acta **48**, 588 (1961).
[*25*] CRESTFIELD, A., K. C. SMITH and F. W. ALLEN: J. Biol. Chem. **216**, 185 (1955).
[*26*] DALY, M. M., and A. E. MIRSKY: J. Biol. Chem. **179**, 981 (1949).
[*27*] DAVIS, F. F., and F. W. ALLEN: J. Biol. Chem. **227**, 907 (1957).
[*28*] DEKKER, C. A., A. M. MICHELSON and A. R. TODD: J. Chem. Soc. **1953**, 947.
[*29*] DISCHE, Z.: Mikrochemie **8**, 4 (1930).
[*30*] — and K. SCHWARZ: Mikrochim. Acta **2**, 13 (1937).
[*31*] EDMAN, P., HAMMERSTEN, E., B. LÖW and P. REICHARD: J. Biol. Chem. **178**, 395 (1949).
[*32*] EULER, H. v., u. L. HAHN: Svensk Kem. Tidskr. **58**, 251 (1946).
[*33*] FRAENKEL-CONRAT, H.: Biochim. et Biophys. Acta **49**, 169 (1961).
[*34*] FUJISAWA, K., u. K. MAKINO: In Papierchromatographie in der Botanik, S. 198, 2. Aufl., herausgegeben von H. F. LINSKENS. Berlin-Göttingen-Heidelberg: Springer-Verlag 1959.
[*35*] GIERER, A., and K. W. MUNDRY: Nature (London) **182**, 1457 (1958).
[*36*] GILHAM, P. T., and H. G. KHORANA: J. Am. Chem. Soc. **80**, 6212 (1958).
[*37*] HANES, C. S., and F. A. ISHERWOOD: Nature (London) **164**, 1107 (1949).
[*38*] HOLIDAY, E. R., and E. A. JOHNSON: Nature **163**, 216 (1949).

[39] Honegger, C. G.: Helv. Chim. Acta **44**, 173 (1961).
[40] Hotchkiss, R. D.: J. Biol. Chem. **175**, 315 (1948).
[41] Jones, W.: Hoppe-Seyler's Z. physiol. Chem. **41**, 101 (1904).
[42] Kay, E. R. M., N. S. Simmons and A. L. Dounce: J. Am. Chem. Soc. **74**, 1724 (1952).
[43] — and A. L. Dounce: J. Am. Chem. Soc. **75**, 4041 (1953).
[44] Khorana, H. G.: Some Recent Developments in the Chemistry of Phosphate Esters of Biological Interest. New York-London: John Wiley & Sons, Inc. 1961.
[45] — and J. P. Vizsolyi: J. Am. Chem. Soc. **83**, 675 (1961).
[46] Khym, J. X., and W. E. Cohn: Biochim. et Biophys. Acta **15**, 139 (1954).
[47] — D. G. Doherty and W. E. Cohn: J. Am. Chem. Soc. **76**, 5523 (1954).
[48] Kiyasu, J. Y.: Private communication, 1962.
[49] Klouwen, H. M., u. H. Weiffenbach: J. Chromatog. **7**, 45 (1962).
[50] Kornberg, A.: Nobel-Lecture 1959, Angew. Chem. **72**, 231 (1960).
[51] Kunitz, M.: J. Gen. Physiol. **24**, 15 (1940).
[52] — J. Gen. Physiol. **33**, 349 (1950).
[53] Levene, P. A., u. H. Mandel: Chem. Ber. **41**, 1905 (1908).
[54] Loring, H. S., J. L. Fairley and H. L. Seagran: J. Biol. Chem. **197**, 823 (1952).
[55] Main, R. K., and L. J. Cole: Arch. Biochem. Biophys. **68**, 186 (1957).
[56] — M. J. Wilkins and L. J. Cole: J. Am. Chem. Soc. **81**, 6490 (1959).
[57] Markham, R., and J. D. Smith: Biochem. J. **45**, 294 (1949).
[58] — — Biochem. J. **49**, 401 (1951).
[59] Marshak, A., and H. J. Vogel: J. Biol. Chem. **189**, 597 (1951).
[60] McCarty, M.: J. Gen. Physiol. **29**, 123 (1946).
[61] Mejbaum, W.: Hoppe-Seyler's Z. physiol. Chem. **258**, 117 (1939).
[62] Michelson, A. M.: J. Chem. Soc. **1959**, 1371.
[63] Nürnberg, E.: Deut. Apotheker Ztg. **101**, 142 (1961).
[64] — Deut. Apotheker Ztg. **101**, 268 (1961).
[65] Ochoa, S.: Federation Proc. **15**, 832 (1956).
[66] — Nobel-Lecture 1959, Angew. Chem. **72**, 225 (1960).
[67] Peterson, E. A., and H. A. Sober: J. Am. Chem. Soc. **78**, 751 (1956).
[68] Randerath, K.: Angew. Chem. **73**, 436 (1961).
[69] — Angew. Chem. **73**, 674 (1961).
[70] — u. H. Stuck: J. Chromatog. **6**, 365 (1961).
[71] — Biochem. Biophys. Res. Comm. **6**, 452 (1961/62).
[72] — Nature **194**, 768 (1962).
[73] — Angew. Chem. **74**, 484 (1962); Internatl. Edition **1**, 435 (1962).
[73a] Randerath, K.: Angew. Chem. **74**, 180 (1962); Internatl. Edition **1**, 553 (1962) and Private communications, 1963.
[73b] — and F. Cramer: Biochim. et biophys. Acta **61**, 346 (1962).
[74] Schmidt, G., and S. J. Thannhauser: J. Biol. Chem. **161**, 83 (1945).
[75] Schuster, H., u. G. Schramm: Z. Naturforsch. **13**b, 697 (1958).
[76] Schwander, H., u. R. Signer: Helv. Chim. Acta **33**, 1521 (1950).
[77] Seiler, H.: Helv. Chim. Acta **44**, 1753 (1961).
[78] Signer, R., u. H. Schwander: Helv. Chim. Acta **32**, 853 (1949).
[79] Smillie, R. M.: Arch. Biochem. Biophys. **85**, 557 (1959).
[80] Sober, H. A., and E. A. Peterson: J. Am. Chem. Soc. **76**, 1711 (1954).
[81] — — In Ion Exchangers in Organic and Biochemistry, p. 318, C. Colman and T. R. E. Kressman, Editors. New York: Interscience Publishers, Inc. 1957.
[82] Staehelin, M., E. A. Peterson and H. A. Sober: Arch. Biochem. Biophys. **85**, 289 (1959).
[83] — Biochim. et Biophys. Acta **49**, 11 (1961).
[84] Stahl, E., u. U. Kaltenbach: J. Chromatog. **5**, 351 (1961).
[85] Steudel, H., u. E. Peiser: Hoppe-Seyler's Z. physiol. Chem. **114**, 201 (1921).
[86] Tamm, C., M. E. Hodes and E. Chargaff: J. Biol. Chem. **195**, 49 (1952).
[87] — H. S. Shapiro and E. Chargaff: J. Biol. Chem. **199**, 313 (1952).

[*88*] Tamm, C., H. S. Shapiro, R. Lipshitz and E. Chargaff: J. Biol. Chem. **203**, 673 (1953).
[*89*] Teichert, K., E. Mutschler u. H. Rochelmeyer: Deut. Apotheker Ztg. **100**, 283 (1960).
[*90*] Tener, G. M., H. G. Khorana, R. Markham and E. H. Pol: J. Am. Chem. Soc. **80**, 6223 (1958).
[*91*] Todd, A. R.: In Perspectives in Organic Chemistry, p. 245, A. R. Todd, Editor. London-New York: Interscience Publishers, Inc. 1956.
[*92*] — Nobel-Lecture 1957, Angew. Chem. **70**, 527 (1958).
[*93*] Vischer, E., and E. Chargaff: J. Biol. Chem. **168**, 781 (1947).
[*94*] — — J. Biol. Chem. **176**, 715 (1948).
[*95*] Volkin, E., and C. E. Carter: J. Am. Chem. Soc. **73**, 1516 (1951).
[*96*] Weber, C. J.: J. Biol. Chem. **86**, 217 (1930).
[*96b*] Weimann, G., and K. Randerath: Experientia (Basel) **19**, 49 (1963).
[*97*] Weissmann, B., P. A. Bromberg and A. B. Gutmann: J. Biol. Chem. **224**, 407 (1957).
[*98*] Wieland, T., u. L. Bauer: Angew. Chem. **63**, 511 (1951).
[*98b*] — G. Lüben, and H. Determann: Experientia (Basel) **18**, 430 (1962).
[*99*] Wyatt, G. R.: Biochem. J. **48**, 581 (1951).
[*100*] — Biochem. J. **48**, 584 (1951).
[*101*] — J. Gen. Physiol. **36**, 201 (1952).
[*102*] —, and S. S. Cohen: Biochem. J. **55**, 774 (1953).

Some recent noteworthy articles

Bové, J. M.: Bull. Soc. Chim. Biol. **45**, 421 (1963). TLC of radioactively labelled adenosine phosphates.

Coffey, R. C., and R. W. Newburgh: J. Chromatog. **11**, 376 (1963). TLC on DEAE and DEAE containing calcium sulfate.

Dyer, T. A.: J. Chromatog. **11**, 414 (193). Separation of RNA hydrolysates by TLC.

Hansbury, E., and D. G. Ott: Los Alamos Scientific Laboratory, Biol. and Med. Res. Group (H-4), Annual Report 1961—1962, LAMS-2780, p. 268. Fractionation of nucleic acid constituents on cellulose and ion exchange layers.

Josefsson, L.: Biochim. et biophys. Acta **72**, 133 (1963). Semiquantitative micro-determination of nucleic acid constituents.

Mahapatra, G. N., and O. M. Friedman: J. Chromatog. **11**, 265 (1963). Separation of isomeric methylated deoxyguanosines on cellulose layers.

Randerath, E., and K. Randerath: J. Chromatogr. **10**, 509 (1963). Preparation of polyphosphate cellulose from polyethyleneimine cellulose.

— — J. Chromatogr. in press. Two-dimensional ion-exchange-TLC of nucleotides.

Randerath, K., and G. Weimann: Biochim. et biophys. Acta **76**, 129 (1963). Detection of complex formation between deoxyribooligonucleotides and polyribonucleotides.

— Biochim. et biophys. Acta **76**, 622 (1963). Separation of deoxyribonucleotides from ribonucleotides.

— and E. Randerath: J. Chromatogr. in press. Ion-exchange-TLC of nucleotides on polyethyleneimine cellulose.

— Experientia (Basel). **20**, 406 (1964). Quantitative ion-exchange-TLC of nucleic acid hydrolysates.

— and E. Randerath: Angew. Chem. **76**, 494 (1964). Enzymatic reactions on ion-exchange layers.

Sanno, Y., M. Honjo and K. Tanaka: Chem. Pharm. Bull. (Tokyo) **9**, 657 (1961). Chromatography of nucleotides on carbon.

Scheig, R. L., R. Annunziata and L. A. Besch: Anal. Biochem. **5**, 291 (1963). Separation of pyridine nucleotides on Silica Gel G layers.

L. Sugars and Derivatives

By

EGON STAHL and U. KALTENBACH

1. Introduction

PARTRIDGE showed, in 1946, that mixtures of sugars can be analyzed by paper chromatography. This observation has been confirmed and extended in over a thousand investigations. Hence, it was of interest to look into the possibility of separating this widespread group of natural products on purely inorganic thin layers. The empirical formula $(\mathrm{H}-\overset{|}{\underset{|}{\mathrm{C}}}-\mathrm{OH})_n$ of the simple carbohydrates explains the markedly hydrophilic character of these compounds. It should be possible to fractionate mixtures of sugars with polar solvent systems on inactive thin layers. Theoretically, it should be of only secondary importance whether the thin layer is already inactive, as is, for example, Kieselguhr G, or whether an active silica gel layer is largely inactivated by the passage of a polar solvent system, e.g., an aqueous solution. In practice, however, it seems that adsorption effects are not altogether precluded on Silica Gel G and Alusil[1] layers. We are inclined to believe that partition processes are mainly responsible for the separations on Kieselguhr G layers.

As with paper chromatography, the speed of migration decreases in the following order: pentoses → hexoses → disaccharides, and → trisaccharides. The furanose forms of hexoses have higher *Rf*-values than the corresponding pyranoses. This regularity, however, does not hold on Silica Gel G or Alusil layers.

One disadvantage of the Kieselguhr G layer is its small capacity (maximum of 5 μg. of sugar per spot). With Silica Gel G and Alusil layers, ten times this amount of sugar can be applied. Table 118 shows, however, that, in practice, only certain mixtures can be separated on Kieselguhr G because the sugars examined so far are not evenly distributed over the whole chromatogram, but lie together in groups (h*Rf*, e.g., 50–70, 25–40, or 30–50).

On non-impregnated Kieselguhr G layers, some sugars separate in two spots [*8*]. This observation has been made also with Silica Gel G layers by PASTUSKA and TRINKS [*4*].

A change of configuration (D, L or α, β) during chromatography was ruled out experimentally [*5*].

It is now supposed that the sugar molecules occur in one spot in the Haworth ring-form and in the other in the open-chain form.

---

[1] See footnote p. 465.

2. Layers and solvent systems

a) Kieselguhr G layers for trace analysis, according to STAHL and KALTENBACH [8]

The preparation and drying of the thin layer, 250 μ, follow the standard procedure. Kieselguhr G "Merck" is mixed with 0.02 M sodium acetate solution in place of water and spread on plates, 20/20 cm. The solvent system (No. 1) of 65 ml. ethyl acetate and 35 ml. isopropanol-water (2:1) proved best. Increasing proportion of the isopropanol-water raises the *Rf*-values and gives better resolutions of di- and tri-saccharides.

This solvent travels about 10 cm. in 25–30 minutes if the trough tank with saturation is used (p. 15). Fig. 183 shows a chromatogram obtained under these conditions; the approximate *Rf*-values are given in Table 118. Higher *Rf*-values are observed when another size of plate (e.g., 5 × 20 cm.) and round tanks are used.

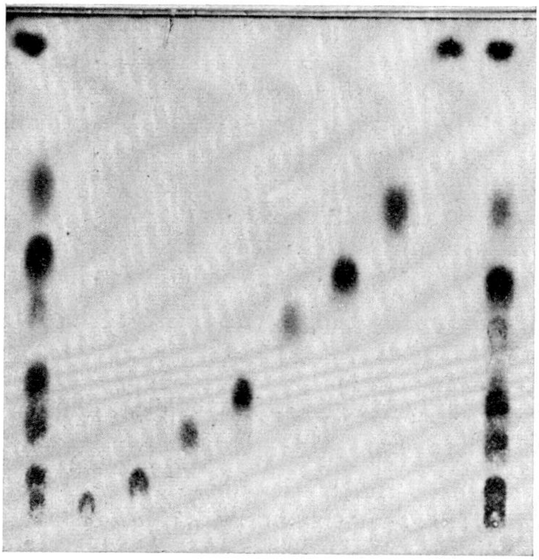

Fig. 183. Thin-layer chromatogram of sugars (0.5 μg. of each) on buffered Kieselguhr G layer. Photograph taken in transmitted light; running time for 10 cm., 25—30 min. [8]

The pure sugar is best dissolved in pyridine, and only 0.5 μg. is applied as standard. Fig. 184 shows that very small amounts of mixtures must be spotted to prevent overlapping of the various fractions.

Products containing sugars in higher concentrations, such as fruit juices, are diluted with pyridine (1:100) to give optimal concentrations (in the 0.1—1 μg. region). Malt extracts and honey are dissolved to give 0.1—0.5% solutions in pyridine.

The method is, as most separations by partition chromatography, susceptible to the presence of extraneous constituents such as inorganic salts, which may occur, for example, in body fluids. Many sugars, e.g., glucose and galactose, are separated into two zones on non-impregnated

Table 118. *Separation of Sugars by TLC under various Conditions*

Layers:	Kieselguhr G impregnated with sodium acetate	Kieselguhr G impregnated with boric acid		Alusil (Silica Gel G + Aluminum Oxide G 1+1) active layers		Alusil (Silica Gel G + Aluminum Oxide G 1+1) inactive layers	
Solvent:	I	II	III	Va	Vc	Vc	VII
hRf	Di 94	Rh 67	Rh 52	Rh 60	Rh 59	Rh 55	Rh 59
	Rh 62	Gl 63	Gl 42	Xy 57	Xy 49	Xy 49	Xy 49
	Ri 49	Sa 63	Ar 42	Gla 54	Gla 48	Ar, Fr 42	Ar, Fr 43
	Xy 39	Ar 62	Xy 39	Ar 49	So 43	Man, So 42	Man 43
	Ar 28	Xy 59	Gls 34	Gaa 49	Ar, Fr 42	Gla 42	So 42
	So 26	Man 58	Man 32	So 47	Man 41	Gl, Sa 39	Gl 39
	Fr 25	La 56	Ga 32	Fr 47	Gaa 40	Gaa 37	Ga 37
	Man 23	Ga 55	Fr 31	Man 47	Gl, Sa 39	Ga 35	Gla 33
	Ga 18	Fr 52	Sa 29	Gl 45	Ga, Tr 35	Tr 31	Sa, Gaa 32
	Gl 17	So 51	La 25	Sa 44	Mal 31	Mal 30	Mal 29
	Sa 8	Gls 48	So 24	Tr 43	Gls 27	Gls 29	La, Tr 26
	Mal 6	Gas 36	Gas 10	Ga 42	Ra 26	La 26	Ra 20
	La 4			Mal 39	La 25	Ra 23	Gls 15
				Gls 38	Gas 19	Gas 19	Gas 13
				La 34			
				Ra 32			
				Gas 27			

Ar = Arabinose
Di = Digitoxose
Fr = Fructose
Ga = Galactose
Gaa = Galactosamine
Gas = Galacturonic acid
Gl = Glucose
Gla = Glucosamine
Gls = Glucuronic acid
La = Lactose
Man = Mannose
Mal = Maltose
Ra = Raffinose
Rh = Rhamnose
Ri = Ribose
Sa = Sucrose
So = Sorbose
Tr = Trehalose
Xy = Xylose

I: Ethyl acetate-isopropanol-water (65+23.5 +11.5)
II: Benzene-glacial acetic acid-methanol (20+20 +60)
III: Methylethyl ketone-glacial acetic acid-methanol (60+20+20)
Va: n-Propanol-ethyl acetate-water-25 % ammonia solution (50+10+30+10)
Vc: as Va, but (60+10+30+10)
VII: n-Butanol-glacial acetic acid-water (60+30+10)

Kieselguhr G layers by the solvent system methyl acetate-isopropanol-water (90 + 5 + 5).

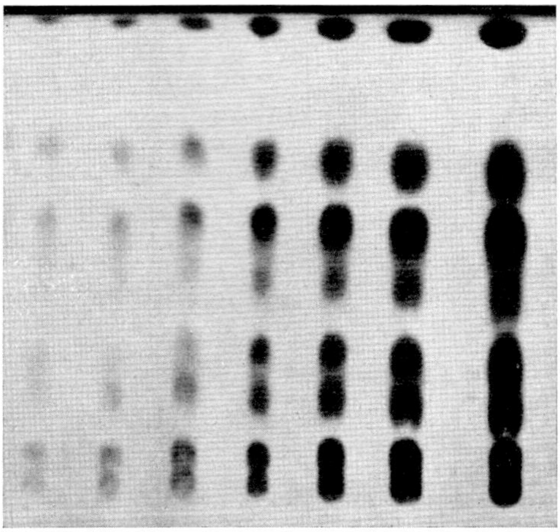

Fig. 184. Thin-layer chromatogram of a sugar mixture (compare Fig. 183). The amount applied per sugar increases from left (0.1 μg.) to right (10 μg.) [*8*]

b) Silica Gel G-boric acid layers, according to PASTUSKA [*4*]

The work so far published shows that Silica Gel G layers, prepared with 0.1 N boric acid solution instead of water, are particularly useful for the separation of mixtures of sugars. PASTUSKA [*4, 5*] used "manually" prepared layers, about 1 mm. thick, whereas PREY et al. [*6*] obtained good separations of certain sugar mixtures on standard plates with various solvents, including those proposed by PASTUSKA. The amounts applied may lie between 50–250 μg., depending on the separation desired and the amount of sugars in the mixture. It is advantageous, however, to apply only between 5 and 50 μg. of substance.

Table 119. *Solvents for TLC of Sugars on Silica Gel G Layers Prepared with 0.1N Boric Acid*

No.	Solvent system	Running time (10 cm.) [*6*]	Remarks	Ref.
II	Benzene-glacial acetic acid-methanol (20 + 20 + 60)	80 min.	Good sepn. of glucose (h*Rf* 63) from fructose (52)	[*4*]
III	Methylethyl ketone-acetic acid-methanol (60 + 20 + 20)	60 min.	Good sepn. of glucose (h*Rf* 48) from sucrose (35)	[*4, 6*]
IV	Butanol-acetone-water (40 + 50 + 10)	60 min.	Good sepn. of sucrose and glucose (both h*Rf* 34) from fructose (14)	[*6*]

Solvent system IV was used also for the separation of di- and trisaccharides. The following h*Rf*-values were reported: raffinose 6, D-melibiose 17, lactose 26, melicitose 33, maltose and sucrose 44–45 [*6*]. The only way of separating a mixture of glucose, fructose, sucrose and raffinose was by working in the two-dimensional technique using solvent systems I and III, consecutively.

c) Silica Gel G- and Alusil[1]-layers, according to WALDI [*9*]

WALDI developed another method for the rapid separation of sugars, sugar-alcohols, sugar-acids, and amino-sugars. He used standard plates, coated either with Silica Gel G or with Alusil[1], i.e., a mixture of Silica Gel G (acidic) and Alumina Oxide G (basic). He distinguished "active" layers, which were dried for 30 min. at 110° C., and "inactive" Alusil layers. The latter were simply air-dried at room temperature. The sugars were applied as 1% aqueous solutions in amounts of 50 μg. The plates were developed in tanks saturated with solvent vapors.

The following solvent systems proved suitable:

α) For "active" and "inactive" Alusil layers

Va = n-Propanol-ethyl acetate-water-25% aqueous ammonia (50 + 10 + 30 + 10);

Vb as Va but 60 + 10 + 20 + 10 by volume;

Vc as Va but 60 + 10 + 30 + 10 by volume.

WEICKER and BROSSMER [*10*] observed at least two spots when they chromatographed pure sugars on Silica Gel G with the basic solvent system n-propanol-conc. ammonia solution-water (60 + 20 + 10). Supposedly, the use of an ammonical solvent system and the catalytic effect of the adsorbent lead to the formation of amino-sugars. The compounds yield positive reactions with ninhydrin (Reagent No. 108) and with the Morgan-Elson reagent (No. 45). This finding should be borne in mind when using the above solvent systems.

β) For "inactive" Alusil layers

VI = n-Propanol-ethyl acetate-water-glacial acetic acid (40 + 10 + 40 + 10);

VII = n-Butanol-glacial acetic acid-water (60 + 30 + 10).

γ) For Silica Gel G layers

VIII = Chloroform-acetone-glacial acetic acid (60 + 30 + 10) *(separation of acetone sorboses)*.

IX = n-Propanol-ethyl acetate-water (70 + 20 + 10) [*6*].

h*Rf*-values of sugars obtained with solvent systems Va, Vc and VII are given in Table 118; those for sugar alcohols are collected in Table 120.

The acetone sorboses can be separated on Silica Gel G plates using solvent system VIII: 1,2-monoacetone sorbose h*Rf* 25–35, 2,3-monoacetone-sorbose h*Rf* 35–45, diacetone sorbose h*Rf* 80.

[1] Alusil = notation for a mixture of Alumina G + Silica Gel G (1 + 1).

Table 120. *Separation of Sugar Alcohols by TLC [9]*

Layers Solvent systems	Rf × 100			
	active Alusil		inactive Alusil	
	Va	Vc	Vc	VII
Sorbitol[1]	41	35	38	37
Mannitol[1]	46	41	42	39
Inositol	29	21	—	—
Dulcitol[1]	45	40	38	36
Adonitol	52	51	46	42
Erythritol	61	57	52	47

[1] WALDI [9] had previously been able to achieve a useful separation of the sugar alcohols dulcitol, mannitol and sorbitol by paper chromatography only. Mannitol is separated from sorbitol and dulcitol on potassium chloride-impregnated S & S paper 2043 b using the continuous flow technique (50 hrs.); solvent system: ethyl acetate-pyridine-water (60 + 30 + 10). Dulcitol is separated from mannitol/sorbitol on non-impregnated paper by the continuous flow technique (48 hrs.) with ethyl acetate-pyridine-water (14 + 5 + 1). Fine-grained cellulose layers were not tried.

Silica Gel G layers and solvent system IX were used to separate mixtures of raffinose and sucrose [6]. It was shown that 1 part of raffinose can be detected in 50 parts of sucrose. The proportion of these compounds is distinctly less favorable, however, in sugar-beet molasses, which contain 1 part in 100 or less. A modification of the wedged-tip technique (p. 35) was found useful for resolving such mixtures. The wedge-path was scraped on the plate as a V-shape with a definite angle (Fig. 185). Chromatography on such shaped areas gave good fractionations of 10–40% molasses solutions.

In a publication [11] dealing with the detection of small amounts of trehalose (h*Rf* 30), it was reported that this compound could be separated from glucose (h*Rf* 50). Water-saturated *sec.* butanol served as the developing solvent. JENSEN mentions the possibility of separating a series of sugars, sugar acids and alginate hydrolyzates [2], and he gives a rather unconvincing illustration of a chromatogram of this type. GÖRLICH [1] used n-butanol-methanol-glacial acetic acid-water (40 + 40 + 5 + 5) for the separation of glucose from oleandrose on Silica Gel G. LICHTI et al. [3a] used TLC successfully in elucidating the structure of the sugar component of digitoxin.

3. Visualization

The many spray reagents known for the detection of sugars in paper chromatography can also be used without difficulty on thin layer chromatograms (Reagents Nos. 7, 8, 9, 10, 18, 45, 67, 110, 116, 120, 135, 137, and 145).

One should, however, choose the most sensitive, and bear in mind that these reagents are not specific for this group of substances. Aniline phthalate (Reagent No. 8) gives good results, for example, with reducing sugars. These compounds are rendered visible after heating as yellow-brown-fluorescing spots under UV-light (365 mμ). PASTUSKA [4] showed that mono- and disaccharides and also uronic acids can be detected with 1,3-dihydroxynaphthalene-(naphthoresorcinol-) sulphuric acid reagent (similar to No. 45). The limit of detection lies below 4 μg. for the aniline phthalate reagent. The distinguishing colors are given in Table 121. The anisaldehyde-sulphuric acid reagent (No. 9) of STAHL and KALTENBACH [7] is considerably more sensitive. As little as 0.05 μg. of sugar can be detected with this reagent. For reaction colors with amounts of 1 μg., see Table 121. This method of detection fails, however, on boric acid-impregnated layers. According to WALDI [9], mono- and diacetone sorboses can be visualized with phosphomolybdic acid reagent (No. 120).

Table 121. *Color-Reactions of some Sugars on Thin-Layer Plates [8, 4]*

Sugar	Anisaldehyde-H_2SO_4 (No. 9) Kieselguhr G layer impregnated with sodium acetate	Naphthoresorcinol-H_2SO_4 (No. 44) Silica Gel G layer impregnated with boric acid
D (+)-Digitoxose	blue	—
L (+)-Rhamnose	green	green
D (—)-Ribose	blue	—
D (+)-Xylose	grey	light blue
L (+)-Arabinose	yellow-green	blue-green
L (—)-Sorbose	violet	red
D (—)-Fructose	violet	red-black
D (+)-Mannose	green	light blue
D (+)-Glucose	light blue-grey	blue-violet
D (+)-Galactose	grey-green	blue-violet
Sucrose	violet	red
Maltose	violet	—
Lactose	greenish	red-violet
D (+)-Glucuronic acid	—	blue
D (+)-Galacturonic acid	—	blue

4. Quantitative determination

PASTUSKA [4] described a procedure for the quantitative titrimetric determination of sugar in zones scraped off chromatograms. The method is based on a wet combustion with potassium dichromate-sulphuric acid and titration of the unused dichromate. The amount of sugar required for a determination is about 0.5 mg. (!).

The zones containing sugar are scraped off the developed and dried chromatogram (layers 0.9 mm. thick) treated with an excess of 0.05 or 0.01 N potassium dichromate solution in 70% sulphuric acid and heated to 90° C. for 60 minutes on a water bath. Then, the solution is cooled and 20 ml. of water and 5 ml. of 5% KI solution are added. After 20 minutes, exactly, the iodine liberated is titrated with 0.01 N sodium thiosulphate. A blank analysis is carried out with an equal area scraped off the plate at the same height as the sample under test.

Standardization of dichromate solution: The titre of the dichromate solution is determined by carrying out the experiments with a 2% glucose solution.

The limits of error for the method are given as below 5%.

5. Thin-layer chromatography of sugar derivatives

The TLC of suitable derivatives is another way of identifying sugars: examples are the hydrazones as mentioned briefly by REICHSTEIN et al. [3]:

Preparation of β-naphthylhydrazones: 1 mg. of sugar is refluxed for 5 minutes with 1 mg. of freshly-sublimed β-naphthylhydrazine in 0.1 ml. methanol. The methanol is removed by distillation with ether. The hydrazone separates on cooling, and after washing with ether and drying, the melting point can be determined.

β-Naphthylhydrazones are applied to the plate as methanol solutions. They are chromatographed on Silica Gel G with ethyl acetate-n-butanol (90 + 10) and the position of the substance is located in UV-light.

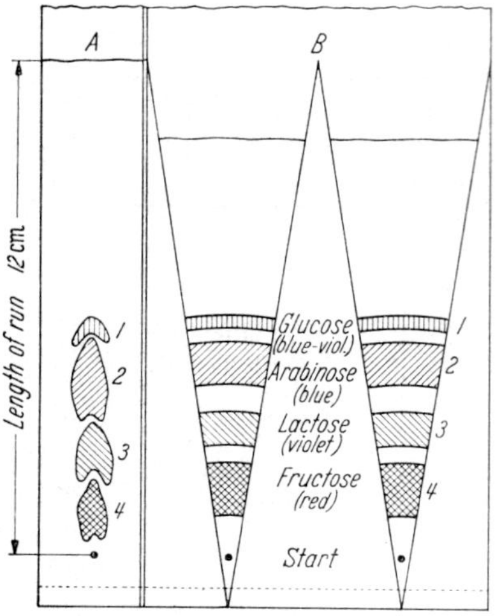

Fig. 185. Thin-layer chromatogram of sugars found in the analysis of urine. A = normal technique, B = wedged-tip technique. For details, see text. (According to RINK and HERMANN [7])

According to TATE and BISHOP [8a], a good micro-preparative separation of sugar acetates can be obtained on Silica Gel G by double development with benzene-methanol (96 + 4) and 17 cm. length of run.

6. Special applications

The method described by STAHL and KALTENBACH [8] was developed for the ultramicro analysis of sugar mixtures with the aim of separating the contents of individual plant or animal cells or small cell-complexes. It is unsuitable for larger amounts of sample, such as 5–100 μg. For analysis of the molasses produced in the beet sugar process, one should use the procedure of PREY et al. [6] which was developed for this purpose. In addition, the method for the quantitative estimation proposed by PASTUSKA [4] may be applied. RINK and HERMANN [7] have worked out a procedure for the rapid determination of sugars occurring, under pathological conditions, in urine.

The layer is prepared by the standard procedure with Silica Gel G and 0.1 M boric acid solution, 1 + 2 w/v, and dried. The solvent system, n-butanol-acetone-0.1 M boric acid, (40 + 50 + 10), is run to a height of 12 cm. A reference mixture of fructose, lactose, arabinose and glucose can be applied alongside the urine sample. The optimal load is about 5 μg. of each sugar. The wedged-tip technique already mentioned proved very advantageous, as shown in Fig. 185. The naphthoresorcinol-sulphuric acid reagent (similar to No. 45) was used for detecting the sugars.

Bibliography to Chapter L. Sugars and Derivatives

[1] GÖRLICH, B.: Planta Medica. **9**, 451 (1961).
[2] JENSEN, A.: TIDSSKR. Kjemi, Bergvesen Metallurgi **1**, 14 (1961).
[3] KOWALEWSKI, Z., O. SCHINDLER, H. JÄGER u. T. REICHSTEIN: Helv. Chim. Acta **43**, 1280 (1960).
[3a] LICHTI, H., M. KUHN u. A. v. WARTBURG: Helv. Chim. Acta **45**, 868 (1962).
[4] PASTUSKA, G.: Z. analyt. Chem. **179**, 427 (1961).
[5] —, u. H. TRINKS: Private communication (1961).

[6] PREY, V., H. BERBALK and M. KAUSZ: Mikrochim. Acta H. 6, 968 (1961).
[6a] — — — Mikrochim. Acta H. 3, 459 (1962).
[7] RINK, M., u. S. HERMANN: Private communication.
[8] STAHL, E., u. U. KALTENBACH: J. Chromatog. **5**, 351 (1961).
[8a] TATE, M. E. and C. T. BISHOP: Can. J. Chem. **40**, 1043 (1962)
[9] WALDI, D.: Private communication.
[10] WEICKER, H., u. R. BROSSMER: Klin. Wochschr. **39**, 1265 (1961).
[11] WYSS-HUBER, M., H. JÄGER u. EK. WEISS: Helv. Chim. Acta **43**, 1010 (1960).

Some recent noteworthy articles

ANET, E. F. L. J.: J. Chromatog. **9**, 291 (1962). Dinitrophenylhydrazones of carbonyl compounds.
DEFERRARI, J. O., R. MUCHNIK DE LEDERKREMER, B. MATSUHIRO and J. F. SPROVIERO: J. Chromatog. **9**, 283 (1962). Acyl derivatives of sugars.
GEE, M.: Anal. Chem. **35**, 350 (1963). Methylated glycosides.
GROSSHOF, H.: Dtsch. Apoth. Ztg. **103**, 1396 (1963). TLC of sugars on magnesium silicate layers.
HAY, G. W., B. A. LEWIS and F. SMITH: J. Chromatog. **11**, 479 (1963). Sugars and their derivatives.
KUHN, R., and H. WIEGANDT: Chem. Ber. **96**, 866 (1963). Gangliosides.
MICHEEL, F., and O. BERENDES: Mikrochim. Acta. **1963**, 519. Sugars and their derivatives.
PREY, V., H. SCHERZ and E. BANCHER: Mikrochim. Acta **1963**, 567. Carbohydrates.
RAGAZZI, E., and G. VERONESE: Il Farmaco **18**, 152 (1963). TLC of sugars on Silica Gel G.
RINK, M., and S. HERRMANN: J. Chromatog. **12**, 416 (1963). Sugars occuring in urine.
TATE, M. E., and C. T. BISHOP: Can. J. Chem. **40**, 1043 (1962). Carbohydrate acetates.
TORE, J. P.: J. Chromatog. **12**, 413 (1963). TLC of sugars on calcium silicate layers.
SCHWEIGER, A.: J. Chromatog. **9**, 374 (1962). TLC of sugars on cellulose layers.
WEILL, C. E., and P. HANKE: Anal. Chem. **34**, 1736 (1962). TLC of malto-oligosaccharides on Kieselguhr G.

M. Thin-Layer Chromatography of Inorganic Ions

By

H. SEILER

I. General

The foregoing chapters describe how many mixtures of organic substances can be rapidly and efficiently separated on thin layers of adsorbent. It was of interest to investigate the possibility of applying thin-layer chromatography to the separation of mixtures of inorganic ions. As early as 1949, MEINHARD and HALL [2] used a technique known as "*Surface Chromatography*" for separating simple mixtures of ferric and zinc salts.

Apprehension that thin layers may flake off when polar solvents are used, thus making inorganic thin-layer chromatography virtually impossible has proved unfounded. It has been shown, in fact, that even aqueous solvents can be used without damaging the layers. This is true whether gypsum, starch, or agar is employed as binder.

Difficulties have arisen, however, from impurities in the commercial silica gel[1]-preparations, which may contain a certain quantity of iron. This can cause considerable distortion on the chromatogram. It has thus proved necessary to purify the adsorbents (see p. 475). This purification procedure, for which hydrochloric acid is used, not only removes unwanted ions but also activates the silica gel as the treatment with acid etches the surface of the gel particles.

II. Theoretical considerations in inorganic thin-layer chromatography [*6*]

In order to obtain criteria relative to the factors governing inorganic thin-layer chromatography, a series of tests was carried out, the results of which are reproduced in Figs. 186a–d and 187 a–b.

These experimental data can be interpreted in the light of the following considerations:

Silica gel acts as a cation exchanger, as UMLAND and KIRCHNER [*13*] were able to show in column chromatography. It is for this reason that, in tests with perchloric acid, the three cations Cu^{2+}, Co^{2+}, and Ni^{2+} give nearly the same *Rf* values. In this case, the rates of migration are determined only by the exchange process, which depends on the pH and ion charge. The influence of the ion charge is evident from tests with Fe^{2+} and Fe^{3+}. *Rf*-Values vary considerably if perchloric acid is added, but when hydrochloric acid is added, there is almost no variation in *Rf*-values. In a hydrochloric acid medium [*3*], chloro-complexes are formed, whereas the perchlorate ion does not form complexes. Hence, the original cations appear as negatively charged anions, and the cation exchange character of silica gel is of no significance in experiments with higher concentrations of hydrochloric acid. Partition takes place according to the order of stabilities of the ionic complexes. As the concentration of Cl-ions in the solvent decreases constantly from bottom to top, this influences the point of equilibrium of the complex formation reaction. Thus, the *Rf* values decrease as follows: – Cu > Co > Ni.

These data also show that two cations may be separated even though they give very similar *Rf* values. If applied separately, they displace each other, with the lower shifting the higher forward. No overlapping occurs, and, in fact, unoccupied zones form between the spots.

This can be explained by the fact that when the exchanger is charged with a cation, H^+-ions are discharged, and this causes a decrease in pH at the border zone of the area occupied by this cation. The free H^+-ions then push forward the cation less firmly bound by the coating material.

Figs. 187a and b show the influence of solvents containing an organic component on the chromatographic behavior of cations. Dioxane, tetrahydrofuran, n-butanol, isopropanol, n-propanol, acetone, ethanol and methanol were each used as solvents with a content of 10% N-hydrochloric acid (Figs. 187a) and with 10% N-perchloric acid (Fig. 187b).

[1] Machery, Nagel & Co., Düren, Germany, is introducing an iron-free preparation.

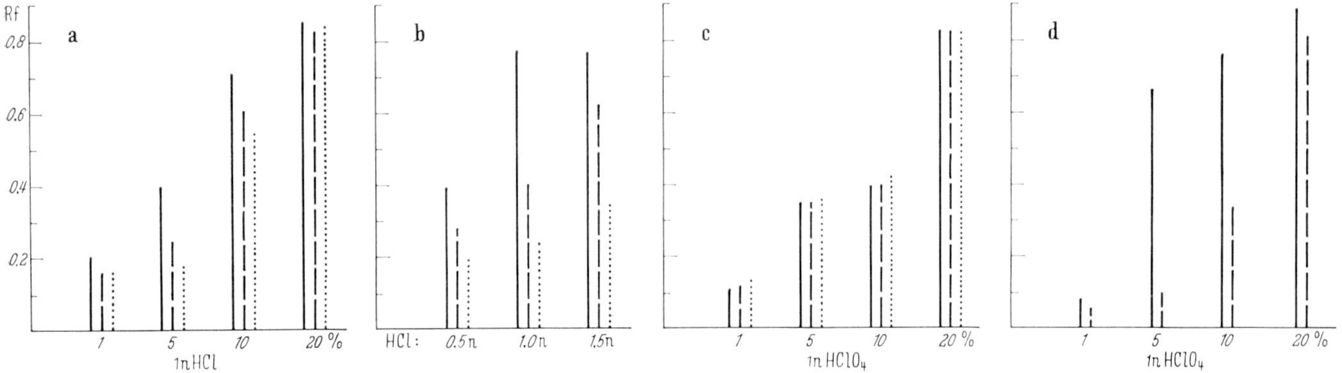

Fig. 186a—c. *Rf* Values of Cu^{2+} (= ——), Co^{2+} (= — —) and Ni^{2+} (= · · · ·) in acetone with addition of acid. Fig. 186d. *Rf* Values of Fe^{2+} (= ——) and Fe^{3+} (= — —) in acetone with increasing quantities of N $HClO_4$

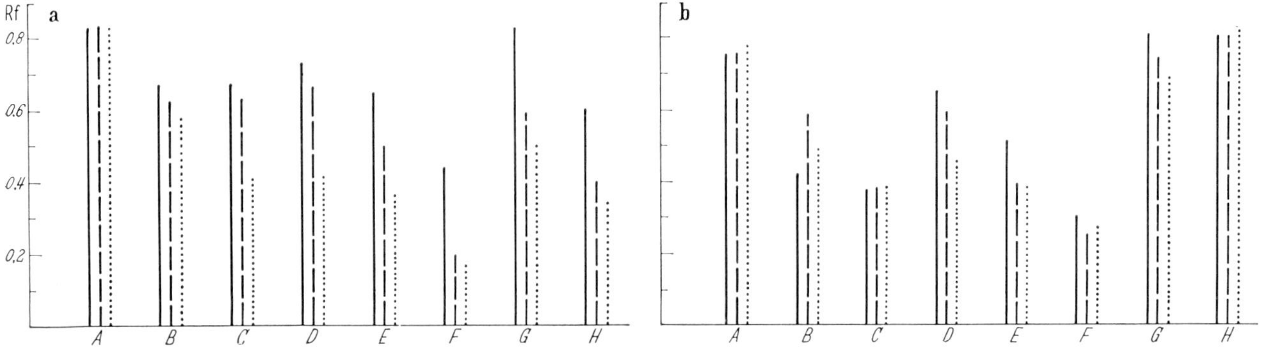

Fig. 187. *Rf* Values of Cu^{2+} (= ——), Co^{2+} (= — —) and Ni^{2+} (= · · · ·) in solvents *A*—*H* with addition of 10% N HCl (Fig. 187a) and 10% N $HClO_4$ (Fig. 187b). *A* = methanol, *B* = ethanol, *C* = acetone, D = n-propanol, *E* = isopropanol, *F* = n-butanol, *G* = tetrahydrofuran, *H* = dioxane

In the experiments with hydrochloric acid, all solvents, with the exception of methanol, gave the same sequence of *Rf* values, i.e., Cu, Co, Ni. In the test with methanol, all three cations showed identical *Rf* values.

A similar sequence was not detectable with perchloric acid and the same organic solvents. Obviously, the selectivity of the ion exchanger can be influenced, to a very marked degree, by the type of solvent used. This has already been found by Sansoni [5] with nonaqueous systems. According to Sansoni, the basicity of the solvent – in Lewis' sense of the term – plays a decisive role. Yet another factor has to be taken into account in the present study relating to the use of aqueous solvents. This factor is the solubility of cation hydrates in the organic component of the solvent. As is evident from Fig. 187 b., selectivity on the part of the ion exchanger and, thus, the ability to separate equivalent ions, can be achieved also with perchloric acid by using a suitable organic component.

In summary, it is obvious, from these experiments, that the ion exchange character of silica gel and the complex formation of cations with the solvents are determining factors in the thin-layer chromatography of inorganic cations.

III. Preparation of sample solutions and plates for thin-layer chromatography

It is generally advisable to pre-separate the mixture to be analyzed into the classical groups of chemical analysis. The following procedure has proved most satisfactory for this purpose.

1. Separation procedure

A solution of 0.1—0.2 g. substance in 3—4 ml. *aqua regia* (3 parts 6 M HCl and 1 part 6 M HNO_3) is diluted with 2—3 ml. of distilled water. If precipitation occurs, the suspension is centrifuged and the deposit washed twice with 0.5 ml. distilled water. The remaining solution and the washing solutions are combined and transferred in a 25 ml. beaker. Five drops of 3% H_2O_2 are added, and the solution is boiled carefully for 2 minutes. After cooling, 6 M NH_3 solution is dropped into the solution until it gives an alkaline reaction. The sample solution is then acidified carefully with 6 M HCl, and a further 6.5 ml. 6 M HCl are added. The solution is transferred to a 25 ml. measuring cylinder, and made up to 19 ml. with distilled water. 1 ml. M thioacetamide solution is added, and the solution is stirred well and heated for 10 minutes in a 50 ml. beaker on a water bath. The precipitate of sulphides is centrifuged off, washed twice with 1 ml. distilled water, and the supernatant and washing water are combined. The precipitate is suspended in 2 ml. distilled water, 2 ml. of a 15 M NH_3 solution is added, and the mixture is heated on a water bath for 10 minutes. One ml. of 2 M thioacetamide solution is added, and the suspension heated for another 10 minutes. It is then centrifuged, washed and the supernatant and washing water discarded. The precipitate containing the cations of the Cu-group is mixed with 2 ml. 6 M HNO_3, boiled, and the resulting solution diluted with distilled water to a concentration of approximately 3% in regard to the sulphides. Between 0.001 and 0.003 ml. of this solution (I) is applied to the plate and chromatographed.

The remaining solution and washing water from the precipitate obtained with thioacetamide, in acidic medium, are concentrated to 2—3 ml., neutralized with 6 M NH_3 solution, and 1 ml. 2 M thioacetamide solution is added.

Diagram Showing Separation into Analytical Groups

Organic Material

Mineralization

Cations, *dissolve mineral substance in aqua regia*

acid precipitation with thioacetamide

Centrifuge: *Treat precipitate with concentrated NH_3 and thioacetamide, dissolve residue in 6 M HNO_3.*
Cu-group = solution I

alkaline precipitation with thioacetamide

Centrifuge: *Dissolve precipitate in 6 M HCl*
$(NH_4)_2S$-group = solution II

precipitation with ammonium carbonate

Centrifuge: *Dissolve precipitate in 6 M acetic acid*
Alkaline earth group = solution III

Evaporate to dryness,
heat residue dry,
dissolve in small portion of 6 M acetic acid.
Alkali group = solution IV

Anions, *mineral substance, fuse with soda, dissolve in distilled water*

Amberlite IR-120 H^+-form

Evaporate until acid vapors evolve, neutralize with NaOH.

Two ml. of 6 M NH_3 solution and 10 ml. distilled water are added and heated for 10 minutes on the water bath. The precipitate containing the cations of the $(NH_4)_2S$-group is centrifuged out, washed twice with 1 ml. distilled water, and dissolved with as little as possible of 6 M HCl. It is heated on the water bath until no more H_2S is given off, and diluted with distilled water until the solution is approximately 3%. About 0.001—0.003 ml. of this solution (II) is applied and chromatographed.

The remaining solution and the washing water from the last precipitation are acidified with 6 M HCl, neutralized with 6 M NH_3 solution, and a further 1 ml. 6 M NH_3 solution is added. Then, 2—3 ml. ammonium carbonate solution is added, heated for 10 minutes on the water bath, cooled, and the solution left for approximately 30 minutes with intermittent shaking. The precipitate is centrifuged and washed twice with 1 ml. of distilled water. The precipitate is dissolved in 6 M acetic acid so as to give an approximate 3% solution. This solution (III) contains the cations of the ammonium carbonate group. Between 0.001 and 0.003 ml. of the solution is applied to a plate and chromatographed.

The supernatant and washing water from the ammonium carbonate precipitation are combined, acidified with 6 M acetic acid, and carefully evaporated to dryness in a porcelain dish. The residue is heated for 10—15 minutes on a clay triangle over an open flame. After cooling, the residue is dissolved by heating in as small amount as possible of 6 M acetic acid, and diluted with enough distilled water to give an approximate 3% solution. This solution (IV) contains the alkali and magnesium ions. From 0.001 to 0.003 ml. of this solution is applied to a plate and chromatographed.

2. Treatment and mineralization

A different treatment is necessary if the substance to be analyzed is present in a form insoluble in *aqua regia*.

If the sample contains fairly large amounts of phosphate or tartrate, for instance, it is essential to remove these anions, which exercise a disturbing influence both on the separation process and on the chromatogram. This is best achieved by means of an ion exchanger in chloride form (e.g. Amberlite IR-400, Cl-form). A slightly acid solution of the sample is fed to a sufficiently large exchanger column and allowed to flow through at a rate of 0.5—1 ml. per minute. The exchanger is then washed with a quantity of distilled water approximately twice the volume of the column.

In many cases where the determination of one single ion is called for, preliminary separation into analytical groups is not necessary. For instance, uranium in rocks can be separated by chromatography with a considerable number of different cations, and identified [*7*], without the accompanying ions interfering. A small sample of rock is melted in a mixture of sodium fluoride and potassium hydrogen sulphate on a platinum wire. The resulting bead is crushed in a few drops of 4.7 N HNO_3, and increasing quantities of the solution are applied to a plate.

If the inorganic ions to be studied are present in biological material, mineralization will be necessary. It is advisable to carry out a test run in order to determine the amount of inorganic material in the sample. A few grams of the substance are heated in a platinum crucible until the residue appears white. It is advantageous to dry the material first for several hours in a drying cabinet at 120—150° C. Mineralization should, if possible, be carried out in a retort furnace at approximately 600° C. This way, the possible formation of an insoluble film in the crucible will be avoided. For the final combustion, sufficient material should be used to give an inorganic residue of 0.1—0.2 g.

There are two methods used for the final combustion: dry combustion, where the material is treated in a platinum crucible without an oxidizing agent being added, and wet mineralization, which involves oxidation with concentrated nitric acid.

For dry combustion, the material is weighed and dried first and then heated in the platinum crucible to approximately 600° C. until it appears almost white. The residue is then mixed with concentrated nitric acid, and if determination of potassium ions is not required, a spatula-pointful of potassium chlorate is added and the mixture is carefully heated. The sample is evaporated to dryness, mixed again with concentrated nitric acid, and the residue is dissolved by heating. The evaporated acid is replaced from time to time. Eventually, the mixture is evaporated to dryness, and the residue dissolved in distilled water. Further treatment can be carried out as described under "C". If traces of heavy metals are to be detected, the acid used for mineralization must first be distilled as it may contain ions from the glass bottle as a result of being left standing for some time. The instruments used for distilling must be boiled out for several days with the corresponding acid.

Wet mineralization should preferably be done either in Pyrex long-necked round-bottom flasks or in a special apparatus [*4*]. All glass equipment should be boiled out for several days before use with concentrated nitric acid. The material is first dried and weighed, then placed in the flasks. Ten ml. of fuming nitric acid per gram dry matter is added, and the whole heated carefully on a sand-bath; reddish-brown fumes evolve. The reaction slows down, but heating is continued until the acid is distilled off. The residue is mixed again with nitric acid. This must be done with care as the material may ignite thus leading to loss of material. Evaporation to dryness is repeated until the residue is pure white. Finally, the sample is dissolved in diluted nitric acid, whereupon group separation (see above) can proceed.

The process of wet combustion can be modified by mixing the material with a few drops of concentrated sulphuric acid before adding the nitric acid; this speeds up the process, though, if alkaline earths are present, sulphates will form that are difficult to dissolve.

Wet mineralization has the advantage over dry combustion of not forming films that are difficult to dissolve. A disadvantage, however, is that stable nitro compounds are formed from proteins that may be present and these have to be destroyed by means of dry combustion. If inorganic ions in material containing proteins are to be determined, it is advisable to precipitate the protein first, preferably with a 10% solution of trichloroacetic acid. The solution is centrifuged, washed with 10% trichloroacetic acid, and the residual solution and washing fluid evaporated to dryness. The residue is dissolved in diluted hydrochloric or nitric acid and further processed.

The preparative methods described so far are mainly suitable for determining cations. If anions are to be determined, the following method is recommended: the material is dried and weighed (for quantity, see above) and heated to approximately 400° C. This does not give complete mineralization. The residue is thoroughly mixed with approximately eight parts of analytical grade Na_2CO_3 and fused for three hours in a torch. The solidified melted material is extracted by boiling in distilled water, and filtered from insoluble matter. As the large quantity of Na^+-ions could affect the separation of anions, it is advisable to exchange the cations for H^+ with a cation exchanger (e.g., Amberlite IR-120). The capacity of the exchanger must be assessed before use. A given quantity of freshly regenerated exchanger is mixed with an excess of NaCl solution. This is shaken for about 30 minutes, filtered free from resin, washed with distilled water, and the H^+-ions are titrated with an NaOH solution using phenolphthalein as an indicator. One-and-a half times the calculated quantity of exchanger is then added to the solution to be analyzed, shaken for approximately 30 minutes, then the resin filtered off is washed two or three times with distilled water. The combined filtrates are carefully concentrated in an evaporation dish using an IR lamp until acid vapors are removed, as indicated by a red coloration of moistened litmus paper when held over the dish. Diluted NaOH is used for exact neutralization. The solution is then made up to volume with distilled water.

3. Purification of Silica Gel G for thin-layer chromatography of inorganic ions [8]

As already mentioned, it has proved essential to purify the adsorbent. We have found that the method described below is the simplest and gives sufficiently clean material for chromatography of inorganic ions.

500 g. "Silica Gel G" (Merck) is stirred with a mixture of 500 ml. concentrated HCl and 500 ml. distilled water in a measuring cylinder. Small amounts of gas evolve at this stage, and the solution turns yellow due to the presence of Fe^{+++}. Dark particles (carbon) may be observed on the surface of the solution. The suspension should be left overnight to allow most of the silica gel to settle. The remaining solution, which is yellow and slightly opaque, is decanted, and the solid phase suspended in 1000 ml. distilled water. When the residual solution is still only slightly opaque, it is decanted again. This washing with distilled water is repeated three times. The silica gel is eventually filtered off by suction and washed with distilled water until the filtrate gives a neutral reaction. It is then further washed with approximately 300 ml. ethanol and finally with approximately 300 ml. benzene. The purified silica gel is dried in an oven at 120° C. for 24 hours.

It is advisable to sieve the silica gel when cleaned and dried, as the purification process may cause nodules which might give rise to difficulties when preparing the plates. For this purpose, we have used a mechanical sieve with 350 mesh = 0.045 mm.

The majority of the calcium sulphate originally present as binder will have passed into solution during the cleaning process. The silica gel prepared in this way can thus be used to obtain plates with various binders.

4. Preparation of layers

28 g. silica gel, cleaned as described above, is thoroughly mixed with 2 g. $CaSO_4 \cdot 2H_2O$, or 2 g. soluble starch, and 50 ml. distilled water and then an additional 10 ml. distilled water is added. It is important to ensure that the suspension is completely homogeneous. The slurry is now applied with the spreader. The amount specified is sufficient for coating five 20 × 20 cm. plates or twenty 5 × 20 cm. plates. The adsorbent-coated plates are dried at room temperature, and then in an oven for 2 hours at 110° C. The plates should be stored over $CaCl_2$ in a desiccator.

IV. Thin-layer chromatography of groups of cations

The following procedures all employ the one-dimensional ascending technique. We have found that it is generally not possible to give a *Rf* value in inorganic thin-layer chromatography as in paper chromatography, because the ratio of migration of the ions to the rate of migration of the solvent front may vary with the moisture content of the adsorbent layer. Relative movements of adjacent individually applied ions are constant. However, in a mixture of several ions, this relative movement is variable, as the different ions often displace each other. These observations have led us to abandon the idea of stating *Rf* values, but to give only the sequence of migration of ions, which remains constant. The individual ions can be readily identified on a chromatogram by the common detection reactions. In the following procedures, the pre-separation of ions into groups by classical methods is assumed.

The best method of application is to spot 2 or 3 different concentrations of the solution to be analyzed. This makes it possible to detect traces of an element as well as to detect occlusion due to excessive concentrations of another element. In any case, it is advisable to run a standard mixture of ions on the same chromatogram.

In all experiments, the chromatography jars are lined with filter paper to saturate them with vapor of the developing solvent.

1. Fractionation of the Cu-group (solution I) [8]

A mixture of 100 ml. n-butanol, 20 ml. 1.5 N HCl and 0.5 ml. acetonylacetone is used as solvent to separate the ions of the Cu-group. Tailing is significantly reduced by adding a weak complexing agent. If a stronger complexing reagent is used, e.g., acetylacetone, the cations are all shifted towards the solvent front, i.e., into the relatively nonpolar part of the solvent partition. In the solvent mixture mentioned, the sequence of ion movement is as follows (Fig. 188): Hg > Bi > Cd > Pb > Cu.

Layer: Silica gel-gypsum (see above).

Standard samples: 0.002 ml. of 0.1 M solutions of $Hg(NO_3)_2 \cdot 2H_2O$; $Cd(CH_3COO)_2 \cdot 2H_2O$; $BiONO_3$; $Pb(NO_3)_2$; $Cu(CH_3COO)_2 \cdot H_2O$; applied as separate spots.

Duration of run: Approximately 2 hours.

Detection: spray with 2% KI solution, dry, hold over conc. ammonia and insert the plate into a tank filled with H_2S gas (Table 122).

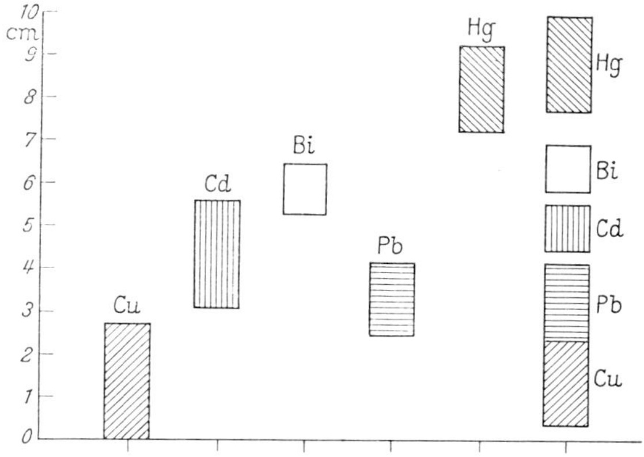

Fig. 188. Fractionation of the Cu-group

Table 122. *Color Reactions of Ions in the Cu-Group*

Ion	KI solution	H_2S
Hg^{II}	red	brown-black
Bi^{III}	brown-yellow	brown-black
Cd^{II}		yellow
Pb^{II}	yellow-brown	brown
Cu^{II}	brown	dark brown

2. Fractionation of the $(NH_4)_2S$-group (solution II) [8]

A mixture of 100 ml. acetone, 1 ml. conc. HCl and 0.5 ml. acetonylacetone is used as solvent to separate the ions of the $(NH_4)_2S$-group. Better separation effects can be achieved if a small crystallizing dish filled with water is put in the tank. The solvent mentioned above gives the following sequence for ion movement (Fig. 189): Fe > Zn > Co > Mn > Cr > Ni > Al.

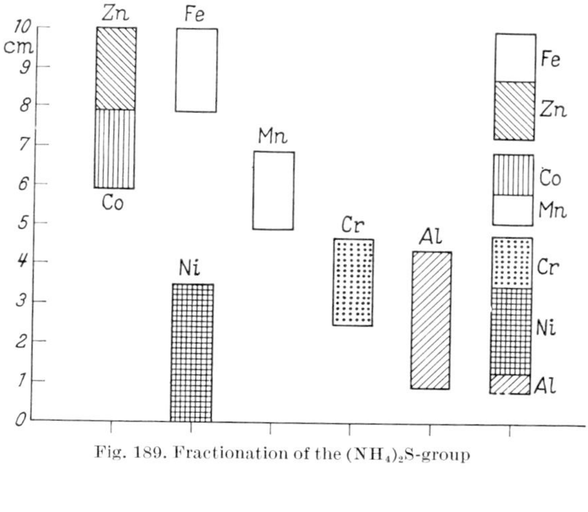

Fig. 189. Fractionation of the $(NH_4)_2S$-group

Layer: Silica gel-gypsum (see p. 476).

Standard samples: 0.002 ml. of 0.1 M solutions of $NiSO_4 \cdot 7\,H_2O$; $Co(NO_3)_2 \cdot 6\,H_2O$;

$ZnSO_4 \cdot 7 H_2O$; $MnSO_4 \cdot H_2O$; $Al(CH_3COO)_3$; $CrCl_3 \cdot 6 H_2O$ and $FeCl_3$, applied as separate spots.

Detection: The plates are held over conc. ammonia and sprayed with a solution of 0.5 g 8-hydroxyquinoline in 100 ml. of 60% ethanol. The resulting colors are evaluated in UV-light (Table 123).

Table 123. *Color Reactions of the Ions in the $(NH_4)_2S$-Group*

Ion	NH_3	Oxin	UV
Fe^{III}		brown	dark
Zn^{II}		pink	yellow
Co^{II}	blue	yellow	dark
Mn^{II}		orange	dark
Cr^{III}	green		dark
Ni^{II}			dark
Al^{III}			light yellow

3. Fractionation of the ammonium carbonate group (solution III) [*9*]

Starch must be used as a binder because the presence of gypsum in the adsorbent layer would lead to the formation of insoluble sulphates[1].

The ions in this group must be present in the form of their acetates. This does not present a problem as the acetates are easily obtained by dissolving the carbonates in 6 N acetic acid.

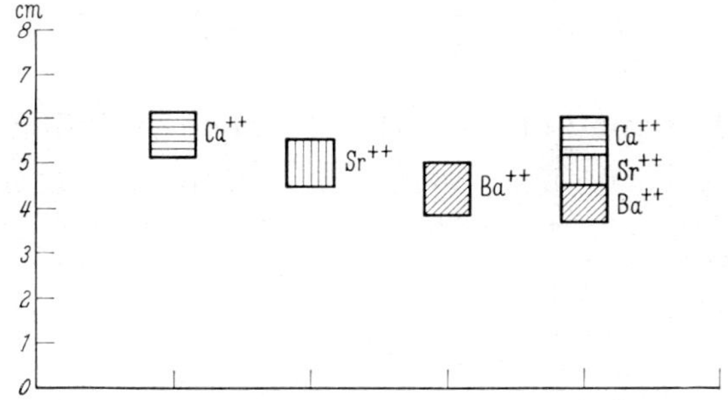

Fig. 190. Fractionation of the ammonium carbonate group

Difficulties do arise, however, when separating this group, due to the fact that marked tail formation occurs when calcium is present. This can be prevented by using an additional 0.001 ml. glacial acetic acid with each application. With a solvent composed of 37.5 ml. ethanol, 37.5 ml. n-propanol, 5 ml. glacial acetic acid, 1 ml. acetylacetone and 20 ml. distilled water, the following sequence of ion movement occurs: Ca > Sr > Ba (Fig. 190).

Layer: Silica gel-starch (see p. 476).

[1] Silica gel with starch as a binder is available from Machery, Nagel & Co., Düren, Germany.

Standard samples: 0.001 ml. of M solutions of Ca-, Sr-, and Ba-acetate, with additional 0.001 ml. glacial acetic acid each time; applied as separate spots.

Duration of run: 80–90 minutes.

Detection: 1.5% solution of violuric acid in distilled water. After spraying, heat in a drying cabinet for 20 minutes at 100° C. to give more distinct coloring (Table 124).

Table 124. *Color Reactions of Ions in the Ammonium Carbonate-Group*

Ion	Ca^{2+}	Sr^{2+}	Ba^{2+}
Color with violuric acid . .	yellow-orange	pink	red-violet

4. Fractionation of the alkali group (solution IV) [*10*]

The presence of gypsum in the layer material has a disturbing effect on the detection of alkali group ions, which is carried out by means of violuric acid. Therefore, starch should be used as a binder. The cations in this group are to be present in the form of their acetates. Alkali salts of strong acids can be treated by means of anion exchangers in their acetate form, but this leads to an increase in volume and makes supplementary concentration necessary.

We attempted, therefore, to carry out the anion exchange on the plate itself. It became evident that barium ions remain at the starting point in the solvent used. The proper procedure is to apply a quantity of barium acetate roughly equivalent to the expected quantity of alkali ions, and then the alkali sulphates on top of this. Salt formation takes place, and the alkali ions can be separated and identified by means of violuric acid. A red stain, caused by reaction with the Ba^{2+}, appears at the starting point. The solvent is allowed to rise higher (15 cm.) than when applying acetates (10 cm.). A solvent composed of 100 ml. absolute ethanol and 1 ml. glacial acetic acid gives the following sequence for ion movement (Fig. 191): Li > Mg > Na > K.

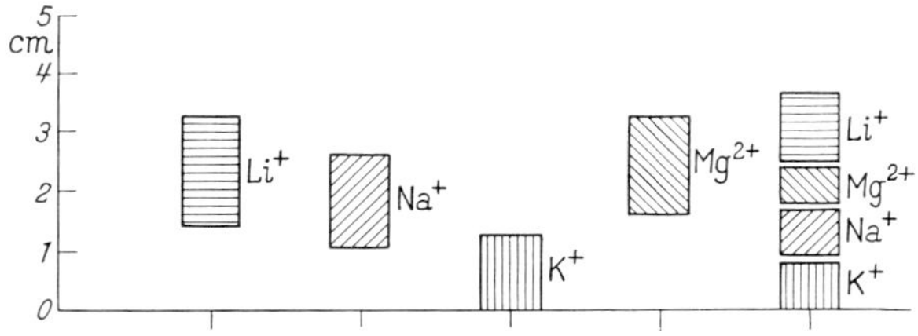

Fig. 191. *Fractionation of the alkali-group*

Layer: Silica gel-starch (see p. 476).

Standard samples: 0.001 ml. of M solutions of Li-, Na-, K-, Mg-acetate, slightly acidified with acetic acid; applied as separate spots.

Duration of run: Acetates 50 minutes, sulphates 70 minutes.

Detection: 1.5% solution of violuric acid in distilled water. The temperature of the water should not rise above 60° C. when dissolving the violuric acid. Spray, then heat for 20 minutes in a drying cabinet at 100° C. for more distinct coloration (Table 125).

Table 125. *Color Reactions of Ions in the Alkali-Group*

Ion	Li^+	Mg^{2+}	Na^+	K^+
Color with violuric acid	light red	yellow-orange	red-violet	blue-violet

5. Separation of U^{VI} and Ga^{III} from cation mixtures [7]

It may be of interest to detect a single ion on its own, without reference to other ions. UO_2^{2+}, for example, can be separated and identified from a mixture of Fe, Cu, Co, Ni, Cr, Al, and Th ions even if the concentration of accompanying ions is considerably greater than that of uranium. Preliminary separation into analytical groups is not necessary. In this case, silica gel-gypsum layers are used for chromatography. A mixture of 50 ml. freshly distilled ethyl acetate and 50 ml. water-saturated ether, with 2 ml. tri-n-butylphosphate added, is used as solvent. The solutions to be applied should be 4.7 N in regard to HNO_3. With this solvent, a uranyl nitrate-(tri-n-butylphosphate) solvate complex forms. Other cations do not develop comparable complexes at this pH. They are, in fact, retained at the starting point, or shortly above it, while the uranium complex is displaced towards the solvent front and can be detected with pyridyl-azo-naphthol (Fig. 192). As little as 1 μg. uranium can be separated and detected satisfactorily.

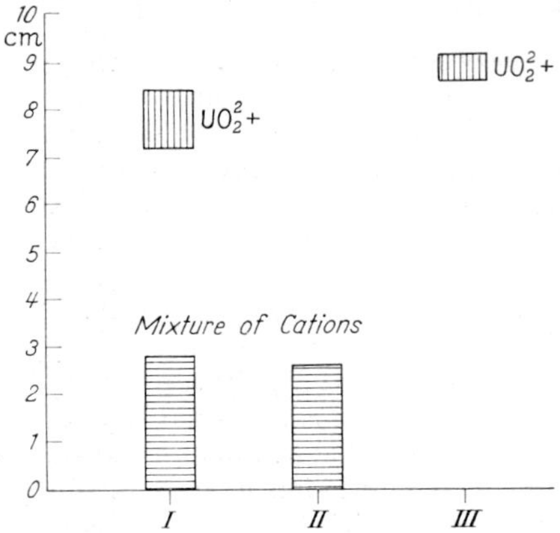

Fig. 192. Separation of UO_2^{2+} from a mixture of cations

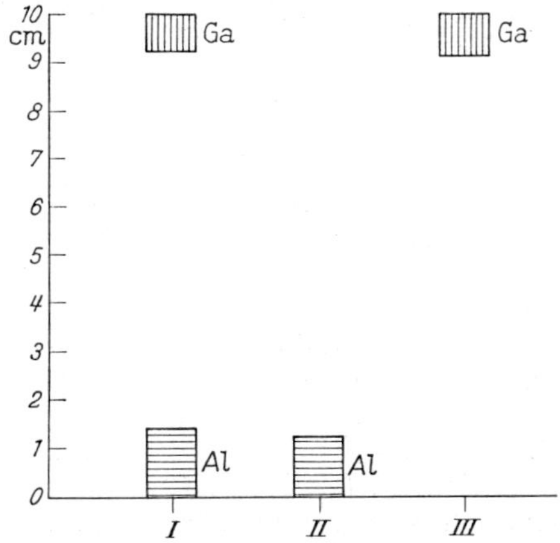

Fig. 193. Separation of Al^{3+} and Ga^{3+}

Layer: Silica gel-gypsum (see p. 476).

Standard samples: 0.001 to 0.004 ml. of a solution of 0.527 g. $UO_2(NO_3)_2 \cdot 6H_2O$ in 50 ml 4.7 N HNO_3.

Duration of run: 10–15 minutes.

Detection: By means of 0.25% solution of pyridyl-azonaphthol in ethanol.

Silica gel-gypsum layers are also used for separating Ga^{3+} from a large excess of Al^{3+} [7]. A mixture of 100 ml. freshly distilled acetone and 0.5 ml. concentrated HCl is used as the developing solvent. $GaCl_3$, which is markedly non-polar, forms in the solvent and it is displaced into the nonpolar section of the solvent partition, while Al^{3+} remains at the starting point. Both are detectable in UV light after spraying with an oxine solution. This method also makes it possible to separate as little as 0.5% Ga^{3+} from Al^{3+} reliably and to identify 1 μg. Ga^{3+} (Fig. 193).

Layer: Silica gel-gypsum (see p. 476).

Duration of run: 10–15 minutes.

Detection: Spray with 0.5% solution of 8-hydroxyquinoline in 60% ethanol, hold over conc. ammonia, and observe in UV light.

V. Thin-layer chromatography of anions

1. Fractionation of halides [11]

Silica Gel G layers are used for fractionating the Na and K salts of the halides. Identification can be affected either by means of ammoniacal $AgNO_3$ and fluorescein and observation in UV light, or by using a pH indicator, bromocresol purple. Zirconium-alizarin is used in detecting fluoride. The solvent for separating halide anions must be more polar than that used for separating cations. This is related to the fact that cations solvate to a marked degree. Cations form solvated systems with solvents other than water, a facility which anions possess to a considerably lesser extent. With iodide ions, migration is also found in nonpolar solvents, and this is due to a partial oxidation of the iodide to iodine – a corresponding yellow stain is visible without the need for detection reagents. In a solvent composed of 65 ml. acetone, 20 ml. n-butanol, 10 ml. conc. ammonia and 5 ml. distilled water, the following sequence of ion movement is found (diag. 194):

$$F' < Cl' < Br' < I'.$$

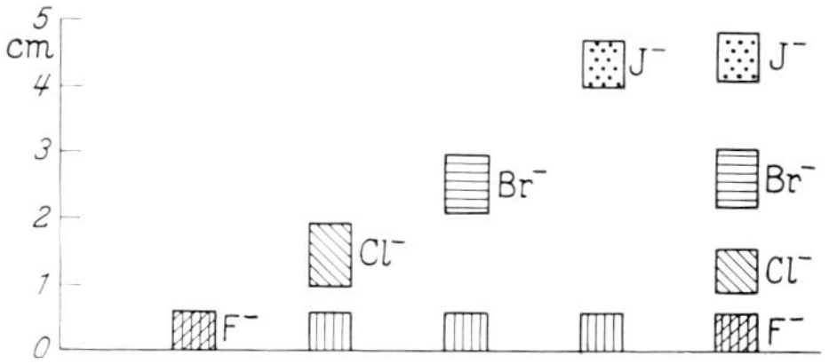

Fig. 194. Separation of halides

When using a pH indicator as detection reagent, it is apparent that the anions run in the form of ammonium salts whereas the alkali ions remain at the starting point. The ammonium salts give light yellow stains and the alkali ions blue zones at the starting point. The fluoride ion remains at the starting point and is detected by means of zirconium alizarin.

Layer: Silica gel-gypsum.

Standard samples: 0.001 ml. of M solutions of NaF, NaCl, KBr, and KI.

Duration of run: 30–40 minutes.

Detection: 1. 0.1% solution of bromocresol purple in ethanol containing a few drops of dilute aqueous ammonia.

2. 1% solution of ammoniacal silver nitrate and 0.1% solution of fluorescein in ethanol.

3. 0.1% solution of zirconium alizarin lake in strong hydrochloric acid.

2. Fractionation of phosphates [*12*]

As with halide ions, the solvent must be significantly more polar when separating anions than when separating cations.

Chromatography on Silica Gel G layers leads to the formation of insoluble calcium phosphate, and this prevents separation of the different phosphoric acid anions. For this reason, silica gel-starch layers are used, as with alkali cations.

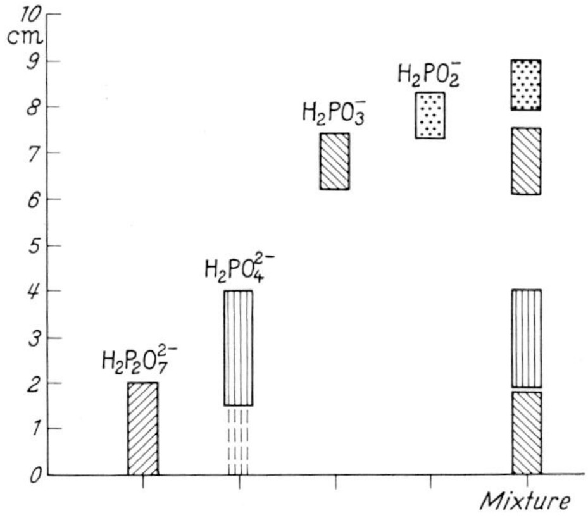

Fig. 195. Separation of phosphates

The sodium salts of pyro- and orthophosphoric acids, as well as of phosphorous and hypophosphorous acids, were used. We investigated both acidic and basic solvent systems. With acid solvents – varying amounts of trichloroacetic acid in various alcohols – a certain tendency to migration was observed, though separation of the anion mixture was unsatisfactory.

Ammoniacal-alcoholic solvents yielded satisfactory separation. Best results were obtained with a solvent containing trichloroacetic acid as a buffer for NH_3. Of the various alcohols used, the systems containing methanol gave the best results, particularly in regard to speed. A mixture of methanol-conc. ammonia-10% trichloroacetic acid-water (50 + 15 + 5 + 30) proved the best solvent (Fig. 195).

Phosphoric acid anions are detected as blue zones by means of ammonium molybdate and stannous chloride.

Layer: Silica gel-starch (see p. 476).

Standard samples: 0.001 ml. of 0.1 M solutions of $Na_2H_2P_2O_7$; NaH_2PO_4; $NaH_2PO_3 \cdot 2.5\ H_2O$ and $NaH_2PO_2 \cdot H_2O$; applied as separate spots.

Duration of run: 50–60 minutes.

Detection: The plates are dried, and sprayed first with a 1% solution of ammonium molybdate in water and then with a 1% solution of stannous chloride in 10% hydrochloric acid. This gives blue spots; but in the case of hypophosphite, the blue coloration may only appear after some time.

I wish to thank my wife, M. SEILER, and also, Prof. H. ERLENMEYER and Dr. B. PRIJS for the great interest they have shown in this work and for their invaluable advice.

Bibliography to Chapter M. Thin-Layer Chromatography of Inorganic Ions

[1] KÜNZI, P.: Dissertation, University of Basel 1962.
[2] MEINHARD, J. E., and N. F. HALL: Analyt. Chemistry **21**, 185 (1949).
[3] MOORE, G. E., and K. A. KRAUS: J. Amer. chem. Soc. **74**, 843 (1952).
[4] PEDRETTI, E.: Ber. schweiz. bot. Ges. **68**, 103 (1958).
[5] SANSONI, B.: Angew. Chem. **66**, 330 (1954); Z. Elektrochem. **57**, 161 (1953); Z. Naturforsch. **11** b, 117 (1956).
[6] SEILER, H.: Helv. chim. Acta **45**, 381 (1962).
[7] SEILER, H. und M.: Helv. chim. Acta **44**, 939 (1961).
[8] — Helv. chim. Acta **43**, 1939 (1960).
[9] SEILER, H.: Unpublished.
[10] —, u. W. ROTHWEILER: Helv. chim. Acta **44**, 941 (1961).
[11] —, u. T. KAFFENBERGER: Helv. chim. Acta **44**, 1282 (1961).
[12] — Helv. chim. Acta **44**, 1753 (1961).
[13] UMLAND, F., u. K. KIRCHNER: Z. anorg. allgem. Chem. **280**, 211 (1955); — G. M. SCHWAB, u. K. JOCKERS: Angew. Chem. **50**, 546 (1937); — R. KUNIN, and R. J. MYERS: "Ion Exchange Resins". New York: John Wiley & Sons Inc.; London: Chapman & Hall. 2nd ed. 1958, S. 59.

Some recent noteworthy articles

MERKUS, F. W. H. M.: Pharm. Weekblad **98**, 947 (1963). TLC of inorganic ions on cellulose.
RÖSSEL, T.: Z. analyt. Chem. **197**, 333 (1963). TLC of condensed phosphates.
SEILER, H.: Helv. Chim. Acta **46**, 2629 (1963). Direct quantitative determination of inorganic cations separated by TLC.
—, B. BIEBRICHER u. H. ERLENMEYER: Helv. Chim. Acta. **46**, 2636 (1963). Separations of *cis-trans* isomeric Co-complexes by TLC.

N. Spray Reagents for Thin-Layer Chromatography

By

D. WALDI

I. Notes on spraying

It is important to apply the reagent solution on to the layer in the finest possible dispersion, i.e., in the form of an aerosol. Home-made sprayers are frequently inadequate for this purpose, since the droplets they produce are too large. Sprayers, such as are now available commercially, for use with ninhydrin (Reagent No. 108), aniline phthalate (Reagent No. 8), bromocresol green (Reagent No. 22), etc., are ideal for this purpose[1]. Another appliance, which appears to be very suitable,

[1] E. Merck, AG, Darmstadt, Germany; Research Specialities Co., Richmond, California, U.S.A.; Mann Research Laboratories, New York 6, N.Y. U.S.A.

is the "Laboratory Spray-Gun"[1]. This sprayer has a propellent-gas container, and the spraying reagent in the glass flask may be changed as required (Fig. 196).

In our work, we apply a specific procedure in order to obtain sprays of even density, which are essential for quantitative estimation. The jet is slowly applied onto the chromatogram three times from a distance of 30–40 cm., in the manner illustrated in Fig. 197. A suitable sprayer has been described by L. REIO [J. Chromatogr. 1, 338 (1958)].

Note: It is difficult to recognize substances on layers of dark adsorbents such as charcoal, ferric oxide etc. The problem was resolved by G. HESSE and M. ALEXANDER who uniformly sprayed a Silica Gel G-ligroin suspension on to the dark, dry layer with an atomizer used for fixatives. A white surface is obtained on which the usual spray reagents can be used.

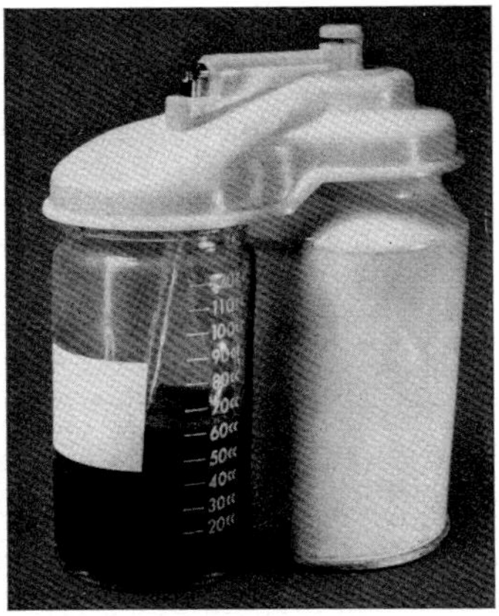

Fig. 196. Propellent-gas sprayer with changeable glass flask for the spraying solution

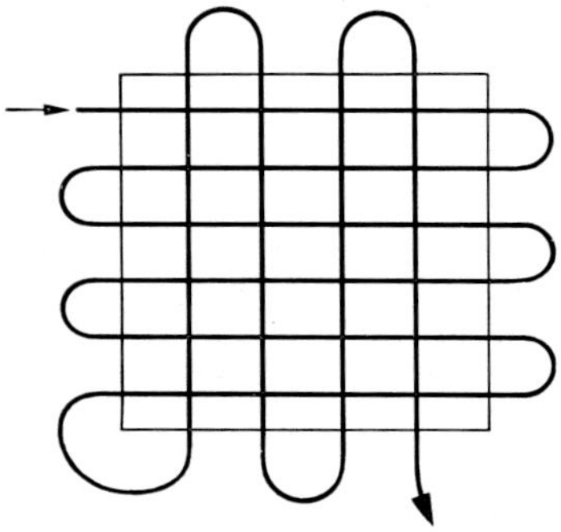

Fig. 197. Spraying diagram. The jet travels over the surface in the direction of the arrow

II. Notes on the preparation of reagents

The reagents used in TLC must satisfy certain standards of chemical purity. It is, therefore, beneficial to employ chemicals specially tested for use in chromatography[2]. The greatest caution must be observed in the preparation of some reagents, and only trained personnel should be allowed to perform this task. As examples, we may point out the danger of poisoning when handling bromocyanogen (Reagent No. 88) and the danger of explosion when handling diazotized sulphanilic acid (No. 37) and Tollens'-reagent (No. 137). It is strongly recommended to wear protective eye-shields when working with highly reactive sprays such as chromic-sulphuric acid solution, *aqua reagia*, antimony chloride, etc., which are often used in TLC. The same precaution should be taken when inspecting chromatograms under filtered UV-light (there is danger of explosion from mercury high-pressure burners).

---

[1] Shandon, London, Great Britain.

[2] See the pamphlet: „Anfärbereagentien für die Dünnschicht- und Papierchromatographie", E. Merck, A. G., Darmstadt, Germany.

Below are given instructions for the preparation and use of recognized spray reagents.

III. Preparation and use of spraying reagents

1. **Alizarin for lithium-, calcium-, magnesium-, aluminum-, thorium-, zirconium-, and ammonium ions, and for selenium.**
Spray I: Saturated alcoholic solution of alizarin.
Spray II: *N*-Sodium hydroxide.
Procedure: Spray chromatogram with I, dry for a short period and then spray with II.
Final treatment: Place chromatograms in a tank saturated with ammonia vapors.

2. **Aluminium chloride for flavonoids.**
Spray: 1% Solution of aluminium chloride in ethanol.
Procedure: Spray chromatogram and inspect under filtered UV-light.

3. **Ammonium molybdate-citrate buffer for vitamin C.**
Solution I: 15% Solution of ammonium molybdate in 1% solution of aqueous ammonia.
Solution II: (Buffer pH 3.8): 48.1 ml 0.1 N hydrochloric acid is mixed with 51.9 ml of a 0.1 M sodium citrate solution (21.008 g. citric acid + 200 ml N-Na OH per liter).
Spray: Mix 15 ml. of I and 10 ml. of II. Add 15 drops sulphuric acid (*d* 1.84) (Stable for not more than 2 days). Reagent No. 42 is more sensitive.

4. **Ammonium molybdate-stannous chloride for phosphoric acids.**
Spray I: Aqueous solution of 1% ammonium molybdate.
Spray II: 1%-Solution of stannous chloride in a 10% solution of HCl.
Procedure: Spray with I and after drying, spray with II; if necessary, heat 3—5 mins. at 105° C.

5. **Ammonium thiocyanate-ferrous sulphate for peroxides.**
Spray: 0.2 g. Ammonium thiocyanate is dissolved in 15 ml. acetone. Immediately before use, add 10 ml. of a freshly prepared 4% solution of aqueous ferrous sulphate.
Note: The presence of peroxide is only indicated if brownish red spots or zones appear immediately. After some time, the entire layer will assume a brownish red hue.

6. **Ammonium thiocyanate for the detection of peroxides.**
Spray I: Dissolve 0.4 g. thiocyanate in 30 ml. acetone.
Spray II: Dissolve 1.2 g. ferrous sulphate in 30 ml. water.
Procedure: Spray first with I, dry for a short time and finally spray with II.

7. **Aniline-phosphoric acid for reducing sugars.**
Spray: 1 Volume of a 2 N aniline solution in water-saturated n-butanol is mixed with 2 volumes of a solution of 2 N o-phosphoric acid in n-butanol.
Treatment: Heat chromatogram to 105° C. for 10 mins.

8. **Aniline phthalate for reducing sugars.**
Spray: Dissolve 0.93 g. aniline and 1.66 g. o-phthalic acid in 100 ml. water-saturated n-butanol.
Treatment: Heat the chromatograms after spraying to 105° C. for 10 mins.

9. **Anisaldehyde-sulphuric acid for steroids, terpenes, sugars, etc.**
Spray: Freshly prepared solution of 5 ml. anisaldehyde in 50 ml. glacial acetic acid, with addition of 1 ml. of sulphuric acid (*d* 1.84).
Treatment: Heat to 100—110° C. for 5—10 mins. The pink background is brightened by treatement with water vapor (from a steam bath). (Lichenous substances, phenols, terpenes, sugars and steroids will stain violet, blue, red, grey, or green.)

Modifications:

a) *Spray:* Add 1 ml. anisaldehyde to 97 ml. glacial acetic acid, then add 2 ml. concentrated sulphuric acid.
Treatment: Heat to 120° C. for 6 mins.

b) Dissolve 0.5 ml. anisaldehyde in a mixture of 10 ml. glacial acetic acid + 85 ml. methanol. Add 5 ml. conc. sulphuric acid. To detect terpene derivatives etc., spray about 10 ml. on the 20 × 20 cm. plate and heat to 100° C. for 10 mins.

c) For the location of sugars: Use a freshly prepared mixture consisting of 0.5 ml. anisaldehyde + 9 ml. ethanol (95%) + 0.5 ml. conc. sulphuric acid + 0.1 ml. glacial acetic acid. After spraying, heat to 90—100° C. for 5—10 mins.

10. Anthrone-reagent for ketoses.
Spray: Dissolve 0.3 g. anthrone in 10 ml. glacial acetic acid. Add 20 ml. ethanol (96%), 3 ml. phosphoric acid (*d* 1.7) and 1 ml. water. The solution will keep for some weeks in a refrigerator.
Procedure: After spraying, heat the chromatogram to about 110° C. for 5—6 mins. Oligosaccharides containing ketoses, as well as ketoses, appear as yellow spots.

11. Antimony trichloride for steroid glycosides and vit. A. (Carr-Price reagent).
Spray: A saturated antimony trichloride solution in chloroform or carbon tetrachloride is freshly prepared (ca. 22%).
Treatment: Heat chromatogram after spraying to 100° C. for 10 mins. and inspect in filtered UV-light.

12. Antimony trichloride-glacial acetic acid for steroids.
Spray: 1 part by weight of antimony trichloride is dissolved in 1 part, by weight, of glacial acetic acid.
Treatment: Heat to 95° C. for 5 mins.

13. Antimony pentachloride for terpenes, essential oils, resins, etc.
Spray: 2 parts by weight of antimony pentachloride are mixed with 8 parts by weight, of carbon tetrachloride.
Procedure: After spraying, the chromatoplates are heated to 120° C. until spots appear.

14. Aurin (Rosolic acid) tricarboxylic acid, ammonium salt. For Al-, Cr-, Li-ions.
Spray: 0.1% Solution of ammonium salt of aurin tricarboxylic acid in a 1% solution of aqueous ammonium acetate.
Treatment: Place chromatogram in a tank saturated with vapors of aqueous ammonia.

15. Benzidine for terpene aldehydes, flavonoids, carbohydrates.
Spray: 0.5 g. Benzidine is dissolved in 20 ml. glacial acetic acid and 80 ml. ethanol.
Treatment: Heat to 100° C. for 15 mins. (Vanillin turns yellow to orange). With some substances, the staining of spots (or fluorescence) is intensified if dilute HCl is sprayed on after heating.

16. Benzidine for the detection of persulphates.
Spray: Dissolve 50 mg. benzidine in 100 ml. N-acetic acid. Persulphates stain blue after spraying.

17. Benzidine-copper sulphate for pyridine monocarboxylic acids.
Spray I: 0.3 g. Copper sulphate is dissolved in 100 ml. of a mixture consisting of 5 vols. water and 4 vols. ethanol.
Spray II: 0.1% Solution of benzidine in ethanol (50%).
Procedure: Spray chromatogram with I, dry at 60° C,, then spray with III (blue spots).

18. Benzidine-sodium metaperiodate for acids, sugars and sugar alcohols.
Spray I: 0.1% Aqueous solution of sodium metaperiodate.

Spray II: To a solution of 2.8 g. benzidine in 80 ml. ethanol (96%), add 70 ml. water, 30 ml. acetone and 1.5 ml. N-HCl.
Procedure: Spray with I, then spray the half-dry chromatogram with II.

19. Benzoyl chloride-zinc chloride for the detection of steroids.
Spray *I:* Solution of 20 g. zinc chloride in 30 ml. glacial acetic acid.
Spray II: 50 g. Benzoyl chloride is dissolved in chloroform. Make up to 100 ml. with chloroform.
Procedure: Spray with I, heat to 90° C. for 5 mins. The dry chromatogram is sprayed with II. Heat to 90° C. for 2—3 mins. (Inspect colors in visible and under filtered UV-light).

20. Basic lead acetate for uronic acids and flavonoids.
Spray: Filter a saturated solution of aqueous lead acetate.
Treatment: Dry at 110° C. for 10 mins.

21. Boric acid-citric acid for quinolines.
Spray: 0.5 g. Boric acid and 0.5 g. citric acid are dissolved in 20 ml. methanol.
Treatment: Heat to 100° C. for 10 mins. (Inspect in filtered UV-light, 8-hydroxyquinoline is yellowish-green).

22. Bromocresol green as indicator-reagent[1].
Spray: 0.04 g. Bromocresol green is dissolved in 100 ml. ethanol (96%). 0.1 N NaOH is added until a blue coloration just appears.

23. Bromocresol purple for halide ions.
Spray: To a 0.1% solution of bromocresol purple in ethanol, add a few drops of dilute ammonia until the color just begins to change.

24. Bromothymol blue for lipids.
Spray: Dissolve 0.04 g. bromothymol blue in 100 ml. of a 0.01 N-NaOH solution.

25. Ceric sulphate-sulphuric acid (modified reagent acc. to Sonnenschein) for alkaloids, iodine containing organic compounds and tocopheryl acetates.
Spray: 0.1 g. Ceric sulphate is soaked in 4 ml. water. Add 1 g. trichloroacetic acid, boil, slowly add (drop by drop) sulphuric acid (*d* 1.84), until the solution clarifies.
Procedure: Heat to 110° C. for a few minutes until spots appear.
Note: The reagent stains the alkaloids apomorphine, brucine, colchicine, papaverine and physostigmine. It may also be used for the detection of organic iodine-containing compounds and for tocopheryl acetates.

26. Chargaff's reagent, phosphomolybdic acid-stannous chloride for choline and choline-containing substances.
Spray I: 1 g. Phosphomolybdic acid is dissolved in 100 ml. of a mixture consisting of equal volumes of ethanol and chloroform.
Spray II: 1 g. Stannous chloride is dissolved in 100 ml. 3 N HCl. Prepare freshly before use.
Procedure: Spray with I, dry for 3 mins., spray with II, dry for 10 mins.

27. Quinidine, general acid reagent.
Spray: 0.3% Solution of quinidine in chloroform.
Pretreatment: If the chromatogram has been developed with a solvent containing volatile acids (non-volatile acids must not be used), it must be heated to 60—80° C. for about 30 mins. in circulating air.
Procedure: After copious spraying, the chromatogram is heated for 10 mins. to 110—120° C. in a drying oven, and then inspected under UV-light.

28. Quinidine-copper sulphate reagent for barbituric and thiobarbituric acids.
Spray: 200 mg. Copper sulphate, 2 ml. pyridine and 20 mg. quinidine are dissolved in 100 ml. of water.
Treatment: The dried chromatogram is exposed to hydrochloric acid vapors. Dark spots in filtered UV-light.

[1] Available in sprayers by E. Merck A. G., Darmstadt, Germany.

29. p-Quinone for detecting ethanolamine.
Spray: 0.5 g. p-Quinone (benzoquinone) is dissolved in a mixture of 10 ml. pyridine and 40 ml. n-butanol.
Note: Red spots of ethanolamine appear immediately after spraying. Choline does not react.

30. Chloramine T for detecting coffeine.
Spray I: Aqueous 10% solution of chloramine T.
Spray II: N-HCl.
Procedure: Spray with I and, after short drying, with II. Heat to 96—98° C. until chlorine odor disappears. Place plate in a humid tank saturated with ammonia (approx. 5 mins.), followed by short heating until maximal appearance of pink-red spots.

31. Chloramine-trichloroacetic acid for digitalis glycosides.
Spray: 10 ml. of a freshly prepared aqueous 3% solution of chloramine is mixed, before use, with 40 ml. of a 25% solution of trichloroacetic acid in ethanol (96%). (Trichloroacetic acid solution keeps for some days).
Procedure: After spraying, heat for 7 mins. at 110° C. in a drying oven. Under filtered UV-light, blue spots appear; glycosides of the A-series yield yellow spots.

32. Chlorine-toluidine test.
Chlorination: Put the plate in a chlorine atmosphere (chlorine from a cylinder: 5—10 mins., or chlorine obtained from equal proportions of 10% HCl and a 1.5% $KMnO_4$ solution: 15—20 mins.) and leave it standing in the air for 3—5 mins., to remove excess chlorine.
Spray: 160 mg. o-Toluidine is dissolved in 30 ml. glacial acetic acid. Make up to 500 ml. with distilled water and add 1 g. KI.
Procedure: First spray cautiously a single corner of the chromatogram. If the background turns blue, wait some time before commencing complete spraying.

33. Chlorosulphonic acid-glacial acetic acid for detecting triterpenes, sterols and steroids.
Spray: 5 ml. Chlorosulphonic acid is dissolved in 10 ml. glacial acetic acid, with cautious cooling.
Treatment: Heat to 130° C. for 5 mins. and inspect the fluorescences in filtered UV-light (365 mμ).

34. Dichromate-sulphuric acid for detecting arylamines.
Spray: 5 g. Potassium dichromate is dissolved in 100 ml. of 40% aqueous sulphuric acid (various colors).

35. α-Cyclodextrin[1] for straight-chain lipids.
Spray: 1% Solution of α-cyclodextrin in 30% aqueous ethanol[1].
Treatment: After spraying, the plate is dried an then placed in a vessel containing iodine vapors.

36. Diethylamine-copper sulphate for thio-acids.
Spray : 0.5 g. Copper sulphate is dissolved in 100 ml. methanol. Add 3 ml. of diethylamine.
Note: Shake before use; keeps only a few days. Thiobarbituric acids yield green spots.

37. Diazotized sulphanilic acid for phenols and for coupling with amines (Pauly's reagent).
Preparation of diazonium salt: Dissolve 25 g. sulphanilic acid in 125 ml. of a 10% solution of NaOH. After cooling, add 100 ml. of a 10% solution of sodium nitrite. The solution is added gradually, drop by drop, through a separatory funnel, into ice-cold HCl (40 ml. HCl (*d* 1.19) in 20 ml. water) while stirring. The reaction temperature must not exceed 8° C. The diazonium salt so formed is filtered by suction and washed successively with iced water, ethanol and ether, and air-dried. The salt, if placed in a brown bottle in a refrigerator,

[1] Applied Science Laboratories, Inc., P.O. Box 140, State College, Pa., U.S.A.

keeps for several months. Owing to its limited keeping quality, however, Fast Blue Salt B is preferred in many cases (Reagent No. 61).
Spray: 0.1 g. Diazonium salt is dissolved, before use, in 20 ml. of an aqueous 10% solution of NaOH.
Note: Attention is drawn to the general safety rules which must be observed during preparation and storage of the explosive diazonium salt.

38. Diazotized p-nitro-aniline solution for phenols.
Stock solution: Dissolve 0.7 g. p-nitro-aniline in 9 ml. concentrated HCl (*d* 1.19) and make up with water to 100 ml.
Spray: 4 ml. of the stock solution is added, drop by drop, to 5 ml. of an ice-cooled aqueous 1% sodium nitrate solution and made up to 100 ml. with iced water.
Note: Prepare solution freshly before use.

39. Diazotization and coupling with β-naphthol to detect sulphonamides.
Spray I: Dissolve 1 g. sodium nitrate to 100 ml. N-HCl.
Spray II: 0.2% Solution of β-naphthol in N-KOH.
Procedure: Spray I is freshly prepared and sprayed. After 1 min., spray with II. Dry chromatogram at 60° C.

40. 2,6-Dibromoquinonechlorimide to detect vitamin B_6.
Spray: 0.4% Solution of 2,6-dibromoquinonechlorimide in methanol.

41. 2,6-Dichloroquinonechlorimide to detect antioxidants as well as cyanamide and its derivatives.
Spray: 1 g. 2,6-Dichloroquinonechlorimide is dissolved in 100 ml. absolute ethanol. If stored in the refrigerator, the solution keeps about 3 weeks. This reagent cannot be used for urea.

42. 2′,7′-Dichlorofluorescein as fluorescent indicator for saturated and unsaturated lipids.
Spray: 0.2% Solution of 2′,7′-dichlorofluorescein in 96% ethanol.

43. 2,6-Dichlorophenol-indophenol sodium salt for organic acids.
Spray: A 0.1% solution of 2,6-dichlorophenol-indophenol sodium salt in ethanol gives red spots on bright-blue background, which soon disappear.

44. 2,6-Dichlorophenol-indophenol silver nitrate for inorganic anions.
Spray: (prepare freshly before use) 0.2 g. 2,6-Dichlorophenol-indophenol sodium salt is dissolved in 100 ml. of 96% ethanol. Add 3 g. silver nitrate, shake and filter.
Remarks: Coloring is disturbed by the presence of collidine and pyridine. Resorcinol, orcinol, phloroglucinol or α-naphthol may be used instead of naphthoresorcinol. 1 Part by volume of trichloroacetic acid may be replaced by 0.1 part by volume of o-phosphoric acid.

45. 1,3-Dihydroxynaphthalene (Naphthoresorcinol)-trichloroacetic acid for ketoses and uronic acids.
Spray: 1 Part by volume of an alcoholic naphthoresorcinol solution (0.2 g. naphthoresorcinol dissolved in 100 ml. ethanol) is mixed with 1 part by volume of an aqueous 20% solution of trichloroacetic acid.
Treatment: For ketoses, heat to 100—105° C. in drying oven for 5—10 mins.; for uronic acids heat to 70—80° C. for 10—15 mins. in a humid atmosphere (water bath).

46. 4-Dimethylaminobenzaldehyde-acetylacetone for amino sugars (Morgan-Elson reagent).
Spray I: 0.5 ml. of a mixture consisting of 5 ml. of a 50% aqueous solution of KOH and 20 ml. ethanol are added immediately before use to a mixture consisting of 0.5 ml. acetylacetone and 50 ml. n-butanol.
Spray II: 1 g. 4-Dimethylaminobenzaldehyde is dissolved in 30 ml. ethanol. Add 30 ml. HCl (*d* 1.19). Before use, dilute with 180 ml. n-butanol.
Procedure: Spray with I, heat to 105° C. for 5 mins. Spray with II and dry for 5 mins. at 90° C.

47. Dimethylaminobenzaldehyde-glacial acetic acid-phosphoric acid for proazulenes and azulenes (EP-reagent).
Spray: 0.25 g. p-Dimethylaminobenzaldehyde (colorless crystals) is dissolved in a mixture of 50 g. glacial acetic acid, analytical grade, and 5 g. o-phosphoric acid, 85%, (high purity) plus 20 ml. water. This solution keeps for months if stored in a brown bottle.
Treatment: Azulene hydrocarbons react after spraying, even at room temperature, giving an intense blue. Proazulenes appear as blue spots, but only after heating to 80° C. for 10 mins. The color fades later and becomes green to yellow. The intensive blue color can be repeatedly reproduced by holding the plate over steam from a water bath.

48. 4-Dimethylaminobenzaldehyde-HCl (van Urk's reagent) for detecting indole derivatives.
Spray: 1 g. 4-Dimethylaminobenzaldehyde is dissolved in 50 ml. HCl (*d* 1.19) and 50 ml. ethanol is added to the solution.
Treatment: Immediately after development, the plates are copiously sprayed until they become transparent (approx. 10 ml. per 20 × 20 $cm.^2$ surface). Plates chromatographed with alkaline volatile solvents are heated to 50° C. for 5 mins. before spraying. Blow vapors of *aqua regia* over the layer. *Aqua regia:* Mix 3 vols. HCl. (*d* 1.19) with 1 vol. nitric acid (*d* 1.4).

49. 4-Dimethylaminobenzaldehyde-HCl (Ehrlich's reagent) for detecting citrulline, urea, tryptamine, tryptophane.
Spray, aqueous: Ehrlich's reagent on urobilinogen (Merck).
Spray, alcoholic: 1% Solution of 4-dimethylaminobenzaldehyde in ethanol (96%).
Procedure: According to requirements, spraying is carried out either with the alcoholic or aqueous solutions.
Treatment: The sprayed chromatogram is placed for 3—5 mins. into a vessel saturated with HCl-vapors.

50. 4-Dimethylaminobenzaldehyde-HCl (modified Ehrlich's reagent) for ergot alkaloids.
Spray: 0.5 g. 4-Dimethylaminobenzaldehyde is dissolved in 100 ml. of hot cyclohexane.
Treatment: After spraying, the chromatogram is placed in a tank saturated with HCl vapors to give a blue coloration.

51. 3,5-Dinitrobenzoic acid for Kedde's reaction with pentacyclic lactones.
Spray: 1 g. 3,5-Dinitrobenzoic acid is dissolved in a mixture consisting of 50 ml. methanol and 50 ml. aqueous 2 N KOH.

52. 2,4-Dinitrophenylhydrazine for detecting free aldehyde- and keto-groups, as well as ketoses.
Spray a: 0.4% Solution of 2,4-dinitrophenylhydrazine in 2 N-HCl.
Spray b: To 1 g. 2,4-Dinitrophenylhydrazine in 1000 ml. ethanol, add 10 ml. HCl (*d* 1.19). Yellow spots are obtained.
Remark: Either "a" or "b" may be used for spraying.

53. Diphenylamine for glycolipids.
Spray: Prepare a mixture consisting of 20 ml. of a 10% ethanolic solution of diphenylamine, 100 ml. conc. HCl and 80 ml. glacial acetic acid.
Treatment: Heat to 100° C. for 5—10 mins.

54. Barium diphenylamine-4-sulphonate[1] as fluorescent indicator.
(For the detection of sugars, barbituric acids, etc.)
Spray: Prepare a saturated (about 0.2%) methanolic solution of barium diphenylaminesulphonate. First dissolve the salt in a few ml. dimethylformamide.
Treatment: If the spots do not fluoresce in short-wave UV-light (254 mμ) after intensive spraying, heat the plate to 110° C. for about 5—10 mins. Many substances may be located in this way.

[1] E. Merck A. G., Darmstadt, Germany.

55. Diphenylboric acid-β-aminoethylester for α- and γ-pyrones.
Spray: 1.0 g. Diphenylboric acid-β-amino-methylester[1] is dissolved in 100 ml. methanol.
Procedure: Spray about 10 ml. of the solution and inspect the fluorescent colors in filtered UV-light (360 mμ).

56. Diphenylcarbazids for Ag-, Pb-, Cu-, Sn, Mn-, Zn-, and Ca-ions.
Spray I: 1% Solution of diphenylcarbazide in ethanol (96%).
Spray II: 25% Solution of ammonia.

57. Diphenylcarbazone for Ag-, Zn-, and Cd-ions and mercuric acetate adducts of unsaturated lipids.
Spray I: Saturated solution of diphenylcarbazone in ethanol (96%).
Spray II: 25% Solution of ammonia or N-NaOH.
Procedure: Spray first with I, then with II.
Note: In the case of Hg-adducts, it is sufficient to spray with a 0.1% ethanolic solution of diphenylcarbazone.

58. Dipicrylamine for the detection of choline.
Spray: Dissolve 0.2 g. dipicrylamine in a mixture consisting of 50 ml. acetone and 50 ml. twice-distilled water. Choline and its derivatives appear as red spots on a yellow background.

59. Dithizon for lead and other heavy metal ions.
Spray I: 0.05% Solution of dithizone in CCl_4.
Spray II: 25% Solution of aqueous ammonia.

60. Dragendorffs-reagent (Bregoff-Delwiche) for quaternary bases (choline).
Stock solution: 8.0 g. of bismuth subnitrate is dissolved in 20 to 25 ml. of 30 per cent HNO_3 (d 1,18). This solution is added slowly and with stirring to a solution containing 28 g. of KI and 1 ml. of 6 N-HCl in approximately 5 ml. of water. The dark precipitate redissolves, giving an orange-red solution which is cooled to 5° C, filtered, and diluted to 100 ml. with water. The orange-red stock solution is stable for a few weeks if kept in a dark bottle in the refrigerator.
The developing solution ($KBiI_4$ reagent) consists of 20 ml. of water, 5 ml. of 6 N-HCl, 2 ml. of Dragendorff stock solution, and 5 ml. of 6 N-NaOH, added in this order. A few drops of 6 N-HCl may be added if all of the $Bi(OH)_3$ does not dissolve on shaking. The $KBiI_4$ reagent is stable for 10 days when kept in the refrigerator.

60a. Dragendorff's-reagent (Thies, Reuther, modified by Vágújfalvi) for the detection of alkaloids.
Stock solution: 2.6 g. Bismuth carbonate and 7.0 g. sodium iodide are boiled for a few minutes with 25 ml. glacial acetic acid. After about 12 hrs., the precipitated sodium acetate crystals are filtered off using a sintered glass funnel. 20 ml. of the clear, red-brown filtrate are mixed with 8 ml. ethyl acetate and stored in a brown bottle.
Spray: 10 ml. stock solution is mixed with 25 ml. glacial acetic acid and 60 ml. ethylacetate. When spraying (using 5—10 ml.), alkaloids and a number of nitrogen-free compounds appear as orange spots.
Treatment: A considerable increase in sensitivity is obtained by spraying with 0.1—0.05 N-sulphuric acid. Ascertain the optimum acid concentration and quantity of spraying reagent by a trial. The background becomes grey and the spots show an intensive orange-red to red color.

61. Fast Blue Salt B for phenols and amines capable of coupling (Diazo-reagent).
Spray: 0.5% Aqueous solution of Fast Blue Salt B, freshly prepared.
Treatment: Spraying with 0.1 N-NaOH.

62. Ferric chloride for hydroxamic acids and phenols.
Spray: 1% Aqueous solution of ferric chloride.

[1] C. Roth, Karlsruhe, Germany.

63. Acetic anhydride-sulphuric acid for triterpenoid glycosides and cholesterol (Liebermann-Burchard Test).
Spray: Mix cautiously, while cooling, 5 ml. acetic anhydride with 5 ml. conc. sulphuric acid. The mixture is added slowly, while cooling, to 50 ml. absolute alcohol. Prepare freshly before use.
Treatment: Heat plates, after spraying, to 110° C. for about 10 mins. Inspect spots under filtered UV-light.

Fluorescent Indicators: see No. 90.

64. Formaldehyde-HCl (Prochazka reagent) for the detection of indole derivatives.
Spray: Freshly prepare a mixture consisting of 10 ml. of a solution of formaldehyde (about 35%), 10 ml. pure HCl, 25%, and 20 ml. ethanol.
Treatment: Heat to 100° C. for 5 mins. The fluorescent colors (yellow-orange-greenish) may be intensified by exposure to *aqua regia* vapors (for preparation of *aqua regia*, see No. 48).

65. Glucose-aniline (Schweppe reagent) for acids.
Spray: 2 g. Glucose is dissolved in 20 ml. water, and also 2 ml. aniline in 20 ml. ethanol. Both solutions are combined in a 100 ml. measuring flask and made up to 100 ml. with n-butanol.
Procedure: After spraying, heat the chromatogram to 125° C. for 5—10 mins. Deep-brown spots appear on a white background.

66. Glucose-phosphoric acid for aromatic amines.
Spray: Dissolve 2 g. glucose in 10 ml. o-phosphoric acid (*d* 1.7) and 40 ml. water; 30 ml. ethanol and 30 ml. n-butanol are added.
Treatment: Heat to 115° C. for about 10 mins.

67. Urea-HCl for the detection of ketoses.
Spray: Dissolve 5 g. urea in 20 ml. 2 N HCl. Mix with 100 ml. ethanol. Ketoses and oligosaccharides containing ketoses stain blue. Short heating promotes the reaction.

68. Hydrazine sulphate for piperonal and vanillin.
Spray: Mix 90 ml. of a saturated aqueous solution of hydrazine sulphate with 10 ml. of 4 N HCl. The wet chromatogram is inspected under a quartz lamp, both before and after exposure to ammonia vapor.

69. Hydroxamic acid-ferric chloride for the detection of acetylcholine and other choline esters.
Solution a: 20 g. Hydroxylammonium chloride is dissolved in 50 ml. water. The solution is made up to 200 ml. with ethanol and stored in the cold.
Solution b: 50 g. KOH is dissolved in as little water as possible and then made up to 500 ml. with ethanol.
Spray I: 1 Volume of solution *a* is mixed with 2 volumes of solution *b*. The precipitated potassium chloride is filtered off. This spraying solution I will keep for about 2 weeks in a refrigerator.
Spray II: 10 g. Very finely powdered ferric chloride ($FeCl_3 \cdot 6H_2O$) is dissolved in 20 ml. 10 N HCl and the solution is agitated with 200 ml. diethyl ether until a homogeneous mixture is obtained. Spraying solution II, if tightly stoppered, will keep for some time.
Procedure: Spray chromatogram with I, dry shortly at room temperature and then spray with II.

70. 8-Hydroxyquinoline for the detection of Ba-, Sr-, and Ca-ions.
Spray: 0.5 g. 8-Hydroxyquinoline is dissolved in 60 ml. ethanol and 40 ml. water.
Treatment: Spray with 25% ammonia solution, or place chromatogram into a humid, ammonia-saturated tank. Inspect in filtered UV-light.

71. 8-Hydroxyquinoline-kojic acid for Mg-, and Al-ions.
Spray I: Solution of 2.5 g. 8-hydroxyquinoline and 0.5 g. kojic acid in 500 ml. ethanol (90%).
Spray II: 25% Ammonia solution. Inspect in filtered UV-light.

72. Iodine as general reagent.
Place the chromatogram in a closed vessel containing a few iodine crystals at the bottom. On placing the vessel in a water bath, great quantities of iodine vapor are given off, which yield a brown color with most organic compounds. *Modification:* Place the chromatogram into a concentrated iodine atmosphere for 5 mins. Excess iodine is removed by letting the chromatogram stand in air for about 5 mins. Spots stain blue when sprayed with a 1% aqueous solution of starch. If there is still too much iodine on the layer, the background stains blue, (make test with a portion of the covered plate).

73. Iodine solution for organic compounds.
Spray: 0.3 g. Iodine is dissolved in 100 ml. aqueous 5% KI solution; better suitable for TLC: 0.5 g. iodine in 100 ml. chloroform.

74. Iodine solution for N-substituted imidazoles.
Spray: 1 g. Iodine is dissolved in 100 ml. ethanol (96%).
Treatment: Heat plate to 100° C. for 30 mins.

75a. Iodine azide solution.
Spray: A solution of 3.5 g. sodium azide in 100 ml. 0.1 N iodine solution is freshly prepared.
Note: Dry iodine azide is explosive,

75b. Iodine azide-reagent.
Spray I: Freshly prepared solution of 1 g. sodium azide in 100 ml. of a 0.005 N iodine solution.
Spray II: 1% Aqueous starch solution.
Procedure: Spray with I, then with II.

76. Iodoplatinate, modified for alkaloids and various nitrogenous heterocyclic compounds.
Spray: 3 ml. of a 10% platinum chloride solution is mixed with 97 ml. water to which are added 100 ml. of a 6% aqueous solution of KI.
Note: Keeps for some time in brown glass bottles.

77. Kalignost®-Rhodamine B for the detection of potassium ions.
Spray I: 0.1 N-NaOH.
Spray II: 1% Ethanolic solution of Kalignost®.
Spray III: 0.5% Ethanolic solution of Rhodamine B.
Procedure: Spray with I, let dry, spray with II, finally spray with III. Strong dark-blue fluorescence in filtered UV-light. With larger amounts of potassium, bright-red spots on dark-red background can be seen in day-light.

78. Potassium ferrocyanide for ferric ions.
Spray: Freshly prepared 2% aqueous solution of potassium ferrocyanide.

79. Potassium ferrocyanide-cobaltous chloride for the detection of choline.
Spray I: Freshly prepared 1% aqueous solution of potassium ferrocyanide.
Spray II: 0.5% Aqueous solution of cobaltous chloride.
Procedure: Spray with I, dry for a short period, and then spray with II. Choline shows up green.

80. Potassium ferricyanide-ferric chloride for the detection of thiosulphates.
Spray I: 1% Aqueous solution of potassium ferricyanide.
Spray II: 1% Aqueous solution of ferric chloride.
Procedure: Spray consecutively with I and II, and inspect in UV-light.

81. Potassium ferricyanide for the detection of vitamin B, (thiochrome-reaction)
Spray: Prior to use, mix 1.5 ml. of a 1% aqueous potassium ferricyanide, 20 ml. distilled water and 10 ml. of a 15% solution of NaOH. After drying, inspect in long-wave UV-light.

82. Potassium hydroxide for coumarins.
Spray: 1% Solution of potassium hydroxide in ethanol.
Procedure: The sprayed chromatograms are dried in a drying oven and inspected under a U.V. lamp.

83. Potassium iodate for sympathicomimetic amines (phenylethylamines).
Spray: 1% Aqueous solution of potassium iodate.
Procedure: After spraying, heat chromatogram to 100—110° C. for 2 mins.

84. Potassium iodine-hydrogen sulphide for heavy metals.
Spray: 2% Aqueous solution of potassium iodide.
Procedure: After spraying, the plate is dried and placed in a vessel containing ammonia vapor. After a few minutes, transfer the plate into another vessel containing hydrogen sulphide gas. (Caution! H_2S is toxic and explosive. It is best to lead it from a Kipp apparatus, under the fume cupboard).

85. Potassium iodide-starch reaction for peroxides
Spray I: 10 ml. of a 4% aqueous solution of potassium iodide is mixed with 40 ml. glacial acetic acid, and a small spatula pointful of zinc powder is added.
Spray II: Freshly prepared 1% aqueous starch solution.
Procedure: After filtering off the zinc dust, spray with I. After 5 mins., spray copiously with II until transparent. Peroxides will appear as blue spots, due to free iodine.

86. Potassium permanganate-acetic acid.
Spray: Mixture of equal volume of 0.1 N potassium permanganate and 2 N acetic acid.

86a. Potassium permanganate-H_2SO_4 (universal reagent)
Spray: 0.5 g. Potassium permanganate is dissolved in 15 ml. conc. H_2SO_4 (Caution! Danger of explosion).
Procedure: Remove solvents from chromatogram. Spray reagent: white spots appear immediately on pink background.

87. Cobaltic nitrate-ammonia for the detection of barbituric acids (Zwikker's reagent).
Spray: 1% Solution of cobaltic nitrate in absolute ethanol.
Treatment: Dry and place in a vessel filled with vapors of aqueous ammonia.
Modifications:
a) *Spray:* 2% Alcoholic solution of cobaltic acetate.
Treatment: Place chromatogram in a vessel containing pyridine vapor.
b) *Spray I:* 0.5% Methanolic solution of cobaltic acetate.
Spray II: 0.5% Methanolic solution of lithium hydroxide.
Procedure: Spray with I, then spray with II.

88. König's reagent for the detection of alkaloids containing a pyridine ring.
Pre-treatment: Before spraying, the chromatogram is placed for 1 hr. in a tank containing a beaker with a bromocyanogen solution (toxic!), which is prepared by adding a 10% sodium cyanide solution to ice-cooled saturated bromine water until all trace of bromine coloration has disappeared.
Spray: 2 g. p-Aminobenzoic acid is dissolved in 75 ml. 0.75 N-HCl. The solution is made up to 100 ml. with ethanol (96%).

89. Kojic acid for metal ions.
Spray: 0.1 g. Kojic acid is dissolved in 100 ml. ethanol (60%).
Procedure: After spraying, inspect fluorescence in UV-light.

90. Fluorescent indicators as general reagents.
a) Barium diphenylamine sulphonate, Reagent No. 54.
b) 2',7'-Dichlorofluorescein, Reagent No. 41.
c) Fluorescein, 0.2% in ethanol.
d) Fluorescent substance "ZS-Super" (Riedel de Haen) added to the adsorbent in amounts of 1%.

e) 0.1% Solution of morin in ethanol.
f) An aqueous 0.04% solution of sodium fluoresceinate for the preparation of layers.
g) Rhodamine B, Reagent No. 129.
h) Zinc silicate, fluorescent substance (P 1, Type 118-2-7)[1] added to the adsorbent in quantities of 0.8%.
i) Methylumbelliferone, Reagent No. 94.

91. Magnesium acetate for the detection of hydroxyanthraquinones.
Spray: 0.5% Methanolic solution of magnesium acetate.
Procedure: After spraying, heat to 90° C. for 5 mins. Orange to violet coloration.

92. Malonic acid-salicylaldehyde for the detection of nitrogenous heterocyclic compounds and amines.
Spray: 0.2 g. Malonic acid and 0.1 g. salicylaldehyde are dissolved in 100 ml. of absolute ethanol.
Treatment: The plates are heated to 120° C. for about 15 mins. and inspected in filtered UV-light. Yellow fluorescent spots.

93. Methyl red with bromothymol blue indicator.
Spray: 0.2 g. Methyl red and 0.2 g. bromothymol blue are dissolved in a mixture consisting of 100 ml. formaldehyde and 400 ml. ethanol (96%). The pH of the solution is adjusted to 5.2 with 0.1 N NaOH.
Treatment: The plates, which have been sprayed several times, are exposed to ammonia vapor.

94. 4-Methylumbelliferone as reagent for nitrogenous heterocyclic compounds (fluorescent indicator).
Spray: 20 mg. Methylumbelliferone is dissolved in 35 ml. ethanol. The solution is made up with water to 100 ml.
Treatment: The chromatograms are placed in a vessel containing ammonia vapor and inspected under a U.V. lamp.

95. Morin for the detection of aluminium ions.
Spray: 1% Solution of morin in glacial acetic acid. Strong bright green fluorescence in U.V. light.

96. β-Naphthoquinone-4-sulphonic acid (sodium salt) for amino acids (Folin reagent).
Spray: 0.2 g. of β-naphthoquinone-4-sulphonic acid (sodium salt) is dissolved in 100 ml. of a 5% solution of Na_2CO_3. Spray about 10—18 mins. after preparation of spray. No further treatment is required.

97. α-Naphthol for arginine and other guanidine derivatives (Sakaguchi-reagent).
Spray I: 0.1% Solution of α-naphthol in N NaOH.
Spray II: Mixture consisting of 100 ml. aqueous 5% NaOH and 2 ml. bromine.
Procedure: Spray with I, then with II.
For the detection of streptomycin, it is recommended to replace spray II by a mixture consisting of 50 ml. of an aqueous solution of sodium hypochlorite (13% active Cl) and 50 ml. ethanol.

98. α-Naphthylamine for the detection of 3,5-dinitrobenzoic acid esters.
Spray I: 1% Ethanolic solution of α-naphthylamine.
Spray II: 10% Methanolic solution of KOH. Red brown spots appear. Coloration may also be carried out with rhodamine B (Reagent No. 129).

99. Sodium dithionite for antimony and bismuth ions.
Spray: 0.1% Aqueous solution of sodium dithionite.

100. Sodium fluoresceinate for the detection of aromatic and heterocyclic compounds.
Spray: 50 mg. Sodium fluoresceine is dissolved in 100 ml. methanol (50%).
Note: If the chromatograms are sprayed evenly, various compounds may be located in U.V. light.

[1] General Electric, Cleveland, Ohio, U.S.A.

101 A. Sodium fluoresceine bromine-test for unsaturated compounds.
The adsorbent slurry is prepared with a 0.04% solution of sodium fluoresceine, instead of water.
After development, the dried plate is exposed to bromine vapor. Due to formation of eosin, quenching of fluorescence is seen in UV-light, except in places where addition compounds have been formed with bromine.

101 B. Sodium fluoresceine-Rhodamine B-sodium carbonate for the detection of chlorinated hydrocarbons, camphors, heterocyclic compounds, etc.
Spray I: 0.5% Ethanolic solution of Rhodamine B.
Spray II: 10% Aqueous solution of Na_2CO_3.
Procedure: Starting with sodium fluoresceinate layers (see Reagent No. 101 A) which after development are sprayed with I. After drying, layers are copiously sprayed with II. In parts, the spots are seen better as fluorescent spots in UV-light. (See also insecticides, p. 360.)

102. Sodium hydroxide for Δ^4-3-ketosteroids.
Spray: 10% Aqueous solution of NaOH.
Procedure: After spraying, dry chromatogram at 80° C. for 10 mins. Δ^4-3 ketosteroids fluoresce yellow in filtered UV-light.

103. Sodium rhodizonate for barium and strontium ions.
Spray I: 1% Aqueous solution of sodium rhodizonate.
Spray II: 25% Solution of ammonia.

104. Sodium sulphide solution for hydrosulphate ions.
See also Reagent No. 84.
Spray: 0.5% Aqueous solution of sodium sulphide, freshly prepared.

105. Sodium thiosulphate-copper acetate for the detection of antimony ions.
Spray I: Saturated aqueous solution of sodium thiosulphate.
Spray II: 0.4 g. Copper acetate is dissolved in a mixture of 2 ml. glacial acetic acid and 48 ml. water.
Procedure: Spray with I, heat briefly, wash off excess sodium thiosulphate with water, then spray with II.

106. Nessler's reagent for the detection of hydroxyamino acids (serine, threonine, hydroxyproline).
Spray I: 1% Aqueous solution of sodium metaperiodate.
Spray II: Nessler's reagent (Merck).
Procedure: Spray with I, dry chromatogram at room temperature, then spray with II.

107. Nessler's reagent for the detection of alkaloids.
Spray: Nessler's reagent (Merck).
Note: Apomorphine, hydrastinine and physostigmine react with Nessler's reagent.

108. Ninhydrin for the detection of amino acids and amines.
Spray: 0.3 g. Ninhydrin is dissolved in 100 ml. n-butanol and mixed with 3 ml. glacial acetic acid.
Treatment: Heat to 60° C. for about 30 mins., or to 110° C. for 10 mins.
Sensitivity: 0.1 μg. proline, 0.001 μg glycine.
Stabilization of ninhydrin-colored spots.
Spray: 1 ml. of a saturated aqueous solution of copper nitrate is dissolved by 0.2 ml. of 10% nitric acid in 100 ml. ethyl alcohol (96%).
Procedure: When the ninhydrin-spots have attained maximal color intensity they are sprayed with the solution, and the plate is placed in a closed vessel, which contains a beaker of concentrated aqueous ammonia. The red copper complex thus obtained is stable only in the absence of free hydrogen ions or of complexing agents.

109. Ninhydrin-copper nitrate for amino acids (polychromatic detection).
Solution I: 0.1 g. Ninhydrin is dissolved in 50 ml. absolute alcohol to which are added 10 ml. glacial acetic acid and 2 ml. 2,4,6-collidine.

Solution II: 0.5 g. Copper nitrate $[Cu(NO_3)_2 \cdot 3\,H_2O]$ is dissolved in 50 ml. absolute alcohol.
Spray: Before use, mix I and II in ratio of 50:3.
Treatment: Heat chromatogram to 110° C. for 4 mins., and note the colors immediately, as some of them begin to change and fade after 10 mins. Sensitivity, about 0.5—1 μg. amino acid.

110. p-Nitroaniline-periodic acid for the detection of desoxysugars.
Spray I: 1 Volume of a saturated aqueous solution of sodium metaperiodate is diluted with 2 volumes of water.
Spray II: 4 Volumes of a 1% ethanolic solution of p-nitroaniline are mixed with one volume of HCl (*d* 1.19).
Procedure: Spray with 1, wait 10 mins., spray with II. Desoxy-sugars and glycols show up as yellow spots, which fluoresce strongly in UV-light. After further spraying with 5% methanolic NaOH, the color changes to green.

111. Sodium nitroprusside for SH-compounds (cysteine), -S-S-compounds (cystine) and cyanamide derivatives (arginine).
Spray I: 1.5 g. Sodium nitroprusside is dissolved in 5 ml. 2 N HCl. After adding 95 ml. methanol and 10 ml. aqueous ammonia solution, filter.
Spray II: 2 g. Sodium cyanide is dissolved in 5 ml. water. Place in a graduated measuring flask and make up to 100 ml. with methanol. Observe all precautions when spraying with sodium cyanide!
Procedure: When spraying with I, SH-compounds are located as red spots. Arginine becomes orange, later turning to grey-blue. When spraying with II, compounds with -S-S- linkages are located as red spots on a yellow background.
Modification for -S-S- compounds:
Spray I: 5 g. Sodium cyanide and 5g. sodium carbonate are dissolved in a graduated 100 ml. measuring flask in ethanol (25%). Make up solution to 100 ml. with 25%-ethanol.
Spray *II:* 2 g. Sodium nitroprusside is dissolved in 100 ml. ethanol (75%).
Procedure: Spray with I, dry briefly in air, spray with II.

112. Sodium nitroprusside for the detection of secondary aliphatic and alicyclic amines.
Spray: 5 g. Sodium nitroprusside is dissolved in 100 ml. of a 10% aqueous solution of acetic aldehyde. 1 volume of this mixture is mixed before use with 1 volume of a 1% solution of sodium carbonate.

113. Sodium nitroprusside-potassium ferricyanide for the detection of cyanamide and its derivatives.
Spray: 1 Volume each of 10% NaOH, 10% sodium nitroprusside and a 10% solution of potassium ferricyanide are mixed with 3 volumes water. The solution is left standing at room temperature before use, for at least 20 mins. If placed in a refrigerator it will keep for several weeks.
Modification:
Spray: 2 ml. of a 5% aqueous solution of sodium nitroprusside are mixed with 1 ml. of 10% NaOH and 5 ml. of a 3% solution of perhydrol, the whole being then diluted with 15 ml. water. The reagent must always be freshly prepared before use.
Cyanamide: violet, dicyanodiamide: carmine red, guanyl-urea: yellow-orange, guanidine: red:orange, arginine: bright-red, creatine: carmine red, creatinine-yellow-brown, agmatine: pink, acetyl guanidine: carmine-red, thiourea: violet, urea: pale pink.

114. Sodium nitroprusside-sodium hydroxide for the detection of α,β-unsaturated lactones (Legal's test).
Spray: 1% Solution of sodium nitroprusside in ethanolic (50%) solution of N-NaOH (red-violet spots).

115. Sodium nitroprusside-hydroxylamine for the detection of thiourea derivatives (Grote's reagent).
Spray: 0.5 g. Sodium nitroprusside is dissolved in 10 ml. water. The solution is mixed with 0.5 g. hydroxylammonium chloride and 1 g. sodium bicarbonate.

After cessation of gas formation, 2 drops of bromine are added. Make up to 25 ml. with water. The reagent keeps about 2 weeks.

116. Sodium nitroprusside-sodium periodate for the detection of desoxy sugars.
Spray I: Mixture of 1 volume saturated aqueous solution of sodium periodate with 2 volumes water.
Spray II: Mixture of 1 volume saturated aqueous solution of sodium nitroprusside, 3 volumes of water and 20 volumes of a saturated ethanolic (96%) solution of piperazine.
Procedure: Spray with I, dry for 10 mins. at room temperature, spray with II.
Note: Maximal blue coloration of desoxy-sugars occurs after 5—10 mins. A more sensitive detection of desoxy-sugars is obtained if spray II is replaced by the following mixture: 4 volumes of an ethanolic solution of 1% p-nitroaniline and 1 volume HCl (36%). Yellow fluorescent spots can be observed in filtered UV-light.

117. Perchloric acid for steroids.
Spray: 2% Aqueous perchloric acid.
Treatment: Heat to 150° C. for 10 mins.
Note: The reaction gives uniform brown spots for comparative quantitative estimations. A more sensitive detection should be obtained by using phosphoric acid reagent, No. 123, and after-spraying with phosphomolybdic acid reagent, No. 120c.

118. Peroxide-ferric chloride for the detection of coumarin.
Spray I: Freshly prepared 0.5% perhydrol.
Spray II: A 2% aqueous solution of ferric chloride.
Procedure: Spray with I, dry at 105° C., then spray with II. Heat in a drying oven until spots appear.

119. o-Phenylenediamine-trichloroacetic acid for the detection of α-keto acids.
Spray: 50 mg. o-Phenylenediamine is dissolved in 100 ml. of a 10% aqueous solution of trichloroacetic acid.
Procedure: After spraying, heat the chromatogram to 100° C. in a drying oven, for not longer than 2 mins, then inspect fluorescent spots under a U.V. lamp.

120. Phosphomolybdic acid, reagent for reducing compounds.
a) *Spray:* Freshly prepared solution of 5 g. phosphomolybdic acid in 100 ml. ethanol.
Treatment: Heat to 80—90° C. for 5 mins.
b) *Spray:* Freshly prepared solution of 10 g. phosphomolybdic acid in 100 ml. ethanol.
Treatment: Heat to 80—90° C. for 5 mins.
c) *Spray:* Freshly prepared solution of 1.5 g. phosphomolybdic acid in 100 ml. ethanol.
Treatment: Heat to 120° C. for 5 mins.
Note: Treatment with ammonia vapors gives a colorless background.

121. Phosphotungstic acid for the detection of lipids.
Spray: 20% Ethanolic solution of phosphotungstic acid.
Treatment: Dry at 70° C. in a drying oven for 20 mins.

122. Phosphomolybdic-tungstic acid for the detection of phenols (Folin-Ciocalteau-reagent).
10 g. Sodium tungstenate and 2.5 g. sodium molybdate are dissolved in 70 ml. water, to which are successively added 5 ml. of phosphoric acid (85%) and 10 ml. concentrated HCl (*d* 1.19). The mixture is boiled under reflux for 10 hrs. Then, 15 g. lithium sulphate, 5 ml. water and 1 drop of bromine are added followed by renewed boiling for 15 mins. The cooled solution is made up to 100 ml. in a measuring flask (stock solution). The solution should not show any green coloration.
Spray I: 20% Aqueous solution of sodium carbonate.

Spraying solution II: Before use, 1 volume of the stock solution is diluted with 3 volumes water.
Procedure: Spray with I, dry for a short period, spray with II.

123. Phosphoric acid for the detection of steroids.
Spray: 1 Volume o-phosphoric acid (*d* 1.7) is diluted with 1 volume water.
Procedure: The plates are sprayed until they are transparently humid and then heated to 120° C. for 10—20 mins. The fluorescent spots are examined under the U.V. lamp. Unsaturated steroids and sterols can be located as blue spots in visible light, when the plates are heated for 10 mins. after spraying with phosphoric acid and are then, while still hot, sprayed with Reagent No. 120c. (freshly prepared). Heating for 2—5 mins. will intensify the color.

124. Picric acid-alkali for the detection of creatinine, glycocyamidine, and lactams of α-guanidine acids (Jaffé-reagent).
Spray I: 1% Ethanolic solution of picric acid.
Spray II: 5% Alcoholic KOH.
Procedure: Spray with I, dry, spray with II. Orange coloration.

125. Pyridylazonaphthol for the detection of uranyl-ions.
Spray: 0.25% Ethanolic solution of pyridylazonaphthol [1-(2-pyridyl-azo)-2-naphthol, Merck, Art. No. 7531].

126. Mercuric acetate-diphenylcarbazone for the detection of purines.
Spray I: 0.25 g. Mercuric acetate is dissolved in 100 ml. of ethanol (96%); add a few drops of glacial acetic acid.
Spray II: 0.05 g. Diphenylcarbazide is dissolved in 100 ml. ethanol (96%).
Procedure: Spray first with I, then with II. The chromatogram turns uniformly violet, but shows shadows at those places where purines are present. When heating to 120° C. in a drying oven provided with a viewing screen, the background color will gradually fade.
Note: The various spots have different degrees of stability and appear gradually during heating. Therefore, continuous observation during heating is necessary. It is best to circle the spots with a pencil immediately after they appear.

127. Mercurous nitrate for the detection of barbituric acids.
Spray: 1% Aqueous solution of mercurous nitrate.

128. Quercetin for the detection of antimony-, copper-, nickel-, iron-, chromium-, manganese-, potassium-, lithium- and beryllium-ions.
Spray: 0.2% Solution of quercetin in ethanol (96%).
Treatment: Spray with 25% solution of ammonia or place into a humid, ammonia-saturated tank. Inspect in filtered UV-light.

129. Rhodamine B as general spraying reagent.
Spray: 0.5 g. Rhodamine B is dissolved in 100 ml. ethanol.

130. Rhodanine = dimethylaminobenzylidine-rhodanone for the detection of cations.
Spray: 1—5% Ethanolic solution of rhodanine in ethanol.
Treatment: Spray with a 25% solution of ammonia, or place in a humid, ammonia-saturated tank. Inspect in filtered UV-light.

131. Dithio-oxamide ($H_2NCS\ CS\ NH_2$) for the detection of copper-, cobalt-, nickel- and manganese-ions.
Spray I: 0.5% Solution of dithio-oxamide in ethanol (96%).
Spray II: 25% Solution of ammonia.
Procedure: Spray with I, dry for a short period, then spray with II, or place chromatograms into a humid, ammonia-saturated tank.

132. Nitric acid for alkaloids and amines.
Spray: 100 ml. Absolute ethanol are mixed with about 50 drops of nitric acid (*d* 1.4). Inspect decomposition products in filtered UV-light.

Note: The spraying solution may be used in this, or at a higher concentration for the detection of other organic compounds in TLC. Fluorescent spots often appear only after a prolonged heating at 120° C.

133. Hydrochloric acid for the detection of glycols.
Spray: 1 Volume HCl (*d* 1.19) is mixed with 4 volumes ethanol
Procedure: Glycols appear as pink spots during heating to 90° C. This mixture can also be used as a general spraying reagent for TLC.

134. Sulphuric acid for alkaloids and amines.
Spray: Mix 100 ml. absolute ethanol with 50 drops sulphuric acid (*d* 1.84).
Note: Due to dehydration, a number of compounds such as glycosides, alkaloids and amines, form decomposition products, which fluoresce in filtered UV-light.

135. Silver nitrate reagent (Dedonder) for the detection of sugars and sugar alcohols.
Spray: Add 1 ml. of a saturated aqueous solution of silver nitrate to 20 ml. acetone, while stirring, then add water, drop by drop, until the silver nitrate has completely dissolved.
Treatment: The plate is put into a dark, ammonia-saturated tank for 15 mins., and is then heated to 80° C., until the dark spots can be clearly located.

136. Silver nitrate-ammonia-fluorescein for the detection of halide ions.
Spray I: 1 g. Silver nitrate is dissolved in 100 ml. of 0.5 N ammonia.
Spray II: 0.1 g. Fluorescein is dissolved in 100 ml. ethanol.
Procedure: Spray with I, after a short drying period spray with II.

137. Silver nitrate-ammonia for the detection of reducing substances (Tollens' or Zaffaroni's reagent).
a) *Spray:* 1 Volume of 0.1 N silver nitrate is mixed, when required, with 1 volume of 5 N ammonia.
Caution is required: when this solution has stood for some time, explosive silver azide is formed!
Treatment: Heat to 105° C. for 5—10 mins. until the dark spots are clearly located.

138. Silver nitrate-ammonia-sodium chloride for the detection of thioacids.
Spray I: When required, 50 ml. of a 0.1 N silver nitrate solution is mixed with 50 ml. 10% ammonia. This solution does not keep.
Spray II: 10% Aqueous sodium chloride.
Procedure: Spray with I, dry, and then spray with II. Expose chromatogram to daylight for some time, until the spots appear.

139. Silver nitrate-bromophenol blue for purines (Wood's reagent).
Spray: 0.2 g. Bromophenol blue is dissolved in 50 ml. acetone. The solution is mixed with 50 ml. of 2% aqueous silver nitrate. This reagent keeps about 1 week.
Procedure: After development in acidic solvents, the chromatogram is dried and exposed to ammonia vapor. The excess ammonia is removed by warm air, then the solution is sprayed onto the layer.

140. Silver nitrate-formaldehyde for the detection of dieldrin, aldrin and lindane.
Spray I: 0.05 N Silver nitrate.
Spray II: 35% Formaldehyde.
Spray III: Methanolic solution of N KOH.
Spray IV: Mixture of equal volumes of perhydrol and nitric acid (*d* 1.4), each freshly prepared.
Procedure: Spray with I, air-dry for 30 mins., spray with II, and again air-dry for 30 mins. After spraying with III, dry for 30 mins. in an oven at 130—133° C. Finally, spray with IV, keep chromatogram in dark for 12 hours, and then expose to sunlight. Dark-brown spots are produced.

141. Silver nitrate-pyrogallol for the detection of acids.
Spray I: 0.17 g. Silver nitrate is dissolved in 1 ml. water. The solution is mixed with 5 ml. of ammonia, and diluted to 200 ml. with ethanol.

Spray II: 6.5 mg. Pyrogallol is dissolved in 100 ml. ethanol.
Procedure: Spray with I, and then with II.

142. p-Toluenesulphonic acid for the detection of steroids and flavonoids.
Spray: 20 g. p-Toluenesulphonic acid is dissolved in 100 ml. absolute ethanol.
Procedure: After spraying, heat a few mins. to 100° C.; inspect chromatogram in filtered UV-light.

143. Trichloroacetic acid for the detection of steroids and glycosides.
Spray: 25 g. Trichloroacetic acid is dissolved in 100 ml. chloroform.
Procedure: After spraying, dry chromatogram to 100° C. in a drying oven and inspect under U.V. lamp.

144. Trifluoroacetic acid for the detection of steroids and glycosides.
Spray: 1% Solution of trifluoroacetic acid in chloroform.
Treatment: Heat to 120° C. for 5 mins.

145. 2,3,5-Triphenyltetrazoliumchloride (TTC) for the detection of reducing sugars, steroids, glycosides and thioacids.
Spray: Before use, mix 1 volume of a 4% methanolic solution of TTC with 1 volume of a methanolic solution of N NaOH.
Treatment: Heat to 110° C. for 5—10 mins. Heating is continued in each case until red spots are clearly located, while the background remains practically colorless.

146. Perchloric acid-ferric chloride for the detection of indole derivatives.
Spray: 100 ml. of a 5% solution of perchloric acid are mixed with 2 ml. of a 0.05 M ferric chloride solution.
Note: Does not react with isatin and other oxyindole derivatives.

147. Vanillin for the detection of amino acids (ornithine, lysine) and amines.
Spray I: 2 g. Vanillin is dissolved in 100 ml. n-propanol.
Spray II: 1% Ethanolic KOH.
Procedure: Spray with I, heat chromatogram to 110° C. in a drying oven for 10 mins. Ornithine fluoresces with an intense yellow-green color in filtered UV-light, but lysine exhibits only a weak green-yellow. After spraying with II, heating is carried out as before. Ornithine stains first salmon-red, and then fades, whereas proline, oxyproline, pipecolinic acid and sarcosine only stain red after a few hours. Glykocoll stains blue-green, the remaining amino acids stain a weak brown.

148. Vanillin-perchloric acid for the detection of pregnanetriol and related compounds.
Spray: 1% Solution of vanillin in a 10% aqueous solution of perchloric acid.
Treatment: Heat to 110° C. for 5—7 mins.
Modification: 1 g. Vanillin, 15 g. toluenesulphonic acid and 15 ml. of a 60% aqueous solution of perchloric acid are made up to 100 ml. with water in a measuring flask. Treatment as above.

149. Vanillin-phosphoric acid for the detection of steroids.
Spray: 1 g. Vanillin is dissolved in 100 ml. of a 50% aqueous solution of o-phosphoric acid.
Treatment: Heat to 120° C. for 10—20 mins.

150. Vanillin-HCl for the detection of catechins.
Spray: 0.5 g. Vanillin is dissolved in 50 ml. HCl (*d* 1.19).
After drying the chromatogram at room temperature, catechols stain red.

151. Vanillin-sulphuric acid for the detection of higher alcohols and ketones.
Spray: 3 g. Vanillin is dissolved in 100 ml. absolute ethanol, and 0.5 ml. sulphuric acid (*d* 1.84) is added to the solution.
Treatment: Heat to 120° C. until green-blue spots are seen.
Modification: Instead of sulphuric acid, one may use 1.5 g. p-toluenesulphonic acid.

152. Violuric acid for the detection of alkali- and alkali earth-ions.
Spray: 1.5% Aqueous solution of violuric acid.
Procedure: When dissolving, violuric acid must not be heated beyond 60° C! After spraying, the chromatogram is heated to 100° C. for 20 mins.
For barium and strontium see also Reagent No. 70.
For lithium and potassium see also Reagent No. 128.
For magnesium and calcium see also Reagent No. 1.

153. Cinnamic aldehyde-acetic anhydride-sulphuric acid for the detection of steroid-saponins.
Spray I: 1 g. Cinnamic aldehyde is dissolved in 100 ml. absolute ethanol.
Spray II: (prepared freshly) Mix 12 volumes acetic anhydride with 1 volume sulphuric acid (*d* 1.84).
Procedure: Spray with I, dry at 90° C. for 5 mins., spray with II. Let the acid act first at room temperature for 1—2 mins., then place chromatogram into a drying oven at 90° C., until colored spots appear.

154. Cinnamic aldehyde-HCl for the detection of indole derivatives.
Spray: 5 ml. Cinnamic aldehyde are dissolved in ethanol (96%). Mix with 5 ml. HCl (*d* 1.19). Must always be freshly prepared.

155. Stannous chloride-potassium iodide for the detection of gold.
Spray: 5.6 g. Stannous chloride is dissolved in 10 ml. HCl (*d* 1.19), after diluting the solution to 100 ml. with water, add 0.2 g. potassium iodide. Black spots appear.

156. Stannic tetrachloride for the detection of triterpenes, sterols, and steroids
Spray: To 160 ml. of a mixture consisting of equal volumes of chloroform and glacial acetic acid add 10 ml. stannic tetrachloride.
Treatment: Heat to 100° C. for 5—10 mins. and inspect plates in filtered UV-light (365 mμ).

157. Zirconium alizarinlake-HCl for the detection of fluoride-ions.
Spray: Dissolve 0.05 g. zirconium chloride ($ZrCl_2 \cdot 8\ H_2O$) and 0.5 g sodium alizarin sulphonate in 100 ml. 2 N-HCl.

O. Terminology of Thin-Layer Chromatography

By

HELMUT K. MANGOLD and M. BRENNER

English	German	French
to scratch off, to scrape off	**abkratzen,** abschaben	détacher
descending	**absteigend**	descendant
to wash, to rinse	**abwaschen,** spülen	laver, rincer
to adsorb; adsorption	**adsorbieren; Adsorption**	adsorber; adsorption
aerosol package	**Aerosolpackung**	cartouche d'aérosol
alumina	**Aluminiumoxid**	l'alumine
sample, mixture	**Analysenmaterial,** Gemisch	l'échantillon, la prise d'essai, le mélange
inorganic	**anorganisch**	minéral
to apply, application	**anwenden, Anwendung**	utiliser, utilisation
aligning tray	**Arbeitsschablone**	le gabarit
to dissolve	**auflösen** (eine Substanz in einem Lösungsmittel)	dissoudre
to resolve, to separate; resolution	**auflösen; Auflösung** (Trennschärfe)	séparer; séparation
ascending	**aufsteigend**	ascendant
to apply, to spot	**auftragen** (eine Substanz)	appliquer
band	**Band**	la bande, la zone
alkaline	**basisch, alkalisch**	basique, alcalin
to coat	**beschichten**	recouvrir d'une couche
to mark	**beschriften, markieren**	marquer
template for marking and sample application	**Beschriftungsschablone**	gabarit de marcage
binder	**Bindemittel**	le liant
BN-chamber, S-chamber	**BN-, S-Trennkammer**	la cuve de séparation BN, S
thickness	**Dicke**	l'épaisseur
thin-layer chromatography, TLC	**Dünnschicht-Chromatographie, DC**	la chromatographie en couches minces, CCM
separated by TLC	**dünnschichtchromatographisch getrennt**	séparé par chromatographie en couches minces
uniform, even	**einheitlich, gleichmäßig**	homogène, régulier
to dip, dipping technique	**eintauchen; Eintauchen**	immerger; l'immersion
to elute; eluate	**eluieren; Eluat**	éluer; l'éluat
eluent	**Elutionsmittel**	l'éluant, l'agent d'élution
demixing	**Entmischen**	le démélange (démixtion)
to develop, developer	**entwickeln; Entwickler**	révéler; le révélateur
desiccator	**Exsiccator**	le dessicateur
colored	**farbig**	coloré
colorless	**farblos**	incoloré
color reagent	**Farbreagens**	le réactif colorant

English	German	French
fixer	**Fixierbad**	le bain de fixation
spot	**Fleck**	la tâche, le "spot"
solvent (mixture)	**Fließmittel**	le solvant
(solvent) front	**Fließmittelfront**	le front du solvant
volatile	**flüchtig**	volatil
gas-liquid chromatography, GLC, gas chromatography	**Gaschromatographie, GC**	la chromatographie en phase gazeuse, CG
jar, tank, trough, vessel	Gefäß zum Entwickeln, **Trennkammer,** Tank	la cuve
calcinated calcium sulphate, gypsum, Plaster of Paris	**Gips (G)**	sulfate de calcium
to identify; identification	**identifizieren;** Identifizierung	identifier; l'identification
ion exchanger	**Ionenaustauscher**	l'échangeur d'ions
wedge-tip technique	**Keilstreifen-Technik**	la technique des bandes en cône
tape, adhesive tape	**Klebeband**	la bande adhésive
silica gel	**Kieselgel**	le gel de silice
diatomaceous earth, kieselguhr	**Kieselgur**	le kieselguhr
column chromatography	**Kolonnen-** oder **Säulenchromatographie**	la chromatographie sur colonnes
grain size, particle size	**Korngröße**	la grosseur des grains ou des particules
speed of migration, travelling speed	**Laufgeschwindigkeit**	la vitesse de migration, de déplacement
length of run	**Laufstrecke**	le parcours
to mix; mixture	**mischen;** Mischung	mélanger; le mélange
mortar and pestle	**Mörser** und **Pistill**	le mortier et le pilon
detection	**Nachweis**	la détection, le révélation, la détermination, le dosage qualitatif
neutral	**neutral**	neutre
paper chromatography, PC	**Papierchromatographie, PC**	la chromatographie sur papier, CSP
the wick for chamber-saturation	**Papierstreifen zur Kammersättigung**	la bande de papier servant à la saturation de la cuve
powder	**Pulver**	la poudre
quantitative evaluation	**quantitative Auswertung**	l'évaluation quantitative
quantitative estimation, determination	**quantitative Bestimmung**	le dosage quantitatif
edge	**Rand**	le bord
X-ray film	**Röntgenfilm**	le film pour rayons X
to saturate, saturated	**sättigen,** gesättigt	saturer, saturé
acidic	**sauer**	acide
layer	**Schicht**	la couche
tail, streak	**Schwanz, Streifen**	la queue de comète, la trainée
tailing, tail formation	**Schwanzbildung**	la formation de trainée, d'une queue de comète

English	German	French
to make visible, to visualize, to indicate, to locate	**sichtbar machen,** kenntlich machen, nachweisen	révéler
sorbent	**Sorptionsmittel**	l'adsorbant, l'agent d'adsorption
syringe	**Spritze**	la seringue
trace	**Spur**	la trace
to spray; atomizer, sprayer, vaporizer, spray gun	**sprühen;** der Sprüher	vaporiser; le vaporisateur
spray reagent, indicator	**Sprühreagens**	le réactif à vaporiser
standard, reference material	**Standard** (Vergleichs-)	l'étalon, le témoin
starting point, point of application (of the sample)	**Start** (-punkt)	le point de départ
(TLC-)spreader, applicator	**Streicher** (Dünnschicht-), **Streichgerät für DC**	le dispositif d'étalement, à étaler, le dispositif d'application
strip	**Streifen**	la bande
streak	**Strich**	la ligne
gradual development (technique)	**Stufentechnik**	technique par étapes
subfractionation	**Subfraktionierung**	le sous-fractionnement
daylight	**Tageslicht**	la lumière du jour
test dyes	**Testfarben**	les couleurs témoins
low-temperature chromatography	**Tieftemperaturchromatographie**	la chromatographie à basse température
the (glass) plate	**Trägerplatte**	le plaque support
resolution, sharpness of separation	**Trennschärfe**	la précision de séparation
to separate, to fractionate, to resolve; separation, fractionation	**trennen; Trennung**	séparer; la séparation
separation-reaction-separation (SRS-) technique	**Trennung-Reaktion-Trennung (TRT-) Technik**	la technique séparation-réaction-séparation (SRS)
drying rack, storage rack	**Trockengestell**	le séchoir
to dry	**trocknen**	sécher
non-polar	**unpolar**, nicht polar	non polaire
to differentiate, to distinguish	**unterscheiden**	différencier, distinguer
UV-light	**UV-Licht**	la lumière U.V., l'U.V.
compound	**Verbindung** (chemische)	le composé chimique
to dilute	**verdünnen**	diluer
standard	**Vergleichssubstanz** (Testsubstanz)	l'étalon, le témoin
ratio	**Verhältnis** (Mischungs-)	le rapport
adjustable	**verstellbar**	réglable
partition	**Verteilung**	le partage
reversed-phase partition	**Verteilung in umgekehrter Phase**	le partage en phase inversée
direction, procedure	**Vorschrift** (Arbeits-)	la mode opératoire
aqueous	**wäßrig**	aqueux
migration rate, speed of migration	**Wanderungsgeschwindigkeit**	la vitesse de migration, de déplacement
zone	**Zone**	la zone, la bande
two-dimensional	**zweidimensional**	bi-dimensionel

P. Commercial Suppliers*

1. Alupharm Chemicals
616 Commercial Pl.
P. O. Box 755
New Orleans, La.
U. S. representative of 72
2. Analabs
Analytical Engineering
Laboratories, Inc.
P. O. Box 5215
Hamden 18, Conn.
U. S. representative of 55
(Adsorbents and accessories)
3. Applied Science Laboratories, Inc.
P. O. Box 140
State College, Pa.
(Plexiglas TLC applicator, adsorbents and accessories)
4. Atomic Accessories Inc.
Subsidiary of Baird-Atomic, Inc.
811 W. Merrick Rd.
Valley Stream, N.Y.
(Scanner for plates containing radioactive substances)
5. W. Bälz & Sohn, K.G.
Heilbronn a. N.
Germany
(UV-lamp: „UVANALYS")
6. BASF, Bad. Anilin & Sodafabriken A.G.
Ludwigshafen, Rh.
Germany
U. S. representative: 15
(Polyethyleneimine and dyestuffs)
7. Becco Chemical Division,
Food, Machinery and Chemical Corp.
Buffalo 7, N.Y.
(Peracetic acid)
8. Becton, Dickinson & Co.
East Rutherford, N.J.
(Disposable self-filling, self-adjustable pipettes)
9. Bio-Rad Laboratories
32nd & Griffin Ave.
Richmond, Calif.
(Coating materials for TLC)
10. Wm. Boekel & Co., Inc.
509 Vine St.
Philadelphia 6, Pa.
(Storage cabinet)
11. C. A. Brinkmann and Comp., Inc.
Cantiague Rd.
Westbury, L. I., N. Y.
U. S. representative of 19, 25, 42, 46
12. Camlab (Glass) Ltd.
Milton Road
Cambridge, Great Britain
Representative of 19, 42, 72
13. Camag A. G.
Muttenz, B. L.
Homburger Str. 24
Switzerland
U. S. representative 47, 67
(Coating apparatus, adsorbents, and accessories for TLC)
14. Chemetron
Milano
Via Sangallo 28
Italy
U. S. representative: 39
(Automatic coating apparatus and accessories)
15. Chemirad Corporation
P. O. Box 187
East Brunswick, N. J
U. S. representative of 6
16. Colab Laboratories, Inc.
Chicago Heights, Ill.
U. S. representative of 66
17. Custom Service Chemicals
New Castle, Del.
(Supply coated plates ready for use)
18. Darco Dept., Atlas Powder Co.
60 E. 42nd St.
New York, N. Y.
(Charcoal as an adsorbent for TLC)
19. C. Desaga, G.m.b.H.
Heidelberg
Hauptstr. 60
Germany
U. S. representative: 11
(Several coating apparatus according to STAHL, adsorbents, densitometer, UV-lamps, accessories for TLC)
20. Despatch Oven Co.
619 S. E. 8th St.
Minneapolis 14, Minn.
(Oven for drying and activating)
21. The Dow Chemical Comp.
Midland, Mich.
(Silicone)
22. Dow Corning Corp.
Midland, Mich.
(Ion Exchangers)
23. Eastman Kodak Comp.
Rochester 3, N.Y.
(X-ray films, developers, fixers, reagents)

* No guarantee for completeness

24. C. Erba, S.p.A.
24, Via Imbonati
Milano, Italy
(Automatic coating apparatus, accessories, UV-lamps)
25. Excorna o.H.G.
Mainz-Rh.
Germany
U.S. representative: 11
(Cellulose powder for analytical chromatography)
26. Farbenfabriken Bayer
Werk Dormagen
Dormagen,
Germany
(Nylon and perlon powders)
27. Farbwerke Hoechst A.G.
Frankfurt-Hoechst a. M.
Germany
(Polyethylene powder)
28. Fisher Scientific Co.
633 Greenwich St.
New York 14, N.Y.
(Pipettes, label glaze, disposable aluminum dishes for weighing)
29. Floridin Co.
P. O. Box 989
Tallahassee, Florida
("Florisil")
30. Fluka A. G.
Buchs S. G.
Switzerland
U. S. representative: 36
(Adsorbents and reagents for TLC)
31. Gallard-Schlesinger
Chemical Manufacturing Corp.
580 Mineola Ave.
Carle Place, L.I., N.Y.
U. S. representative of 64
32. General Electric
X-Ray Department
Milwaukee 1, Wisc.
("Supermix" photogr. developer)
33. J. Haltermann
Hamburg
Germany
(Undecane)
34. Hamilton Company, Inc.
P. O. Box 307
Whittier, Calif.
(Microsyringes)
35. L. Hormuth, Inh. W. E. Vetter
Wiesloch/Baden, Germany
(Accessories for TLC; spray-technique)
36. International Chem. and Nuclear Corp.
13332 E. Torch St.
City of Industry, Calif.
U. S. representative of 30.
37. Johns Manville Corp.
Celite Division
270 Madison Ave.
New York 16, N.Y.
(Celite)
38. Joyce, Loebl and Comp.
Gateshead on Tyne
Great Britain
U. S. representative: 48
(Densitometer "Chromoscan" for TLC-plates)
39. Kensington Scientific Corp.
1717 Fifth St.
Berkeley 10, Calif.
U. S. representative of 14.
40. Kopp Laboratory Supplies, Inc.
70—13 35th Rd.
Jackson Heights 72, N. Y.
(Glass apparatus for collecting fractions and eluting them)
41. Laboratorium Prof. Dr. Berthold
Wildbad im Schwarzwald
Postfach 160
Germany
(Radiochromatogram scanner)
42. Macherey, Nagel & Co.
Düren, Rhld.
Germany
U. S. representative: 11
("MN"-Coating materials for TLC according to STAHL*)*
43. Mallinckrodt Chemical Works
2nd & Mallinckrodt Sts.
St. Louis 7, Mo.
(Silicic acid)
44. Mann Research Laboratories, Inc.
136 Liberty St.
New York 6, N. Y.
(Coating materials and spray reagents for TLC)
45. K. Markgraf
Berlin,
Germany
(Apparatus for TLC-electrophoresis)
46. E. Merck, A .G.
Chemische Fabrik
Darmstadt,
Germany
U. S. representative: 11
(Adsorbents for TLC according to STAHL, *spray reagents)*
47. Microchemical Specialties Co.
1825 Eastshore Hwy.
Berkeley 10, Calif.
U. S. representative of 13
48. National Instrument Laboratories, Inc.
12300 Parklawn Dr.
Rockville, Md.
U. S. representative of 38

49. Nuchar Industrial Chemical Sales
230 Park Ave.
New York, N. Y.
(Charcoal)
50. Nuclear Chicago Corp.
351 E. Howard Ave.
Des Plaines, Ill.
(Scanner for strips containing radioactive substances)
51. Packard Instrument Comp., Inc.
Box 428
La Grange, Ill.
(Radiochromatogram scanner and chemicals for liquid scintillation counting)
52. Pharmacia
Uppsala,
Sweden
U. S. representative: 53
("Sephadex" for gel filtration)
53. Pharmacia Fine Chemicals, Inc.
501 Fifth Ave.
New York 17, N.Y.
U. S. representative of 52.
54. Photovolt Corp.
1115 Broadway
New York 10, N. Y.
(Densitometer for TLC)
55. G. Pleuger, S. A.
511, Turnhoutsebaan
Wijnegem,
Belgium
U. S. representative: 2
(Coating apparatus and accessories)
56. E. I. du Pont de Nemours & Co. (Inc.)
Photo Products Department
Wilmington 98, Del.
(Fluorescent minerals)
57. H. Reeve Angel & Co., Ltd.
9, Bridewell Place
London, E. C. 4
Great Britain
U. S. representative: 58
(Whatman "Chromedia" material)
58. H. Reeve Angel & Co., Inc.
52, Duane St.
New York 7, N. Y.
U. S. representative of 57
59. Research Specialties Co.
200 S. Garrard Blvd.
Richmond, Calif.
(Coating apparatus, accessories, and spray reagents, apparatus for thin-layer electrophoresis)
60. Riedel-de Haën A. G.
Seelze, Hann.
Germany
(Fluorescent minerals, adsorbents)
61. C. Roth
Karlsruhe,
Germany
Representative of 13, 30
62. C. Schleicher & Schüll
Dassel, Krs. Einbeck
Germany
U. S. representative: 63
(Cellulose powder, ion-exchangers)
63. C. Schleicher & Schuell Co.
543 Washington St.
Keene, N. H.
U. S. representative of 62.
64. Scientifica
P. O. Box 1084
Clifton, N. J.
(Coating materials for TLC)
65. Serva Entwicklungslabor
Heidelberg
Römerstr. 118
Germany
U. S. representative: 31
(Spreading rod, coating materials)
66. Severoceske chemicke zavody Lovosice, n. p.
Factory "Rudnik"
Lovosice,
Czechoslovakia
(Silone)
67. Shandon Scientific Co.
65 Pound Lane
London N. W. 10
Great Britain
U. S. representative: 16
(Coating apparatus and accessories)
68. A. H. Thomas Comp.
Vine St. at Third
Philadelphia 5, Pa.
U. S. representative of 13
(Multiple spot applicator)
69. Toyo Rayon Co.
Nakano-shima, Kita-ku,
Osaka, Japan
(Polyamide)
70. Ultra-Violet Products, Inc.
5114 Walnut Grove Ave.
San Gabriel, Calif.
("Mineralight" UV-lamps)
71. Wacker-Chemie
München,
Germany
(Silicones)
72. Wako Pure Chemicals Co.
Tokyo, Japan
(Silica Gel for TLC)
73. M. Woelm
Eschwege,
Germany,
U. S. representative: 1
(Coating materials for TLC)

Conversion table for *Rf* into *Rm* and vice versa

100 · Rf	0	1	2	3	4	5	6	7	8	9		
00	∞	3,000	2,698	2,522	2,396	2,299	2,219	2,152	2,093	2,042	*—1,996*	***99***
01	1,996	1,954	1,916	1,881	1,848	1,817	1,789	1,762	1,737	1,713	*—1,690*	***98***
02	1,690	1,669	1,648	1,628	1,609	1,591	1,574	1,557	1,540	1,525	*—1,510*	***97***
03	1,510	1,495	1,481	1,467	1,453	1,440	1,428	1,415	1,403	1,392	*—1,380*	***96***
04	1,380	1,369	1,358	1,347	1,337	1,327	1,317	1,307	1,297	1,288	*—1,279*	***95***
05	1,279	1,270	1,261	1,252	1,243	1,235	1,227	1,219	1,211	1,203	*—1,195*	***94***
06	1,195	1,187	1,180	1,172	1,165	1,158	1,151	1,144	1,137	1,130	*—1,123*	***93***
07	1,123	1,117	1,110	1,104	1,097	1,091	1,085	1,079	1,073	1,067	*—1,061*	***92***
08	1,061	1,055	1,049	1,043	1,038	1,032	1,026	1,021	1,015	1,010	*—1,005*	***91***
09	1,005	1,000	0,994	0,989	0,984	0,979	0,974	0,969	0,964	0,959	*—0,954*	***90***
10	0,954	0,949	0,945	0,940	0,935	0,931	0,926	0,922	0,917	0,913	*—0,908*	***89***
11	0,908	0,904	0,899	0,895	0,891	0,886	0,882	0,878	0,874	0,869	*—0,865*	***88***
12	0,865	0,861	0,857	0,853	0,849	0,845	0,841	0,837	0,833	0,829	*—0,826*	***87***
13	0,826	0,822	0,818	0,814	0,810	0,807	0,803	0,799	0,796	0,792	*—0,788*	***86***
14	0,788	0,785	0,781	0,778	0,774	0,770	0,767	0,764	0,760	0,757	*—0,753*	***85***
15	0,753	0,750	0,747	0,743	0,740	0,736	0,733	0,730	0,727	0,723	*—0,720*	***84***
16	0,720	0,717	0,714	0,711	0,707	0,704	0,701	0,698	0,695	0,692	*—0,689*	***83***
17	0,689	0,685	0,682	0,679	0,676	0,673	0,670	0,667	0,665	0,662	*—0,659*	***82***
18	0,659	0,656	0,653	0,650	0,647	0,644	0,641	0,638	0,635	0,633	*—0,630*	***81***
19	0,630	0,627	0,624	0,621	0,619	0,616	0,613	0,610	0,607	0,605	*—0,602*	***80***
20	0,602	0,599	0,597	0,594	0,591	0,589	0,586	0,583	0,580	0,578	*—0,575*	***79***
21	0,575	0,572	0,570	0,567	0,565	0,562	0,560	0,557	0,555	0,552	*—0,550*	***78***
22	0,550	0,547	0,545	0,542	0,540	0,537	0,535	0,532	0,530	0,527	*—0,525*	***77***
23	0,525	0,523	0,520	0,518	0,515	0,513	0,511	0,508	0,506	0,503	*—0,501*	***76***
24	0,501	0,499	0,496	0,494	0,491	0,489	0,487	0,484	0,482	0,479	*—0,477*	***75***
25	0,477	0,475	0,472	0,470	0,468	0,465	0,463	0,461	0,459	0,456	*—0,454*	***74***
26	0,454	0,452	0,450	0,447	0,445	0,443	0,441	0,439	0,436	0,434	*—0,432*	***73***
27	0,432	0,430	0,428	0,425	0,423	0,421	0,419	0,417	0,414	0,412	*—0,410*	***72***
28	0,410	0,408	0,406	0,404	0,402	0,399	0,397	0,395	0,393	0,391	*—0,389*	***71***
29	0,389	0,387	0,385	0,383	0,381	0,378	0,376	0,374	0,372	0,370	*—0,368*	***70***
30	0,368	0,366	0,364	0,362	0,360	0,357	0,355	0,353	0,351	0,349	*—0,347*	***69***
31	0,347	0,345	0,343	0,341	0,339	0,337	0,335	0,333	0,331	0,329	*—0,327*	***68***
32	0,327	0,325	0,323	0,321	0,319	0,317	0,316	0,314	0,312	0,310	*—0,308*	***67***
33	0,308	0,306	0,304	0,302	0,300	0,298	0,296	0,294	0,292	0,290	*—0,288*	***66***
34	0,288	0,286	0,284	0,282	0,280	0,278	0,277	0,275	0,273	0,271	*—0,269*	***65***
35	0,269	0,267	0,265	0,263	0,261	0,259	0,258	0,256	0,254	0,252	*—0,250*	***64***
36	0,250	0,248	0,246	0,244	0,242	0,240	0,239	0,237	0,235	0,233	*—0,231*	***63***
37	0,231	0,229	0,227	0,225	0,224	0,222	0,220	0,218	0,217	0,215	*—0,213*	***62***
38	0,213	0,211	0,209	0,207	0,205	0,203	0,202	0,200	0,198	0,196	*—0,194*	***61***
39	0,194	0,192	0,190	0,189	0,187	0,185	0,183	0,181	0,180	0,178	*—0,176*	***60***
		9	*8*	*7*	*6*	*5*	*4*	*3*	*2*	*1*	*0*	100 · Rf

Continuation

100 · Rf	0	1	2	3	4	5	6	7	8	9		
40	0,176	0,174	0,172	0,170	0,169	0,167	0,165	0,163	0,162	0,160	*—0,158*	***59***
41	0,158	0,156	0,154	0,153	0,151	0,149	0,147	0,145	0,144	0,142	*—0,140*	***58***
42	0,140	0,138	0,136	0,135	0,133	0,131	0,129	0,127	0,126	0,124	*—0,122*	***57***
43	0,122	0,120	0,119	0,117	0,115	0,113	0,112	0,110	0,108	0,107	*—0,105*	***56***
44	0,105	0,103	0,101	0,100	0,098	0,096	0,094	0,092	0,090	0,089	*—0,087*	***55***
45	0,087	0,085	0,084	0,082	0,080	0,078	0,077	0,075	0,073	0,072	*—0,070*	***54***
46	0,070	0,068	0,066	0,065	0,063	0,661	0,059	0,057	0,056	0,054	*—0,052*	***53***
47	0,052	0,050	0,049	0,047	0,045	0,043	0,042	0,040	0,038	0,037	*—0,035*	***52***
48	0,035	0,033	0,031	0,030	0,028	0,026	0,024	0,022	0,020	0,019	*—0,017*	***51***
49	0,017	0,015	0,014	0,012	0,010	0,008	0,007	0,005	0,003	0,002	*±0,000*	***50***
		9	*8*	*7*	*6*	*5*	*4*	*3*	*2*	*1*	*0*	100 · Rf

N. B. Rf-values from 0—0,5 : Rm positive
Rf-values from 0,5—1,0 (*Italics*): Rm negative

Examples:

Rf	0,040	0,346	0,510[1]	0,654[1]
Rm	1,380	0,277	—0,017	—0,277[2]

[1] Use column of *Rf*-values in *Italics*.

[2] Rm-values which have been found via the column written in *Italics*, are negative.

Author Index

Numbers in *italics* refer to the bibliography; *italic* numbers in brackets [ ] refer to the numbering of references in the particular chapter.

Subject Index

Printed by Brühlsche Universitätsdruckerei, Gießen